T4-AHW-006

The Crystal Chemistry and Physics of Metals and Alloys

WILEY SERIES ON THE SCIENCE AND TECHNOLOGY OF MATERIALS

Advisory Editors: E. Burke, B. Chalmers, James A. Krumhansl

THE CRYSTAL CHEMISTRY AND PHYSICS OF METALS AND ALLOYS
W. B. Pearson

X-RAY DIFFRACTION METHODS IN POLYMER SCIENCE
L. E. Alexander

PHYSICAL PROPERTIES OF MOLECULAR CRYSTALS, LIQUIDS, AND GLASSES
A. Bondi

FRACTURE OF STRUCTURAL MATERIALS
A. S. Tetelman and A. J. McEvily, Jr.

ORGANIC SEMICONDUCTORS
F. Gutmann and L. E. Lyons

INTERMETALLIC COMPOUNDS
J. H. Westbrook, editor

THE PHYSICAL PRINCIPLES OF MAGNETISM
Allan H. Morrish

FRICTION AND WEAR OF MATERIALS
Ernest Rabinowicz

HANDBOOK OF ELECTRON BEAM WELDING
R. Bakish and S. S. White

PHYSICS OF MAGNETISM
Sōshin Chikazumi

PHYSICS OF III-V COMPOUNDS
Otfried Madelung (translation by D. Meyerhofer)

PRINCIPLES OF SOLIDIFICATION
Bruce Chalmers

APPLIED SUPERCONDUCTIVITY
Vernon L. Newhouse

THE MECHANICAL PROPERTIES OF MATTER
A. H. Cottrell

THE ART AND SCIENCE OF GROWING CRYSTALS
J. J. Gilman, editor

SELECTED VALUES OF THERMODYNAMIC PROPERTIES OF METALS AND ALLOYS
Ralph Hultgren, Raymond L. Orr, Philip D. Anderson and Kenneth K. Kelly

PROCESSES OF CREEP AND FATIGUE IN METALS
A. J. Kennedy

COLUMBIUM AND TANTALUM
Frank T. Sisco and Edward Epremian, editors

MECHANICAL PROPERTIES OF METALS
D. McLean

THE METALLURGY OF WELDING
D. Séférian (translation by E. Bishop)

THERMODYNAMICS OF SOLIDS
Richard A. Swalin

TRANSMISSION ELECTRON MICROSCOPY OF METALS
Gareth Thomas

PLASTICITY AND CREEP OF METALS
J. D. Lubahn and R. P. Felgar

INTRODUCTION TO CERAMICS
W. D. Kingery

PROPERTIES AND STRUCTURE OF POLYMERS
Arthur V. Tobolsky

PHYSICAL METALLURGY
Bruce Chalmers

ZONE MELTING, SECOND EDITION
William G. Pfann

The Crystal Chemistry and Physics of Metals and Alloys

W. B. Pearson

Dean of Science,
University of Waterloo,
Waterloo, Ontario, Canada

Wiley – Interscience

A Division of John Wiley & Sons, Inc.
New York · London · Sydney · Toronto

Library of Congress Catalogue Card Number: 70-176284

ISBN 0-471-67540-7

Printed in the United States of America.

10 9 8 7 6 5 4 3 2 1

Preface

I am very conscious that a book which nowadays presumes to be called *The Crystal Chemistry and Physics of Metals and Alloys* will be found to have many omissions in the eyes of its various readers. This results mainly from the enormous field of information to be covered in a finite time, and from the limitations of the author, since the literature and ideas enlarge at a rate greater than the expansion of my comprehension. Indeed, somebody once told me that he had fitted a mathematical function to the relationship between the thickness and year of publication of the volumes of the three editions of Wells' *Structural Inorganic Chemistry*. This indicated that a fourth edition, if published in a year which I now forget, should have a thickness greater than four feet!

This situation weights the subject matter of this book in partisan fashion along the lines of interest of the author, and here I am conscious of having made a choice that eliminates almost all discussion of what structures are formed by what particular combinations of elements, and of examination of the limitations in the occurrence of particular structures in terms of the Periodic Table, with analysis of these causes. Instead, I have chosen to concentrate on description of the particular crystal structures, the general factors which control their occurrence and on atomic dimensions in the structures. Insofar as some choice had to be made, I feel that this is logical, since I have written the *Handbook of the Lattice Spacings and Structure of Metals and Alloys* so that the present book ranks as a sequel to it. Had I but written Hansen's *Aufbau der Zweistofflegierungen*, then it would have been more logical to have discussed metals structures from the viewpoint of the Periodic Table! Indeed, the title of the book has been a matter of no little concern, since how can a book presume to be called *Crystal Chemistry* and ignore so large and important segment of the subject? Furthermore, the book delves into physics and physical principles when these are appropriate for discussion of structural stability; yet *Crystal Physics* has quite a different connotation. Since, however, any features apparent in energy band considerations must have their counterpart in chemical bond considerations and *vice versa*, it seems that the most satisfactory compromise is achieved by calling it *The Crystal Chemistry and Physics of Metals and Alloys*.

While writing this book I have found that ideas which I held about the structures of metals have largely disappeared and been replaced by new values. These earlier ideas about structural relationships evolved mainly from the principle of derivative structures which was useful in relation to physical properties. In this regime metals were, of course, the realm of close packings in their own right, but otherwise tended to be the rump that was left when things were not insulators, ionics, or semiconductors. In this feeling I was not alone, since this has been the position accorded to metals in general books on crystal chemistry. My appreciation that the crystal chemistry of metals is something that can be approached in its own right, rather than as an apology concerning what is left, comes largely from considering the results of experimental Fermi surface studies on metals, and from the confidence that these results give in discussing electron energy relationships in reciprocal space. Second, it comes from appreciating how useful it is to consider the structures of metals to be made up of layer networks of atoms whenever possible.

Of course, I might have attempted to extend the concept of sphere packing which was used so successfully by Laves in 1956, but I have rejected this as a basis of my thinking for two reasons. In the first place the sphere model seems inappropriate for metals whose atoms are compressible and readily accept different distances to ligands in different directions, but more important, it seemed at that time impossible to achieve any quantitative assessment or measurement of the strength of a "geometrical factor." Furthermore, even though we may define a structure type on a strictly geometrical basis (p. 15), our thinking is generally prejudiced by a subjection of geometrical realities to chemical conventions and habits. Thus, when occasion arises, and we consider structures containing close packed layers of atoms, we like to consider one particular component as forming a close packed framework and the other component(s) as occupying tetrahedral or octahedral holes therein. In this way we have come to regard the NiAs structure as built up of a close packed array of As atoms with the Ni atoms occupying the octahedral holes therein. Rarely do we regard it as a simple hexagonal array of Ni atoms with alternate trigonal prisms centered by As. Neither do we regard the Ni_2In structure as a superstructure of the AlB_2 type, made up of a simple hexagonal array of Ni atoms in which alternate trigonal prisms are centered by Ni and In in ordered fashion. This historical limitation of our way of thinking which derives from early classifications of structures on an ionic model, must be abandoned when we come to consider specifically the structures of metals, since there is no *a priori* difference between one component atom and another. In eventually freeing myself from this conventional avenue of thought, I am now rather surprised to find that I have completed a book in which close packing of triangular nets of atoms is nothing special *per se*, being only a particular case in a wider group of structures built up of triangular, hexagonal, and kagomé networks of atoms which can be stacked together according to the sequences normally considered in close packings.

Indeed, the escape from chemical premptions to a new freedom of thought based purely on geometry, leads to new groupings of structures and structural relationships, when atomic coordination is considered. As noted above, one immediate consequence of this is the relegation of structures containing close packed layers of atoms from their position of paramount importance, merely to a special case in the much wider group of structures made up by packing together morphologically triangular, hexagonal, and kagomé nets of atoms. Another is the greatly increased importance of the triangular prism in developing relationships between structure types. It is, for instance, interesting to recall that the triangular prism was not mentioned at all during a week-long symposium on "the Geometry and Systematics of Crystal Structures" in Lexington in 1969, whereas in this book it plays a major role in relating together some 80 structure types!

Although in this book I have attempted to make energy band considerations the prime structural consideration, they are a weak feature of the overall free energy of a phase and so are readily dominated by chemical bond and other factors. For this reason full use is also made of the derivative structure concept and geometrical concepts where they are of advantage. Indeed, one of the features of this is the discovery of a way of assessing quantitatively, at least to some extent, the importance of geometrical factors in structures.

These thoughts lead to two principles which, although largely self-evident, do not appear to have been explicitly stated and applied systematically in considerations of structural architecture and the stability of metallic phases. The first is that the size of a metal atom is an essentially adjustable parameter to the extent that the metal may generally be compressed readily in the direction of certain ligands in order to achieve particular contacts which lead to structural stability for chemical bond or geometrical reasons (see pp. 52–68). The second which follows from experimental Fermi surface studies and energy band calculations, is that the effective electron concentration (*s*, *p*, *d* valence electrons) in partly filled energy bands is likely to be only one or two electrons per atom in a metallic alloy whose structure is controlled by geometrical or energy band factors. If there are other "valence electrons" they occupy filled energy bands below the Fermi level. If the effective electron concentration is larger (2–4 electrons per atom), then structures are probably controlled by specific chemical bond (M–N) interactions, and when it approaches or reaches 4 electrons per atom, then all energy bands are probably filled and lie below the Fermi level so that semiconductors or insulators are obtained. The first of these concepts permits to some extent a quantitative evaluation of the influence of geometrical effects in the structures of metals, which heretofore have been largely only a conversation piece. The immediate consequence of the second precept is in providing a net outlook on the effective electron concentrations of transition metal alloys, and the realization that many features of the alloys of Cu, Ag, and Au that repeat themselves in the transition metals series are probably occurring at the same effective electron concentrations.

I shall undoubtedly be criticized, as I have been in the past, for not *wholeheartedly* embracing the latest theoretical developments to account for relative structural dimensions and stability, still adhering mainly to the basic principles annunciated by Jones. However, these still appear to be correct some forty years later and so provide an acceptable framework for discussing alloy stability. On the other hand, the pseudopotential theory, although offering attractive qualitative explanations of the stability and relative cell dimensions of the structures of a number of metals and alloys, still requires, I believe, acceptable quantitative developments in this field, before it is adopted as the basis of our thinking.

It may be noticed that much of the material in this book is drawn from papers that I have published in the last few years; but I must remark that it was the preparation of the book which in fact led to the writing of the papers as one or other aspect of the subject matter was studied in detail, rather than *vice versa*. Not only is this, as far as I am aware, only the second book on the crystal chemistry of metals and alloys (following Schubert's *Kristallstrukturen zweikomponentiger Phasen*), but much of the material presented in the first five chapters seeks to establish a new and unified point of view, rather than paraphrasing discussions of alloy chemistry to be found in reviews and papers in the literature. Nevertheless, it is obvious that many of the concepts set forth find their origins and inspiration in one place or another in the existing literature.

One of the greatest problems in organizing this book has been to decide how best to arrange and group some seven hundred different structure types found in metals and alloys, in order that they may be described coherently. This I feel has now in the main been handled successfully by taking layer networks of atoms as the basis of the arrangement and atomic coordination, which follows from the layer arrangements, as the secondary criterion. Generally this pattern leads to the conventional association of structure types, although in some cases where triangular prismatic and/or octahedral coordinations are involved, the switch from chemical premptions to purely geometrical logic establishes some new bedfellows as indicated above. In describing many crystal structure types, the accuracy and reliability of the structural determinations comes in question. All structures described without comment on the accuracy of the determination have at least been examined by single-crystal methods, even though they may not have been refined according to modern standards. Other structures which may have been determined only by powder diffraction methods and need confirmation have this fact recorded. Structures of hydride, boride, carbide, or nitride phases generally have the method of determination recorded so that their reliability can be assessed by the reader. It would have been convenient to present the descriptions of less important structures and ideas in smaller print so as to emphasize the more important, but unfortunately this has proved impossible.

The assembly of a book of this type by one person is clearly impossible without the support and assistance of many people. I am most grateful, first, to the National Research Council of Canada, whose constant support through this work made it indeed possible. Second, I value greatly the collaboration and discussions with Drs. D. T. Cromer and A. L. Larson of the Los Alamos Laboratory of the University of California A.E.C., who have graciously made available to me the contents of their metals structure data bank; with Dr. A. L. Loeb of the Lexington Laboratory of the Kennecott Copper Corporation and Dr. J. Lima de Faria and his collaborators of the Laboratório de Técnicas Físico-Químicas Aplicadas á Mineralogia e Petrologia, Lisbon, Portugal, for many happy discussions regarding structural relationships; and with Drs. L. D. Calvert and J.-P. Jan of the National Research Council for guidance in the more subtle points of crystallography and physics. I am particularly grateful to Mr. J. K. Byron, Mr. A. C. Parshad, Miss Eileen De Jong, Mrs. Olga Lauber, and my wife who have helped me through many difficulties and technical aspects of handling crystal data, and to Mr. S. Martin, Miss Celia Clyde, and other members of the Drafting Office at the National Research Council for their efforts in producing drawings of crystal structures.

Finally, handling a large amount of information in a descriptive manner demands the use of jargon as a means of achieving the necessary brevity, and this book is no exception. Nevertheless, the indulgence therein is not excessive and it mainly follows well established patterns. In order that the reader may readily comprehend such jargon as is used, it is collected and defined altogether in a section at the beginning of the text.

University of Waterloo
Waterloo, Ontario
February, 1972

W. B. PEARSON

Contents

Jargon

One of the limitations of communication between those who describe and discuss crystal structures is the lack of a common system of abbreviated description or jargon. This is perhaps not surprizing, because not all systems are applicable to all situations, and in a particular circumstance one may offer advantages over another. Knowledge of certain systems, such as the Ždanov–Beck and Jagodzinski methods of describing the stacking of close packed layers, is mandatory, since they have come into wide-spread use.

In this book a certain amount of jargon has to be used in order to eliminate much unnecessary verbal description, and certain accepted conventions are followed. This material is here assembled for easy reference and study, since its appreciation is required in order to make full use of the book. Two special systems call for attention: the Schläfli symbols describing layer networks of atoms, which are adopted for the primary classification of crystal structure types in this book, and the extension of the use of A, B, C symbols for describing the stacking of close packed layers of atoms to include a, b, c describing the stacking of graphite-like hexagonal nets, and α, β, γ describing the stacking of kagomé nets of atoms. The stacking of close packed layers of atoms is but a special case of a more general situation which includes the mixed stacking together of morphologically triangular and hexagonal nets and kagomé nets of atoms.

In the jargon assembled below, the use or absence of italics and the particular use of brackets has significance which should be noted.

General

Component atoms: M, N, T (transition metal), X (metalloid), Ln (lanthanon)

Stacking sequences:	A, B, C, etc.
Crystallographic planes:	(001), *ab* plane, etc. {001} family of equivalent planes
Crystallographic directions:	[001], *c* direction, etc. ⟨001⟩ family of equivalent directions
Atomic or other radii:	R
Atomic or other diameters:	D
Interatomic distances:	d_1, d_2 etc.
"Anion":	A
"Cation":	C
CN	Coordination number
Zintl border	Line in the Periodic Table which separates Group III from Group IV nontransitional elements.
Superstructure	An ordered phase, frequently formed from a solid solution. The general usage nowadays does not necessarily imply a multiplication of the edges of the fundamental (disordered) cell.
Antisymmetrically	One polygon is said to cover another similar polygon antisymmetrically when it is rotated about their mutual centers by approximately half of the angle subtended at the center by the polygon edge (Figure J-1).
Siteset	A set of crystallographically equivalent atomic positions. These are listed for the 230-space groups in the *International Tables for Crystallography*, Volume I.†

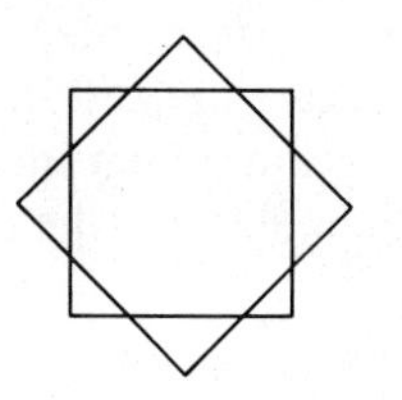
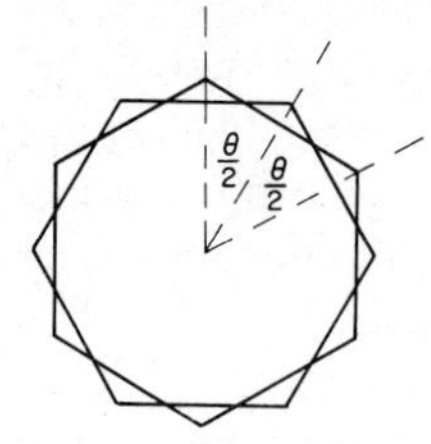

Figure J-1

†*International Tables for Crystallography*. International Union of Crystallography, Birmingham: Kynoch Press.

Schläfli Symbols to Describe a Network of Atoms

A planar network of atoms is described by recording in order the size and number of polygons surrounding each nonequivalent node in the network: 3 specifies a triangle, 4 a square or rectangle, 5 a pentagon, 6 a hexagon, etc. Multiplicities are recorded by an index super. When a network has nodes which are nonequivalent in polygonal surrounding, the net is described by listing successively the different corners. Thus the net shown in Figure J-2 is a $3^2434 + 3^34^2(2:1)$ net. The ratio following the network description indicates, in order, the relative frequency of occurrence of the nodes described. [It should be pointed out that although a close packed layer of atoms forms a 3^6 net composed of equilateral triangles, not all 3^6 nets of atoms are necessarily close packed layers. For instance, the triangles of 3^6 nets of b.c. cubic {110} layers have angles of 55, 55 and 70° approximately.]

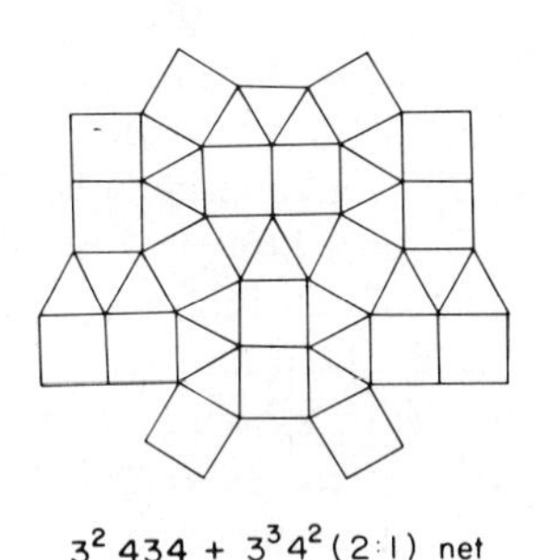

$3^2\,434 + 3^34^2(2:1)$ net

Figure J-2

Orientation of Square Nets; 4^4

It is frequently convenient to be able to refer to the orientation of a 4^4 net relative to the unit cell axes if they lie parallel to the same plane, whether or not they happen to be orthogonal. This is done as follows: Let the net, transposed parallel if necessary, pass through an origin at the left hand bottom corner of the unit cell. One edge of the net must then intersect one of the two opposite sides of the unit cell from bottom right to top left corner (Figure J-3). The orientation is then specified as that fraction of the cell edges (starting from the bottom right hand corner of the cell) where the intersection with a 4^4 net occurs, counting the full length of each cell edge as $\frac{1}{2}$, regardless of their relative lengths. Thus, a 4^4 net which is oriented parallel to orthogonal cell edges is a 1.4^4 net. Other common orientations are designated in the figure.

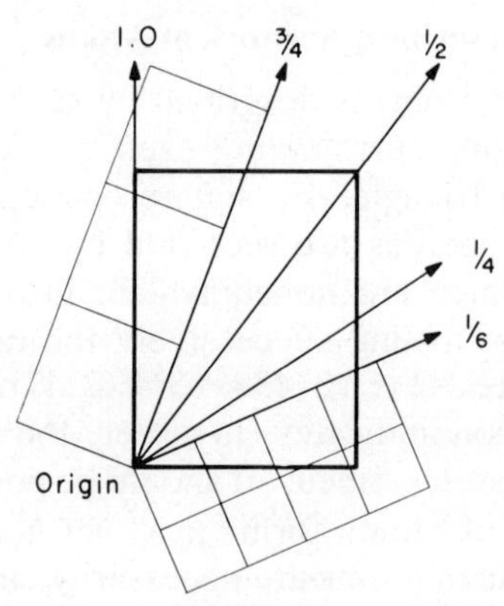

Figure J-3

Stacking of Layer Nets

The morphologically triangular 3^6 (close packed), hexagonal 6^3 and kagomé 3636 nets are of frequent structural occurrence, and since a 3^6 net can be subdivided into a 6^3 and a larger 3^6 net (ratio of number of sites 2:1), or into a kagomé net and a larger 3^6 net (ratio of number of sites 3:1), a compatible sequence of stacking symbols for the three nets can be chosen which has a wide application in describing the atomic arrangements in structures. The stacking symbols A, B, C; a, b, c; α, β, γ for 3^6, 6^3 and 3636 nets respectively relate the positions of the nodes of the nets to a hexagonal cell which contains respectively 1, 2, or 3 net points as shown in Figure J-4*a*. Sometimes it is convenient for structural descriptions also to use a larger cell which contains respectively 3, 6, and 9 nodes, the

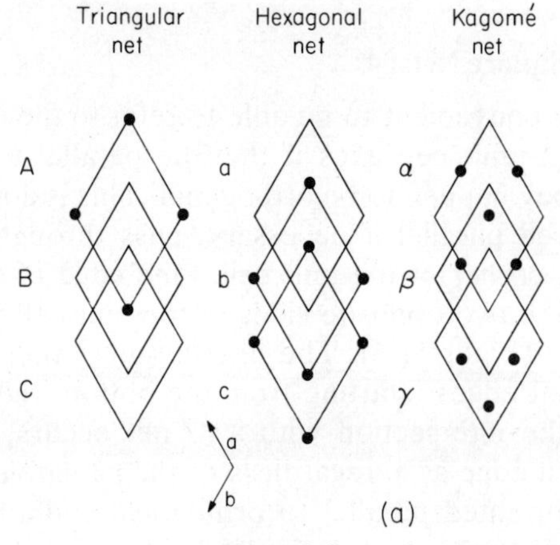

Figure J-4

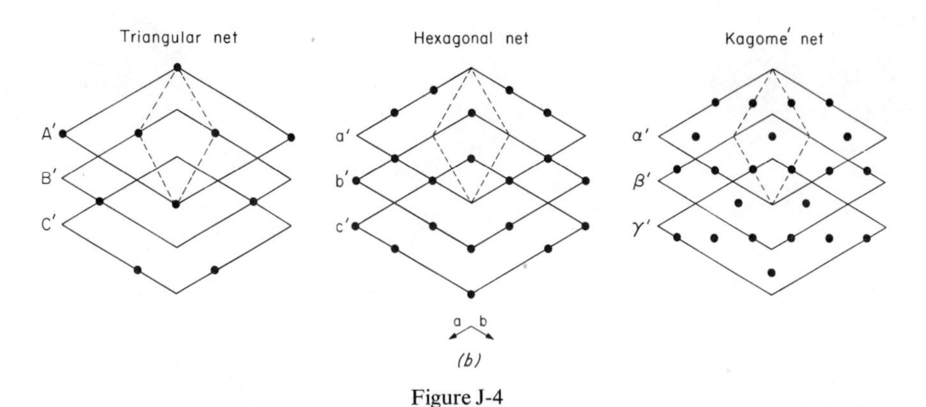

Figure J-4

stacking positions being referred to as A′, B′, C′; a′, b′, c′ and α', β', γ' as shown in Figure J-4*b*. [These 18 symbols and the information that they give about the number and arrangement of atoms relative to a hexagonal cell, and the stacking sequence that their successive listing implies for a structure, permit the simple description of atomic positions in several hundred structure types, including those of close packing].

"Ždanov–Beck" Notation for Describing Stacking of Close Packed 3^6 Nets

Ždanov derived symbols for describing successive arrangements of close packed layers; much the same symbols have been used by Wyckoff. Beck more recently has generalized the use of Ždanov symbolism. In this book the symbolism used to describe stacking sequences of 3^6 nets is referred to as the "Ždanov–Beck" notation, without making full use of the results of Beck's analysis.

Ždanov indicated the stacking sequences in polytypes of silicon carbide and similar substances by listing the number of moves in one direction in $(11\bar{2}0)$ planes of a hexagonal cell, that successive layers of one of the components in the structure make. Thus a layer stacking sequence ABAC is described as 2,2; ABABCBCAC is 212121 (Figure J-5). Multiple-repetitive sequences can be described with a sub or superscript index. Thus 333332 is $3_5 2$, or 3332333212133323332 could be shortened to $(3_3 2)_2 121 (3_3 2)_2$.

The part of Beck's analysis adopted here is to differentiate movements in one or other direction in the $(11\bar{2}0)$ plane calling one positive and the other negative (symbol with bar underneath); thus the Ždanov 22 symbol becomes $2\underline{2}$ in Ždanov–Beck notation. Furthermore, Beck gives only

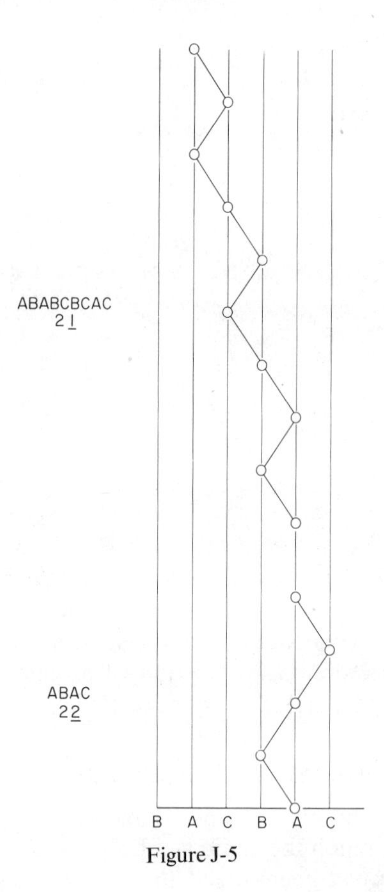

Figure J-5

the simplest stacking sequence describing the structure, rather than repeating the description for all of the layers in the unit cell. Thus the Ždanov–Beck description of the rhombohedral structure with layers 212121 becomes just $2\underline{1}$. The cubic repeat sequence when successive layers are all stacked with movements in the same direction, which is sometimes described by the symbol ∞, becomes $1\underline{0}$ in Ždanov–Beck notation.

Jagodzinski Symbols

Close packed layers which are surrounded in cubic stacking sequence $A\overline{B}C$ are described by the symbol c and those which are hexagonally surrounded $A\underline{B}A$ are given the symbol h. Thus the double hexagonal sequence $AB\overline{A}C$ is described by the symbols chch.

Ramsdel Symbols

Ramsdel described hexagonal or rhombohedral structures involving the successive stacking of (ordered) close packed layers by the letters H or R, respectively, preceded by a number giving the number of close packed layers in the repeat sequence. Thus the double hexagonal sequence ABAC is 4H. The nine-layer Sm structure is 9R.

In this book any of these symbolisms for describing the arrangements of close packed layers may be used depending on which is the most convenient in a particular context.

Periodic Table

The form of the Periodic Table of the elements invariably used throughout this book, and the designation of the A and B subgroups, is shown in Figure J-6.

1 H																	2 He
3 Li	4 Be			A Group								5 B	6 C	7 N	8 O	9 F	10 Ne
11 Na	12 Mg											13 Al	14 Si	15 P	16 S	17 Cl	18 A
19 K	20 Ca	21 Sc	22 Ti	23 V	24 Cr	25 Mn	26 Fe	27 Co	28 Ni	29 Cu	30 Zn	31 Ga	32 Ge	33 As	34 Se	35 Br	36 Kr
37 Rb	38 Sr	39 Y	40 Zr	41 Nb	42 Mo	43 Tc	44 Ru	45 Rh	46 Pd	47 Ag	48 Cd	49 In	50 Sn	51 Sb	52 Te	53 I	54 Xe
55 Cs	56 Ba	57† La	72 Hf	73 Ta	74 W	75 Re	76 Os	77 Ir	78 Pt	79 Au	80 Hg	81 Tl	82 Pb	83 Bi	84 Po	85 At	86 Rn
87 Fr	88 Ra	89 Ac	‡	Transition metals								B Group					

†Lanthanides (rare earths)	58 Ce	59 Pr	60 Nd	61 Pm	62 Sm	63 Eu	64 Gd	65 Tb	66 Dy	67 Ho	68 Er	69 Tm	70 Yb	71 Lu
‡Thorides	90 Th	91 Pa	92 U	93 Np	94 Pu	95 Am	96 Cm	97 Bk	98 Cf	99 E	100 Fm	101 Mv		

Figure J-6

1

Introduction: Crystal Chemistry

The fundamental problem of the solid state is to understand how the electronic energy levels of the component atoms of a solid whose energies and occupation are well known for the free atoms from spectroscopic data, interact and are modified when the atoms condense to form the solid. Whatever our *a priori* understanding of this problem may be, there is one experiment which shows the resultant of all of the complex electronic interactions in the solid, and that is the solution of the crystal structure. The crystal structure therefore provides the first and foremost experiment of the solid state, and by gaining an understanding of the principles involved in the formation of various types of structures, and of those which lead to distortions in structures, it may be possible to infer something about the electronic interactions that have occurred. Such studies are called crystal chemistry. It is a subject of great fascination both through its architectural aspects and because of the possibility of correlating the dependence of crystal structure and physical properties, such as electrical conductivity and magnetic behavior, with common atomic properties and interactions.

The type of crystal structure results from the character, and differences thereof, of the component atoms. Characterless atoms which, like billiard balls, have only uniform diameter and an incompressible exterior, would be expected to pack together in a crystal structure according to Laves' principles of the highest symmetry, highest coordination, and the densest packing of the atoms. These conditions are well met in the close packed cubic or hexagonal structures of many metals. When two components are present in the structure and differences of atomic character occur, the observed structures may differ in lesser or greater extent from those based on Laves' principles. In metals, atomic size, electron concentration, and directional chemical bonds may influence the crystal structure, but the dominant influence that can outweigh all others in controlling phase formation and composition, is the difference of ionization potential and electron affinity of the component atoms, commonly characterized by the electronegativity difference. Indeed, when this factor is large, ionic rather than metallic substances are obtained. In ionic structures, since the anions are generally larger than the cations, the structure type resulting from a given stoichiometry satisfying valence rules, is determined principally by the maximum number of anions that can be accommodated without touching each other in the coordination polyhedra about the cations. If they come in contact, repulsion between the anions makes the structure unstable, so that the relative cation and anion sizes, or the radius ratio, is the primary factor controlling ionic structure type; thereafter symmetry requirements and packing considerations generally lead to the choice of the most symmetric coordination polyhedron possible. Ultimately, the composition and electrostatic interactions control the arrangement of the coordination polyhedra which may share corners, edges, and/or faces. These arrangements are important to electron transport and magnetic properties in transition metal compounds. In general, an ionic compound adopts a structure with as high a Madelung constant as possible, provided that the relative sizes of the ions do not lead to anion–anion contacts.

One of the main objectives of studies of crystal chemistry is to classify and arrange the various observed structure types so as to show relationships which may exist between them. Such classifications provide a means of greatly condensing descriptive details of crystal structures, thus permitting much structural information to be comprehended and borne in mind. Crystal chemical classification of structures has in the past[1] generally been based on an ionic model where the ratio of the cationic to anionic radii is the most important parameter in assessing structure type. This type of classification does not recognize the fact that frequently both metals and ionic substances have the same types of structure for the underlying reason that in each case the bonds between the atoms are nondirectional in character. For some purposes such as discussing the structures of semiconductors, a more useful criterion is *the amount of directional character of the bonds*, starting with diamond as the prototype for fully directional bonds, and examining how structures change with decreasing directional character of the bonds, either as metallic properties or as ionic properties increase. The atomic parameters adopted as gauges for these considerations are also important for predicting the type of electrical conductivity of a substance and thus a connection between crystal chemistry and conductivity type may be established.

In developing a crystal chemistry of alloys one of the problems is to choose criteria for describing structures that show to the best advantage the relationships between the greatest numbers of structures. A classification of alloy structures that is based on coordination polyhedra *alone* does not appear to be a good method, or, for example in a limited region of alloy structures, the classification of *tetrahedral structures*[2] which, because of its restricted character has no physical properties fully identified with it, must be less useful than a classification based on close packing of the anions and the occupation of octahedral and/or tetrahedral sites by the cations, because semiconductivity broadly identifies with it. There can be no doubt that primary classification of the structures of metals and alloys by the stacking of (nearly) planar layers of atoms is the most effective way to regard them. In the first place there are the structures formed by the stacking of morphologically triangular, hexagonal, or kagomé (hexagon-triangle) layer networks of atoms one over the other in sequence. Each of these layers has the property that it can be located about one of three equivalent sites, A, B, or C, giving the possibility of variation in the stacking sequence as well as in the succession of the net types. In addition, planes of atoms may be made up of a combination of different layer networks such as triangular plus hexagonal, each of which is occupied by a different chemical component. An unlimited number of structure types can be derived from permutation of the stacking and net sequences. When the permutation involves only stacking sequences of the equilateral triangular net, *close packed* structures are obtained, and by ordering the component atoms on the triangular nets (generating sub-networks), the number of possible structure types is further increased. In such arrangements the atoms either surround each other in cubo-octahedra or twinned cubo-octahedra and they form tetrahedra, octahedra, and/or right trigonal prisms, which themselves may provide the coordination polyhedra of atoms located at their centers. Such atoms and those centering the coordination polyhedra severally form interpenetrating close packed cubic, hexagonal, or simple hexagonal arrays.

Other structures can be formed by stacking together layer networks of atoms containing squares alone, or together with triangles, pentagons, and/or hexagons; the squares, pentagons, or hexagons of one net may be centered by atoms of nets above and below, or they may not. Two of the characteristic coordination polyhedra found in these structures are the cube and the archimedian antiprism, but many structural variations are covered by this group and they are hard to classify on a simple system. Another group of structures which lends itself much more readily to simple classification is the Frank–Kasper structures which contain *only* interpenetrating triangulated CN 12, 14, 15, or 16 polyhedra. These structures are formed by interleaving pentagon-triangle, pentagon-hexagon-triangle, or hexagon-triangle primary layers with triangular, square-triangle, or square secondary nets of atoms, so that all pentagons and hexagons of the primary net are centered by atoms of the secondary net. The structures are examples of (somewhat distorted) tetrahedral close packing of metal atoms.

Classifications such as these embrace very many of the structures of metals, but numerous structures still remain which have to be described individually, or which can at best be grouped together according to some minor coordination feature. There are also small groups of structures possessing some common feature, such as those containing close pairs of atoms or those in which one component occupies a β-Sn like array of sites.

The second most important objective of a crystal chemistry of metals is to study the influence of various factors in controlling the type of crystal structure that is formed, and also the composition at which phases occur. Here, electron concentration has long been recognized as exerting an influence in particular cases, and the present experimental knowledge of Fermi surfaces together with theoretical energy-band calculations must be examined in relation to crystal structure. Questions of the importance of directional chemical bonding and the influence of the electronegativity difference of the component atoms must be studied, and some effort must be made to examine in a quantitative manner the importance of geometrical features in controlling structure and the structural dimensions of phases. The influence of temperature on crystal structure and ordering has to be considered, as well as that of high pressure, since an increasing number of investigations are being carried out using the pressure variable.

REFERENCES

1. Cf, e.g., R. C. Evans, 1937, *Crystal Chemistry*, 1st ed, Cambridge: Cambridge University Press.
2. E. Parthé, 1964, *Crystal Chemistry of Tetrahedral Structures*, New York: Gordon and Breach.

2

Crystal Structure Relationships in Metals and Alloys

One of the functions of crystal chemistry is the establishment of systematic relationships between different crystal structures and particularly between families of related structures. This task is made easier by establishing broad classifications of structure type according to some structural features which are easily recognized and of frequent occurrence. In this way structures with similar characteristics are grouped together so that more detailed similarities can easily be recognized.

From the viewpoint of practical crystallography, when a new phase has had its crystal system, space group, and the number of atoms per unit cell characterized, the first question that arises is whether the structure type has been determined previously. At this stage of structure analysis a classification of known structure types according to Bravais lattice and number of atoms per unit cell proves most valuable, since it may only be necessary to compare the formulas and space group of two or three compounds with the same classification in order to answer the question, thus saving a time-consuming search of the literature. To this end listings of the crystal structures of metals by Bravais lattice and number of atoms per cell are given in Pearson[1] and the same classification symbols are used throughout this book in order to specify briefly the crystal system. The symbols give successively the crystal system, lattice symbol, and the number of atoms per crystallographic cell according to the designation in Table 2-1. Thus, for example, the fluorite structure is described as *cF*12. Another advantage of describing structures in this manner is that it provides a simple means of identifying in terms of crystallographic information structure types that are named after some compound or phase

Table 2-1 Symbols Used for the Fourteen Bravais Lattices

Symbol	System	Lattice Symbol
aP	Triclinic (anorthic)	*P*
mP	Monoclinic	P^a
mC		C^a
oP	Orthorhombic	*P*
oC		*C*
oF		*F*
oI		*I*
tP	Tetragonal	*P*
tI		*I*
hP	Hexagonal (and trigonal, *P*)	*P*
hR	Rhombohedral	*R*
cP	Cubic	*P*
cF		*F*
cI		*I*

[a]Second setting, *y* axis unique.

characteristic of the type. Thus Al_7Cu_2Fe (*tP*40) tells immediately that the Al_7Cu_2Fe structure has a primitive tetragonal cell containing 40 atoms. In this book the old *Strukturbericht* names for structure types which were unique, but noninformative, have generally been rejected in favor of chemical names followed by the structure symbols, as above.

The most promising way of classifying and describing the known structures of metals and alloys so as to demonstrate structural relationships on a broad basis might at first sight appear to be the *derivative structure* method or the *lattice complex* method. But neither of these methods has yet been applied extensively to metals' structures, although the principles governing the former have been set out by Buerger (p. 17) and applied in the inorganic field, and symbolism for lattice complexes in all systems has now been completed by Hellner and collaborators.[2] A crystal algebra method which has been developed by Loeb[3] for the cubic and hexagonal classes appears to be essentially an off-shoot of the lattice complex analysis. Broad classification based on coordination polyhedra alone does not appear to be satisfactory, although it is useful in picking out specific families of structures. In this book, since full lattice complex symbolism has not been available, structural classification and grouping is based primarily on the consideration that the structures of metals are built up of essentially planar layer networks of atoms.

Systematic consideration of these layers leads naturally to grouping structures according to coordination polyhedra where this is an advantage, as for example in the case of Frank–Kasper structures (p. 31) or those dominated by triangular prisms, and it readily permits the recognition of families of polytypic structures which are important in considering factors governing alloy phase stability. The derivative structure method assumes a role of much less importance than was originally imagined, appearing only incidentally in the relationship of one group of structures to another, except that the principle is inherent in the divisions of layer networks into subnetworks, and in the introduction of extra layers of atoms, say in the octahedral holes of a close packed array.

Today when there is so much available structural information, a new problem has arisen in condensing and tabulating the information so that the characteristics of a large number of structure types can be comprehended and borne in mind. If such shorthand descriptions as are devised can be typed on-line for tabulation, can reveal structural relationships, and can give information on the relative atom positions and coordination, so much the better. One useful method of devising short coded descriptions of structural properties is again to regard the structures of metals and alloys as formed by the stacking of layers of atoms, since this is a principle of wide applicability embracing a large number of structure types, and one that is readily given to coding.

DEFINITION OF CRYSTAL STRUCTURE TYPE

One seemingly trivial point that has bemused crystallographers for many years is how to define the conditions under which two substances are said to have the same crystal structure and, to a lesser extent, what to call the relationship between two substances that are agreed to have the same structure! The seat of difficulty in the first point is that in crystal structures with degrees of freedom (atomic parameters, x, y, z, axial ratios, and interaxial angles), the atomic coordination or the relative atomic arrangement may alter at a specific value of one of the free parameters. Thus it is possible for two substances having the same space group, occupation of the same sitesets, and values of a free atomic parameter which differ by only 1% of the cell edge, to have a formally different layer stacking sequence as, for example, in the $NaHF_2$ and $CsICl_2$ forms of the *Strukturbericht* $F5_1$ structure type (see p. 430 for full details). The pyrite (FeS_2) and CO_2 structures provide another often discussed example of the same kind, in which the atoms occupy the same sitesets of the same space group, but have different coordinations.

Formally it would seem that substances with occupation of the same sitesets in the same space group should be regarded as having the same structure, but in practice this is an insufficient condition because of the different coordination that may obtain with different values of the free parameters, which could lead to quite different chemical and physical properties of the substances. This problem has been discussed in committees of the International Union of Crystallography and particularly at a meeting called for the purpose in Lisbon in 1968. It was agreed among participants at this meeting that some distinction must be made between substances which have the same sitesets and space group, yet different atomic arrangements; that is to say, pyrite and CO_2 can not be regarded as having the same type of crystal structure. Substances only have the same structure type when the atoms occupy the same sitesets in the same space group, and the atoms on a given siteset have the same coordination.

Turning to the second point of what to call two substances that are regarded as having the same crystal structure, there are three possible adjectives: *isomorphic* (isomorphism), *isotypic* (isotypism), and *isostructural*, each of which is variously used. The first is a word of confused meaning from historical definitions and it should not be used in a crystallographic sense. The second and third are acceptable from a crystallographic point of view and they are free of historical connotations. They have been used interchangeably, and are so used in this book to denote structures with the same occupied sitesets and the same coordination, although of the two words isotypic is preferred and used where possible. Structures in which the same sitesets and the same space group are occupied, but which have different atomic coordinations, are said to be *isopoint* structures. Thus pyrite and CO_2 are *isopuntal* structures, but they are *not isotypic*. Isotype embraces the whole range of structural properties, whereas isopoint only embraces the formal aspect of space group and sitesets, without regard to the consequences of the structural degrees of freedom that these may leave unspecified.

Even with these points settled, there is still the further problem of whether substances with different numbers of component atoms can be regarded as isotypic. From a crystallographic point of view it is the number of different sitesets which characterize a structure type, rather than the type of atoms that occupy the different sets. Therefore substances satisfying the isotypic or isopuntal conditions given above, are to be regarded as isotypic or isopuntal regardless of whether there are one, two, three, etc. different component atoms occupying the different sitesets, provided that these do not degenerate or further subdivide. Thus for example $Mg_6Si_7Cu_{16}$ and Th_6Mn_{23} (each with space group $Fm3m$, occupied sitesets 4(*b*), 24(*d*), 24(*e*), 32(*f*), and 32(*f*)) are isostructural or isotypic, as are also α-Mn and $Re_{24}Ti_5$. On the other hand $NiTi_2$ is not isostructural with CFe_3W_3, even though they both have space group *Fd*3*m*, and occupied sitesets 16(*d*), 32(*e*), and 48(*f*); siteset 16(*c*) of the CFe_3W_3 structure is unoccupied in the $NiTi_2$ structure.

Sometimes an important distinction must be made between a structure and its so-called *antitype*. For example, if a nonmetallic or metalloid component in the antitype occupies the sitesets of the metal in the normal structure, the different coordination that it has compared to that in the normal structure may cause significant differences in physical properties of the two arrangements. Thus in AmO_2 with the fluorite structure, O is four-coordinated by Am and also surrounded octahedrally by six O atoms, whereas in the antifluorite structure of Li_2O it is surrounded by a cube of Li atoms. This distinction which is important in semiconductors, ionics, and insulators and generally unimportant in metals, introduces chemical considerations that are not pertinent to the definition of a structure type. Structure type, as seen above, is defined purely according to geometrical principles of space group, occupied sitesets, and atomic coordination.

CLASSIFICATION OF STRUCTURE TYPES AND RELATIONSHIPS BETWEEN STRUCTURES

Derivative and Degenerate Structures

Since, speaking in general terms, a set of equipoints may be subdivided into two or more subgroups (with increase of unit cell size if necessary), or two sets of equipoints occurring in a structure may under certain circumstances be combined in a single set giving a structure of higher symmetry, a concept of *derivative* structures (the former) and *degenerate* structures (the latter) can be formulated which is useful in classifying and predicting structure types and in showing structural relationships. Thus for example sphalerite (ZnS) is a derivative structure of diamond and a degenerate structure of chalcopyrite ($CuFeS_2$). The method of derivative structures has been used for predicting semiconductivity in compounds,[4] and in describing "tetrahedral" structures,[5] whereas Buerger[6] has discussed derivative structure types systematically from the fundamental crystallographic aspect of the relationship of their symmetries to those of the basic structures. Starting with some simple structure as a basic type, all possible derived structures can be predicted with their cell multiplicity and space group. This leads to an essentially unlimited number of substitutional derivative structures, and so it is only profitable to conduct analyses for basic structures of common interest, and to proceed only within the range where some of the predicted types are recognized among known structures.

A derived structure has fewer symmetry operations than the basic structure and a degenerate structure has more. A derived structure must therefore have either a larger cell or a lower symmetry (or both) than the basic structure from which it is derived. Derived structures are obtained by ordered *substitution* of components in the basic structure, or by *distortion* of the crystal structure, or both. When derivative structures are formed by ordered substitution of two or more components for a single component occupying one siteset in the basic structure, it may involve (i) partial replacement of the component of the basic structure, (ii) complete replacement of the component by two or more new components, and/or (iii) introduction of vacant sites (symbol □) as a "component" in i or ii above. Structural degeneracy may be achieved by replacing two or more ordered components by a single component,† or by disordering them on the lattice sites.

Since disordering produces a degenerate structure, it follows that an ordering process, as in $AuCu_3$, results in a derived structure. Such a process is referred to as forming a *superstructure*, because of the original observation that it leads to a multiple cell. However, such ordering need not lead to a multiple cell, if the symmetry of the ordered structure is reduced relative to the disordered structure. Nevertheless, the name superstructure has now come to refer to the formation of an ordered structure from a disordered solid solution regardless of whether there is multiplication of the cell edge(s).

Derived structures which are formed with the ordered introduction of vacant sites, generally result from the need to satisfy valence rules. In deriving chalcopyrite ($CuFeS_2$) from sphalerite (ZnS) no vacant sites are introduced, but the Hahn phases ($CdAl_2S_4$), also derived from sphalerite, include one ordered vacant site per formula which might be written as Cd□Al_2S_4. Since the ordered introduction of vacancies is permitted in derived structures, the concept might also be extended to include *stuffed* or filled-up structures in which extra atoms are added on sites unoccupied in the basic structure. Thus if the NiAs structure is regarded as a basic type, the Ni_2In structure in which extra atoms are introduced at

†Buerger's (*loc cit.*) original use of the term degenerate appears to have referred to the components only, and not the combining of two sitesets to give a supergroup, as here. In Buerger's use the two (or more) sitesets have different symmetry in both the original and the degenerate structure, as for example in the spinels $MgAl_2O_4$ and Fe_3O_4. In the sense that degenerate structures are considered here, the different sitesets assume the same symmetry properties (form a supergroup) when they become occupied by a single component as in the following series of structures, each of which is a degenerate derivative of the former: $Cu_2FeSnS_4 \rightarrow CuFeS_2 \rightarrow ZnS \rightarrow C$.

sites $\pm(\frac{2}{3}, \frac{1}{3}, \frac{1}{4})$, is a stuffed derivative structure, whereas pyrrhotite and defect ordered structures with compositions such as M_7N_8, M_6N_5, M_4N_3, and the CdI_2 structure, can be regarded as defect derivative forms of the NiAs structure. A structure such as the $BiLi_3$ type can be regarded as a stuffed fluorite, sphalerite, rocksalt, or even a stuffed Cu structure, but such extended use of the concept of derived structures is generally unnecessary. Frequently structural distortion accompanies the formation of a substitutional derived structure because of the ordered disposition of atoms of different sizes, or because of the ordered distribution of vacant sites. Distorted derived structures can, however, also be obtained without substitution, as in the case of the MnP structure which is a distorted form of the NiAs structure, the SnS structure which is a distorted form of rocksalt, or the rhombohedrally distorted forms of the rocksalt and cesium chloride structures (SnSe and PbLi structures respectively).

Although these principles are not those on which the arrangement of structures in Chapters 7 to 16 is primarily based, it will be observed that they are frequently followed in the separation of structure types. Lima-de-Faria and Figueiredo[7] have published tables of derived close packed inorganic structure types.

Lattice Complexes

A lattice complex is defined as an arrangement of equivalent points that are related by space group symmetry operations including lattice translations.[8] A given lattice complex may therefore occur in more than one space group and may have more than one location in regard to any chosen origin for a unit cell. Listing the lattice complexes occupied by the crystallographically different atoms in a structure therefore provides a means of both describing and classifying structures, and of demonstrating relationships between different structure types, particularly as lattice complexes themselves exhibit derivative characteristics. Such descriptions are very convenient for relatively simple structures in the cubic system as Hellner[2] has demonstrated, but when all crystal systems are considered, there are over four hundred different lattice complexes for which symbolism is required. Furthermore, lattice complexes described by atomic parameters x, y, z have degrees of freedom which may lead to changes of coordination about the lattice points as the values of the parameter(s) change. Thus the atomic coordination must also be specified and described by appropriate symbols for lattice complexes with degrees of freedom. Finally, in crystal systems with symmetry lower than cubic, the geometrical configuration of the lattice complexes may change significantly with axial ratios and angles between the crystal axes.

Because the structures of metallic phases generally involve atoms on numerous sitesets with degrees of freedom, requiring the specification of coordination, the lattice complex description is likely to be too long and cumbersome for practical use, and the attractive features of the description of simple structures in the cubic system are no longer apparent.

Crystal Algebra Descriptions of Structure Types

The crystal algebra descriptions of structure types developed by Loeb[3] are essentially lattice complex descriptions in which the atomic distribution in a structure is displayed in the form of a distribution matrix. Such matrices provide an attractive means of demonstrating structural relationships and showing atomic coordination, but as the matrices usually occupy several lines,[9] they are awkward to tabulate, and it is not apparent that they offer advantages over the shorter lattice complex descriptions of structures, although the symbolism is less involved. Crystal algebra descriptions can be given for most hexagonal and rhombohedral structure types and for some 55 out of 75 cubic structure types considered, so that they cover about a quarter of the known structures of metals and alloys, but their use for more complex structures is limited.

Coordination as a Basis for Classification

(Packing of Right Triangular Prisms). Kripjakevič[10] has discussed the classification of structures according to coordination polyhedra, but it does not appear that primary classification by this means is as effective or informative as other methods, although it may be useful to know structures in which the simpler coordination polyhedra such as tetrahedra, octahedra, and trigonal prisms occur, and particularly how they are interconnected. The method of interconnection of these polyhedra is likely to be more important in mineral structures than in metals, except for transition metal compounds and alloys where the sharing of faces, edges, or corners has an important influence on physical properties, because it is one of the main factors determining the possibility and extent of interaction between the transition metal atoms. Nevertheless, coordination has been chosen as the second factor upon which is based the arrangement of structure types set forth in Chapter 7 and subsequent chapters, since the primary classification of structures according to layer networks of atoms and their stacking sequence also specifies coordination arrangements.

Structures can be grouped according to the number and kind of connections between a given type of coordination polyhedron which they contain. Consider for example structures containing right triangular prisms which may be (body) centered, or uncentered.

When these prisms pack together to fill space, they share three side faces and both basal end faces, and therefore the three c' edges and six a' edges also.† Transpositions and rotations of neighboring prisms or layers of prisms may reduce the number of connections between them, and, if only the connections between centered prisms are considered, the number of connections also depends on the distribution of the centered prisms. Structures containing triangular prismatic arrangements of atoms can therefore be partitioned systematically according to the number and type of face and edge connections that are made between the centered prisms. Transpositions or rotations from a space filling prismatic arrangement can be tabulated, together with the stacking sequence describing the prism centering, as well as the direction of the prism axes in relation to the crystallographic axes of the unit cell, and the prism axial ratios.[11] In this way structures containing a triangular prismatic arrangement of atoms can be listed and much of their relevant structural information condensed and presented in tabular form (see Table 2-2) so that structural comparisons can be made systematically.

Including the case of both completely filled and completely uncentered prisms, there are five stacking sequences that are useful for describing the relative arrangements of filled and empty *next-neighbor* triangular prisms which share 3 side + 2 basal faces and 3 side edges + 6 basal edges with prisms above and below (that is to say, arrangements in 11 of the 38 prisms that touch a central triangular prism at faces, edges, or corners). The stacking sequences are described in the same notation as that adopted for the stacking of triangular (3^6) and hexagonal (6^3) nets (p. 4). Thus in Table 2-3, $\underline{A}$ represents the stacking of the atoms that form the prisms and B, C, or 'a' give the arrangement of the atoms that center the prisms.

A partition according to the type of connections between filled prisms is shown in Table 2-2 for some structures containing triangular prisms. The structure type is indicated under the appropriate column giving the stacking sequence describing the arrangement of filled or empty prisms in the structure. Transpositions and rotations refer to movements from a space filling arrangement of triangular prisms. When counting the connections between filled prisms (cols. 1–4 of Table 2-2) *only the basal edge connections which occur between prisms in layers above and below the prisms considered are counted.* Thus no a' edges are counted as shared if the prisms each share 2 side faces, even though this results in the actual sharing of 4 a' edges per prism. The reason for this is that the information is already contained in the statement that 2 side faces and 3 c' edges are shared, and the specification of a' edge connections can therefore be

†Basal edges are referred to as a' edges and edges parallel to the prism axis are referred to as c' edges.

Table 2-2 Some Examples of Structures Containing Right Triangular Prisms None of the Prisms Are Centered, then Connections between all Prisms Are

Connections							Stacking Sequence	
c' face	c' edge	a' face	a' edge	Transpositions	Rotations	Structure Symbol	A̲a	A̲B
3	3	2	6			$hP6$	$CaIn_2$	
						$hP6$	NbS_2(H.T.)	
						$hP3$	ω-Cr-Ti	
						$hP3$	AlB_2	
						$hP4$	NOTaZr	
						$hP3$	δ-UZr_2	
						$hP8$	β-Na_2S_2	
						$hP6$	$InNi_2$	
2	3	2	6	*a,b,c*		$oC4$	Ga	
				a,b,c		$oC12$	$ZrSi_2$	
				a,b,c		$oC8$	CrB	
				b,c		$oC4$	β'-Cu_3Ti	
				a/2,*b*/2,*c* on one side, *c* on other	alt. layers 90° ⊥ *c* about center of *ab* face	$tI16$	MoB	
				b/2,*c*		$oP8$	β-Cu_3Ti	
				a		$oC12$	Ga_2Zr	
				a/2,*b*.*c*/2		$oC16$	$BCMo_2$	
1	3	2	6	*a,b*	alt. pairs prisms, rot 90° about *c* direction	$tP20$	Al_2Gd_3	
0	3	2	6			$hP2$		LiRh
						$hP2$		WC
						$hP2$		Mg
						$oP8$		SiTi
1	2	2	6	*a,b,c*		$mC12$	As_2Nb	
0	2	2	6	*a,b,c*		$oC12$	($Ag_{0.93}Cu_{1.07}S$)	
				a,b,c		$oP12$	Co_2Si (anti-$PbCl_2$)	
				a,b,c		$mC16$	CoGe	
0	0	2	6	*a,b,c*		$oC16$	Al_2CuMg	
				b,c		$oC16$	$CoPu_3$	
				a,b,c		$oC16$	BRe_3	

[a]Although sharing of a prisms side face involves also the sharing of basal (*a*) edges, they are not so counted, as this information is already contained in the number of side faces and *c* edges shared. Counting of *a* edges is reserved for giving information on connections to filled

Arranged According to Connections between Neighboring Centered Prisms (When Counted).[a]

Prism Formed by	Prism Centered by	Prism Axis Along Crystal Direction	Prism Axial Radio c'/a'	Remarks
Ca	In slightly o.c.	*c*	0.79	
Nb	S o.c. alt. up and down	*c*	1.80	
disordered		*c*	0.61	
Al	B	*c*	1.08	
Zr	N + Ta alt. in columns ‖ *c*	*c*	1.06	Alt. prisms in columns ‖ *c* end-centered by O.
Zr	(Zr + U) randomly	*c*	0.61	
Na(1)	Na(2) and 2S o.c. alt. in $\underline{A}B\underline{A}C$ arrangement	*c*	1.14	S also form prisms centered by Na(2).
Ni(1)	In + Ni(2) alt. in $\underline{A}B\underline{A}C$ arrangement	*c*	0.61	
Ga	—	*a*	1.08, 0.92	
Zr	Si	*a*	1.01	Layers sep. by square net of Si.
Cr	B	*a*	1.09 1.01	
disordered	B	*a*	0.68, 0.60	
Mo		*a*,*b* in alt. layers	1.12 1.00	Prism a' = prism c'.
Cu(2)	Cu(1) + Ti alt. in $\underline{A}B\underline{A}C$ arrangements	*a*	1.36 1.19	
Zr	Ga(3) o.c.	c		Layers sep. by square net Ga(1 & 2)
Mo(1)	B	*a*	1.07 1.00	Double layers Mo(2) & C in 4^4 nets between prism layers.
Gd(1&2)	Al slightly o.c.	*c*	~0.9	Distorted prisms. Gd(1) center cubes separating prisms.
Li	Rh	*c*	1.64	
W	C	*c*	0.98	
Mg	Mg	*c*	1.62	Half Mg form prism about other and *vice versa*.
Ti	Si slightly o.c.	*b*	1.06, 0.90	
As	Nb	*b*	1.14, 0.88 0.84	
Ag + Cu	S o.c. toward Cu edge	*a*	1.12 1.02	Ag–Ag edges shared, Cu–Cu unshared. Cu of one prism lies just outside AgAgAgAg face of other prism.
Co	Si	*b*	1.49, 1.46 1.40	
Co	Ge(2)	*b*	1.54 1.49 1.44	Ge(1) in spaces between prisms, themselves form prisms with Co at basal edge centers.
Al + Mg	Cu	*a*	1.44, 1.35	
Pu	Co	*a*	0.97, 0.95	
Re	B	*a*	1.06, 0.95	

prisms in layers above and below the prism concerned. *a*, *b*, and *c* always refer to the unit cell of the structure; a' and c' refer to the basal and vertical edges of the prisms. Abbreviations: alt. alternate, o.c. off-center, sep. separated.

Table 2-3 Stacking Sequence in Prisms[a]

Stacking Sequence	Shared Edges and Faces	Structural Arrangement
AA	All side and basal faces.	Empty prisms filling space.
AaAa	All side and basal faces.	Filled prisms filling space.
ABAB	Filled prisms share each c' edge with two other filled prisms, basal faces with other filled prisms.	Alternate columns of filled and empty prisms running in c' direction.
ABAC	Filled prisms share each c' edge with two other filled prisms, each basal edge a' with one other filled prism basal edge.	Layers of alternate filled and empty prisms. Filled prisms sit between empty prisms of layers above and below.
ABABACAC	Combination of the two above arrangements	Filled and empty prisms alternate in each layer. Filled prism has filled prism below it in layer below and empty prism above it in layer above, and *vice versa*.

[a]Underlined symbol A represents the stacking of the atoms which form the prism layers; symbols not underlined indicate the stacking of layers of atoms that center the prisms.

reserved to give explicit information regarding connections to prisms in layers above and below the one considered. Crystal axes a, b, and c in the various columns always refer to the crystallographic unit cell of the compound; only in the first four columns are the a' and c' edges of the prisms considered and in the penultimate column where the axial ratio, c'/a', refers to the prism edges.

The list of structures containing triangular prisms (Table 2-2) is by no means complete; it just serves as an example of how structures may be systematically compared on the basis of a coordination polyhedron.

TABULAR REDUCTION OF STRUCTURAL DATA: STRUCTURES CONSIDERED AS STACKED LAYERS OF ATOMS

A broad coverage of half of the known structure types of metals and semiconductors can be achieved by considering the structures to be formed by stacking together morphologically triangular (3^6), hexagonal

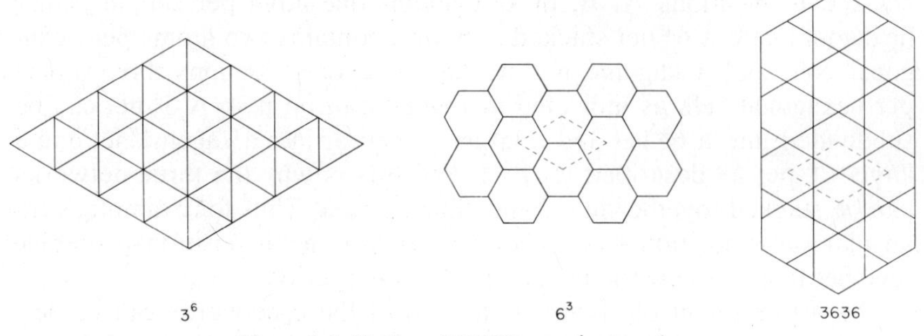

Figure 2-1. 3^6, 6^3, and 3636 kagomé networks.

(6^3), or kagomé (3636) networks of atoms (Figure 2-1), and networks containing squares. The close packed structures formed by packing together 3^6 nets of atoms, although the most important, are only a part of those made up of 3^6, 6^3, and kagomé layers which form a compatible group, since each of them can be stacked in any of three equivalent positions A, B, C at the corners of an equilateral triangle. This group of structures, together with those made up of networks of atoms containing squares, and more complex structures of the Frank–Kasper type (p. 31) can be readily described in terms of codes. Such codes allow tabular presentation of a large quantity of structural information, so that similar structural features can be recognized and compared directly from the codes without having to generate the structures themselves. One of the main reasons why these groups of layer structures embrace so many structure types, is that the stacking of the layers generates solids containing the simplest basic coordination polyhedra: tetrahedra, right trigonal prisms, and octahedra (antiprisms), cubes and archimedean antiprisms (anticubes), pentagonal prisms and antiprisms, and hexagonal prisms and antiprisms.

Structures Formed by the Stacking of Morphologically Triangular or Hexagonal Nets, or Kagomé Nets of Atoms

Structures that can be regarded as formed by the successive stacking of 3^6, 6^3, and/or kagomé nets of atoms can be simply described by reference to a code based on the symbols used to describe the stacking positions of the layers, which are given on p. 4. Structures involving only 3^6 nets of atoms (equilateral triangles) located so that the nodes of one net center the triangles of nets above and below, are of the well known close packed type, provided that the spacing between the layers is correct. With atoms of equal size and ideal packing, the normal distance between layers is 0.817 of the distance between the atoms. A 3^6 net stacked in positions A, B, or C contains one atom per simple planar hexagonal cell; a 6^3 net stacked a, b, or c contains two atoms per hexagonal cell; and a kagomé net stacked α, β or γ contains three atoms per hexagonal cell, as indicated in Figure J-4*a* on p. 4. A 3^6 net can be subdivided into a 6^3 net and a larger 3^6 net, or into a kagomé net and a larger 3^6 net as described on p. 4, and this is why the three networks can be stacked together in a compatible manner. Thus for example, a triangular net in position A and a coplanar kagomé net in position α combine together to give a new triangular net of shorter period.

The arrangement of atoms in any one of these networks can be described relative to the basal hexagonal cell by the appropriate code letter. Thus two similar atoms at $\frac{1}{3}, \frac{2}{3}, 0$ and $\frac{2}{3}, \frac{1}{3}, 0$ in a hexagonal unit cell are part of a 6^3 network whose position relative to the cell requires the code letter "a." Similarly, atoms at $\frac{1}{3}, \frac{2}{3}, \frac{1}{4}$ and $\frac{2}{3}, \frac{1}{3}, \frac{3}{4}$ in a hexagonal unit cell represent two 3^6 networks of atoms whose stacking positions require the code letters B and C. If atoms of two (or more) different kinds lie in the same plane in ordered sites, each can be described by a different subnetwork. Thus if one component atom is at $\frac{1}{3}, \frac{2}{3}, 0$ and a different component at $\frac{2}{3}, \frac{1}{3}, 0$, the net is described as [BC], where the square brackets indicate that the atoms lie in the same plane. Indeed, some layers of atoms may require two network symbols to describe them even though only one component is present, if the atoms occupy different sitesets. Thus an array of atoms at $0, 0, 0$; $\frac{1}{3}, \frac{2}{3}, 0$; $\frac{2}{3}, \frac{1}{3}, 0$ would be described by [Aa]. In some structures several different layers of atoms occur together as a group which, although not coplanar, are considerably closer to each other than the average interlayer distance observed throughout the structure. Such groups of layers are enclosed in round brackets; for example, (BC).

When a structure contains only two different types of component atom, it is easy to distinguish between layers belonging to each component by underlining the layer symbol pertaining to the first component in the formula name of the structure, the component with the smallest number of atoms in the formula being placed first. Thus the symbols describing the structure of the Laves phase $MgZn_2$ (p. 657) are $\mathrm{A\underline{B})\beta(\underline{B}A\underline{C})\gamma(\underline{C}}$, and of the NiAs structure (p. 452) are $\mathrm{\underline{A}B\underline{A}C}$. When layers of atoms are incompletely occupied, they can be so designated by an asterisk in the position super following the symbol. Thus the defect $Nb_{1.4}S_2$ structure (p. 468) is described by the symbols $\mathrm{\underline{A}^{*}B\underline{A}B\underline{A}^{*}C\underline{A}C}$. Such symbols are highly informative, and for this structure type they tell the stacking sequence and occupation of the layers, as well as those Nb layers that are only partially occupied. The symbols also indicate that the unit cell contains only one atom per layer and give its approximate coordinates, so that the structural arrangement can be regenerated from the code symbols.

This coding is most useful for describing hexagonal and rhombohedral structures (hexagonal coordinates), since these are mainly layer structures, and there are very few cases among the simpler structures where the arrangements of atoms in the layers can not be described by the symbols. One of the convenient features of the notation is that it specifies the atomic arrangement in relation to the unit cell, rather than just specifying the type of net arrangement. In describing structures whose unit cell contains many atoms, it is sometimes necessary to consider a larger basal hexagonal cell having three times the area and rotated 90° to the simplest basal cell considered above. The larger cell contains respectively 3, 6, and 9 atoms of the 3^6, 6^3, and kagomé nets. The larger cell and the three stacking arrangements corresponding to those for the smaller hexagonal cell are shown in Figure J-4*b* on p. 5. They are described by the primed symbols A′, B′, C′; a′, b′, c′, and α', β', γ' for the 3^6, 6^3, and kagomé nets respectively.

Close packed or other arrangements with cubic symmetry can also be described in terms of these symbols if an appropriate hexagonal cell is chosen with the layers of atoms normal to the *c* axis. Sometimes this description offers advantages, as for example in comparing the cubic Laves phase $MgCu_2$ with the other Laves phases. Structures with lower symmetry can also be described with these symbols if the atom positions and axial ratios are such as to result in triangular, hexagonal, and/or kagomé networks of atoms that are not too greatly distorted.

It should be noted that a 3^6 network need not be made up of equilateral triangles; b.c. cubic {110} layers, for example, are 3^6 nets with triangle angles approximately 55°, 55°, and 70°, but the ⟨110⟩ stacking of these nets does not result in the nodes of one net lying over the centers of the triangles of nets above and below, as required in close packing of triangular nets.

Structural relationships are readily recognized from the stacking symbols by observing sequences describing simple structures which repeat in part, or with site permutation in more complex structures. Thus it is unnecessary to make a drawing of the whole structure in order to appreciate such relationships. For example, the stacking unit which characterizes the Laves phases is of the type $\beta(\underline{B}A\underline{C})$, but as γ must follow C, the simplest fragment of this which can repeat continuously and generate a structure is the two units found in $MgZn_2$ $(\underline{C}A\underline{B})\beta(\underline{B}A\underline{C})\gamma$, Nevertheless, the single Laves stacking unit is found combined with $CaCu_5$ structural units of the type $[\underline{A}a]\alpha$ in structures such as the Ce_2Ni_7 and Er_2Co_7 types. Here, one Laves unit is followed by two $CaCu_5$ units giving a complex unit of the type (CAB) β[Bb]β[Bb]β which occurs twice in BC stacking in the Ce_2Ni_7 structure $\{A\underline{B})\beta[\underline{B}b]\beta[\underline{B}b]\beta(\underline{B}A\underline{C})\gamma[\underline{C}c]\gamma[\underline{C}c]\gamma(\underline{C}\}$ and three times in ACB stacking in the Er_2Co_7 structure $\{\alpha[\underline{A}a]\alpha(\underline{A}B\underline{C})\gamma[\underline{C}c]\gamma[\underline{C}c]\gamma(\underline{C}A\underline{B})\beta[\underline{B}b]\beta[\underline{B}b]\beta(\underline{B}C\underline{A})\alpha[\underline{A}a]\}$.

The simplest stacking sequence giving 1:3 stoichiometry, $[\underline{A}\alpha]$, appears to be unknown structurally because of the unfavorable arrangement of the layers one above the other, but combinations of this stacking unit are known in a range of structures from $SnNi_3$ (p. 323), where they are arranged in hexagonal sequence, $[\underline{B}\beta][\underline{C}\gamma]$, through the $BaPb_3$, $TiNi_3$, $HoAl_3$, and $PuAl_3$ structures, where they are arranged in mixed hexagonal and cubic stacking, to the $AuCu_3$ structure (p. 322) where they are arranged in cubic stacking $[\underline{A}\alpha][\underline{C}\gamma][\underline{B}\beta]$.

For most purposes it is sufficient to describe structures in terms of the stacking symbols A, B, C, α, β, γ, etc., but in order to give a detailed condensation of crystal structure information it is necessary to specify the components occupying the layers, and the relative spacings between the layers, so that the structures can be regenerated with reasonable accuracy from the coded data. In the full symbol that is adopted, the component atom occupying the layer is typed on-line with the stacking symbol following it super, and the layer spacing given sub as the fractional height (decimal with preceding 0 omitted) of the layer along the *c* axis of the hexagonal cell. Table 2-4 gives a few examples of condensed

Table 2-4 Structures Formed by Stacking Triangular, Hexagonal, and/or Kagomé Nets of Atoms[a]

Structure		Layer Stacking Sequence
		HEXAGONAL STRUCTURES
hP 2	WC	$W_0^A C_5^B$
	Mg	$Mg_{25}^B Mg_{75}^C$
hP 2.7	ϵ-$PtZn_{1.7}$	$Pt_0^A Zn_{45}^{B*} Zn_{50}^{\alpha*} Zn_{55}^{C*}$
hP 3	CdI_2	$Cd_0^A I_{25}^B I_{75}^C$
	ω-Cr-Ti	$M_0^A M_5^a$
	Se	Chain structure
	AlB_2	$Al_0^A B_5^a$
	δ-UZr_2	$Zr_0^A M_5^a$
	NFe_2	$N_0^{A*} Fe_{25}^B N_5^{A*} Fe_{75}^C$
hP 3.75	Cr_7S_8	$Cr_0^A S_{25}^B Cr_5^{A*} S_{75}^C$
hP 4	NOTaZr	$Zr_0^A O_0^B Ta_5^B N_5^C$
	NLi_3	$N_0^A Li_0^a Li_5^A$
	NiAs	$Ni_0^A As_{25}^B Ni_5^A As_{75}^C$
	La	$La_0^A La_{25}^B La_5^A La_{75}^C$
	ZnS	$S_0^B Zn_{37}^B S_5^C Zn_{87}^C$
hP 5	Ni_2Al_3	$Al_0^A Ni_{15}^B Al_{35}^C Al_{65}^B Ni_{85}^C$
	La_2O_3	$O_0^A La_{24}^B O_{37}^C O_{63}^B La_{76}^C$

Table 2-4 (*continued*)

Structure		Layer Stacking Sequence
hP 6	HgS	Chain structure
	$InNi_2$	$Ni_0^A Ni_{25}^C In_{25}^B Ni_5^A Ni_{75}^B In_{75}^C$
	$CaCu_5$	$Ca_0^A Cu_0^a Cu_5^{\alpha}$
	CoSn	$Sn_0^A Co_0^{\alpha} Sn_5^a$
	$TeCu_2$	$Cu_{16}^a Te_{31}^A Te_{69}^A Cu_{84}^a$
	ReB_2	$B_{05}^C Re_{25}^B B_{45}^C B_{55}^B Re_{75}^C B_{95}^B$
	$CaIn_2$	$In_{05}^B Ca_{25}^A In_{45}^B In_{55}^C Ca_{75}^A In_{95}^C$
	MoS_2	$S_{13}^C Mo_{25}^B S_{37}^C S_{63}^B Mo_{75}^C S_{87}^B$
	NbS_2(H.T.)	$S_{13}^B Nb_{25}^A S_{37}^B S_{63}^C Nb_{75}^A S_{87}^C$
hP 7	Al_3Zr_4	$Al_0^{\beta} Zr_{25}^B Zr_5^b Zr_{75}^B$
	$Nb_{1.4}S_2$	$Nb_0^{A*} S_{13}^B Nb_{25}^A S_{37}^B Nb_5^{A*} S_{63}^C Nb_{75}^A S_{87}^C$
hP~7.4	$Mo_{\sim 0.84}N$	$Mo_0^{A*} N_{13}^B Mo_{25}^A N_{37}^B Mo_5^{A*} N_{63}^C Mo_{75}^A N_{87}^C$
hP 8	$AsNa_3$	$Na_{08}^C Na_{25}^A As_{25}^B Na_{42}^C Na_{58}^B Na_{75}^A As_{75}^C Na_{92}^B$
	$AlCCr_2$	$C_0^A Cr_{09}^B Al_{25}^C Cr_{41}^B C_5^A Cr_{59}^C Al_{75}^B Cr_{91}^C$
	Th_3Pd_5	$Pd_0^{C'} Pd_0^a Th_5^{B'}$ (distorted)
	NNi_3	$N_0^B Ni_{25}^{B'} N_5^C Ni_{75}^{C'}$
	AsTi	$As_0^A Ti_{13}^B As_{25}^C Ti_{37}^B As_5^A Ti_{63}^C As_{75}^B Ti_{87}^C$
	$MnTa_3N_4$	$M_0^A N_{13}^B Ta_{25}^C N_{37}^B M_5^A N_{63}^C Ta_{75}^B N_{87}^C$
	ReB_3	$B_0^A B_{05}^C Re_{25}^B B_{45}^C B_5^A B_{55}^B Re_{75}^C B_{95}^B$
	$Cu_{0.65}NbS_2$	$Cu_{06}^{C*} S_{13}^B Nb_{25}^C S_{37}^B Cu_{44}^{C*} Cu_{56}^{B*} S_{63}^C Nb_{75}^B S_{87}^C Cu_{94}^{B*}$
	GaS	$S_{10}^C Ga_{17}^B Ga_{33}^B S_{40}^C S_{60}^B Ga_{67}^C Ga_{83}^C S_{90}^B$
	β-Na_2S_2	$Na_0^A S_{15}^C Na_{25}^B S_{35}^C Na_5^A S_{65}^B Na_{75}^C S_{85}^B$
	$SnNi_3$	$Sn_{25}^B Ni_{25}^{\beta} Sn_{75}^C Ni_{75}^{\gamma}$
hP 9	δ_H^{II}-NW_2	$W_{06}^A N_{15}^B W_{27}^B W_{39}^B N_5^A W_{61}^C W_{73}^C N_{85}^C W_{94}^A$
	Cr_2N	$N_0^c Cr_{25}^{B'} N_5^C Cr_{75}^{C'}$
	δ-AgZn	$Zn_0^A M_{25}^{C'} Zn_{25}^C M_{75}^{B'} Zn_{75}^B$
	Pb_2Li_7	$Li_0^A Li_{08}^C Pb_{25}^B Li_{33}^A Li_{42}^C Li_{58}^B Li_{67}^A Pb_{75}^C Li_{92}^B$
	$CrSi_2$	—
	PFe_2	$P_0^{B'} Fe_0^a P_5^{C'} Fe_5^A$
	$InMg_2$	$In_0^{B'} Mg_0^a In_5^{C'} Mg_5^A$
hP 10	N_2Be_3	$N_0^A Be_{08}^B Be_{25}^A N_{25}^B Be_{42}^B N_5^A Be_{58}^C Be_{75}^A N_{75}^C Be_{92}^C$
	C_2Mo_3	$C_0^{A*} C_{08}^{B*} Mo_{17}^C Mo_{25}^A Mo_{33}^C C_{42}^{B*} C_5^{A*} C_{58}^{C*} Mo_{67}^B Mo_{75}^A Mo_{83}^B C_{92}^{C*}$
	Pt_2Sn_3	$Sn_{07}^C Pt_{14}^B Sn_{25}^A Pt_{36}^B Sn_{43}^C Sn_{57}^B Pt_{64}^C Sn_{75}^A Pt_{86}^C Sn_{93}^B$
hP 12	$AgBiSe_2$	$Ag_0^A Se_{07}^C Bi_{16}^B Se_{25}^A Ag_{33}^C Se_{41}^B Bi_5^A Se_{59}^C Ag_{67}^B Se_{75}^A Bi_{84}^C Se_{93}^B$
	WAl_5	$Al_0^c W_0^C Al_{25}^{C'} Al_5^b W_5^B Al_{75}^{B'}$
	K_2S_2	$K_0^{B'} S_{18}^a S_{32}^A K_5^{C'} S_{68}^A S_{82}^a$
	CuS	$S_{06}^A Cu_{11}^B Cu_{25}^C S_{25}^B Cu_{39}^B S_{44}^A S_{56}^A Cu_{61}^C Cu_{75}^B S_{75}^C Cu_{89}^C S_{94}^A$
	$MgZn_2$	$Zn_0^A Mg_{06}^B Zn_{25}^{\beta} Mg_{44}^B Zn_5^A Mg_{56}^C Zn_{75}^{\gamma} Mg_{94}^C$
	$NbSe_2(4H)$	$Nb_0^A Se_{07}^B Se_{19}^C Nb_{25}^A Se_{32}^C Se_{43}^A Nb_5^B Se_{57}^A Se_{68}^C Nb_{75}^A Se_{81}^C Se_{93}^B$
hP 14	W_2B_5	$B_0^A B_{03}^C W_{14}^B B_{25}^b W_{36}^B B_{47}^C B_5^A B_{53}^B W_{64}^C B_{75}^C W_{86}^C B_{97}^B$
	β-Si_3N_4	$N_{25}^{B'} N_{25}^B Si_{25}^{\gamma} N_{75}^{C'} N_{75}^C Si_{75}^{\beta}$
	Ni_3Te_4	—
hP 16	Si_3Mn_5	$Mn_0^a Mn_{25}^{B'} Si_{25}^{C'} Mn_5^a Mn_{75}^{C'} Si_{75}^{B'}$ (distorted)
	$TiNi_3$	$Ti_0^A Ni_0^{\alpha} Ti_{25}^B Ni_{25}^{\beta} Ti_5^A Ni_5^{\alpha} Ti_{75}^C Ni_{75}^{\gamma}$

Table 2-4 (*continued*)

Structure		Layer Stacking Sequence
hP 18	$\mathrm{FeMg_3Si_6Al_8}$	$\mathrm{Al_0^{B'}Fe_0^{A}Si_{22}^{C'}Al_{23}^{a}Mg_5^{B'}Al_5^{A}Al_{77}^{a}Si_{78}^{C'}}$ (distorted)
	$\mathrm{Ga_4Ti_5}$	$\mathrm{Ga_0^{A}Ti_0^{a}Ga_{25}^{C'}Ti_{25}^{B'}Ga_5^{A}Ti_5^{a}Ga_{75}^{B'}Ti_{75}^{C'}}$
hP 19	$\mathrm{Th_7S_{12}}$	—
hP 20	α-$\mathrm{Cr_2S_3}$	$\mathrm{Cr_0^{A'}S_{13}^{C'}Cr_{25}^{B}S_{38}^{B'}Cr_5^{A'}S_{63}^{C'}Cr_{75}^{C}S_{88}^{B'}}$
	$\mathrm{Fe_3Th_7}$	—
hP 22	$\mathrm{Sn_5Ti_6}$	$\mathrm{Ti_0^{\alpha}Ti_{25}^{\gamma}Sn_{25}^{c}Sn_{25}^{\beta}Ti_5^{\alpha}Ti_{75}^{\beta}Sn_{75}^{b}Sn_{75}^{\gamma}}$
	$\mathrm{Cr_5S_6}$	$\mathrm{Cr_0^{A'}S_{13}^{C'}Cr_{25}^{b}S_{38}^{B'}Cr_5^{A'}S_{63}^{C'}Cr_{75}^{c}S_{88}^{B'}}$
hP 24	$\mathrm{HoD_3}$	$\mathrm{D_{10}^{B'}D_{17}^{B}Ho_{25}^{C'}D_{25}^{A}D_{33}^{C}D_{40}^{B'}D_{60}^{C'}D_{67}^{B}Ho_{75}^{B'}D_{75}^{A}D_{83}^{C}D_{90}^{C'}}$
	$\mathrm{PuAl_3}$	$\mathrm{Al_{08}^{\beta}Pu_{09}^{B}Al_{25}^{\alpha}Pu_{25}^{A}Pu_{41}^{B}Al_{42}^{\beta}Al_{58}^{\gamma}Pu_{59}^{C}Al_{75}^{\alpha}Pu_{75}^{A}Pu_{91}^{C}Al_{92}^{\gamma}}$
	$\mathrm{CeNi_3}$	$\mathrm{Ni_0^{A}Ce_{04}^{B}Ni_{13}^{\beta}Ce_{25}^{B}Ni_{25}^{b}Ni_{37}^{\beta}Ce_{46}^{B}Ni_5^{A}Ce_{54}^{C}Ni_{63}^{\gamma}Ce_{75}^{C}Ni_{75}^{c}Ni_{87}^{\gamma}Ce_{96}^{C}}$
	$\mathrm{VCo_3}$	$\mathrm{V_0^{A}Co_0^{\alpha}V_{17}^{C}Co_{17}^{\gamma}V_{33}^{B}Co_{33}^{\beta}V_5^{A}Co_5^{\alpha}V_{67}^{B}Co_{67}^{\beta}V_{83}^{C}Co_{83}^{\gamma}}$
	$\mathrm{PCu_3}$	$\mathrm{Cu_0^{A}Cu_{08}^{C'}Cu_{17}^{B}P_{25}^{B'}Cu_{33}^{C}Cu_{42}^{C'}Cu_5^{A}Cu_{58}^{B'}Cu_{67}^{B}P_{75}^{C'}Cu_{83}^{C}Cu_{92}^{B'}}$
	FeS	$\mathrm{S_0^{A}S_{02}^{B}Fe_{13}^{B'}S_{25}^{C'}Fe_{38}^{B'}S_{48}^{B}S_5^{A}S_{52}^{C}Fe_{63}^{B'}S_{75}^{C'}Fe_{88}^{B'}S_{98}^{C}}$
	$\mathrm{MgNi_2}$	$\mathrm{Ni_0^{\alpha}Mg_{09}^{A}Ni_{13}^{B}Mg_{16}^{C}Ni_{25}^{\gamma}Mg_{34}^{C}Ni_{37}^{B}Mg_{41}^{A}Ni_5^{\alpha}Mg_{59}^{A}Ni_{63}^{C}}$ $\mathrm{Mg_{66}^{B}Ni_{75}^{\beta}Mg_{84}^{B}Ni_{87}^{C}Mg_{91}^{A}}$
		CUBIC STRUCTURES
cP 2	CsCl	$\mathrm{Cs_0^{A}Cl_{17}^{B}Cs_{33}^{C}Cl_5^{A}Cs_{67}^{B}Cl_{83}^{C}}$
cP 4	$\mathrm{AuCu_3}$	$\mathrm{Au_0^{A}Cu_0^{\alpha}Au_{33}^{C}Cu_{33}^{\gamma}Au_{67}^{B}Cu_{67}^{\beta}}$
	$\mathrm{ReO_3}$	$\mathrm{Re_0^{A}O_{17}^{\beta}Re_{33}^{C}O_5^{\alpha}Re_{67}^{B}O_{83}^{\gamma}}$
cP 5	$\mathrm{CFe_4}$	$\mathrm{C_0^{A}Fe_{25}^{A+\alpha}C_{33}^{C}Fe_{58}^{C+\gamma}C_{67}^{B}Fe_{92}^{B+\beta}}$
	$\mathrm{CaTiO_3}$	$\mathrm{Ca_0^{A}O_0^{\alpha}Ti_{17}^{B}Ca_{33}^{C}O_{33}^{\gamma}Ti_5^{A}Ca_{67}^{B}O_{67}^{\beta}Ti_{83}^{C}}$
cP 6	$\mathrm{OCu_2}$	$\mathrm{O_0^{A}O_{17}^{B}Cu_{25}^{A+\alpha}O_{33}^{C}O_5^{A}Cu_{58}^{C+\gamma}O_{67}^{B}O_{83}^{C}Cu_{92}^{B+\beta}}$
cP 8	$\mathrm{VCu_3S_4}$	$\mathrm{V_0^{A}Cu_{17}^{\beta}S_{24}^{A+\alpha}V_{33}^{C}Cu_5^{\alpha}S_{59}^{C+\gamma}V_{67}^{B}Cu_{83}^{\gamma}S_{92}^{B+\beta}}$
cF 4	Cu	$\mathrm{Cu_0^{A}Cu_{33}^{B}Cu_{67}^{C}}$
cF 8	C	$\mathrm{C_0^{A}C_{25}^{A}C_{33}^{B}C_{58}^{B}C_{67}^{C}C_{92}^{C}}$
	ZnS	$\mathrm{Zn_0^{A}Zn_{25}^{A}Zn_{33}^{B}S_{58}^{B}Zn_{67}^{C}S_{92}^{C}}$
	NaCl	$\mathrm{Na_0^{A}Cl_{17}^{C}Na_{33}^{B}Cl_5^{A}Na_{67}^{C}Cl_{83}^{B}}$
cF 12	$\mathrm{CaF_2}$	$\mathrm{Ca_0^{A}F_{08}^{B}F_{25}^{A}Ca_{33}^{B}F_{42}^{C}F_{58}^{B}Ca_{67}^{C}F_{75}^{A}F_{92}^{C}}$
	AsAgMg	$\mathrm{As_0^{A}Mg_{08}^{B}Ag_{25}^{A}As_{33}^{B}Mg_{42}^{C}Ag_{58}^{B}As_{67}^{C}Mg_{75}^{A}Ag_{92}^{C}}$
cF 16	$\mathrm{BiF_3}$	$\mathrm{Bi_0^{A}F_{08}^{B}F_{17}^{C}F_{25}^{A}Bi_{33}^{B}F_{42}^{C}F_5^{A}F_{58}^{B}Bi_{67}^{C}F_{75}^{A}F_{83}^{B}F_{92}^{C}}$
	NaTl	$\mathrm{Na_0^{A}Tl_{08}^{B}Tl_{17}^{C}Na_{25}^{A}Na_{33}^{B}Tl_{42}^{C}Tl_5^{A}Na_{58}^{B}Na_{67}^{C}Tl_{75}^{A}Tl_{83}^{B}Na_{92}^{C}}$
	$\mathrm{AlMgCu_2}$	$\mathrm{Al_0^{A}Cu_{08}^{B}Mn_{17}^{C}Cu_{25}^{A}Al_{33}^{B}Cu_{42}^{C}Mn_5^{A}Cu_{58}^{B}Al_{67}^{C}Cu_{75}^{A}Mn_{83}^{B}Cu_{92}^{C}}$
cF 24	$\mathrm{MgCu_2}$	$\mathrm{Mg_0^{A}Cu_{13}^{\alpha}Mg_{25}^{A}Cu_{29}^{C}Mg_{33}^{B}Cu_{46}^{\beta}Mg_{58}^{B}Cu_{63}^{A}Mg_{67}^{C}Cu_{79}^{\gamma}Mg_{92}^{C}Cu_{96}^{B}}$
	$\mathrm{AuBe_5}$	$\mathrm{Au_0^{A}Be_{13}^{\alpha}Be_{25}^{A}Be_{29}^{C}Au_{33}^{B}Be_{46}^{\beta}Be_{58}^{B}Be_{63}^{A}Au_{67}^{C}Be_{79}^{\gamma}Be_{92}^{C}Be_{96}^{B}}$
	$\mathrm{MgSnCu_4}$	$\mathrm{Mg_0^{A}Cu_{13}^{\alpha}Sn_{25}^{A}Cu_{29}^{C}Mg_{33}^{B}Cu_{46}^{\beta}Sn_{58}^{B}Cu_{63}^{A}Mg_{67}^{C}Cu_{79}^{\gamma}Sn_{92}^{C}Cu_{96}^{B}}$
cF 32	$\mathrm{GeCa_7}$	$\mathrm{Ge_0^{A}Ca_0^{\alpha}Ca_{17}^{C+\gamma}Ge_{33}^{B}Ca_{33}^{\beta}Ca_5^{A+\alpha}Ge_{67}^{C}Ca_{67}^{\gamma}Ca_{83}^{B+\beta}}$
cF 56	$\mathrm{MgAl_2O_4}$	$\mathrm{O_0^{A+\alpha}Al_{08}^{\beta}O_{17}^{C+\gamma}Mg_{21}^{B}Al_{25}^{A}Mg_{29}^{C}O_{33}^{B+\beta}Al_{42}^{\gamma}O_5^{A+\alpha}Mg_{54}^{C}Al_{58}^{B}Mg_{63}^{A}}$ $\mathrm{O_{67}^{C+\gamma}Al_{75}^{\alpha}O_{83}^{B+\beta}Mg_{88}^{A}Al_{92}^{C}Mg_{96}^{B}}$
cI 10	$\mathrm{PtHg_4}$	$\mathrm{Pt_0^{A}Hg_{08}^{C+\gamma}Pt_{17}^{B}Hg_{25}^{A+\alpha}Pt_{33}^{C}Hg_{42}^{B+\beta}Pt_5^{A}Hg_{58}^{C+\gamma}Pt_{67}^{B}Hg_{75}^{A+\alpha}Pt_{83}^{C}Hg_{92}^{B+\beta}}$
		RHOMBOHEDRAL STRUCTURES
hR 1	Hg	$\mathrm{Hg_0^{A}Hg_{33}^{C}Hg_{67}^{B}}$
hR 2	As	$\mathrm{As_{11}^{C}As_{23}^{A}As_{44}^{B}As_{56}^{C}As_{77}^{A}As_{89}^{B}}$

Table 2-4 (*continued*)

Structure		Layer Stacking Sequence
hR 3	α-MoS_2	$Mo_0^A S_{08}^B S_{25}^A Mo_{33}^C S_{41}^A S_{59}^C Mo_{67}^B S_{75}^C S_{92}^B$
	$CdCl_2$ ($TaSe_2$)	$Ta_0^A Se_{08}^C Se_{25}^A Ta_{33}^C Se_{42}^B Se_{58}^C Ta_{67}^B Se_{75}^A Se_{92}^B$
	Sm	$Sm_0^A Sm_{11}^C Sm_{22}^A Sm_{33}^C Sm_{45}^B Sm_{55}^C Sm_{67}^B Sm_{78}^A Sm_{89}^B$
	δ_R^V WN_2	$W_0^A N_{15}^C N_{18}^A W_{33}^C N_{49}^B N_{51}^C W_{67}^B N_{82}^A N_{85}^B$
hR 4	GaSe	$Ga_{05}^A Se_{10}^C Se_{24}^B Ga_{28}^C Ga_{38}^C Se_{44}^B Se_{57}^A Ga_{62}^B Ga_{72}^B Se_{77}^A Se_{90}^C Ga_{95}^A$
	$CrCuS_2$	$Cu_0^A S_{13}^A Cr_{19}^C S_{26}^A Cu_{33}^C S_{46}^C Cr_{53}^B S_{59}^C Cu_{67}^B S_{80}^B Cr_{86}^A S_{93}^B$
	β-NiTe	$Te_{08}^C Ni_{13}^A Ni_{20}^C Te_{26}^A Te_{41}^B Ni_{46}^C Ni_{54}^B Te_{59}^C Te_{74}^A Ni_{80}^B Ni_{87}^A Te_{92}^B$
	$CrNaS_2$	$Cr_0^A S_{07}^C Na_{17}^B S_{26}^A Cr_{33}^C S_4^B Na_5^A S_6^C Cr_{67}^B S_{74}^A Na_{83}^C S_{93}^B$
hR 5	S_2Ni_3	$S_{08}^C Ni_{17}^{B'} S_{25}^A S_{42}^B Ni_5^{B'} S_{58}^C S_{75}^A Ni_{83}^{B'} S_{92}^B$ (distorted)

[a]Subscript numbers give height of atom plane as fraction of c edge of cell (decimal points are excluded; thus 5 means 0.5). An asterisk (*) indicates a partially occupied layer. M indicates two components disordered on sites.

crystal chemical data for hexagonal, rhombohedral, and cubic structures.[11] As all of the essential crystal chemical information on such structures can be condensed into very little space in a table such as this, individual verbal description of the various structure types becomes largely redundant; it is simpler and more comprehensive to think of the structures and compare them in terms of the code symbols. Crystal structure relationships can be read directly from the symbols, or the structural arrangement of the atoms can be regenerated from the code when required. Atomic coordination can also be deduced from the coding, and Table 2-5 lists this for certain layer sequences.

The importance of specifying the relative spacing between the layers can be demonstrated by considering close packed structures. These are represented by the stacking symbols A, B, and C (A′, B′, and C′) only, but *not all* structures represented by combinations of the symbols A, B, and C are close packed; only those with the appropriate spacing between the layers are. Thus the sphalerite structure represented by the symbols $Zn_0^A S_{25}^A Zn_{33}^B S_{58}^B Zn_{67}^C S_{92}^C$, is not an overall close packed structure even though the Zn atoms and the S atoms severally form close packed cubic arrays (ABC); the Zn and S layers are much too close together.

Tetrahedral Close Packing: Frank–Kasper Structures

There is an essentially unlimited family of possible structures which contain only interpenetrating coordination polyhedra with 12 (icosahedron) 14, 15, or 16 vertices and triangulated faces. These interpenetrat-

Table 2-5 Coordination of Atoms in Various Layer Sequences

Layer Sequence	Polyhedra Formed	Coordination	CN
ABC---- (cubic close packing)	Tetrahedra and octahedra	Each atom by cubo-octahedron	12
AB-- (AC) (hexagonal close packing)	Tetrahedra and octahedra A(B) form trigonal prisms	Each atom by twinned cubo-octahedron	12
AA	Trigonal prisms		
aa	Hexagonal prisms		
$\alpha\alpha$	Trigonal and hexagonal prisms		
Aa----	All 'a' hexagonal prisms centered by A	A atoms by 'a' hexagonal prisms	12
	All A trigonal prisms centered by 'a'	'a' atoms by A trigonal prisms	6
Aα	All α hexagonal prisms centered by A	A atoms by α hexagonal prisms	12
	Each edge face of A prisms centered by α	α atoms by 4 planar A atoms	4
Bα-- (Cα--)	Half α trigonal prisms centered by B(C)	B(C) by α trigonal prisms	6
	Half B trigonal prisms contain 3 α atoms, other half empty	α by B(C) atoms	2
aα--	All α trigonal prisms centered by 'a' α hexagonal prisms empty	'a' by α trigonal prisms	6
	'a' hexagons covered antisymmetrically by α hexagons	α by 4 planar 'a' atoms	4
bα-- (cα--)	All α hexagonal prisms and half α trigonal prism centered by b(c).	Half b(c) by α hexagonal prisms	12
		Half b(c) by α trigonal prisms	6
	b(c) hexagonal prisms have triangle of α atoms located about center	α by b(c) atoms	2

ing coordination polyhedra result in space filling by a somewhat distorted tetrahedral packing of the atoms, and although ideal tetrahedra themselves can not pack together to fill space completely, as regards volume, at least, the misfit is relatively small in the case of the icosahedron. The structures are in most cases generated by the alternate stacking of primary layers of pentagon-triangle, pentagon-hexagon-triangle, or hexagon-triangle nets of atoms with secondary layers of triangular, square-triangle, or square nets of atoms such that all pentagons of successive primary nets are covered antisymmetrically by pentagons of neighboring primary nets, and similarly for the hexagons. The atoms of the secondary nets center all and only the pentagons and/or hexagons of the primary nets. The rather remarkable geometrical properties of these structures make them easy to classify and describe by means of a code which is described after some discussion of how these properties arise.

The maximum number of spheres that can surround a central sphere of equal size without compression of the atomic contacts is 12, and these can be distributed in three different symmetrical arrangements. Compared to the ideal close packed cubic or hexagonal arrangement of 12 equal spheres around a central sphere, all of which are in mutual contact, the icosahedral arrangement in which the 12 spheres contacting the central sphere are not in mutual contact, has a probable advantage of a lower free energy. This is because slight compression of the twelve contacts to the central atom so that the surrounding atoms also come mutually in contact, results in the 13 atoms occupying a smaller volume than 13 close packed atoms, and compression of some of the contacts is a feature of common occurrence in the structures of metals.

The icosahedron with 12 vertices each connected to 5 surrounding surface atoms has 20 triangular faces and is thus made up of 20 tetrahedra surrounding the central atom. In the ideal icosahedron where spheres of equal radius surround a central atom of the same radius, the tetrahedra are but little distorted from the ideal. Ideal icosahedra can not pack together to fill space completely, and furthermore structures containing only interpenetrating ideal icosahedra are impossible. In fact, it appeared that even when the ideal criterion is relaxed in practical structures this situation is never realized, although interpenetrating icosahedra and CN 14, 15, or 16 triangulated coordination shells are frequently found (however, now see γ brasses, p. 583). The systematic treatment of such structures is due to Frank and Kasper[12] who argue that a combination of 5 and 6 surface coordination,† rather than say 5 and 4, is likely to be most favored for interpenetrating coordination polyhedra in the structures of alloys containing atoms of "approximately" (within 25%) the same size. They make use of a theorem of Euler ($V-E+F=2$, where V is the number of vertices, E of edges, and F of faces) for any polyhedron to show that there are only 3 possible symmetrical polyhedra with triangulated coordination shells containing both five-fold and six-fold surface coordinated vertices, if the six-fold surface coordinated atoms are not contiguous. These are the CN 14, 15, and 16 polyhedra illustrated together with the icosahedron in Figure 2-2. There is no CN 13 coordination shell and no shells with CN $>$ 16 which satisfy these conditions, and

†Surface coordination is defined as the number of neighbors, n, surrounding an atom which lies on the surface of the specified polyhedron. The atom is briefly referred to as an S_n atom.

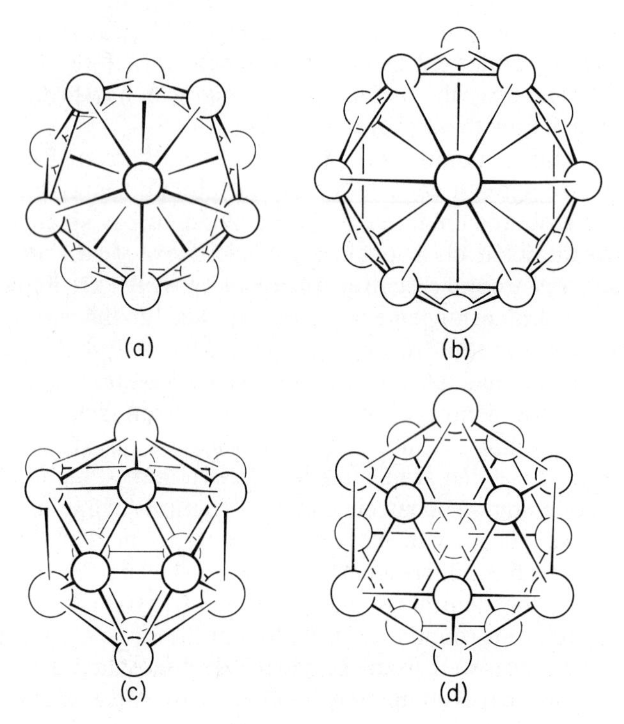

Figure 2-2. Pictorial diagrams of the CN 12 icosahedron (*a*) and CN 14, 15, and 16 (*b*, *c*, and *d*, respectively) coordination polyhedra of Frank and Kasper. Central atoms are omitted.

the 3 permitted polyhedra are uniquely prescribed. Naturally, as the size of the coordination shell increases, so the distortion from an ideal tetrahedral arrangement of the atoms increases since 24, 26, and 28 "tetrahedra" respectively have to be packed about the central atom to provide the required number of triangular faces for the CN 14, 15, and 16 polyhedra. There are of course other structures that can be regarded as belonging to the distorted tetrahedrally close packed type besides those containing only interpenetrating CN 12, 14, 15, or 16 polyhedra, and some structures such as those of the R and δ phases contain volumes in which the atoms are arranged in plane layers and other regions where the planarity is disturbed. Only the layered structures containing these four coordination polyhedra will be further discussed and these are referred to as Frank–Kasper structures.

Frank and Kasper demonstrated that structures formed by the interpenetration of the four coordination polyhedra contain planar or approximately planar layers of atoms. They called those made up of triangles with

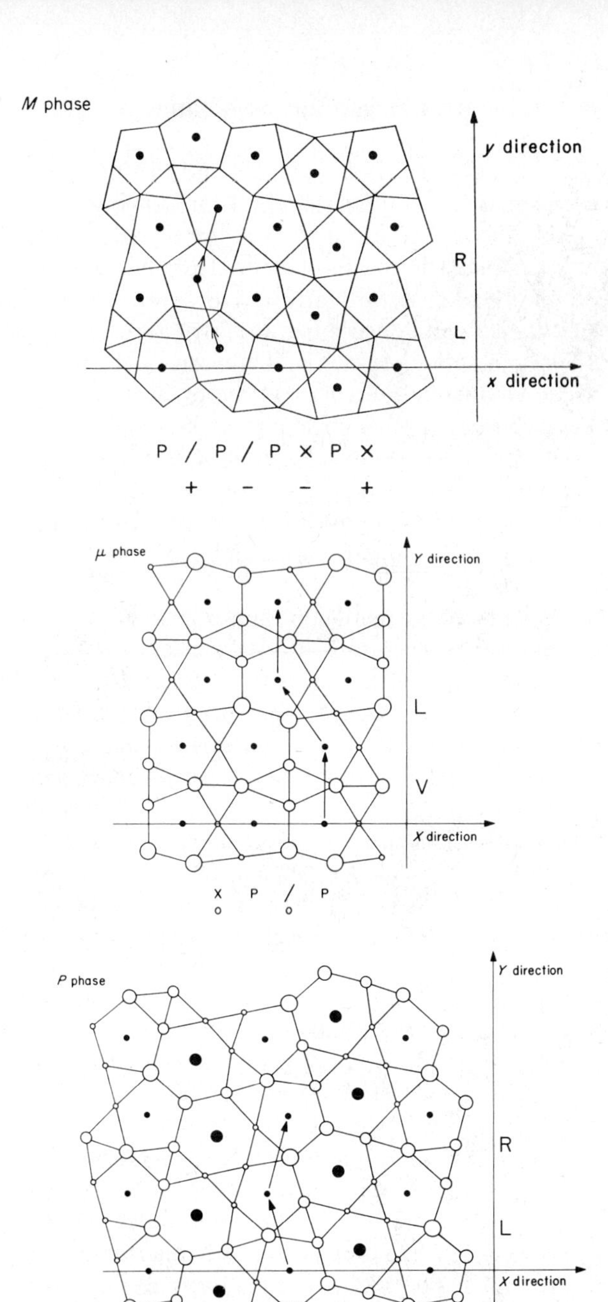

Figure 2-3. Arrangements of pentagons and triangles in the main layers of the M and μ phases, and the hexagons, pentagons, and triangles in the P phase, together with the repeat sequence along the X and Y directions.

hexagons and/or pentagons *primary* layers and the intervening layers of triangles and/or squares *secondary* layers. Lines joining points of six-fold surface coordination were referred to as *major* ligands and in the $MgCu_2$ structure, for example, where CN 12 and 16 polyhedra interpenetrate, these form a diamond-structure array. However, Frank and Kasper proceeded heuristically in describing the structures and used several different models and sets of symbols. Thus the Laves phases were described by three triangular secondary layers separating the primary kagomé layers, whereas the σ and P phases were described by a single secondary layer between primary layers. Only later did it become apparent[13] that it was possible to describe all of the structures in terms of a single secondary layer of atoms which interleaves the primary layers.

Layered structures obeying the conditions set out in the first paragraph of this section fall into two groups: The first is characterized by a linear repeat sequence of the pentagons and/or hexagons along a direction (X) called the basal repeat row; repetition of the repeat sequence along the row and about a direction (Y) normal to it, generates the whole of the primary net (Figure 2-3). In structures of this group the secondary net is generated by sets of parallel (generally zigzag) lines of atoms (Figure 2-4). The second group comprises those structures whose secondary nets can not be generated by sets of parallel (zigzag) lines. These structures are frequently characterized by having a two-dimensional tile of pentagons and/or hexagons which by repetition generates the whole of the primary net as in the case shown in Figure 2-5. In both groups of structures the primary nets whose pentagons and/or hexagons cover those of the primary nets above and below antisymmetrically, may or may not be equivalent. Examples of the second group of structures are scarcely known at present, and so the coding and structural description set forth below applies only to the first group, which is well known in such structures as the σ, μ, P, and Laves types described on pp. 654–675.

Coding of Structures with Primary Nets Having Linear (Zigzag) Repeat Units.[14] Pentagons can be represented by the letter P and hexagons by H and the method of joining these along the basal repeat row by / when they join edge to edge, and X when they join apex to apex with two intervening triangles (Figure 2-6*a*). The basal repeat row is chosen so that along it hexagons can only share opposite edges or apices, or an edge and next-but-one apex. Pentagons can only share an edge and next-but-one apex, or next-but-one edges or apices with neighboring polygons. The nets are oriented relative to two orthogonal axes, X and Y, so that the basal repeat row of polygons sharing edges or apices lies along the X direction, and the direction in which the basal row repeats itself in the same orienta-

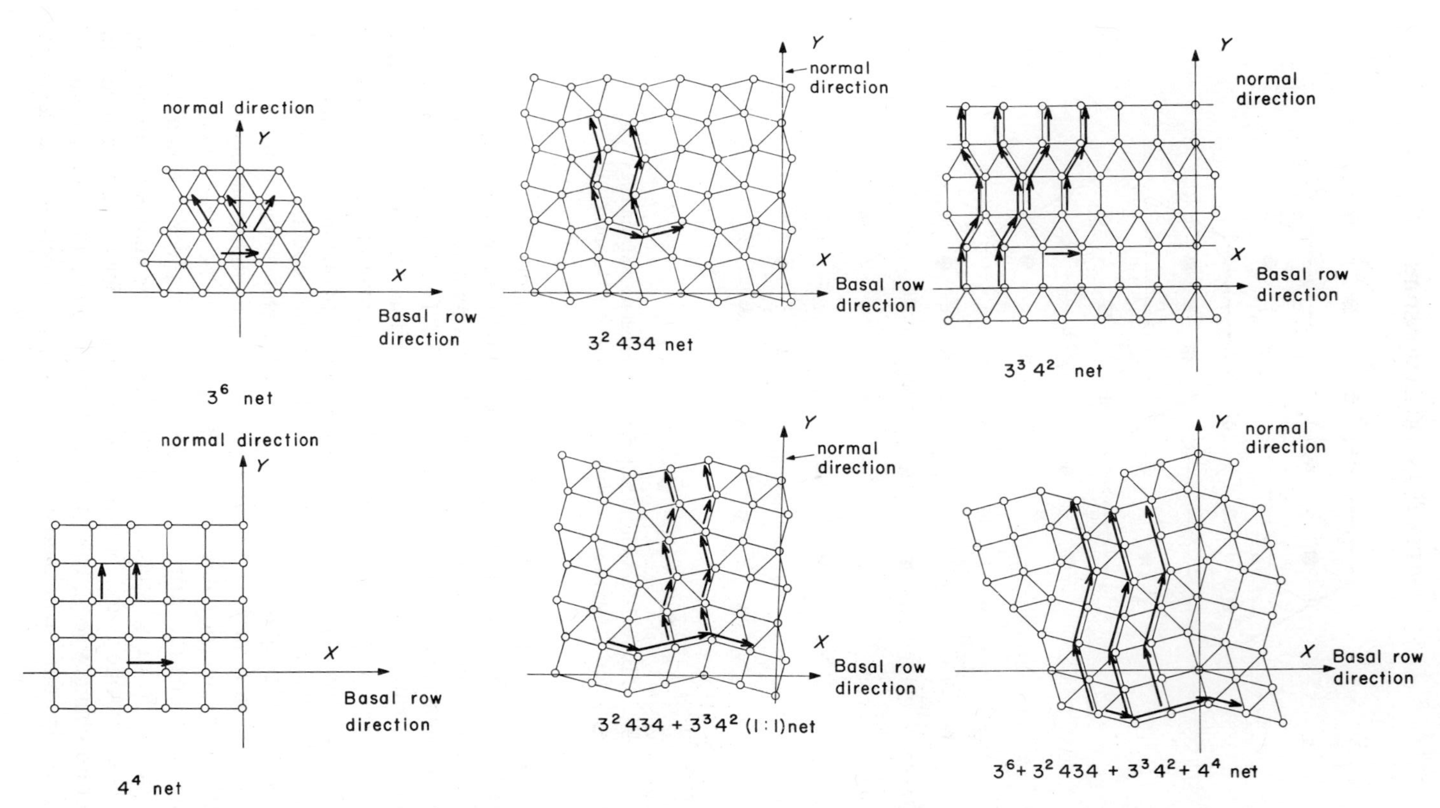

Figure 2-4. Secondary nets for Frank–Kasper structures, showing zigzag repeat lines about Y.

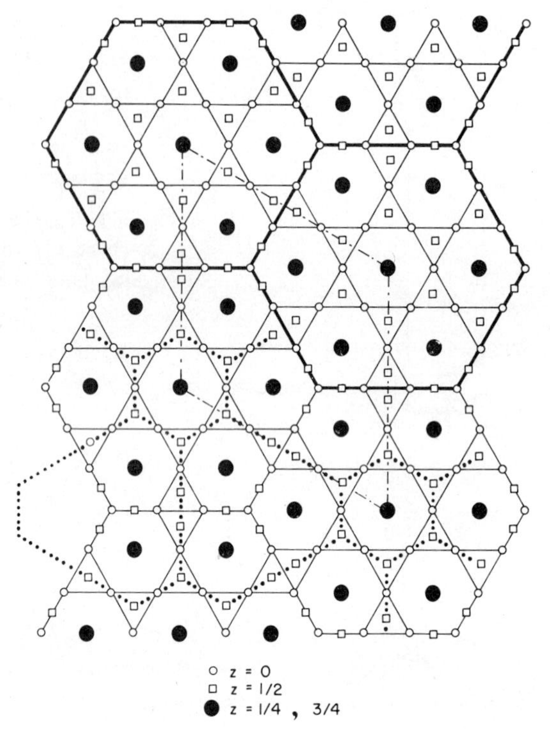

Figure 2-5. Main repeat nets of hexagons and triangles for a hypothetical structure. (Frank and Kasper[12]).

tion makes an angle between 0° and approximately 30° with the *Y* axis. The next row in the *Y* direction is not necessarily a crystallographic repeat of the first basal row, but it contains pentagons and/or hexagons in the same sequence and orientation (Figure 2-3). The basal row can repeat along the *Y* direction with shift movements either left (symbol L) or right (symbol R) or none (symbol V), so that the pentagons or hexagons form strips along this direction. Along these strips pentagons must join at an apex and have one intervening triangle on one side (see Figure 2-6*b*); this condition establishes the direction of the shift movements along the *Y* direction in a structure whose primary net contains pentagons, since there is only one possible choice. Hexagons on the other hand may be connected along the strips by opposite edges or apices in which case the direction L, R, or V is continued; or they may be connected by an edge and next-but-one apex, or two next-but-one apices or edges when the direction changes (e.g., from L to R, L to V, etc.). The repeat sequence of pentagons and/or hexagons in the primary net can now be described in terms of these symbols. For example, along the basal row for the *M* phase it is P/P/PXPX and along the *Y* direction LR (Figure 2-3). For the *P* phase it is P/HXP/HX; LR (Figure 2-3).

In the basal row the connections between the pentagons and/or hexagons may be such as to cause the row to continue in a direction up (coded +) or down (coded −) relative to the direction of the *X* axis, or in a direction parallel to it (coded 0). When the direction changes (i.e. + to −), the kink introduced in the basal row direction must always result in a turn *toward* the *X* axis. The basal row repeat unit must also contain an even number of kinks (+− or −+ sequences). These symbols may be added to those previously given to describe the primary nets. Thus the *M* phase becomes $P\underset{+}{/}P\underset{-}{/}P\underset{-}{X}P\underset{+}{X}$; LR (see Figure 2-3). They are however redundant, since they give no more information than is already contained in the sequence P/P/PXPX; they do nevertheless provide a convenient

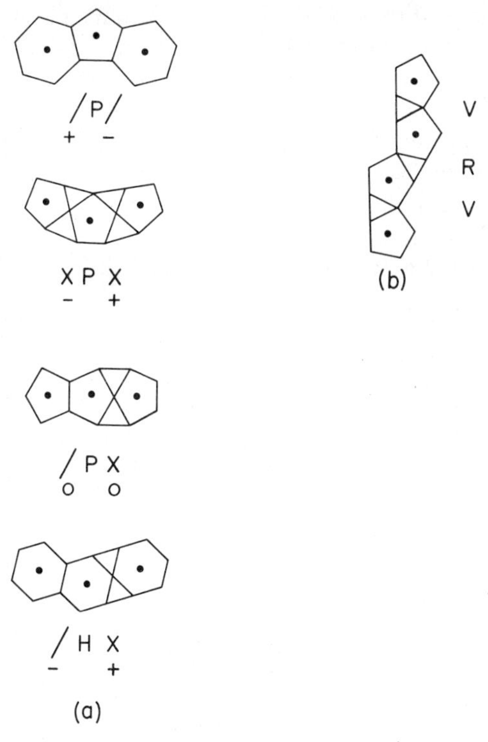

Figure 2-6. (*a*) Joining of pentagons and/or hexagons along the basal row. (*b*) Joining of pentagons along the *Y* direction, showing the intervening triangles which determine whether movements are L, R, or V.

aid to visualizing the sequence of connections along the basal repeat row, and in ensuring that the correct sequence, which contains an even number of kinks, has been chosen for any given structure.

Description of one primary net of a Frank–Kasper structure in terms of these symbols permits the whole net to be generated by repetition of the basic repeat unit along the *X* and *Y* axes. Since pentagons and hexagons of one primary net cover those of the primary nets above and below antisymmetrically, symbols describing these are obtained merely by interchanging *X* for / and *vice versa*; the repeat sequence along the *Y* axis remains unchanged. Thus the primary nets above and below $\text{P}\underset{+}{/}\text{P}\underset{-}{/}\text{P}\underset{-}{\text{X}}\text{P}\underset{+}{\text{X}}$; LR of the *M* phase are $\text{P}\underset{+}{\text{X}}\text{P}\underset{-}{\text{X}}\text{P}\underset{-}{/}\text{P}\underset{+}{/}$; LR. The designation of the repeat sequence of the primary net also specifies the secondary net, since all of the atoms of the secondary net center all of the pentagons and/or hexagons of the primary nets (not the triangles), and therefore the sequence and coordination of all of the atoms in the structure. The secondary net is triangular (3^6) if the primary net sequence is (0: LR), square (4^4) if it is (0: V) and square-triangle with sequences such as (+−; LR), (0; VL), etc.

With this coding any structures of the first group having a linear (zigzag) basal repeat unit can be described with a great condensation of structural information (Table 2-6) and the atomic arrangement can be regenerated

Table 2-6 Coding for Some Group 1 Structures

Structure	Planes in Which Nets Lie	Coding for Structure		Schläfli Symbols for Secondary Net and Ratio of Numbers of Different Corners
A. *σ* phase	001	$\text{H}\underset{+}{\text{X}}\text{H}\underset{-}{/}$	LR	3^2434
M phase	001	$\text{P}\underset{+}{/}\text{P}\underset{-}{/}\text{P}\underset{-}{\text{X}}\text{P}\underset{+}{\text{X}}$	LR	$3^2434+3^34^2$ (1/1)
P phase	001	$\text{P}\underset{+}{/}\text{H}\underset{-}{\text{X}}\text{P}\underset{-}{/}\text{H}\underset{+}{\text{X}}$	LR	$3^2434+3^34^2$ (1/1)
B. *β*-W	001	$\text{H}\underset{0}{\text{X}}$	V	4^4
Zr_4Al_3	110	$\text{P}\underset{0}{\text{X}}\text{P}\underset{0}{/}$	V	4^4
Zr_4Al_3	001	$\text{H}\underset{0}{\text{X}}$	L,R	3^6
$MgZn_2$	110	$\text{P}\underset{0}{\text{X}}\text{P}\underset{0}{/}$	LR	3^6
$MgCu_2$	$1\bar{1}0$	$\text{P}\underset{0}{\text{X}}\text{P}\underset{0}{/}$	L^{3a}	3^6
$MgNi_2$	110	$\text{P}\underset{0}{\text{X}}\text{P}\underset{0}{/}$	LLRR	3^6
MgAlCu (9-layer Laves)	110	$\text{P}\underset{0}{\text{X}}\text{P}\underset{0}{/}$	$(\text{LLR})^{3a}$	3^6
μ phase	110	$\text{P}\underset{0}{\text{X}}\text{P}\underset{0}{/}$	$(\text{VL})^{3a}$	3^34^2
Idealized planar *R*		$\text{P}\underset{0}{\text{X}}\text{H}\underset{0}{\text{X}}\text{P}\underset{0}{/}\text{H}\underset{0}{/}$	$(\text{VL})^{3a}$	3^34^2

[a]Tripling to achieve rectangular repeat.

readily from the symbols. Possible structures can be predicted systematically, and the coordination of the atoms can be determined by inspection, or with a little scribbling, since it is uniquely prescribed by the type of connections between the pentagons and/or hexagons. Figure 2-7 shows all possible coordination arrangements for group one structures.

In order to be able systematically to predict new structure types it is convenient to group Frank–Kasper structures as follows.

A, Table 2-6: Structures with + and − along X, and L and/or R along Y. The secondary net is square-triangle and kinks along the parallel lines are of about 150°. The σ, P, and M phases belong to this class. New structures are obtained by changing the +− sequence and by changing H and

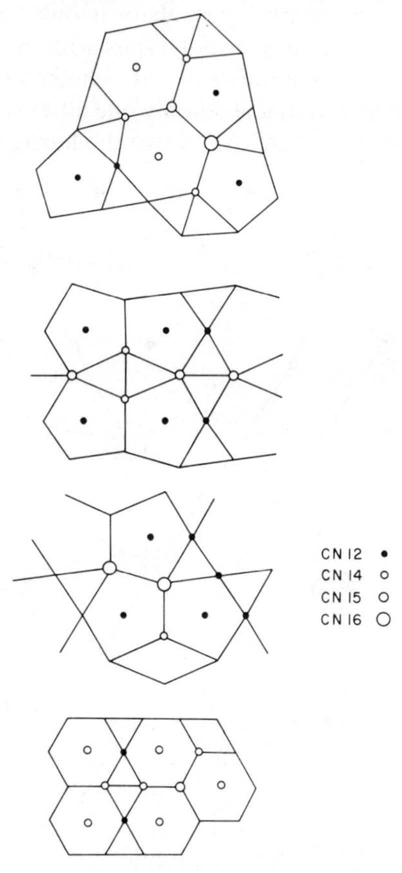

Figure 2-7. All possible coordination arrangements in Frank–Kasper structures which can be described by linear (zigzag) repeats of the secondary nets along the Y direction. See text.

P for a given +− sequence, maintaining an even number of kinks in the basal repeat sequence. For any given +− HP sequence, further structures are obtained systematically by changing the LR repeat sequence along the *Y* axis.

B, Table 2-6: Structures with 0 along the *X* axis and L, R, or V along *Y*. L followed by R now produces 120° kinks along *Y*, and V followed by L or R 150° kinks. The secondary net is rectangular (4^4) with only V, and triangular with only L followed by R and *vice versa*. The Laves phase polytypes, the μ, Al_3Zr_4, and β-W structures are found in this class.

All-H and all-P structures differ fundamentally because of the unique way that the pentagons must be connected along the strips in the *Y* direction. With triangular secondary layers there is only one all-H structure, although there is a whole family of polytypes with the all-P structure. In the all-H structure there is nothing to say whether the strips are L or R; either strip can be selected, and so only one all-H structure arises with a 3^6 triangular secondary net (Zr_4Al_3, down the hexagonal axis). L and R

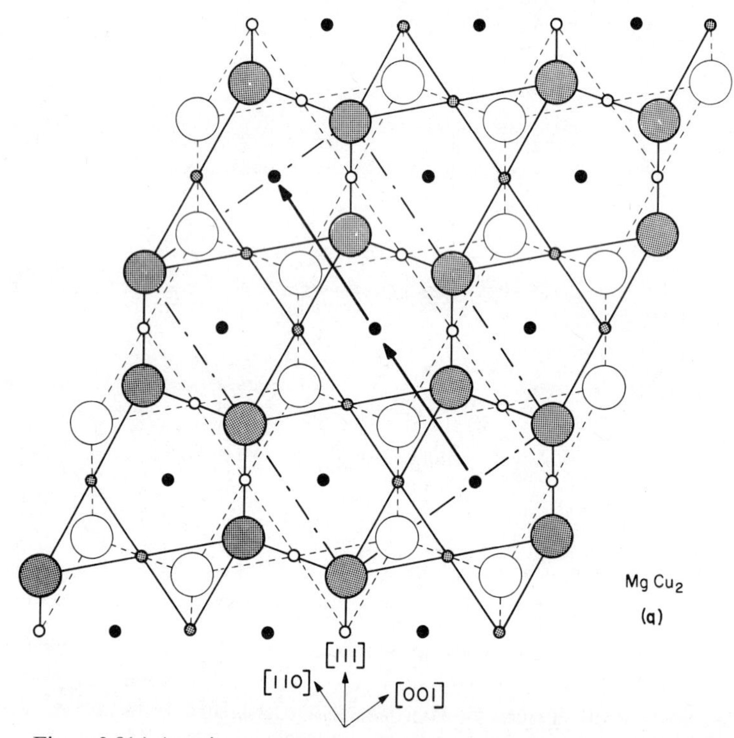

Figure 2-8(*a*) Atomic arrangement in the $MgCu_2$ (*cF*24) structure projected on the ($1\bar{1}0$) plane of a hexagonal cell. (b) Atomic arrangement of the $MgZn_2$ (*hP*12) structure projected onto the (110) plane. Large circles

are however distinct in the all-P structures because that direction is chosen which connects the pentagons along the strips apex to apex with an intervening triangle on one side (Figure 2-6*b*). Thus a pentagon of one basal row is adjacent to two pentagons in the row above; it shares an edge or apex with two intervening triangles with one, and an apex with single intervening triangle with the other which defines the direction of the pentagon strip. The Laves phases are the all-P structures with various combinations of L and R repeats along the *Y* axis. Thus it is seen that all of the Laves phases are merely stacking polytypes of the cubic $MgCu_2$ PXP/; L^3 structure (Figure 2-8 *a* and *b*), being analagous to the polytypes

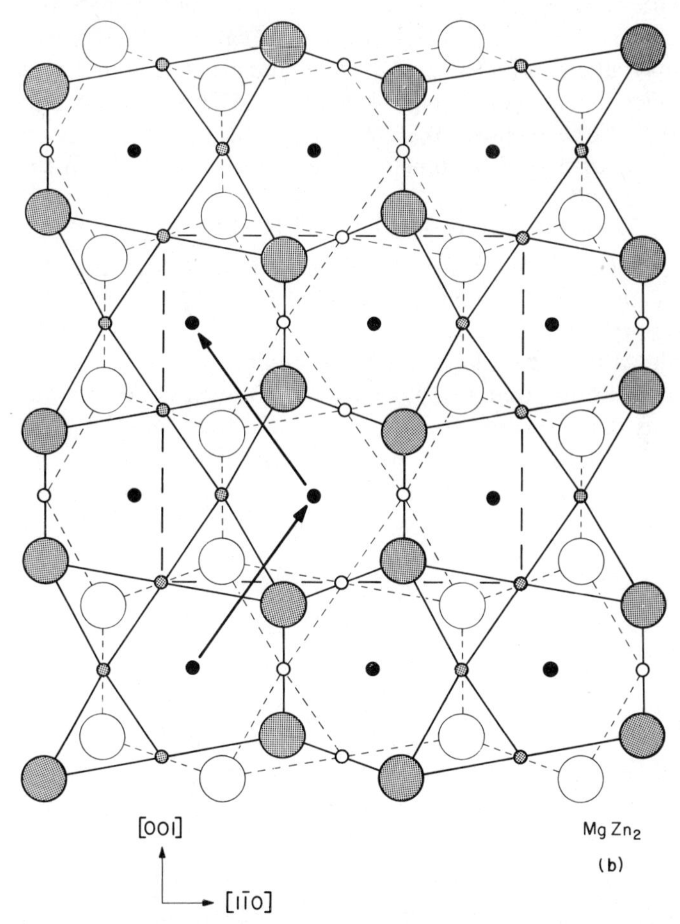

indicate CN 16 sites and small circles icosahedrally surrounded sites. Arrows indicate movements L or R of successive rows of pentagons. (Shoemaker and Shoemaker[13, 15].)

of the SiC structure. In the cubic structure and in rhombohedral polytypes, the unit cell contains three times the repeat sequence along the *Y* direction. For the same reasons L and R must differ in H–P structures and new structural types can be generated by varying or extending the H–P repeat sequence along *X* and varying the L, R, V repeat sequence along *Y*.

Several features of this scheme of coding are not uniquely defined, so that in some cases different descriptions can be given for a particular structure. This does not impair the usefulness of the coding, since the regeneration of the structure from the symbols adopted is indeed unique.

(1) The possible choice of primary and secondary nets may not be unique. Thus the Al_3Zr_4 structure (p. 664) can be described either with pentagon-triangle primary and 4^4 secondary nets parallel to the (110) plane, or with hexagon-triangle primary and 3^6 secondary nets parallel to the (001) plane (Table 2-6).

(2) Either of two orthogonal axes can be selected as the *X* direction in hexagon-triangle primary nets. Thus the hypothetical hexagon-μ phase (Figure 2-9) can be described either as H/$\underset{-}{\mathrm{H}}\underset{+}{\mathrm{X}}$; L, H$\underset{0}{/}$; VL, or H$\underset{0}{/}$; VR.

(3) With triangular 3^6 secondary nets there are two possible choices of the *X* direction in pentagon-triangle primary nets and three in the hexagonal-triangle primary net. In the descriptions which have been given, the conditions adopted (e.g., of joining P and/or H along the basal repeat row) ensure that a single basal repeat row can be obtained. If these conditions are relaxed, then other basal rows may be selected that give different repeat sequences, and in particular two alternate paths (about a mirror plane lying along the *X* axis) may occur for the basal row (coded $\pm$ or $\mp$). Such methods allow the possibility of describing

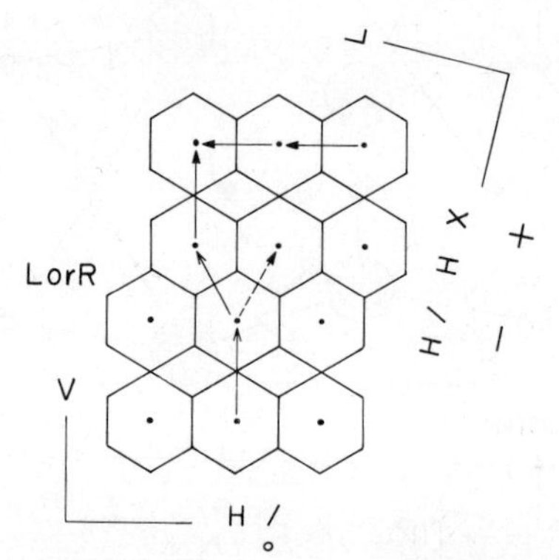

Figure 2-9. Diagram showing different possible descriptions of a hypothetical hexagon μ-phase structure.

group two structures and, for example, the hypothetical structure shown in Figure 2-5, p. 38 could be coded $H\underset{\pm}{X}H\underset{0}{/}H\underset{\mp}{X}$, hexagonal. However, there is little point in pursuing this further at present, since few if any group two type structures are known.

These tetrahedrally close packed structures will presumably become increasingly important as our knowledge of intermetallic phases, particularly those of the transition metals, increases, In observed structures the hexagons and pentagons, here treated ideally, may be somewhat distorted from their regular shapes as they accommodate atoms of different sizes and the layers of atoms may not be exactly planar.

Structures Generated by Layer Networks Containing Squares

These structures are generated from primary nets containing squares and other polygons which are generally interleaved with 4^4 secondary nets, or from 4^4 networks of atoms alone. The presence of squares gives rise to interpenetrating octahedra or centered cubes when the nodes of successive 4^4 nets are displaced by half of the diagonal of the squares, and to cubes or archimedean antiprisms when the squares overlie each other directly or antisymmetrically, although these may not be recognized as the ultimate coordination polyhedra in the structure. Thus squares covering each other antisymmetrically with interleaving 4^4 nets would be regarded as forming CN 10 polyhedra, rather than CN 8 archimedean antiprisms. However, as the cube is a relatively small coordination polyhedron compared to those of the Frank–Kasper structures, structures can be expected in which some or all of the cubes or anticubes are not centered. Centering of cubes in some square-triangle nets may be achieved by displacing the nodes of subsequent networks of the same type by one-half of the diagonal of the squares, or it may be achieved by interleaving a subsidiary layer of atoms between the primary networks. There are many structures built up of nets containing pentagons, hexagons, and larger polygons together with squares, but the examples of coding structure types given here are confined to structures containing 4^4 nets of atoms alone, and to those with primary 3^2434 nets.

Structures Generated by 4^4 Nets of Atoms. These structures are made up of 4^4 nets of atoms aligned parallel or at 45° to the cell edge. The nets are occupied by one component only and there are no vacant sites so that the chemical repeat unit of the net is equal to the net period. The important parameters for describing the structures are the separation of the nets normal to their plane, and the net origin (node) and orientation relative to one fixed net (unit cell). If the net orientation is parallel to the cell edge, its origin may be located either over the cell origin or transposed by $a/2$,

$b/2$, or $(a+b)/2$. If the net is rotated 45° to the cell edge, its origin may be transposed by $a/2$, $b/2$, $(a+b)/4$, or $(3a+b)/4$ from the cell origin. These conditions are illustrated at the top of Table 2-7[11] which summarizes information on some structures within this class. By numbering these net origins, information on the components occupying the nets, the net origin and orientation, and the spacing between the layers can be given in a linear set of symbols similar to those for structures based on

Table 2-7 Structures with 4^4 Nets Only (Length of Net Edge a or $a/\sqrt{2}$)

Sub numbers give height of atom plane as fraction of cell edge (decimal with 0. omitted). Super numbers locate origin and orientation of 4^4 net as follows:

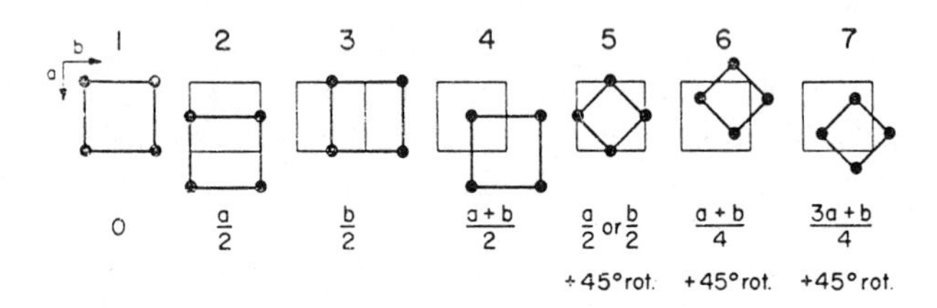

Structure	Description of structure
Po (cP 1)	Po^{1}_{0}
CsCl (cP 2)	$Cs^{1}_{0}Cl^{4}_{5}$
$AuCu_3$ (cP 4)	$Au^{1}_{0}Cu^{4}_{0}Cu^{5}_{5}$
ReO_3 (cP 4)	$Re^{1}_{0}O^{5}_{0}O^{1}_{5}$
CFe_4 (cP 5)	$C^{1}_{0}Fe^{6}_{27}Fe^{7}_{73}$
$CaTiO_3$ (cP 5)	$Ca^{1}_{0}O^{4}_{0}Ti^{4}_{5}O^{5}_{5}$
Cu_2O (cP 6)	$O^{1}_{0}Cu^{6}_{25}O^{4}_{5}Cu^{7}_{75}$
VCu_3S_4 (cP 8)	$V^{1}_{0}Cu^{2+3}_{0}S^{6}_{24}Cu^{1}_{5}S^{7}_{74}$
Cu (cF 4)	$Cu^{1+4}_{0}Cu^{5}_{5}$
ZnS (cF 8)	$Zn^{1+4}_{0}S^{6}_{25}Zn^{5}_{5}S^{7}_{75}$
C (cF 8)	$C^{1+4}_{0}C^{6}_{25}C^{5}_{5}C^{7}_{75}$
NaCl (cF 8)	$Na^{1+4}_{0}Cl^{5}_{0}Cl^{1+4}_{5}Na^{5}_{5}$
AgAsMg (cF 12)	$As^{1+4}_{0}Ag^{6}_{25}Mg^{7}_{25}As^{5}_{5}Mg^{6}_{75}Ag^{7}_{75}$
CaF_2 (cF 12)	$Ca^{1+4}_{0}F^{6+7}_{25}Ca^{5}_{5}F^{6+7}_{75}$
$AlCu_2Mn$ (cF 16)	$Al^{1+4}_{0}Mn^{5}_{0}Cu^{6+7}_{25}Mn^{1+4}_{5}Al^{5}_{5}Cu^{6+7}_{75}$
$BiLi_3$ (cF 16)	$Bi^{1+4}_{0}Li^{5}_{0}Li^{6+7}_{25}Li^{1+4}_{5}Bi^{5}_{5}Li^{6+7}_{75}$
NaTl (cF 16)	$Na^{1+4}_{0}Tl^{5}_{0}Na^{6}_{25}Tl^{7}_{25}Tl^{1+4}_{5}Na^{5}_{5}Tl^{6}_{75}Na^{7}_{75}$
W (cI 2)	$W^{1}_{0}W^{4}_{5}$
$PtHg_4$ (cI 10)	$Pt^{1}_{0}Hg^{6+7}_{25}Pt^{4}_{5}Hg^{6+7}_{75}$

Table 2-7 (*continued*)

Structure	Crystal plane	Description of structure
FeNiN (*tP* 3)	*ab*	$\mathrm{Fe_0^1N_0^4Ni_5^4}$
$\mathrm{FeSi_2}$ (*tP* 3)	*ab*	$\mathrm{Fe_0^1Si_{27}^4Si_{73}^4}$
AuCuI (*tP* 4)	*ab*	$\mathrm{Au_0^{1+4}Cu_5^5}$
γCuTi (*tP* 4)	*ab*	$\mathrm{Cu_{10}^1Ti_{35}^4Ti_{65}^1Cu_9^4}$
$\mathrm{CuTi_3}$ (*tP* 4)	*ab*	$\mathrm{Cu_0^1Ti_0^4Ti_5^5}$
βNp (*tP* 4)	*ab*	$\mathrm{Np_0^{1+4}Np_{38}^3Np_{63}^2}$
PbO (*tP* 4)	*ab*	$\mathrm{O_0^{1+4}Pb_{24}^3Pb_{76}^2}$
PtS (*tP* 4)	*ab*	$\mathrm{Pt_0^3S_{25}^1Pt_5^2S_{75}^1}$
$\mathrm{CCo_2Mn_2}$ (*tP* 5)	*ab*	$\mathrm{Mn_0^{1+4}C_5^4Co_5^5}$
$\gamma\mathrm{H_{0.5}Zr}$ (*tP* 6)	*ab*	$\mathrm{Zr_0^{1+4}H_{25}^6Zr_5^5H_{75}^7}$
$\mathrm{Cu_2Sb}$ (*tP* 6)	*ab*	$\mathrm{Cu_0^{1+4}}\mathrm{S}\underset{\sim}{\mathrm{b}}{}_{28}^{2}\mathrm{C}\underset{\sim}{\mathrm{u}}{}_{28}^{3}\mathrm{C}\underset{\sim}{\mathrm{u}}{}_{72}^{2}\mathrm{S}\underset{\sim}{\mathrm{b}}{}_{72}^{3}$
SeSiZr (*tP* 6)	*ab*	$\mathrm{Si_0^{1+4}Zr_{25}^3Se_{38}^2Se_{63}^3Zr_{75}^2}$
$\mathrm{CdIn_2Se_4}$ (*tP* 7)	*ab*	$\mathrm{Cd_0^1Se_{25}^6In_5^5Se_{75}^7}$
$\mathrm{BBe_{4-5}}$ (*tP* 10)	*ab*	$\mathrm{Be_0^{1+4}B_{17}^3Be_3^{1+4}Be_5^5Be_7^{1+4}B_{83}^2}$
Pa (In) (*tI* 2)	*ab*	$\mathrm{Pa_0^1Pa_5^4}$
Sn (*tI* 4)	*ab*	$\mathrm{Sn_0^1Sn_{25}^3Sn_5^4Sn_{75}^2}$
$\mathrm{H_2Th}$ (*tI* 6)	*ab*	$\mathrm{Th_0^1H_{25}^5Th_5^4H_{75}^5}$
$\mathrm{MoSi_2}$ (*tI* 6)	*ab*	$\mathrm{Mo_0^1Si_{17}^4Si_{33}^1Mo_5^4Si_{67}^1Si_{83}^4}$
$\mathrm{CaC_2I}$ (*tI* 6)	*ab*	$\mathrm{Ca_0^1C_{10}^4C_4^1Ca_5^4C_6^1C_9^4}$
AsNb (*tI* 8)	*ab*	$\mathrm{Nb_0^1As_{17}^2Nb_{25}^3As_{42}^1Nb_5^4As_{67}^3Nb_{75}^2As_{92}^3}$
$\mathrm{TiAl_3}$ (*tI* 8)	*ab*	$\mathrm{Ti_0^1Al_0^4Al_{25}^5Al_5^1Ti_5^4Al_{75}^5}$
$\mathrm{AgTlTe_2}$ (*tI* 8)	*ab*	$\mathrm{Tl_0^1Te_{13}^4Ag_{25}^3Te_{38}^1Tl_5^4Te_{63}^1Ag_{75}^2Te_{88}^4}$
$\mathrm{ThCu_2Si_2}$ (*tI* 10)	*ab*	$\mathrm{Th_0^1Si_{13}^4Cu_{25}^5Si_{38}^1Th_5^4Si_{63}^1Cu_{75}^5Si_{88}^4}$
$\alpha\mathrm{ThSi_2}$ (*tI* 12)	*ab*	$\mathrm{Th_0^1Si_{08}^4Si_{17}^2Th_{25}^3Si_{33}^2Si_{42}^1Th_5^4Si_{58}^1Si_{67}^3Th_{75}^2Si_{83}^3Si_{92}^4}$
$\mathrm{BaAl_4}$ (*tI* 10)	*ab*	$\mathrm{Ba_0^1Al_{13}^4Al_{25}^5Al_{38}^1Ba_5^4Al_{63}^1Al_{75}^5Al_{88}^4}$
$\mathrm{CdAl_2S_4}$ (*tI* 14)	*ab*	$\mathrm{Al_0^{1+4}S_{13}^6Cd_{25}^3S_{38}^7Al_5^{1+4}S_{63}^6Cd_{75}^2S_{88}^7}$
$\mathrm{ZrAl_3}$ (*tI* 16)	*ab*	$\mathrm{Al_0^5Zr_{12}^1Al_{14}^4Al_{25}^5Al_{36}^1Zr_{38}^4Al_5^5Zr_{62}^4Al_{64}^1Al_{75}^2Al_{86}^4Zr_{88}^1}$
$\mathrm{HfGa_2}$ (*tI* 24)	*ab*	$\mathrm{Ga_0^5Hf_{07}^1Ga_{08}^4Ga_{16}^2Hf_{18}^3Ga_{25}^{1+4}Hf_{32}^3Ga_{34}^2Ga_{41}^1Hf_{43}^4Ga_5^5Hf_{57}^4Ga_{59}^1}$
		$\mathrm{Ga_{66}^3Hf_{68}^2Ga_{75}^{1+4}Hf_{82}^2Ga_{84}^3Ga_{91}^4Hf_{93}^1}$
CuTe (*oP* 4)	*ba*	$\mathrm{Te_{22}^1Cu_{45}^3Cu_{55}^2Te_{78}^4}$
$\mathrm{Co_2N}$ (*oP* 6)	*ba*	$\mathrm{N_0^1Co_0^6N_5^4Co_5^7}$ (Co atoms off-site somewhat)
$\mathrm{Sb_2Tl_2Se_4}$ (*oP* 8)	*ba*	$\mathrm{Sb_0^1Tl_0^4Se_{23}^4Se_{27}^3Sb_5^2Tl_5^3Se_{73}^3Se_{77}^4}$
Ga (*oC* 4)	*ac*	$\mathrm{Ga_{13}^1Ga_{38}^4Ga_{63}^3Ga_{88}^2}$
BCr (*oC* 8)	*ac*	$\mathrm{B_{06}^4Cr_{15}^1Cr_{35}^4B_{44}^1B_{56}^2Cr_{65}^3Cr_{85}^2B_{94}^3}$
$\mathrm{ZrGa_2}$ (*oC* 12)	*bc*	$\mathrm{Ga_0^{1+4}Zr_{15}^2Ga_{18}^3Ga_{32}^4Zr_{35}^1Ga_5^5Zr_{65}^1Ga_{68}^4Ga_{82}^3Zr_{85}^2}$
$\mathrm{VAu_2}$ (*oC* 12)	*ac*	$\mathrm{V_0^{1+4}Au_{17}^5Au_{33}^{1+4}V_5^5Au_{67}^{1+4}Au_{83}^5}$
$\mathrm{ZrSi_2}$ (*oC* 12)	*ac*	$\mathrm{Si_{06}^4Zr_{10}^1Si_{25}^5Zr_{40}^4Si_{45}^1Si_{55}^2Zr_{60}^3Si_{75}^{1+4}Zr_{90}^2Si_{95}^3}$
BCU (*oC* 12)	*ac*	$\mathrm{B_{04}^4U_{13}^1C_{17}^4C_{33}^1U_{37}^4B_{47}^1B_{53}^2U_{63}^3C_{67}^2C_{83}^3U_{87}^2B_{96}^3}$
$\mathrm{TaPt_2}$ (*oC* 12)	*bc*	$\mathrm{Ta_0^6Pt_{17}^7Pt_{33}^6Ta_5^7Pt_{67}^6Pt_{83}^7}$
$\mathrm{BCMo_2}$ (*oC* 16)	*ac*	$\mathrm{B_{03}^4Mo_{07}^1C_{19}^1Mo_{19}^4Mo_{31}^1C_{31}^4Mo_{43}^4B_{47}^1B_{53}^1Mo_{57}^3Mo_{69}^2C_{69}^3}$
		$\mathrm{C_{81}^2Mo_{81}^3Mo_{93}^1B_{97}^3}$

triangular, etc. networks of atoms. A great condensation of structural information is obtained for the structures listed and structural details can be regenerated from the symbols, coordination determined, and structure types compared. However, many of the structures which can be so described are more conveniently considered under other headings. Thus the Cu ($cF4$) or $AuCu_3$ ($cP4$) structures which can be described in terms of 4^4 nets stacked along {100}, are more effectively considered in terms of close packed layers stacked along {111}.

Structures with Primary Square-Triangle 3^2434, Nets of Atoms. Numerous structures are known in which primary 3^2434 nets of atoms occur; frequently these are interleaved with 4^4 nets of atoms that center the cubes or cubic antiprisms formed by the 3^2434 nets. Details of these structures can be described by a code that gives the stacking and orientation of the 3^2434 nets, centering of the major prisms, and whether the squares of the nets lie over the cell corners, base center, or midpoint of the basal edges of the cell.

The squares of the 3^2434 net either lie over the cell corners and cell base centers (stacking symbol A), or over the midpoints of the cell edges (stacking symbol B). The nets are oriented so that the direction normal to the top of the square over the cell base center, or over the midpoint of the base side at the top of the cell (looking towards the cell origin), points either toward 11 o'clock or toward 1 o'clock. In the first case the stacking symbol is given a prime (A′, B′); in the second case it is used unprimed (Figure 2-10). Thus a primed-unprimed (A′A) stacking sequence gives cubic antiprisms, a primed-primed (B′B′) or unprimed-unprimed (BB) sequence gives cubes, and a mixed stacking sequence (AB) gives an eight-cornered polyhedron that is greatly distorted from a cube. Centering of the eight-cornered polyhedra by interleaved 4^4 nets of atoms is indicated by 1 or $\frac{1}{2}$ between the symbols describing the 3^2434 layers, depending on whether all or half of the polyhedra are centered. When the centering atoms lie in the same plane as the 3^2434 net, the fraction of squares centered is given sub to the symbol describing the layer. Thus symbols describing respectively all, half and no polyhedra filled would be A1A, A$\frac{1}{2}$A, and AA. A'_1 would indicate centering of the squares in the A′ layer. Finally some of these structures contain square 4^4 nets of atoms parallel to the basal cell edges and with their nodes located over $\frac{1}{4}$, $\frac{1}{4}$; etc. of the basal plane, giving squares over the cell corners, base center, and mid-points of the basal cell edges. Such nets are described by the stacking symbol C (Figure 2-10). Table 2-8 provides examples of some structures containing 3^2434 nets of atoms which can be described by these symbols.

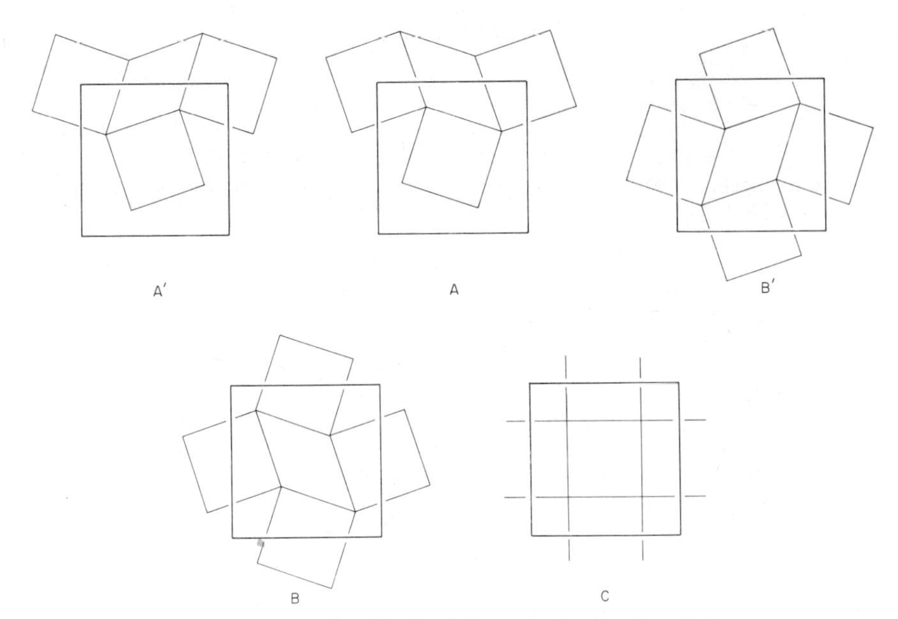

Figure 2-10. Symbols for 3^2434 and 4^4 nets variously oriented. See text.

Table 2-8 Coding for Some Structures Built Up of 3^2434 Nets of Atoms

Structure Name	Structure Symbol	Stacking Sequence†
$CuAl_2$	*tI*12	$A'\underline{1}A\underline{1}$
$SeTl_2$	*tP*30	$\underline{1}^{*}B\underline{B}_1'B\underline{1}^{*}B'\underline{B}_1B'$
$CoGe_2$	*oC*23	$\underline{1}A'\underline{1}C\underline{1}B\underline{1}C$
$PdSn_2$	*tI*48	$A\underline{1}C\underline{1}B\underline{1}C\underline{1}A'\underline{1}C\underline{1}B'\underline{1}C\underline{1}$
$SiPt_3$	*mC*16	$B'1B1$ + Si (centers diamonds)
$CoGa_3$	*tP*16	$A1A1$ + Co (centers $\frac{1}{2}$ triangles)
$PdSn_3$	*oC*32	$A_{\underline{1}}A'B'_{\underline{1}}B_1B'A'_{\underline{1}}$
$PtPb_4$	*tP*10	$\underline{1}A'A$
$NbTe_4$	*tP*10	$A\frac{1}{2}A'\frac{1}{2}$
$PtSn_4$	*oC*20	$\underline{1}AB'\underline{1}BA'$
$PdGa_5$	*tI*24	$1A'\underline{1}A1A\underline{1}A'$
Si_2U_3	*tP*10	$\underline{A}\underline{1}A'$
Al_2Gd_3	*tP*20	$\underline{B}_1'B\underline{B}_1'B$
B_3Cr_5	*tI*32	$\underline{A}_1A'\underline{1}A\underline{A}_1'A\underline{1}A'$
Pb_3Ba_5	*tI*32	$\underline{A}_1'A\underline{1}A'\underline{A}_1A'\underline{1}A$

†First component in formula is underlined. *Half occupied randomly.

REFERENCES

1. W. B. Pearson, 1967, *Handbook of Lattice Spacings and Structures of Metals*, Vol. 2, New York: Pergamon Press pp. 5–78.
2. Unpublished, and also E. Hellner, 1965, *Acta Cryst.*, **19**, 703. H. Burzlaff, W. Fischer, and E. Hellner, 1968, *Ibid.*, **A24**, 57. W. Fischer, 1968, *Ibid.*, **A24**, 67.
3. A. L. Loeb, 1958, *Acta Cryst.*, **11**, 469. 1962, *Ibid.*, **15**, 219. 1964, *Ibid.*, **17**, 179. 1967, TR-130 Ledgemont Laboratory; I. L. Morris and A. L. Loeb, 1960, *Acta Cryst.*, **13**, 434.
4. E. Mooser and W. B. Pearson, 1957, *J. Chem. Phys.*, **26**, 893. C. H. L. Goodman, 1958, *J. Phys. Chem. Solids*, **6**, 305.
5. E. Parthé, 1963, *Z. Kristallogr.*, **119**, 204.
6. M. J. Buerger, 1969, private communication.
7. J. Lima-de-Faria and M. O. Figueiredo, 1969, *Z. Kristallogr.*, **130**, 41, 54.
8. P. Niggli, 1919, *Geometrische Kristallographie des Diskontinuums*, Gebr. Bornträger: Leipzig. P. Niggli, 1928, *Kristallographische und strukturtheoretische Grundbegriffe, Handbuch d. Experimental Physik*, Akademische Verlagsgesellschaft: Leipzig. P. Niggli, 1941, *Lehrbuch der Mineralogie*, Part 1, Gebr. Bornträger: Berlin. C. Hermann, 1960, *Z. Kristallogr.*, **113**, 142. G. Mentzer, 1960, *Z. Kristallogr.*, **113**, 178. E. Hellner, 1965, *Acta Cryst.*, **19**, 703. H. Burzlaff, W. Fischer, and E. Hellner, 1968, *Acta Cryst.*, **A24**, 57. W. Fischer, 1968, *Ibid.*, **A24**, 67.
9. A condensed notation can be devised: A. L. Loeb, 1968, Technical Memorandum 18, Ledgemont Laboratory, Kennecott Copper Corporation.
10. P. I. Kripjakevič, 1963, *Ž. Strukt. Khim.*, **4**, 117, 282 (Translation, Consultants Bureau, New York). Review, 1965, *Acta Cryst.*, **19**, 692.
11. W. B. Pearson, 1968, *Helv. Phys. Acta*, **41**, 1070.
12. F. C. Frank and J. S. Kasper, 1958, *Acta Cryst.*, **11**, 184; 1959, *Ibid.*, **12**, 483.
13. C. B. Shoemaker and D. P. Shoemaker, 1967, *Acta Cryst.*, **23**, 231.
14. W. B. Pearson and C. B. Shoemaker, 1969, *Acta Cryst.*, **B25**, 1178.
15. C. B. Shoemaker and D. P. Shoemaker, 1969, *Developments in the Structural Chemistry of Alloy Phases*. New York: Plenum Press, p. 107.

3

Factors Affecting Crystal Structure

In an alloy system if a stable phase and crystal structure exists at a given temperature, pressure, and composition, it is that with the lowest Gibbs free energy, G, given by the relationship $G = H - TS$, where H is the enthalpy, T the temperature in degrees absolute, and S the entropy. If in a binary system at a certain composition, no phase has a uniquely low Gibbs free energy under given conditions of temperature and pressure, two phases exist in equilibrium and these are determined by the tangent to the free-energy-composition curves of two single phases which lie at the lowest free energy. Among the factors controlling the crystal structure that a phase of given composition may adopt are the environmental factors, temperature and pressure, which are associated with the entropy term, TS, and the geometrical, energy band, chemical bond, and electrochemical factors resulting from the (difference in) properties of the component atoms, which relate mainly to the enthalpy term, H. The main contribution to the enthalpy in metals comes from nearest-neighbor interactions which might also be referred to as the chemical bonding. Interactions between next nearest and further neighbors, may be expected to make a contribution which is at least an order of magnitude smaller. On the energy band picture the nearest-neighbor interactions or chemical bonds may be expected to correlate mainly with energy bands of valence electrons lying below the Fermi level, particularly in phases containing transition metals, whereas interactions involving next nearest neighbors probably correlate more with the electrons at the Fermi level It is for this reason that, if energy band effects are to be observed in metals and alloys (e.g., effects of electron concentration in controlling alloy phase structure), they are only likely to be observed in experimental situations where (i) the main contribution to the enthalpy, the near-neighbor interactions, remain largely unchanged, or (ii) all of the outer valence electrons occupy sheets of Fermi surface in partly filled Brillouin zones, so that the effects of interaction between Fermi surface and Brillouin zone planes on the electron energies make a relatively large contribution to the enthalpy.†

The influence of atomic size on the crystal structure of metals and alloys is complex; to some extent it may be thought of as giving rise in part to the geometrical factor, although in another sense it may be more realistic to consider that geometrical and chemical bond effects themselves control the near-neighbor distances between at least some of the ligands in many of the structures of metallic phases. Thus metallic atoms generally appear to be readily "compressible," frequently approaching each other in some directions more closely than expected for unit bond distances, so that their structural behavior is not the same as that of atoms or ions in valence compounds which are insulators. In order to emphasize this difference, geometrical and chemical bond factors are first discussed.

GEOMETRICAL FACTORS

Laves[1] pointed out that, in the absence of other dominant factors, the structures of metals should exhibit the highest degree of space filling, the highest symmetry, and the greatest number of connections between the atoms. These geometrical principles have since become a conversation piece, but the difficulty of explicitly analyzing their importance in structures is that they are not readily ammenable to quantitative measurement. Although Laves himself introduced a space filling model for the structures of binary phases and this has been developed further by Parthé[2], it was based on an incompressible sphere model of the atom and so is unsuitable for discussing metals because compressibility of the atoms is one of their main properties; it can much better be applied to ionic substances. As the space filling model of Laves and Parthé only *compares structures* and does not examine the *dimensional behavior of phases with given structures*, it is also unsuited for quantitative assessment of the relative importance of geometrical or other effects in controlling the occurrence of structures and their dimensions. Furthermore, their model has obvious difficulties in its application to complex structures.

A more useful model for metals[3] is one which allows the atoms of a

†For fundamental discussion of Brillouin zones in relation to alloys, see references given on p. 80.

binary alloy M_xN_y to be compressed until successively, according to the structural geometry, M–M, M–N, N–N, etc. contacts are established. These are considered to occur when the interatomic distances, d_M, d_{MN}, d_N are equal respectively to $2R_M = D_M$, $R_M + R_N = \frac{1}{2}(D_M + D_N)$, $2R_N = D_N$, where R_M, R_N are the atomic radii and D_M, D_N the atomic diameters of the two components†.

By expressing the distances between all of the close interatomic contacts in the structure in terms of the cell and atomic parameters, and thence in terms of d_M, the distance between an arbitrarily selected set of M–M contacts, lines on a diagram of a reduced strain parameter $(D_M - d_M)/D_N$ *vs.* D_M/D_N can be drawn to represent each of the interatomic contacts resulting from the geometry of the structure, and such are straight lines (Figure 3-1). The value of $(D_M - d_m)/D_N$ for the arbitrarily selected set of M–M contacts (in complex structures there may be several different sets of M–M contacts) is set at zero for all D_M/D_N values in order to locate the diagram on the strain parameter scale.

Such diagrams are called near-neighbor diagrams (n.n.d.). It does not matter whether, say, $(D_M - d_M)/D_N$ or $(D_N - d_N)/D_N$ is taken as the reduced strain parameter since the latter only rotates the whole diagram to place the line for the chosen N–N rather than M–M contact along the zero value of the strain parameter; the relative geometry of the diagram remains otherwise the same and no new information is obtained. The complexity of the structure presents no more difficulty in creating n.n.d. than the time involved in calculations for the various interatomic distances given by the structural geometry. Since the d_M values for known phases with the given structure can be determined from the lattice (and atomic) parameters of the phases, points can be plotted on these diagrams representing the occurrence of actual phases. The distribution of these can be compared with the lines for contacts between the various atoms to determine whether chemical bond or geometrical factors are important in controlling the occurrence and cell dimensions of phases with the particular structure. When a point representing a phase has a larger value of the strain parameter $(D_M - d_M)/D_N$ than a particular contact line, that is to say, lies above it on the n.n.d., then the contacts represented by the line are compressed in the structure of the phase; if it lies below the line, then those contacts have not been established in terms of the D_M and D_N values assumed for the component atoms of the phase.

†The metallic radii chosen are those appropriate for the coordination of the atoms in the phase; see p. 146, 151. In constructing n.n.d. for phases MN_x, the component N is always selected so that $x > 1$ unless $x = 1$, when N is the second component in the formula as written.

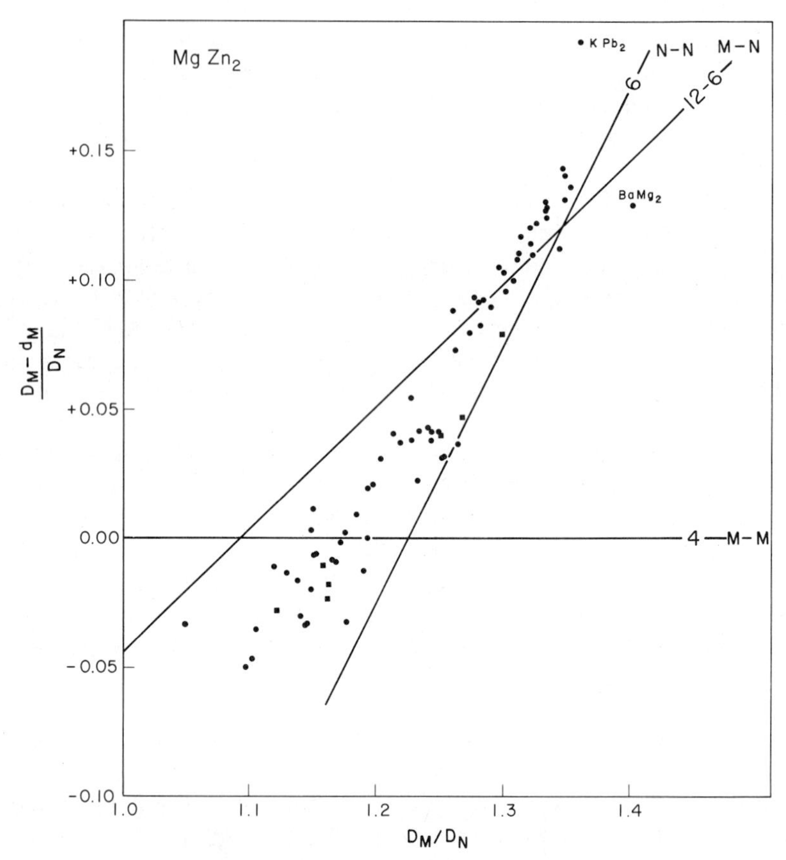

Figure 3-1. Near-neighbor diagram for phases with the $MgZn_2$ ($hP12$)(MN_2) structure constructed for the ideal axial ratio, $c/a = 1{\cdot}63$, and for the ideal atomic parameters $z_{Mg} = \frac{1}{16}$, $x_{Zn} = \frac{5}{6}$. Circular points represent actual phases with the $MgZn_2$ structure. Square points represent phases containing Mn with appropriately chosen valency (i.e., atomic radius). 6 N–N indicates N atoms have 6 N neighbors. 12–6 M–N indicates M has 12 N neighbors and N has 6 M neighbors.

When the structure has variable axial ratios or variable atomic parameters, the lines for various contacts may change their positions on the diagram as the structural parameters change. Strong evidence for the importance of a particular contact may be obtained if the points for observed phases follow the position of the contact line on the diagram as it changes with change in axial ratio or atomic parameters. The near-neighbor diagrams given in this book for structures with variable atomic parameters or variable axial ratios are drawn for representative values of these parameters, or if these have a large spread in known phases with the structures, then the contact lines are drawn for several representative (or extreme) values of these parameters as in Figure 3-9, for example.

The n.n.d. model emphasises the precise experimentally measured structural parameters (i) the atomic coordination in the structure (represented by the diagram) and (ii) the cell dimensions and atomic parameters of phases with the structure (represented by points on the diagram), and places least significance on the uncertain parameters—the atomic diameters (by virtue of considering structures adopted by many phases).

Since evidence of a chemical bond factor on a near-neighbor diagram is obviously to be shown by a distribution of points for observed phases along the contact line for the particular chemical bond, the question arises how evidence of geometrical effects is to be recognized. This leads to the conclusion that there are two distinct types of geometrical effect. One is the attainment of high coordination alone and is referred to as the *coordination factor*. This would be the controlling factor when points for phases with the particular structure are distributed only about the region where two or more contact lines giving high coordination intersect. The attainment of high coordination then controls the radius ratio at which the phases occur, and the structural dimensions (value of the strain parameter $(D_M - d_M)/D_N$) so as to give the high coordination. An example of this is found in the σ phase ($tP30$) (Figure 3-2) which is one of

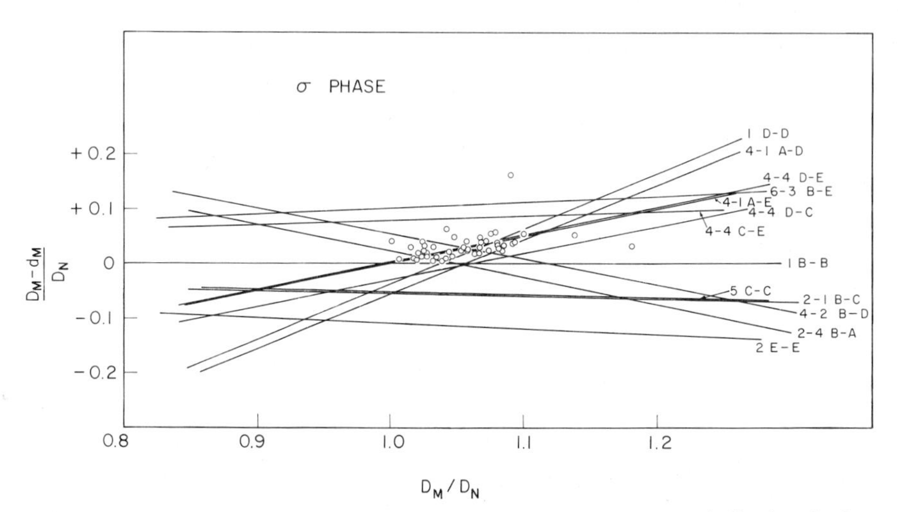

Figure 3-2. Near-neighbor diagram for the σ phase ($tP30$) structure, indicating the importance of the coordination factor. A to E represent the five different sitesets found in the structure in the generally accepted notation, where A refers to positions 2(a), B to 4(g), C to 8(i), D to 8(i), and E to 8(j). Constructed for the axial ratio and atomic parameters found in the Cr–Fe σ phase. The line for the single B–B contact is arbitrarily assumed for the zero value of the strain parameter. Numbers indicate numbers of D–D, A–D, etc. contacts. M and N refer to the two component atoms forming phases with the σ structure.

the Frank–Kasper structures containing interpenetrating triangulated coordination polyhedra, where the attainment of the high coordination is an obvious factor of importance. The n.n.d. for the β-W ($cP8$) structure (Figure 3-3), another Frank–Kasper phase, shows that the main body of the phases with the structure are located about the line for 12–4 M–N contacts† and between the lines for $8+2$ N–N contacts, in the most favor-

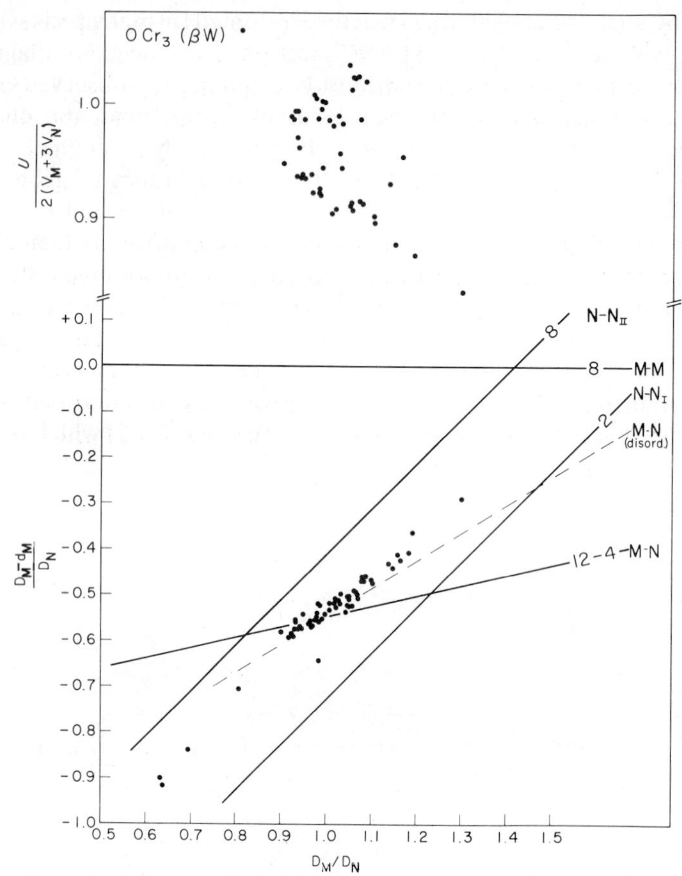

Figure 3-3. (Lower) Near-neighbor diagram for the β-W ($cP8$)(MN_3) structure indicating the importance of the coordination factor. Numbers indicate number of N–$N_{\rm I}$, N–$N_{\rm II}$, etc. contacts. (Upper) Comparison of observed unit cell volume, U, with the elemental atomic volumes of component atoms $2(V_M+3V_N)$ as a function of the radius ratios of the component atoms for phases with the β-W structure.

†12–4 M–N contacts indicates that M is surrounded by 12 N atoms and that N is surrounded by 4 M atoms.

able location to achieve the overall 12–14 coordination of the structure. Figure 3-4 shows an example where a possible strong coordination factor exerts no influence whatsoever on the occurrence of phases with the structure. Lines for various M–M, M–N, and N–N contacts in the α-$ThSi_2$ ($tI12$) structure severally intersect in a region covered by radius ratios from 1.8 to 2.1, and phases with these radius ratios and structural

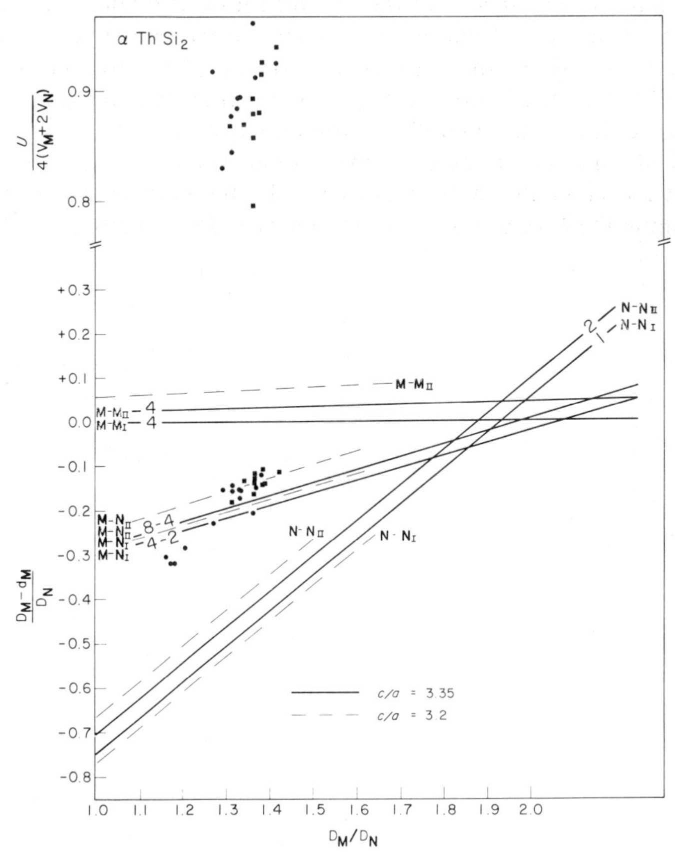

Figure 3-4. (Lower) Near-neighbor diagram for phases with the α-$ThSi_2(tI12)(MN_2)$ structure, indicating the importance of a chemical bond factor. Squares indicate phases with defect $ThSi_2$ structures. Constructed for the axial ratios indicated and the ideal value of the atomic parameter, $x_{Si} = \frac{5}{12}$. Numbers indicate the number of M–M_I, M–M_{II}, M–N_I, etc. contacts. (Upper) Comparison of unit cell volume, U, with the sum of the elemental atomic volumes of the component atoms $4(V_M + 2V_N)$ as a function of the radius ratio of the component atoms. Squares indicate phases with defect $ThSi_2$ structures.

dimensions to locate them in the region of intersection of these lines, would achieve the high coordination of 20–9. However, no observed phases with the α-$ThSi_2$ structure meet these conditions; instead they have radius ratios from 1.15 to 1.45 and are located about the lines for M–N contacts, showing that the M–N bond factor controls the structural dimensions and not the coordination factor.

The other geometrical effect that is recognized is regarded as the true *geometrical factor*, since here the attainment of high coordination stabilizes the occurrence of phases *over a wide range of radius ratios of the component atoms*. In this case points for the phases on the n.n.d. are expected to be distributed close to or between two or more lines of contacts giving high overall coordination, both about the crossing point of the lines (as in the coordination effect) and over a wide range of D_M/D_N values on either or both sides of it. In these circumstances, such as are found in the $MgCu_2$ ($cF24$) Laves phase (MN_2, Figure 3-5) when, for

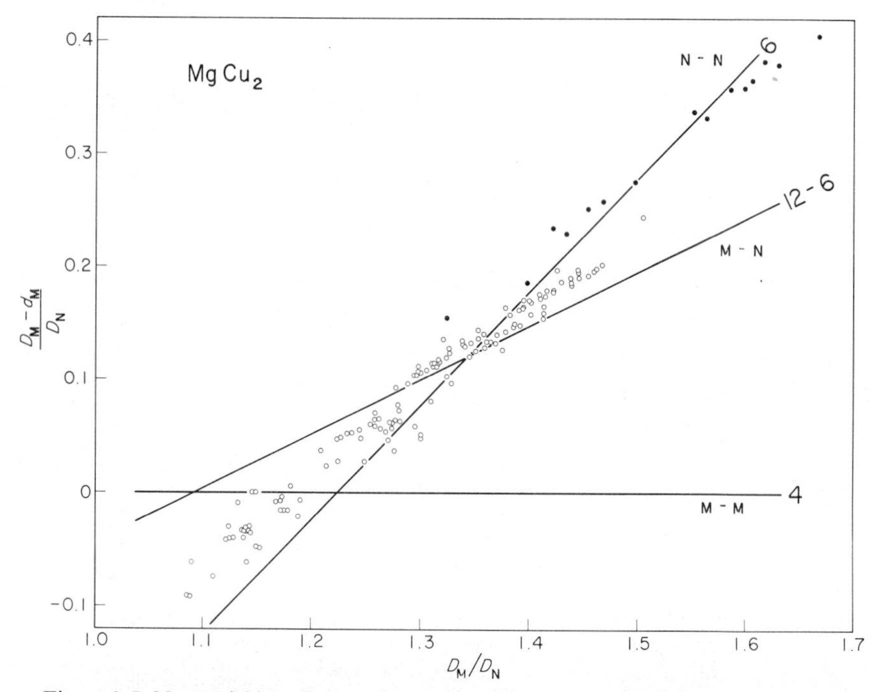

Figure 3-5. Near-neighbor diagram for phases with the $MgCu_2$ ($cF24$)(MN_2) structure showing the influence of the geometrical factor. Filled circles indicate phases with electronegativity difference of the component atoms, $\Delta x \geqslant 1$. Numbers indicate numbers of N–N, M–N, etc. contacts.

example, the radius ratio of the component atoms D_M/D_N is larger than 1.225, the energy gained by the formation of the 12–6 M–N and 6 N–N contacts, permits the compression of the 4 M–M contacts and results in the phase having a sufficiently low free energy to be stable over a wide range of radius ratios. Note that at the radius ratio 1.225 generally referred to in the literature as "ideal," a phase exactly forms 4 M–M and 6 N–N contacts if the value of the strain parameter, $(D_M - d_M)/D_N$, is correct. But this value does not appear to be much more significant† than a radius ratio of 1.347, where with the correct value of the strain parameter, a phase forms 6 N–N and 12–6 M–N contacts without strain, as well as the 4M–M contacts which are somewhat compressed. $SmRu_2$ with $D_M/D_N = 1.346$ is an example of such behavior as the following data show:

$D_M = 3.604$ Å $\quad d_{M\,\mathrm{obs}} = 3.252$ Å: M–M contact with compression
$D_N = 2.678$ Å $\quad d_{N\,\mathrm{obs}} = 2.680$ Å: N–N contact without compression
$D_{MN} = 3.142$ Å $\quad d_{MN\,\mathrm{obs}} = 3.141$ Å: M–N contact without compression

Note that in phases with the largest electronegativity differences (which are found at high values of the radius ratio), not only are the M–M contacts greatly compressed, but also the M–N contacts are compressed until N–N contacts definitely occur. This is indeed strange since it would be imagined that when the electronegativity difference is large, the structural dimensions would be such that points for the phases would lie on or close to the line for M–N contacts. Figure 3-5 shows one other interesting effect which tends to confirm the account of the geometrical factor given above. The so-called ideal radius ratio, 1.225, is that at which compression of M–M contacts to give M–N and N–N contacts changes to compression of N–N contacts to give M–M and M–N contacts on proceeding to lower radius ratios. There is, however, a notable asymmetry in the distribution of phases about this radius ratio in Figure 3-5; many phases occur at higher values extending up to 1.67, whereas fewer phases occur below and they extend only to a value of 1.08. This does not result from a lack of pairs of suitable elements to give lower radius ratios, but because at values lower than 1.225, two-thirds of the atoms in the structures (the N atoms) have to be compressed in order to achieve 12–16 coordination with the formation of M–M and M–N contacts also. Above 1.225 only one-third (the M atoms) have to be compressed in order to achieve a similar result giving structural stability. Clearly it takes more

†This is not quite true since at a radius ratio $R_M/R_N = 1.225$, hard spheres can form the $MgCu_2$ and ideal $MgZn_2$ structures and their volume ratio V_M/V_N has a value of 2.

energy to compress two-thirds of the atoms in the structure than is regained by achieving 12–16 coordination. Figure 3-5 shows that on going to radius ratios below 1.225, the compression of the N atoms is soon insufficient to establish M–M and M–N contacts, and structural stability disappears at radius ratios a little below 1.10. On the other hand, above 1.225 when only one-third of the atoms have to be compressed in order to attain structural stability resulting from 12–16 coordination, the energy balance is favorable, and $MgCu_2$ phases are found over a wide range of radius ratios extending to 1.67. Data for the hexagonal $MgZn_2$ (*hP*12) phase are similar to those for $MgCu_2$ as shown in Figure 3-1.

In superstructures formed by ordering close packed solid solutions, the geometrical factor must play an important role. Although n.n.d. give no information on the preference of ordered structures to disordered solid solutions, they do demonstrate the occurrence of the geometrical factor, since the ordered phases over a range of radius ratios are distributed between lines for contacts giving the 12–12 coordination. Such is the situation for phases with the $AuCu_3$ (*cP*4) structure shown in Figure 3-6.

Thus, from the use of near-neighbor diagrams and in the light of experience gained in their study, it is possible to understand the influence that geometrical effects can exert on the structural dimensions of phases and on their range of occurrence in terms of the different sizes of the component atoms, and so in some sense to make a quantitative appraisal of the geometrical factor. Perhaps the most revealing feature of the near-neighbor diagram is in relation to atomic size. Once it is established that two elements form a phase with a given structure, the value of the radius ratio is fixed, but the value of the strain parameter is entirely free, being determined by the unit cell dimensions (and atomic parameters) adopted by the phase. Indeed, phases are found to adopt cell dimensions and structural parameters to give a value of the strain parameter which satisfies a chemical bond or geometrical factor that is responsible for the structural stability. In so doing, various atomic contacts may be compressed compared to the assumed elemental atomic diameters (see, e.g. Figures 3-1, 3-2, 3-4, 3-5, and 3-8). The apparent facility with which compression of contacts to ligands in certain directions can occur is one of the characteristic features of metals. Of course the radius ratio and the actual value of the strain parameter depend on the assumption of an atomic diameter for each of the component atoms. These values are derived from the structures of the elements[4] and depend on the numbers of ligands and valence electrons (see p. 146). Nevertheless, whereas details may depend on the relative values of the atomic diameters assumed (movements of

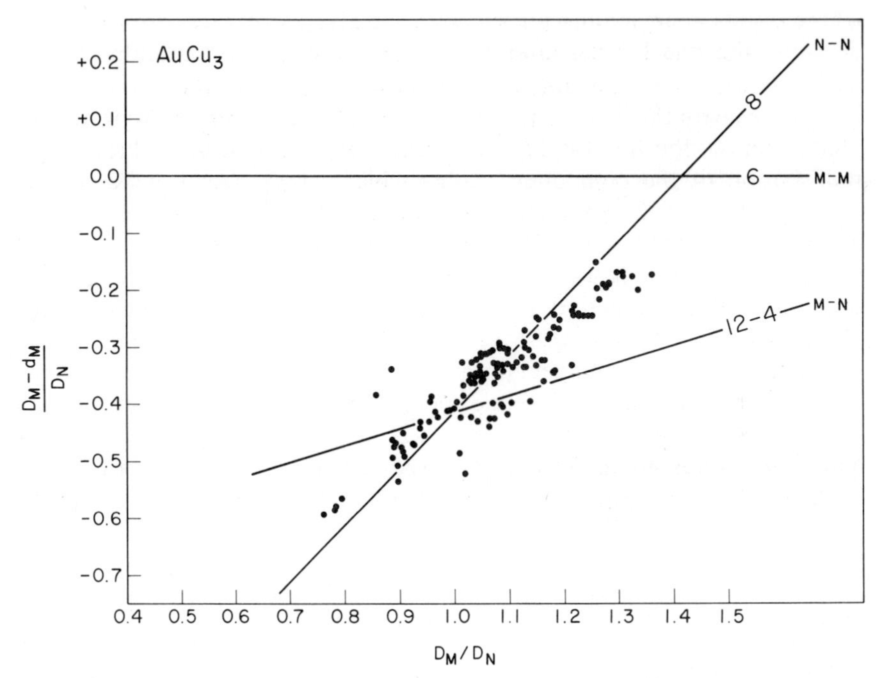

Figure 3-6. Near-neighbor diagram for phases with the $AuCu_3$ (MN_3) ($cP4$) structure. Numbers indicate number of N–N, M–M, etc. contacts.

the radius ratio and strain parameter on the n.n.d. are coupled), the main conclusions regarding the compressibility of metal atoms in certain directions remain unaffected. The influence of the relative sizes of the component atoms and their geometrical arrangement on the unit cell dimensions of specific structures are discussed in Chapter 4.

CHEMICAL BOND FACTOR

A chemical bond factor can be said to control a structure when a particular set of chemical bonds determines the unit cell dimensions (and atomic parameters) of phases with the structure, so that the observed interatomic distances equal those expected for the particular chemical bond. This control of the structural dimensions may occur regardless of the close approach of certain atoms in the structure as in the case of the $AsNa_3$ structure discussed below. The existence of a chemical bond factor is particularly well demonstrated by near-neighbor diagrams (p. 53) where points representing phases with the given structures are found to lie along the line for the interatomic separation of the particular bond. Thus Figure 3-7 demonstrates the well-known *M–N* bond factor controlling phases with the sphalerite structure (*cF*8). Points representing actual phases follow the line for *M–N* contacts regardless of the value of the radius ratio of the component atoms which varies over a wide range, and regardless of the large electronegativity difference of the component atoms in some compounds which may be expected to modify the assumed values of the atomic radii. Figure 3-8 shows the importance of the 6–3 *M–N*(2) coordination in controlling the structural dimensions of phases with the $AsNa_3$ (*hP*8) type structure, despite the apparent compression of the atoms along numerous other *M–N* and *N–N* contacts. The 6 As–Na(2) bonds and 2 As–Na(2) bonds which are slightly compressed, surround As by two tetrahedra of Na(2) atoms and resonance of tetrahedral sp^3 bonds about As may be expected to occur in similar fashion to that found in the semiconductors $MgSi_2$ with the antifluorite structure

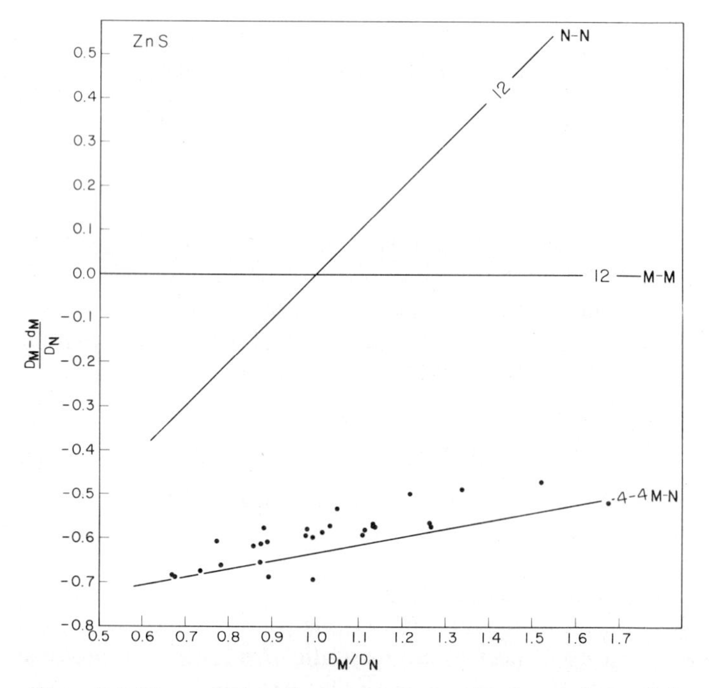

Figure 3-7. Near-neighbor diagram for phases with the sphalerite, ZnS (*cF*8) (*MN*) structure, indicating the importance of the chemical bond factor. Numbers indicate number of *N–N*, *M–M*, etc. contacts.

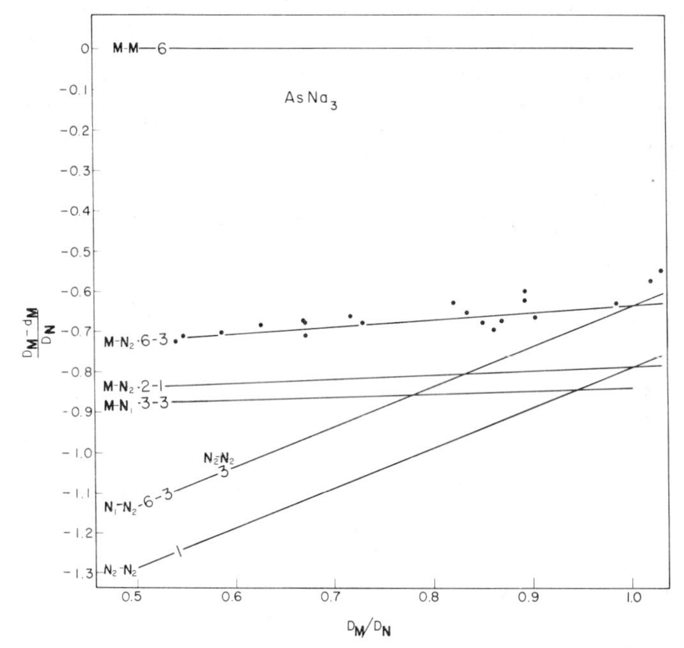

Figure 3-8. Near-neighbor diagram for phases with the $AsNa_3$ (*hP*8) (MN_3) structure, indicating the importance of the chemical bond factor. Constructed for the axial ratio $c/a = 1.78$ and the ideal value of the atomic parameter $x_{Na} = 5/12$. Numbers indicate number of M–N_2, M–N_1, etc. contacts.

(*cF*12) and $BiLi_3$ (*cF*16). Indeed, this bonding had already been recognized as important in some compounds with the $AsNa_3$ structure following reports that they exhibited semiconducting properties. Most significant, however, is that it accounts for the very constant value of the unit cell axial ratio ($c/a = 1.79 \pm 0.03$) found in some twenty phases with this structure, since the rigidity of strong tetrahedral sp^3 bonds does not permit much variation in bond angles. This is seen by comparing the constancy of the axial ratio of phases with the wurtzite structure (*hP*4), where the coordination is tetrahedral, with the great range of values that is observed in the NiAs structure, where the coordination is trigonal prismatic and octahedral. A M–N chemical bond factor also appears to be important in the CdI_2 structure (*hP*3), since points representing phases with this structure lie along the line for M–N contacts and follow the position of this line as it varies on the n.n.d. with the value of the axial ratio of the hexagonal cell, when it departs from the ideal value of 1.63.

Thus, phases having axial ratios close to 1.33 lie close to the *M–N* line for this axial ratio, whereas those with c/a values close to 1·75 lie near the *M–N* line for that ratio (Figure 3-9). The α-$ThSi_2$ (*tI*12) structure already referred to (p. 57) provides a further example of the definite

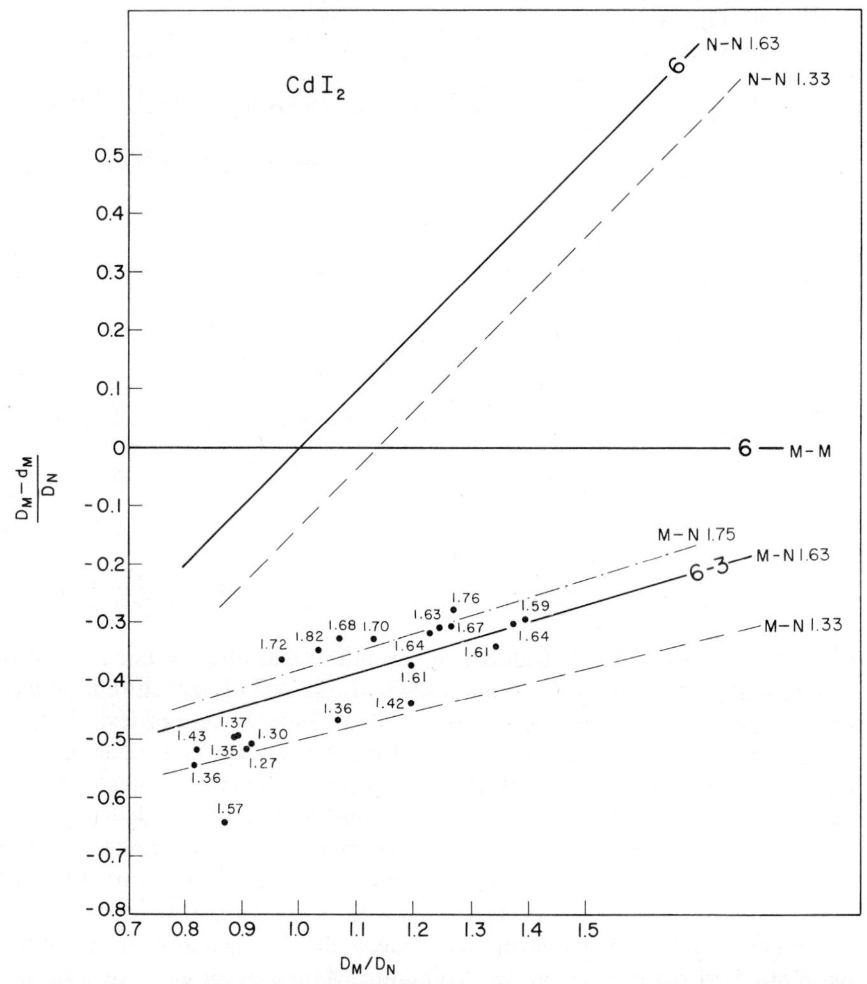

Figure 3-9. Near-neighbor diagram for phases with the CdI_2 (*hP*3)(MN_2) structure. Full lines for axial ratio, $c/a = 1.63$, broken lines for $c/a = 1.33$, and chain line for 6–3 *M–N* contacts for $c/a = 1.75$. Constructed for $z_I = 0.25$. Numbers on lines indicate numbers of *N–N*, *M–M*, etc. contacts. Small numbers give the axial ratios, c/a, of phases with the CdI_2 structure.

occurrence of a chemical bond factor. Here (Figure 3-4, p. 57), known phases are distributed about the lines for 8–4 and 4–2 M–N contacts, showing that these are the dominant feature controlling the structural dimensions. However, the points do not follow the slope of these M–N lines, but they lie parallel to the lines for $2+1$ N–N contacts indicating a secondary importance of the three-dimensional, three-connected net of N atoms that runs throughout the structure.

Clear evidence of the structural influence of bond factors can be provided by n.n.d. In cases of rather more complex structures with many different interatomic contacts, or of structures with very many known phases all of which might not occur for the same reasons, interpretation may be somewhat less certain, although the general principle controlling the structural dimensions should be apparent. Thus, comparing Figures 3-10 and 3-11, it is clear that the Si_3Mn_5 ($hP16$) structure forms for geometrical reasons since the phases are distributed between the lines for 6–4 and 6 N–N contacts and 1–1, 4–6, and 2–2 M–N contacts giving high overall coordination, whereas the dominant role of the M–N bond factor in the NaCl ($cF8$) structure is apparent despite the rather wide scatter of phases about the M–N contact line. Finally, it may be noted that the distribution of phases on n.n.d. of structures of MN_x phases where $x > 3$, naturally follows the lines for N–N contacts, because this component constitutes more than 75% of the atoms in the structure.

Two different situations can be distinguished in considering chemical bond factors. If the bonds have a high ionic character, then they are largely nondirectional and the only condition to be satisfied apart from interatomic distance, is for the larger anions to form as large a symmetrical coordination polyhedron as possible, subject to the limitation provided by relative cation and anion sizes, since the anions cannot come in material contact. When the chemical bonds are largely covalent, their directional character must be suitably accounted for by the coordination in the structural arrangement. It is for this reason that so many semiconductors adopt structures based on a close packed array of the anions, because the tetrahedral and octahedral surroundings that these provide for the cations satisfy the symmetry requirements for cation–anion bonds formed by s and p valence electrons.

The characteristic of metallic phases generally is that the bonds between metal atoms are essentially nondirectional so that high symmetrical coordination is found according to Laves' principles. In certain circumstances however, such as in some phases formed with the lighter elements or the transition metals, or when the electronegativity difference between the components is not small, evidence for the occurrence

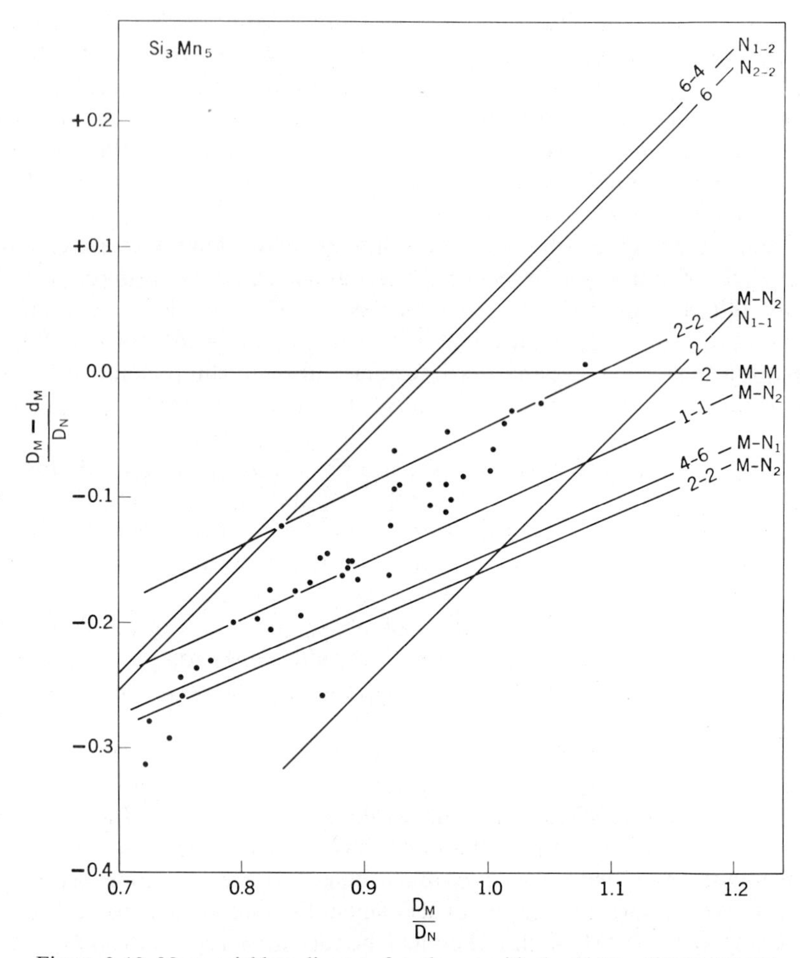

Figure 3-10. Near-neighbor diagram for phases with the Si_3Mn_5 ($hP16$)(M_3N_5) structure. Constructed for the axial ratio $c/a = 0.70$ and atomic parameters $x_{Si} = 0.60$, $x_{Mn} = 0.236$. Numbers indicate number of N–N, M–N, etc. contacts.

of directional chemical bonds may be detected through the appearance of specific coordination features, or of irregular coordination polyhedra. Thus in the metallic AlB_2 ($hP3$) phases the appearance of the graphite-like net of B atoms, or in the metallic $ThSi_2$ structure the open three-connected net of Si atoms indicates directional character in the bonding of these components.

In the structures of metals there may be some problem in distinguishing between contacts that occur for geometrical reasons between atoms

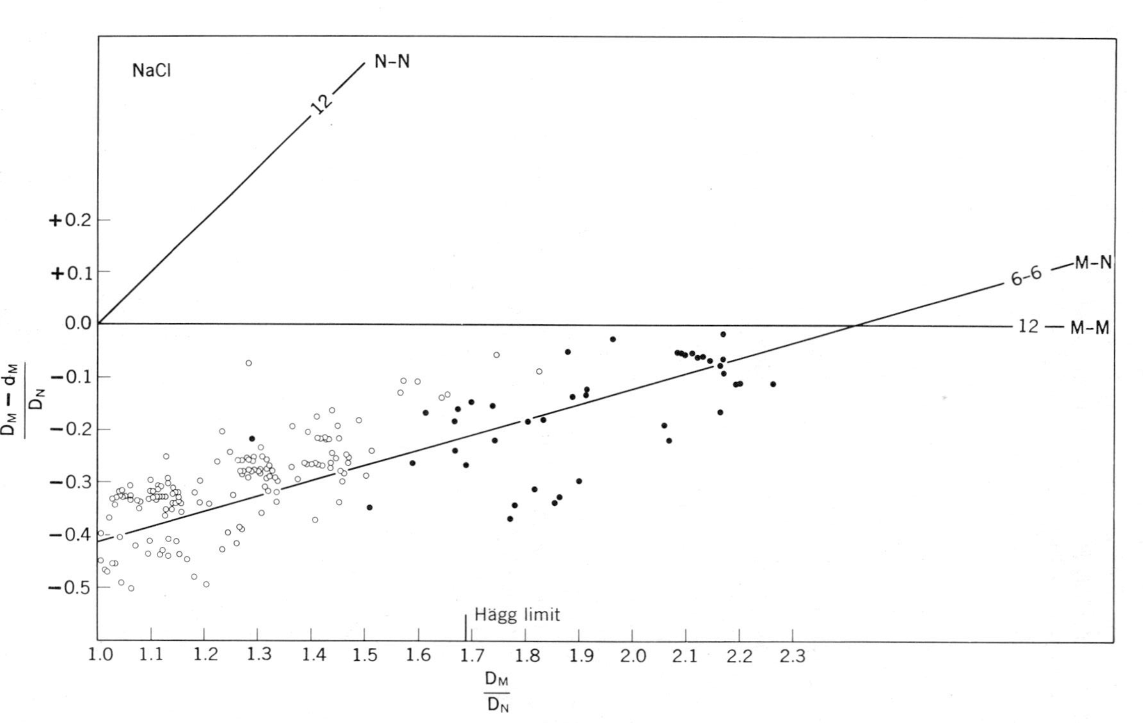

Figure 3-11. Near-neighbor diagram for substances with the NaCl($cF8$) (MN) structure. Oxides and hydrides are omitted. Dark circles represent carbides and nitrides. Numbers indicate number of N–N, M–N, etc. contacts.

without directional bonding characteristics, and those between atoms that are connected by strong directional bonding, so that an active bond factor controls the structure. The general principle governing these considerations is that as the electronegativity differences between the components gets larger, and as the average atomic number of the components becomes smaller, so the importance of the chemical bond factor is likely to increase. Chemical bond factors control structures between components for which there is a large electrochemical factor, and when the effective number of valence electrons per atom (see p. 109) exceeds 2 (i.e., is 3 or 4), especially if one component attains a filled valence subshell, either by electron sharing or transfer.

ELECTROCHEMICAL FACTOR: ELECTRONEGATIVITY

When one component in a binary alloy is very electropositive relative to the other, there is a strong tendency for them to form compounds of considerable stability in which valence rules are satisfied. Such alloys are said to exhibit a strong electrochemical factor and this is the strongest effect in determining the constitution of alloys, and one which dominates all other effects such as energy band or geometrical factors. When the electrochemical factor is strong, both terminal solid solubilities and the range of composition over which intermediate phases are homogeneous, tend to be very restricted; random atomic arrangements have little stability and there is a corresponding increase in stability of ordered structural arrangements in which valence rules are satisfied. The size of the electrochemical factor is determined by the difference between the electronegativities of the component atoms.

The electronegativity value, introduced by Pauling, is itself hard to define. It can best be regarded as a parameter expressing the tendency of an atom to attract electrons to itself in a particular solid; it is therefore a measure relative to the other atoms in the solid, and electronegativity differences of two components M and N, $\Delta x = |x_M - x_N|$, are normally discussed. Perhaps the most fundamental expression of the electronegativity value is Mulliken's[5] representation of it as the arithmetic mean of the first ionization energy and the electron affinity of the atom, expressed in volts. Pauling,[6] however, derived the first electronegativity values from thermochemical data on the basis of a relationship between the bond energy of the compound E_{MN}, the homopolar bond energies of the elements E_{MM}, E_{NN}, and x, k being a proportionality constant,

$$E_{MN} = \tfrac{1}{2}(E_{MM} + E_{NN}) + k(x_M - x_N)^2 \qquad (1)$$

assuming (later) a value of 2.05 for the electronegativity of hydrogen.

Mulliken[5] gave a theoretical justification for Pauling's empirical equation (1) using the LCAO molecular orbital method, but pointed out that the x values are "inherently incapable of representing exactly definable properties of the atoms"; they are thus rough averages, and the value x_M is not in general entirely independent of the properties of atom N.

An element does not have a unique electronegativity value, since this depends on its oxidation state, x increasing rapidly with the valence of the ion (about 0.3 to 0.5 per unit valence increase for the cation or decrease for the anion). On the other hand the electronegativity values depend very little on bond type and details of crystal structure, coordination, etc. For this reason they are a general parameter of atoms and so are useful for discussing and comparing the properties of solids.

There have been many papers written on the subject of electronegativity and values have been derived from several physical properties. For further information, reference can be made to Pauling's book[6] or to Gordy and Thomas,[7] who have reviewed data on electronegativity values and given selected values which are listed in Table 3-1 and compared with the thermochemical values of Pauling or of Haissinskij.[8] Figure 3-12 shows the variation of electronegativity values according to the periodic arrangement of the elements.

In addition to depending on the valence state of an atom, the electronegativity value also depends on the charge on the atom, and conversely the charge separation in a compound or alloy depends on the electronegativity difference of the component atoms. Indeed, it may be imagined that there is a redistribution of atomic charge until the electronegativity difference has been reduced to zero. Thus any attempt to make rigorous use of electronegativity values is likely to be fraught with corrections and adjustments sufficient to reduce the exercise to a state of impotence. Nevertheless, electronegativity values taken at some face value provide a useful guide for estimating how strong the electrochemical factor may be in a given system of two components, and in assessing the related problem of the charge distribution, which is the difference between the nuclear charge and that of the valence electrons surrounding or associated with the nucleus. However, specification of the atomic boundaries introduces difficulties of definition. The problem affects metals in questions of electron screening, semiconductors where it influences energy band gaps and charge-carrier mobilities, energy band calculations for two-component systems in the derivation of the crystal potential, and chemical bond discussions where it appears in the extent of the ionic contribution to the bonding.

The influence that a large electronegativity difference has on the constitution of an alloy system depends on the effective number of valence

Table 3-1 Electronegativity Values of the Elements

		Electronegativity		
		Bond energies		Selected value
Element	Atomic Number	Pauling[6] or Haissinskij[8]		Gordy & Thomas[7]
H	1		2.05	2.1_5
Li	3		1.0	0.95
Na	11		0.9	0.9
K	19		0.8	0.80
Rb	37		0.8	0.8
Cs	55		0.7	0.75
Fa	87		~0.7	0.7
Be	4		1.5	1.5
Mg	12		1.2	1.2
Ca	20		1.0	1.0
Sr	38		1.0	1.0
Ba	56		0.85 (0.9)	0.9
Ra	88		~0.8	0.9
B	5		2.0	2.0
Al	13		1.5	1.5
Sc	21		1.3	1.3
Y	39		1.2 (1.3)	1.2
La	57		1.1	1.1
Ac	89		1.0	1.1
Ce	58	Ce^{III}	1.05	1.1
Pr	59		1.1	1.1
Nd	60			~1.2
Pm	61			~1.2
Sm	62			~1.2
Eu	63			~1.1
Gd	64			~1.2
Tb	65			~1.2
Dy	66			~1.2
Ho	67			~1.2
Er	68			~1.2
Tm	69			~1.2
Yb	70			~1.1
Lu	71			~1.2
Ti	22		1.6	1.6
Zr	40		1.4 (1.6)	1.5
Hf	72		~1.3	1.4
Th	90		1.1	1.0
Pa	91	Pa^{III}	~1.4	1.3
		Pa^{V}		1.7
U	92	U^{IV}	1.3	1.4

Table 3-1 (*continued*)

Element	Atomic Number	Electronegativity: Bond energies, Pauling[6] or Haissinskij[8]		Electronegativity: Selected value, Gordy & Thomas[7]
		U^{VI}		1.9
Np	93			~1.1
Pu	94			~1.3
Am	95			~1.3
Cm	96			~1.3
Bk	97			~1.3
Cf	98			~1.3
V	23	V^{III}	1.35	1.4
		V^{IV}	1.65	1.7
		V^{V}	~1.8	1.9
Nb	41		~1.6	1.7
Ta	73	Ta^{III}	~1.4	1.3
Cr	24	Cr^{II}	1.5	1.4
		Cr^{III}	1.6	1.6
		Cr^{IV}	~2.1	2.1
Mo	42	Mo^{IV}	~1.6	1.6
		Mo^{VI}	~2.1	
W	74	W^{IV}	~1.6	1.6
		W^{VI}	~2.1	2.0
Mn	25	Mn^{II}	1.4	1.4
		Mn^{III}	~1.5	1.5
		Mn^{VII}	~2.3	2.5
Tc	43	Tc^{V}		1.9
		Tc^{VII}		2.3
Re	75	Re^{V}		1.8
		Re^{VII}		2.2
Fe	26	Fe^{II}	1.65	1.7
		Fe^{III}	1.8	1.8
Ru	44	Ru^{III}	2.05	2.0
Os	76		2.1	2.0
Co	27		1.7	1.7
Rh	45		2.1	2.1
Ir	77		2.1	2.1
Ni	28		1.7	1.8
Pd	46		2.0	2.0
Pt	78		2.1	2.1
Cu	29	Cu^{I}	1.8	1.8
		Cu^{II}	2.0	2.0
Ag	47		1.8	1.8

Table 3-1 (*continued*)

		Electronegativity		
	Atomic	Bond energies		Selected value
Element	Number	Pauling[6] or Haissinskij[8]		Gordy & Thomas[7]
Au	79		2.3	2.3
Zn	30	Zn^{II}	1.5	1.5
Cd	48	Cd^{II}	1.5	1.5
Hg	80	Hg^{I}	1.8	
		Hg^{II}	1.9	1.8
Ga	31		1.6	1.5
In	49		1.6	1.5
Tl	81	Tl^{I}	1.5	1.5
		Tl^{III}	1.9	1.9
C	6		2.5	2.5
Si	14		1.8	1.8
Ge	32		1.7	1.8
Sn	50	Sn^{II}	1.65 (1.7)	1.7
		Sn^{IV}	1.8	1.8
Pb	82	Pb^{II}	1.6	1.6
		Pb^{IV}	1.8	1.8
N	7		3.0	3.0
P	15		2.1	2.1
As	33		2.0	2.0
Sb	51	Sb^{III}	1.8 (1.8)	1.8
		Sb^{V}	2.1	
Bi	83		1.8	1.8
O	8		3.5	3.5
S	16		2.5	2.5
Se	34		2.3 (2.4)	2.4
Te	52		2.1	2.1
Po	84		2.0	2.0
F	9		4.0	3.9_5
Cl	17		3.0	3.0
Br	35		2.8	2.8
I	53		2.6 (2.5)	2.5_5

electrons. If the electron concentration is high so that one component can attain a filled valence shell by electron sharing or transfer, it results in the formation of compounds of great stability which generally have narrow ranges of composition. Terminal solid solutions are generally very restricted in this case. If on the other hand the effective electron concen-

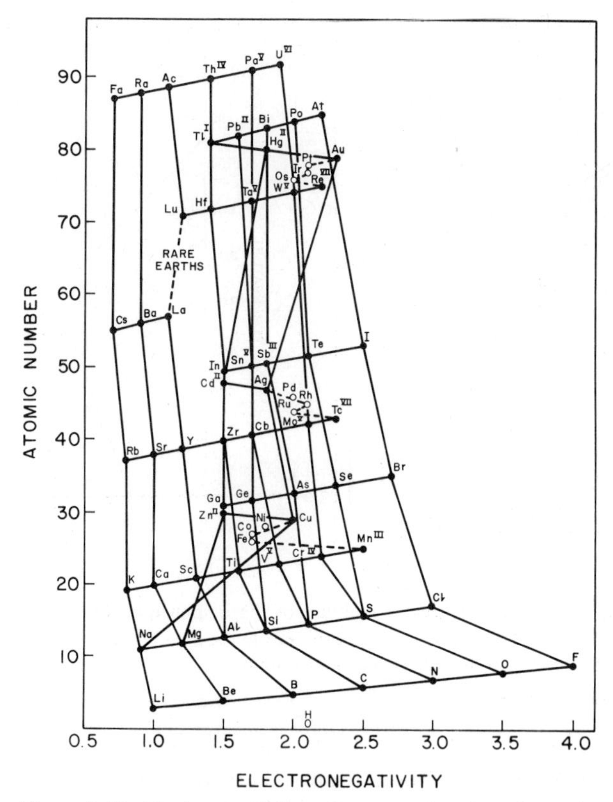

Figure 3-12. Electronegativity values as a function of atomic number. (Gordy and Thomas [7].)

tration is low and no intermediate "*compounds*" occur, an increase in electronegativity difference may increase the stability and range of homogeneity of intermediate phases as in the example shown in Figure 3-13 where an increase of Δx by 0.5 greatly increases the temperature of formation and width of the homogeneous β' phase. A large electronegativity difference also restricts terminal solid solutions in alloys of low electron concentration. The combined effects of electronegativity difference and relative atomic size on terminal solid solubility are further discussed on p. 76.

SIZE FACTOR

In the present context a size factor may be defined as the influence exerted on alloy constitution, structure, or structural dimensions by the

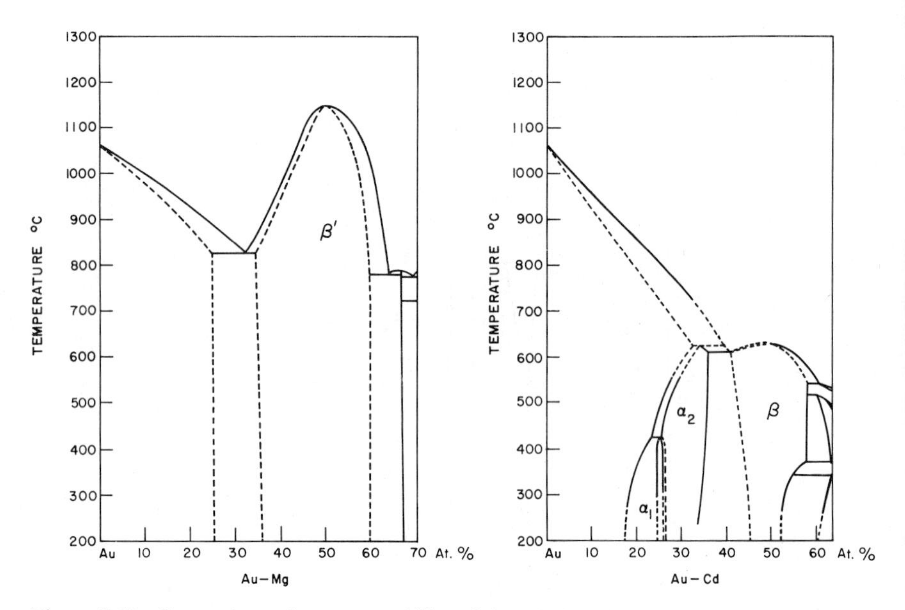

Figure 3-13. Comparison of greater stability of the β' phase in the Au–Mg system (left) than in the Au–Cd system (right), which results from the greater electronegativity difference between the component atoms in the Au–Mg system.

relative sizes of the component atoms. The sizes of atoms, themselves, in various situations are the subject of Chapter 4, and it has already been demonstrated on pp. 52–68 how geometrical and chemical bond factors can determine the interatomic distances between atoms in structures that they control. In the context of near-neighbor diagrams, the size factor appears as the radius ratio which determines whether a particular coordination can be realized in a given structure type, and it indicates to some extent the amount of strain that may be involved in satisfying a geometrical or chemical bond factor for a particular pair of elements. Generally, however, relative atomic size has little significance in relation to the chemical bond factor, except in ionic structures where the relative sizes of the cations and anions prescribe the maximum number of anions that can surround the (smaller) cations. Atomic size (interatomic distance) may also be influenced by the energy band factor, as indicated by the axial ratios and interatomic distances of the hexagonal metals (p. 115), by the influence of electron concentration on the axial ratios of hexagonal phases (p. 121), and by the interatomic distances in long-period superstructures.

The undisputed province of the size factor in controlling the structures of alloys is somewhat limited, being confined to determining permissible coordinations, the extent of terminal solid solutions, and certain geometrical influences. Nevertheless, even in the structures of metals, the importance of relative atomic size in determining permitted or preferred coordination must not be underestimated. The distribution of phases as a function of radius ratio is, for example, shown in Figure 3-14 for the structures of MN_x ($x > 10$) phases containing one large component atom, and it is apparent that relative atomic size does influence the choice of structure. Relative atomic size is also the controlling factor for the occurrence of Hägg interstitial phase structures (c.p. cubic, c.p. hexagonal, or simple hexagonal arrangement of the metal atoms) formed by alloys of the metalloids B, C, and N and the transition metals, when the metalloid-metal radius ratio does not exceed 0.59 (see p. 294). The distribution of the component atoms on the various equivalent sites in transition metal phases of the σ, P, R, μ, and α-Mn types, where only coordination or geometrical factors are important, is determined mainly by the relative atomic sizes; the largest atoms always prefer sites with CN 16 and 15 and the smallest atoms take sites with CN 12.

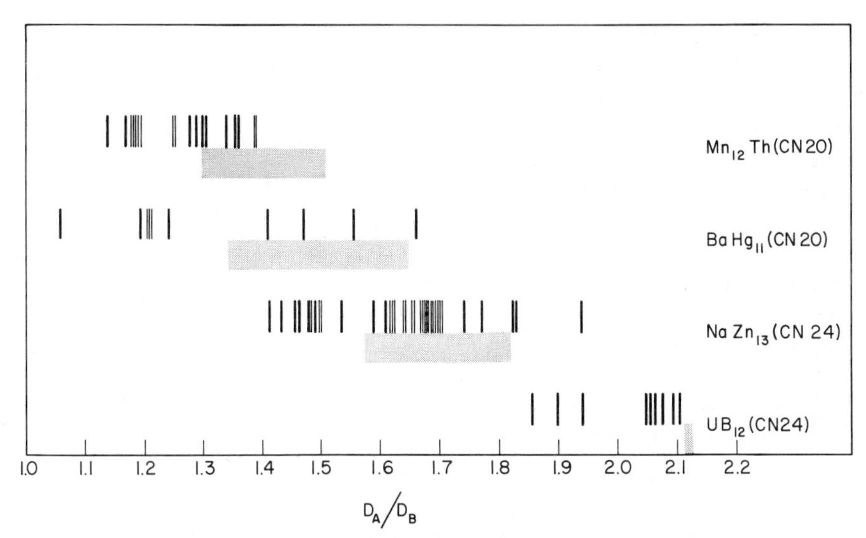

Figure 3-14. Distribution of phases with the $ThMn_{12}$ ($tI26$); $BaHg_{11}$ ($cP36$), $NaZn_{13}$ ($cF112$), and UB_{12} ($cF52$) structures according to radius ratio of the component atoms. See text. The shaded portion under each subject indicates the range of radius ratios of the component atoms, which would satisfy a high coordination factor in the structure.

Size Factor and Terminal Solid Solubility

Hume-Rothery and coworkers[9] first pointed out that if the closest distance of approach in the elemental structures of a solvent and solute differed by more than 13 or 14%, the terminal solid solubility would be very restricted, whereas if it was less, extended solutions may form. This has come to be recognized as the 15% rule with respect to the atomic diameters for the coordination concerned. Although the original proposal probably gives better agreement with experimental observations, the latter is generally more convenient. Further restrictive factors such as a large valency difference and a large electronegativity difference were also recognized as acting to reduce terminal solubility even when the size factor was not exceeded, and these are, for example, discussed by Hume-Rothery and Raynor.[10] The size factor rule is recognized as a negative rule. Certainly when the 15% limit is exceeded, it is very unlikely that extended solid solution can occur, but when the difference is less it is only permissive; wide solid solutions still may not form. Darken and Gurry[11] took the size and electronegativity factors into account simultaneously by plotting diagrams of electronegativity *vs.* atomic diameter for various elements, and drawing an ellipse about the element selected as solvent with diameters ± 0.4 units of electronegativity difference and $\pm 15\%$ size difference for solvent and solute. Solutes falling within the elipse may form extended solid solutions with the solvent, those falling without should form only very restricted solid solutions. Waber *et al.*[12] have applied the Darken-Gurry method to the study of 1455 terminal solid solutions in 850 alloy systems where data on terminal solid solubility were available. The plots were made for the CN(12) radii and electronegativity values of Teatum *et al.*[4] and 5 at.% was arbitrarily chosen as the boundary between restricted and extended solid solubility. Although not explicitly stated, 5% appears to refer to the maximum solid solubility at any temperature and not the solubility at room temperature. Of systems predicted to have extended solubility 61.7% were correct, and 84.8% of predictions of restricted solid solubility were correct, the overall percentage of correct predictions on the basis of their criterion was 76.6%, which seems rather reasonable. By comparison, the size-factor-alone criterion did much better at predicting restricted solid solubility (90.3%), but was only about 50% reliable in predicting extended solubility. Details of their results are summarized in Table 3-2 and, as an example, their Darken-Gurry plot for silver is shown in Figure 3-15. Regarding common solvents, it is worth noting that for aluminum, 3 out of 10 correct predictions of wide solubility were made, and 36 out of 38 correct predictions of restricted solid solubility. For iron the figures are respectively 19 out of 20 and 28 out of 36;

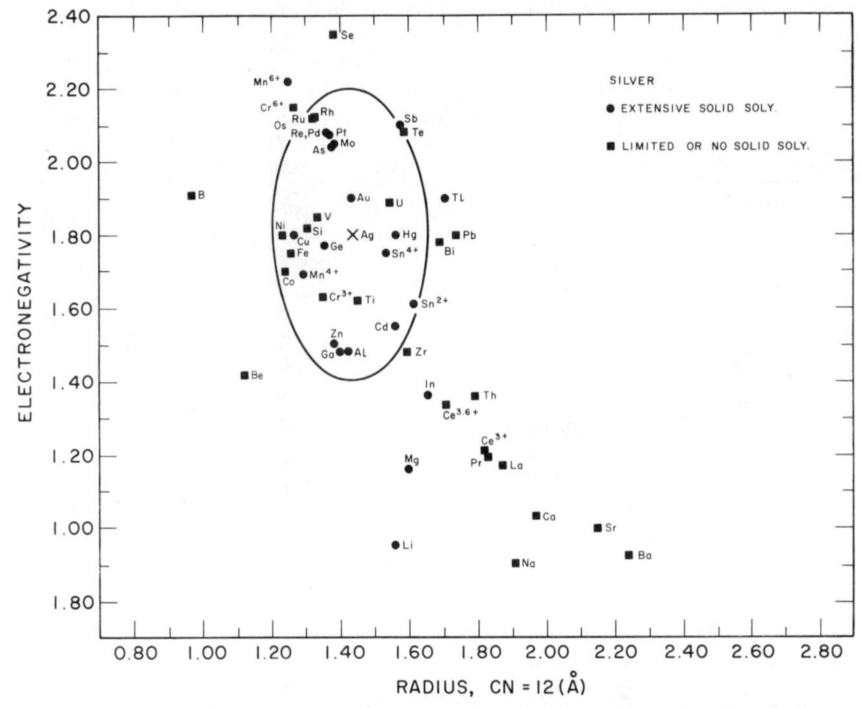

Figure 3-15. Darken–Gurry plot for various solutes dissolved in silver. The ellipse indicates the limiting range of electronegativity differences and atomic radii for the expected occurence of extensive solid solutions in silver.[12] Circles indicate alloys in which extensive solid solutions are found; squares indicate alloys in which limited or no solid solubility is found.

for copper 11 out of 15 and 20 out of 28; for gold 12 out of 22 and 14 out of 22; for tin 4 out of 15 and 26 out of 27; and for zinc, surprisingly, 0 out of 10 and 26 out of 26!†

Figures such as these justify the point made by Hume-Rothery and co-workers, that terminal solid solubility depends mainly on the relative sizes of solute and solvent so that, if the 15% difference limit is exceeded, wide solid solubility is very unlikely. However, when it is less than 15%, several other factors may also intervene and prevent the expected wide solid solubility, so that the rule is permissive below 15% and compulsive in its predictions above that limit. Hume-Rothery,[13] considering common solvents and solutes, points out that expressing the size factor

†The data for zinc tend to confirm the point made on p. 142 that the size of Zn determined from the elemental structure where the axial ratio is greatly distorted from the ideal value of 1.63, may not be appropriate for Zn in its alloys.

Table 3-2 Results of the Darken-Gurry Analysis for 62 Elements[12]

	No. of Systems in Which Solute Considered Soluble			No. of Systems in Which Solute Considered Insoluble			Total Number		
Solvent	Pred	Obs	Pct	Pred	Obs	Pct	Pred	Obs	Pct
Li	1	1	100.0	13	11	84.5	14	12	85.6
Be	—	—	—	17	16	94.1	17	16	94.1
B	—	—	—	2	2	100.0	2	2	100.0
C	—	—	—	3	3	100.0	3	3	100.0
Na	1	0	0.0	19	19	100.0	20	19	95.0
Mg	8	3	37.5	27	24	88.9	35	27	77.2
Al	10	3	30.0	38	36	94.7	48	39	83.1
Si	19	1	5.3	17	16	94.1	36	17	47.2
P	1	1	100.0	—	—	—	1	1	100.0
S	1	1	100.0	12	11	91.7	13	12	92.3
K	1	1	100.0	11	10	91.0	12	11	91.7
Ca	4	2	50.0	15	15	100.0	19	17	89.5
Ti	15	15	100.0	17	9	52.9	32	24	75.0
V	12	11	91.7	11	7	63.6	23	18	78.2
Cr	14	13	92.8	18	16	88.8	32	29	90.6
Mn	13	13	100.0	18	14	77.7	31	27	87.1
Fe	20	19	95.0	36	28	77.8	56	47	83.0
Co	10	10	100.0	32	21	65.7	42	31	73.8
Ni	12	11	91.6	31	18	58.1	43	29	67.5
Cu	15	11	73.3	28	20	71.4	43	31	72.1
Zn	10	0	0.0	26	26	100.0	36	26	72.2
Ga	6	0	0.0	17	17	100.0	23	17	73.9
Ge	11	1	9.1	11	11	100.0	22	12	54.5
As	11	4	36.4	9	9	100.0	20	13	65.0
Se	1	1	100.0	12	11	91.7	13	12	92.3
Rb	2	2	100.0	2	2	100.0	4	4	100.0
Sr	2	2	100.0	2	1	50.0	4	3	75.0
Y	5	5	100.0	15	15	100.0	20	20	100.0
Zr	7	6	85.7	19	10	52.6	26	16	61.5
Nb	9	9	100.0	7	6	85.7	16	15	94.2
Mo	10	7	70.0	8	5	65.5	18	12	66.7
Ru	5	5	100.0	3	1	33.3	8	6	75.0
Rh	10	8	80.0	1	0	0.0	11	8	72.7
Pd	13	11	84.6	13	3	23.1	26	14	53.8
Ag	24	14	58.3	25	20	80.0	49	34	69.4
Cd	9	3	33.3	20	20	100.0	29	23	79.3
In	4	2	50.0	17	12	70.6	21	14	66.7
Sn	15	4	26.7	27	26	96.3	42	30	71.4
Sb	10	3	30.0	25	25	100.0	35	28	80.0
Te	8	0	0.0	12	10	83.3	20	10	50.0
Cs	2	2	100.0	4	4	100.0	6	6	100.0
Ba	2	2	100.0	8	8	100.0	10	10	100.0

Table 3-2 (*continued*)

Solvent	No. of Systems in Which Solute Considered Soluble			No. of Systems in Which Solute Considered Insoluble			Total Number		
	Pred	Obs	Pct	Pred	Obs	Pct	Pred	Obs	Pct
La	6	6	100.0	15	15	100.0	21	21	100.0
Ce	8	7	87.5	15	15	100.0	23	22	95.6
Pr	1	0	0.0	9	9	100.0	10	9	90.0
Nd	1	1	100.0	—	—	—	1	1	100.0
Gd	2	2	100.0	—	—	—	2	2	100.0
Hf	4	3	75.0	2	2	100.0	6	5	83.3
Ta	9	6	66.7	4	3	75.0	13	9	69.2
W	8	3	37.6	8	6	75.0	16	9	57.2
Re	6	6	100.0	4	3	75.0	10	9	90.0
Os	7	7	100.0	1	0	0.0	8	7	87.5
Ir	5	5	100.0	1	0	0.0	6	5	83.3
Pt	18	17	94.3	10	6	60.0	28	23	82.1
Au	22	12	54.5	22	14	63.6	44	26	59.1
Hg	14	0	0.0	18	18	100.0	32	18	56.3
Tl	6	5	83.3	27	25	92.6	33	30	90.8
Pb	9	6	66.7	31	26	83.9	40	32	80.0
Bi	9	1	11.1	33	32	97.0	42	33	78.6
Th	5	4	80.0	17	14	82.3	22	18	81.7
U	15	8	53.3	20	16	80.0	35	24	68.6
αPu	8	1	12.5	20	20	100.0	28	21	75.0
δPu	10	3	30.0	18	18	100.0	28	21	75.0
εPu	10	7	70.0	18	18	100.0	28	25	89.3
Total	514	317	61.67	941	798	84.80	1455	1115	76.63

rule in terms of atomic volumes (15% in diameters is equivalent to 52% in volumes) results in disaster. This is due to the observation (p. 143) that elemental atomic volumes of the Group V–VII nontransitional elements and Group IV elements with the diamond structure are not meaningful for alloys. The use of atomic volumes also introduces a change in degree because of the magnification given to the difference between the closest distance of approach, specified in the original formulation of the rule, and the atomic diameter implied by the use of atomic volumes. In the linear case this difference between the closest distance of approach and the atomic diameter generally used, may not be very significant in regard to a borderline case close to 14–15%, whereas in the volume case it could locate the solute definitely within the unfavorable region.

ENERGY BAND FACTOR: ELECTRON CONCENTRATION

This section is written with the presumption that the reader has some knowledge of the electron theory of metals insofar as it involves energy-reciprocal-lattice-vector relationships, the consequences of crystal potential, Fermi surface, energy bands, density of states, Brillouin zones, etc. An introduction to these topics is available in Barrett and Massalski[14] and they are considered in more detail in Jones.[15]

Historical Introduction

Hume-Rothery's[16,17] realization of the importance of electron concentration in controlling the composition limits of certain metallic phases and Jones'[18] interpretation of this in terms of the Bloch theory of metals set a stage which has remained largely unchanged for 40 years, although it is now being re-examined in the light of direct experimental knowledge of the Fermi surfaces of metals.

One of the features of metals, which is particularly true of alloys of the noble metals Cu, Ag, and Au with the succeeding B Group II, III, and IV elements, is their ability to form phases and solid solutions with apparent disregard of normal valence rules. In 1926 Hume-Rothery noticed that, although valence rules were not followed, there were regularities in the occurrence of the structures of intermediate phases in these alloys in terms of the average number of valence electrons per atom (called electron concentration, e/a). The b.c. cubic, β, or the primitive cubic CsCl, β', phases occurred close to an electron atom ratio of $\frac{3}{2}$, the γ brasses at $\frac{21}{13}$ and the c.p. hexagonal ϵ phases at $\frac{7}{4}$. Later Hume-Rothery and co-workers[17] showed that there were also regularities in the limits of the solidus for α f.c. cubic solid solutions of B Group solutes in Cu and Ag, when they were expressed in terms of electron concentration, and also in the solidus curves as a function of temperature. In Cu and Ag alloys the α solid solution boundary occurs at e/a values close to 1.4 and at rather lower values in Au alloys. These limits are lowered by a large electrochemical factor such as occurs when Group V solutes are dissolved in the noble metals. Furthermore, it was noted that the upper limit of the β phase boundaries lay close to an e/a value of 1.5. Jones[18] accounted for these observations in terms of particular E *vs.* k relationships resulting from the interaction of Fermi surface and Brillouin zone boundaries. It was shown that the α, β, and γ phase limits occurred sometime after contact of appropriate Brillouin zone planes by the Fermi surface, leading to a decreasing density of states curve, and hence the probability that some other structure would be more stable than the one considered. Some thirty years later Jones' explanation was confirmed by the demonstration from specific heat measurements of a decreasing density of states curve at compositions where the β and γ phases became unstable in Cu–Zn alloys,[19] and energy band calculations[20] and measurements of the Fermi surface[21] of β brasses further demonstrated the plausibility of Jones' arguments.

In the years following Hume-Rothery's and Jones' work of 1934, the compositions of many alloy phases, including those of the transition metals, were found to exhibit at least a partial dependence on electron concentration, although the simple regularities of the B group alloys were never again realized. This led to a period in which attempts were made to account for the electron concentration at which various phases occurred in terms of "zones" derived from strong X-ray reflections of the phases concerned (cf. review by Taylor[22]). This idea followed from Jones' incomplete zone used to discuss the γ-brass structure, but the idea was based on the belief that the strength of the X-ray scattering from crystal planes and the size of the energy gaps across Brillouin zone faces should vary in magnitude in similar fashion, whereas in fact they are each determined by quite different atomic factors.† There is no particular reason for an *a priori* parallel in the two effects; only when all the geometrical contributions to the structure factor give zero intensity, may zero X-ray intensity and an identically zero band gap be expected.‡ Secondly, the so-called "zones" made up of planes giving these strong X-ray reflections did not in general contain an integral number of *Brillouin zones* which have exactly 2 electrons per primitive cell. Furthermore, there have been misconceptions in the properties of these incomplete zones. In his treatment in 1960 Jones[15] is careful to distinguish between those which are true Brillouin zones (complete) and those which are not (incomplete), since the latter must "leak" and the electrons spill over beyond their bounding planes before they are filled up. At present it is uncertain whether consideration of the so-called large zones makes any significant contribution to an understanding of phase stability. For example, the large zone for the γ-brass structure which Jones constructed from {330} and {114} planes that give very strong X-ray reflections, contains 90 electron states per unit cell. This was held responsible for γ phases being stable at electron concentrations of less than 90 per

†The pseudopotentials appropriate here can have either a negative or positive sign like the atomic factors for neutron scattering, whereas those for X-ray scattering are always positive.

‡In practice the situation is more complicated and spin-orbit splitting and other effects may result in a band gap, except in magnetic fields large enough to cause magnetic breakdown. Furthermore, the size of the energy band gap is not constant across a Brillouin zone plane.

cell and for the formation of defect structures so that this electron concentration should not be exceeded.† Newer work on some of these defect structures indicates an error in the earlier experimental observations and Dodd and co-workers,[23] for example, find no defects in the Au–Zn γ phase which is stable at electron concentrations as high as 92 per cell. Furthermore, it is uncertain whether the $\{330\}$ and $\{114\}$ planes do indeed both have significantly larger energy gaps across them than other planes of similar interatomic spacing. (Pseudopotential coefficients such as those given by Harrison[24] are not reliable for calculating the band gaps across planes for the γ phases because of the presence of d electrons in the metals, Cu, Zn, etc.)

Since the direct experimental determination of the Fermi surface of pure metals and some alloys has become a reality, the importance of the older concepts must be assessed in relation to the experimental results, and also in relation to the great wealth of structural information on alloys which has become available during the last few decades. In addition to this, the increasing amount of knowledge derived from energy band structure calculations of metals and alloys and semiconductors has to be considered. Here, however, it is realized that even if the crystal potential problems can be handled satisfactorily with the possibility of carrying the calculations through to self-consistency, the problem of the exchange potential is by no means solved, and without experimental knowledge of the Fermi surface in metals or of energy band gaps in semiconductors on which to base theoretical energy band calculations, they can not at present necessarily be considered as a reliable and authoritative statement of E *vs.* k relationships. Furthermore, if E is an approximately constant function of k for several energy bands located about the energy of the Fermi level, it may be quite impossible to prescribe from theoretical considerations alone, whether a particular energy band lies just above or just below the Fermi level for a great part of its path in E as a function of k. In such conditions the form of the Fermi surface cannot be prescribed theoretically with any authority; only as a result of prior experimental knowledge could the Fermi level be adjusted to give the correct situation. At the same time, however, it must also be realized that experiments only give properties such as extremal cross-sectional areas or caliper diameters of the Fermi surface, and a reliable model is required to interpret the results. Therefore, except perhaps in the simplest cases, neither theoretical calculation nor experimental observation can be complete without the other. Energy band calculations for rare earth

†The actual recorded limit was about 88 electrons per cell, in reasonable agreement for a rapidly decreasing $N(E)$ curve as the zone was being filled up.

compounds are even more dependent on experimental knowledge of Fermi surface or optical and electrical energy band gaps, because of the great sensitivity of the *f* bands to the exchange potential, as Cho[25] has shown. It is essential to have experimental data with which to gauge the best value of the exchange potential to give the correct results in such energy band calculations.

Experimental determinations of Fermi surfaces suggest that with a reasonable number of conduction electrons per primitive cell (say less than 20), there will be sheets of Fermi surface in no more than four or five partly filled Brillouin zones, and of these, the distribution in perhaps only two is likely to exert a significant influence on crystal structure. Although the appropriate Bravais lattice for a particular structure prescribes the succession of possible Brillouin zones, the component atoms in the structure determine the number of electrons that have to be accommodated in the Brillouin zones, and the crystal potential, which prescribes the energy band gaps across the Brillouin zone faces and their variation over the faces and edges. Therefore, in achieving the lowest electron energy for any particular metallic phase, both the geometrically possible structures and the relative character of the component atoms are important. Energy band calculations also indicate that the symmetry of the energy bands at the Fermi level (relative *s*, *p*, *d*, *f*, character) is likely to be important in determining crystal structure, since the coordination of the atoms may have to be consistent with this. This amounts to an equivalent statement in the chemical bonding approach to crystal structures (cf. e.g. Altmann *et al.*,[26]) which would maintain that the symmetry of the possible hybridized bond orbitals must be consistent with the spatial distribution of the atoms in the structure. This conclusion is quite obvious in the case of many semiconductors whose bonding characteristics demand either tetrahedral or octahedral coordination (cf. p. 206).

Thus it is seen that electron concentration is not the only factor involved in the E *vs.* k relationships that can influence crystal structure, and so in referring to this effect we prefer to call it the *energy band factor*, realizing that it embraces both the effects of electron concentration and of symmetry of the energy bands at the Fermi level. The energy band factor may be expected to show its influence in determining crystal structure and in limiting the electron concentration at which a phase is stable, as well as in influencing the unit cell dimensions of phases with a given structure. As Fermi surface is a property of all metals, the energy band factor must play a role in determining the structures of all metals. However, it is generally a very weak effect readily dominated by stronger effects such as the chemical bond factor and it is only in rather special circumstances, when other strong effects are largely absent, that unambiguous evidence of its control of crystal structure can be observed. Such situations are found in the alloys of Group I B noble metals with the elements of Groups II to IV B, in families of ordered polytypically related structures where the near-neighbor coordination is not changed in going from one polytype to another, and in ordered alloys where the introduction of antiphase domain boundaries may provide a means of placing planes of energy discontinuity close to large areas of Fermi surface, thus reducing the electron energy (p. 90).

Mechanisms for Achieving Structural Stability

There are two main ways by which interaction of Fermi surface with planes of energy discontinuity has been observed to control structural stability as a function of electron concentration. One is the general effect of Fermi surface touching any plane of energy discontinuity in lowering the electron energies, and the other is concerned with the relative filling of electron states in specific Brillouin zones and its effect on the density of states for a phase in one crystal structure or another.

The principle upon which a structure may be stabilized by the presence of planes of energy discontinuity at or close to the Fermi surface is illustrated in Figure 3-16, which shows the parabolic E *vs.* k relationship for free electrons. Introduction of a plane of energy discontinuity lowers the energies of electrons at k vector values less than that of the plane, and raises them at higher k values. If therefore the Fermi surface extends to, or nearly to a plane of energy discontinuity, the electron energies are lowered, and if the Fermi surface contains large flat areas which are surrounded by planes of energy discontinuity, the electron energies are lowered over a large fraction of the solid angle. Thus the overall electron energies can be lowered by an appreciable amount so that considerable stabilization of the phase is achieved. It is not necessary for the energy discontinuity planes involved in this process to constitute a specific Brillouin zone since the electron energy lowering is a general property of any plane of energy discontinuity.

If, as a Fermi surface expands or contracts with changing electron concentration in a series of alloys, planes of energy discontinuity can be introduced which lie close to large relatively flat areas of the Fermi surface, by such processes as the formation of low energy antiphase domain boundaries, or the introduction of planar stacking faults in ordered close packed structures, structural stability may be maintained over a considerable range of electron concentration. Introduction of antiphase domain boundaries (of the "first kind"; see p. 90) or planar stacking faults in ordered close packed structures of compositions MN or MN_3 satisfies

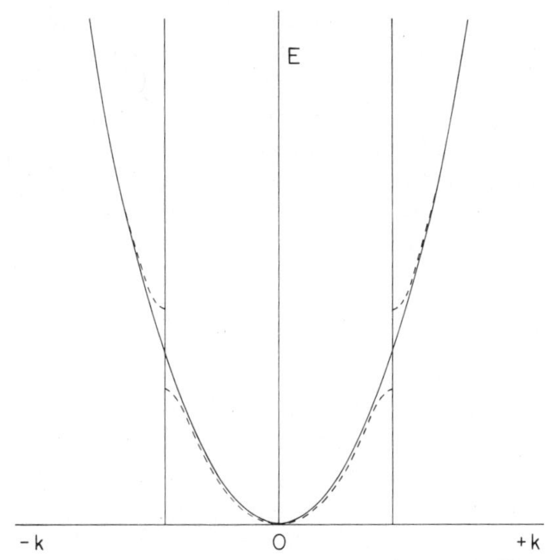

Figure 3-16. Parabolic *E vs. k* relationship for free electrons (full line), showing the effect of a plane of energy discontinuity (vertical lines) on the electron energies (broken lines).

the condition for observing energy band effects discussed on p. 51 (that the near-neighbor environment of the atoms, and therefore the main contribution to the enthalpy remains essentially unchanged between one structure and another) so that differences in the smaller enthalpy contributions arising from interaction of electrons and planes of energy discontinuity are effective in controlling the crystal structure. Quantitative evidence of this behavior which has been found by Sato, Toth, and coworkers[27] is discussed on pp. 90–102.

Superstructures themselves may derive stability due to multiplication of the cell edge introducing planes of energy discontinuity which lie close to large flat areas of Fermi surface, so that the electron energies are lower than they would be in a disordered solid solution. Although it is difficult to see that introduction of the cubic first zone influences the development of the ordered β'CsCl structure from the disordered b.c. cubic structure, the suggestion of Lomer[28] illustrates how the large γ-brass cell may develop from β' and be stabilized by multiplication of the cell edge. In the second zone the Fermi surface of β' brass contains large rather flat electron surfaces parallel to the cubic zone faces and at about one-third of the zone radius from the boundary. A superlattice vector one-third of the β' reciprocal lattice vector could couple two such areas across the zone boundaries and thus stabilize a supercell with a tripled cell edge in real space, as in γ-brass. Omission of two atoms from the supercell (27 CsCl type cells) would give a good value of the required matrix element for the larger cell. Thus the γ-brass structure is seen as one which is stabilized by the introduction of Brillouin zone boundaries at the Fermi surface with consequent lowering of the electron energy.

The density of states, $N(E)$, is important to the consideration of whether a particular alloy phase might be stable in a particular crystal structure. A high density of states would suggest that the phase could be stable in the structure, whereas a low density of states would suggest that some other structure permitting a higher density of states (and of course, lower overall electron energies, E) would be stable in its stead. The density of states for a series of alloys is still increasing with electron concentration if the Fermi surface has not made contact with planes of the Brillouin zone for the structure. If, however, the electron concentration is such that the Brillouin zone planes are touched but not overlapped by the Fermi surface, so that most of the electron states in the zone are filled, the $N(E)$ curve is falling. As the electron concentration is further increased by alloying, $N(E)$ falls to lower values until some other crystal structure with a larger density of states can accommodate the valence electrons at a lower overall energy, and change of structure occurs. This is a general principle of phase stability which may or may not be important in specific cases. It can only be recognized as controlling phase stability if a number of phases with given structures are found to have common limits of stability in terms of electron concentration. This is the situation which was recognized by Hume-Rothery in the α, β, γ, and ϵ phases of the Group II to IV B metals with Cu, Ag, or Au. Its recognition was possible because the second condition discussed on p. 51 is satisfied: all of the outer valence electrons occupy sheets of Fermi surface in partly filled Brillouin zones so that the difference of interaction between Fermi surface and Brillouin zone planes in one structure or another has a relatively large influence on the total enthalpy and therefore the free energy of the phase. Coupled with this, the comparative similarity of the Group I to IV B metals does not lead to such large changes in the near-neighbor contribution to the enthalpy on changing alloy composition across the different phase fields, so as to dominate and obscure the energy band effect in controlling structures as a function of electron concentration. Support for this concept derives from the beginning of the breakdown of Hume-Rothery's rules observed in alloys of Cu, Ag, or Au with Group V B metals, because the large electrochemical difference of the components increases the importance of near-neighbor interactions and the chemical bond factor.

Dependence of the Structures of Ordered Close Packed Alloys on Electron Concentration

Qualitative evidence. Ordered MN_3 alloys of the noble metals with some succeeding B Group metals have structures which depend in a general way on electron concentration as shown in Figure 3-17. The ordering of the M component on the close packed planes is such as to give nets composed of triangles and/or rectangles arranged in rows (p. 327). In order of increasing electron concentration, superstructures with M nets composed of one row of rectangles to one of triangles (TR) are followed by those with two of rectangles to one of triangle (TR_2) and one structure is shown with four rows of rectangles to one of triangles (TR_4). In the Au–Cd and Au–Mg systems, superstructures with a more complex arrangement of triangles and rectangles of the M component in the close packed planes (i.e., not arranged in rows) are also found in the electron concentration range where TR structures occur.

Electron concentration also appears to play a definite role in determining the structures adopted by ordered transition metal MN_3 type

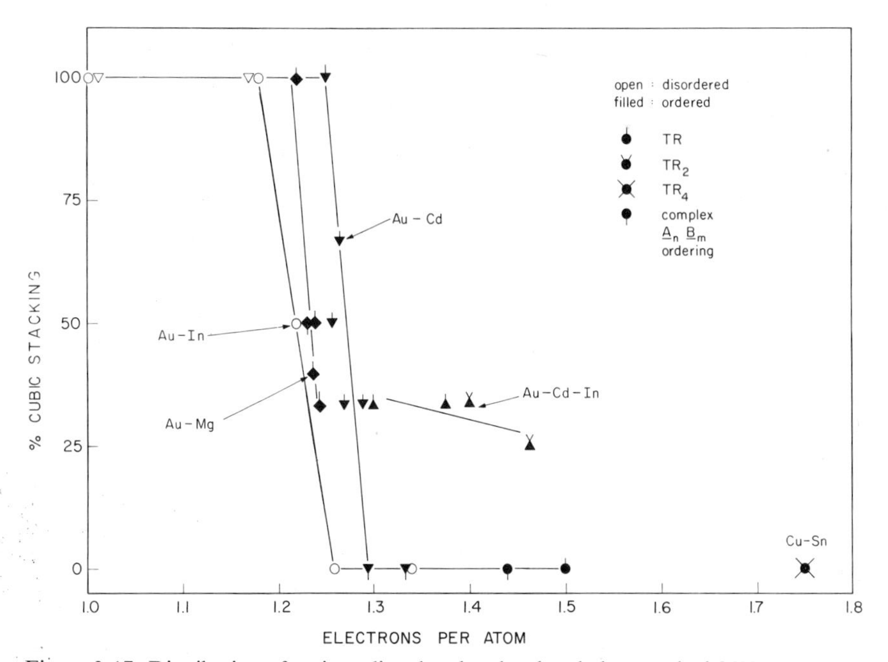

Figure 3-17. Distribution of various disordered and ordered close packed MN_3 structures as a function of electron concentration and percentage cubic stacking of the layers in Au–Mg◊, Au–Cd▽, Au–In○, Au–Cd–In△, and Cu–Sn○ alloys, T, TR, TR_2, TR_4, and complex ordering of the layers indicated by symbols shown on diagram.

compounds. Up to an electron concentration of about 8 *s*, *p*, and *d* electrons per atom, these phases adopt the $AuCu_3$ (*tP*4) structure. Thereafter they pass through various polytypes of $AuCu_3$ with decreasing proportion of cubically surrounded layers, possibly achieving the $SnNi_3$ (*hP*8) polytype with hexagonally surrounded layers at an electron number of about 8.5. With further increase of electron number, the structures change from the triangular 3^6 arrangement (*T*) of the *M* component to the rectangular 4^4 (*R*) arrangement and the proportion of cubically stacked layers increases again in polytypes of the $TiAl_3$ (*tI*8) structure until all layers are cubically surrounded, and the $TiAl_3$ structure itself is found at an electron number of about 8.75. Several examples of this behavior are shown in Figure 3-18. This figure is composed from data in

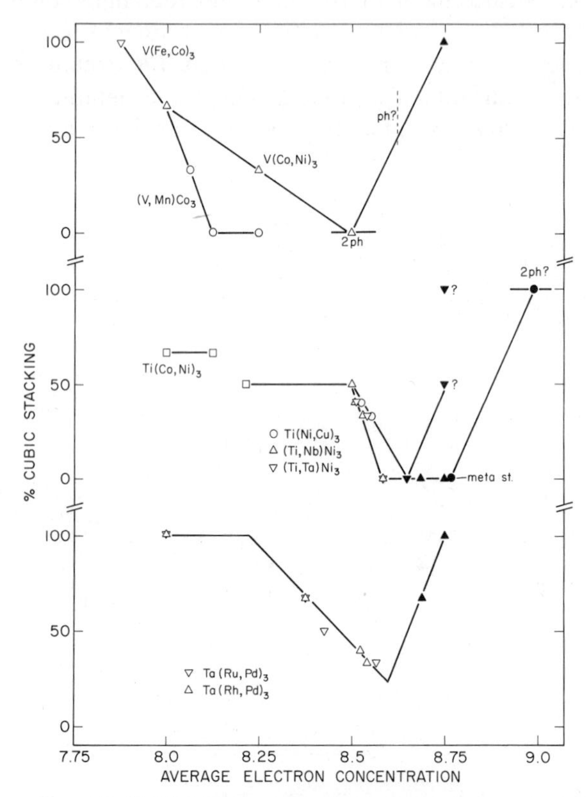

Figure 3-18. Distribution of various ordered close packed MN_3 structures of the transition metals as a function of average electron concentration (*s*, *p*, and *d* electrons) and percentage cubic stacking of the layers: Ph = Phase; St, Stable.

the published literature, but Dr. B. C. Giessen, who has examined such alloys extensively, informs me that it is a general pattern of behavior. Furthermore, in the region of electron concentration where the hexagonally close packed $SnNi_3$ structure occurs, it appears that alloy stability is maintained through change of axial ratio of the hexagonal phase, since this first decreases and then subsequently increases with increasing electron concentration.

Although electron concentration is the major factor controlling the structures of such phases, relative atomic size may not be without influence, as is shown by the change of structure from $TiNi_3$ (*hP*8) (or $TiPd_3$ with the same structure, which has 50% cubically surrounded layers) to the $AuCu_3$ structure of $TiPt_3$, whereas in the alloys $Ti(Ni_{0.11}Pt_{0.89})_3$ and $Ti(Pd_{0.17}Pt_{0.83})_3$ a polytype with 74% cubically surrounded layers is found. An additional and weaker influence of size could well account for these differences in detail of structure type at a given electron concentration; nevertheless the main pattern of structural behavior as a function of electron concentration is clear. van Vucht[29] shows from space filling diagrams that the packing density increases in the structural order $AuCu_3$, $SnNi_3$, $TiCu_3$ (hexagonal polytype of $TiAl_3$) and that the difference between these structure types increases with increasing radius ratio R_M/R_N of the component atoms (Figure 3-19). Since the packing density of the $AuCu_3$ and $SnNi_3$ structures, particularly, decreases with increasing radius ratio, it may be argued that the more disparate the sizes of the two components, the more the $TiAl_3$ polytypes are to be favored in order to give the best packing density. Furthermore, partial replace-

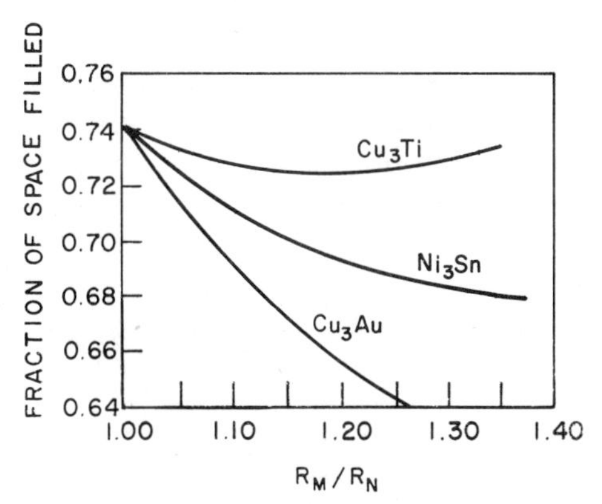

Figure 3-19. Fraction of space filled as a function of radius ratio R_M/R_N for phases with the $TiCu_3$, $SnNi_3$, and $AuCu_3$ structures, based on a hard sphere model. (van Vucht [29].)

ment of the M component by atoms of larger size should induce structural changes in the direction of the hexagonal $TiAl_3$ polytypes. Although in some alloys shown in Figure 3-18 structural changes are in accord with predictions from packing density considerations, it is by no means always the case, and electron concentration remains the only feature with which all changes can be correlated.

Quantitative Evidence: Antiphase Domain Structures. Ordering of the f.c. cubic solid solution in the AuCuI ($tP4$) structure distributes Au and Cu atoms alternately on layers normal to the c axis of the tetragonal cell. Shift of one AuCuI type cell by $\frac{1}{2}(\mathbf{a}+\mathbf{c})$ in the (010) plane (Figure 3-20) introduces an out-of-step shift, or antiphase boundary. In the AuCuII structure ($oP40$) this shift occurs every five cells along the b axis of the crystal and the resulting structure has orthorhombic symmetry and contains 10 AuCuI type pseudocells lying along the b direction. Such is a one-dimensional long period superstructure where the period M is defined as the number of AuCuI type pseudocells within the antiphase domain (here 5). This type of superstructure results in splitting of (110) type superstructure reflections which are obtained by electron diffraction from thin films; the splitting is $x = 1/(2\text{M})$ in terms of $1/a$, the unit reciprocal lattice parameter. In addition to the distortion ($c/a < 1$) of the AuCuI pseudocell, there is also a further small distortion of the cell along the direction b of the period so that $b'/10a > 1$.

One-dimensional long period superstructures such as this are found in MN and MN_3 alloys, and two-dimensional long period structures are also observed in MN_3 alloys; these are characterized by two different domain periods M_1 and M_2 and two step-shifts. Four sublattices are required to describe the possible arrangements of domain boundaries in MN_3 structures, whereas only two are required for MN structures. This leads to two possible kinds of antiphase domain boundaries $\{hkl\}$ in MN_3 structures. If the out-of-step vector is $<uvw>$, then $hu+kv+lw = 0$ expresses the "first kind" of boundary and $hu+kv+lw \neq 0$ the

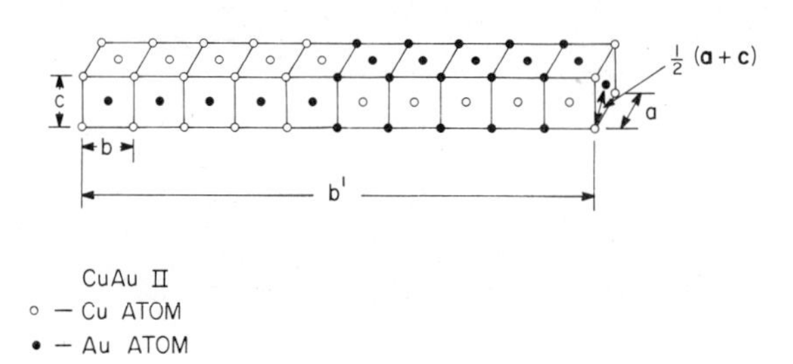

Figure 3-20. Diagram of the CuAu II antiphase-domain structure.

"second kind." This is because the M atoms on sublattice site 1 can replace N atoms on any of sublattice sites 2, 3, or 4 at the domain boundary, the out-of-step vectors being $\frac{1}{2}(\mathbf{a}_1+\mathbf{a}_2)$, $\frac{1}{2}(\mathbf{a}_2+\mathbf{a}_3)$, and $\frac{1}{2}(\mathbf{a}_3+\mathbf{a}_1)$. Only one vector, say $\frac{1}{2}(\mathbf{a}_1+\mathbf{a}_2)$, lies in the plane of the antiphase boundary, (010) of Figure 3-21*a*, and the boundary is formed by a shift along this plane giving an antiphase boundary of the "first kind." The other two vectors, which are equivalent, do not lie in the antiphase boundary (Figure 3-21*b* and *c*) and in order to create the domain boundary by shifting the crystal along these out-of-step vectors, it is necessary to remove a plane of atoms. These are referred to as boundaries of the "second kind," and they introduce wrong nearest neighbors because of the removal of a plane of atoms, whereas boundaries of the first kind formed by movement in the plane of the boundary do not introduce wrong nearest neighbors. There is thus a significant difference in energy of formation of boundaries of the two kinds, and only boundaries of the first kind are found in one-dimensional long period superstructures.

Formation of long period superstructures introduces a small distortion of the fundamental ordered cell in the direction along the period, and the distortion may either increase or decrease the cell edge according to circumstances discussed below. Note that these distortions are quite small (about 1%), distinctly smaller than those which may occur in the

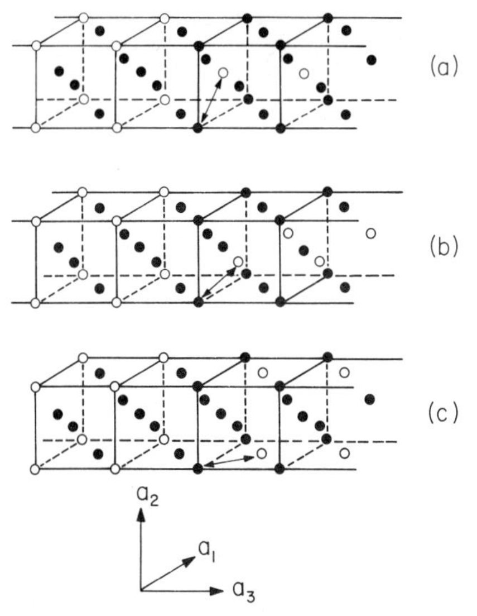

Figure 3-21. Creation of an antiphase-domain boundary by three different out-of-step vectors. See text.

ordering of the fundamental cell. Thus in AuCu I, the tetragonal distortion due to ordering is about 8% ($c/a = 0.92$), whereas that due to the introduction of antiphase domains in AuCu II is less than 1% ($b/a = 1.0035$). The fundamental cell of the MN_3 type structure which remains cubic on ordering, becomes slightly tetragonally distorted on forming the one-dimensional long period structure, the distortion being orthorhombic when two-dimensional long period superstructures occur.

With this introduction, the work of Sato and Toth showing the influence of electron concentration on ordering in alloys between the noble metals can be discussed.

The reduced Brillouin zone of the ordered AuCu I structure is a cuboid formed by (001) and {110} planes, with square cross-section normal to the k_z axis (Figure 3-22). Electrons overlap the (001) faces and the Fermi surface probably touches the {110} planes. When the long-period AuCu II superstructure forms so that the {110} reciprocal lattice points split in the direction of the period, the {110} faces of the zone become doubled and tilted since they are constructed as the planes normal to, and bisecting the line joining the origin to the {110} reciprocal lattice points (Figure 3-23). Selection of four planes inside the original AuCu I Brillouin zone seen in Figure 3-23 in cross section through the origin and parallel to (001), gives a new zone of smaller volume than the original AuCu I zone, and selecting the four planes outside gives a new zone of

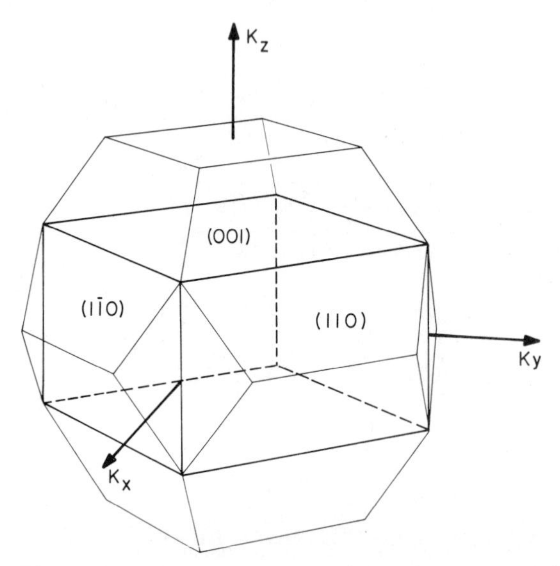

Figure 3-22. First and second extended Brillouin zones for the AuCu I structure. Planes refer to the first zone.

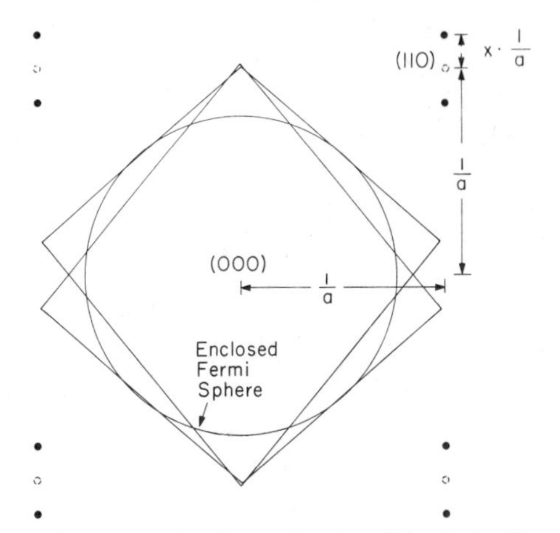

Figure 3-23. Reciprocal lattice of the CuAu II structure in a plane through the origin parallel to (001), showing the Fermi sphere and planes of energy discontinuity arising from the splitting of the superstructure reflections. (Sato and Toth [27].)

larger volume than the AuCu I zone. Sato and co-workers show that the electron concentration, e/a, corresponding to a Fermi sphere that just touches either the inner or outer set of {110} planes can be calculated in terms of the splitting, x, of the reciprocal lattice {110} points. The volume of the inscribed sphere expressed as the electron to atom ratio of the states which it contains is $(\pi/12)(2 \pm 2x + x^2)^{3/2} = (\pi/12)[2 \pm 1/M + 1/(2M)^2]^{3/2}$ and the electron concentration can be represented by $e/a = (\pi/t^3 12)(2 \pm 2x + x^2)^{3/2}$ where t is an empirical factor (generally about 0.95) recognizing that the actual Fermi surface is distorted from a spherical shape.† The + and − signs correspond to the cases of the outer and inner sets of {110} planes.

The electron concentration in the AuCu II phase can be varied by substituting in it other elements of different valency, and it is found that this changes the period of the superstructure. Indeed it has been shown[30] that the variation of period M with electron concentration follows closely the form of the above relationship, showing that the size of the long period superstructure is controlled by the electron concentration according to the explanation given above. The data are given in Figure 3-24. The full lines represent the expression with $t = 0.95$. The upper one indicates

†It may also be defined as the ratio of the distance from the origin to the {110} planes, to the radius of the free electron Fermi sphere in reciprocal space.

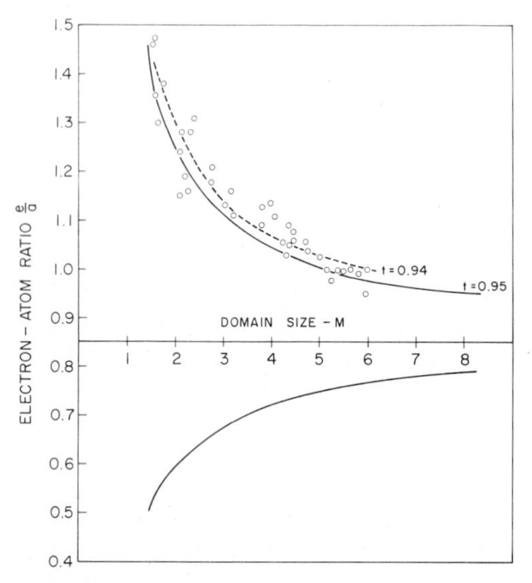

Figure 3-24. CuAu II structure: Curves for electron atom ratio *vs.* domain size, M, derived from the theoretical equation on p. 93, for $t = 0.95$ (full line) and $t = 0.94$ (broken line), showing agreement with experimental data obtained from alloys with electron-atom ratio greater than unity. (Sato and Toth [27].)

that the new outer $\{110\}$ energy discontinuity planes ($e/a > 1$) stabilize and control the size of the long period superstructure. The lower full curve refers to the inner $\{110\}$ planes which could only influence structure when the electron atom ratio is less than about 0.8.

The expansion of the fundamental cell edge in the direction of the super-period ($b/a > 1$) is in agreement with predictions. In cases where the inner $\{110\}$ planes stabilize structures, a contraction of the fundamental cell edge is expected in the direction of the super-period.

The first (reduced) Brillouin zone for ordered cubic MN_3 alloys is a cube with volume $1/a^3$ which can accommodate 0.5 electrons per atom of MN_3. The second zone bounded by $\{110\}$ planes is a rhombic dodecahedron, and as the first two zones together can accommodate one electron per atom, the planes bounding the second zone may play an important role in MN_3 superstructures. A one-dimensional long period superstructure would result in doubling of 8 of the 12 $\{110\}$ planes, and a two-dimensional long period superstructure would influence all 12 $\{110\}$ planes, so that the period and stabilization of the long period superstructure according to electron concentration, can be treated and discussed in terms of the splitting of 110 reflections in much the same way as for the MN superstructures. In fact, for one-dimensional long period superstructures in MN_3 alloys, the relationship between electron-atom ratio and the separation of the superstructure spots is the same as for MN alloys.

Furthermore, it has been shown that conditions for stabilization of antiphase domains of the second kind are unfavorable compared to those for domains of the first kind, and therefore the latter should always be found in MN_3 one-dimensional long period superstructures, as indeed, they are. The direction of the super-period in MN_3 one-dimensional structures is always in one of the [100] directions (and antiphase boundaries are in {100} planes) as this gives the maximum change of the {110} energy discontinuity planes for a given separation of the {110} reciprocal lattice points.

One-dimensional long period superstructures have been observed in Ag_3Mg, Au_3Zn, Au_3Cu, and Pd_3Mn[31] alloys where $e/a > 1$, and the outer {110} planes control the super-period, and for Cu_3Pt and Cu_3Pd where $e/a < 1$ and the inner {110} planes control the super-period. When the outer planes control the structural properties ($e/a > 1$), the tetragonal cell is observed to expand slightly in the direction of the super-period ($c/a > 1$), whereas when the inner planes control the structure, the cell contracts in the direction of the super-period ($c/a < 1$) as predicted by the theory. Figure 3-25 gives an example of the observed variation of electron-

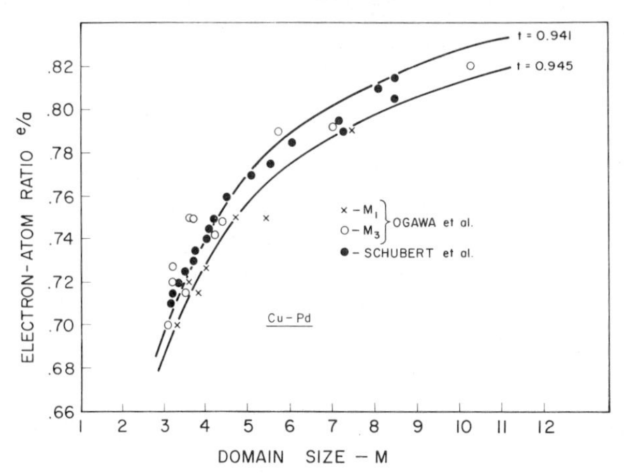

Figure 3-25. $AuCu_3$ antiphase domain type structure: Theoretical curves for electron atom ratio *vs.* domain size, M, for $t = 0.945$ and 0.941 showing agreement with experimental data obtained from Cu–Pd alloys with electron atom ratios less than unity. (Sato and Toth [31].)

atom ratio with domain size, and the theoretically predicted relationship for a long period superstructure stabilized by the inner {110} planes. The excellent agreement leaves little doubt that the electron concentration does indeed control the structural dimensions and arrangement.

Two-dimensional long period superstructures have been observed in alloys such as Cu_3Pd, Au_4Zn, and Au_3Mn[31] and here boundaries of the second kind can occur.

General considerations suggest that stabilization of long period superstructures results from the Fermi surface of the alloy just touching the {110} energy discontinuity planes modified by the introduction of antiphase domains. It is to be noted that the Fermi surfaces of the noble metals themselves are very flat in the [110] direction, and if the Fermi surfaces of the solid solutions or ordered alloys are not greatly different, considerable stabilization of the superstructures can be achieved because of the large solid angle over which the electron energy is lowered by the presence of the {110} superstructure planes.

Apparent nonintegral values of the super-period are generally found in long period superstructures and these presumably result from an admixture of supercells with integral periods, giving the best overall fit to the Fermi surface resulting from the particular electron concentration. However, when the period is short ($M = 2$) there is a definite tendency for the period not to respond to small changes of e/a, because of the relatively large disturbance of the periodic potential that would result from a random admixture of $M = 2$ and $M = 3$ cells. Also, when M has small values, the change of angle of the energy discontinuity planes arising from the split 110 reflections is large, and the simple arguments given above for large values of M, scarcely apply. Indeed, when $M = 1$ and 2 characteristic crystal structures arise ($TiAl_3$ (*tI*8) and $ZrAl_3$ (*tI*16) types respectively), which have to be considered in terms of their own specific Brillouin zone structure. Such structures may therefore be expected to exist as discrete phases stable over a fairly narrow range of electron concentration, unless the Brillouin zone boundaries can be kept at the Fermi surface by some other mechanism such as structural distortion.[27]

Long Period Structures Stabilized by Distortion. If a one-dimensional long period superstructure with $M = 2$ does not respond to increasing electron concentration with increase of period, it may still be able to keep the Brillouin zone planes at the Fermi surface by considerable distortion of the unit cell.[32] Figure 3-26 shows a section through the origin parallel to the (010) plane of a Brillouin zone formed by {110} planes appropriate to a MN_3 phase with one-dimensional long period superstructure. It indicates how a larger Fermi sphere, can be accommodated by a tetra-

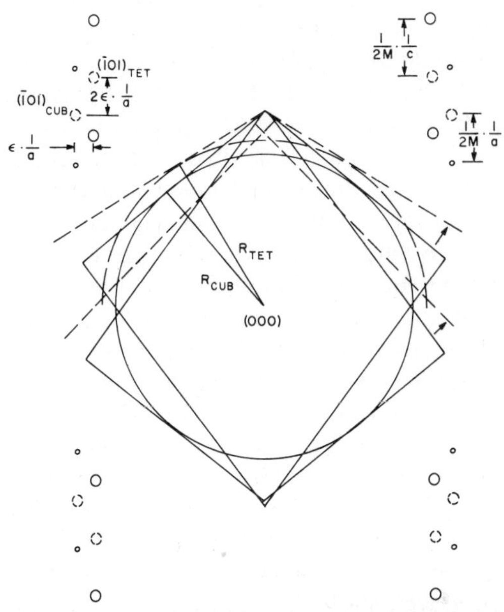

Figure 3-26. Reciprocal lattice for a tetragonally distorted one-dimensional long period superstructure MN_3 with $M = 2$, in a plane through the origin and parallel to the (010) plane, showing accommodation of a larger Fermi sphere by energy discontinuity planes formed by the split {110} reflections. (Sato and Toth [27].)

gonal distortion of the structure in the direction of the super-period, while keeping the atomic volume fixed. For the positive case where the electron concentration is increased above the value corresponding to a long period cell with M = 2, the relationship between the tetragonal distortion parameter ϵ (Figure 3-26) and the electron-atom ratio is $e/a = (\pi/12t^3)(2.5625 + 4.25\epsilon + 21.75\epsilon^2)^{3/2}$ where $c/a = 1 - 3\epsilon$. The tetragonal distortion takes place in the direction of the period with $c/a < 1$, which is in the opposite sense to that accompanying the formation of antiphase domains for the positive case ($c/a > 1$). Furthermore, the extent of the distortion required to accommodate a significantly larger Fermi sphere (several per cent) is notably larger than that which occurs when antiphase domains form (less than one per cent).

An example of this type of behavior has been reported in the Pd–Mn system.[32] In alloys containing less than 25 at.% Mn, Pd_3MnI has the $AuCu_3$ type structure; above 25 at.% Mn as e/a increases above 1.20 (assuming e for Mn = 3 and for Pd = 0.6), the Pd_3MnII structure occurs which is a tetragonal long period superstructure with a fixed value of M = 2 (it has essentially the Al_3Zr structure type). The tetragonal distortion increases with increasing Mn content (e/a) reaching $c/a = 0.88$ by about 32 at. % Mn, when the structure appears to have changed homogeneously into the CuAuI type. The observed tetragonal distortion is in the direction of the super-period and much larger than that associated with the formation of the antiphase domains and has $c/a < 1$, as predicted above. However, the variation of c/a with e/a does not follow exactly the form predicted by the above equation, and it appears that the amount of distortion is not controlled by the electron concentration alone. The tetragonal distortion removes the degeneracy of the three sublattices occupied by Pd atoms so that excess Mn can occupy one of them preferentially allowing homogeneous transformation to the AuCuI structure.† This ordering also reduces the energy required for the tetragonal distortion because of the relative size effect as the structure approaches more and more closely to that with alternate layers of Pd and Mn atoms. Thus, as the electron concentration increases with addition of Mn over 25 at. %, the ordered distribution of the excess Mn atoms facilitates the tetragonal distortion which is believed to keep the Brillouin zone planes at the Fermi surface, permitting structural stability over a considerable range of composition.

Structures Stabilized by Changes of Layer Stacking Sequence. Periodic modulation of a regular close packed structure by low energy boundaries can also be achieved by the systematic introduction of a stacking fault in a layer of atoms. Thus, a close packed cubic sequence of layers could, for example, be modulated by the introduction of a stacking fault on every fifth and sixth layers giving a hexagonal structure which would be described as 41 in Ždanov–Beck notation. These principles apply to any close packed array of atoms regardless of the ordering within the layers, and so they apply to ordered structures such as the $AuCu_3$, Al_3Ti (M = 1), and Al_3Zr (M = 2) types. If, therefore, long period superstructures are fixed at M = 1 or M = 2 and do not respond in period of antiphase boundary to changes of electron concentration, there is the further possibility of introducing low energy stacking faults which change the

†At temperatures below about 450°C in the region about $Pd_{21}Mn_{11}$, an ordered superstructure based on a block of 8 f.c. cubic cells can be obtained. (A. Kjekshus, R. Møllerud, A. F. Andresen, and W. B. Pearson, 1967, *Phil. Mag.*, **16**, 1064).

structure and alter the energy discontinuity planes, so as to enclose a larger Fermi sphere and give structural stability with increasing *e/a* in a series of alloys.[27,33] Any close packed arrangement of atoms can be described on hexagonal or orthorhombic axes with the close packed planes corresponding to the basal plane and the *c* axis perpendicular to them. In the orthorhombic reciprocal lattice the separation between the closest spot to the origin in the *c* direction, which corresponds to 111 of the f.c. cubic crystal, and the origin itself gives the reciprocal distance between neighboring close packed layers. In the notation of Sato and Toth a periodic modulation of a close packed structure, L, (corresponding to M of the antiphase domain structures) is expressed as the number of close packed layers (or groups of similarly oriented layers; *vide infra*) in the repeat sequence, and it corresponds to the number of spots occurring in an electron diffraction pattern between the origin and 111 of the f.c. cubic array. Modulations with odd numbers of layers L = 1, 3, etc. differ from those with even numbers L = 2, 4, etc., since in the latter, diffraction spots occur at the same level normal to the reciprocal *c* axis, whereas in the former they do not, being shifted by $\pm\frac{1}{3}$L in the *c* direction (Figure 3-27). L even situations correspond to hexagonal symmetry (H^s) and L odd to rhombohedral (R^s). Structural arrangements are referred to by the L number and H^s or R^s. Thus the cases shown in Figure 3-27 are respectively $1R^s$, $2H^s$, $3R^s$, and $4H^s$. However, the number of close packed layers does not necessarily have a one to one correspondence to the period L. In the R^s case it corresponds to 3L, or an integral multiple thereof, consistent with $1R^s$ representing cubic close packing ABC, since groups of three similarly oriented layers form units with a stacking shift between each. In the hexagonal case it normally corresponds to L, but particulars of ordering in the layers may result in an integral multiple of L, just as it might result in an integral multiple of 3L in the R^s case. For this reason the symbols R^s and H^s with superscript *s* are used to differentiate them from those of Ramsdel's earlier notation (see p. 7). For example the 9-layer Sm structure is described as $3R^s$, whereas in Ramsdel's notation it is 9R.

Examples of stacking order modulation to shift the energy zone boundaries as the size of the Fermi sphere increases are found in Au-Mn alloys with 22 to 28 at.% Mn, where the basic structure is a MN_3 long period superstructure with M = 1 (the Al_3Ti, *tI*8 type) and in alloys about Au_3Mg and Au_3Cd where the basic structure is a long period superstructure with M = 2. In Au–Mn alloys the sequence of structures (which may coexist) with increasing *e/a* or Mn content is $1R^s$, $5H^s$, $3R^s$, and $6H^s_1 + 6H^s_2$. The Ramsdel descriptions giving the number of layers in the structural periods are respectively 6R, 10H, 18R, and $6H_1 + 6H_2$. The

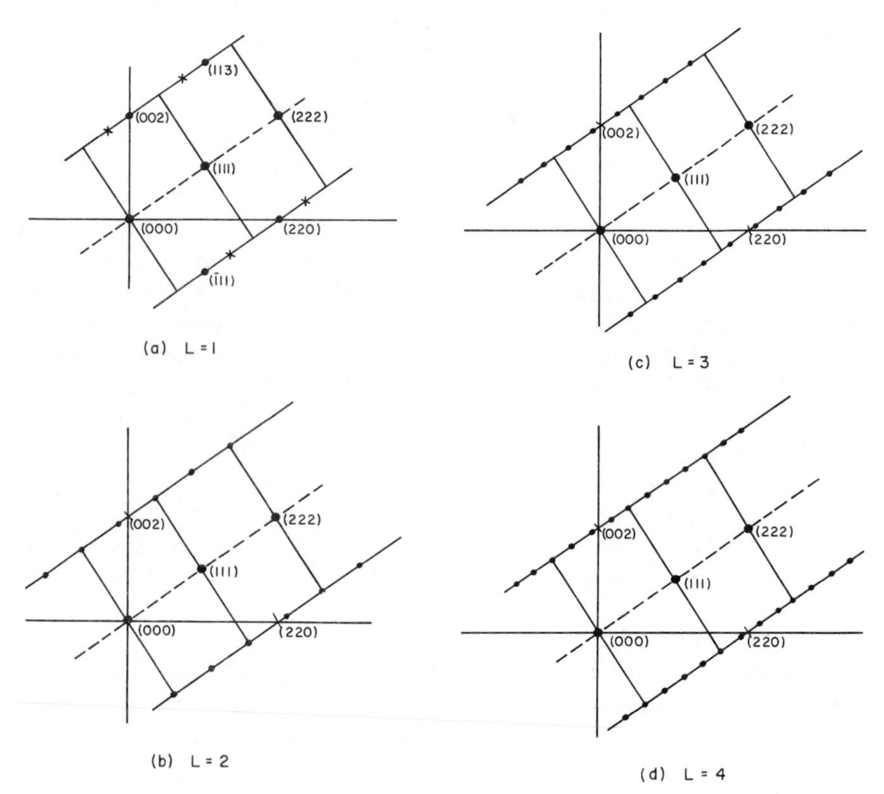

Fig. 3-27. Reciprocal lattices for structures formed by the close packing of close packed layers of atoms in different stacking sequences (L = 1 is 1 R^S f.c. cubic, L = 2 is $2H^S$, L = 3 is $3R^S$, and L = 4 is $4H^S$). In cubic notation the planes shown are ($1\bar{1}0$) including the [111] axis. (Sato and Toth [27].)

equation $e/a = (\pi/12t^3)(2 \pm 2x + x^2)^{3/2}$ (p. 93) indicates that the M = 1 structure should be stable at an e/a value of about 1.45 corresponding, with trivalent Mn, to a composition of about 22 at.% Mn, where the M = 1 structure does indeed occur. Since M does not increase with e/a, stacking modulations L = 5, 3, and 6 are introduced in order, successively, to shift the planes of energy discontinuity to accommodate the increasing Fermi sphere.

In the M = 1 structure the Fermi surface contacts $\{1, 0, \frac{3}{2}\}$ type boundaries (Figure 3-28). In the orthorhombic modulated structures the $\{1, 0, \frac{3}{2}\}$ type spots separate into two groups with different shifts relative to the origin. Analysis of the shifts of the spots with the strongest intensity for each modulated structure ($1R^s$, $5H^s$, $3R^s$, $6H^s$) (Sato *et al.*[33]) indicates that the volumes which would be included within the Brillouin

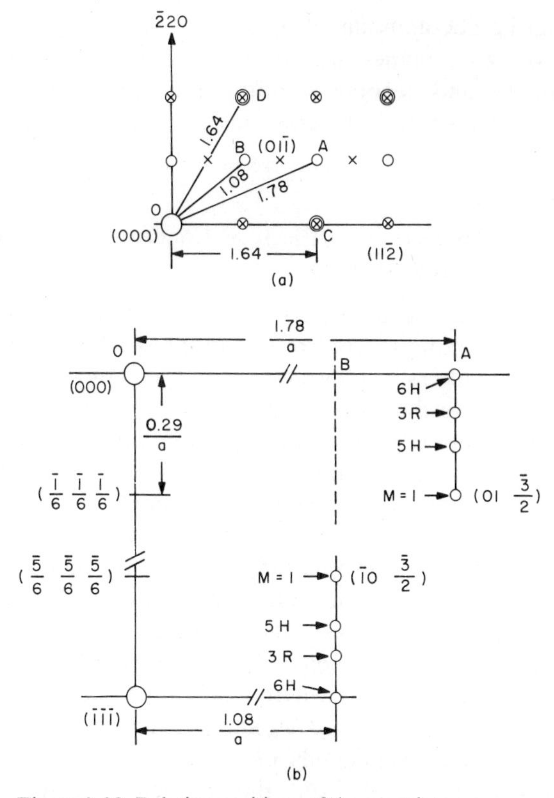

Figure 3-28. Relative positions of the most intense superstructure reflections of modulated structures which are related to the $\{1, 0, \frac{3}{2}\}$ reflections of the M = 1 structure for Au–Mn alloys. (*a*) Basal plane of the orthorhombic reciprocal lattice showing the projections of the locations of the shift axes. (*b*) shows the actual shifts of the reflections corresponding to $\{1, 0, \frac{3}{2}\}$ along the OA and OB axes in the reciprocal lattice for the various modulated structures. (Sato and Toth [27].)

zones corresponding to them increase in the observed order of their occurrence as e/a increases with the Mn content of the alloys (i.e., $1R^s \rightarrow 5H^s \rightarrow 3R^s \rightarrow 6H^s$).

In Au–Cd alloys about Au_3Cd (M = 2),[34] the observed order of structures modulated by layer stacking sequences is $1R^s$, $4H^s$, $6H^s$, $3R^s$ (Ramsdel's 12R, 4H, 6H, and 36R structures), and in Au–Mg alloys about Au_3Mg[35] it is $1R^s$, $4H^s$, $7R^s$, $10H^s$ with increasing electron concentration (Ramsdel's 12R, 4H, 84R, and 10H structures). These structural sequences are also in the order of increasing volume of Fermi sphere that can be accommodated in the energy zone as e/a increases, provided that the zone planes are estimated from a weighted average of both the strongest and secondary diffraction spots for the particular structures (strongest spots only were used for the analysis for $M = 1$ structures above).

Dependence of Composition on Electron Concentration in a Group of Phases with Tetragonal Structures

In the cases of ordered alloys so far described where electron concentration has been found to control the antiphase domain period or the structure among members of a family of polytypically related structures, the near-neighbor coordination of the atoms remains essentially unchanged between one structural modification and another. There is, however, a family of some 40 known tetragonal structures formed between transition metals, T, and Group III or IV elements, X, whose composition and structure depend in some way on electron concentration, and in which the near-neighbor coordination appears to change significantly between one structure and another. These structures, which are discussed in detail on pp. 594 to 602, were discovered mainly by Nowotny and co-workers[36] and reviewed by Jeitschko and Parthé,[37] and by Parthé.[38] The phases have formulas T_nX_m where n and m are integers and $2 > m/n \geqslant 1.25$. The unit cells contain n β-Sn like pseudocells of T metal atoms and m interpenetrating pseudocells of X atoms along the c axis. The β-Sn like pseudocells result from the T metals forming equilateral triangular 3^6 nets which are stacked in $TiSi_2$ like sequence, A°C°B°D°, on planes parallel to the (110) planes of the tetragonal structures (p. 594). The X atoms are present to the extent that they satisfy certain electron concentration requirements as discussed below, their m pseudocells being stretched up to equal the height along [001] of the n β-Sn pseudocells whose relative dimensions are fixed in providing equilateral triangular nets of the T atoms. This stretching of m pseudocells of X atoms to the height of n pseudocells of T atoms results in a "sinusoidal" like variation of the $T-X$ distances on proceeding along [001] from one T atom to the next, as noted by Völlenkle *et al.*[39] and shown in Figure 3-29. The effect of this regularity may well be to give comparable average near-neighbor environments to the atoms in the different tetragonal structures, and thus comparable contributions to the enthalpy, so that there is little change of free energy on going from one structure to another, as regards the contributions arising from near-neighbor interactions.

Nowotny and co-workers[36] and Jeitschko and Parthé[37] noted the importance of electron concentration in controlling the composition of

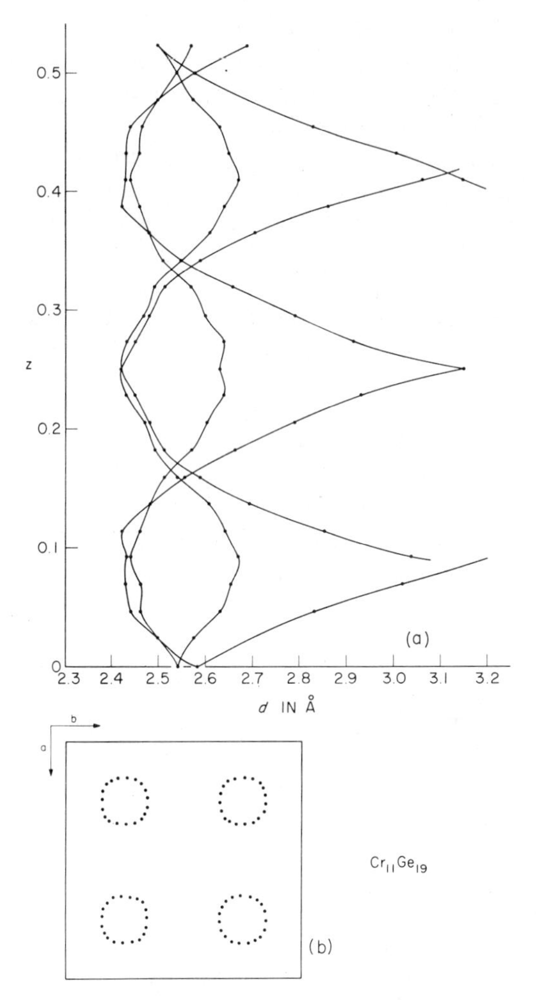

Figure 3-29. $Cr_{11}Ge_{19}$ (*tP*120) structure. (*a*) Cr–Ge interatomic distances for Cr atoms at fractional heights along [001] indicated by the ordinate. (*b*) Positions of the Ge atoms in the $Cr_{11}Ge_{19}$ structure projected on to the (001) plane.

these phases. When $m/n = 2$, the T and X atoms lie on the same planes and tetragonal symmetry with A°C°B°D° stacking is impossible; indeed $TiSi_2$ is orthorhombic; $CrSi_2$ is hexagonal with A°B°C° stacking, and $MoSi_2$, although tetragonal, has A°B° stacking of the layers. The possible value(s) of m/n for any two components appear to be controlled by a unique relationship between the valency of the T atom, V_T, and the ratio

mV_X/nV_T of the number of valence electrons contributed per formula by X to that contributed by T[40] (counting outer s, p, and d electrons) as shown in Figure 3-30a. Although the number of electrons per atom in the structures is uniquely prescribed according to the valency of the X and T atoms, there is no limiting value on its magnitude, as indicated in Figure 3-30(b). The electron concentration per T atom or per β-Sn pseudocell of T atoms is sensibly limited to 14 (or 56) as shown in Figure 3-30c; a fact also noted by Parthé. Indeed, the inflection in the curve of Figure 3-30a results from this limitation, since with the further limitation $m/n < 2$, it is not possible to maintain 14 electrons per T atom much below a V_T value of 7 (i.e., Mn), as indicated in Figure 3-30c. The unique curve of Figure 3-30a at V_T values above 7 results from structural stability associated with 14 electrons per T atom, but why a unique curve still appears to control the compositions of structures at V_T values between 7 and 4 as the number of electrons per T metal atom in the structures decreases from 14 to 12, is not at present clear. Experience seems to show that any m/n ratio that gives a point lying sufficiently close to the line of Figure 3-30a, results in a new tetragonal structure of this general type, and structures containing as many as 600 atoms in the unit cell have been reported.

Parthé[38] points out that the superstructure containing a given T metal requires that a certain number of electrons be contributed by the X atoms in order to achieve stability. Although his analysis does not require such an assumption, it would agree rather exactly with the requirement of 14 electrons (T metals contributing all of their s, p, and d electrons) per T metal atom of Groups VIII_2, VIII_1, and VII, the number then decreasing to 12 at Group IV (Figure 3-31). Facts such as these leave little doubt that the compositions and therefore the heights of the supercells of this family of structures are controlled by electron concentration. Since it can be demonstrated (p. 595) that the equilateral triangular arrangement of the T atoms in these structures controls the relative structural dimensions, and that there is no correlation between the relative dimensions of the β-Sn pseudocells and the composition ratio m/n, it cannot be argued that this pseudocell is anything more than a by-product of the A°C°B°D° stacking sequence of the T layers parallel to the (110) plane in the structures. Nevertheless, structural control by electron concentration of necessity involves some geometrical feature, since it results from the action of planes of energy discontinuity in reciprocal space. The one constant geometrical feature common to all of these 40 tetragonal structures is the β-Sn like pseudocell of T metal atoms with $c/na\sqrt{2} \sim 0.577 = 1\sqrt{3}$ (the value for equilateral triangles in the nets of T atoms), so it would seem that structural stability must result from the achievement of some constant electron concentration

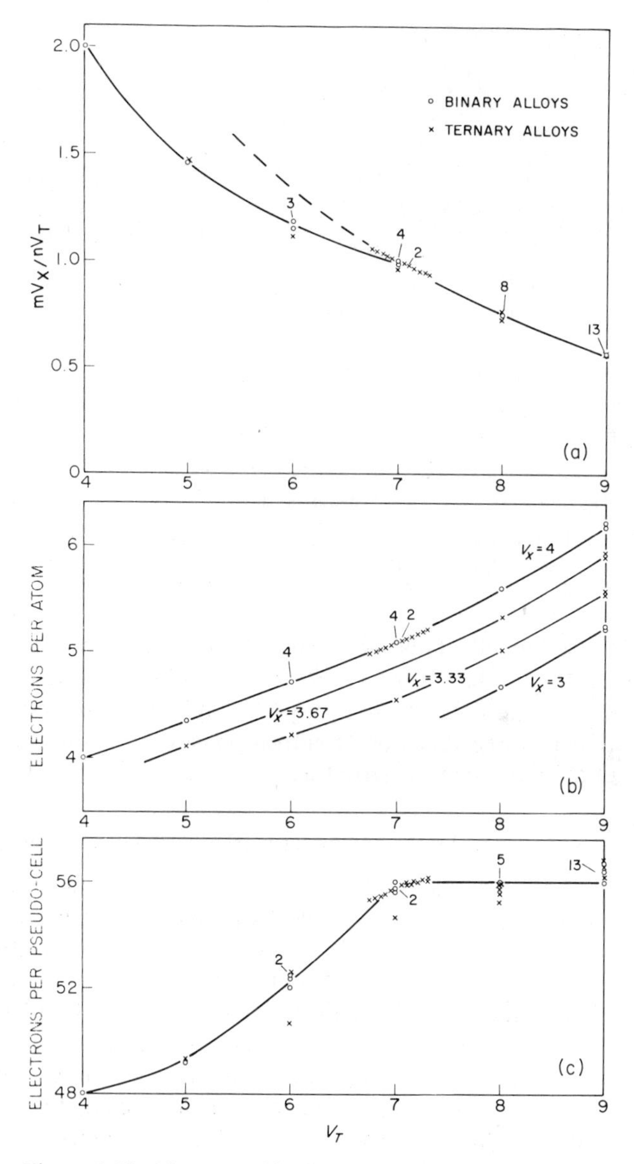

Figure 3-30. Nowotny chimney and ladder phases (*a*) The ratio mV_X/nV_T. (*b*) The number of electrons per atom. (*c*) The number of electrons per β-Sn type T metal pseudocell as a function of the average number of electrons of the transition metal atoms (counting outer *s*, *p*, and *d* electrons). The numbers on the diagram indicate the number of different crystal structure types represented by a single point. (Pearson [40].)

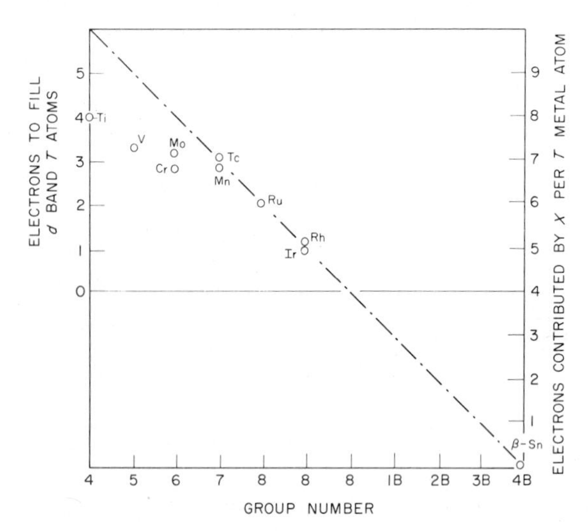

Figure 3-31. Number of electrons contributed by the X atom per T metal atom in Nowotny chimney and ladder phases. (Parthé [38].)

within this pseudocell. This conclusion is entirely consistent with the observed fact that (at least for $V_T \geqslant 7$) the controlling feature is a constant electron concentration *per T atom* or per β-Sn like pseudocell, and not as in all other recognized cases of "electron phases," a controlling electron concentration *per atom* in the structure.

Further Discussion of Electron Concentration and Alloy Structures

The relative importance of the energy band factor compared to other factors in controlling the stability of alloy phases is well illustrated by considering phases with the CsCl structure, characteristic of the Hume-Rothery β' phases whose composition is controlled by an upper limiting electron concentration of 3.0 per primitive cell (loosely 1.5 electrons per atom). Figure 3-32 shows a histogram for the occurrence of metallic phases with this structure at different apparent valence electron concentrations (i.e., outer s, p, and d electrons of the component atoms). The most striking feature of this is that the envelope bounding the counts at intervals of a half electron per atom shows no maximum corresponding to preferential occurence of phases with the CsCl structure at an electron concentration of 1.5 e/a, and by far the largest number of phases with this structure occur at electron concentrations of 2.0 and 2.5 e/a. Admittedly, this results from the large number of rare earth components with a valency of three, but even if the rare earth compounds are excluded (dotted line in Figure 3-32), there still appears to be no particular significance attached to the electron concentration of 1.5 *e*/*a* as regards the number of phases with the structure. Indeed, phases with the CsCl structure are primarily controlled not by electron concentration, but by the interaction between the *M* and *N* atoms as is indicated by the near-neighbor diagram in Figure 10-9, p. 569. Only when the chemical bond factor is relatively weak, as in phases formed between Cu, Ag, or Au and the Group II to IV B metals, does the electron concentration control the composition of the phases.

The sharp cut-off in the occurrence of phases with the CsCl structure above 3 *e*/*a* is also striking. Reference to the empty-lattice model for

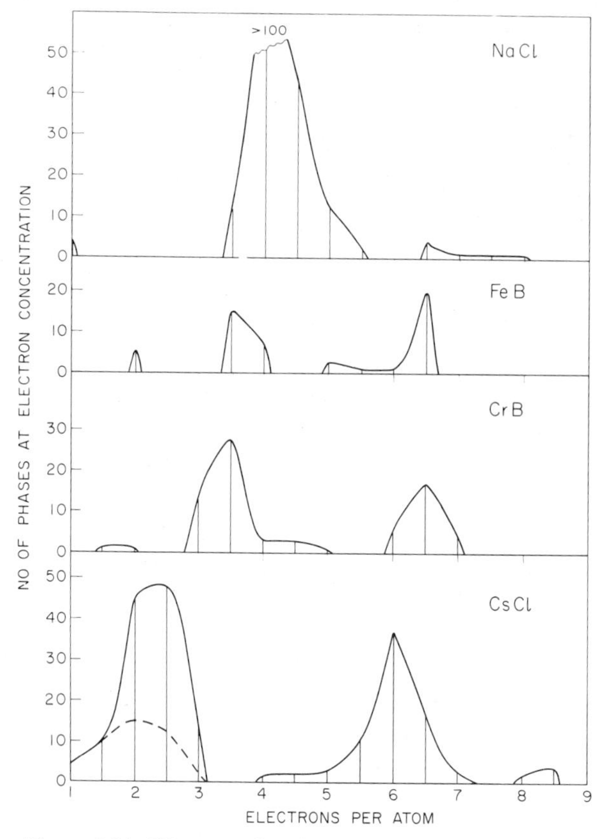

Figure 3-32. Histogram for the occurrence of phases with the NaCl, FeB, CrB, and CsCl structures at intervals of 0.5 outer *s*, *p*, and *d* electrons per atom. The broken line for CsCl structures indicates the numbers of phases which do not contain rare earths.

calculating the Fermi surface suggests no reason why this electron concentration should exert any limitation on the occurrence of the structure. However, it is also apparent from Figure 3-32, that in phases dominated by M–N interactions, the FeB and CrB types are preferred to the CsCl type at electron concentrations over three, and that the NaCl type is the preferred structure at an electron concentration of 4 when one of the components can attain a filled valence subshell by electron sharing and/or transfer.

The behavior of β-Al–Cu–Ni alloys provides a useful example to dispel possible arguments that an apparent limitation of the β phase boundary by an electron concentration of 1.5 e/a, results not from decreasing inherent instability of the β phase as this value is approached, but trivially from the much greater stability of the γ phase at a composition slightly richer in the B Group metal. As shown in Figure 3-33,[41] the Al–Ni β phase extends to a composition well above 50 at. % Al which would correspond to an e/a ratio considerably greater than the limiting value of 1.5, if the solid solution were substitutional. However, a defect solid solution is formed with omission of Ni atoms in alloys containing more than 50% Al so that the electron concentration does not increase above 3 electrons per primitive cell, thus firmly establishing the dependence of the alloys on electron concentration. It is, however, possible that the omission of Ni atoms occurs for spatial reasons because of the large

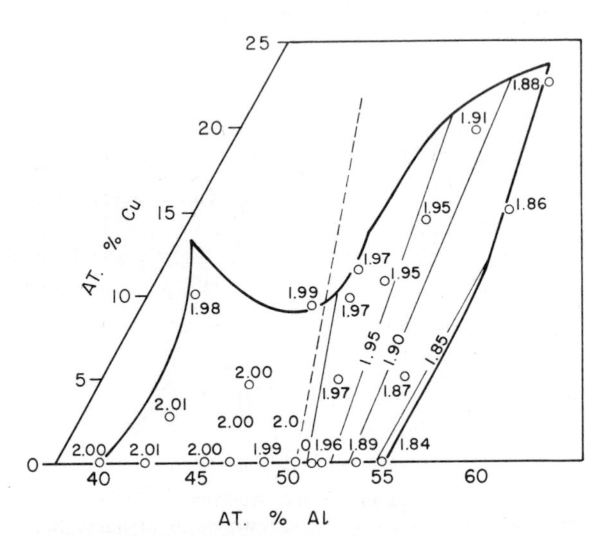

Figure 3-33. Boundaries of the β' phase (CsCl type) in Al–Cu–Ni alloys. Numbers give the number of atoms in the CsCl type cell. See text. (Lipson and Taylor[41].)

Al atom, but the results for the ternary β-Al–Cu–Ni alloys dispel this interpretation, since atoms start to be omitted when the electron concentration would exceed 3 per cell (broken line of Figure 3-33) rather than along the line in the ternary diagram representing 50 at.% Al. Just why the ternary alloys extend so far into the Cu field is uncertain, since in the alloys richest in Cu the electron concentration does greatly exceed 3 per cell.

A further maximum is found about an electron concentration of 6 e/a, in the histogram showing the occurrence of the CsCl structure as a function of electron concentration (Figure 3-32), and this results from phases containing transition metals. Our knowledge of the Fermi surfaces and band structures of metallic phases indicates that this is only an "apparent" electron concentration, resulting from two or three filled energy bands of valence electrons per atom lying below the Fermi level, and the effective electron concentration in partly filled valence bands is probably not different from that shown by the nontransition metal group of phases with the CsCl structure (i.e., 1 to 3). Herein lies an important point stemming from our knowledge of Fermi surfaces and energy band structures, that in phases with truly metallic structural characteristics, the effective electron concentration in partly filled energy bands is likely to be only one or two electrons per atom; all other apparent valence electrons are in filled energy bands lying below the Fermi level. If the effective electron concentration has a higher value, say 2.5 to 4 e/a, then structures are likely to be obtained in which M–N interactions (chemical bonds) play a dominant role. This is certainly the case when the electron concentration is high enough for one of the components to achieve a filled valence shell, and it generally results in a situation where all energy bands are filled and lie below the Fermi level at the absolute zero of temperature, giving semiconductor-insulator properties. Thus transition metal phases such as σ or α-Mn, whose occurrence appears to be somewhat influenced by an apparent electron concentration in the range of 6.5 to 7 electrons per atom, are seen as having an effective electron concentration in partly filled energy bands of 1 or 2 e/a, the remainder of the apparent electron concentration being present in filled energy bands below the Fermi level. Studies of the σ and other ternary transition metal phases by Beck and co-workers[42] reveal that the phases frequently exist over narrow ranges of composition across the ternary diagrams indicating a rather sharp dependence on electron number. Figures 3-34 and 3-35 give examples in the Cr–Mo–Co and Cr–Mo–Ni systems for comparison, and also in the Mn–Mo–Fe and Mn–Mo–Co systems where the extent of the σ phase is particularly striking.

In many Hume-Rothery electron phases, β, γ, and ϵ, it appears that

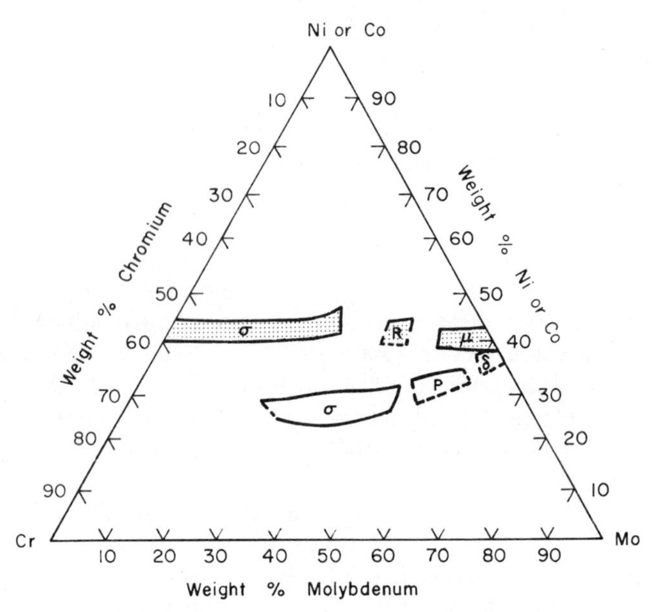

Figure 3-34. Location of the σ, R, μ, P, and δ phases in the Cr–Mo–Co (shaded) and Cr–Mo–Ni systems at 1200°C. (Beck and coworkers [42].)

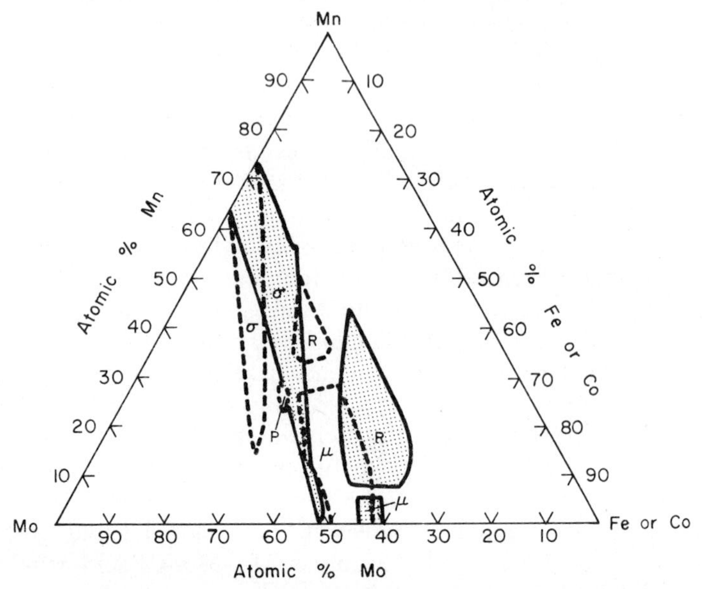

Figure 3-35. Location of the σ, R, μ, and P phases in the Mn–Mo–Fe (shaded) and Mn–Mo–Co systems at 1240°C. (Beck and co-workers [42].)

transition metals frequently contribute nothing to the "valence" electron concentration, since the rules are often obeyed with zero valency for the T metal. Fermi surface studies[43] confirm this behavior for β-PdIn, for example, since the observed de Haas–van Alphen periods are almost identical with those found for β-AgZn, implying $e/a = 1.5$ and hence zero valence electrons contributed by Pd. In other alloys, however, the T metals do contribute electrons to the valence band and Haworth and Hume-Rothery[44] have used the slope of the $\alpha/(\alpha+\beta)$ and $(\alpha+\beta)/\beta$ phase boundaries in ternary Cu–Zn–T and Cu–Al–T alloys to estimate the apparent valence electron contributions from the T metals, assuming that the Hume-Rothery rules are obeyed. Figure 3-36 summarizes some of their results and shows the derived T metal "valencies." Constitution is influenced by electron concentration in both of these examples—the occurrence of Hume-Rothery phases containing transition metals on the basis of their apparent zero valence, and the dependence of α and β phase boundaries in ternary alloys on the valence contribution of the added element.

Laves and Witte[45] have demonstrated that in ternary Laves phases, MN_2, of the elements Cu, Zn, Al, Si, Co, or Ag with Mg, electron concentration controls the choice between the cubic $MgCu_2$ and hexagonal $MgZn_2$ structures, the more complex hexagonal $MgNi_2$ modification occurring, if it is found, in the region between the other two (Figure 3-37).

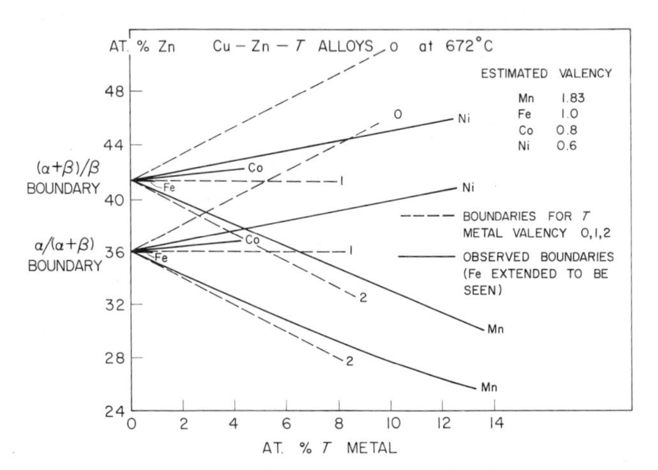

Figure 3-36. Observed $\alpha/(\alpha+\beta)$ and $(\alpha+\beta)/\beta$ phase boundaries (full lines) for Cu–Zn–T alloys at 672°C, and T metal valencies estimated therefrom. Broken lines indicate phase boundaries expected for T metal valencies of 0, 1, and 2. (Haworth and Hume-Rothery [44].)

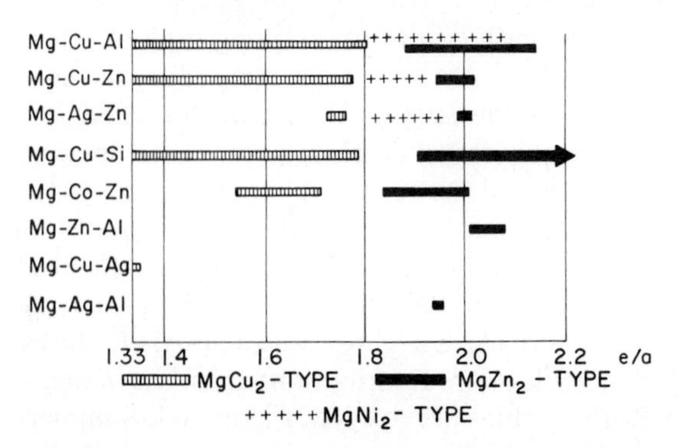

Figure 3-37. Distribution of $MgCu_2$, $MgZn_2$, and $MgNi_2$ structures as a function of electron concentration in ternary systems formed by Mg and other metals. (Laves and Witte [45].)

They observed that the $MgCu_2$ structure occurred in the range from about 1.3 to 1.8 e/a, thereafter the $MgNi_2$ structure might be found, and the $MgZn_2$ structure occurred generally in the range from 1.85 to 2.2 e/a. In more recent work, Komura and co-workers[46] found that in the $Mg(Cu, Zn)_2$ and $Mg(Ag, Zn)_2$ systems various $MgCu_2$ polytypes (p. 654) occur in the range of electron concentration between about 1.98 and 1.90 electrons per atom. Figure 3-38 shows approximately their findings in the $Mg(Cu, Zn)_2$ system, where the $MgZn_2$ structure, occurring at 2.0 e/a, soon gives way to other polytypes as the electron concentration is reduced. Specific heat data for alloys such as Mg–Cu–Al and Mg–Cu–Si[47] confirm the apparent influence of electron concentration on stability, since they indicate a rapidly decreasing N(E) curve as the boundary of the cubic phase is approached. When, however, the occurrence of all

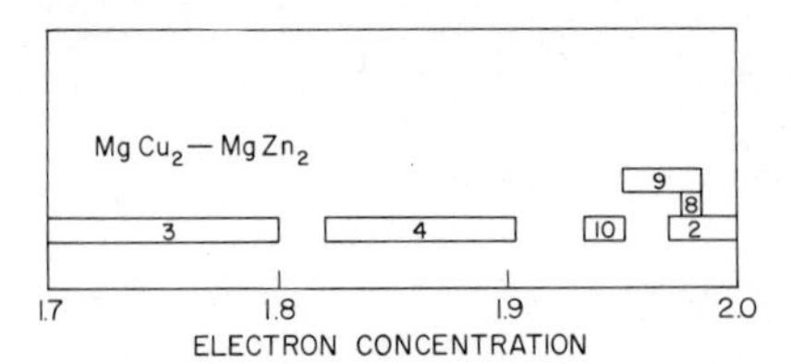

Figure 3-38. Diagram indicating very approximately the $MgCu_2$ polytypes found in the $Mg(Cu, Zn)_2$ system by Komura and co-workers [16]. The numbers indicate the number of $\alpha(ABC)$ groups in the repeat sequence of the polytype according to the description given on p. 654.

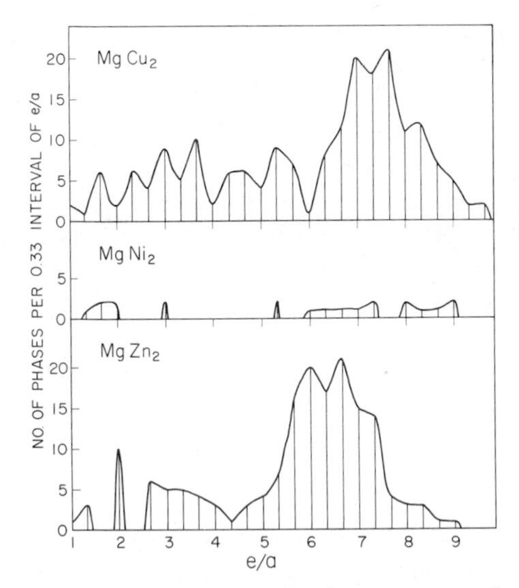

Figure 3-39. Histogram for the occurrence of $MgCu_2$, $MgNi_2$, and $MgZn_2$ Laves phases at intervals of 0.33 electron concentration per atom (number of outer *s*, *p*, and *d* electrons).

binary Laves phases as a function of the average valency electron concentration per atom is examined (Figure 3-39), it is seen that they cover the whole range of electron concentrations, and that there is a preference for the $MgCu_2$ structure at an apparent electron concentration about 7, and for the $MgZn_2$ structure at slightly lower values. Further analysis of the influence of the energy band factor, other than at low electron concentrations where it can be recognized, is impossible without a detailed knowledge of the energy band structures of the various phases, and furthermore, it is apparent (cf. p. 58) that the occurrence and structural dimensions of phases with these structures are generally controlled by geometrical factors.

Symmetry of the Electron States at the Fermi Level

The influence on crystal structure of the symmetry of the electron states at the Fermi level is more difficult to demonstrate than the influence of electron concentration. This is because when the character of the electrons is strongly pronounced (e.g. in sp^3 bonds), the situation is considered as the chemical bond factor and generally the Fermi level lies above filled energy bands. Nevertheless, the change of crystal structure on proceeding down a Group from, say, the diamond structure of C,

Si, Ge, and α-Sn to the distorted structure of β-Sn and then the f.c. cubic structure of Pb, or from the tetrahedrally coordinated structures of SiS_2 and GeS_2 to the CdI_2 structure of SnS_2, is evidence for change from $s-p$ to predominantly p symmetry of the electron states at the Fermi level as the atomic number of the elements increases.

The elemental rare earths and their alloys appear to provide some evidence which might be interpreted as showing the influence of the $4f$ electrons on structure, since it is difficult to account logically for both their structures (Table 3-3) and for changes of their structures with pressure (Figure 3-40) on the basis of atomic size, whereas the changes can be explained if the relative estimated sizes of the $4f$ shells are considered.[48]

Although it seems improbable that alone, the form of the pseudopotential could be used to account for the systematic changes through the family of polytypic structures that occur in the elemental rare earths[49] and some of their alloys, even though it has been used in arguments explaining why certain metals have f.c. cubic, c.p. hexagonal or other structures,[50] a more sophisticated development by Havinga *et al.*[51] appears to account

Table 3-3 Structures of Trivalent Rare Earths

Metals	Structure Type	Layer Stacking Sequence
La and Ce	f.c. cubic	c
La, Ce, Pr, Nd and Pm	Double hexagonal c.p.	$(ch)_2$
Sm	Rhombohedral	$(chh)_3$
Gd to Lu	Hexagonal c.p.	h

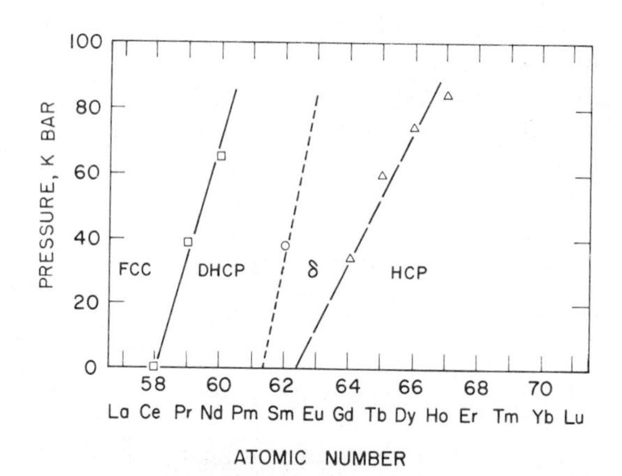

Figure 3-40. Structural changes in rare earths as a function of pressure. (Gschneidner and Valletta [48].)

well on a qualitative basis for these polytypic structure sequences both in rare earth elements and alloys, and in other non-transition metal alloys. This development includes "(1) a screening, due not only to the finite free paths of the Fermi electrons, but also due to the finite values of the pseudopotentials, (2) relativistic effects, which for many metals are equally important as the nonrelativistic scattering of the Fermi electrons and (3) a volume effect in ordered MN_3 compounds, depending on the radius ratio of the atoms M and N." The stability of the close packed polytypes is found to depend mainly on the relative importance of three factors: normal scattering-matrix elements in the band-structure energy, relativistic scattering terms, and "singular" effects due to interaction of at least two Bragg planes simultaneously. In MN_3 compounds a volume effect is also included. These authors also point out that explanations of polytypic structure sequences in the elemental rare earths and their alloys, based on the influence of the $4f$ electrons, are very improbable, although they do not discuss the influence of pressure on the structures of the elemental rare earths, which this proposition sought to explain.[48]

Influence of Energy Band Factor on Relative Structural Dimensions

For many years Jones' theories[52] of the interaction of Brillouin zone planes and Fermi surface were held to account for the relative cell dimensions of hexagonal metals and alloys. More recently the results of Fermi surface studies of the Group II metals, although not generally inconsistent with Jones' explanation of their relative dimensions, have shown the account of the striking variation of the lattice parameters of the Mg–Cd solid solution[53] to be untenable, since Γ overlap (*vide infra*) has already occurred in elemental Mg, thus destroying one of the most pleasing explanations resulting from the theory. More recently still pseudopotential theory, which adopts a spherical Fermi surface and is based on quite different assumptions to Jones' model, has been able to account qualitatively for the relative cell dimensions of the Group II hexagonal metals, the distortion of the Hg structure, and for the dimensional behavior of the Mg–Cd solid solution.[54] Full details of the pseudopotential method are discussed in Vol. 24 of *Solid State Physics*;[55] it is not possible to go into details of it here, especially since further quantitative developments are required to established thoroughly the credence of the new explanations of the relative cell dimensions of hexagonal metals and alloys. In these circumstances the developments of the Jones' theory are discussed historically in the following section in order to present the experimental data and to reveal the problems inherent in their interpretation.

According to Jones[52] the interaction of Brillouin zone planes and the valence electrons of noncubic metals such as those with c.p. hexagonal structures may introduce shear strains causing the relative cell dimensions (axial ratios) to depart from values that might be regarded as "ideal" (i.e., 1.63 for c.p. hexagonal metals). Such strains are particularly expected to develop when a nearly filled zone is overlapped into succeeding zones, as in the case of the Group II hexagonal metals where electrons from zones 1 and 2 overlap into zones 3 and 4. On this basis Jones attempted to account for the relative cell dimensions of these Group II metals and of solid solutions in Mg which result in changes of electron concentration. Although this and other work referred to below has doubtful significance today, since the model assumed for the Fermi surface of Mg differs in important aspects from that determined experimentally at a later date (*vide infra*), it is necessary to discuss it in some detail to show how data on the Fermi surfaces of these metals raise problems on dimensional changes observed in their alloys. Furthermore, Jones' conclusions concerning the effects of the interactions of valence electrons and Brillouin zone planes on relative cell dimensions are independent of the model to which they were actually applied.

Hexagonal Metals. Hume-Rothery and Raynor[53] accounted for the dimensional changes in magnesium alloys in terms of postulated overlaps of various planes of the first and second hexagonal Brillouin zones, and the striking lattice parameter changes in the Mg–Cd solid solution (Figure 3-41) were accepted as one of the stronger experimental confirmations of Jones' theory. In the literature (e.g., Jones,[52] Hume-Rothery and Raynor,[53]) this discussion has been given in terms of the extended zone shown in Figure 3-42(i) which is the first and second Brillouin zones for a primitive hexagonal Bravais lattice, or in terms of an incomplete zone composed of part of it, and overlaps were referred to as A, B, or Q (now M, Γ, or K *vide infra*). In this book all discussion is based on the primitive hexagonal Brillouin zone shown in Figure 3-42(ii) where symmetry points are designated in order to describe the zone origin and faces and edges of interest. The figure also shows the second Brillouin zone mapped in the reduced zone.† Briefly, following Jones (1934,[52]), Hume-Rothery and Raynor contended that for Mg the first Brillouin zone was nearly full, heavy overlaps into the second zone occurring on the faces at A and M, and the second zone overlapped into the third zone about K, whereas for Cd the second zone overlapped into the third zone across

†Translation of the portions of the second extended zone by reciprocal lattice vectors normal to the faces of the first or reduced zone which they cover, packs them into the reduced zone.

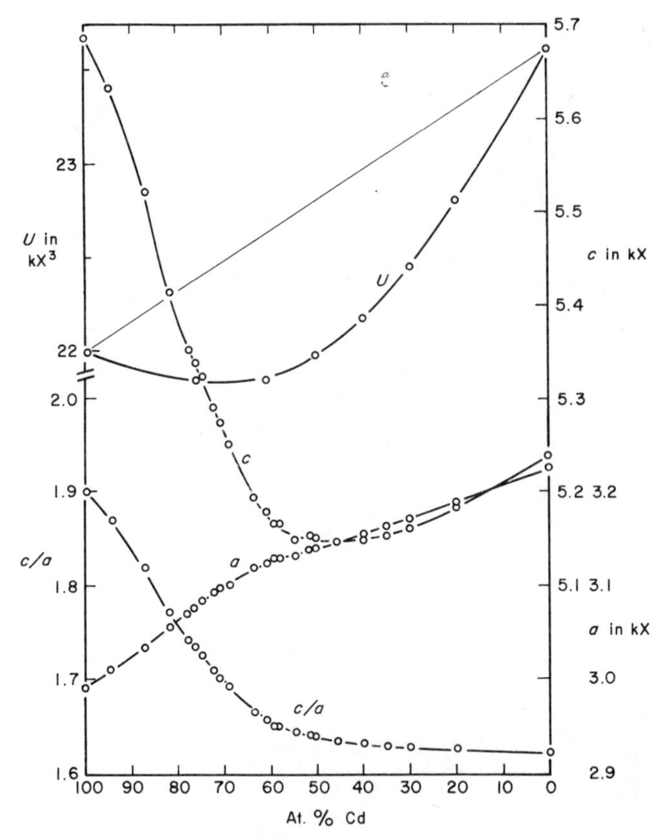

Figure 3-41. Variation of unit cell volume, *a*, *c*, and *c*/*a* with composition for the Mg–Cd solid solution at 310°C. From data of W. Hume-Rothery G. V. and Raynor[53].

planes centered at Γ, although there was no overlap at K. The Γ overlap accounted for the large *c*/*a* value (~ 1.89) in Cd (and Zn) compared to Mg (which has an almost ideal value), because in Cd the shear strain which it causes contracts the zone normal to the plane, or expands the crystal along the *c* axis in real space. When Cd is added to Mg, both *a* and *c* contract (Figure 3-41) consistent with the smaller size of Cd, until in the region of 40 to 60 at.% Cd, the Γ overlap was thought to begin and the K overlaps to decrease until they finally disappeared. This results, as the Cd content of the alloy increases, first in a fairly constant *c*, then a rapid increase of *c* because of compression of the Brillouin zone in the [001] direction of real space. The more rapid decrease of the *a* parameter

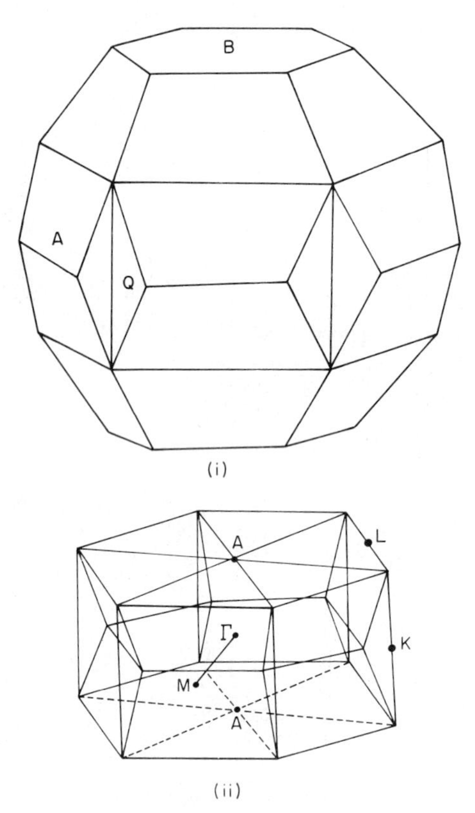

Figure 3-42. (i) Extended first and second Brillouin zones for the hexagonal c.p. ($A3$) structure. (ii) Reduced Brillouin zone with the second zone mapped therein for the hexagonal c.p. ($A3$) structure. Designation of symmetry points is indicated.

beyond about 65 at.% Cd was held to result from decreased overlaps of the first Brillouin zone at M.

Secondly, Raynor reported[53] sharp changes of slope in the c lattice parameter (but not a) of Mg in its solid solutions with solutes that increase the electron concentration (In, Tl, Sn, and Pb), but not with solutes Ag and Cd, that do not increase e/a. These changes, which resulted in an increase of the rate of change of c with solute concentration, were then incorrectly interpreted as the result of the onset of the Γ overlap in Mg at an electron concentration slightly above 2.0 e/a. Some of Raynor's experimental data on which this interpretation is based, have subsequently been disputed, but it may be observed that the absolute thermoelectric power of Mg alloys shows exactly analogous effects suggesting some sharp change in magnesium Fermi surface in alloys with In, Al, Sn, and Pb which increase the electron concentration, but not in alloys with Ag and Li that decrease it.

Experimental Fermi surface studies of the Group II metals have now shown (Figure 3-43[56]) (1) hole surfaces in zones 1 and 2 which imply A overlaps of zone 1. (2) Electron "lenses" (Γ overlaps of zone 2 into 3) in Cd, Zn, and Mg, decreasing in size in this order. (3) Electron "cigars" (K overlaps of zone 2 into 3), absent in Cd, very small in Zn, and substantial in Mg and Be. (4) V shaped surfaces (L overlaps from zone 2

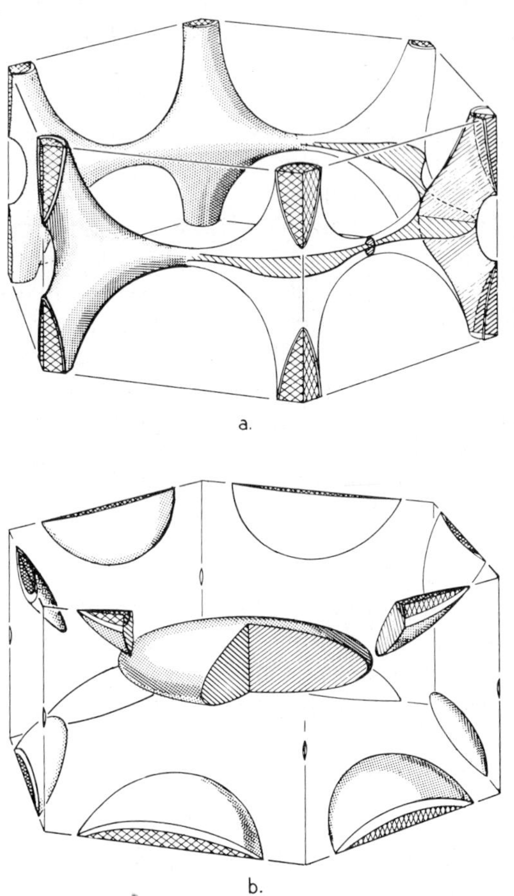

Figure 3-43 (*a*) Fermi surface in first and second reduced Brillouin zones for Zn. (*b*) Fermi surface in third and fourth reduced Brillouin zones for Zn. (Gibbons and Falicov[54].) *Note*: Minor details may by now be revised.

to zones 3 and 4). Only the zone 1 and 2 hole surface and zone 3 cigars are found for Be. These observations are consistent with the explanations given above for the large axial ratio of Cd and Zn (1.886 and 1.856, respectively) and they are consistent with Be having an axial ratio (1.568) less than the ideal value, if the K overlaps results in an expansion of the cell edges in the basal plane (i.e., along a). It could also be argued that the presence of both "lenses" and "cigars" (Γ and K overlaps of the second zone) in Mg resulted in simultaneous expansion of a and c with little change of c/a from the ideal value (c/a for Mg is 1.624). However, the occurrence of substantial Γ overlaps already in Mg appears to demolish completely the explanation of the Mg–Cd alloy c lattice parameter variation and the bends in Mg alloy c parameter curves at an electron concentration slightly above $2e/a$. If the Γ overlap has already taken place in Mg, it can not account for the rapid increase of c with composition above 60 at.% Cd in Mg–Cd alloys or above $2\ e/a$ in Mg alloys. These alloys are further discussed below.

Including the terminal Zn and Cd solid solutions, η, three independent c.p. hexagonal phases, δ, ϵ, and η, are found in alloys of the noble metals with the following B Group metals. The axial ratios of these phases show a characteristic variation with electron concentration[57,58,59] which is in large measure independent of the component metals (Figures 3-44 and 3-45). The relative changes in cell dimensions with e/a ratio have also been interpreted in terms of Brillouin zone overlaps that occur as the electron concentration increases.[14,52,57] The main features of the axial ratio changes are, first, the onset of decreasing c/a with increasing electron concentration in the region of the δ phase at an electron concentration of about 1.40, and secondly, the continued decrease of c/a throughout the ϵ phase field until at an e/a ratio of about 1.85, it starts increasing rapidly. At or just beyond this composition, the ϵ phase ceases to be stable and it is separated by a heterogeneous region from the terminal Zn or Cd solid solution (η phase), which has a much larger c/a value that continues to rise rapidly until the pure metals are reached.

At the ideal axial ratio the (100) planes of the close packed hexagonal structure are closer to the origin in reciprocal space than the (002) planes, and it is therefore reasonable to assume that M overlap of the first into the second zone (Figure 3-42 p. 118) starts at an e/a value of about 1.4 causing a more rapid expansion of the unit cell in the basal plane and the decrease of c/a with increasing e/a.† This process continues until in the

†Following this e/a ratio, the c parameter remains approximately constant or decreases somewhat (Figure 3-45), presumably in order to keep the unit cell volume and therefore the energy from increasing excessively.

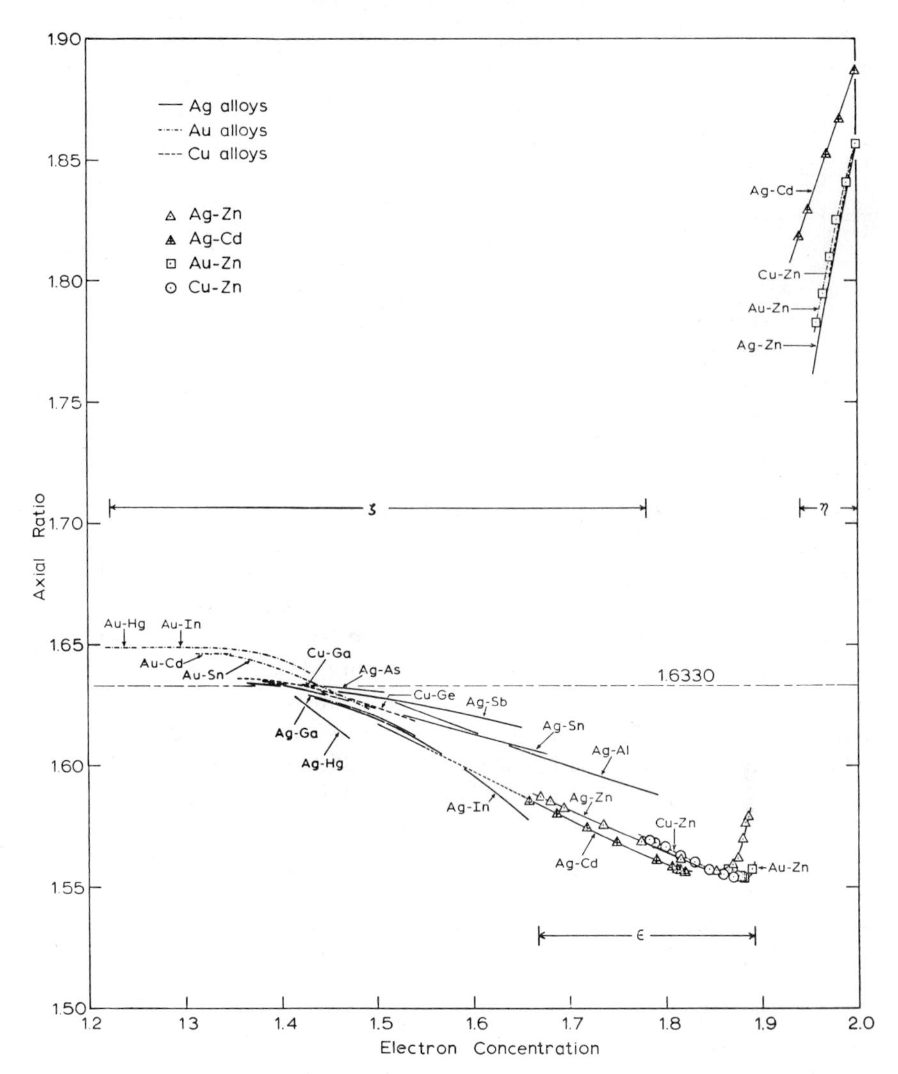

Figure 3-44. Axial ratio, c/a, as a function of electron concentration per atom in the ζ, ϵ, and η phases of Cu, Ag, and Au alloys which have the hexagonal c.p. ($A3$) structure. (Massalski and King [56].)

ϵ phase at $e/a \sim 1.85$, the onset of Γ overlap causes a rapid expansion of c and increase of c/a. The question also arises whether the K overlap might have occurred in ϵ phase alloys with c/a values less than 1.60; this also would tend to decrease c/a.

The onset of Γ overlap causes such a rapid change of structural

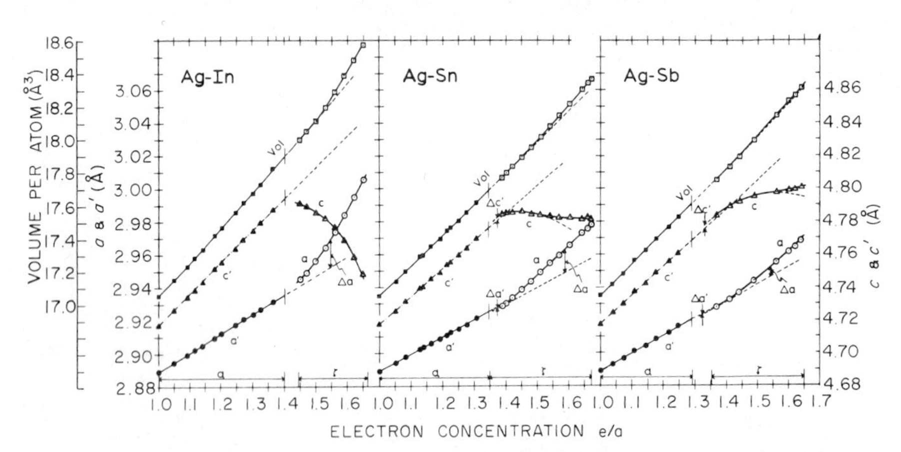

Figure 3-45. Variation of atomic volume, a and c with electron concentration in the α f.c. cubic solid solution (expressed in terms of a hexagonal cell) and the c.p. hexagonal phase of Ag–In, Ag–Sn, and Ag–Sb alloys (Massalski[57].)

dimensions that no homogeneous alloys are found between e/a values of about 1.9 and 1.95. The η solid solutions first occur at e/a values of 1.95, or higher, where the unit cell axial ratios have values greater than 1.77. Some appreciation of the structural disruption that is caused by the Γ overlap can be seen from considering the c lattice parameter of Ag–Zn alloys. At the ϵ phase limit at 89 at.% Ag the c parameter is 4.407 Å, whereas at the η phase boundary at about 95.6 at.% Ag, c has a value of 4.773 Å and 4.947 Å for pure Zn. The required change in c would therefore be 0.54 Å over 11 at. % in composition, or at a rate of more than 100% over a 100% range of composition. Such a change is clearly much too large to be achieved by homogeneous solid solution, and a heterogeneous region must separate the two c.p. hexagonal phases. It is also interesting to note that no pure c.p. hexagonal metals are known with axial ratio values between 1.65 and 1.85 and the only homogeneous c.p. hexagonal solid solution covering this range of axial ratios is that in the Mg–Cd system discussed above.

Solid solutions of the Group II metals do not extend very far to higher electron concentrations, heterogeneous phase fields occurring. Nevertheless such evidence as there is, indicates that the axial ratio in Group II metal solid solutions continues to increase rapidly at electron concentrations greater than 2.0.

Mg–Cd alloys are certainly remarkable since c changes homogeneously from about 5.18 Å at 60 at. % Cd to 5.68 Å at pure Cd, corresponding to a change of about 24% over a range of 100% in composition. Since 15% size difference is the limit normally found for extended solid solutions, let alone for continuous solid solution, the homogeneous change of *c* parameter at a rate of 24% over the whole composition range is indeed surprising, even though it is facilitated by a simultaneous contraction of *a*. Furthermore, the atomic volume (Figure 3-41, p. 117) contracts markedly compared to a line joining the atomic volumes of Mg and Cd. Although Hume-Rothery and Raynor suggested that the contraction of *a* and *c* in the Mg-rich solid solution was the normal contraction due to the smaller size of Cd, it is clear from Figure 3-41 that the contraction is very much greater. Nevertheless, the axial ratio remains almost constant at the value for Mg in this region suggesting that there is no great redistribution of electrons overlapping Brillouin zone planes, and the contraction must presumably be attributed to the introduction of *d* electron states with Cd. Considering ϵ and η alloys, it is apparent that the only reason why this solid solution can form, is the similarity of both Mg and Cd in having Γ overlap. Indeed, the nature of the solid solution which readily forms Mg_3Cd, MgCd, and $MgCd_3$ superstructures at low temperatures, suggests great electronic similarity of the two components, with superstructures forming through slight adjustments of the energy. Even so, this picture is at variance with the behavior of the unit cell dimensions which indicate a dramatic redistribution of crystal energies. The 9.7% increase in *c* between 60 and 100 at.% Cd is accompanied by a 2% decrease in *a* in the same composition range, but it seems most unlikely that the disappearance of the K overlap could cause the contraction of *a*, with the expansion of *c* occurring only to maintain an appropriate atomic volume in the alloys. Such considerations do not therefore appear to provide a satisfactory explanation of the behavior of the Mg–Cd alloy system. The difference between Zn and Cd on the one hand and Mg on the other, certainly arises in part from the presence of occupied *d* states in the former and their absence in the latter, although it is not obvious how this difference might account for details of the dramatic variation of lattice parameters in Mg–Cd alloys. On the other hand it is noted that pseudopotential theory has been able to account, at least qualitatively for the axial ratios of the hexagonal metals and in particular for the variation of c/a across the Mg–Cd solid solution.[54]

One further feature is of interest in Mg–Cd alloys; the superstructures $MgCd_3$ and Mg_3Cd are formed with doubling of the *a* axis of the unit cell† and therefore with introduction of inner Brillouin zone boundaries in reciprocal space within those of the disordered solid solution, the boundaries corresponding to (100) planes in real space. Such adjustments

†These have the $SnNi_3$ (*hP*8) type structure which is an ordering of a c.p. hexagonal solid solution analogous to the $AuCu_3$ ordering of a f.c. cubic solid solution.

would not be expected to influence the Γ overlap greatly, and it is perhaps significant in considering that the changes of lattice parameters of the disordered solid solution are not related to Γ overlap, to observe that 2 *c*/*a* for Mg_3Cd is almost identical with the value for $MgCd_3$, the values being 1.608 and 1.618 respectively. The approximate *c*/*a* values for the corresponding disordered alloys are respectively 1.63 and 1.72.

Regular variation of the axial ratio of c.p. hexagonal phases with average valence electron concentration (*s*, *p*, and *d* electrons) is also found in the transition metal series as pointed out by Rudman.[60] Regardless of the component atoms, a minimum in *c*/*a* occurs at a valence electron number of about 8, and it is probable that there is a maximum in the region of 6 valence electrons, low axial ratios again being found in the region of 4 or 3 valence electrons (Figure 3-46). Since, however, the number of apparent valence electrons occupying filled energy bands lying below the Fermi level is unknown, it is impossible to compare these results with the variation of *c*/*a* and *e*/*a* found in alloys of the noble and B Group metals.

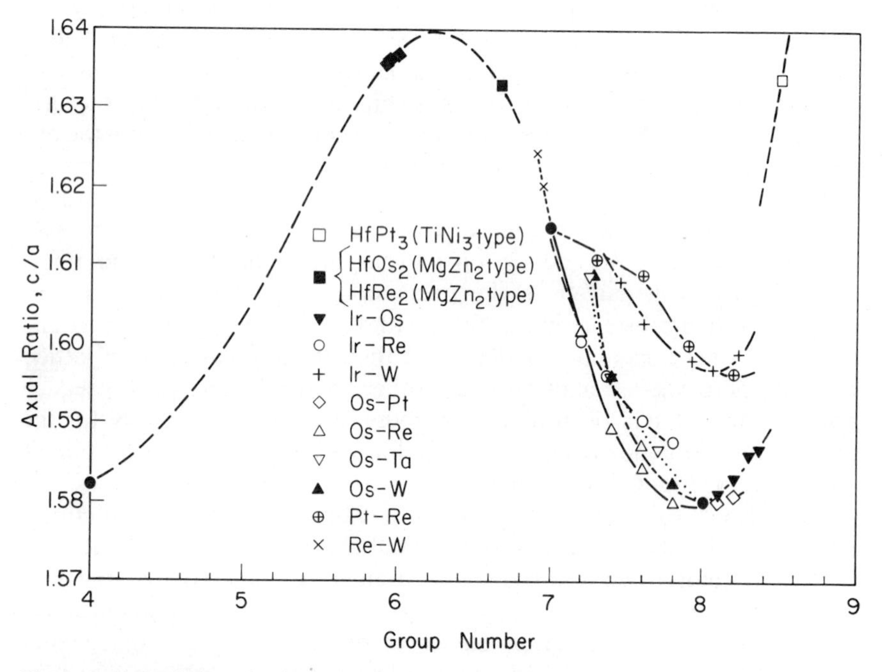

Figure 3-46. Variation of axial ratio, *c*/*a*, with average Group number in hexagonal phases of transition metal alloys which have the c.p. hexagonal (*A*3) structure, unless otherwise noted. (P. S. Rudman, private communication.)

Cubic Metals. It can also be demonstrated that the cell dimensions of cubic metals in their solid solutions are subject to energy band effects because of differences in the potential fields of the solvent and solute ions, particularly when they are of different valencies. Lattice parameters of terminal solid solutions of B Group solutes in the noble metals show a dependence on the valency difference of solute and solvent, although they also depend on other factors such as relative atomic size, since the lattice spacing curves are not superimposed when plotted as a function of e/a. Thus Owen[61] has shown that there is generally a linear dependence of the percentage lattice distortion $\Delta a \times 100/a$ of the solid solution on the solute valency, although there are several exceptions to this rule. Raynor[62] showed that for these solvents and solutes $\Delta R/(e-1) = K(V_{\text{solute}} - V_{\text{solvent}} + C)$ where K and C are constants ($C = 0$ for elements from the same Period); e is the electron-atom ratio, V the valency and ΔR is the difference between the observed atomic radius and that calculated for size effect, taking the closest distance of approach in the elemental structures of Group I, II, and IV B elements for their relative sizes. ΔR is thus the lattice distortion due to valence difference after elimination of size effect.

Since for a monovalent solvent $(e-1) = [n_M/(n_M + n_N)](V_M - 1)$ where $n_M/(n_M + n_N)$ is the atomic fraction of the solute M, $\Delta R(n_M + n_N)/n_M = K(V_M - 1)^2$ for the case where $C = 0$ and both solvent and solute belong to the same Period. A plot of ΔR for constant solute concentration, such as one atomic percent, is linear with $(V_M - 1)^2$ for solvent and solute of the same Period. The relationship does not hold with Group V solutes such as As and Sb, and it is apparent that the large electrochemical factor here present also influences the lattice parameter variation. Raynor's study serves to demonstrate the regular dependence of lattice parameters on valence differences of solvent and solute when other influential factors are absent, but, as with other energy band effects, it is readily masked by more important factors such as relative atomic size, relative size of ions and atom cores, and the electrochemical factor so that no useful quantitative calculations of lattice parameter variation can be made which are based on valence differences alone.

Summary

It is now becoming clear that there are many cases in which energy band effects control the stability of alloy structures. The first definite recognition of this was in the nineteen thirties with Hume-Rothery's work on the alloys of Cu, Ag, or Au with the subsequent B Group metals, and Laves and Witte's work on the Laves phases. In each of these cases other strong influences which might have dominated the energy band effects were absent, and the influence of electron concentration could be unequivocally demonstrated. In later years the work of Sato on the one hand and of Giessen on the other, has shown that electron concentration can control the choice of structure in a family of polytypic structures, where change of structure from one member to another does not alter the near-neighbor coordination. These ideas have been extended to show how structural distortion may maintain phase stability when further change of structure along a series is impossible. Much of the newer work demonstrates that the important factor is the occurrence of planes of energy discontinuity at the Fermi surface which lower the overall electron energies and thus give structural stability. The main criterion for finding energy band effects in a series of related alloy structures is that the near-neighbor coordination does not change significantly from one structure to the next so that, in the field of study, the major contribution to the enthalpy and thus to the stability of the alloys remains unchanged, since changes in this are likely to dominate completely the smaller changes in electron energies associated with energy band effects.

On the theoretical side a new approach—pseudopotential theory—is having increasing success in accounting for the stability of structures adopted by elemental metals and for the relative cell dimensions of noncubic crystals. In this treatment the energy of a metal crystal is computed as the sum of the Ewald energy and a band structure energy, calculated by the pseudopotential method.

THE INFLUENCE OF TEMPERATURE ON CRYSTAL STRUCTURE

The influence of temperature on the structures of metals arises principally because it changes the balance between the binding energy at the absolute zero and the entropy term which determine the stability of a phase according to the expression $G = H - TS$, involving the Gibbs free energy, G, the heat capacity, H, and the entropy, S. At low temperatures where TS is generally small, the stable structure is that with the lowest heat capacity or the greatest binding energy. As the temperature is increased another phase with higher entropy due to uncertainty of positional or electronic parameters may achieve a lower free energy and become stable leading to a phase transition.

Positional disorder, which increases the entropy, results from thermal vibrations that cause uncertainty in the location of atoms about their equilibrium positions, and also from random occupation of lattice sites in binary or multicomponent alloys, or from the random introduction of vacant sites. Vibrational disorder can, for example, be considered responsible for the $\alpha \rightarrow \beta$ Sn transition. Thermal vibrations are greater in β-Sn ($tI4$) than in α-Sn ($cF8$) which has the larger binding energy, and this results in $T\Delta S$ exceeding ΔH (cubic $\rightarrow$ tetragonal) at 292°K, the equilibrium transformation temperature,[63] as shown in Figure 3-47. The increased entropy due to random site occupation is responsible for the disordering of ordered alloys or superstructures as the temperature is raised. The ordered state may be stabilized at the lowest temperatures by the energy band factor (see p. 85), or by the increase of binding energy resulting from size effects if atomic repulsions are reduced by ions with larger cores surrounding themselves by those with smaller cores, or from chemical effects if better chemical bonding results when M is surrounded by N. As the temperature increases so does positional disorder, and this decreases the ordering force, cooperatively creating more disorder until a temperature is reached where the long-range order abruptly disappears. Thus the increased entropy due to disorder induces disorder as the temperature is raised, and disordered structures are stabilized relative to ordered structures at high temperatures because of the entropy term TS. Such considerations are pertinent to the stability of superstructures like Cu_3Au.

The vibrational entropy term which is larger in simple loosely packed structures than in complex highly involved structures, is held responsible for the transformation of the α and β forms of Mn to the simpler f.c. and b.c. cubic forms at high temperatures. The vibrational entropy term has also been held responsible for transformations from c.p. hexagonal or f.c. cubic to b.c. cubic structures at high temperatures, on the basis that

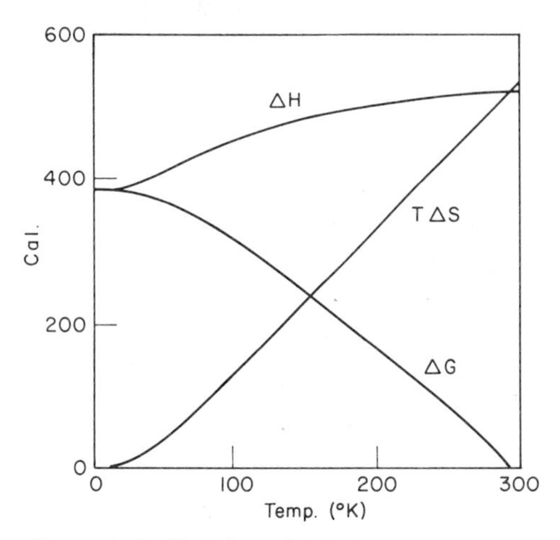

Figure 3-47. Variation of ΔH, $T\Delta S$ and ΔG with temperature for Sn. (Lumsden [63].)

the b.c. cubic structure is less closely packed and therefore has a higher vibrational entropy. However, this explanation now appears doubtful since it has been shown (see p. 139) that the c.p. structures are not necessarily more closely packed than the b.c. cubic structure, and the c.p. hexagonal structure is definitely less closely packed when its axial ratio decreases by more than about 0.05 from the ideal value. Furthermore, it seems uncertain whether the 8 close neighbors in the b.c. cubic structure may not offer more vibrational constraint than the 12 more distant neighbors in the c.p. structures.

Electronic entropy arising from the disorientation of magnetic spins in transition metal alloys is analogous to positional entropy in ordered alloys. At high temperatures the electronic entropy term leads to a disordering of the magnetic spins, whereas at lower temperatures depending on the strength of the magnetic coupling, the spins may order ferro- or antiferromagnetically, and this in turn may lead to structural changes as the temperature is lowered. For instance, antiferromagnetic spin ordering in cubic structures such as NiO, or quenched f.c. cubic Mn-rich Mn–Cu alloys causes a lowering of the crystal symmetry from cubic to rhombohedral in the first case and to tetragonal in the second. Antiferromagnetic spin ordering in hexagonal crystals generally only results in a change of the relative cell dimensions, as for example in MnTe (Figure 3-48) where the stresses due to ordering are absorbed by a change of axial ratio.[64]

Disordered structures which are stabilized by entropy at high temperatures may become mechanically unstable at lower temperatures if they have not ordered, because a shear motion may lower the interaction energy between the atom cores and result in a structure with a lower heat capacity. Such transformations which also involve some mechanical

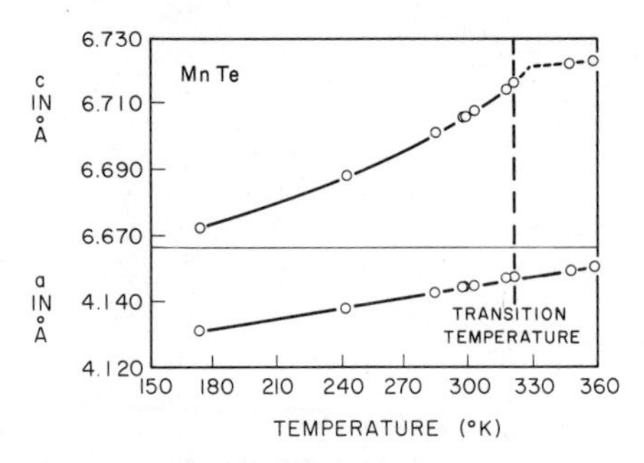

Figure 3-48. Variation of the lattice parameters of MnTe (NiAs structure) with temperature. (Greenwald [64].)

deformation of the parent or daughter structure, are known as martensitic transformations. Numerous martensitic phases may occur in a given system and in the β-brass region, for example, at compositions richer in Cu than the equiatomic, both disordered β and ordered β' brass may undergo martensitic transformations. The rate of cooling or quenching may be important in influencing the structure obtained, since it may be necessary to prevent decomposition of a phase into two other stable phases, or to prevent ordering on cooling in order to preserve the phase to low temperatures where it becomes mechanically unstable.

The most important effect of the increasing amplitude of vibration of the atoms about their equilibrium sites with increasing temperature, is the ultimate melting of the crystal. At temperatures above about 90% of the melting point, measurable numbers of equilibrium vacancies (vacant lattice sites) appear, and the number of these increases to the melting point where the percentage of vacant sites may be typically 10^{-3}%. Vibration amplitudes are greatest between atoms feebly bonded together, and this may result in the melting of a metal or semiconductor into a liquid containing molecules. Thus the melting of hexagonal Se (p. 239) results from the breakdown of the very feeble forces holding the chains together, although the Se chains remain intact in the liquid for many hundred degrees above the melting point of the solid. In other cases such as complex Al alloys containing Mg and Si, evidence has been found for the existence of $SiMg_2$ molecules in the molten alloys.

Thus it is seen that change of temperature can influence the structures of metals in many ways, and therefore studies of structure or constitution as a function of temperature are inherently important. Furthermore, from the relative change of structural dimensions with temperature (expansion coefficients), information can be derived concerning interatomic forces.

THE INFLUENCE OF PRESSURE ON CRYSTAL STRUCTURE

Experimental interest in studying the effect of the pressure variable on metals and alloys has led to the development of techniques for the simultaneous application of high pressures and high temperatures, and also for studying structural changes *in situ* at high pressure. The application of high pressure causes changes of structure to those with a higher degree of space filling, and these changes are generally facilitated by raising the temperature so that diffusion can occur more readily. In some cases the high pressure structures can be retained by pressure quenching and also by cooling the product below room temperature, and various apparently metastable structures may be obtained on removing the pressure, or on annealing samples in which the equilibrium high-pressure structure has been retained by quenching.

The type of changes found at high pressure resemble very much those that are found as a function of temperature. Thus phases may disproportionate forming two new phases, or there may be changes of crystal structure at essentially constant composition. In some cases the products are related to the initial structure, appearing as distortions of it, as in changes from the rocksalt to the GeS (*oP*8) type structure. In other cases where plausible distortions reducing the free space in the structure might be expected, entirely new structural types are found as in the case of $TaSe_2$,[65] for example.

In simple inorganic solids, there seems to be a general pattern of structural change, either from a regular structure with given coordination to a distorted form thereof, or from a structure of low coordination to a regular structure of higher coordination. In each case the tendency is for space filling and coordination to increase with increasing pressure. These series of changes are portrayed diagramatically in the few examples shown in Figure 3-49. In some instances where a change of crystal structure might have been anticipated, as in the case of Zn and Cd, the only effect of pressure appeared to be a change of axial ratio which decreased markedly, reaching a value of about 1.733 at 180 kb for Zn and 1.68 at 300 kb for Cd.[66] However, later work, has suggested that the structure of Cd might change from c.p. hexagonal to double hexagonal c.p. at pressures above about 135 kb.[67]

The effect of pressure on the crystal structures of the elements of Groups IV, V, and VI B is generally to induce changes such as are found on proceeding down the Group increasing the principal quantum number, *n* (Figure 3-50). This is not surprising since high pressure, in bringing the atoms closer together, must increase the relative metallic character

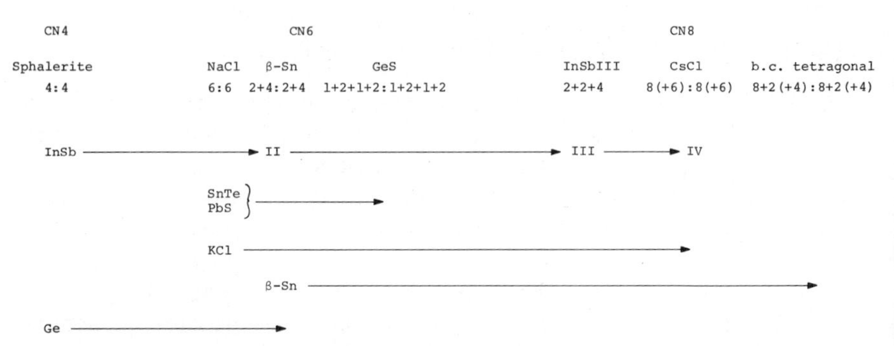

Figure 3-49. Changes of crystal structure type with increasing pressure for various substances.

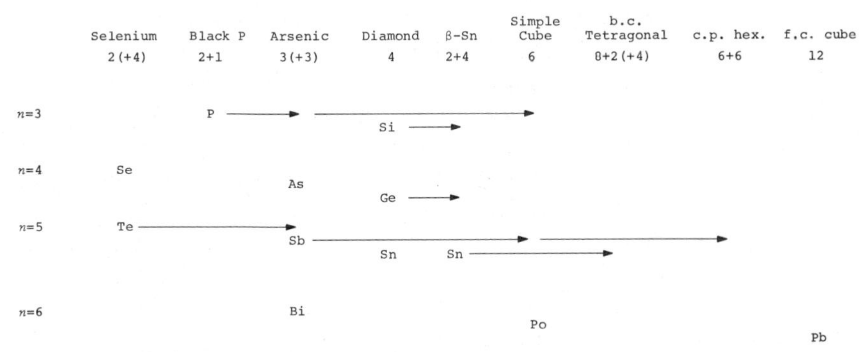

Figure 3-50. Crystal structures of elements to the right of the Zintl border, and changes thereof reported with increasing pressure; however, some of these, e.g. for Te and Sb, are disputed.

of these elements to the right of the Zintl border which have valence structures of relatively high atomic volume. Phosphorus best exhibits this behavior. At normal pressures black P has a two-layer structure with 2 + 1 coordination; between 50 and 83 kb this transforms to the arsenic type structure with 3 (+3) coordination and at 120 kb it takes the simple cubic structure which is not known at normal pressures in Group V, but occurs for Po in Group VI. The sequence of changes of Sb with increasing pressure are from the rhombohedral structure to a simple cubic form at about 50 kb, to an even more characteristically metallic structure, c.p. hexagonal with $c/a = 1.58$ at about 85 kb. Te is reported to change from its hexagonal chain structure with CN 2 (+4) to the rhombohedral As structure with increase of coordination to 3 (+3) at about 15 kb, and thence to a further undetermined structure at about 45 kb, but see comment in Figure 3-50 caption.

In Group IV, the tetrahedral surroundings of the Si diamond structure are greatly distorted in the Si III (*cI*16) form obtained at high pressures, although the coordination of Si remains 4. Another modification, Si II, which has the β-Sn type structure has also been reported. Ge transforms to the β-Sn structure under pressure, and another dense tetragonal form with CN 4 is formed on release of pressures greater than 120 kb. It has now been shown that this structure is a stable form under high pressure. Above about 39 kb at 314°C or 115 kb at room temperature, white tin transforms to a b.c. tetragonal structure with $c/a = 0.91$, which is a distortion of the b.c. cubic (*cI*2) structure, so that the atoms are arranged intermediately between close packing and b.c. cubic {110} packing. The report that it also forms a b.c. cubic structure at high pressures requires confirmation.

The effect of pressure on the close packed structures of metals is in some cases to induce changes from one member of the family of polytypic close packed structures to another. This behaviour is observed in the rare earth series where changes in the direction from c.p. hexagonal to Sm, to double hexagonal, to cubic are engendered by increasing pressure (Figure 3-40, p. 114). Fe changes to a c.p. hexagonal form at pressures above 130 kb, the axial ratio being 1.643 at 150 kb and 25°C.[68] Magnesium (and possibly Cd at 135 kb) changes from c.p. hexagonal to double hexagonal c.p. at pressures above about 50 kb (Perez-Albuerne *et al.*[67]).

Finally it is noted that there is much that is reminiscent in the structures induced by high pressures, to the metastable structures obtained by exceedingly rapid quenching experiments (p. 287).

REFERENCES

1. F. Laves, 1956, *Theory of Alloy Phases*, Cleveland: American Socicty of Metals, p. 124.
2. E. Parthé, 1961, *Z. Kristallogr.*, **115**, 52.
3. W. B. Pearson, 1968, *Acta Cryst.*, **B24**, 7; **B24**, 1415.
4. E. Teatum, K. Gschneidner, and J. Waber, 1960, U.S. AEC, LA-2345.
5. R. S. Mulliken, 1934, *J. Chem. Phys.*, **2**, 782; 1935, *Ibid.*, **3**, 573.
6. L. Pauling, 1932, *J. Amer. Chem. Soc.*, **54**, 3570; See also Pauling 1960, *The Nature of the Chemical Bond*, 3rd. ed, Ithaca: Cornell University Press.
7. W. Gordy and W. J. O. Thomas, 1956, *J. Chem. Phys.*, **24**, 439.
8. M. Haissinskij, 1946, *J. Phys. Radium*, **7**, 7.
9. W. Hume-Rothery, G. W. Mabbott, and K. M. Channel-Evans, 1934, *Phil. Trans. Roy. Soc.*, **A223**, 1.
10. W. Hume-Rothery and G. V. Raynor, 1954, *Structure of Metals and Alloys*, London: Institute of Metals.
11. L. Darken and R. W. Gurry, 1953, *Physical Chemistry of Metals*, New York: McGraw-Hill, pp. 86–89.
12. J. T. Waber, K. Gschneidner, A. C. Larson, and M. Y. Prince, 1963, *Trans. Met. Soc., AIME*, **227**, 717.
13. W. Hume-Rothery, 1966, *Acta Met.*, **14**, 21.
14. C. S. Barrett and T. B. Massalaski, 1966, *Structure of Metals*, 3rd ed, New York: McGraw-Hill, Chapters 12 and 13.
15. H. Jones, 1960. *The Theory of Brillouin Zones and Electronic States in Crystals*, Amsterdam: North Holland Publishing Co.,
16. W. Hume-Rothery, 1926, *J. Inst. Met.*, **35**, 295, 307.
17. W. Hume-Rothery, G. W. Mabbott, and K. M. Channel-Evans, 1934, *Phil. Trans. Roy. Soc.*, **A233**, 1. For later summary see W. Hume-Rothery and G. V. Raynor, 1962, *The Structure of Metals and Alloys*, 4th ed, London: Institute of Metals.
18. H. Jones, 1934, *Proc. Roy. Soc., Lond.*, **A144**, 225; *Ibid.*, **A147**, 396; 1937, *Proc. Phys. Soc., Lond.*, **49**, 250.
19. B. W. Veal and J. A. Rayne, 1962, *Phys. Rev.*, **128**, 551; 1963, *Ibid.*, **132**, 1617.
20. K. H. Johnson and H. Amar, 1965, *Phys. Rev.*, **139**, A 760; 1966, H. Amar, K. H. Johnson, and K. P. Wang, *Ibid.*, **148**, 672; F. Arlinghaus, 1967, *Phys. Rev.*, **157**, 491.

21. J.-P. Jan, W. B. Pearson, and Y. Saito, 1967, *Proc. Roy. Soc., Lond.*, **A297**, 275.
22. W. H. Taylor, 1954, *Acta Met.*, **2**, 684.
23. R. B. Maciolek, J. A. Mullendore, and R. A. Dodd, 1967, *Acta Met.*, **15**, 259.
24. W. A. Harrison, 1963, *Phys. Rev.*, **131**, 2433; 1964, *Rev. Mod. Phys.*, **36**, 256.
25. S. J. Cho, 1969, International Colloquium on Rare Earth Elements, Paris-Grenoble, May 5–10, CNRS.
26. S. L. Altmann, C. A. Coulson, and W. Hume-Rothery, 1957, *Proc. Roy. Soc., Lond.*, **A240**, 145.
27. See, e.g., H. Sato and R. H. Toth, 1968, *Bull. Soc. Fr. Minér. Crist.*, **91**. 557; 1965, in *Alloying Behavior and Effects in Concentrated Solid Solutions*, New York: Gordon and Breach, p. 295.
28. W. Lomer, 1967, in *Phase Stability in Metals and Alloys*, New York: McGraw-Hill, p. 280.
29. J. H. N. van Vucht, 1966, *J. Less-Common Metals*, **11**, 308.
30. H. Sato and R. S. Toth, 1961, *Phys. Rev.*, **124**, 1833.
31. H. Sato and R. S. Toth, 1962, *Phys. Rev.*, **127**, 469.
32. H. Sato and R. S. Toth, 1965, *Phys. Rev.*, **139**A, 1581.
33. H. Sato, R. S. Toth, and G. Honjo, 1967. *J. Phys. Chem. Solids*, **28**, 137; H. Sato, R. S. Toth, G. Shirane, and D. E. Cox, 1966. *Ibid.*, **27**, 413.
34. H. Sato and R. S. Toth, 1968, *J. Phys. Chem. Solids*, **29**, 1562.
35. M. Hirabayashi, N. Ino, and K. Higara, 1967, *J. Phys. Soc. Japan*, **22**, 1009.
36. For references see Jeitschko and Parthé.[37] See also 1966. *Mh. Chem.*, **97**, 506; 1967, *Z. Kristallogr.*, **124**, 9; 1968, *Mh. Chem.*, **99**, 2048.
37. W. Jeitschko and E. Parthé, 1967, *Acta Cryst.*, **22**, 417.
38. E. Parthé, 1969. in *Developments in the Structural Chemistry of Alloy Phases*, New York: Plenum Press, p. 49.
39. H. Völlenkle, A. Preisinger, H. Nowotny, and A. Wittmann, 1967, *Z. Kristallogr.*, **124**, 9.
40. W. B. Pearson, 1970, *Acta Cryst.*, B**26**, 1044.
41. H. Lipson and A. Taylor, 1939, *Proc. Roy. Soc., Lond.*, **A173**, 232.
42. P. A. Beck and co-workers, 1951, *Trans. AIME*, **191**, 872; 1952, *Ibid.*, **194**, 1071; 1955, *Ibid.*, **203**, 765; 1957, *Ibid.*, **209**, 69; 1960. *Ibid.*, **218**, 733, 617.
43. W. B. Pearson, J.-P. Jan, and Y. Saito, 1967, in *Phase Stability in Metals and Alloys*, New York: McGraw-Hill, p. 273.
44. J. B. Haworth and W. Hume-Rothery, 1952, *Phil. Mag.*, **43**, 613; W. Hume-Rothery, 1948, *Ibid.*, **39**, 89.
45. F. Laves and H. Witte, 1936, *Metallwirtschaft*, **15**, 840.
46. Y. Komura, 1969, Eighth International Congress of Crystallography, Stony Brook, N.Y., August; Y. Komura, E. Kishida, and M. Inoue, 1967. *J. Phys. Soc. Japan*, **23**, 398.
47. See for example the review by W. E. Wallace and R. S. Craig, 1967, in *Phase Stability in Metals and Alloys*, New York: McGraw-Hill, p. 255.
48. K. A. Gschneidner and R. M. Valletta, 1968, *Acta Met.*, **16**, 477.
49. C. H. Hodges, 1967, *Acta Met.*, **15**, 1787 (see p. 1793).
50. V. Heine and D. Wearie, 1966, *Phys. Rev.*, **152**, 603.
51. E. E. Havinga, J. H. N. van Vucht, and K. H. J. Buschow, 1969, *Philips Res. Repts.*, **24**, 407.
52. H. Jones, 1950, *Phil. Mag.*, **41**, 663; 1934, *Proc. Roy. Soc., Lond.*, **A147**, 400.
53. W. Hume-Rothery and G. V. Raynor, 1940, *Proc. Roy. Soc., Lond.*, **A174**, 471; G. V. Raynor, 1940, *Ibid.*, **A174**, 457; 1942, *Ibid.*, **A180**, 107.
54. D. Weaire, 1968, *J. Phys. C (Proc. Phys. Soc.)*, Ser. 2, **1**, 210.

55. Ehrenreich, Seitz, and Turnbull, Ed, *Solid State Physics*, Vol. 24, 1970, New York: Academic Press.
56. D. F. Gibbons and L. M. Falicov, 1963, *Phil. Mag.*, **8**, 177.
57. T. B. Massalski and H. W. King, 1961, *Proc. Mater. Sci.*, Vol. 10, p. 1; T. B. Massalski, 1962, *J. Phys. Radium*, **23**, 647.
58. T. B. Massalski and H. W. King, 1962, *Acta Met.*, **10**, 1174.
59. T. B. Massalski, 1963, *Science and Human Progress*, Mellon Institute, p. 145.
60. P. S. Rudman, 1966, Symposium, AIME Annual Meeting, New York, March 1.
61. E. A. Owen, 1947, *J. Inst. Met.*, **73**, 471.
62. G. V. Raynor, 1949, *Trans. Faraday Soc.*, **45**, 698.
63. J. Lumsden, 1952, *Thermodynamics of Alloys*, London: The Institute of Metals, p. 94.
64. S. Greenwald, 1953, *Acta Cryst.*, **6**, 396.
65. E. Bjerkelund, A. Kjekshus, and V. Meisalo, 1968, *Acta Chem. Scand.*, **22**, 3336.
66. R. W. Lynch and H. G. Drickamer, 1965, *J. Phys. Chem. Solids*, **26**, 63.
67. E. A. Perez-Albuerne, R. L. Clendenen, R. W. Lynch, and H. G. Drickamer, 1966, *Phys. Rev.*, **142**, 392.
68. R. L. Clendenen and H. G. Drickamer, 1964, *J. Phys. Chem. Solids.*, **25**, 865.

4

Atomic Size

THE CONCEPT OF ATOMIC SIZE

The nature of the electronic distribution about an atom precludes rigorous definition of atomic size either in the gaseous or solid state, but it has long been known (cf. Bragg[1]) that a size can be attributed to atoms which by addition will reproduce observed interatomic distances in solids to within a few percent. Furthermore, if particular series of compounds with similar crystal structures or similar coordination are considered, systems of radii can be devised which reproduce interatomic distances rather effectively within the series (cf. Pauling and Huggins' tetrahedral radii,[2] or Geller's[3] radii for compounds with the β-W structure). Specific systems of radii for ionic crystals have been derived by Wasastjerna,[4] Goldschmidt,[5] Pauling,[6] Zachariasen,[7] and Shannon and Prewitt,[8] for covalent-metallic solids by Pauling,[9,10] and for metals by Bokij,[11] whereas Slater[12] has given a system of general radii applicable to all solids. The more detailed treatments provide methods of applying corrections for such factors as coordination, ionic repulsion, type of valence bonds, partial ionicity and so forth, with considerable success in some areas, and rather less in others. The great amount of labor that has been applied to the problem of atomic sizes has only served to show that no overall system of "corrected" radii can be expected to provide exact interatomic distances in all cases, or conversely, no single system of radii can be used generally as a criterion of bond strength or type. Therefore for general discussion, a system of radii such as those proposed by Slater for application without correction, may well be the most useful.

Two distinctly different electronic situations influencing atomic radii can be considered. In the first, which concerns the approach of uncharged atoms with filled valence subshells, repulsive forces become important and outweigh the weak attractive forces when the tails of the wave functions of the electrons of the outermost filled shell begin to overlap. This situation is met in the structures of the inert gases and between nonbonded atoms in compounds where they have attained filled valence subshells through electron sharing. The sizes of uncharged atoms with filled valence subshells are known as van der Waals diameters. The ions in ionic compounds also have filled valence subshells, but the coulombic attractions between the charges on unlike ions cause ionic radii to be somewhat smaller than van der Waals radii. The second situation concerns the approach of covalently bonded atoms where the overlap of the wave functions results in the electron sharing characteristic of the covalent bond. Slater has pointed out that there appears to be a definite correlation between the radius of the maximum radial charge density of the outermost shell and the atomic or covalent radius. Sums of covalent radii are smaller than the sums of ionic radii for the same elements. Covalent bonds can be either saturated, when the electron sharing (including nominal transfer) results in unit or multiple bond strengths and filled valence subshells on the atoms bonded together, or unsaturated when there are not enough valence electrons for all atoms bonded together to attain filled valence subshells, and either pivotal or uninhibited resonance of the bonds occurs. Bond strengths in the second case, which includes metals and some semiconductors, are less than unity and the "covalent" sizes of atoms may vary according to the situation in which the atom occurs.

Such considerations lead to the recognition of three important sets of radii: the van der Waals radii of the metalloid and nonmetallic atoms lying to the right of the Zintl border which can attain filled valence subshells, the ionic radii, and the covalent and/or metallic radii resulting from bonding through overlap of atomic wave functions. Since in practically all metallic substances the number of ligands exceeds the number of valence electrons available for bond formation, interatomic distances are longer than those expected for covalent bonds of unit strength, upon which covalent radii are based. The covalent radii have therefore to be adjusted for metallic valency and coordination in order to reproduce observed interatomic distances in metals, but the situation is complicated by the manner in which metallic atoms readily surround themselves with neighbors at various interatomic distances. Although both the ionic and the covalent systems of atomic radii are purely empirical scales, conceptually they are associated with the two limiting types of chemical bond. Either one or other scale may be appropriate to a given substance, but there is a primary difficulty in dealing with those compounds in which the bonding is a mixture of both the ionic and covalent types, and in determining the influence of this on the atomic sizes, whichever of the two scales is chosen.

Atomic sizes influence the crystal structure of ionic compounds because, when the larger anions come in contact, anionic repulsion renders that particular structure unstable. Therefore, for any given number of anions surrounding a cation, there is a limiting low value of the ratio of the radius of the cation to the radius of the anion, R_c/R_a, at which the anions come in contact and the structure becomes unstable. These limiting radius ratios for the stability of simple symmetrical coordination polyhedra are given in Table 5-19, p. 246.

Since in general, metallic atoms can be readily compressed by ligands in various directions, and ionic charges are essentially absent in metals, size restrictions on structural stability are not pronounced as in the ionic case, and the influences of size on structure and *vice versa* are much more subtle. Only in the case of the most disparate atomic sizes, as in the Hägg interstitial compounds, or in phases with large coordination polyhedra (CN $\geqslant$ 20) surrounding a large atom, is structure type obviously influenced by relative atomic size. Indeed, a strong case can be advanced to show that the chemical bond, coordination, or geometrical factors that control the majority of metallic structures, themselves influence structural dimensions and therefore the relative distances between the various ligands in alloys, rather than that the formal sizes of the atoms themselves influence the structure type (see pp. 74). This suggests certain new ideas concerning metallic structures. For instance, it has always been held that the Laves phases form because of the ease with which atoms in the proportion MN_2 with radii in the ratio $R_M/R_N = 1.225$ or atomic volumes in the ratio of 2 to 1, pack together to fill space, whereas the discussion on p. 59, on which the above thesis is in part based, shows that this is by no means necessarily the case. The geometrical feature of 16–12 coordination is satisfied through control of interatomic distances over a very wide range of, and almost regardless of, the relative atomic sizes of the components forming the Laves structures. There are many other examples in addition to that of the Laves phases, which show the control of interatomic distances by the factors that give structural stability in metallic phases. Indeed, the important question may well be, what does a structure do to the interatomic distances of the components in the phases that adopt it, rather than what is the influence of the relative sizes of the metallic atoms in controlling the structures of metallic phases? Relative atomic size does, however, control the distribution of the component atoms on the various lattice sites in alloys with Frank–Kasper type and related structures such as the σ phase, where the larger atoms occupy the 15- or 16-fold coordinated sites and the smaller atoms the sites with CN 12 or 14.

That the energy band factor also controls interatomic distances is apparent from the dimensions of the Group II elements Mg, Zn, and Cd, and from various superstructures based on close packing where either structural change, or change of relative structural dimensions maintains phase stability with changing electron concentration (see pp. 116 and 96). Hexagonal phases generally have relative cell dimensions which depend on electron concentration or average group number (see pp. 121 and 124). The extent of terminal metallic solid solutions is influenced mainly by relative atomic size provided that the electrochemical factor is small. The empirical rule of Hume-Rothery and co-workers[13] that extended solid solutions are not to be expected in alloys when the difference in the sizes of the components exceeds about 14%, is well established (cf. Waber *et al.*[14]), as discussed on pp. 76 to 79.

Thus in metals, interatomic distance is generally a feature resulting from chemical bond, geometrical or energy band factors that control structure, rather than a feature which itself controls structure. Compression of atoms in the directions of some atomic contacts is a common feature of the structures of most metals, and the average atomic volume in intermetallic phases is generally less (often considerably) than the appropriate average of the atomic volumes of the component atoms. In alloys which are semiconductors, in very many of whose structures the anions form a close packed array with the cations located at the centers of the tetrahedral and/or octahedral interstices therein, no apparent relative size restrictions exist, such as are found in ionic compounds with similar structures. Interatomic distances generally follow those for the expected bond numbers.

ATOMIC VOLUME

In metals the atoms generally have far more near neighbors than valence electrons required to form bonds of unit strength, and so metallic-covalent radii for unit bond strength are inappropriate for metallic alloys, although they would be suitable for many semiconducting compounds where the metal forms bonds of unit strength with the anions. Such radii have to be corrected according to the number of valence electrons furnished by the metal. After correction, it might reasonably be expected that the radii would reproduce observed interatomic distances between nearest neighbors, or alternatively that observed interatomic distances could be used to determine fractional bond strengths. However, experience with near-neighbor diagrams, suggests that these accepted views must be modified somewhat as discussed on p. 60.

Since the valence or conduction electrons in a metal are more profitably regarded as forming a quasi-free "electron gas" rather than directed chemical bonds, it would seem that atomic volume which "averages" the whole environment of an atom is a more sensible size parameter than interatomic distance which depends specifically on the number of valence electrons and the coordination. Atomic diameters derived from interatomic distances in the structures of the elements are structure-dependent parameters, whereas atomic volumes are much more nearly independent of structure, and this applies even in the face centered cubic ($A1$) and body centered cubic ($A2$) structures. For example, by equating the atomic volumes V of an element in the $A1$ (CN 12) and $A2$ (CN 8) structures

$$V = a^3_{(A1)}/4 = a^3_{(A2)}/2, \tag{1}$$

it is seen that the near-neighbor interatomic distance d, is about 3% larger in the $A1$ structure

$$\frac{d_{(A1)}}{d_{(A2)}} = \frac{a_{(A1)}/\sqrt{2}}{\sqrt{3}\cdot a_{(A2)}/2} = \frac{\sqrt[3]{2}\cdot a_{(A2)}/\sqrt{2}}{\sqrt{3}\cdot a_{(A2)}/2} = 1.03 \tag{2}$$

and empirically it is indeed found to be some $2\frac{1}{2}$ to 3% larger in $A1$ structures, implying that atomic volumes in the two structures are nearly equal. Thus, the average difference in atomic volume (ignoring sign) between 19 elemental allotropes with the $A2$ structure and the $A1$ (or c.p. hexagonal, $A3$) structure at the same temperature is only 0.79% as Rudman[15] has shown (see Table 4-1). In more complex structures interatomic distance is bound to be even more structure-dependent than in simple structures. One of the interesting features shown by Table 4-1 is that the "close packed" $A1$ and $A3$ structures are not necessarily more close packed in terms of atomic volumes than the $A2$ structure. This, of course, is not surprising for although on a hard-sphere space-filling model (Laves,[16] Parthé[17]), the space-filling factor of the $A1$ structure ($\sim$ 0.74) is about 9% larger than that of the $A2$ structure ($\sim$ 0.68), the radius of the atom in the CN 12 $A1$ structure is about 3% larger than in the $A2$ structure, so that little volume change is to be expected on transformation of one to the other, agreeing generally with experimental observations.

As the axial ratio of the c.p. hexagonal structure departs from 1.63, the atomic volume must increase compared to that of the ideal structure, unless some of the atomic contacts are compressed. Therefore it is not surprising to find a decrease of volume on going from $A3 \rightarrow A2$ in those metals such as Be whose axial ratio is less than 1.63. When the

Table 4-1 Volume Changes of Metals in Allotropic Transformation from *A*3 or *A*1 Structures to *A*2 at the Equilibrium Transformation Temperature (After Rudman[15])

Metal	c/a or $c/2a$	$\Delta V/V$(%) $A3 \rightarrow A2$ Transformation	$A1 \rightarrow A2$ Transformation
Li	1.637	−0.4	
Na	1.634	−0.4 [0.28]	
Be	1.568	−3.58	
Ca			0.20
Sr			−0.30
Ti	1.587	−0.55	
Zr	1.593	−0.66	
Hf	1.581	−1.05	
Mn			0.8
Fe			1.06 ($\gamma-\alpha$) 0.48 ($\gamma-\delta$)
Tl	1.598	−0.15	
La			1.3
Ce			0.1 ($\gamma-\delta$)
Pr	1.611[a]	0.5[a]	
Nd	1.613[a]	0.1[a]	
Yb			−1.25
Th			−0.1
Pu			−2

[a]Double hexagonal close packing

volume change is examined as a function of axial ratio (*A*1 has c/a equivalent to 1.63) it appears that the ideal close packed structures have a slightly better space filling than the *A*2 structure (Figure 4-1). Transformations involving Li, Na, Ca, Sr, and Th are exceptions where the *A*2 structure has about the same degree of space filling as the *A*1 or *A*3 structure. These metals are either relatively compressible, or those in which the transformation takes place at a high temperature close to the melting point where the atoms are probably quite compressible. Experience with near-neighbor diagrams suggests that in the *A*2 structure of compressible metals, the 8 closest contacts are compressed somewhat so as to reduce the expansion of the 6 other slightly longer contacts, the observed nearest-neighbor distance being slightly shorter than the "intrinsic" diameter of the atoms. This results in the atoms having a smaller than expected atomic volume in the *A*2 structure. In the *A*1 or ideal *A*3 structure, where the atoms are surrounded by 12 nearest neighbors

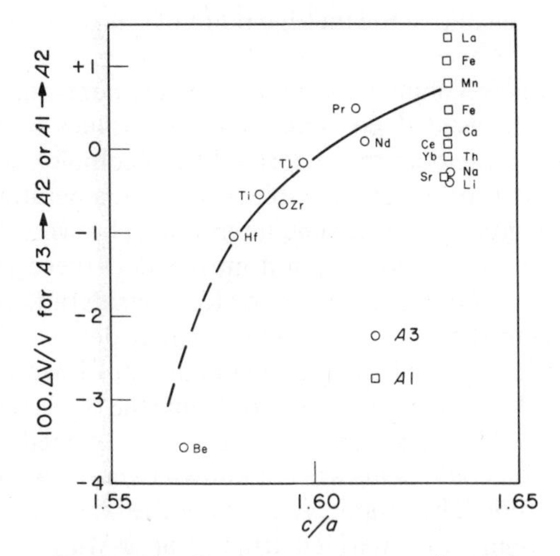

Figure 4-1. Percentage change of atomic volume on going from the c.p. hexagonal (*A*3) or c.p. cubic (*A*1) structures to the b.c. cubic structure (*A*2) as a function of the axial ratio of the hexagonal cell.

and the next nearest neighbors are at a much larger distance away, there is no reason for compression of the contacts and the atoms are separated by their normal "intrinsic" interatomic distance giving the normal atomic volume. The apparent contraction of the interatomic distance in the *A*2 structure of relatively compressible metals results in a better space filling in the structure of these elements so that there is little volume change, or even an expansion on going from *A*2 to *A*1 or ideal *A*3, whereas in the structures of noncompressible metals the atomic volumes in *A*1 or *A*3 structures are some 0.5 to 1.0% smaller than in the *A*2 structures. This account appears to be confirmed by the observation that the volume of the high temperature *A*2 Fe structure is only 0.48% larger than the *A*1 structure, whereas the low temperature *A*2 structure is 1.06% larger.

The whole concept of size factor and its application to problems of phase occurrence and stability is bound, at best, to be a very approximate consideration. Not only is there some structural dependence of the derived atomic volumes of the elements, but Rudman[15] points out that atomic volumes of metals increase by some 6% up to their melting point, so that the room temperature atomic volumes are themselves not strictly comparable because of the different melting points of the metals. Secondly, any particular alloy phase which is considered in terms of size factor may be stable at room temperature, at high temperatures only, or indeed at temperatures well above the melting point of one (or both) of its components.

In deriving atomic volumes of the metallic elements, the $A1$, $A2$, or $A3$ type structures are used if possible, since the values obtained must be much more compatible than those derived from complex structures, such as the α-Mn structure, which is stable at room temperature. This structure results from complex electronic interactions between the Mn atoms, and an atomic volume derived from it must reflect these special Mn–Mn interactions and be quite inappropriate for considerations of Mn in its general alloy chemistry. Carrying the particular details of these interactions into considerations of the alloying behavior of Mn can be avoided if an atomic volume can be derived from the high-temperature $A1$ or $A2$ allotropes, but these phases cannot be retained by quenching, and their expansion coefficients are not known over the whole range down to room temperature. The atomic volumes of the γ and δ forms of Mn are certainly larger than those derived from α or β-Mn (12.2 and 12.5 $Å^3$, respectively), but the volume of about 13.85 $Å^3$ estimated for δ-Mn at room temperature by Rudman[15] appears to be too large. When Mn-rich γ-Mn alloys are quenched, the f.c. cubic γ structure distorts tetragonally. The presence of this tetragonal phase, which is retained at room temperature, makes extrapolation of the atomic volumes of γ phase solid solutions to 100% Mn to obtain the atomic volume of pure γ-Mn, rather objectionable. Nevertheless, extrapolation of the atomic volumes of 4 different γ-Mn solid solutions points to an atomic volume of 12.8 to 12.9 $Å^3$ for γ-Mn, which volume we adopt. It is certain that the α-Mn volume of 12.2 $Å^3$ given in reviews by Rudman[15] and King[18] should not be used in size factor considerations.

Similar considerations must apply to the structures of U, Np and Pu, which are complex at room temperature, and some attempt should also be made to derive atomic volumes from the high temperature $A2$ or $A1$ modifications. For Ga, which only exists in complex orthorhombic modifications, atomic volume is the only sensible method of obtaining an atomic size from the structure of the element itself. Giessen and coworkers[19] have derived a value of the atomic volume of 12 coordinated Ga (18.27 $Å^3$ at -190°C) by quadratic extrapolation of the volume of α ($A1$) and α' (tetragonally distorted $A1$) metastable Al–Ga alloys to 100% Ga, which can be compared with the value of 19.6 $Å^3$ obtained from the orthorhombic structure at room temperature.

Zn and Cd with the $A3$ type of structure have much larger axial ratios (1.86 and 1.89, respectively) than the ideal value of 1.63 for true hexagonal close packing, so that the derived atomic volumes may differ considerably from those exhibited by Zn and Cd when they engage in forming alloys with other components. There appears to be little that can be done about this problem except to realize that it contributes further to the approximate nature of size factor considerations in alloy chemistry. When the axial ratio of the hexagonal form is close to the ideal value as in Co, the atomic volume (11.10 Å^3) is almost identical to that obtained for the cubic close packed *A*1 form (11.13 Å^3), as expected. The atomic volume of *A*1 γ-Ce (34.35 Å^3) is slightly smaller than that of double hexagonal close packed β-Ce (34.47 Å^3) with an axial ratio, $c/2a = 1.607$, which is smaller than the ideal value.

Atomic volume is not a meaningful parameter for Group V–VII nontransition metal elements which have more valence electrons than available bonding orbitals so that some electrons must enter nonbonding or antibonding orbitals, unless electron transfer occurs. Consider for example Se, which in its hexagonal structure forms two unit strength bonds that link the atoms together in chains. Each Se atom also has four other neighbors in neighboring chains where it sees the filled valence subshells of the nonbonding electrons, and so it is separated from them by distances corresponding approximately to van der Waals diameters. The atomic volume of Se in this structure gives a rather meaningless average of two bonded and four essentially nonbonded contacts (but see p. 237 for further details), and its value is quite different to the atomic volume of Se in, say, $PtSe_2$ with the CdI_2 type structure where it forms three bonds and has three nonbonded contacts, or of Se in $NiSe_2$ with the pyrite structure, where it has four bonded ligands. Nevertheless in the 2, 3, and 4 unit-strength bonds formed, respectively, in elemental Se, in compounds with the CdI_2 type structure and the pyrite structure, Se exhibits a diameter which is essentially independent of structure and the number of bonds of unit strength that are formed. Thus, in these elements near-neighbor interatomic distances have to be considered rather than atomic volumes. This is apparent on comparing the variation of atomic volume and of interatomic distance with atomic number on proceeding across a (nontransition) Period. The balance between the attractive and repulsive potentials results in a monotonic decrease of interatomic distance with increasing atomic number across the Period as shown in Figure 4-2. Atomic volume, however, increases again after Group IV when there are more valence electrons than bonding orbitals available for them.

In addition to the effect of electrons in nonbonding orbitals on atomic volumes of the elements in semiconducting compounds, the attainment of filled valence subshells by electron sharing in forming directed covalent bonds can lead to a large increase in atomic volume, as is apparent from comparison of the dimensions of the gray (CN 4) and white (CN 4+2) forms of tin. The atomic volume of diamond–tin is about 23%

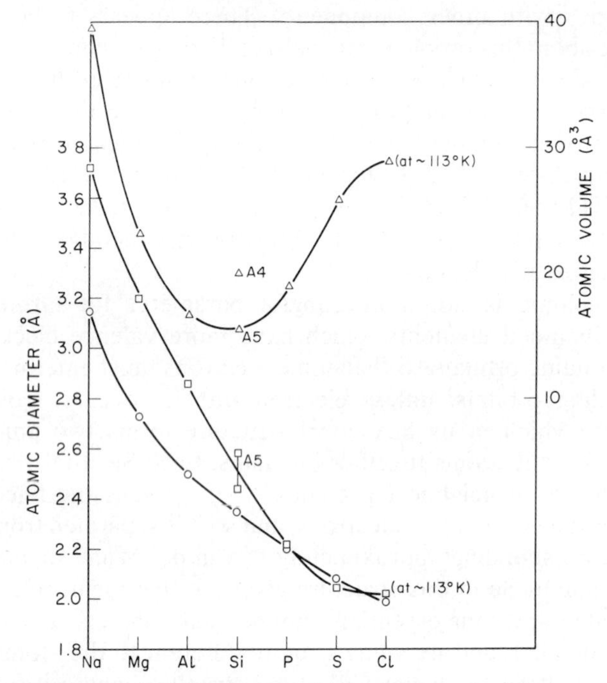

Figure 4-2. atomic volume, △; atomic diameter, □, and Pauling single bond diameter, ○, of the atoms of the third Period as a function of the Group number. *A*4 refers to the diamond structure of Si and *A*5 to the white tin structure.

larger than that of white tin, although the interatomic distance in diamond-tin is 0.21 and 0.37 Å shorter than the two distances in white tin! Nevertheless, it is reasonable to suppose that 4 valence electrons are involved in cohesion in each of the two modifications.

Table 4-2 lists the atomic volumes derived for the metallic elements, including the Group IV elements Si, Ge, and Sn, whose atomic volumes are Rudman's values[15] taken from the metallic forms with the white tin (*A*5 type) structure. Although there are many arguments why atomic volume should be used for examining size effects in metals, in practice it appears to be a somewhat disappointing parameter, and more useful information can generally be gained from considering interatomic distances. It is most useful for comparing the average atomic volumes of intermetallic phases with those of the elements in order to assess what degree of atomic expansion or contraction occurs. The Hume-Rothery solid solubility rule and Vegards' Law (p. 174) can also be examined

Table 4-2 Atomic Volumes of Metals

Element	Structure	Atomic Vol. (Å³)	Element	Structure	Atomic Vol. (Å³)
Li	*A*2	21.622	V	*A*2	13.814
Na	*A*2	39.49	Nb	*A*2	18.077
K	*A*2	75.3	Ta	*A*2	~18.0
Rb	*A*2	92.6			
Cs	*A*2	~116.0	Cr	*A*2	12.001
			Mo	*A*2	15.580
Be	*A*3	8.110	W	*A*2	15.855
Mg	*A*3	23.240			
Ca	*A*1	43.632	Mn	*A*1	12.8
Sr	*A*1	56.325	Tc	*A*3	14.213
Ba	*A*2	62.99	Re	*A*3	14.705
Al	*A*1	16.603	Fe	*A*2	11.776
Sc	*A*3	25.002	Ru	*A*3	13.574
Y	*A*3	33.012	Os	*A*3	13.993
La	double *A*3	37.415			
	*A*1	37.14	Co	*A*1	11.128
				*A*3	11.077
Ce	double *A*3	34.472	Rh	*A*1	13.766
	*A*1	34.349	Ir	*A*1	14.144
Pr	double *A*3	34.560			
Nd	double *A*3	34.181	Ni	*A*1	10.939
Sm	rhombohedral	33.12	Pd	*A*1	14.724
Eu	*A*2	48.099	Pt	*A*1	15.104
Gd	*A*3	33.103	Cu	*A*1	11.807
Tb	*A*3	31.969	Ag	*A*1	17.057
Dy	*A*3	31.522	Au	*A*1	16.961
Ho	*A*3	31.119			
Er	*A*3	30.642	Zn	*A*3	15.212
Tm	*A*3	30.099	Cd	*A*3	21.581
Yb	*A*1	41.281	Hg[b]	*A*10	23.4
Lu	*A*3	29.496			
			Ga	Complex	19.595
Ti	*A*3	17.665	In	*A*6	26.144
Zr	*A*3	23.272	T1	*A*3	28.583
Hf	*A*3	22.321			
			Si	*A*5[a]	15.5
			Ge	*A*5[a]	18.0
			Sn	*A*5	27.047
			Pb	*A*1	30.326

[a]High pressure forms
[b]At low temperature

in terms of atomic volumes, but there appears to be little advantage in this and, as observed above, atomic volumes can not be employed in examining size effects in phases formed with elements to the right of the Zintl border.

COVALENT AND METALLIC RADII

Although it may be inherently desirable to consider size effects in metals in terms of atomic volumes, it appears that consideration of interatomic distances can be much more fruitful. It has already been pointed out that in metals, interatomic distances depend primarily on the number of ligands and on the number of valence electrons of an atom, and other factors such as the type of bond only exert a secondary influence. Pauling's rule[9], $R_{(1)} - R_{(n)} = 0.30 \log n$, relating radii for bonds of strength n (number of valence electrons per ligand) to those of unit strength, provides a means of adjusting radii for coordination (or for estimating effective valencies). No matter what may be the limitations of any particular system of metallic radii or valencies that is adopted, Pauling's rule appears to be most realiable, providing a sound basis for comparing interatomic distances in metals.

The second problem in dealing with interatomic distances in metals is to select some system of atomic radii upon which discussion of interatomic distances can be based. That of Slater[12] is too general and not accurate enough, whereas those of Pauling[9, 10] which prove satisfactory for elements to the right of the Zintl border, are inappropriate for transition metals, and the 1947 radii are based on a system of atomic valencies of B Group metals which must today be considered unacceptable, although in 1949 radii were also given for normally accepted valencies. The radii given by Teatum and co-workers[20] for 12 coordination (p. 150) which are based directly on interatomic distances in metals, appear to be the most useful for discussing metallic alloys and these are used in this book for calculating near-neighbor diagrams, etc., whereas those of Pauling[9, 10] are used for discussing semiconductors generally. Pauling's empirical equation[9] is used to adjust the radii to the particular coordination for which they are required. Thus, once a system of radii for given valency and coordination is accepted, the adjustments of the radii for other coordinations are prescribed and independent of the system.

Pauling's Single-Bond Radii

In Table 4-3 the Pauling[9] single-bond radii, $R_{(1)}$, and valencies to which they refer are listed for the A group elements, excluding the transition metals, and for the Group V–VII B and rare-earth elements. The radii for any other coordination number, as for example $R_{(\mathrm{CN}\,12)}$, can be calculated from the equation $R_{(1)} - R_{(n)} = 0.30 \log n$ by setting $n = V/12$, where V is the elemental valency.† The Group I–IV B element radii

†Pauling found that the correction calculated from this equation for the change from CN 12 to CN 8 was too large when compared to the lattice parameters of the b.c. cubic forms of Fe, Ti, Zr, and Tl. This was attributed to the bond number being less than $V/8$ in the b.c.

Table 4-3 Pauling's Metallic and Covalent Radii of Elements

Element	Valency V	$R(1)$
Li	1	1.225
Na	1	1.572
K	1	2.025
Rb	1	2.16
Cs	1	2.35
Be	2	0.889
Mg	2	1.364
Ca	2	1.736
Sr	2	1.914
Ba	2	1.981
B	3	0.80
Al	3	1.248
Sc	3	1.439
Y	3	1.616
La	3	1.690
Ce	3.2	1.646
Pr	3.1	1.648
Nd	3.1	1.642
Pm		
Sm	2.8	1.66
Eu	2	1.850
Gd	3	1.614
Tb	3	1.592
Dy	3	1.589
Ho	3	1.580
Er	3	1.567
Tm	3	1.562
Yb	2	1.699
Lu	3	1.557
Cu	1	1.352
Ag	1	1.528
Au	1	1.520
Zn	2	1.309
Cd	2	1.485
Hg	2	1.490
Ga	3	1.266
In	3	1.442
Tl	3	1.460
	1	1.540
C	4	0.771
Si	4	1.173
Ge	4	1.223
Sn	4	1.399
Pb	4	1.430
	2	1.540
N	3	0.74[28]
P	3	1.10
As	3	1.21
Sb	3	1.41
Bi	3	1.52
O	2	0.74[28]
S	2	1.04
Se	2	1.17
Te	2	1.37
Po	2	1.53
F	1	0.72[28]
Cl	1	0.994
Br	1	1.142
I	1	1.334

cubic structure, since there was an appreciable amount of bonding to the 6 next-nearest neighbors. An empirical correction from CN 12 to CN 8 based on the radii of the b.c. cubic forms of these metals was therefore proposed by Pauling, However, Thewlis[21] pointed out that in his treatment Pauling had compared lattice parameters for the hexagonal and cubic forms of Ti, Zr, and Tl which were determined at different temperatures. Furthermore, in comparing the b.c. cubic and f.c. cubic radii of iron, Pauling had accepted a value of the lattice parameter of γ-Fe which was obtained by extrapolating the lattice parameters of the γFe-Ni solid solution to zero solute content—a process which may be unsatisfactory (cf. Axon and Hume-Rothery;[22] Hume-Rothery and Raynor[23]). When the high-temperature lattice parameters of the b.c. cubic forms of Fe, Ti, Zr, and Tl are reduced to 20°C with the known expansion coefficients, the atomic radii in structures with CN 8 and CN 12 can be compared on an equitable basis. Thewlis then showed that there is no systematic discrepancy between the calculated and observed ΔR values, so that the equation requires no empirical modification of the type employed by Pauling.

based on Pauling's system of valencies – 5.44, 4.44, 3.44, and 2.44 respectively – are not reproduced here; instead the radii of Pauling[10] are given, since these were derived for the normal valencies 1, 2, 3, and 4 respectively. Neither Pauling's 1947 nor 1949 transition metal radii which recognize the influence of bond orbital type, are given here since the radii of Teatum *et al.* are preferred for alloys, and as explained on pp. 160 to 167, crystal field effects appear to influence the radii of valence compounds.

Details of the methods of obtaining the radii of the metallic elements are given in Pauling.[9,10] For the most part data on interatomic distances were taken from Neuberger's 1936 summary.[24] Although the more recent collections of data[25,26] do not change the interatomic distances by very much, except in a few cases such as Rb and V, a number of the elemental modifications discussed by Pauling (e.g., hexagonal Cr and Ni or the *A*15 structure of β-W) have since been shown to be compounds formed with hydrogen, carbon, nitrogen or oxygen. The radii of the nonmetallic elements of Groups IV to VII are the normal covalent radii and, with the exception of nitrogen, oxygen, and fluorine, they are the interatomic distances between singly-bonded atoms chosen as described in Pauling (1940).[27] Radii of 0.70 Å for nitrogen and 0.66 Å for oxygen were obtained by interpolation between the well established radius of carbon, and a radius of 0.64 Å derived for fluorine from spectroscopic data, which is now known to correspond to an excited state of the molecule (Schomaker and Stevenson,[28] Wells[29]). Although these radii were included in Pauling's (1947) table, they have been omitted from Table 4-3 where only the Schomaker-Stevenson[28] radii are given, which were obtained from the N–N and O–O bond distances in hydrazine and hydrogen peroxide, respectively. The radius of the fluorine atom was obtained from the interatomic distance in the gaseous fluorine molecule which was determined by electron diffraction.

When the Schomaker–Stevenson radii for N, O, and F are adopted, the Pauling additivity rule is no longer obeyed for many of the bonds formed between atoms with low atomic number, and these elements. For instance, there are serious discrepancies between the observed and calculated bond lengths for the following pairs of atoms in compounds of the type $X(CH_3)_n$ and XCl_n: N–C, O–C, F–C, O–Cl, F–Cl, Si–Cl, P–Cl, and S–Cl.† These discrepancies were found particularly between atoms having an electronegativity difference of 0.5 or greater. This led Schomaker and Stevenson[28] to propose an empirical equation (referred to below as the S–S equation) to correct radii sums for the ionicity from

†The observed bond energies for such bonds also depart from the values expected from the principle of additivity, being in each case larger than expected.

the electronegativity difference $|x_M - x_N|$ between the component atoms:

$$R_{(MN)} = R_{(M)} + R_{(N)} - 0.09\,|x_M - x_N| \tag{3}$$

Since it was so arranged, the equation makes satisfactory corrections for the calculated lengths of the bonds listed above. However, in other bonds such as C–Cl where the Pauling $R_{(M)} + R_{(N)}$ sum agrees exactly with the observed value, use of the S–S equation causes discrepancies between observed and calculated interatomic distances, and this is particularly so with bonds between an element of high atomic number and carbon or chlorine. Wells[30] shows convincingly that no equation of the S–S type can give a satisfactory correction for ionicity which would make the rule for the additivity of covalent radii hold rigorously.

Pauling's principle of the additivity of covalent bond radii has been much discussed particularly by Schomaker and Stevenson, Buroway[31] and Wells.[29,30] It appears that the original assumption that partial ionicity was unlikely to modify greatly the bond distances calculated from covalent radii is not true for bonds between the light elements and N, O and F, although it is probably satisfactory for bonds formed between elements of higher atomic number. Secondly, the earlier implication that the bond length in a compound having mixed covalent-ionic bonding would lie between the sums of the covalent and the ionic radii does not seem to be correct; the bonds are frequently shorter than either of the sums. However, attempts to correct the sums of covalent radii for ionicity with equations of the S–S type fail to correct all bond distances, because ionicity does not appear to be the only factor influencing the radius sums. The references cited above, in comparing observed interatomic distances with those calculated with the S–S equation, show that the S–S correction becomes too large as the average atomic number (or interatomic distance) gets large. If the partial ionicity of a covalent compound is responsible for the shortening of interatomic distances and for the increase in bond energies, then coulombic as well as covalent forces must contribute to this process. Since coulombic forces decrease as $1/R^2$ it would be reasonable to modify the S–S equation to take account of the decrease of these forces with interatomic distance, normalizing it to give the right correction, say for the length of the O–C bond. This reduces the size of the S–S correction for bonds formed between atoms of higher atomic number and therefore improves it. Indeed, Pauling[32] in order to improve the S–S relationship, empirically reduces the constant 0.09 of equation (3) with increasing average atomic number of the component atoms forming bonds. Further adjustments of the S–S equation have been proposed which take account of the effect of relative atomic sizes in modifying the ionicity estimated from Δx values, and these can give improved agreement between calculated and observed interatomic distances.[33]

The agreement between observed and calculated interatomic distances in many semiconducting valence compounds is not good when the Pauling radii are used. In view of the uncertainties about bond number and the effects of different orbitals forming bonds, attempts to draw detailed conclusions from comparison of calculated and observed interatomic distances should be very cautious.

Metallic Radii of Teatum and Co-Workers

Teatum and co-workers'[20] metallic radii for CN 12 (given in Table 4-4) were taken whenever possible from the observed interatomic distances in the f.c. cubic *A*1 structure, the hexagonal c.p. *A*3 structure (by averaging d_1 and d_2) or the b.c. cubic *A*2 structure. Since the near-neighbor coordination is 8 in the *A*2 structure, the observed radii were increased by the value indicated (as a function of radius for CN 8) in Figure 4-3. This relationship was derived by comparing observed radii for *A*1 or *A*3 structures with *A*2 structures at the same temperature (corrected by known expansion coefficients where necessary). The lattice parameters for this compilation were taken from Pearson[25] and for the rare earths, Sc and Y from Gschneidner.[34] Radii for elements with other than the *A*1, *A*2, or *A*3 structures were taken from Pauling[9] and converted to Å units and to CN 12, except for those of Pa, α-U, α-Np,

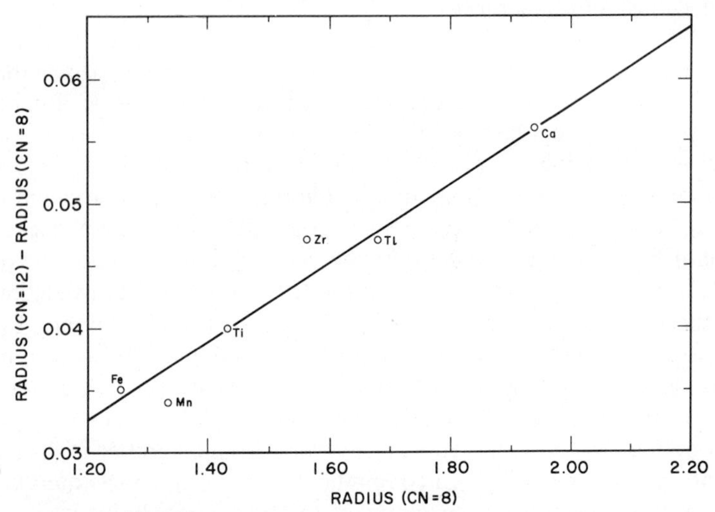

Figure 4-3. Difference of radii for CN 12 and 8 as a function of the CN 8 radius, used in adjustment of atomic diameters by Teatum and coworkers [20].

Table 4-4 Radii of the Elements for CN 12 (After Teatum, Gschneidner, and Waber[20])

Atomic Number	Element	Valency	Radius (CN = 12)	Atomic Number	Element	Valency	Radius (CN = 12)
1	H	−1	0.78	51	Sb	5	1.59
3	Li	1	1.562	52	Te	6	1.60
4	Be	2	1.128	55	Cs	1	2.731
5	B	3	0.98	56	Ba	2	2.243
6	C	4	0.916	57	La	3	1.877
7	N	−3	0.88	58	α-Ce	3.6	1.715
8	O	−2	0.89	58	γ-Ce	3.1	1.825
11	Na	1	1.911	59	Pr	3	1.828
12	Mg	2	1.602	60	Nd	3	1.821
13	Al	3	1.432	61	Pm	3	1.810
14	Si	4	1.319	62	Sm	2.9	1.802
15	P	−3	1.28	63	Eu	2.1	2.042
16	S	−2	1.27	63	Eu	3	1.799
19	K	1	2.376	64	Gd	3	1.802
20	Ca	2	1.974	65	Tb	3	1.782
21	Sc	3	1.641	66	Dy	3	1.773
22	Ti	4	1.462	67	Ho	3	1.766
23	V	5	1.346	68	Er	3	1.757
24	Cr	3	1.360	69	Tm	3	1.746
24	Cr	6	1.282	70	Yb	2	1.940
25	Mn	4	1.304	70	Yb	3	1.740
25	Mn	6	1.264	71	Lu	3	1.734
26	Fe	6	1.274	72	Hf	4	1.580
27	Co	6	1.252	73	Ta	5	1.467
28	Ni	6	1.246	74	W	6	1.408
29	Cu	1	1.278	75	Re	6	1.375
30	Zn	2	1.394	76	Os	6	1.353
31	Ga	3	1.411	77	Ir	6	1.357
32	Ge	4	1.369	78	Pt	6	1.387
33	As	5	1.39	79	Au	1	1.442
34	Se	6	1.40	80	Hg	2	1.573
37	Rb	1	2.546	81	Tl	3	1.716
38	Sr	2	2.151	82	Pb	4	1.750
39	Y	3	1.801	83	Bi	5	1.70
40	Zr	4	1.602	84	Po	6	1.76
41	Nb	5	1.468	87	Fr	1	2.80
42	Mo	6	1.400	88	Ra	2	2.26
43	Tc	6	1.360	89	Ac	3	1.878
44	Ru	6	1.339	90	Th	4	1.798
45	Rh	6	1.345	91	Pa	5	1.63
46	Pd	6	1.376	92	U	6	1.56
47	Ag	1	1.445	93	Np	6	1.555
48	Cd	2	1.568	94	α-Pu	5.0	1.58
49	In	3	1.663	94	δ-Pu	4.5	1.644
50	Sn	2	1.623	94	ϵ-Pu	4.8	1.591
50	Sn	4	1.545	95	Am	3	1.81

and α-Pu which were derived from data of Zachariasen and Ellinger,[35] hydrogen taken from Laves,[36] and Eu^{3+} and Yb^{3+} which were derived from the radii of the other trivalent rare earths by interpolation. Table 4-4 also lists the valencies of the elements to which the radii refer.

These radii appear to be the most suitable set to use for metallic alloys together with Pauling's bond number relationship for adjusting them to coordinations other than 12, and they are used in this book unless otherwise stated.

IONIC RADII

Ionic radii are of some importance in the crystal chemistry of alloys, since it may be necessary to compare ionic and covalent radii sums in order to assess the character of a particular phase. In other cases the ionic scale may even be intrinsically appropriate for the atomic sizes in a metallic conductor such as TiO, where the metallic conductivity arises from an overlap of the *d* electron wave functions, the valence electrons otherwise completing closed ionic shells. For this reason we give a system of ionic radii and discuss briefly their derivation and application.

In ionic crystals the equilibrium separation of the ions depends mainly on the balance of the attractive and repulsive forces between unlike ions, and also to some extent on the repulsion between like ions, particularly the larger anions. For this reason interionic distances depend on the coordination number of the cations by the larger anions and on the relative sizes of the cations and anions. Therefore a system of derived ionic sizes which is intended to reproduce by addition the observed interionic spacings in crystals, must be suitably corrected for coordination number and, in some cases also, for the ratio of the radii of the cations and anions.

The early systems of ionic radii derived by Wasastjerna[4] and by Goldschmidt[5] take 1.32 Å for the radius of the O^{2-} ion, whereas Wyckoff[37] in a system based, like that of Goldschmidt, entirely on interatomic distance measurements, takes $R_{O^{2-}} = 1.40$ Å as do Pauling[6] and Shannon and Prewitt[8] whose systems of ionic radii are discussed here.

Shannon and Prewitt's effective ionic radii for ions in different valence states and coordinations are listed in *Acta Crystallographica*, Volume **B25**, pp. 928 and 929, with revisions in Volume **B26**, p. 1046. These were derived using the approximately linear relationships between atomic volume and unit cell volume for more than 60 isotypic series of oxides and fluorides and experimental data for more than 1000 interatomic distances. Pauling's[6] crystal radii, listed in Table 4-5, are less detailed, being determined for a standard coordination of 6 and a standard radius ratio of about 0.75. Details of their derivation can be found in the refer-

Table 4-5 Pauling's Crystal Ionic Radii for Certain Ions

Element	Valency	Crystal Radii (Å)		Element	Valency	Crystal Radii Å
Li	1+	0.60	(0.68)[a]	C	4+	0.15
Na	1+	0.95	(0.97)	Si	4+	0.41
K	1+	1.33		Ti	4+	0.68
Rb	1+	1.48		Zr	4+	0.80
Cs	1+	1.69	(1.67)	Ce	4+	1.01
Cu	1+	0.96		Ge	4+	0.53
Ag	1+	1.26		Sn	4+	0.71
Au	1+	1.37		Pb	4+	0.84
Be	2+	0.31	(0.35)	H	1−	2.08
Mg	2+	0.65		F	1−	1.36
Ca	2+	0.99		Cl	1−	1.81
Sr	2+	1.13		Br	1−	1.95
Ba	2+	1.35		I	1−	2.16
Zn	2+	0.74		O	2−	1.40
Cd	2+	0.97		S	2−	1.84
Hg	2+	1.10		Se	2−	1.98
				Te	2−	2.21
B	3+	0.20	(0.23)			
Al	3+	0.50		N	3−	1.71
Sc	3+	0.81		P	3−	2.12
Y	3+	0.93		As	3−	2.22
La	3+	1.15		Sb	3−	2.45
Ga	3+	0.62				
In	3+	0.81		C	4−	2.60
Tl	3+	0.95		Si	4−	2.71
				Ge	4−	2.72
				Sn	4−	2.94

[a]Amended values given by Ahrens.[38] Only values differing by 0.02 Å or more are given.

ence cited. Radii given in parentheses in Table 4-5 are changes recommended by Ahrens[38] in reviewing scales of ionic radii in relation to ionization potentials. Adjustments of the radii for coordination other than 6 are calculated from the expression $[B/6]^{1/(n-1)}$, where B is the ligancy of the cation in the structure considered, and n is the appropriate average value of the Born exponent (Table 4-6) for the anions and cations in the structure. Values for the correction factor for different CN and values of the Born exponent, are listed in Table 4-7.

Pauling's radii were derived from consideration of attractive and repulsive forces between unlike ions only, although other factors also contribute to the lattice energy of an ionic crystal. When the radius ratio of the ions is less than the standard value of 0.75, but *especially* when it approaches the lower limit for the stability of a given structure (see p. 246),

Table 4-6 Values of the Born Exponent

Ion Type (inert gas shell)[a]	n
He	5
Ne	7
Ar and Cu^+	9
Kr and Ag^+	10
Xe and Au^+	12

[a] Cu^+, Ag^+, and Au^+ are the ions with 18 electrons in their outer shells.

Table 4-7 Correction Factor for Ionic Crystal Radii for Change of Ligancy from Standard Value of 6

Ligancy	$n =$ 6	7	8	9	10	11	12
12	1.149	1.122	1.104	1.091	1.080	1.072	1.065
9	1.085	1.070	1.060	1.052	1.046	1.041	1.038
8	1.059	1.049	1.042	1.037	1.032	1.029	1.026
7	1.031	1.026	1.022	1.019	1.017	1.016	1.014
6	1.000	1.000	1.000	1.000	1.000	1.000	1.000
5	0.964	0.970	0.974	0.978	0.980	0.982	0.984
4	0.922	0.935	0.944	0.951	0.956	0.960	0.964

repulsive forces between like ions (particularly the larger anions) become important in addition to the normal repulsion between ions of opposite charge which approach each other too closely. Corrections for this effect have been applied successfully to radii sums for the alkali halides,[39] and the treatment can be extended to ionic crystals of other types,[40] but it is said that in most cases, sums of crystal radii corrected for change of coordination from 6, reproduce the observed interatomic distances in ionic crystals rather well.

Another factor contributing to the lattice energy of an ionic crystal is the crystal field stabilization energy found in many transition metal compounds. This depends on the coordination and the spectroscopic ground state of the transition metal ions, and it may reduce the transition metal radii by more than 20% of the values expected from the coulombic interactions discussed above. Since this stabilization energy depends mainly on the crystal structure (cation coordination) and on the number and configuration of the *d* electrons, giving a large number of possible situations, transition metal ions sizes are not included in Table 4-5. For further discussion of ionic radii see also Wells.[41]

VAN DER WAALS RADII

In the solid state the inert gases such as argon, which have filled valence subshells, are held together by electrostatic quadrupolar or higher order van der Waals forces, and the interatomic distances in these solids are therefore known as van der Waals diameters. In semiconducting compounds nonbonded atoms which achieve filled valence subshells by electron sharing in forming covalent bonds, are expected to be separated in crystals by distances which correspond approximately to sums of the van der Waals radii of the atoms. Closer approach must result in overlapping of the tails of the electron wave functions and the establishment of repulsive forces which hold the atoms apart. Knowledge of the van der Waals radii of anions lying to the right of the Zintl border is important in the crystal chemistry of semiconductors, because significantly closer approaches between anions which are supposed to have filled valence subshells, suggests either that they do not have filled subshells, and/or that there is some direct shared electron pair (or ionic) bonding between them. Weak intermolecular, interlayer or interchain bonding of this type, which is found in the structures of I, Se, Te, As, and Sb and various compounds, may give rise to unexpected electrical and other physical properties which can be understood on the basis of crystal chemistry (see pp. 233 to 243).

Table 4-8 gives the van der Waals radii for a few elements from the data of Pauling[42] and von Hippel[43] (Se, Te).

Table 4-8 van der Waals Radii of Atoms

		H	1.20		
N	1.5	O	1.40	F	1.35
P	1.9	S	1.85	Cl	1.80
As	2.0	Se	2.00	Br	1.95
Sb	2.2	Te	2.20	I	2.15

SIZE OF TRANSITION METAL ATOMS

In the transition series the atomic volume decreases rapidly with increase in the number of bonding *d* electrons due to the increased cohesion; finally it increases again as more and more *d* electrons enter antibonding orbitals (Figure 4.4). Consideration of heats of atomization of the noble metals Cu, Ag, and Au which follow the transition metals, reveals about 20–30 Kcal more cohesive energy than can be accounted for by a single bonding electron (see, e.g., Kambe,[44] Mott,[45] and Brewer[46]), indicating that the *d* electrons still contribute to the cohesion, even though the *d*

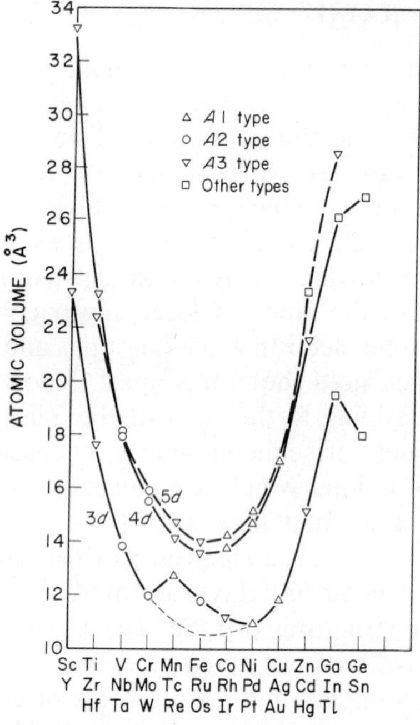

Figure 4-4. Atomic volumes of the transitions metals of the first, second and third long Periods as a function of group number: △, *A*1 type is the f.c. cubic structure; ○, *A*2 type is the b.c. cubic structure; ▽, *A*3 type is the c.p. hexagonal structure. (Pearson [47].)

shells are filled. This poses a difficult problem in view of the size of the energy involved, and it is one for which no very satisfactory quantitative account seems yet to have been given. In view of the contribution of the *d* electrons to cohesion in the Group I B metals, it is natural that the atomic volume should increase further on proceeding from Group I to the Group II B metals, Zn, Cd, and Hg, whose heats of atomization at room temperature have relatively low values of about 30 Kcal/gm atom, showing that there is no further cohesion from the *d* electrons in Group II B. However, atomic volumes increase still further on proceeding to the Group III B metals and even to the Group IV metals (Figure 4-4), instead of decreasing steadily as would be expected at the start of a (sub) period. The only explanation of this behavior seems to be that the *d* electrons still influence the atomic sizes even though it is demonstrated that they do not contribute noticeably to the cohesive energy. Nevertheless, it should be noted that this influence is in the opposite sense to that expected for a *d* shell which contracts with increasing atomic number. Of course, it might be argued that this increasing atomic volume running through Groups I B to III or IV B is just that to be expected for the Pauling[9] series of valencies which attributed 5.44, 4.44, 3.44, and 2.44, respectively to the Group I B to IV B metals, but it seems unlikely that these valencies can be accounted for by the observed heats of atomization, and their meaning is certainly obscure in relation to the experimentally determined Fermi surfaces of the Group I and II B metals.

Returning to consideration of Figure 4-4, the atomic volumes of the 4*d* and 5*d* transition metals are seen to pass smoothly through a minimum at the Fe Group, whereas those of the 3*d* series do not. From the smooth variation of the atomic volumes of the 4*d* and 5*d* *T* metals, that expected for the 3*d* metals can be reliably predicted and is indicated by the broken line in Figure 4-4. This reveals that not only is the size of Mn anomalous, as has long been recognized, but that the sizes of Fe and Co are also.[47] Indeed, the increased atomic volume which appears as a step at Mn (*A*1 or *A*2 modifications, *vide infra*) continues through Fe and Co. As no anomaly is found in the 4*d* and 5*d* Periods, it seems improbable that this anomaly can be attributed to entry of electrons into the upper part of the *d* bands; nor does it appear possible to attribute it to the particular collective electronic distribution now held responsible for the atomic moments and/or ferromagnetic properties of these 3*d* T metals. Consider α-iron for example; the lattice parameters, which have been measured continuously from room temperature up through the Curie point to 910°C[48,49] (Figure 4-5), give no indication of a volume change of more than 1 Å^3 (such as we are discussing) associated with the magnetic disordering, and furthermore Basinski and co-workers[48] have shown that the difference in atomic volume of b.c. cubic and f.c. cubic iron at 916°C is only 0.12 Å^3. At one time it was believed that the magnetic properties of these metals resulted from the presence of nonbonding *d* electrons, although this concept is now abandoned in the face of various difficulties (cf. Mott[50]). Nevertheless, it could well account for the increased atomic volume of these metals because of the decreased cohesion that would result from nonbonding *d* electrons, and indeed, if the electron number of Mn, Fe, and Co is reduced by about 2, then their atomic volumes are observed to lie close to the line for the variation of atomic volume predicted from the behavior of the 4*d* and 5*d* series, where there is no ferromagnetism (Figure 4-4). The atomic volume of Mn, which calls for special comment, has already been discussed on p. 142. In Figures 4-4 and 4-6 the rather conservative atomic volume of 12.8 Å^3 is adopted for γ-Mn. The γ and δ modifications of Mn appear to have magnetic moments corresponding to 1 to 2 free spins.

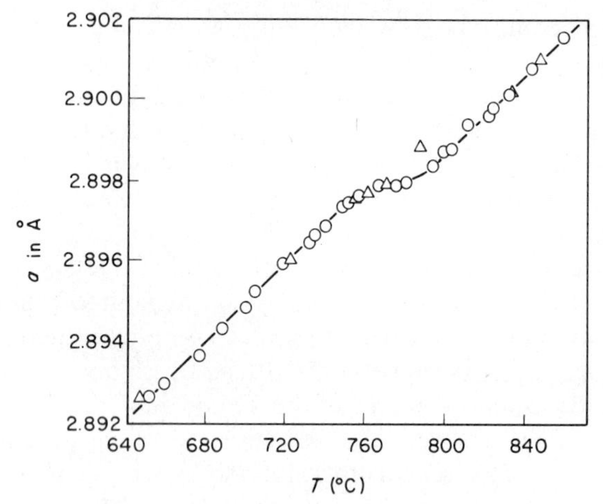

Figure 4-5. Lattice parameters of α-iron on passing through the Curie point: ○ measurements by Ridley and Stuart[49]; △ measurements by Basinski and co-workers.[48]

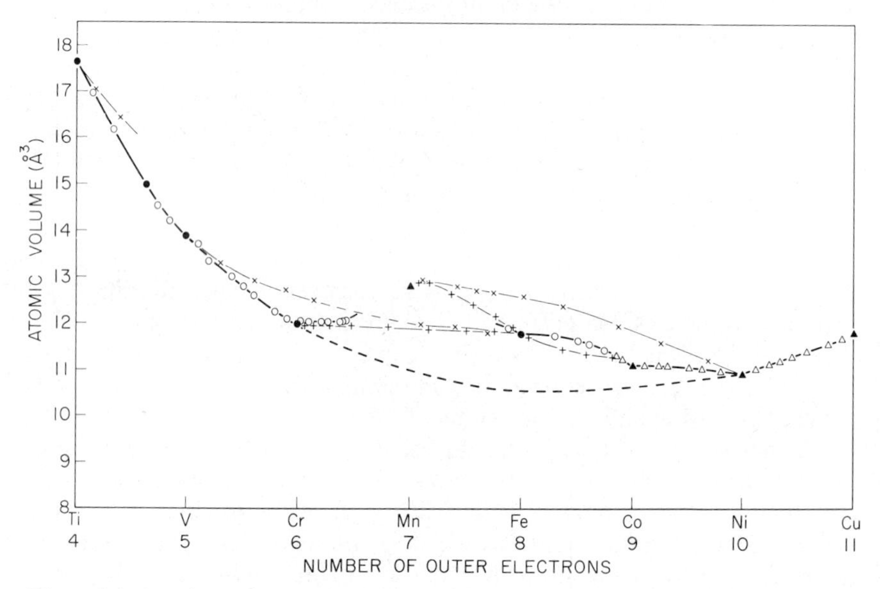

Figure 4-6. Atomic volumes of solid solutions between various transition metals of the 3*d* long Period. (Pearson[47].)

△, body-centered cubic, *A*2 type ⎫ atomic volumes of alloys between neighboring metals
○, face-centered cubic, *A*1 type ⎭

+, Atomic volumes of alloys between metals separated by two atomic-number units.

×, Atomic volumes of alloys between metals separated by three atomic-number units.

---, Expected variation for the 3*d* series.

Solid Solutions between Transition Metals

The changes of atomic volume with average valence $(s+d)$ electron number can be traced out in greater detail by examining solid solutions formed between the transition metals. Figures 4-6 and 4-7 show data for the 3*d* and 4*d* series.

In the 3*d* series the smooth decrease of atomic volume with increasing electron number ceases at about Cr, but a definite increase of atomic volume towards that of Mn is only becoming apparent at 40 at.% Mn where the lattice parameter measurements of Cr–Mn alloys cease. In the small range of solid solution obtained by dissolving Mn in Fe, the atomic volume is increasing straightaway towards that of Mn. On the other hand, in the continuous solid solution formed between Cr and Fe, there is no sign of increased atomic volume near electron number 7 corresponding to Mn. Such behavior is quite different from that found by Beck and co-workers for the electronic specific heat of b.c. cubic *T* metal solid solutions where, for example, rather similar data as a function of electron number are obtained for Cr–Mn, Cr–Fe, and V–Fe solid solutions.[51] The specific heat data suggest the suitability of a rigid

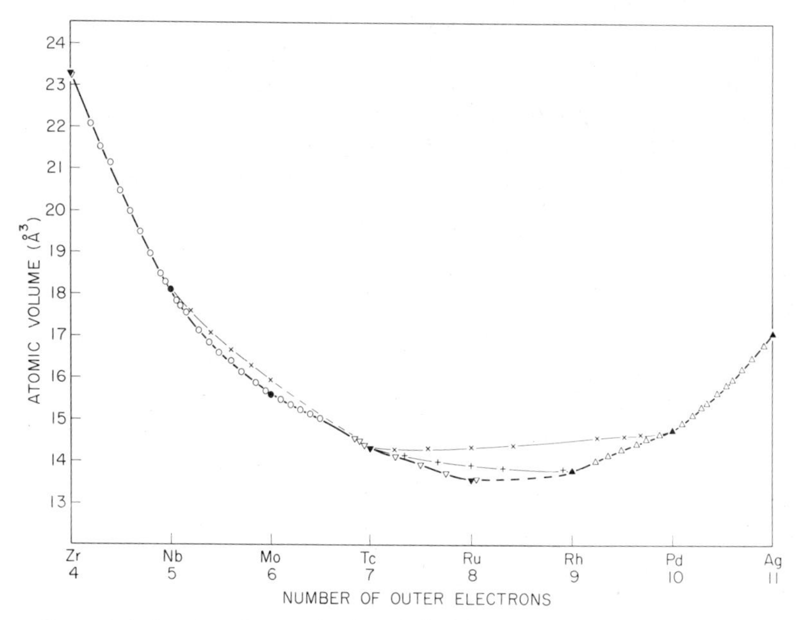

Figure 4-7. Atomic volumes of solid solutions between various transition metals of the 4*d* long Period. Symbols as for Figure 4-6 with the addition of ▽ for the hexagonal close packed structure. (Pearson[47].)

band model for these alloys, but the atomic volume data appear largely independent of the details of the density of states curve. The data of Figures 4-6 and 4-7 show generally that the atomic volumes of solid solutions of metals separated by two or more atomic number units change smoothly between the volumes of the two end components. There is no evidence of atomic volumes following either the maxima or minima suggested by the atomic volumes of elements of intermediate atomic number.

Apparent atomic volumes of metals in solid solution, determined by extrapolating the initial atomic volume variation (with at. %) in the solid solution to 100% solute, can be usefully compared with the elemental atomic volumes in some cases. In Figure 4-8, for example, available data are shown for the 3*d* transition metal solutes in Ti, Cr, Fe, Ni, and Cu as solvents. The prominence of the step occurring at Mn (absent in the first solvent, Ti) increases with increasing atomic number of the solvent as would be expected from various considerations. The elemental atomic volumes of the solutes are also indicated in Figure 4-8, and from this it can be seen by appropriately transposing the scales, that Ni is almost an ideal solvent for the metals from Cr to Cu, since the apparent atomic volumes of the solutes in nickel are almost identical to their elemental atomic volumes.

A phenomenon of frequent occurrence in purely metallic phases with identical structures, is the smaller apparent volume of Co than Ni, in spite of the fact that elemental Co has a larger atomic volume than Ni. Figure 4-9 taken from Jeitschko *et al.*[52] provides examples of this in the structures of binary disilicides, in the ternary *E* and *V* phases, and the ternary U*M*Si and ThM_2Si_2 forms of the Fe_2P and $BaAl_4$ structures. The most probable explanation of this effect involves Figure 4-4, p. 156, where it was pointed out that elemental Mn, Fe, and Co have anomalously large atomic volumes and a curve representing the expected "normal" atomic volumes derived from the behavior of the 4*d* and 5*d* *T* metals is shown. The minimum in this lies between Fe and Co, and Ni has a distinctly larger "normal" atomic volume than Co. If then in alloys Co behaves as a "normal" 3*d* metal, it is no longer surprising to find that it has a smaller apparent volume than Ni; it is only in the elemental form that it assumes an anomalously large volume.

Transition Metal Size in Compounds with Elements of Groups V and VI

The sizes of *T* metal atoms in compounds with the Groups V to VII elements are particularly influenced by the relative energies of crystal field and spin orbit coupling effects and of coulombic interactions between

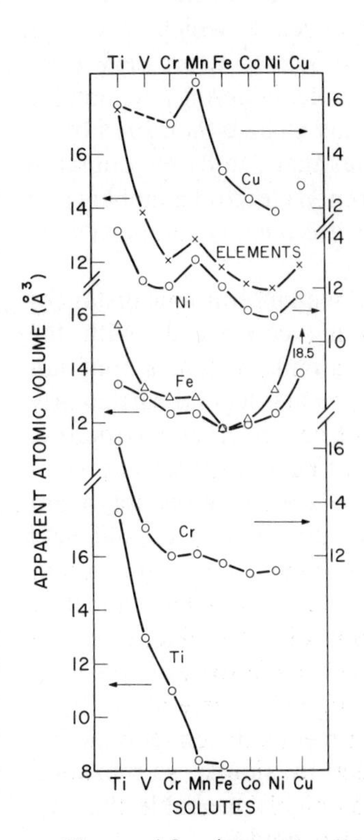

Figure 4-8. Apparent atomic volume of various 3*d* transition metal solutes in the solvents indicated on the diagram. *Elements* (×) refers to the variation of the atomic volumes of the elements themselves. (Pearson [47].)

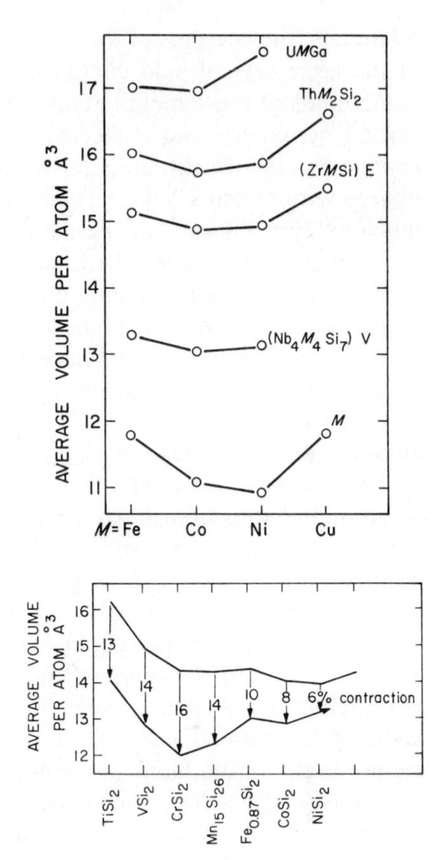

Figure 4-9. (Upper) Average volume per atom of ternary phases with the U*M*Ga (*hP*9, ordered Fe_2P type), ThM_2Si_2 (*tI*10), Zr*M*Si *E* phase and $Nb_4M_4Si_7$ *V* phase structures, compared with the atomic volumes of the elements. M = Fe, Co, Ni, and Cu. (Lower) Upper line gives the average volume per atom in the silicides calculated from the elemental atomic volumes taking 15·5 Å for Si, obtained from its H.P. white Sn structure. Lower line gives the measured atomic volume in the silicides. (Jeitschko and co-workers [52].)

the electrons of the *T* metal atoms. Generally speaking, in the 3*d* metals the energies of coulombic interactions are greatest, and crystal field effects dominate those of spin orbit coupling,[53] so that high-spin states may be realized for electron numbers between d^4 and d^7. In the 4*d* and

$5d$ metals, the energies of these three effects are comparable, and because of the large crystal field energies, only low-spin configurations occur in compounds of these metals. These competing influences which depend on crystal symmetry and d electron number, are able to exert very large energies on the transition metal atoms in the semiconducting and metallic phases with Groups V to VII elements that may change their radii by as much as 25%. In view of the many symmetries that can be encountered in anion coordination polyhedra and the different d electron numbers and configurations, it is practically impossible to prescribe any simple system of radii for the T elements in these compounds.

It has already been shown necessary to deal with interatomic distances rather than atomic volumes when considering compounds with the elements of Groups V to VI, because the atomic volumes (but not atomic radii for closest neighbors) of these elements depend particularly on the number of nonbonding electrons, and so on crystal structure. In addition, transition metal ions, which do not have a spherically symmetric S ground state, and are surrounded by coordination polyhedra of polar groups or electronegative ions, are subject to a crystal field stabilization energy that reduces the sizes of the ions. This reduction in ion size, whose magnitude depends on the type of ligand, crystal symmetry, coordination, and the number and configuration of the d electrons, can in principle be calculated, and Hush and Pryce[54] have for example estimated it for monoxides of the first transition Period, where all of the T metals occur in the high-spin state. Transition metals in semiconducting pnictides and chalcogenides are also subject to similar stabilization energies which affect their sizes, and it seems highly probable that the same effect occurs in metallic pnictides and chalcogenides.

The monoxides of the transition metals have the rocksalt structure in which the T metal ions are surrounded octahedrally by O^{2-} ions. There is a smooth monotonic decrease in T metal radius with increasing atomic number for the d^0, high spin d^5, and d^{10} oxides CaO, MnO, and ZnO† which do not experience crystal field stabilization energy (Figure 4-10). Interpolation of this curve allows the sizes of the intermediate d^1, d^2, etc. T^{2+} ions in octahedral coordination to be estimated in the absence of crystal field stabilization energy which makes the observed values smaller. The most striking feature of Figure 4-10 is the magnitude of the contraction that can occur; for example the radius of vanadium is decreased by about 25% by the crystal field stabilization energy!

†ZnO has the 4 coordinated wurtzite structure. The interionic separation for hypothetical ZnO with the rocksalt structure is taken as the sum of the radius 1.40 Å of the oxygen ion and a radius of 0.74 Å estimated by Ahrens[38] for Zn in six coordination.

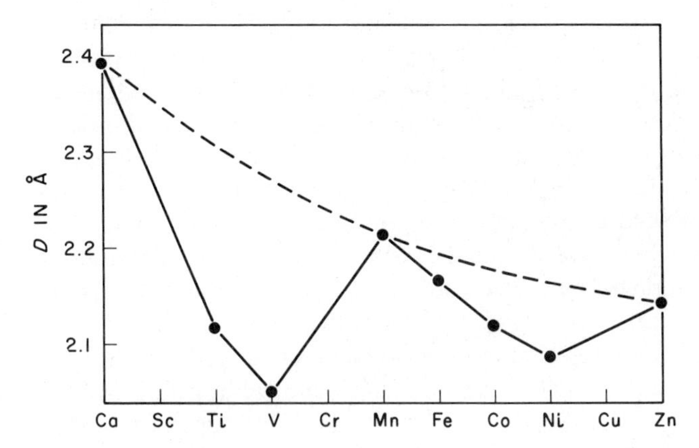

Figure 4.10. Atomic diameter of the transition metal atoms in the monoxides TO.

Hush and Pryce showed that the crystal field stabilization energy per ion of T metal in a symmetrical octahedral ionic field is proportional (among other things) to $\xi = 4n_{(t2g)} - 6n_{(eg)}$, where $n_{(t2g)}$ and $n_{(eg)}$ are respectively the number of electrons in the $t2g$ and eg orbitals of the T-metal ion. ξ has the characteristic variation with the number of d electrons shown in Figure 4-11. It has two possible values between d^4 and d^7 because of the high- and low-spin configurations, and it is appropriately zero

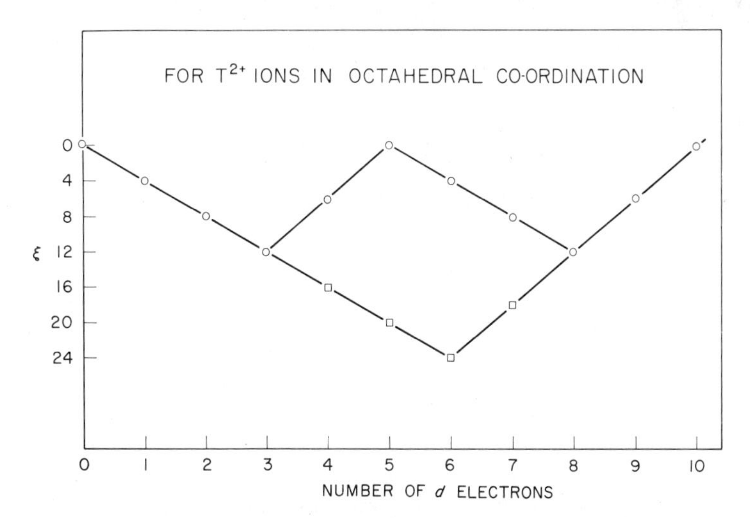

Figure 4-11. The function $\xi = 4n_{(t2g)} - 6n_{(eg)}$ for different numbers of transition metal d electrons. See text.

for d^0, high-spin d^5 and d^{10} configurations. All of the ions considered by van Santen and van Wieringen[55] and Hush and Pryce were in the high-spin state, but both high and low spin states can be recognized (from magnetic properties) in transition metal chalcogenides and pnictides, and this results in two different sizes for the atoms corresponding to the high and low-spin states. Figure 4-12 shows as a function of d electron concentration, the radii of divalent $3d$ T-metal atoms derived from various chalcogenides where the T metal is surrounded (approximately) octahedrally by the anions. The d electron number can be determined from either a covalent or ionic bond model, since both give the same results. The T metal radii in Figures 4-12 and 4-13 were determined from the (average) interatomic distances in the compounds by subtracting the Pauling[9] single-bond radii for S, Se, Te, P, As, or Sb. The difference in radii between the high and low-spin states of Fe is striking, and Blasse[56] has also demonstrated similar differences for Co^{III} in oxides. More striking still is the way that the radii follow the ξ function scaled between $t2g^3\ eg^2$ MnS_2 and $t2g^6$ FeS_2, and shown as a full line in Figure 4-12. This indicates a strong depen-

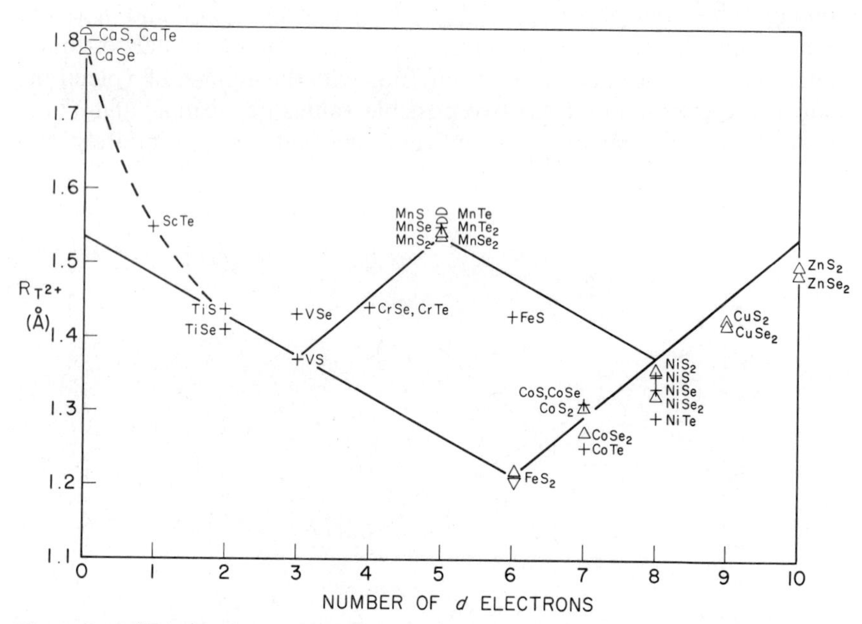

Figure 4-12. "Divalent" $3d$ transition metal radii for octahedral coordination in the compounds indicated, as a function of the number of d electrons on the transition metal atom. The full line represents the function ξ appropriately scaled. Radii were derived from measured interatomic distances and Pauling's radii for S, Se, and Te. Structure types: ●, NaCl; +, NiAs; △, pyrite; ▽, marcasite.

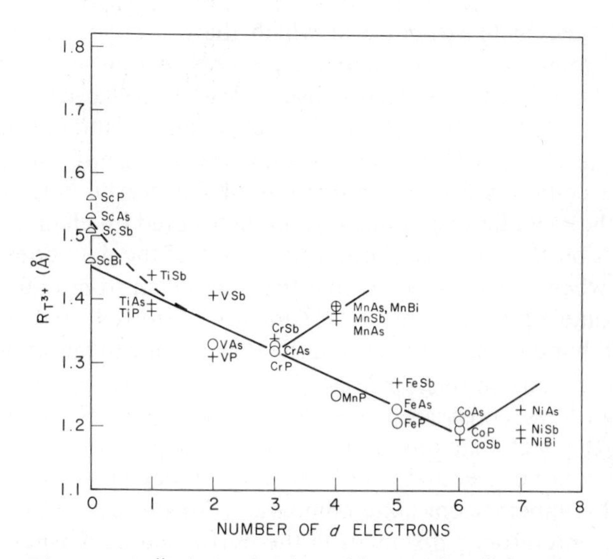

Figure 4-13. "Trivalent" 3*d* transition metal radii for octahedral coordination in the compounds indicated, as a function of the number of *d* electrons on the transition metal atom. The full line represents the function ξ appropriately scaled. Radii were derived from the measured interatomic distances and Pauling's radii for P, As, Sb, and Bi. Structure types: ●, NaCl; +, NiAs; ○, MnP.

dence of the atomic size on the number and configuration of the *d* electrons in these compounds, that include both metals and semiconductors, and which resembles that already found for ionic compounds. The observed radii (Figure 4-12) do not follow the ξ function at the start of the transition Period where its values are small, and it is only to be expected that the normal coulombic contraction of atomic radius with increasing electron concentration, found at the start of a Period, should dominate the crystal field effects there.

Figure 4-13 shows similar data for trivalent 3*d* metals in octahedral coordination, together with the ξ function suitably scaled. Only Mn compounds are found in the high-spin state, although MnP which had already been attributed to a low-spin state by Goodenough,[57] so behaves here. Such data as are available for 4*d* and 5*d* *T* metals, which only occur in low-spin states, also suggest a similar dependence of *T* metal radius on *d* electron number. The apparent radii of tetravalent *T* metals in the 3*d*, 4*d* and 5*d* series all appear to pass through a minimum at a *d* electron number of 4 or 5, insteady of 6 as expected. Such data are derived from the d^5 $CoSb_2$ structure where there is a close *T*–*T* approach, and the d^4 marcasite structure in which there are close approaches in chains of T atoms, $-T-T-T-$, and it is possible that these T atoms should be regarded as d^6, not d^4 or d^5, because of d electron sharing.

The data of Figures 4-12 and 4-13 reveal a fair amount of scatter and it could be questioned whether indeed they are significant. Data for semiconductors could follow the predictions of the crystal field theory and those for the metallic compounds be so distributed by chance as to give the impression that they also do. The scatter of the data does not seem excessive when it is considered that they are taken from different structures in some of which the octahedra are distorted, and in some of which $T-T$ bonding may occur, and the d electron configuration may not correspond exactly to high or low-spin states. If only one structure type such as pyrite is considered, the data show much less scatter. More serious is the fact that interatomic distance depends on bond number (average number of electrons per bond) and although in the data of Figure 4-13, bond numbers seem quite comparable, this is less certain in Figure 4-12. Here, interatomic distances in the pyrite and marcasite structures, where the bonds on a covalent bond model have unit strength, are compared with those in the NiAs structure, where the bond strength may be less. Nevertheless, where both structure types appear together in several compounds of Mn, Co, and Ni it is quite impossible to attribute any systematic difference in interatomic distance to difference in bond number in the two structures. Finally, the significance of applying a function such as ξ which refers to localized d electrons, to metals must be questioned, but this seems to be permissible since it is well known that transition metal atoms can have spin moments in metallic phases, and Mn with a moment of 4 spins in AuMn, PtMn, and NiMn provides a good example.

Thus it appears that in both semiconducting and metallic compounds of the T metals with the anions of Groups V and VI generally, there is likely to be a strong dependence of the T metal size on the d electron number and configuration, and the symmetry of the anion coordination polyhedron surrounding the T metal. Therefore attempts to derive a general set of transition metal radii for use in compounds with the Group V and VI elements are impossible, because crystal field stabilization energy can decrease radii by up to 25%, and the difference between radii of high- and low-spin states of a T metal may be as large as 15 or 20%. There must be so many special cases depending on the type of T metal coordination, d electron number (valency) and configuration, as to make the derivation of detailed tables of radii futile. Formulas such as those of Pauling[10] for determining T metal radii as a function of the percentage d character of the bonds cannot be satisfactory for these compounds, since they predict a monotonic decrease in $3d$ T-metal radius with increasing atomic number in, say, a series of TX_2 compounds such as pyrites with d^2sp^3 hybrid T metal bonds.

SIZES OF RARE-EARTH ATOMS

Anomalies are found in the relative sizes (interatomic distances or atomic volumes) of the elemental trivalent rare earths and to a slight extent in some of their alloy phases and intermetallic compounds. These anomalies take the form of departures from the expected smooth mono-

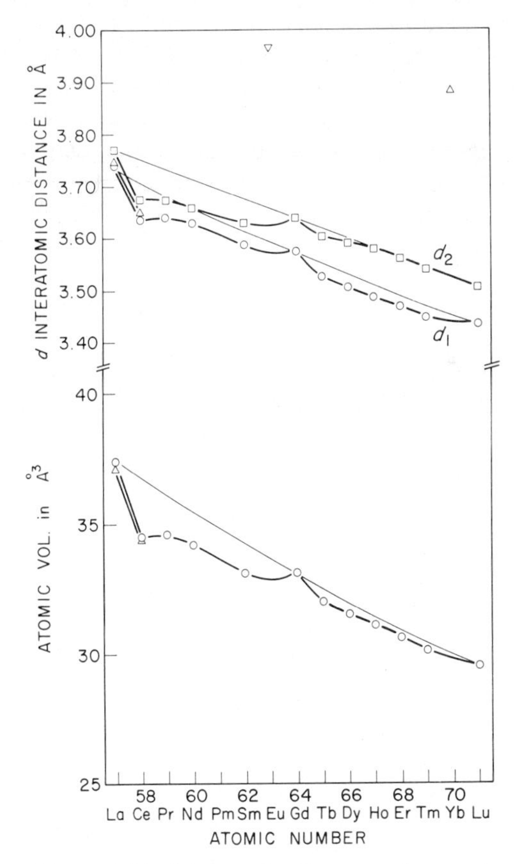

Figure 4-14. Interatomic distances d_1 and d_2 (basal plane) and atomic volume of the rare earths. △ indicates results for the f.c. cubic, ▽ for the b.c. cubic structure, and □ and ○ for other structures.

tonic decrease of atomic size on proceeding across the rare-earth Period, and resemble those found for the 3*d* transition metal monoxides where crystal field stabilization energies decrease the size of the metals which do not have $3d^0$, high spin $3d^5$, or $3d^{10}$ spherical *S* ground states. Thus the $4f^0$, high spin $4f^7$, and $4f^{14}$ trivalent metals La, Gd, and Lu have larger sizes than the intermediate rare earths as shown in Figure 4-14. It seems most probable that these anomalies are to be attributed to crystal field effects, since even if there is some band mixing of *f* states, the nature of the 4*f* levels which have a maximum radial density at a radius of about 0.5 Å, is predominantly shell-like.

The divalent elemental rare earths $4f^7$ Eu and $4f^{14}$ Yb have structures consistent with those of the Group II A metals Ca, Sr, and Ba, and accordingly much larger atomic volumes than the trivalent rare earths. Not infrequently Eu and Yb form phases with the same crystal structures as the trivalent rare earths, but with a very much larger atomic volume as indicated in Figures 4-15 and 4-16. Indeed, the relative atomic volume provides a very good criterion for the valency of the rare earth in any series of alloys. For example, from this and other criteria it is estimated

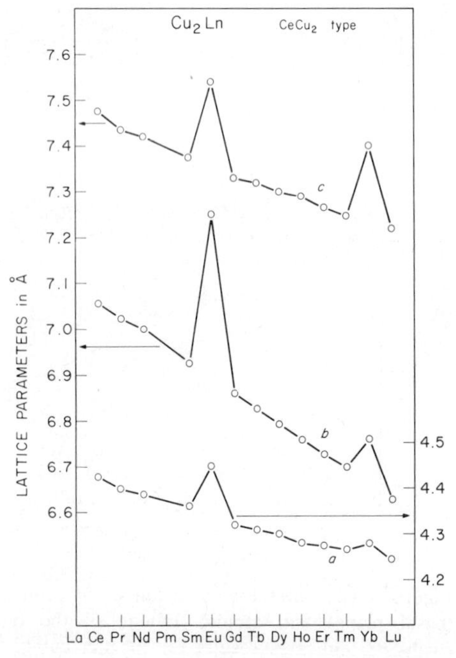

Figure 4-15. Lattice parameters of rare earth–copper alloys with the $CeCu_2$ (*oI*12) structure.

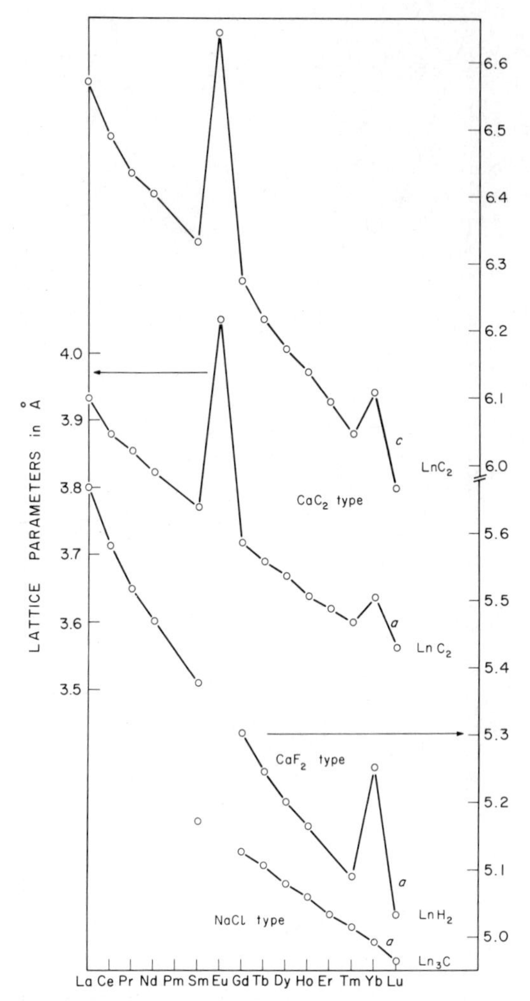

Figure 4-16. Lattice parameters of REC_2 phases with the CaC_2 (*tI*6) structure, REH_2 alloys with the CaF_2 (*cF*12) structure, and RE_3C alloys with the NaCl (*cF*8) type structure.

that the valency of Ce in elemental form is about 3.5,[58] and frequently in alloys it is observed that the valency of Ce is larger than 3, as indicated for rare-earth sesquicarbides in Figure 4-17. Sm is normally only found with valency 3; however, the dimensions of its chalcogenides SmS, SmSe, and SmTe with the NaCl structure (Figure 4-18) suggest that in these it has a valency close to 2. Indeed, with this value, there would be 8 valence

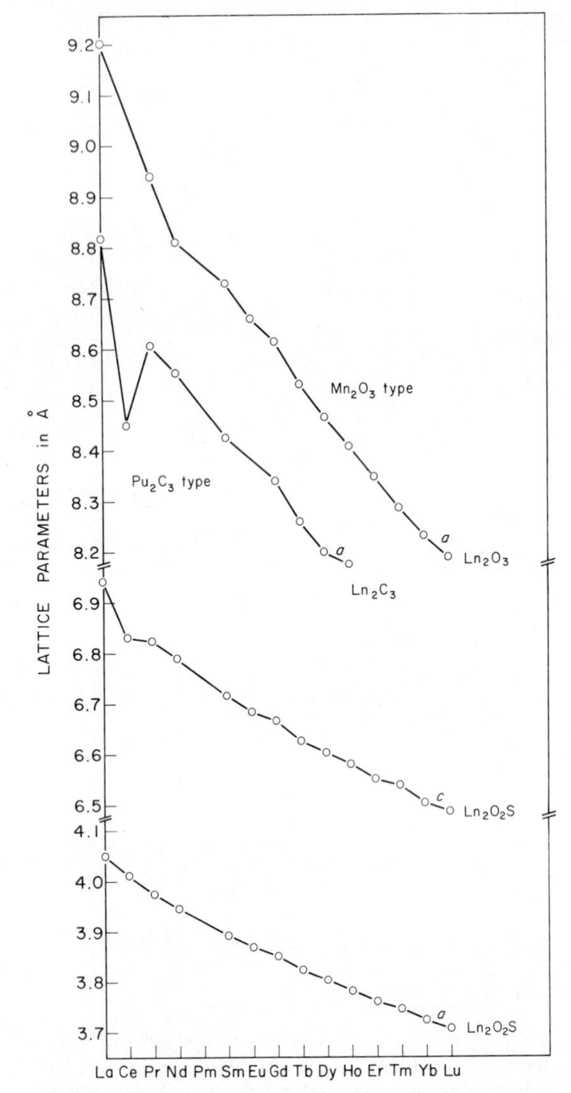

Figure 4-17. Lattice parameters of RE_2O_3 alloys with the Mn_2O_3 ($cI80$) structure, RE_2C_3 with the Pu_2C_3 ($cI40$) structure, RE_2O_2S with the ordered La_2O_3 ($hP5$) structure.

electrons per chalcogen atom so that they would attain filled valence subshells, giving a more stable state than that arising from trivalent Sm and 9 valence electrons per chalcogen atom.

In some alloys or compounds Yb, and much more rarely Eu, are present

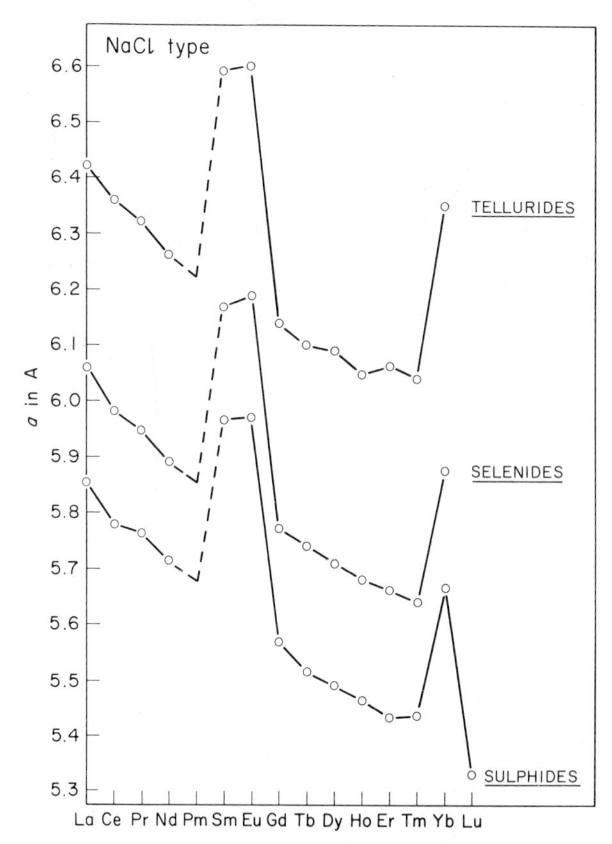

Figure 4-18. Lattice parameters of *RE*Te, *RE*Se, and *RE*S phases with the NaCl ($cF8$) structure.

in the trivalent state, as shown in examples to be found in Figures 4-17 and 4-19. In other cases they just do not form the same phases as the other rare earths (Figure 4-20). For example, in the pnictides with the rocksalt structure (Figure 4-19), where the trivalent state is favored so that the Group V element achieves a filled valence subshell, the nitride, phosphide, arsenide, and antimonide of Yb^{3+} have been obtained, but only the nitride of Eu^{3+}. On the other hand all rare earths with the exception of Eu and Yb form with the Group I and II metals, phases which have the CsCl structure (Figure 4-20). In this example, the trivalent state favors the formation of phases with the CsCl structure, but it seems that the result does not provide enough energy to excite Eu or Yb to the trivalent state, so that their compounds do not form. Combination with a Group V element which can attain a filled valence subshell in the rocksalt structure, may,

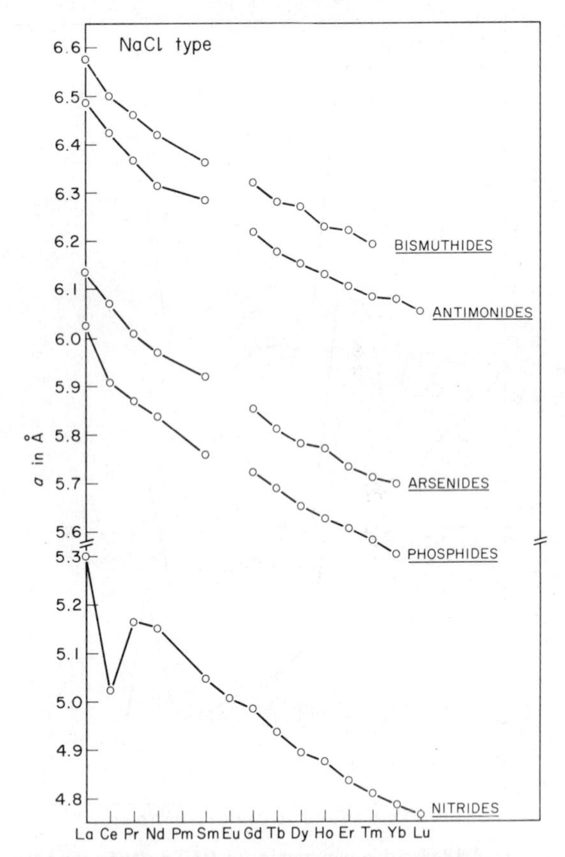

Figure 4-19. Lattice parameters of *RE*Bi, *RE*Sb, *RE*As, *RE*P, and *RE*N phases with the NaCl (*cF*8) structure.

however, provide sufficient energy for excitation as shown above. For similar reasons, Eu and Yb form oxides with the Mn_2O_3 (*cI*80) and RE_2O_2S structures (Figure 4-17), in which the trivalent state attains a sufficiently low free energy to be stable.

The rare earths have sizes that are large relative to many of the elements with which they form alloys. One consequence of this is that they form many structures controlled by the coordination factor in which the *RE* atom is surrounded by a large coordination polyhedron of the other components. Another consequence is that they do not form a number of structures common to the Group III metals Al, Ga, and In, presumably because of their large size. They do not form tetrahedrally close packed Frank–Kasper phases such as σ, μ, P, and R, whose structures are controlled by the coordination factor, because the atoms are too large, but

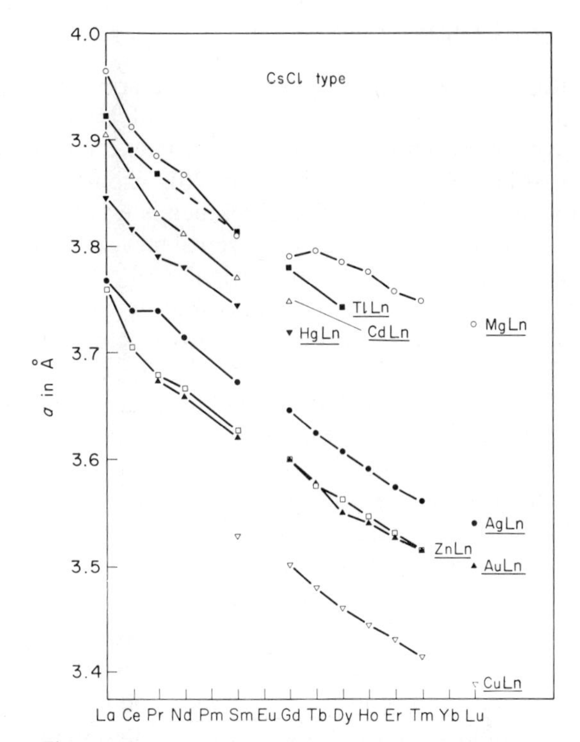

Figure 4-20. Lattice parameters of rare-earth alloys with the CsCl ($cP2$) structure: ○, *RE*Mg; ■, *RE*Tl; △, *RE*Cd; ▼, *RE*Hg; ●, *RE*Ag; □, *RE*Zn; ▲, *RE*Au; ▽, *RE*Cu.

they readily take the Laves phase structures. Although these are also tetrahedrally close packed Frank–Kasper structures, their stability is due to the geometrical factor which maintains the $16-12$ coordination over a wide range of radius ratios (up to 1.67 for the $MgCu_2$ structure; see p. 58) and hence the structure can be adopted by *RE* phases despite the large size of the *RE* atoms.

Also because of their large size, the *RE* atoms generally only form structures based on a close packed array of *RE* atoms with other close packed layers inserted at $\frac{1}{4}$, $\frac{1}{2}$, or $\frac{3}{4}$ spacing (i.e., in tetrahedral and octahedral holes) when one component is in defect, as in the RE_2X_3 phases with the sphalerite structure, and RE_2O_3 phases with the Mn_2O_3, Sc_2S_3, or spinel structures. The only exceptions to this rule seem to be the common occurrence of *RE* phases with the NaCl structure and distorted forms, and also *RE* hydrides with the CaF_2 structure; hydrides with the BiF_3 structure are always defective, REH_{3-x}.

TERMINAL SOLID SOLUTIONS

Relative atomic size is the major factor determining the extent of primary solid solutions, since a size difference of 15% between solvent and solute provides a limit beyond which it is exceedingly improbable that extended solid solutions occur (see p. 76). The permissivity for extended solid solutions when the size difference is less than 15%, may be dominated by other factors such as relative valency, energy band characteristics, or electronegativity difference which still prevent the formation of a wide range of solid solubility. The variation of lattice parameters across a solid solution range between two isostructural elements, might reasonably be expected to be linear and this expectation has been expressed in Vegard's Law,[59] or if atomic volumes rather than lattice spacings are considered fundamental, the equivalent consideration is expressed in Zen's Law.[60] In general, neither of these laws are obeyed (they are of course not identical—if one is obeyed the other cannot be), and so it is of practical interest to have information on the variation of lattice parameters with composition, and also to examine the basis of departures of observed data from either Vegard's or Zen's Laws.

Vegard's Law

Vegard's Law[59] calls for a linear variation of the lattice parameter (a), as a function of atomic concentration (x) in solid solution between two components (1, 2) of similar structure.

$$a = x_1 a_1 + x_2 a_2$$

Rarely can a "law" which is so seldom obeyed, have enjoyed so long a vogue, evoked so much discussion, or inspired so many serious attempts and calculations to explain its failure; interest in it in the last few years has increased rather than diminished! Examples in which Vegard's Law is strictly obeyed in metallic systems are few, and one might guess that they amount at most to a few percent of those cases that can reasonably be tested. It has often been considered that an expression of the law in terms of atomic volumes (V) might be more satisfactory and in 1956 Zen[60] published this, maintaining that Vegard's Law itself could not be expected to hold when the atomic volumes of the components differed significantly. However, Zen's Law

$$V = x_1 V_1 + x_2 V_2$$

is not more successful than Vegard's Law (see e.g. column 4 of Table 4-10, p. 178).

parameters in solid solutions with certainty, and so there is some value in experimentally determining the apparent size factors in solid solutions. King[18] has determined these in terms of atomic volumes for 469 substitutional solid solutions from lattice parameter data, and his values for four common solvents are given in Table 4-10. The volume size factor Ω_{sf} expressed in percent, is defined as $100\ (\Omega_N^* - \Omega_M)/\Omega_M$, where Ω_M is the atomic volume of the solvent and Ω_N^* is the effective atomic volume of the solute, N, obtained by linear extrapolation of the initial atomic volume *vs.* concentration relationship to 100% solute. Ω_{sf} can also be expressed as $(100/\Omega_M)(\partial\Omega/\partial x)$ where $\partial\Omega/\partial x$ is the rate of change of atomic volume with concentration. Table 4-10 also lists values of a linear size factor, l_{sf}, which is derived from Ω_{sf} on the basis of the Seitz radius,

$$R_0 = \left(\frac{3\Omega}{4\pi}\right)^{1/3}, \text{ where } l_{sf} = \frac{R_{0N}^* - R_{0M}}{R_{0M}} = \frac{1}{R_{0M}} \cdot \frac{\partial R_0}{\partial x} \quad \text{or} \quad l_{sf} = \left(\frac{\Omega_N^*}{\Omega_M}\right)^{1/3} - 1,$$

and values of l_{sf} have been calculated with this relationship.

Only if Zen's Law for solid solution is obeyed, will the initial linear variation of atomic volume with concentration give an apparent atomic volume of the solute N on extrapolation to 100% solute, that is the same as the elemental atomic volume of N. In general their values differ and a measure of this difference expressed as a percentage, given by $100\ (\Omega_N^* - \Omega_N)/\Omega_N$, is listed ($VLF$) for all of the solid solutions in Table 4-10. King also lists atomic volumes of the elements, which were presumably used to calculate the values of $100\,(\Omega_N^* - \Omega_N)/\Omega_N$; several of these (notably for the elements Ba, Eu, Hf, Pr, Sc, Sn, Tb, and Yb) differ appreciably from the accepted values (see Table 4-2). Secondly, the values given for Mn (α-Mn structure), Si and Ge (diamond structure), and elements such as P and S of Groups V and VI, are not expected to be comparable to their sizes in solid solution in normal metals with $A\,1$, $A\,2$, or $A\,3$ structures, and so the comparison of Ω_N^* with Ω_N for these solutes has no very great significance.

The desirability of considering atomic volumes rather than lattice parameters or interatomic distances has already been discussed on p. 138 (see also King[68]). Is is in determining size factors of terminal solid solutions that atomic volumes have the greatest advantage, since difficulties of dealing with noncubic solid solutions and differences in coordination between solvent and solute are avoided, and furthermore, compared to determining interatomic distances, details of the structure and atomic parameters do not have to be known. The disadvantage is that if for some reasons "linear" size factors are required, these have to be calculated on the basis of the Seitz radius, and although this results in a

Table 4-10 Volume Size Factors (Ω_{sf}), Linear Size Factors (l_{sf}), Vegard's Law Factors (VLF) for Some Metallic Solid Solutions

Solution	Ω_{sf} (%)	l_{sf} (%)	VLF (%)
Ag–Al	−9.18	−3.16	−8.89
–As	+10.35	+3.33	−12.61
–Au	−1.78	−0.60	−1.22
–Bi	+70.92	+19.56	−17.59
–Cd	+14.84	+4.71	−9.10
–Cu	−27.75	−10.27	+43.09
–Ga	−5.09	−1.73	−17.38
–Ge	+1.66	+0.53	−23.42
–Hg	+14.00	+4.46	−17.06
–In	+23.50	+7.28	−51.84
–Mg	+7.13	+2.32	−21.38
–Mn	+0.09	+0.02	−86.02
–Pb	+54.52	+15.60	−13.09
–Pd	−17.21	−6.10	−4.15
–Pt	−20.05	−7.19	−9.40
–Sb	+44.93	+13.16	+18.21
–Sn	+32.40	+9.81	−16.53
–Tl	+39.42	+11.71	−16.85
–Zn	−13.74	−4.80	−3.27
Al–Ag	+0.12	+0.03	−2.55
–Ca	+177.10	+40.46	+6.12
–Cr	−57.23	−24.66	−40.82
–Cu	−37.77	−14.62	−12.52
–Ga	+4.94	+1.62	−11.06
–Ge	+13.13	+4.19	−17.03
–Li	−2.10	−0.70	−24.82
–Mg	+40.82	+12.08	+0.58
–Mn	−46.81	−18.98	−27.70
–Pb	−53.63	−22.66	−84.67
–Si	−15.78	−5.56	−30.17
–Sn	+24.09	+7.45	−23.84
–Th	+156.61	+36.91	+29.65
–Ti	−15.06	−5.22	−20.17
–V	−41.42	−16.37	−30.04
–Zn	−5.74	−1.95	+2.89
Au–Ag	−0.64	−0.21	−1.20
–Al	−10.17	−3.51	−8.26
–As	+17.69	+5.57	−7.31
–Cd	+13.14	+4.20	−10.94
–Co	−25.22	−9.23	−11.68
–Cr	−16.45	−5.82	+18.08
–Cu	−27.81	−10.29	+3.66
–Fe	−19.87	−7.12	+10.16
–Ga	−4.32	−1.46	−17.17
Au–Ge	+5.54	+1.81	−20.94
–Hf	−3.30	−1.11	−26.01
–Hg	+18.90	+5.94	−13.95
–In	+20.57	+6.43	−21.86
–Li	−19.24	−6.88	−36.62
–Mn	−5.35	−1.82	+31.46
–Mo	−14.86	−5.22	−7.30
–Ni	−21.92	−7.92	+21.03
–Pd	−14.20	−4.98	−1.07
–Pt	−12.62	−4.40	−1.87
–Sb	+34.62	+10.41	−24.44
–Sn	+28.78	+8.80	−19.27
–Ta	+3.40	+1.11	−2.68
–Ti	−7.76	−2.66	−11.46
–Tl	+23.82	−7.43	−26.45
–V	−8.94	−3.08	+11.42
–Zn	−13.82	−4.84	−3.93
–Zr	+13.19	+4.21	−17.48
Cu–Ag	+43.52	+12.79	−0.60
–Al	+19.99	+6.26	−14.66
–As	+38.77	+11.53	−23.87
–Au	+47.59	+13.85	+2.79
–Be	−26.45	−9.73	+7.10
–Cd	+67.40	+18.74	−8.22
–Co	−3.78	−1.28	+1.84
–Cr	+19.72	+6.18	+17.85
–Fe	+4.57	+1.50	+4.87
–Ga	+24.11	+7.46	−25.18
–Ge	+27.77	+8.51	−33.34
–Hg	+5.44	+1.78	−47.10
–In	+79.03	+21.42	−19.21
–Mg	+50.80	+14.67	−23.37
–Mn	+34.19	+10.30	+28.69
–Ni	−8.45	−2.90	−1.22
–P	+16.51	+5.22	−28.12
–Pd	+27.96	+8.56	+2.59
–Pt	+31.19	+9.47	+2.58
–Sb	+91.87	+24.25	−24.99
–Si	+5.08	+1.68	−38.02
–Sn	+83.40	+22.41	−19.92
–Th	+49.12	+14.25	−46.44
–Ti	+25.74	+7.93	−15.95
–Tl	+129.16	+31.84	−5.33
–Zn	+17.10	+5.40	−9.09

usuable convention, the l_{sf} values obtained in practice are not quite the same for cubic solvents and solutes, as those that would be derived from considerations of lattice parameters.

PRACTICAL USE OF INTERATOMIC DISTANCES

Introduction

From the discussions of Chapter 3 and the present chapter, it might appear on the one hand that interatomic distances are the most powerful tool for diagnosing the factors controlling phase stability, and for predicting physical properties such as the type of electrical conductivity and magnetic properties. On the other hand, it might appear that atomic sizes are a matter of such uncertainty and subject to so many influences that can not be measured, or their effects predicted, that arguments based on observed interatomic distances could at best be a futile pastime. In relation to the physicists' rigid quantitative approach, the latter is most certainly true, although in the chemists' realm of working hypotheses, the former assessment is correct since consideration of cell dimensions and interatomic distances may permit the prediction of properties, and furthermore, relative interatomic distances must be demonstrably consistent with bonding or other models advanced to account for known properties. For example, the discussion of near-neighbor diagrams on pp. 52–68 indicates how cell dimensions or interatomic distances can demonstrate or confirm whether chemical bond or geometrical effects control structural stability, or they may demonstrate the existence of crystal field stabilization energies, or particular *d* electron configurations in transition metal compounds (p. 160). They may also yield information on probable Jahn–Teller distortions and thus *d* electron configuration in transition metal or other compounds[69] (see p. 270). The physicist is also interested in interatomic distances of simple structures as a scaling parameter in reciprocal space for the band structure or Fermi surface of metals, and also in relation to thermal expansion and magnetic interactions. In intermediate alloy phases the metallurgist may be interested in relating the axial ratios and changes of cell dimensions with composition to possible models of Fermi surface and energy-band structure, and to general questions of phase stability. As noted, many of these cases have already been discussed in Chapters 3 and 4; here only a few additional examples are given to show what measure of success may be obtained, and to emphasize some of the difficulties of making practical use of observed interatomic distances or atomic volumes.

Interatomic Distances in Substances with Mixed Covalent and Ionic Bonding

Although there is unlikely to be any very significant charge separation in metallic alloys, this is not the case with semiconductors, and in substances with mixed covalent and ionic bonding, there is uncertainty over the appropriate scales of radii and possible corrections thereto. Covalent and ionic bonds are not by definition opposites, and although incomplete charge transfer in an ionic compound implies a corresponding admixture of covalent bonding, the reverse is not necessarily true and an ionic component in the bonding of a covalent substance implies nothing about the strength of the covalent bonds.[70] A compound MX can well form covalent bonds of unit strength and still have an ionic component $M^{\delta+}X^{\delta-}$ in the bonding and, as Cochran[71] has pointed out, both definition and attempts to measure charge separation can be pitfalls for the unwary. First therefore, the extent of charge separation is likely to be unknown. Secondly, it will be uncertain whether the covalent or ionic scale of radii is appropriate. Although sums of ionic radii are normally larger than covalent radii sums, the effect of an ionic component in the bonding of a covalently bonded substance is to *decrease* the interatomic distances. Much has been written about corrections for this, and as explained on p. 148 the situation is generally unsatisfactory. Little, however, appears to have been said about the effects on interatomic distances of a covalent component in the ionic bonding of a compound, and it may be significant that the observed interatomic distances in compounds such as NaCl and MgS, where ionicity is incomplete, are *larger* than the sums of the calculated ionic radii (Table 4-12). Thus it is certain that there is not a monotonic variation of interatomic distance between ionic and covalent radii sums as the ionicity in a series of similar substances changes (hypothetically) from 100 to zero percent. For example, an oxygen atom may have its ionic diameter of 2.80 Å in a compound of sufficient ionicity, and its covalent size (1.44 Å), slightly perturbed, if the ionicity is insufficient. Such behavior could greatly influence alloy phase diagrams, and it is interesting here to compare compounds of Ti with oxygen and with nitrogen. There is clear evidence that oxygen combined with Ti has its ionic diameter of 2.8 Å, whereas nitrogen has its covalent diameter of 1.48 Å. For example, one of the forms of TiO_2 has the rutile type of structure (p. 415), in which the small Ti ion sits at the center of an octahedron of the larger oxygen atoms. The limiting ratio of $R_{cation}/R_{anion} > 0.414$ for stability of this arrangement is satisfied. There is no TiN_2 compound, but the existence of Ti_2N with the antirutile type of structure immediately suggests Ti with its large, and N with its small covalent diameter. On the other hand Ti_2O with large O atom has the anti-CdI_2 type of structure. Another example is provided by the system TiO–TiN. Both TiO and TiN have the rocksalt structure and their lattice parameters, respectively $a = 4.1822$ Å and 4.246 Å, differing only by 1.5%, would readily suggest that they should form a continuous solid solution. Instead, it is reported[72] that whereas TiO dissolves about 60 wt.% TiN, TiN itself dissolves practically no TiO. The electronic structures of these two compounds would suggest, if anything, the opposite effect: that TiN being normally metallic would dissolve TiO without difficulty, whereas TiO having a filled valence band and a partly filled conducting *d* band might not dissolve TiN. In fact it appears that TiO having oxygen in its large ionic diameter can dissolve TiN without difficulty, no matter whether N adopts its large ionic diameter or its small covalent diameter. If TiN, on the other hand, is largely covalent having N with its small covalent diameter, then, as observed, solution of TiO would be impossible if oxygen in the Ti environment insisted on adopting its large ionic size.

The oxides, nitrides, and carbides of Ti and V all have the rocksalt structure, but comparison of observed and calculated interionic distances (Table 4-11) indicates that only the oxides can have a basically ionic structure. Nevertheless, VO and VN are reported to form a continuous solid solution, and this perhaps calls in question the interpretation of TiO–TiN alloys given above, although there are differences between VO and TiO. All of these compounds have been reported to have a range of solid solubility extending up to the equiatomic composition and perhaps beyond. TiC, TiN, VC, VN, and VO all show an increase in unit cell edge with increasing content of nonmetallic component on the metal-rich side of the equiatomic composition, whereas TiO shows a definite *decrease*. VO, thus behaves more like the other interstitial compounds in this respect than like TiO. However, TiO is known to contain

Table 4-11

Compound	M–X d_{obs} (Å)	Ionic d_{calc} (Å)
TiO	2.090	2.08
TiN	2.123	2.39
TiC	2.164	3.28
VO	2.047	1.99
VN	2.085	2.30
VC	2.091	3.19

Table 4-12

Compound	Structure	Observed Inter-atomic Distance, d_{obs} (Å)	Δd	Sum of Ionic Radii for Appropriate C.N. (Å)	Sum of Covalent Radii Assuming $n = 1$[a]	Electronegativity Difference Δx
NaF	rocksalt	2.32	0.21_6	2.31		3.05
MgO	rocksalt	2.104		2.05	2.10	2.3
AlN	wurtzite	1.885 (1.917 along c)		2.07	1.99	1.5
SiC	sphalerite	1.887		2.81	1.94	0.7
NaCl	rocksalt	2.814	0.219	2.76	2.57	2.1
MgS	rocksalt	2.595	0.244	2.49	2.40	1.3
AlP	sphalerite	2.351	0.001	2.47	2.35	0.6
Si	diamond	2.352		—	2.35	0
CuBr	sphalerite	2.46	0.01	2.78	2.49	1.0
ZnSe	sphalerite	2.45	0.01	2.59	2.48	0.9
GaAs	sphalerite	2.44	0.01	2.71	2.48	0.5
Ge	diamond	2.45		—	2.45	0
AgI	sphalerite	2.808	0.003	3.28	2.86	0.75
CdTe	sphalerite	2.805	0.001	3.05	2.86	0.6
InSb	sphalerite	2.806	0.004	3.13	2.87	0.3
Sn	diamond	2.810		—	2.80	0
ZnS	sphalerite	2.351		2.44	2.35	
CsCl	CsCl	3.566		3.78	3.34	

[a] n is the bond order

more than 10% of vacancies on both the cationic and anionic sites and little further can be said at present until the concentration of vacant sites as a function of composition in these compounds, is known in some detail.

Table 4-12 compares observed interatomic distances with calculated interionic distances for series of isoelectronic compounds with the rocksalt, wurtzite, sphalerite or diamond structures. The strong contraction of the interionic distance on passing from the monovalent NaF to divalent MgO results from the greatly increased coulombic contraction of doubly charged ions. On the other hand there is practically no change in interatomic distance on going from AlN to SiC, and the bonding in these compounds is predominantly covalent because of the very high polarizing power of three and four valent ions. Interatomic distances are essentially constant in the isoelectronic series from Ge to CuBr and from Sn to AgI; although the electronegativity differences and energy band gaps increase significantly throughout the series, it is apparent that the increasing ionicity has little influence on interatomic distances, unless it is perhaps to cause them to contract slightly, and that the covalent radii scale is appropriate if n is assumed to be 1.

The evidence presented in this section does little to show how to treat *a priori* the sizes of atoms in compounds of intermediate ionicities, since it seems to confirm the view that there are only two limiting scales of atomic size that can be considered, and that there is not a continuous transition of sizes between them. It is therefore necessary to choose whichever scale appears to be most appropriate to a particular compound, and then to treat the increase or decrease of ionicity as a perturbation, endeavoring to correct for it if necessary. The situation is therefore an unhappy one.

A further trouble arises in determining whether the bond order, n, in valence compounds should be determined by the satisfaction of valency in the formula, or by the coordination found in the crystal structure. For instance, in AlP with the sphalerite structure, interatomic distances agree on the assumption that $n = 1$, but since each atom has valency three and four neighbors, it is not *a priori* certain whether $n = 1$ or $\frac{3}{4}$. Indeed, in the case of BP agreement is obtained with $n = \frac{3}{4}$ rather than 1. The difficulty increases with divalent compounds with the sphalerite structure (d_{obs} and d_{calc} agree with $n = 1$ for ZnS, but with $n = \frac{1}{2}$ for BeS, BeSe or BeTe) and with valence compounds such as MgS with the rocksalt structure where atomic coordinations are 6. On the other hand, in applying the radii to metallic phases, it is clear that n is to be taken as the ratio of the elemental valency to the coordination.

Use of Interatomic Distance to Diagnose Bond Character

Although the covalent or ionic radii sums may not precisely reproduce the observed interatomic distances in crystals, they are nevertheless often sufficiently accurate to diagnose gross changes of bond character. For instance, there is frequently a considerable difference between the sums of ionic and covalent radii in compounds, as Table 4-12 shows, and there is little doubt that ZnS and CdTe are predominantly covalent whereas NaCl and CsCl appear to be highly ionic. On the other hand, it is not possible to say anything much about the character of MgO from comparison of interatomic distances as there is very little difference between covalent and ionic radii sums. Another example is provided in Table 4-11, p. 181 where interatomic distances and calculated ionic radii of the carbides, nitrides, and oxides of Ti and V are compared, with the conclusion that the valence bonding involving *s* and *p* electrons in TiO and VO is basically ionic, whereas in the nitrides and carbides it is not.

One of the more important applications of atomic radii to semiconductors is to determine whether there is direct chemical bonding or only van der Waals bonding between certain atoms. This may influence the electrical properties of layer and chain type compounds as indicated in Chapter 5, where several examples are discussed.

In compound semiconductors with complex structures and various different interatomic distances, Pauling's bond order equation[9] may be used to calculate the bond number sums for the various atoms to see whether they exhibit their normal chemical valencies. For example, Tideswell and co-workers[73] calculating the bond numbers for the atoms in Sb_2Se_3 (*oP*20) using Pauling's distance of 2.58 Å for Sb–Se bonds of unit strength, find a constant valency for different atoms of the same kind, and that the ratio of the valencies of Sb to Se is 3 : 2. However, it is necessary to assume a larger unit bond distance of 2.63 Å in order to obtain the normal chemical valencies of 3 and 2 for Sb and Se respectively in the bond summation. Data taken from their paper are given in Table 4-13.

In other cases calculation of interatomic distances may reveal direct anion–anion or cation–cation bonding which is important to understanding electrical properties. Sometimes such observations are unexpected. For example, of phases with the Ir_3Ge_7 (*cI*40) structure, only Re_3As_7 and Tc_3As_7 with 56 outer *s*, *p* and *d* electrons per formula unit have been reported to be semiconductors,[74] and these would appear to be perfectly normal valence compounds since 3 Re atoms and 7 As atoms can each provide or accept 21 valence electrons. However, when interatomic distances in the structure are examined,[75] it is discovered that there are bonds between the Re atoms and also bonds between the As atoms, in

Table 4-13 Bond Distances, Bond Numbers, and Valencies in Sb_2Se_3[73]

Central Atom	Bonded Atom	Observed Distance and Multiplicity (Å)	Bond Number $d_1 = 2.58$ Å	Bond Number $d_1 = 2.63$ Å
Sb (1)	Se (2)	2.66 (2)	0.74 (2)	0.89 (2)
	Se (3)	2.66 (1)	0.74 (1)	0.89 (1)
	Se (1)	3.22 (2)	0.09 (2)	0.10 (2)
	Se (2)	3.26 (1)	0.08 (1)	0.09 (1)
	Se (3)	3.74 (1)	0.01 (1)	0.01 (1)
			$\Sigma = 2.49$	$\Sigma = 2.97$
Sb (2)	Se (1)	2.58 (1)	1.00 (1)	1.21 (1)
	Se (3)	2.78 (2)	0.47 (2)	0.56 (2)
	Se (1)	2.98 (2)	0.21 (2)	0.26 (2)
	Se (2)	3.46 (2)	0.03 (2)	0.04 (2)
			$\Sigma = 2.42$	$\Sigma = 2.93$
Se (1)	Sb (2)	2.58 (1)	1.00 (1)	1.21 (1)
	Sb (2)	2.98 (2)	0.21 (2)	0.26 (2)
	Sb (1)	3.22 (2)	0.09 (2)	0.10 (2)
			$\Sigma = 1.60$	$\Sigma = 1.93$
Se (2)	Sb (1)	2.66 (2)	0.74 (2)	0.89 (2)
	Sb (1)	3.26 (1)	0.08 (1)	0.09 (1)
	Sb (2)	3.46 (2)	0.03 (2)	0.04 (2)
			$\Sigma = 1.62$	$\Sigma = 1.95$
Se (3)	Sb (1)	2.66 (1)	0.74 (1)	0.89 (1)
	Sb (2)	2.78 (2)	0.47 (2)	0.56 (2)
	Sb (1)	3.74 (1)	0.01 (1)	0.01 (1)
			$\Sigma = 1.69$	$\Sigma = 2.02$

addition to the Re–As bonds, and instead of a straightforward case of semiconductivity in a normal valence compound, semiconductivity in a substance containing a complex array of cation–cation and anion–anion bonds, must be explained. Taking the single bond radius of As ($R_1 = 1.21$ Å) and a single bond radius of Re derived from the elemental structure ($R_1 = 1.285$ Å), Pauling's bond order equation can be used to determine the strengths of the various bonds. If then it is assumed that the As(1) and As(2) atoms have a filled valence subshell of 8 valence electrons consistent with the reported semiconductivity of Re_3As_7, the general valence equation, $(n_a + n_c + b_a - b_c)/N_a = 8$ (p. 202), can be satisfied by placing the remaining 4.5 electrons in d orbitals of the Re atoms. However, this does not satisfy the reported diamagnetism of the compounds.[75] If, on the other hand, a single bond radius of 1.235 Å is assumed for As, the bond numbers are then as listed in Table 4-14. These are consistent with the Re atoms each sharing 7 electrons in bonding, As(1) 3.5 and As(2) 4. The 3 Re atoms per formula devote one electron to Re–Re bonds, and the 7 As atoms use 7 electrons in As–As bonding. Since each As atom has a filled valence subshell to account for the semiconductivity each As(1) has half a non-bonded electron. The number of valence electrons used (Table 4-14) is thus 50 leaving 2 electrons in the *d* orbitals on each of the 3 Re atoms. The valence equation is satisfied as shown in Table 4-14, and so by assuming a slightly larger size for the As atom, both the reported semiconductivity and diamagnetism are accounted for.

Compounds with the Ir_3Ge_7 structures which do not have 56 outer *s*, *p*, and *d* electrons are not expected to be semiconductors. Owing to the uncertainty in the exact values of the atomic radii it may not be possible to predict this for Ir_3Ge_7 with 55 electrons, unless perhaps the magnetic properties are known accurately, but metallic properties can certainly be predicted from the interatomic distances and the general valence formula for compounds such as Ru_3Sn_7 which has only 52 outer electrons. Attention to precise details arising from bond number calculations in semiconducting compounds cannot be justified in view of all the uncertainties already discussed, although when the valence sums agree with the normal chemical valencies as in the example of Sb_2Se_3, more confidence can be placed in the calculation of relative bond strengths within the compound.

Interatomic distances between transition metals are also important in determining whether or not the *d* electrons are likely to be localized, and this has a direct influence on magnetic and electrical properties and also in structural considerations. The critical radius R_c for change from localized to collective behavior of the T metal *d* electrons depends on the T metal and the number of *d* electrons, as well as the component atoms surrounding the T atom and the symmetry of its coordination polyhedron. Goodenough[76] has derived expressions for R_c from correlation of conductivity and magnetic data of oxides and of metals. From consideration of sesquioxides of the first Period transition metals, he postulates that the critical radius R_C (3*d*) for cations in octahedral sites of a close-packed oxygen sublattice is given by

$$R_c(3\text{d}) \sim 3.05 - 0.03(Z - Z_{\text{Ti}}) - 0.04\Delta(J(J+1))\ \text{Å}$$

where Z and Z_{Ti} are, respectively, the atomic numbers of the particular T-metal cation and of Ti, and $\Delta(J(J+1)) = J_l(J_l+1) - J_c(J_c+1)$ where J_l and J_c are the total quantum numbers for localized *vs.* collective

Table 4-14 Bond Numbers Calculated for Re_3As_7 on the Assumption that R_1 for As = 1.235 Å and R_1 for Re = 1.285 Å

	Interatomic Distances[75] (Å)	Calculated Bond Numbers	Electrons Used in Bonding, per Atom	Electrons Used in Forming As–As Bonds per Formula	Electrons Used in Bonding per Formula	Nonbonding Electrons Assuming As Have Filled Valence Subshells	General Valence Equation (see p. 202)
Re–1 Re	2.779	0.45	0.45				$n_a = 7 \times 5 = 35$
–4 As(1)	2.584	0.78	3.13				$n_c = 3 \times 7 - 6 = 15$
–4 As(2)	2.559	0.86	3.43				$b_a = 7$
			7.0		21	6	$b_c \sim 1$
							$N_a = 7$
As(1)–4 Re	2.584	0.78	3.13				$\dfrac{n_a + n_c + b_a - b_c}{N_a} = \dfrac{35 + 15 + 7 - 1}{7} = 8$
–4 As(1)	3.082	0.10	0.40	1.2			
			3.53		10.5	2.5	
As(2)–3 Re	2.559	0.86	2.57				
–1 As(2)	2.482	0.95	0.95 }	5.88			
–3 As(2)	2.925	0.175	0.52 }				
			4.04		16	0	
				7	47.5	8.5	
					56.0 (47.5 + 8.5)		

electrons. The rule may be applied to T-metal cations in the octahedral ligand field of other anions, although if these are more polarizable than oxygen it appears that R_c is somewhat larger. For metals themselves where the critical radii are largest, Goodenough finds $R_c(3d) \sim 1.53$ Å, $R_c(4d) \sim 1.97$ Å, $R_c(5d) \sim 2.21$ Å and comparison with the results for oxides shows $R_c(3d)_{\text{metals}} \sim (R_c(3d)_{\text{oxides}} + 0.1)$ Å and $R_c(4d) \sim (R_c(3d) + 0.44)$ Å, $R_c(5d) \sim (R_c(3d) + 0.68)$ Å.

These relationships, which were advanced tentatively by Goodenough, are not sufficient to describe all cases, as is apparent from consideration of NiMn with the AuCu I structure. In this structure Mn is surrounded by 4 Mn atoms at 2.63 Å and 8 Ni atoms at 2.56 Å, yet has a spin moment of $2S = 4$, thus behaving more like a rare earth than a transition metal.

Interatomic Distances in Complex Metallic Phases

In alloys the concept of metallic radius becomes strained when the distances between different ligands are not the same. When this happens in valence compounds, Pauling's rule relating observed radius $R(n)$, $R_{(1)}$ and bond number n generally leads to a sensible interpretation of the situation in terms of fractional bond numbers, if the bond numbers add up to show that a component is exerting its expected valency as, for example, in Sb_2Se_3 discussed on p. 184. Many metallic alloys present immediate problems because of uncertainty of atomic size, and even in the case of the c.p. hexagonal elemental structures when the axial ratio is not ideal, it is uncertain whether the closest distance of approach or an average of the two distances represents the best value of the diameter of the atoms. In alloys such as Laves phases with $R_M/R_N > 1.225$ where the M atom has CN 16 and the 4 M–M contacts require a smaller M radius than that which describes the 12 M–M contacts (p. 59), the chemical concept relating valency, coordination, and radius is probably irrelevant, since the energies involved are probably not the same as those concerned in the formation of chemical bonds *per se*. In particular, Shoemaker and Shoemaker[77] point out that the radius of M depends not only on its valency and coordination, but also on the size of the constituents with which it is alloyed [although this may be nothing more than the geometrical requirements of the structure that must be satisfied to achieve the conditions for structural stability, and the relative sizes of the component atoms]. They find that by defining two radii each for atoms in CN polyhedra 14, 15, and 16 in Frank–Kasper type transition metal phases $\{R^*_{14}$, R^*_{15}, and R^*_{16} referring to the distances to the major ligands (atoms with surface coordination number 6) and R_{14}, R_{15}, and R_{16} referring to distances to minor ligands with surface coordination number 5$\}$

and a single R_{12} radius for icosahedra, interatomic distances in these complex phases can be predicted to within 0.06 Å, even though the observed values have a spread of up to 0.9 Å. The effect of the average sizes of the component atoms on these radii is taken into account by establishing relationships between observed values and $\bar{R}$, the average radius of the component atoms for CN 12, suitably weighted according to composition.

The success of this work in predicting interatomic distances in complex phases of transition metals further emphasizes that no single set of metallic radii can be used for predictions of interatomic distances if sufficient accuracy is sought. In this case the interatomic distances depend not only on the component atoms, but on the type of ligand, the number of neighbors and on the average sizes of the constituent atoms, which are incidentally all factors that are involved in satisfying coordination factor requirements for phase stability.

LATTICE PARAMETER VARIATION

Changes of Slope of Lattice Parameter Curves about Crystallographically Stoichiometric Compositions of Intermediate Phases

Addition of a solute to a pure metal leads to lattice parameter variation with composition that depends on the relative sizes of the two components and other factors. Lattice parameter-composition curves for various solutes in a given solvent thus proceed in many different directions as shown for example in Figure 59 of Pearson (1958).[78] Similar behavior should therefore be expected for ordered intermediate phases with a range of homogeneity on either side of a stoichiometric composition which corresponds to full ordered occupation of the various sitesets for the structure. Thus, the rate of change of lattice parameters with composition may not be the same when component M replaces N in one siteset on approaching the stoichiometric composition, as that found beyond the stoichiometric composition where M replaces N in another siteset.

The simplest situation to find evidence of this behavior is, of course, in a fully ordered structure such as the CsCl type occurring at the equiatomic composition. A phase MN with the CsCl structure with solution of M on proceeding to M-rich compositions and of N on proceeding to N-rich compositions can be regarded as analogous to a pure metal taking solutes M or N into solid solution. If M and N have distinctly different sizes, the lattice parameter variation on either side of the stoichiometric composition is no more likely to be the same than it is for two solutes of disparate sizes forming solid solutions in a pure metal, and a change of slope of the lattice parameter curve is to be expected at the equiatomic composition. In many cases structural complexities, incomplete ordering or nonspecific replacement if several sitesets are available for the substituting atom, as well as other relative properties of the component atoms may tend to obscure such an effect. On the other hand, large electronegativity difference, high degree of ordering and unique siteset requirements on the position of the substituting atoms would tend to favor its observation. Figure 4-22 gives an example of this effect in AgMg with the CsCl structure.[79] Density measurements (Figure 4-22) show that the change of slope in the lattice parameter curve at the equiatomic composition can not be attributed to the introduction of vacant lattice sites on either side of the stoichiometric composition.

Rather more specific examples of this effect can be observed in uniaxial crystals such as those of the ordered CuAu I structure, which can be regarded as generated by stacking square nets of Cu and Au atoms alternately along [001]. Phases such as PtMn, PdMn, NiMn, and IrMn with this structure, and CuAu I itself, have ranges of homogeneity which extend on either side of the equiatomic composition, and they exhibit a reasonably high degree of positional ordering. On proceeding towards the equiatomic composition by increasing the concentration of the larger component M, M atoms replace the N atoms on one set of 4^4 nets until at the composition MN, alternate 4^4 nets are composed completely of M atoms. Since the net axes lie parallel to the basal cell edges and the atoms touch along this direction, the cell expands along a until at the equiatomic composition, the a parameter is approximately equal to the M–M separation in its elemental structure. Proceeding therefore from a hypothetically similar atomic arrangement of pure N atoms to 50 at.% M when the structure contains alternate layers of M atoms, the a parameter expands from the elemental N–N spacing to the elemental M–M spacing, or twice as fast as predicted by Vegard's law for continuous N–M solid solution. Accompanying this compulsory expansion of a, there is a contraction of the c axis to prevent the cell volume, and therefore the free energy, from increasing excessively.

Beyond the equiatomic composition, further addition of the larger M atoms results in substitution for N in the other layers. Since the a axis of the crystal is fully expanded to the elemental M–M spacing, the new interlayer M–M contacts now cause a compulsory expansion of the crystal along its c axis, and an accompanying contraction of a to balance the increase of cell volume. Since in this case 50 at.% M represents M–N contacts and hypothetical structures with 0 and 100 at.% M represent N–N and M–M contacts, the expansion of c is expected to be in accord with the normal predictions of Vegard's Law. Figure 4-23 shows these

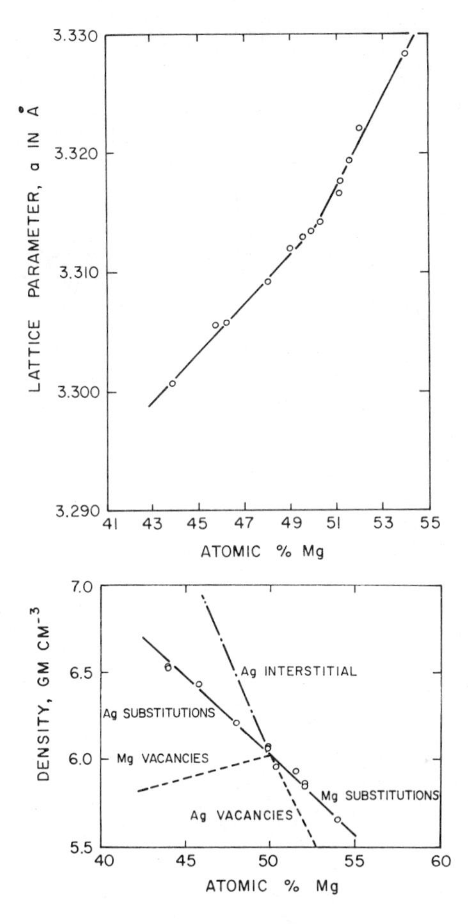

Figure 4-22. (Upper) Lattice parameter variation across the homogeneous range of the AgMg phase with the CsCl ($cP2$) structure. (Lower) Density measurements for the solid solution indicating that the method of composition change is by substitution. (Hagel and Westbrook[79].)

changes of unit cell edges for PtMn[80] with sharp changes of slope occurring at the equiatomic composition. The compulsory expansion of a below 50 at.% of the larger component does essentially have a rate of twice Vegard's Law prediction, and of c above 50 at.% an expansion equal to Vegard's Law prediction in some of these alloy systems such as PtMn, IrMn, and AuCu. The contraction of c below 50 at.% of the larger component and of a above, which occur to prevent the cell volume

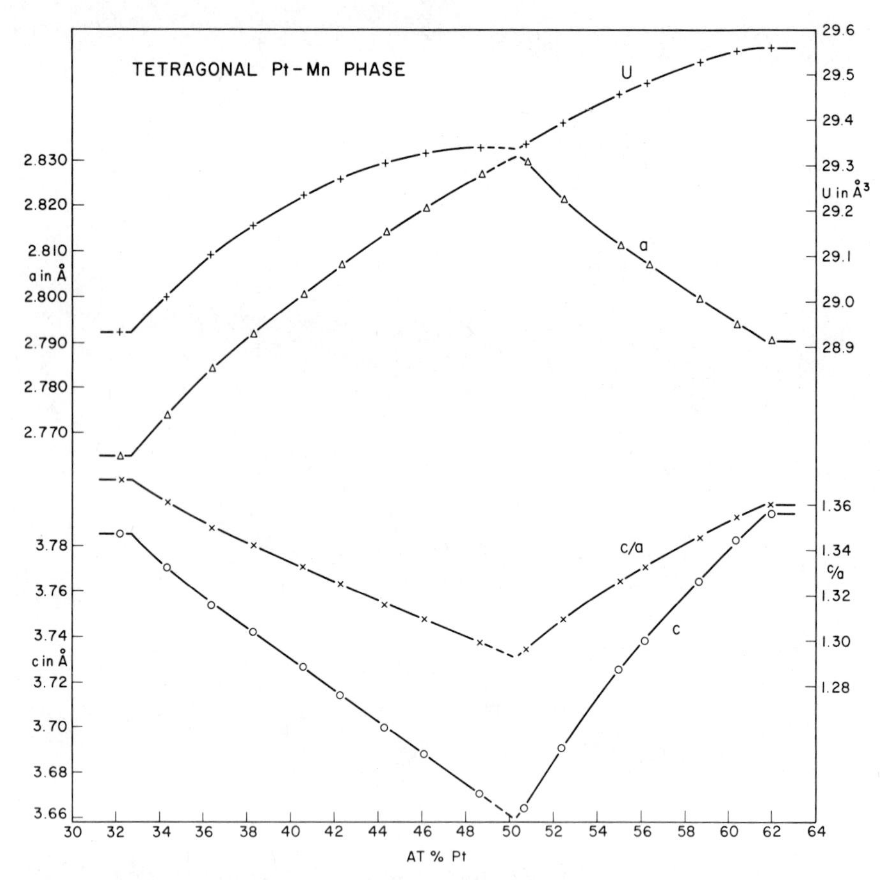

Figure 4-23. Lattice parameter variation across the homogeneous range of the PtMn phase with the AuCuI (*tP*2) structure.[80] Lattice parameters are given for the primitive cell containing two atoms.

expanding too rapidly, follow no such predictions as is to be expected. Some phases, however, do not follow these predictions closely if at all, PdMn being a notable example. In this case it appears that magnetic effects also influence the variation of lattice parameters, and it is to be noted that the magnetic interactions in the Pd alloy are significantly different to those in the Pt, Ir, and Ni phases.

Lattice Parameter Anomalies and Vacant Sites

Numerous cases of lattice spacing anomalies in solid solutions have been cited in the literature and attributed, from density measurements, to the occurrence of vacant lattice sites, often over rather narrow concentration ranges. More recent repetitions of such work in the Au–Ni, Sn–In, and Sn–Cd solid solutions, for example, show that either the lattice spacing anomalies, or density anomalies suggesting vacant sites, do not exist. For this reason the subject is not further pursued here. In some cases, such as the AlNi and AlCo phases[81] with the CsCl structure, however, the changes of slope in lattice parameter composition curves due to the appearance of vacant sites are well established. Bradley and Jay's[82] analysis of the lattice parameter variation across the Fe_3Al phase field provides another well known example, details of which may be changed by later work.

Lattice Parameter Changes Resulting from Magnetic Ordering

Magnetic-ordering introduces stresses in the crystal lattice and generally causes lattice parameter changes as a function of composition

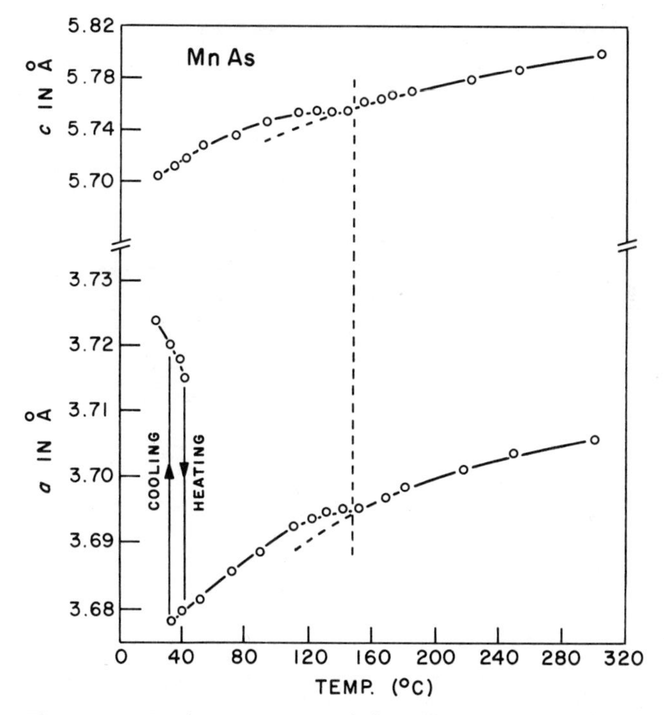

Figure 4-24. Lattice parameter variation of MnAs as a function of temperature. MnAs has the NiAs (*hP*4) structure at high temperatures, becomes orthorhombically distorted (*oP*8) below about 125°C, and undergoes a first order change back to the NiAs structure at about 40°C. (Willis and Rooksby [83].)

or temperature, and discontinuous changes when it accompanies a first-order crystal structure change; the rare earths provide some striking examples of these. The changes accompanying ferromagnetic ordering are generally relatively small, as in iron (Figure 4-5, p. 158) and nickel, although when there is a simultaneous first-order structure transformation, as for example in MnAs,[83] they can be quite large (Figure 4-24). Much larger changes are generally found to accompany second-order antiferromagnetic ordering. In cubic crystals this generally results in structural distortion and lowering of the symmetry of the "chemical" cell, as for example in the case of CoO with the rocksalt structure which becomes tetragonal below T_N. Uniaxial crystal structures generally absorb the stresses of antiferromagnetic ordering by change of the relative axial lengths of the unit cell, as in the case of the PdMn sample[84] (CuAu I structure) shown in Figure 4-25, or of MnTe shown in Figure 3-48, p. 128.

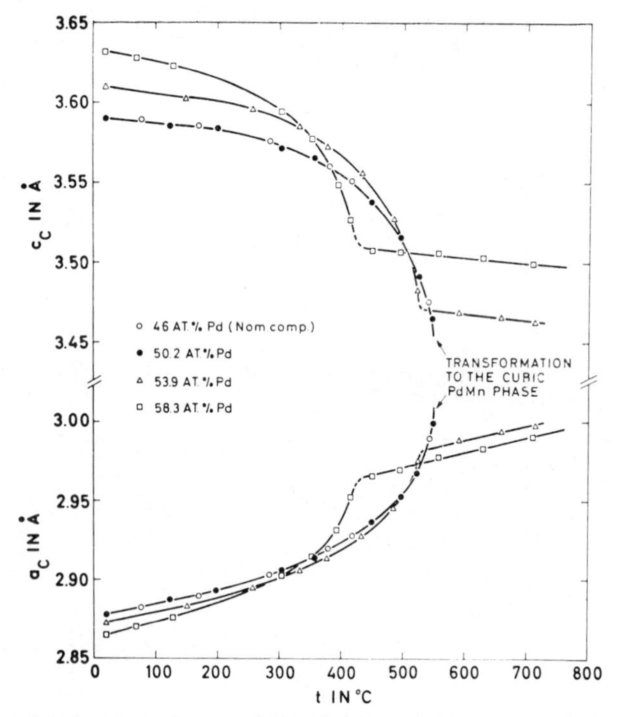

Figure 4-25. Lattice parameters of PdMn alloys with the AuCu I (*tP*2) structure, as a function of temperature. The observed changes of slope occur at the Néel point, which in one of the cases probably coincides with the transformation of the structure from tetragonal to cubic. (Kjekshus *et al.* [84].)

REFERENCES

1. W. L. Bragg, 1920, *Phil. Mag.*, **40**, 169.
2. L. Pauling and M. L. Huggins, 1934, *Z. Kristallogr.*, **87**, 205.
3. S. Geller, 1956, *Acta Cryst.*, **9**, 885 (see also *Ibid.*, **10**, 374, 380, 685, 687 for discussion).
4. J. A. Wasastjerna, 1923, *Soc. Sci. Fenn. Comm. Phys. Math.*, 1, **38**, 22.
5. V. M. Goldschmidt, T. Barth, G. Lunde, and W. Zachariasen, 1926, "Geochemische Verteilungsgesetze der Elemente", *Skr. Norsk. Videns-Akad. Oslo, I Mat.-Nat. Kl.* No. 2.
6. L. Pauling, 1927, *J. Amer. Chem. Soc.*, **49**, 765; now see *Idem*, 1960, *The Nature of the Chemical Bond*, 3rd ed, Ithaca: Cornell University Press, p. 511.
7. W. H. Zachariasen, 1931, *Z. Kristallogr.*, **80**, 137.
8. R. D. Shannon and C. T. Prewitt, 1969, *Acta Cryst.*, **B25**, 925; *Idem*, 1970. *Ibid.*, **B26**, 1046.
9. L. Pauling, 1947, *J. Amer. Chem. Soc.*, **69**, 542.
10. L. Pauling, 1949, *Proc. Roy. Soc. Lond.*, **A196**, 343.
11. G. B. Bokij, 1953, *Dokl. Akad. Nauk, SSSR*, **89**, 459.
12. J. C. Slater, 1962, *Solid State and Molecular Theory Group*, M.I.T. Q.P.R. No. 46, Oct., p. 6.
13. W. Hume-Rothery, G. W. Mabbott, and K.M. Channel-Evans, 1934, *Phil. Trans. Roy. Soc.*, **A223**, 1.
14. J. T. Waber, K. Gschneidner, A. C. Larson, and M. Y. Prince, 1963, *Trans. Met. Soc. AIME*, **227**, 717.
15. P. S. Rudman, 1965, *Trans. Met. Soc., AIME*, **233**, 864.
16. F. Laves, 1965, *Theory of Alloy Phases*, Cleveland, Ohio: American Society for Metals, p. 124.
17. E. Parthé, 1961, *Z. Kristallogr.*, **115**, 52.
18. H. W. King, 1966, *J. Mater. Sci.*, **1**, 79.
19. B. C. Giessen, U. Wolff, and N. J. Grant, 1968, *J. Appl. Cryst.*, **1**, 30.
20. E. Teatum, K. Gschneidner, and J. Waber, 1960, LA-2345, U.S. Department of Commerce, Washington, D. C.
21. J. Thewlis, 1953, *J. Amer. Chem. Soc.*, **75**, 2279.
22. H. J. Axon and W. Hume-Rothery, 1948, *Proc. Roy. Soc.*, **A193**, 1.
23. W. Hume-Rothery and G. V. Raynor, 1954, *The Structure of Metals and Alloys*, London: Institute of Metals.
24. M. C. Neuberger, 1936, *Z. Kristallogr.*, **93**, 1.
25. W. B. Pearson, 1958, *Handbook of Lattice Spacings and Structures of Metals and Alloys*, London: Pergamon Press, Volume 1.
26. *Idem*, 1967, *Ibid*, Volume 2.
27. L. Pauling, 1940, *Nature of the Chemical Bond*, 2nd ed, Ithaca: Cornell University Press.
28. V. Schomaker and D. P. Stevenson, 1941, *J. Amer. Chem. Soc.*, **63**, 37.
29. A. F. Wells, 1950. *Structural Inorganic Chemistry*, 2nd ed, Oxford: Clarendon Press, p. 50.
30. A. F. Wells, 1949, *J. Chem. Soc.* **55**, pp. 56–61.
31. A. Buroway, 1943, *Trans. Faraday Soc.*, **39**, 79.
32. L. Pauling, 1960, *Nature of the Chemical Bond*, 3rd ed, Ithaca: Cornell University Press. p. 229.
33. W. B. Pearson, 1962, *J. Phys. Chem. Solids*, **23**, 103.
34. K. A. Gschneidner, 1961, *Rare Earth Alloys*, Princeton: Van Nostrand.
35. W. H. Zachariasen and F. H. Ellinger, 1957, *J. Chem. Phys.* **27**, 811.

36. F. Laves, 1956, *Theory of Alloy Phases*, Cleveland: American Society for Metals, p. 124.
37. R. W. G. Wyckoff, 1951, *Crystal Structures*, New York: Interscience.
38. L. H. Ahrens, 1952, *Geochim. et Cosmochim. Acta*, **2**, 155.
39. L. Pauling, 1960, *Loc. cit.* [32] 3rd ed, pp. 523–526.
40. L. Pauling, 1960, *Loc. cit.* [32] 3rd ed, p. 526.
41. A. F. Wells, 1962, *Structural Inorganic Chemistry*, 3rd ed, Oxford: Clarendon Press, p. 66.
42. L. Pauling, 1960, *Loc. cit.* [32] 3rd ed, p. 260.
43. A. von Hippel, 1948, *J. Chem. Phys.*, **16**, 372.
44. H. Kambe, 1955, *Phys. Rev.*, **99**, 419.
45. L. Brewer, 1963, *USAEC Publication*, UCRL-10701, p. 13.
46. N. F. Mott, 1962, *Rep. Progr. Phys.*, **25**, 218 (p. 234).
47. W. B. Pearson, 1968, *Z. Kristallogr.*, **126**, 362.
48. Z. S. Basinski, W. Hume-Rothery, and A. L. Sutton, 1955, *Proc. Roy. Soc.*, **A229**, 459.
49. N. Ridley and H. Stuart, 1968, *Brit. J. Appl. Phys.*, Ser. 2, **1**, 1291.
50. N. F. Mott, 1962, *loc. cit.* [46], pp. 239–241.
51. C. H. Cheng, C. T. Wei, and P. A. Beck, 1960, *Phys. Rev.*, **120**, 426, (See, e.g., Figures 13 and 14).
52. W. Jeitschko, A. G. Jordan, and P. A. Beck, 1969, *Trans. Met. Soc. AIME*, **245**, 335.
53. D. H. Martin, 1967, *Magnetism in Solids*, London: Iliffe Press, pp. 128–143.
54. N. Hush and M. H. L. Pryce, 1958. *J. Chem. Phys.*, **28**, 244; **26**, 143.
55. J. H. van Santen and J. S. van Wieringen, 1952, *Recueil*, **71**, 420.
56. G. Blasse, 1965, *J. Inorg. Nucl. Chem.*, **27**, 748.
57. J. B. Goodenough, 1964, in *Transition Metal Compounds*, ed, E. R. Schatz, New York: Gordon and Breach, p. 65.
58. K. A. Gschneidner and R. Smoluchowski, 1963, *J. Less-Common Metals*, **5**, 374.
59. L. Vegard, 1921, *Z. Phys.*, **5**, 17.
60. E-an Zen, 1956, *Amer. Min.*, **41**, 523.
61. K. A. Gschneidner, 1966, Symposium, *The Met. Soc., AIME.*
62. B. J. Pines, 1940, *J. Phys., USSR*, **3**, 309.
63. J. Friedel, 1955, *Phil. Mag.*, **46**, 514.
64. K. A. Gschneidner and G. H. Vinegard, 1962, *J. Appl. Phys.*, **33**, 3444.
65. E. S. Sarkisov, 1960, *Ž. Fiz. Khim.*, **34**, 432 [*J. Phys. Chem.*, **34**, 202].
66. M. Simerska, 1962, *Czechosl. J. Phys.*, **12**, 54.
67. N. Ridley, H. Stuart, and L. Zwell, 1969, *Trans. Met. Soc.*, **245**, 1834.
68. H. W. King, 1966, *Alloying Behavior and Effects in Concentrated Solid Solutions*, New York: Gordon and Breach, p. 85.
69. L. E. Orgel, 1960, *An Introduction to Transition-Metal Chemistry: Ligand-Field Theory*, London, New York: Methuen-John Wiley; J. D. Dunitz and L. E. Orgel, 1960, *Adv. in Inorg. Chem. Radiochem.*, **2**, 1.
70. E. Mooser and W. B. Pearson, 1961, *Nature, Lond.*, **190**, 406.
71. W. Cochran, 1961, *Nature, Lond.*, **191**, 60.
72. O. Schmitz-Dumont and K. Steinberg, 1954, *Naturwiss.*, **41**, 117.
73. N. W. Tideswell, F. H. Kruse, and D. J. McCullough, 1957, *Acta Cryst.*, **10**, 99.
74. F. Hulliger, 1966, *Nature, Lond.*, **209**, 500.
75. P. Jensen, A. Kjekshus, and T. Skansen, 1969, *J. Less-Common Metals*, **17**, 455.
76. J. B. Goodenough, 1963, *Magnetism and the Chemical Bond*, New York: Interscience, pp. 26–28, 265–266, 295–297. Note that in Goodenough's work R_c *is the critical diameter*, not radius as here.

77. C. B. Shoemaker and D. P. Shoemaker, 1969, *Developments in the Structural Chemistry of Alloy Phases*, New York: Plenum Press, p. 107.
78. W. B. Pearson, 1958, *loc. cit.* [25] p. 346.
79. W. C. Hagel and J. H. Westbrook, 1960, *Trans. Met. Soc. AIME*, **221**, 951.
80. K. Brun, A. Kjekshus, and W. B. Pearson, 1964, *Phil. Mag.*, **10**, 291.
81. M. J. Cooper, 1963, *Phil. Mag.*, **8**, 805.
82. A. J. Bradley and A. H. Jay, 1932, *J. Iron and Steel Inst.*, **125**, 339; 1932, *Proc. Roy. Soc.*, **A136**, 210.
83. B. T. M. Willis, and H. P. Rooksby, 1954, *Proc. Phys. Soc.*, **B67**, 290.
84. A. Kjekshus, R. Møllerud, A. F. Andresen, and W. B. Pearson, 1967, *Phil. Mag.*, **16**, 1063.

5

Valence Compounds of Metalloids: Crystal Chemistry of Semiconductors

METALLIC AND SEMICONDUCTOR BONDS

Since Wilson's[1] physical definition of a semiconductor as a substance which, at the absolute zero of temperature, has a filled valence band separated by a range of forbidden energies (energy gap) from higher conduction bands, permitted no means of saying *a priori* what substances may be semiconductors, there was good reason for seeking an alternative chemical statement of the definition which would allow separation of metals on the one hand and semiconductors, ionics, and insulators on the other, and thus permit the prediction of semiconductivity. The somewhat belated discovery of a chemical definition of semiconductivity followed a few years after Welker's[2] empirical discovery of semiconductivity in the III–V compounds such as InSb. It is interesting today to reflect that had Pauling in the nineteen thirties but thought about the relationship of electrical conductivity and his valence bond theory, he must have been led to predict semiconductivity in many of the several hundred semiconductors which we now recognize! Indeed, the history of the development of knowledge of semiconductors has many fascinating aspects. The property was recognized and mainly characterized in the interval between 125 and 75 years ago. In 1931 Wilson brought forward his physical definition of a semiconductor which has resulted in a rather thorough understanding of the electrical properties of known semiconductors. Yet on the eve of Welker's discovery, about 20 years after Wilson's publication, there were still only a handfull of known semiconductors. Welker's discovery of semiconductivity in the III–V compounds lifted the lid from Pandora's box and led to the rapid recognition of semiconductivity in several hundred substances. This remarkable outcome resulted from the realization of several obvious factors relating chemical bonding and electrical conductivity which had long remained overlooked.

One of the continuing difficulties in dealing theoretically with the properties of substances in the solid state which contain two or more components, is to decide on the extent of separation of ionic charge, since it is a property which does not lend itself to direct and precise measurement. Therefore in developing a chemical equivalent of Wilson's physical definition of a semiconductor, it is best to ensure that considerations of the degree of ionicity do not enter the discussion. This can be achieved completely within the framework of known definitions of chemical bonds by distinguishing between *electron configuration in bonds* and the *filling of valence orbitals*, and the results are satisfying because they lead to simultaneous consideration of both crystal chemistry and electrical properties in terms of the same useful atomic parameters.

First, it is considered that there are three fundamental types of chemical bonds in which electron configuration is specified, and that the chemical bonding in any known substance can be simulated by a suitable admixture of these three bond types:

(i) covalent or shared-electron-pair bonds,

(ii) ionic bonds resulting from the coulombic attraction between charged atoms with filled valence subshells,

(iii) weak van der Waals bonds resulting from interactions between uncharged atoms with filled valence subshells.

Next, two *secondary* bond types are defined in which the *degree of filling of the valence subshells* rather than the electron configuration is specified (the actual bonds in any substance being composed of a suitable admixture of the three types detailed above):

(i) *Metallic bonds*[3] result when bonds between atoms that do not have filled valence subshells run continuously throughout the crystal structure. Under these conditions when an electric field is applied, charge carriers are free to move throughout the lattice because of the vacant orbitals on atoms which are bonded together.

(ii) *Semiconductor bonds*[4] result when atoms attain filled valence subshells by electron sharing or transfer, so that there are no bond paths running continuously throughout the structure between any atoms which have vacant orbitals in their valence shells (the presence of such atoms in the structure does not destroy the semiconductor bonds provided that there are no direct bond paths between them which run throughout the structure). Under these conditions when an electric field is applied, the filled valence orbitals prevent the free movement of charge carriers throughout the lattice and electrons require an activation energy to excite them into higher conducting orbitals before they can move throughout the lattice carrying an electric current.

The relationship between the secondary metallic and semiconducting bond types is shown diagrammatically in Figure 5-1. Semiconductor bonds occur in valence compounds, which may be semiconductors, ionic conductors, or insulators at a specific temperature depending on the size of the energy band gap. When the electronegativity difference between the anions and cations is unity or less, corresponding to an energy band gap of

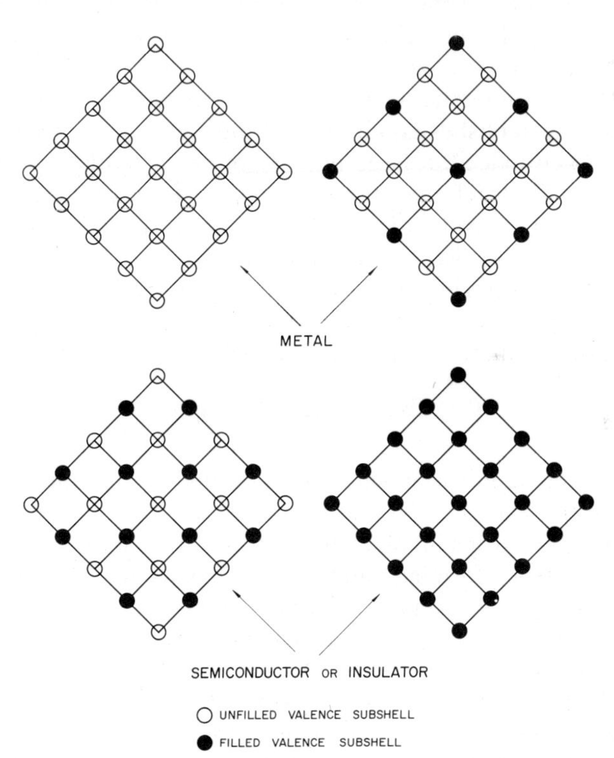

Figure 5-1. Schematic diagram of two-dimensional atomic arrays illustrating the difference between metallic and semiconductor bonds. Bond paths are indicated by lines joining atoms. Metallic bonds occur when there are bond paths running throughout the structure between atoms having vacant orbitals in their valence subshells. Semiconductor bonds occur when atoms having filled valence subshells interrupt these bond paths, or when all atoms attain filled valence subshells by electron sharing.

2.5 eV or less, semiconductivity occurs. When it is larger, the conductivity may be ionic under conditions of thermal excitation since, whatever the structure, thermal activation energy of about 2.5 to 3 eV is generally sufficient to free conducting ions. Nevertheless if the activation energy for electronic conduction in a compound is larger than this and ionic conductivity is observed under conditions of thermal excitation, it may still be possible to excite electronic conduction specifically at low temperatures by other methods of excitation such as bombardment by X-rays. In the case of most organic substances, thermal stability is so low that decomposition occurs before either ionic or electronic conductivity can be excited thermally.

The definition of a semiconducting bond becomes of interest in crystal chemistry when it is appreciated that location of the anions on the sites of a close packed array, and the cations at the centers of the tetrahedral and/or octahedral interstices thereof, to the extent that they satisfy the anion valency, establishes conditions for the occurrence of semiconducting bonds. Since indeed, a large number of semiconducting compounds have structures in which the anions lie on a close packed array, or a slight distortion thereof, consideration of these provides a good basis for discussion of the crystal chemistry of semiconductors.

Since substances having semiconductor bonds are valence compounds, they must adhere to a general valence rule that can be expressed in the following form for anions whose valence shell contains 8 s and p electrons,[4]

$$\frac{n_a+n_c+b_a-b_c}{N_a}=8 \tag{1}$$

where n_a is the number of valence electrons on the anions, n_c is the number on the cations, *less* any unshared valence electrons, b_a is the number of electrons involved in forming anion–anion bonds, b_c is the number of electrons forming cation–cation bonds and N_a is the number of anions, *all of these numbers being calculated per formula unit of the compound.* Thus for example for tetragonal ZnP_2 (see Figure 5-9, p. 221, $n_a=10$, $n_c=2$, $b_a=4$, $b_c=0$, and $N_a=2$; for PbSe, where Pb has two unshared electrons, $n_a=6$, $n_c=2$, $b_a=b_c=0$, and $N_a=1$.

Since this expression is a statement of the valence rule, it is obeyed by valence compounds and not by metals which are not valence compounds. If it is assumed that any anion–anion or cation–cation bonds formed are unit-strength shared electron pair bonds, the formula can be expressed in terms of the total coordination numbers of anions (Γ_a) and cations (Γ_c) in a compound.[5] The number of anion–anion bonds formed by each anion is then b_a/N_a and cation–cation bonds formed by each cation is b_c/N_c. The anions form $N_a(\Gamma_a-b_a/N_a)$ bonds with the cations which can be equated to $N_c(\Gamma_c - b_c/N_c)$, the number of bonds formed by the cations with the anions:

$$N_a(\Gamma_a - b_a/N_a) = N_c(\Gamma_c - b_c/N_c)$$

Therefore
$$b_a - b_c = N_a\Gamma_a - N_c\Gamma_c$$

and
$$\frac{n_a + n_c + N_a\Gamma_a - N_c\Gamma_c}{N_a} = 8 \qquad (2)$$

which is a general form of various expressions that have been used in discussing diamond type compounds. If the structures of elements lying to the right of the Zintl border only are considered ($n_c = N_c = 0$), equation (2) reduces to the $8 - N$ rule of Hume-Rothery

$$n_a + \Gamma_a = 8 \qquad (3)$$

since n_a is identical with the group number. Thus, for example, for selenium $n_a = 6$ ($= N$), whence $\Gamma_a = 2$.

The structural use of formula (1) based on the assumption that any anion–anion or cation–cation bonds are unit-strength shared pair bonds applies only to valence compounds. When the formula is to be used as a criterion of semiconductivity to separate semiconductors from metals, the *actual* numbers of electrons forming any anion–anion and cation–cation bonds, must be considered. Thus, although some substances such as As or BiIn obey the approximate form (2), they do not obey the exact form (1).

Understanding the metallic conductivity of InBi (*tP*4) is an important example since it demonstrates that the anions must attain a filled valence subshell by electron sharing or transfer if the substance is to be a semiconductor, emphasizing that if and when they do attain this, they must (unless b_a or $b_a - b_c \equiv 1, 2, \ldots$) be separated from each other by distances at least equal to their van der Waals diameters. If the anions approach each other significantly closer than this distance, then it is unlikely that they do in fact obtain a filled valence subshell so that

$$\frac{n_a + n_c + b_a - b_c}{N_a} \neq 8$$

and there is probably some direct valence bonding between them, causing either modification of normal semiconducting properties or metallic conductivity, depending on the strength of the bonding. In InBi the Bi atoms approach each other more closely (Bi–Bi = 3.68 Å) than the Bi van der Waals diameter (~ 4.4 Å) suggesting that there is some direct valence bonding between them ($b_a \neq 0$ in formula (1) making it invalid) so that conditions for semiconducting bonds are not satisfied, in keeping with the observed metallic conductivity of the compound. In contrast, the Sb–Sb distances of 4.58 Å in the semiconductor InSb ($cF8$), which has the sphalerite structure, are even larger than the van der Waals diameter of Sb, allowing no possibility of chemical bonds between the anions. Careful study of the data set out in Table 5-1 is revealing since it shows that metallic properties need not arise through close approach of the cations, if their valence electrons are all used in forming shared pair bonds with the anions or (to greater or lesser extent) ionically transferred to complete the anions' shells. However, the possibility of ionic transfer occurring to complete the anion valence shell ("ionic criterion for semiconductivity") is itself an insufficient condition for semiconductivity, since CuMgSb (Sb on c.p. cubic lattice, Cu in tetrahedral, Mg in octahedral holes thereof) which satisfies the ionic criterion is metallic, whereas LiMgSb (Sb on c.p. cubic sites, both Li and Mg in tetrahedral holes) shows the expected semiconducting properties. It appears that metallic properties generally result when the cations occupy the octahedral and *one* set of tetrahedral holes in a close packed array of anions.

There is widespread evidence that equation (1) holds generally as a criterion for predicting semiconductivity, and that the few apparent exceptions such as InBi or some transition metal compounds which are metallic, can well be accounted for by the equation when details of the distances between the component atoms are examined. The equality with 8 in equation (1), of course, only applies to anions with a filled valence shell of 8 electrons, and not to boron which may have a filled valence subshell of 6 or H and Au that can have a filled valence subshell of 2 electrons.

Busmann[6] has pointed out that the $8-N$ coordination relationships of the elements lying to the right of the Zintl border are also found among the anion subarrays of semiconducting compounds containing anion–anion bonds, which are called *polyanionic* compounds. Such behavior is to be expected from equation (1); if $b_c = 0$, then

$$\frac{b_a}{N_a} = 8 - \frac{n_a + n_c}{N_a} \tag{4}$$

and the number of anion-anion bonds formed by each anion (b_a/N_a) equals eight minus the total number of valence electrons available per anion, which is an expression of the $8-N$ rule. Numerous examples of polyanionic semiconducting compounds are discussed on pp. 214 to 222. When both b_a and b_c are zero, normal semiconductors are obtained in which there are only anion–cation bonds and the anions are separated by van der Waals distances, corresponding to the case, $8-N=0$, of the

Table 5-1

Compound	Conductivity	Anion–Anion Distance (Å)	Anion van der Waals Diameter (Å)	Cation–Cation Distance (Å)	Cation Diameter in Pure Metal (Å)
InBi	Metallic	3.68	~ 4.4	3.54	3.24, 3.37
InSb (sphalerite type)	Semiconductor	4.58	~ 4.4	4.58	3.24, 3.37
Mg_2Si[a] (fluorite type)	Semiconductor	4.49		3.18	3.19, 3.20
Mg_2Ge[a]	Semiconductor	4.52		3.19	3.19, 3.20
Mg_2Sn	Semiconductor	4.78		3.38	3.19, 3.20

[a]Note that in Mg_2Si and Mg_2Ge the Mg atoms approach each other as close or closer than in Mg metal, whereas in InBi the In–In separation is significantly larger than in metallic In.

inert gases, where there are only central forces between the atoms in the solid state. It is interesting to note the very large number of normal semiconducting compounds that have structures with the anions lying on close packed cubic or hexagonal arrays corresponding to the common structures of the inert gases. In these structures there are only very weak interactions between the nonbonded close packed anions, and the main reason why semiconductors adopt such structures is that they allow the directional requirements of the chemical bonds to be satisfied, while maintaining the densest possible packing of the atoms, in accordance with Laves' principles for structure formation. The requirements of the common tetrahedral and octahedral symmetries of bond orbitals formed from atomic *s* and *p* electrons are satisfied if the cations are located at the centers of the tetrahedral and octahedral interstices in the close packed anion array.

Polycationic compounds ($b_c \neq 0$) satisfying equation (1) are semiconductors, and the presence of semiconducting bonds implies that the cation–cation bonds do not run continuously throughout the crystal lattice. Several such compounds are known and their structures are discussed on p. 223.

NORMAL SEMICONDUCTING COMPOUNDS WITH ANIONS FORMING CLOSE PACKED ARRAYS

Systematic derivation of normal semiconducting compound structures based on close packing of the anions (b_a and $b_c = 0$) follows the method of derivative structures (p. 17), and all possible derivative structures can be predicted by a partition of possible cation valencies satisfying the valency of the anions in the formula bearing in mind that each anion site provides three possible cation sites at the centers of the tetrahedral and octahedral holes. When the many possible anion close packed sequences are considered, and when structural distortion is permitted, the number of possible structures rapidly becomes innumerable. It is generally convenient to relate derivative structures to basic structural types. For example starting with a close packed cubic array of anions (*cF*4 structure) filling of the octahedral holes, one tetrahedral hole, two tetrahedral holes, one tetrahedral and one octahedral hole, or all tetrahedral and octahedral holes gives respectively the rocksalt (NaCl), sphalerite (ZnS), fluorite (CaF_2), CuMgSb, and Li_3Bi structures (Figure 5-2), which can be regarded as the five basic types derived from cubic close packing of anions, and to which further derivative types can be related by four processes:

(1) by substituting in an *ordered* manner different cations of the same valency or same average valency per atom, thus forming a superstructure of the original binary compound;

(2) by substituting cations of higher average valency so that some of the interstitial sites remain vacant in an *ordered* fashion. If the vacant cation sites occur in a random fashion then no change of crystal structure occurs, as for instance in one form of In_2Te_3 which has the sphalerite type of structure;

(3) by substituting cations of lower average valency, filled up derivatives are obtained;

(4) through a slight displacement of the atoms from the face-centered cubic sublattice sites, any number of structures can be obtained directly from the five basic types or from any of the structures obtained by methods (1), (2), and (3) above.

The symmetry of structures containing atoms slightly displaced from the face-centered lattice sites is generally, although not invariably, lower than cubic. The symmetry of the ordered structures of cases (1), (2), and (3) is also generally lower than cubic because of the layering of vacancies or atoms of different sizes which occurs. This ordering frequently causes displacement of some of the atoms (the anions) from their

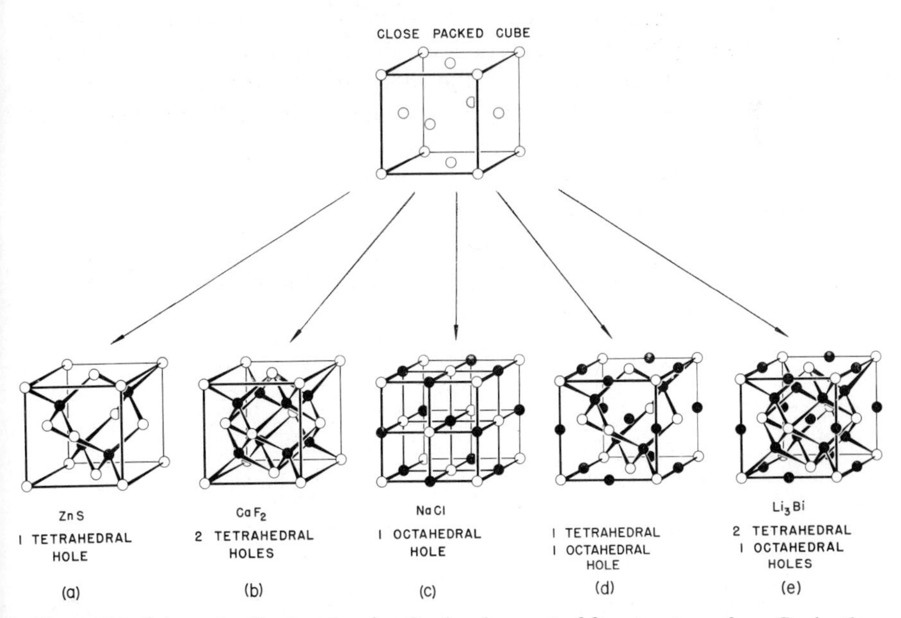

Figure 5-2. Schematic diagram showing the development of five structures from Cu, by the addition of a further one, two or three interpenetrating *F* complexes.

exactly close-packed lattice sites. The five basic structural types, sphalerite, rocksalt, fluorite, CuMgSb, and Li_3Bi, all have cubic symmetry and the stacking and population of the component atoms in layers is identical in each of the three directions of the crystal axes.

Organization of structures of semiconductors derived from a close packed hexagonal array of anions is less certain than in the cubic case. In the first place, the close approach of the two sets of anion tetrahedra in pairs sharing faces, so that only one set can be centered by a cation, results in the formation of only two basic derived structures from a c.p. hexagonal array of anions: the NiAs type with occupied octahedral holes and the wurtzite (ZnS) type with one set of occupied tetrahedral holes. The two tetrahedral holes combine to give a trigonal bipyramidal hole, and simultaneous occupation of octahedral holes and this, or a tetrahedral hole, probably leads to metallic properties as in the cubic case. Secondly, atoms at the centers of the octahedral holes in the NiAs structure form a simple hexagonal array rather than a c.p. hexagonal array, and therefore come close together along [001], particularly when the axial ratio of the unit cell is considerably less than the ideal value of 1.63. Since the NiAs structure is formed by transition metals whose *d* orbitals may overlap along [001], such compounds are generally metallic, although semiconductors are known, for example, in MnTe and FeS.

The structure types adopted by most of these semiconducting compounds are described in Chapters 7 to 16 according to their systematic structural relationships; a few of the more simple derivative types are listed for convenience in Tables 5-2 and 5-3. In addition there are many normal semiconducting compounds whose structures are not based on a close packed anionic array and the structures of these are also described in these chapters.

Table 5-2 Summary of Some Derived Structures of Compounds with Close-Packed-Cubic Anion Arrays (Often Distorted)

Structure	Kind and Fraction of Holes Occupied by Cations	
	Tetrahedral	Octahedral
Derived from ZnS	1, 0	0
$ZnCdSe_2$ disordered	1, 0	0
$CuFeS_2$	1, 0	0
Cu_2FeSnS_4	1, 0	0
α-Ag_2HgI_4 disordered	$\frac{3}{4}$, 0	0
$CdAl_2S_4$	$\frac{3}{4}$, 0	0
$CdIn_2Se_4$	$\frac{3}{4}$, 0	0
In_2Te_3 disordered	$\frac{2}{3}$, 0	0

Table 5-2 (*continued*)

Structure	Kind and Fraction of Holes Occupied by Cations	
	Tetrahedral	Octahedral
Derived from CaF_2 or Mg_2Sn	1, 1	0
LiMgSb	1, 1	0
Li_3AlN_2	1, 1	0
Li_5GeP_3 disordered	1, 1	0
Mg_3As_2 (anti-Mn_2O_3)	$\frac{3}{4}, \frac{3}{4}$	0
Zn_3P_2	$\frac{3}{4}, \frac{3}{4}$	0
Cu_3VS_4	$\frac{3}{4}, \frac{1}{4}$	0
CuTe	$\frac{1}{2}, \frac{1}{2}$	0
HgI_2	$\frac{1}{4}, \frac{1}{4}$	0
SiS_2	$\frac{1}{4}, \frac{1}{4}$	0
SnI_4	$\frac{1}{8}, \frac{1}{8}$	0
Derived from NaCl	0, 0	1
$AgBiSe_2$ disordered	0, 0	1
$AgBiSe_2$ 287°–120°C } $NaCrS_2$ }	0, 0	1
$AgBiSe_2$ trigonal 120°C	0, 0	1
$PbSnS_2$	0, 0	1
$BiNaSe_2$	0, 0	1
La_2O_3	0, 0	$\frac{2}{3}$
TiO_2 anatase	0, 0	$\frac{1}{2}$
$CdCl_2$	0, 0	$\frac{1}{2}$
$CrCl_3$	0 0	$\frac{1}{3}$
α-UF_5	0, 0	$\frac{1}{5}$
Derived from Li_3Bi	1, 1	1
Cu_2Te[a]	$\frac{1}{2}$, 0	1
Cu_2Sb[a]	$\frac{1}{2}$,n	1
Li_2MgSn	1, 1	1
$MgAl_2O_4$	$\frac{1}{8}, \frac{1}{8}$	$\frac{1}{2}$
γ-Al_2O_3	Defect spinel ($21\frac{1}{3}$ metal atoms randomly on spinel sites)	
Cu_4Te_3	$\frac{1}{2}, \frac{1}{2}$	$\frac{1}{3}$
Co_9S_8[a]	$\frac{1}{2}, \frac{1}{2}$	$\frac{1}{8}$
MCo_8S_8	$\frac{1}{2}, \frac{1}{2}$	$\frac{1}{8}$

[a]Compounds having these structures, are not necessarily expected to be semiconductors; they may well be metallic because of the close approach of cations occupying both octahedral and tetrahedral holes.

Table 5-3 Summary of Some Derived Structures of Compounds with Close-Packed-Hexagonal Anion Arrays

Structure	Kind and Fraction of Holes Occupied by Cations	
	Tetrahedral	Octahedral
Derived from wurtzite, ZnS	1, 0	0
$AgInS_2$ (H.T. form)	1, 0	0
Cu_3AsS_4 enargite	1, 0	0
$ZnAl_2S_4$ (H.T. form)	$\frac{3}{4}$, 0	0
β-Ga_2S_3 disordered	$\frac{2}{3}$, 0	0
Al_2Br_6	$\frac{1}{3}$, 0	0
Derived from NiAs	0, 0	1
MnP	0, 0	1
CrS	0, 0	1
TiAs	0, 0	1
α-Al_2O_3	0, 0	$\frac{2}{3}$
$FeTiO_3$ (ilmenite)	0, 0	$\frac{2}{3}$
$LiSbO_3$	0, 0	$\frac{2}{3}$
CdI_2	0, 0	$\frac{1}{2}$
$AuTe_2$	0, 0	$\frac{1}{2}$
$HgBr_2$	0, 0	$\frac{1}{2}$
TiO_2 brookite	0, 0	$\frac{1}{2}$
$FeCl_3$ or BiI_3	0, 0	$\frac{1}{3}$
α-WCl_6	0, 0	$\frac{1}{6}$
UCl_6	0, 0	$\frac{1}{6}$
Derived from unknown structure	1, 0	1
Mg_3Bi_2 (anti-La_2O_3)	1, 0	$\frac{1}{2}$
Mg_2SiO_4	$\frac{1}{8}$, $\frac{1}{8}$	$\frac{1}{2}$

Tetrahedral Structures

Tetrahedral structures alone can be considered in what amounts to a full development of the Grimm–Sommerfeld[7] rule in much the same way as the valence relationship (1), p. 202, is a full development of the octet rule. The treatment developed by Parthé[8] defines a tetrahedral structure as one in which every atom has four neighbors surrounding it tetrahedrally, and so excludes structures such as the fluorite (CaF_2) type. Since its field is much more limited than that of the general octet rule, it is less useful for considering the structure of valence compounds as a whole, although in recognizing "defect tetrahedral structures" as those with vacant coordination sites, it has the advantage of offering a good nomenclature for describing the family of tetrahedral structures.

If a compound M_nN_m has a tetrahedral structure, then $(ne_M + me_N +$

$\cdots)/(n+m+\cdots)=4$, where e_M, e_N, $\cdots$ are the numbers of valence electrons on atoms M, N, $\cdots$. Putting $Z=n/m$ gives

$$Z=\frac{(e_N-4)}{(4-e_M)} \tag{5}$$

If e_M has integral values 1 to 4 and e_N 4 to 7, the solutions of Z for the different combinations of the integral values of e_M and e_N are those given in Table 5-4. These ten solutions for $Z=1$ to 4 can be expressed in terms of general composition formulas where integers 1 to 7 represent an element of that Group number and subscripts refer to the number of such atoms in the formula of the compounds. The ten solutions for Z are thus 4, 35, 26, 17, 3_26, 3_37, 25_2, 2_37_2, 15_3, and 1_26_3.

The advantage of this notation is that 0 can be used to represent a vacant tetrahedral site. Since each of the 4 atoms surrounding a vacant site needs one extra electron to fill a nonbonding orbital directed towards the site, the symbol 0 represents 4 extra valence electrons in the structure, 0_2 represents 8 and so on. For the compound $M_nN_m0_p$, $(ne_M+me_N+\cdots)/(n+m+\cdots p)=4$ and the three solutions of Z with $e_M=0$ in Table 5-4 above, take account of possible defect tetrahedral structures, the three binary defect compounds being 05_4, 96_2, and 0_37_4.

These $10+3=13$ solutions for Z can be added together severally to give compositions of ternary and quaternary compounds which are also tetrahedral or defect tetrahedral structures. Thus, for example, $1_26_3+3_26=136_2$ is the chalcopyrite structure, or $26+3_26+06_2=23_206_4$ is the structure of the Hahn phases represented by $ZnAl_2S_4$. The solution 4 represents any combination of Group IV elements such as Ge, Si, SiC, or a Ge–Si solid solution. $4+0_37_4=40_37_4$ is known in the compound SnI_4 which, because of the number of defect sites, is a molecular compound; $4+06_2=406_2$ is a chain structure like that found in SiS_2 (see p. 388); and $2_37_2+0_37_4=207_2$ is represented by the layer structure of HgI_2. Table 5-5 lists normal and defect tetrahedral structures classified

Table 5-4

e_M	e_N			
	4	5	6	7
4	All compositions			
3		1	2	3
2		$\frac{1}{2}$	1	$\frac{3}{2}$
1		$\frac{1}{3}$	$\frac{2}{3}$	1
0		$\frac{1}{4}$	$\frac{1}{2}$	$\frac{3}{4}$

Table 5-5 Tetrahedral Structures[8]

Normal Tetrahedral Structures	
4 or 44	C, Si, Ge, Sn, SiC
35	BN, BP, BAs, AlN, AlP, AlAs, AlSb, GaN, GaP, GaAs, GaSb, InN, InP, InAs, InSb
26	BeO, BeS, BeSe, BeTe, BePo, MgTe, ZnO, ZnS, ZnSe, ZnTe, ZnPo, CdS, CdSe, CdTe, CdPo, HgS, HgSe, HgTe, MnS, MnSe
17	CuF?, CuCl, CuBr, CuI, AgI, NH_4F
3_26	Al_2O
25_2	ZnP_2, CdP_2, $ZnAs_2$
3_246	Al_2CO
245_2	$BeSiN_2$, $MgGeP_2$, $ZnSiP_2$, $ZnSiAs_2$, $ZnGeP_2$, $ZnGeAs_2$, $ZnSnAs_2$, $CdGeP_2$, $CdGeAs_2$, $CdSnAs_2$
14_25_3	$CuSi_2P_3$, $CuGe_2P_3$, $ZnGe_2As_3$
1_246_3	Cu_2SiTe_3, Cu_2GeS_3, Cu_2GeSe_3, Cu_2GeTe_3, Cu_2SnS_3, Cu_2SnSe_3, Cu_2SnTe_3
136_2	$CuAlS_2$, $CuAlSe_2$, $CuAlTe_2$, $CuGaS_2$, $CuGaSe_2$, $CuGaTe_2$, $CuInS_2$, $CuInSe_2$, $CuInTe_2$, $CuTlS_2$, $CuTlSe_2$, $CuFeS_2$, $CuFeSe_2$, $AgAlS_2$, $AgAlSe_2$, $AgAlTe_2$, $AgGaS_2$, $AgGaSe_2$, $AgGaTe_2$, $AgInS_2$, $AgInSe_2$, $AgInTe_2$, $AgFeS_2$
156	CuAsS
1_356_4	Li_3PO_4, Li_3AsO_4, Cu_3PS_4, Cu_3AsS_4, Cu_3AsSe_4, Cu_3SbS_4, Cu_3SbSe_4
23_245_4	$ZnIn_2GeAs_4$, $ZnIn_2SnAs_4$, $CdIn_2GeAs_4$
134_25_4	$CuGaGe_2P_4$
1_2246_4	Cu_2FeSnS_4, $Cu_2FeGeSe_4$, $Cu_2FeSnSe_4$
12_236_4	$AgCd_2InTe_4$
Defect Tetrahedral Structures	
40_37_4	SiI_4, GeI_4, SnI_4, SiF_4
406_2	GeS_2
207_2	$ZnCl_2$, $ZnBr_2$, ZnI_2, HgI_2
3_206_3	Al_2S_3, Al_2Se_3, Ga_2S_3, Ga_2Se_3, Ga_2Te_3, In_2Se_3, In_2Te_3
$3_605_46_3$	$3_605_46_3$, $Ga_6As_4Se_3$, $In_6As_4Te_3$
23_206_4	$ZnAl_2S_4$, $ZnAl_2Se_4$, $ZnAl_2Te_4$, $ZnGa_2S_4$, $ZnGa_2Se_4$, $ZnGa_2Te_4$, $ZnIn_2Se_4$, $ZnIn_2Te_4$, $CdAl_2S_4$, $CdAl_2Se_4$, $CdAl_2Te_4$, $CdGa_2S_4$, $CdGa_2Se_4$, $CdGa_2Te_4$, $CdIn_2Se_4$, $CdIn_2Te_4$, $HgAl_2S_4$, $HgAl_2Se_4$, $HgAl_2Te_4$, $HgGa_2S_4$, $HgGa_2Se_4$, $HgGa_2Te_4$, $HgIn_2Se_4$, $HgIn_2Te_4$
1_2207_4	Cu_2HgI_4, Ag_2HgI_4
$1_25_206_4$	$CuSbS_2$, $CuBiS_2$, $CuSbSe_2$?
13_206_37	$CuIn_2Se_3Br$, $CuIn_2Se_3I$, $AgIn_2Se_3I$

by Parthé. All binary and ternary tetrahedral structures can be represented in a diagram such as that of Figure 5-3. Ternary compounds are located according to atomic proportion in the appropriate triangle given by the Group numbers of the component atoms, and binary

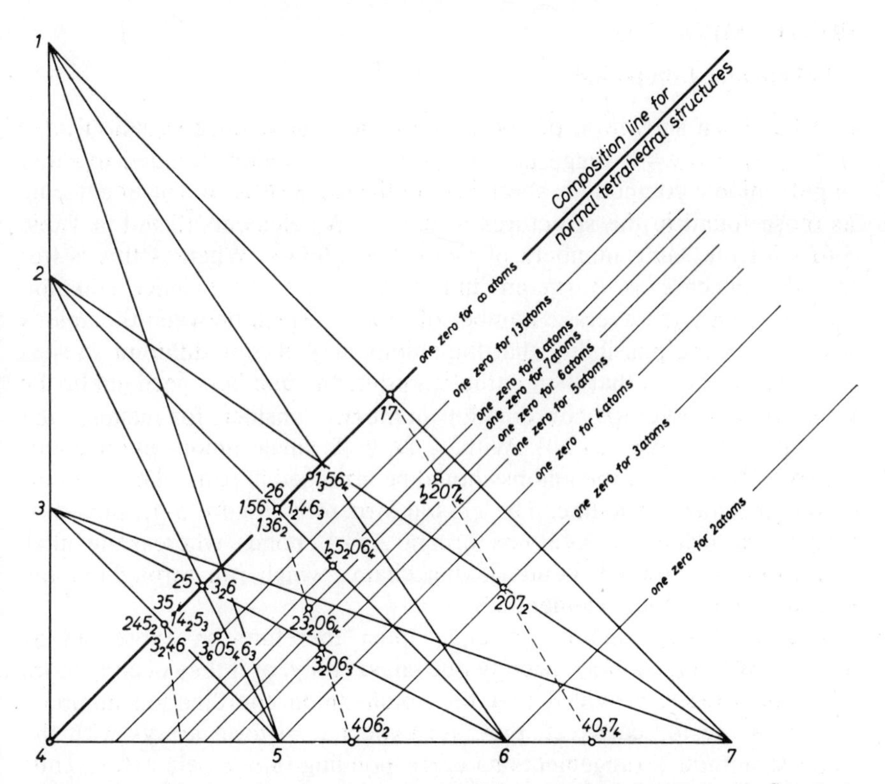

Figure 5-3. Diagram showing normal and defect tetrahedral structures. See text. (Parthé [8].)

compounds are located along lines giving the Group numbers of the two components. Normal tetrahedral compounds lie on a unique line, and defect tetrahedral compounds lie on lines parallel to the unique line. The location of these lines is determined by the number of atoms $(n+m+\cdots)$ per defect (p) in the structural formula of the compound, $y = (n+m+\cdots)/p$. Thus for the structure represented by the formula $1_2 5_2 06_4$, $y = (2+2+4)/1 = 8$. Study of Figure 5-3 taken from the work of Parthé[8] reveals many other interesting relationships between the various tetrahedral structures. A compound which cannot be appropriately placed in this diagram because its formula cannot be obtained by combinations of the 13 solutions from equation (5), cannot have a tetrahedral structure as defined above. Conversely, if the composition of a new compound of unknown structure fits appropriately in the diagram of Figure 5-3, it can confidently be predicted that it has a tetrahedral structure and the number of vacant sites, if any, can be specified.

POLYCOMPOUNDS

Polyanionic Compounds

It has been shown on p. 204 how the general valence equation $(n_a + n_c + b_a - b_c)/N_a = 8$ suggests that the arrays of anions bonded together in polyanionic compounds should adopt the same structural arrangements as those found in the structures of the $(8-N)$ elements (listed in Table 5-6) with the same numbers of nearest neighbors. Whereas this is frequently the case as Busmann[6] first pointed out, the valence equation considers only the average number of bonds formed between the anions and ignores the possibility that the anions may occupy different sitesets in the structure, so that their actual coordination numbers need not be the same as their average coordination number. Consider, for instance the structure of $GeAs_2$ (*oP*24). With $b_a/N_a = 1$, single anion–anion bonds are expected so that the anions should be arranged in pairs like the atoms in the structure of iodine. The crystal structure (Figure 5-4), however, reveals that half of the As atoms form no As–As bonds, whereas the other half form two As–As bonds giving chains which run throughout the structure like those in elemental Se.

The structure building principle must therefore be expressed as follows: When the anions occupy one siteset only, or if they occupy more than one siteset, provided that the anion–anion coordination numbers are the same for each set, they are expected to form arrays with the same structural arrangements as corresponding $(8-N)$ elements. Thus

Table 5-6 Correlation of Anion Arrays in Polyanionic Compounds with the (8–*N*) Structures of the Elements

b_a/N_a for Compound	8–*N* for Element	Elements	C.N. of Element	Structural Arrays
4	4	C, Si, Ge, α-Sn	4	Three-dimensional: tetrahedral coordination
		Graphite	3	Hexagonal-net layers: multiple bonds
3	3	White P	3	Tetrahedra
		Black P, As, Sb	3	Double layers
		N_2	1	Pairs: multiple bonds
2	2	O_2	1	Pairs: multiple bonds
		S, Se, Te	2	Rings, spirals, chains
1	1	Br, I	1	Pairs

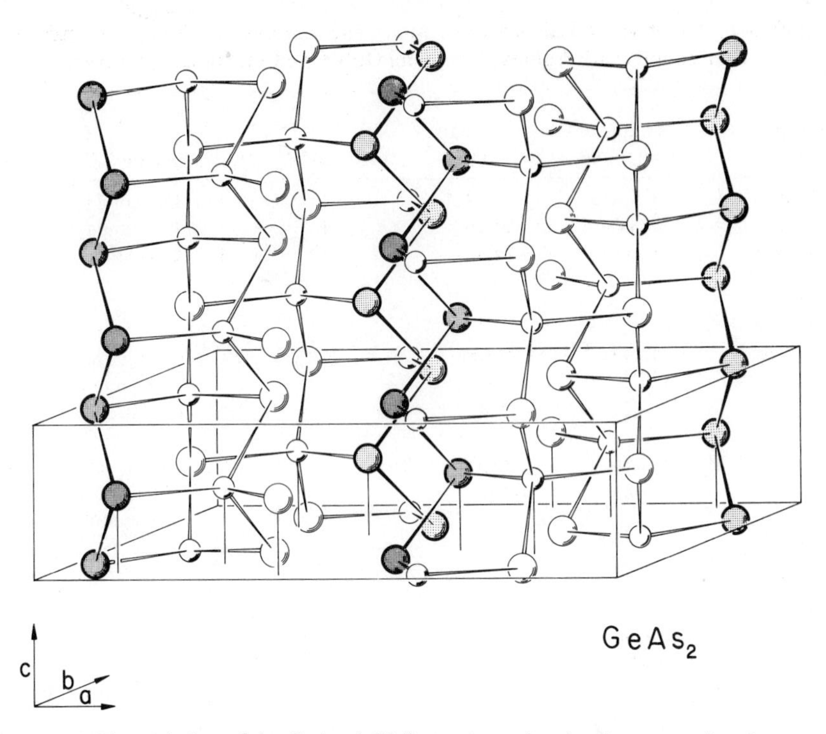

Figure 5-4. Pictorial view of the $GeAs_2$ ($oP24$) structure, showing the connections between the As(2) and As(3) atoms and the absence of As–As connection between As(1) and As(4).

the valence rule only says what arrangements *are possible* in the structure of a polyanionic compound, it is not a means of predicting *a priori* whether the anions occupy one or more sitesets in a structure, or whether the anion–anion coordination numbers are the same for anions on different sitesets. However, if valence considerations indicate a fractional value for the ratio b_a/N_a, it seems certain that the anions must occupy more than one siteset in the structure.

Polyanionic Compounds in Which the Anions All Form the Same Number of Anion–Anion Bonds. The formulas of all possible polyanionic compounds obeying the general valence rule and having anions on one siteset only in the structure, can be determined by partitioning the valencies of the anions and cations as shown in Table 5-7. Details of some examples of polyanionic compounds and the observed anionic subarrays in their structures are also included in the table.

There are several known polycompounds with $b_a/N_a = 3$ and anion arrays corresponding to those of the Group V ($8-N=3$) elements. In the

Table 5-7 Partition of Valencies of Anions and Cations for Possible Polyanionic Compounds with Anions Occupying Only One Crystallographic Site

Partition of Valencies	Example of Compound	Structure Type	$(n_a+n_c)/N_a$	b_a/N_a	Anion X Sub Arrays
$M_2^{I}X^{II}$	Cu_2Mg	$cF24$	4	4	Diamond type
$M^{I}X^{III}$	NaTl	$cF16$	4	4	Diamond type
$M^{II}X_2^{III}$	$CaGa_2$	$hP3$	4	4	Graphite type
$M_2^{I}X^{III}$			5		
$M^{II}X^{III}$			5		
$M^{I}X^{IV}$	KGe	$cP64$[a]	5	3	Ge_4 tetrahedra as in white P
	NaPb	$tI64$	5	3	Pb_4 tetrahedra as in white P
$M^{II}X_2^{IV}$	$CaSi_2$	$hR6$[a]	5	3	Double-layer As type
	CaC_2	$tI6$	5	3	C_2 pairs as in N_2 structure
$M^{III}X_3^{IV}$			5		
$M_3^{I}X^{III}$			6		
$M_3^{II}X_2^{III}$			6		
$M_2^{I}X^{IV}$	Li_2Si Li_2Sn		6		
$M^{II}X^{IV}$	CaSi	$oC8$	6	2	Planar zigzag chains
$M_2^{III}X_3^{IV}$			6		
$M^{I}X^{V}$	LiAs	$mP16$[a]	6	2	Spiral chains
$M^{II}X_2^{V}$	ZnP_2	$tP24$[a]	6	2	Spiral chains
$M^{III}X_3^{V}$	$CoAs_3$	$cI32$	6	2	Squares: 4-fold rings possibly as in γ–O_2 (ignoring rotations)
$M^{IV}X_4^{V}$			6		
$M_3^{I}X^{IV}$			7		
$M^{III}X^{IV}$			7		
$M_3^{II}X_2^{IV}$			7		
$M_2^{I}X^{V}$	Na_2As?		7		
$M^{II}X^{V}$	ZnSb	$oP16$	7	1	Pairs
$M^{IV}X_2$	$SiAs_2$?		7		
$M_2^{III}X_3^{V}$			7		
$M^{I}X^{VI}$	NaS?		7		
$M^{II}X_2^{VI}$	FeS_2	$cP12$, $oP6$	7	1	Pairs
$M^{III}X_3^{VI}$	$GaTe_3$? $InTe_3$		7		
$M^{IV}X_4^{VI}$			7		
$M^{V}X_5^{VI}$			7		

[a]In these structures the anions actually lie on more than one crystallographic site, but b_a/N_a has the same value for each.

ionic acetilides such as CaC_2 ($tI6$) (Figure 8-44a, p. 410) there are triple bonds between the pairs of carbon atoms, which thus resemble the multiply-bonded N_2 pairs in the structure of solid nitrogen. A number of polycompounds formed between an alkali metal and Si or Ge have the KGe ($cP64$) structure type shown in plan on (001) in Figure 5-5. In these compounds $b_a/N_a = 3$ and the anion subarrays are Si_4 or Ge_4 tetrahedra resembling the P_4 tetrahedra found in the structure of white phosphorus $(8-N=3)$. Pb_4 tetrahedra are also found in the structure of the metallic phase NaPb ($tI64$) for which b_a/N_a could be imagined to have a value of 3. Anionic subarrays with a double-layer arsenic-type of arrangement $(8-N=3)$ are found in the structure of $CaSi_2$ ($hR6$) shown in Figure 5-6. Each Si atom is bonded to three more Si atoms in the same double layer and the general valence rule is satisfied with $b_a/N_a = 3$. Si occupies

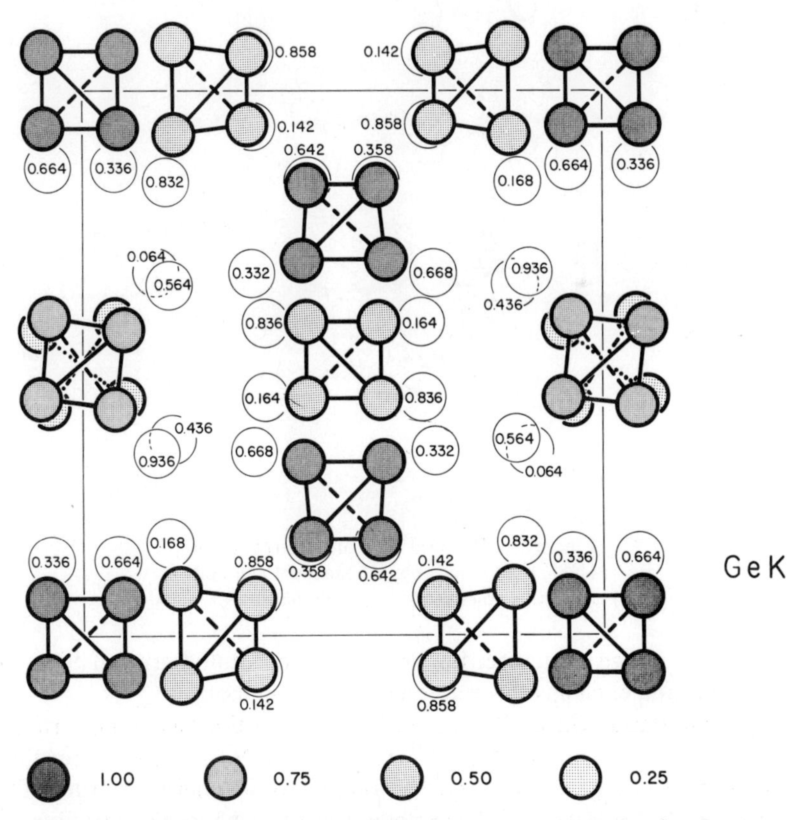

Figure 5-5. The structure of KGe ($cP64$) viewed down [001] showing Ge tetrahedra. Figures indicate fractional heights of the atoms. K, open circles. (Busmann [6].)

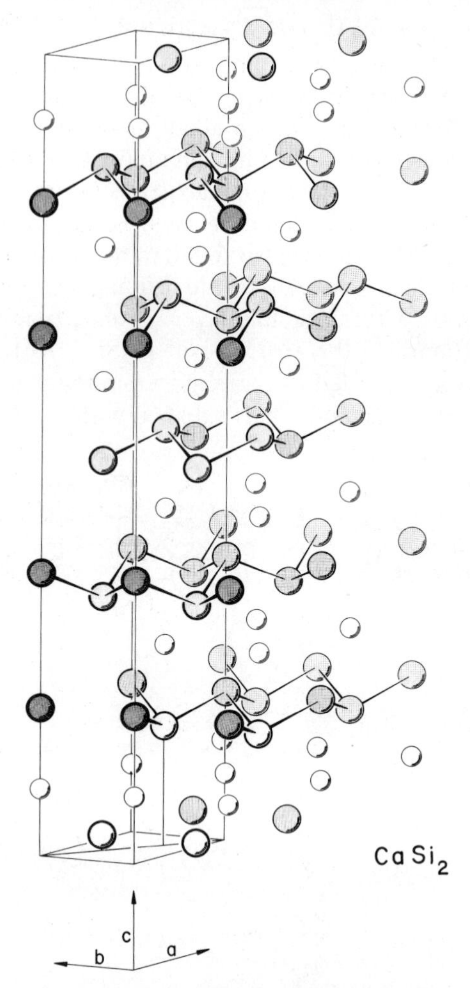

Figure 5-6. Pictorial view of the structure of $CaSi_2$ (*hR*6), showing a hexagonal cell and the arrangement of Si in arsenic-like arrays.

two sitesets in the structure, but there are no bonds between Si(1) and Si(2) atoms and each has three anion nearest neighbors.

The compounds CaSi, CaGe, and CaSn with the CrB type structure (*oC*8) satisfy the general valence rule with $b_a/N_a = 2$ and accordingly the anions form zigzag arrays corresponding to the chain structure of the Group VI elements Se and Te as shown in Figure 5-7. The structures of LiAs, NaSb, and KSb (*mP*16) contain spiral chains of As or Sb atoms (Figure 5-8) as expected from the value of $b_a/N_a = 2$ for the compounds.

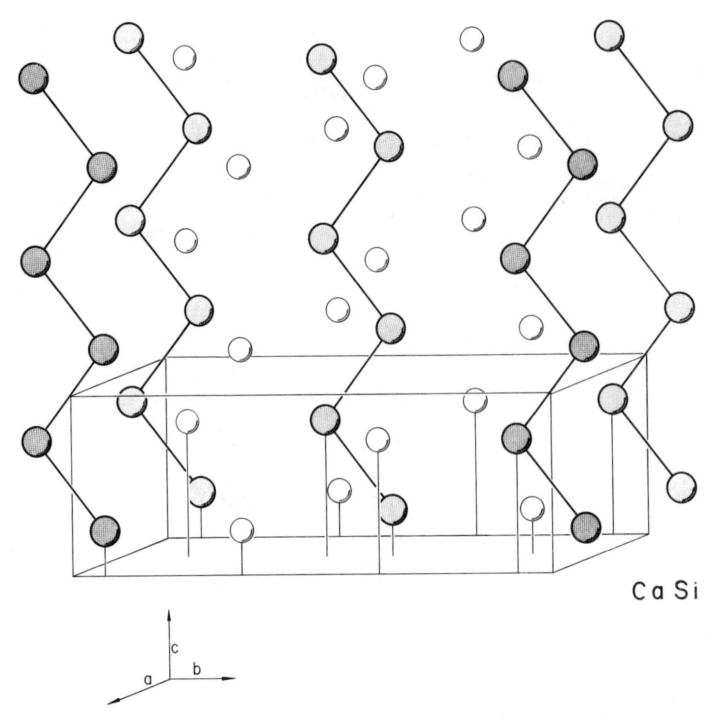

Figure 5-7. Pictorial view of the CrB type structure of CaSi (*oC*8), showing selenium-like chains of Si atoms.

Although the As atoms lie on two sitesets in the LiAs structure and bonds occur between As(1) and As(2), the expected chain-like arrays of As atoms are obtained since both As(1) and As(2) have two anion nearest neighbors. In the tetragonal modification of ZnP_2 (*tP*24) shown in Figure 5-9, and also in the structure of $ZnAs_2$ and PdP_2 (*mC*12), shown in Figure 5-10, spiral chains of P or As atoms are found, corresponding to $b_a/N_a = 2$ in each case.

The semiconductor skutterudite, $CoAs_3$ (*cI*32), is a polyanionic compound satisfying the general valence rule with $b_a/N_a = 2$, if Co is assumed to provide three valence electrons. The As atoms, as shown in Figure 5-11, form square rings which may possibly correspond to the structural arrangement in γ-oxygen.

Few of the possible polyanionic compounds with $b_a/N_a = 1$ indicated in Table 5-7, have so far been recognized. In the structures of ZnSb and CdSb (*oP*16) where $b_a/N_a = 1$, Sb atoms are bonded together in pairs (Figure 5-12) corresponding to the structural arrangement in crystalline

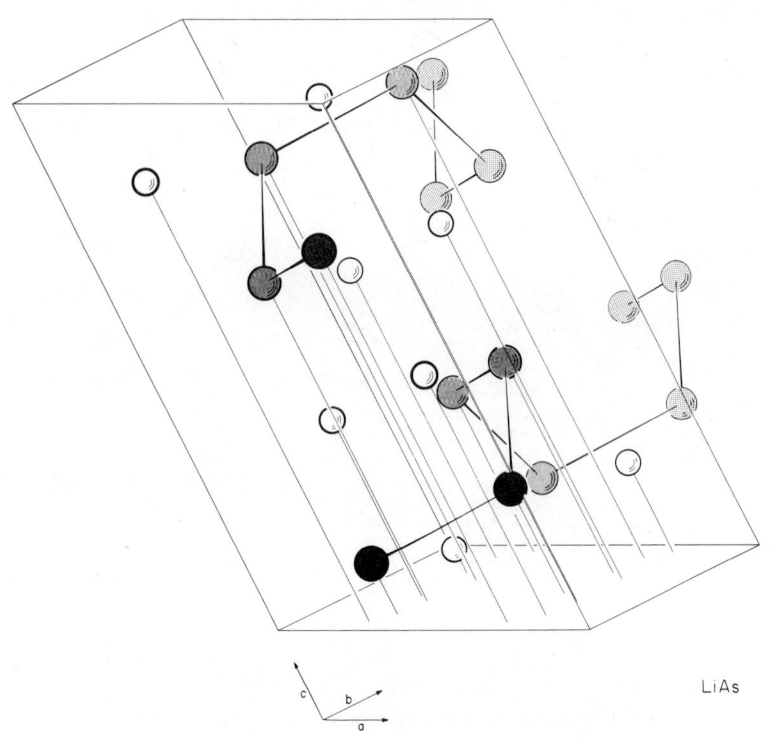

Figure 5-8. Pictorial view of the LiAs (*mP*16) structure, showing spiral arrays of As atoms.

bromine and iodine ($8-N=1$). FeS_2 with the pyrite (*cP*12) or marcasite (*oP*6) structures (Figure 5-13) also satisfies the general valence rule with $b_a/N_a=1$ and a valency of two for Fe. S_2 atom pairs occur corresponding to the pairs of atoms in the structures of the Group VII elements.

Polyanionic Compounds with Anions on More Than One Siteset and Different Numbers of Anion Nearest Neighbors. When the anions in the structure of a polyanionic compound occupy more than one siteset having different anion–anion coordination numbers, and particularly when the average b_a/N_a value for the compound is nonintegral, as for example in CdP_4, it is impossible to predict *a priori* what type of anionic subarray is to be expected. Nevertheless, it does appear that the anionic arrays bear systematic similarities to those already described for compounds whose anions all have the same number of anion nearest neighbors. Thus in the monoclinic structures of CdP_4 (*mP*10) and ZnP_2 (*mP*24) some P atoms

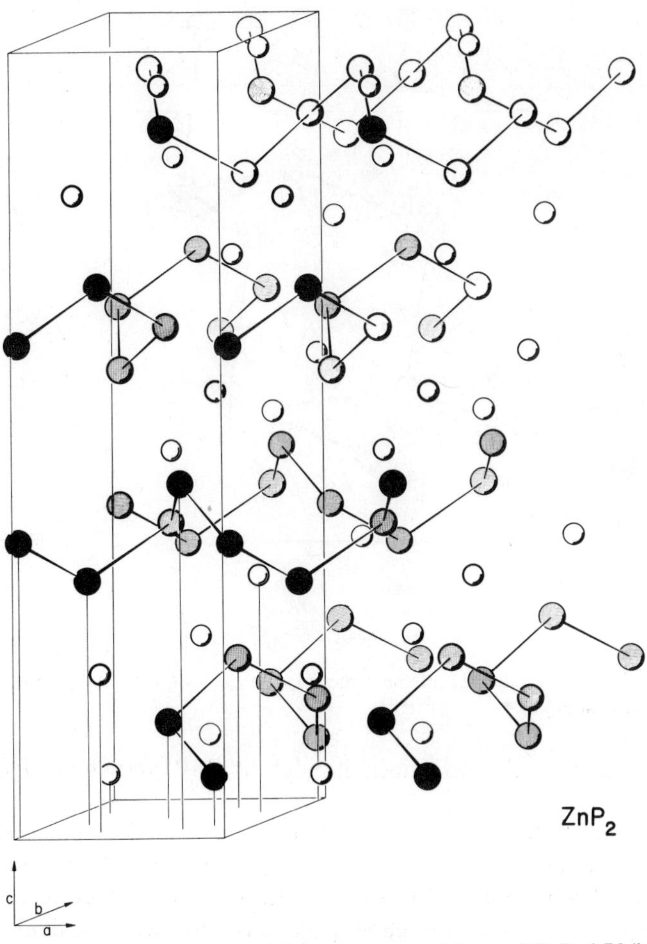

Figure 5-9. Pictorial view of the red tetragonal form of ZnP_2 (*tP*24), showing the spiral arrays of P atoms.

form bonds with two P neighbors and the remainder with three. Both types of P atoms are bonded together to form chains in which the anions each have two neighboring anions. These chains are linked in a three-dimensional array by bonds between the P atoms having three anion nearest neighbors. In the CdP_4 structure shown in Figure 5-14, bonds between the P(1) and P(2) atoms form spiral chains. The P(2)–P(2) bonds which join the chains in a three-dimensional array lie in layers extending throughout (001) planes of the structure, the general arrange-

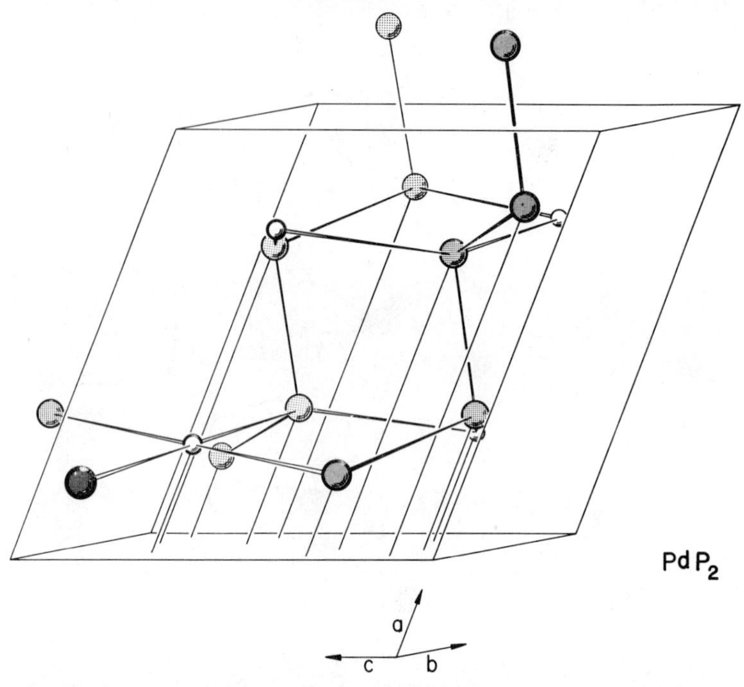

Figure 5-10. Pictorial view of the structure of PdP_2 ($mC12$) showing the spiral chains of P atoms.

ment being very similar to that in monoclinic ZnP_2 shown in Figure 5-18, p. 228.

Another polyanionic compound whose structure has been examined and found to contain anions on different sitesets with different numbers of anion nearest neighbors is $GeAs_2$ ($oP24$), already referred to on p. 214. This compound satisfies the general valence rule with an overall b_a/N_a value of unity, but the As atoms lie on four different sitesets. The anion–anion coordination is zero for As(1) and As(4) and two for As(2) and As (3) which are bonded together, giving zigzag chains of atoms running throughout the structure as shown in Figure 5-4. These zigzag chains of As atoms correspond to the structures of $(8-N)$ elements of Group VI, as would be expected for anions with $b_a/N_a = 2$.

Organic compounds can also be regarded as polyanionic or polycationic compounds, since they obey the general valence rule, but the "anions" (carbon atoms in a paraffin) or "cations" (carbon and hydrogen atoms in compounds such as aldehydes, amides, or acids) generally occupy many sitesets in the structure, and as the structures are usually of the molecular type, they are not considered here.

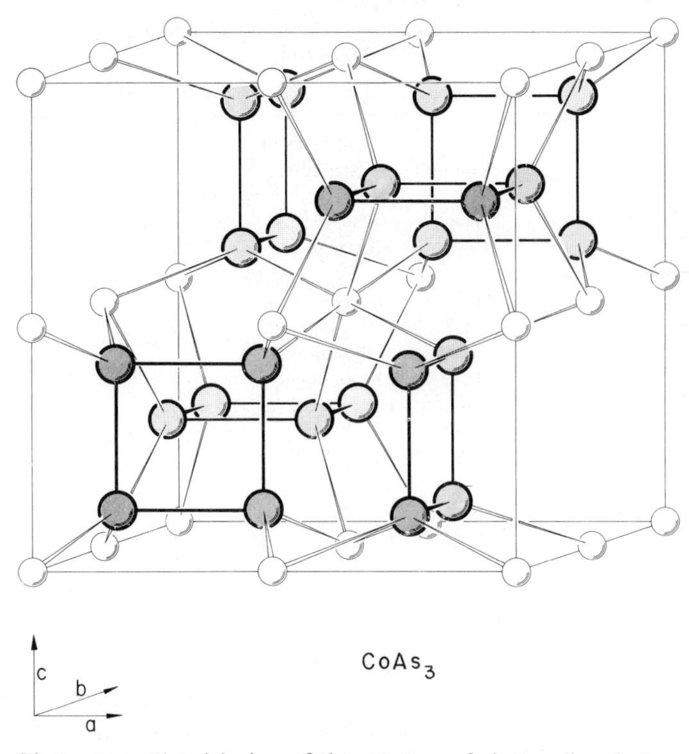

Figure 5-11. Pictorial view of the structure of skutterudite, $CoAs_3$ ($cI32$), showing the square planar arrangements of As atoms. Note the origin of the cell is transposed from that normally chosen.

Polycationic and Polycationic–Polyanionic Compounds

A few examples of known semiconducting polycationic compounds in which the cation–cation bonds do not run continuously throughout the structure are given in Table 5-8. For these compounds $(n_a+n_c)/N_a > 8$ so that $b_c/N_a > 0$. As $b_a/N_a = 0$ in polycationic compounds, there are no bonds between the anions and the anion subarrays might therefore be expected to resemble the close-packed structures of the inert gases. However, such structural arrangements are probably incompatible with the condition that the cation–cation bonds do not run continuously throughout the structure, since cations bonded together must be isolated from other cation-pairs by the arrays of anions.

The polycationic compounds GaS, α-GaSe ($hP8$), and β-GaSe ($hR4$) with four layers of atoms arranged in the sequence ABBA are interesting semiconductors, since the pairs of metal atoms bonded together do not cause metallic conductivity, being isolated from other cation pairs by

Table 5-8 Some Polycationic (Semiconducting) Compounds

$(n_a + n_c)/N_a > 8$ $b_c/N_a \neq 0$

	Compound	Symmetry of Structure	$(n_a + n_c)/N_a$	b_c/N_a
$M^{IV}X^{V}$	SiAs		9	1
	GeAs	Monoclinic	9	1
$M^{III}X^{VI}$	GaS	Hexagonal	9	1
	GaSe	Hexagonal or rhombohedral	9	1
	GaTe	Monoclinic	9	1
	InS	Orthorhombic	9	1
	InSe		9	1
	AsSe	Amorphous	9	
$M^{II}X^{VII}$	HgCl		9	1
	HgBr		9	1
	HgI		9	1

bonds to the anions. The structure of β-GaSe is shown in Figure 5-15. Figure 7-48, p. 257 compares the stacking sequence in these three structures with that in the more complex monoclinic structure of the polycationic compounds GaTe and GeAs (*mC*24) (Figure 5-16), which contains groups of atoms with a similar four-layer motif. The fragments of four-layer structure extend indefinitely in the [010] direction as shown in Figure 5-16, and the overall result is a complex layer structure.

β-InS (*oP*8) with an orthorhombic structure is also a polycationic compound in which the general valence rule is satisfied with $b_c/N_a = 1$, but in contrast to the layer structures discussed above, the bonds form a three-dimensional array as indicated in Figure 5-17.

The most interesting polycompound of all is the monoclinic form of ZnP_2 (*mP*24) which is both polyanionic and polycationic, the general valence rule being satisfied as follows:

$$\frac{n_a + n_c + b_a - b_c}{N_a} = \frac{10 + 2 + 4\frac{1}{2} - \frac{1}{2}}{2} = 8.$$

There are four different phosphorus and two different zinc sites in the structure. One quarter of the P atoms (P(1)) form three P–P bonds and one P–Zn bond; the remaining three quarters (P(2), P(3), and P(4)) form two P–P and two P–Zn bonds. The Zn(1) atoms form four Zn–P bonds,

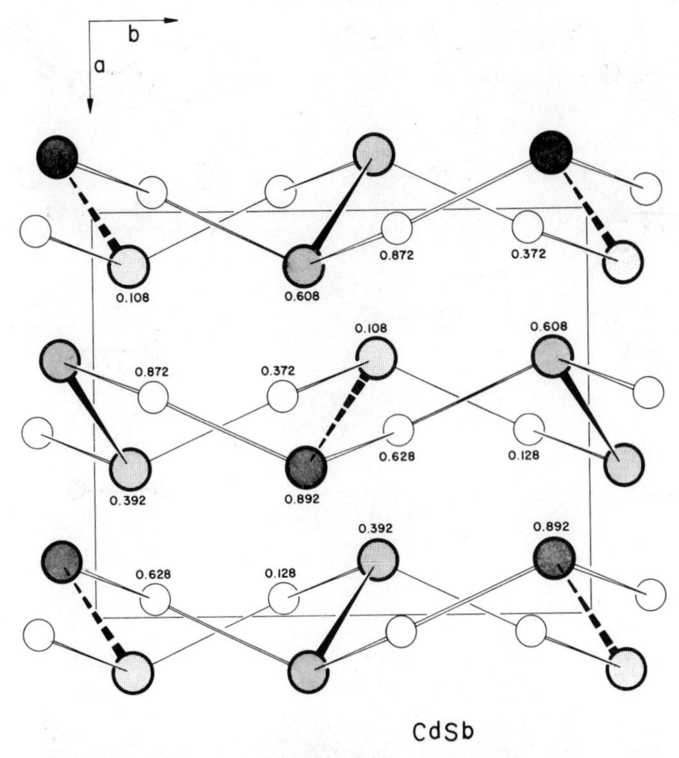

Figure 5-12. View of the CdSb ($oP16$) structure projected down [001] showing arrangement of pairs of Sb atoms.

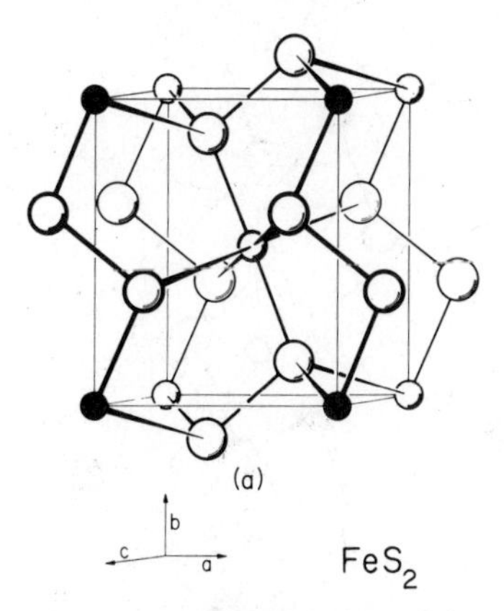

Figure 5-13. Pictorial view of the marcasite structure ($oP6$), showing S pairs (large circles).

Figure 5-14. Pictorial view of the CdP_4 (*mP*10) structure showing the arrangement of the connections between P atoms, and the octahedral surrounding of Cd.

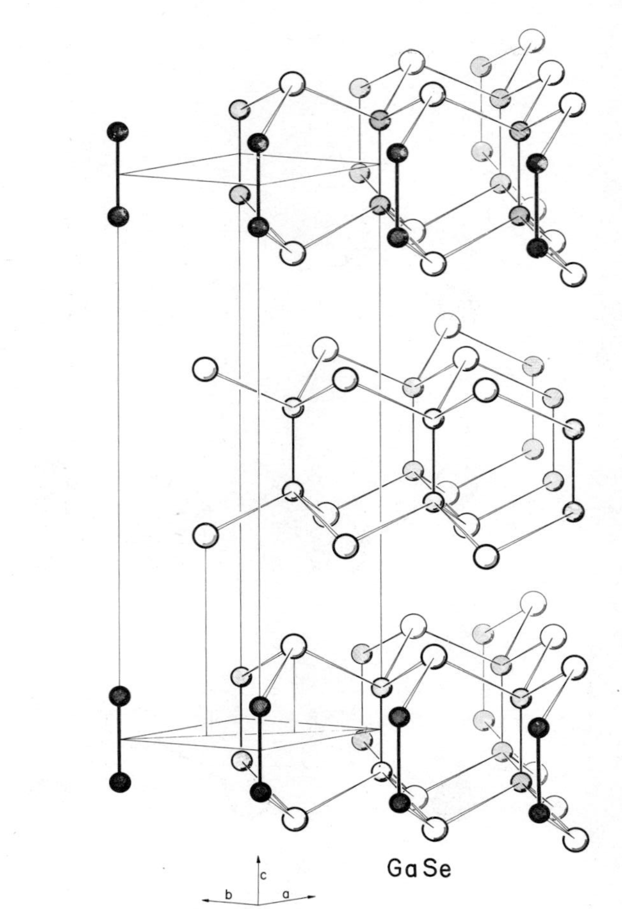

Figure 5-15. Pictorial view of the structure of β-GaSe (*hR*4), showing the hexagonal cell. Se large circles.

and Zn(2) atoms form one Zn–Zn and three Zn–P bonds. As Figure 5-18 shows, the phosphorus atoms are bonded together in chains with a "pitch" of eight atoms, but they are not wound in spirals. These chains are linked together in a three-dimensional array by P(1)–P(1) bonds which lie in bands extending throughout (100) planes in the structure, and the metal–metal bonds between Zn(2) atoms also lie in the same bands.

PARTIAL IONIC STRUCTURES

The semiconductivity of a number of substances suggests that one of the components is present in the structure as an ion, having transferred its electrons to a two- or three-dimensional framework of covalently bonded atoms in which the anion achieves a filled valence shell. The structures of TlS and TlSe (*tI*16) belong to this class. The Tl atoms occupy two different sitesets and the compounds appear to have formulas of the type $Tl^{I}Tl^{III}Se_2$, the T^{III} and the two Se atoms forming chains:

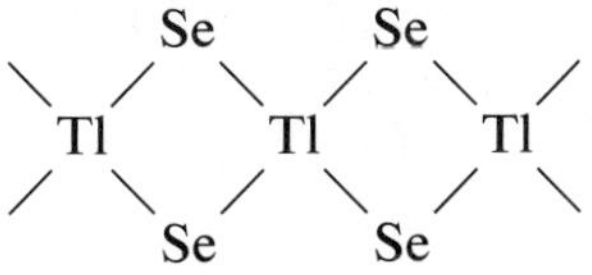

```
      Se      Se
 \  /    \  /    \  /
  Tl      Tl      Tl
 /  \    /  \    /  \
      Se      Se
```

The Tl^{I} atoms are packed between the chains as shown in Fig. 5-19 and it appears that they are present as Tl^{+} ions permitting the Tl^{III} atom to assume the character of a Group IV atom; thus the $Tl^{III}Se_2$ chains resemble those in the SiS_2 structure (Figure 8-26, p. 390). The constitution of $FeKS_2$ (*mC*16), if it is a semiconductor, is probably similar, since FeS_2Fe chains run throughout the structure, and Fe with valency three requires one more electron for the S atoms to achieve filled valence subshells.

Comparing the structure of the semiconductor Li_3Bi (*cF*16) (Figure 5-2, p. 207) with the anti-fluorite structure of Mg_2Sn (*cF* 12), it appears that a similar electron transfer involves the unique Li atoms in the octahedral holes of the array of Bi atoms, so that shared electron-pair covalent bonds only occur between Bi and the Li atoms in the tetrahedral holes. This accounts for the semiconductivity of the compound, for it appears that otherwise most intermetallic compounds have metallic properties when there is a simultaneous occupation of octahedral and tetrahedral holes in a close-packed cubic array of ions. Similar considerations probably also apply to the structure of Na_2Sb (*hP*8) in which the Sb atoms occupy a close-packed hexagonal array of sites and form a bonded array with the Na(2) atoms (see the near-neighbor diagram,

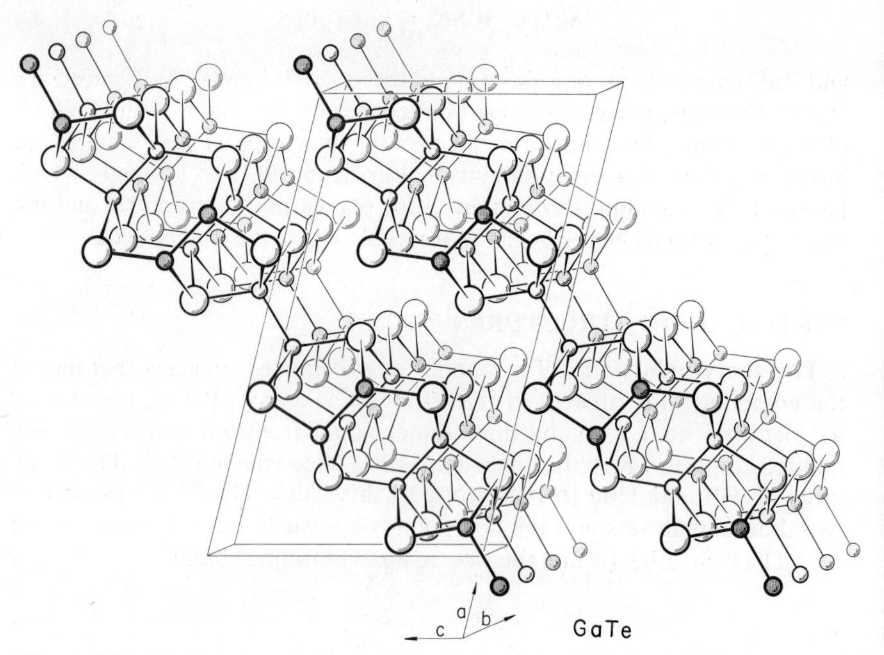

Figure 5-16. Pictorial view of the atomic arrangement in the GaTe ($mC24$) structure. Te large circles.

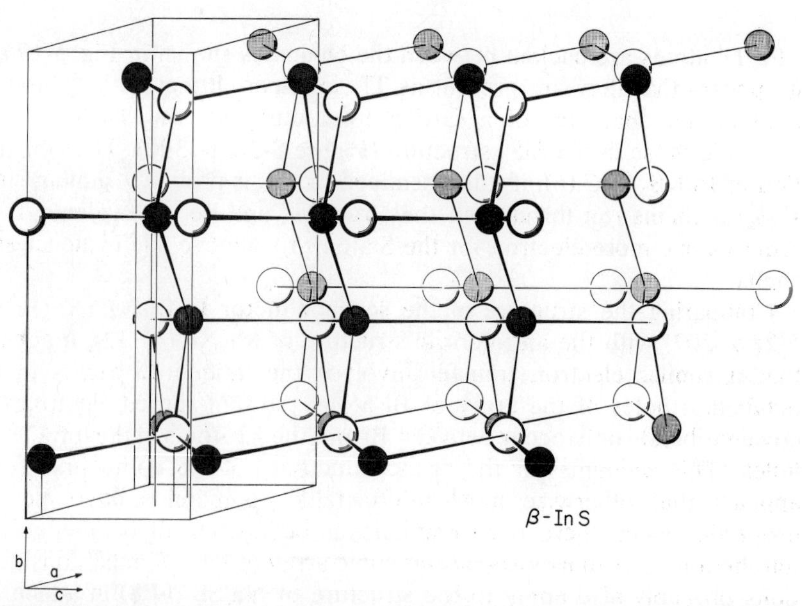

Figure 5-17. Pictorial view of the structure of β-InS ($oP8$), indicating In–In pairs (shaded circles).

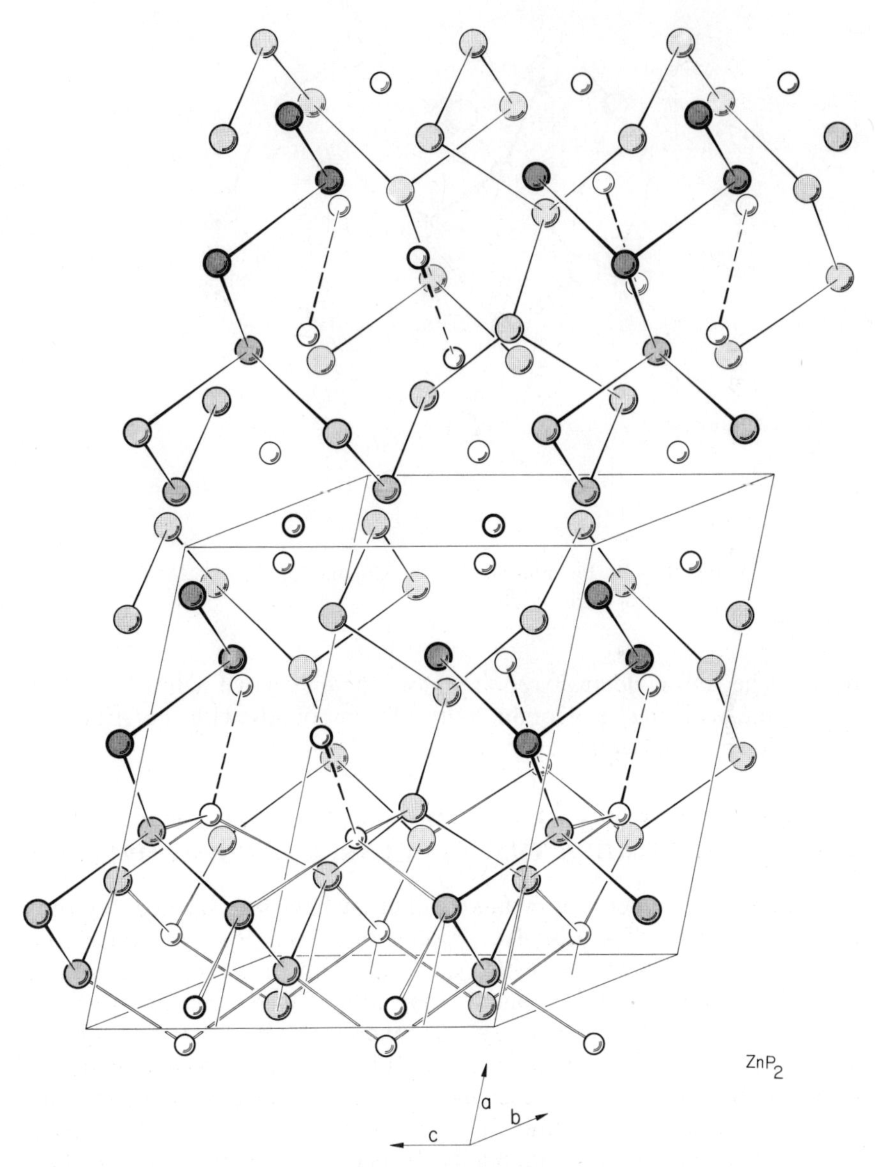

Figure 5-18. Pictorial view of the black monoclinic form of the ZnP_2 (*mP*24) structure, showing the connections between the P atoms, and the Zn atoms connected in pairs.

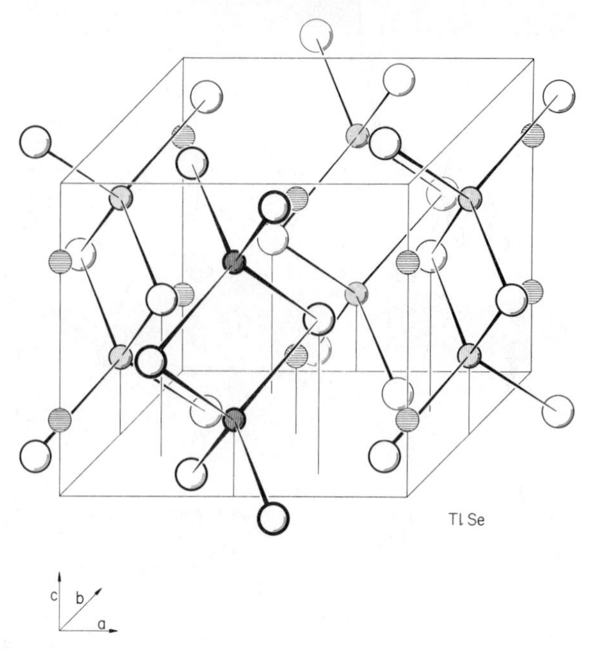

Figure 5-19. Pictorial view of the structure of TlSe (*tI*16). Te(1) small hatched balls.

p. 63. The extra electron required for Sb to achieve a filled subshell in keeping with the semiconductivity of compounds with the structure, is supplied by Na(1) ions.

STRUCTURES OF ELEMENTS AND COMPOUNDS WITH MORE VALENCE ELECTRONS THAN AVAILABLE BONDING ORBITALS

One obvious property of metals is that they have less valence electrons than available bonding orbitals; the semiconductors that have been considered provide the anions with exactly as many electrons as they have available bonding orbitals so that they attain filled valence subshells, although the cations present may and generally do have bonding orbitals available. There is a further class of "alloys" that are semiconductors in which the components have more valence electrons than there are bonding orbitals available for electron sharing, and these are the Group V to VII elements and compounds formed between them. Since their physical properties depend strikingly on the relative distances between the atoms of the most electronegative component, they are worth special study. The main structural characteristic of this group of substances is the arrangement of the atoms in chains, layers, or columns. These result from the inefficient use of the available bonding electrons which have to be retained on the atoms as unshared pairs in nonbonding orbitals, since there are insufficient *s* and *p* orbitals available for each atom to form four bonds. According to Pauling's postulate,[3] bonds from atoms with a filled valence subshell cannot resonate to various positions to increase the rather restricted coordination, unless vacant orbitals can be made available in the valence shell. Indeed, if four or more bonds are to be formed, the surplus valence electrons must either be promoted into antibonding orbitals or into higher bonding orbitals which may be lowered by hybridization with the other valence orbitals. There is a fundamental difference between the Group IV elements and those of Groups V–VII. By maintaining a nonbonding pair of *s* electrons, the heavier Group IV elements obtain a vacant orbital in their valence shell which permits pivotal resonance of the bonds. Thus, octahedral coordination can be achieved without the participation of higher valence states that appear to be necessary when the Group V–VII elements are octahedrally coordinated. If higher *d* states participate with the *s* and *p* valence states of Group V or VI elements to form sp^3d^2 or p^3d^3 bonds suitable for octahedral coordination, the transference of valence electrons from nonbonding to bonding orbitals not only makes more efficient use of the available electrons, but also provides the energy required for the excitation. Bi_2Te_3 discussed on p. 240, appears to be a good example where admixture of higher states with those derived from atomic *s* and *p* states occurs, since all atoms are approximately octahedrally coordinated in its structure.

The type of crystal structure and the physical properties of the Group IV to VII elements and compounds formed between them (summarized in Tables 5-10 to 5-12) can be entirely accounted for in terms of the chemical bonds that are formed, and the degree of filling of the bond orbitals by valence electrons. The structures of these elements and their compounds reflect the dehybridization of the sp^3 bonds with increasing Period, or principal quantum number. This is a consequence of the increased separation of *s* and *p* atomic levels with increasing principal quantum number. However, the separation of the *s* and *p* levels is not so great in the fourth and fifth Periods that hybridization of *s*, *p*, and *d* levels cannot occur, and these bonds play a role in the structure of elements in these Periods and of the compounds which they form. In the sixth Period, at least for Pb, the $6s^2$ level has become so stable, that *s*, *p*, *d* hybridization does not occur, and octahedral coordination appears to be achieved through *p* bonds alone.

Table 5-9 summarizes the distribution of 4, 5, and 6 valence electrons in possible bond orbitals. When vacant orbitals (indicated by an asterisk in

Table 5-9 Distribution of 4, 5, and 6 Valence Electrons in Possible Bond Orbitals

	Electrons Per Atom		
Type of Bond Orbital	4	5	6
nonbonding + p orbitals	s^2p^{2*}	s^2p^3	s^2p^4
(sp^3) hybrid	(sp^3)	(s^2p^3)	(s^2p^4)
(sp^3d^2) hybrid		(sp^3d^{1*})	(sp^3d^2)
(p^3d^3) hybrid		$s^2(p^{3***})$	$s^2(p^3d^{1**})$

*Asterisk indicates vacant orbital.

Table 5-10 Elements and Compounds with 4 Electrons Per Atom

Si, Ge, α-Sn, (SiC)	β-Sn, Si(H.P.), Ge(H.P.)	Pb
Diamond structure (sphalerite, wurtzite etc.) Semiconductor Bonds (sp^3)	β-Sn structure ($tP4$) Metal Bonds $s^2p^{2*}+(sp^3)$[a]	f.c. cubic ($cF4$) Metal Bonds s^2p^{2*}[b]

[a]C. A. Coulson, 1969. *Valence*, 2nd ed, Oxford: Clarendon Press, p. 335.
[b]In agreement with the results of Fermi surface studies which are analyzed to show that a band containing two states lies considerably lower in energy than bands containing the other two electrons (J. R. Anderson and A. V. Gold, 1965, *Phys. Rev.*, **139**, A, 1459).

Table 5-9) occur on an element, or on both components in a compound, metallic properties result.

Group IV Elements, and their Compounds with Group VI Elements

The stability of the $6s^2$ level in Pb results in PbS, PbSe, and PbTe with the rocksalt structure being semiconductors as the anions achieve a filled valence subshell, whereas the metallic conductivity of SnTe with the same structure probably results from the formation of sp^3d^2 bonds in orbitals which are deficient of one electron (sp^3d^{1*}). The distorted rock-salt structures of the semiconducting compounds SnS, SnSe, GeS, and GeSe come presumably from admixture of some sp^3 hybrid bonding with the resonance bonds formed by the Group IV elements. The inability of Pb to form s, p, d hybrid bonds also accounts for the probable absence of PbS_2, $PbSe_2$, and $PbTe_2$ compounds, whereas SnS_2 and $SnSe_2$ are semi-conductors with the CdI_2 structure. Here it appears that Sn forms sp^3d^2 bonds (with the transference of two electrons) and the anions achieve a filled valence subshell accounting for the semiconductivity.

The structures and electrical properties of the IV–VI compounds are consistent with the following assumptions: (i) that in Pb compounds

Table 5-11 Compounds with an Average of 5 Electrons Per Atom[a]

As	GeS GeSe	GeTe
(i) Two-layer structure, distorted NaCl type Metallic Bonds s^2p^3 layer and (sp^3d^{1*}) interlayer (ii) Amorphous Semiconductor Bonds s^2p^3 layer	Orthorhombic, ($oP8$) distorted NaCl type, 3-dimensional structure Semiconductor Bonds Ge, s^2p^{2*}; S, s^2p^4, some (s^2p^4) admixed	*High Temp.* NaCl structure Metallic Bonds (sp^3d^{1*}) — *Low Temp.* As-like 2-layer structure Metallic Bonds s^2p^3 and (sp^3d^{1*})
Sb	SnS SnSe	SnTe
		NaCl structure Metallic Bonds (sp^3d^{1*})
Bi	PbS PbSe	PbTe
Two-layer structure distorted NaCl type Metallic Bonds s^2p^3 layer and (sp^3d^{1*}) interlayer	NaCl structure Semiconductor Bonds Pb, s^2p^{2*}, S, s^2p^4	

[a]Descriptions apply to all substances enclosed within a box.

the tendency to maintain a nonbonding $6s^2$ level and form pivotally resonating p bonds is always stronger than the tendency to admit higher d orbitals to the bonds; (ii) in Sn compounds the tendency to combine higher d orbitals with those derived from atomic s and p states may be stronger than the tendency to maintain a nonbonding $5s^2$ pair; and (iii) the tendency to form hybrid sp^3 bond orbitals increases in compounds of Sn, Ge, and Si in this order.

Group V and VI Elements and Compounds Formed between Them

As noted on p. 230, the distribution of valence electrons in nonbonding orbitals in Group V–VII elements and compounds formed between them, gives rise to crystal structures in which the atoms form chains, columns, or layers creating one- or two-dimensional macromolecules. Provided that only intramolecular bonds (nearest-neighbor bonds) are considered, these elements and compounds formed between them contain semiconductor bonds and should show normal semiconducting properties.

Table 5-12 Compounds with 6 Electrons on Either Anion or Cation (by Transfer)[a]

	SiS_2 $oI12$, chain structure Semiconductor Bonds Si, (sp^3), S s^2p^4 or (s^2p^4)	$SiSe_2$	$SiTe_2$ CdI_2 layer structure Semiconductor Bonds Si, (sp^3d^2), Te, s^2p^3 or (s^2p^3)
Se Chain structure distorted NaCl. Semiconductor Bonds s^2p^4 chain and (sp^3d^2) weak inter-chain	GeS_2 $oF72$,3-dimensional structure. Semiconductor Bonds Ge, (sp^3), S s^2p^4 or (s^2p^4)	$GeSe_2$ Distorted layer structure (?) Semiconductor Bonds Ge, (sp^3d^2), S, s^2p^3 or (s^2p^3)(?)	$GeTe_2$ Not formed (?)
Te	SnS_2 CdI_2 layer structure Semiconductor Bonds Sn, (sp^3d^2). S, s^2p^3 or (s^2p^3)	$SnSe_2$	$SnTe_2$ Not formed
Po Simple cubic Metallic Bonds (sp^3d^2), $s^2(p^3d^{1*})$ (?)	PdS_2 (occurrence reported?) Not formed because of strong preference of Pb to form bonds of type s^2p^2 rather than (sp^3d^2)	$PbSe_2$	$PbTe_2$

[a]Descriptions apply to all substances enclosed within a box.

As the anions attain filled valence subshells by electron sharing in forming the intramolecular bonds, pivotal resonance of these bonds to take account of any intermolecular bonding which may occur between the anions is impossible according to Pauling's postulate. Direct anion–anion intermolecular bonding, which is an empirical fact in many of these substances, therefore requires the participation of higher atomic (d) orbitals in the bonds. Hybridized sp^3d^2 or p^3d^3 bonds so formed are suitable for the observed distorted octahedral coordination. If these bond orbitals on the anions are incompletely filled, the electrons therein must have metallic properties. It is therefore postulated that if direct chemical intermolecular bonding occurs between the atoms in the structures of the Group V–VII elements, or between *anions* in the structures of compounds formed between them, some modification of the normal semiconducting properties must be expected. On the other hand, if direct intermolecular bonding occurs between anions and cations or only between cations, no modification is required because vacant orbitals on the cations can account for resonance bonding without destroying the filled valence subshells on the anions. Naturally, if the intermolecular contacts are no closer than van der Waals contacts, such bonds are not involved and normal semiconducting properties occur. Examination of interatomic distances and hence intermolecular bonding in chain, layer, and columnar structures of the Group V–VII elements and compounds formed between them, shows that the physical and electrical properties of these substances are indeed in accord with the above postulates.

Arsenic, Antimony, and Bismuth. The Group V elements As, Sb, and Bi which have five valence electrons, each form double-layer structures in their normal crystalline forms so that every atom has three neighbors in the same double layer and three more neighbors at rather larger distances in neighboring layers, as shown in Figure 5-20. Table 5-13 shows that the second-nearest neighbor distances are considerably shorter than the van der Waals diameters of these elements, indicating some interlayer valence bonding admixed with the intralayer bonding. Since filled valence subshells are obtained in forming the intralayer bonds, permitting no

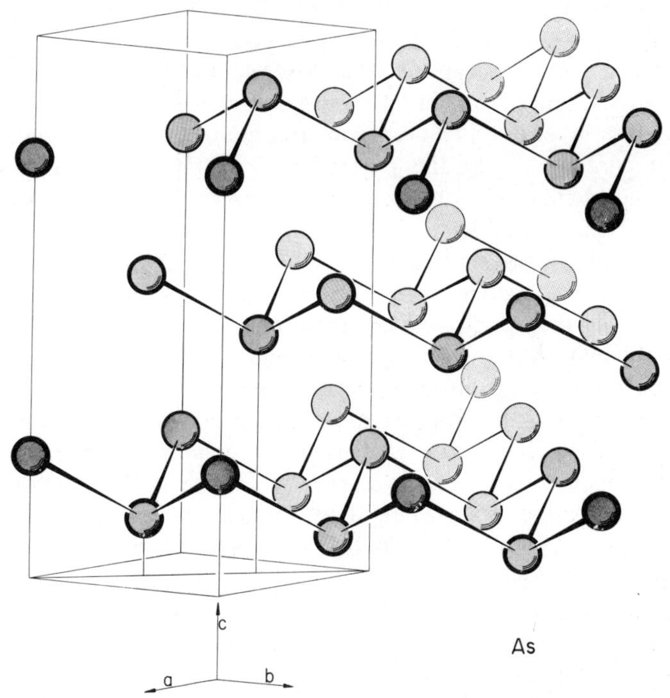

Figure 5-20. Pictorial view of the As ($hR2$) structure showing the hexagonal cell.

Table 5-13

Element	d_1 Intralayer Distance (Å)	d_2 Interlayer Distance (Å)	van der Waals Diameter (Å)	d_2/d_1
As	2.51	3.15	4.0	1.25
Sb	2.87	3.37	4.4	1.17
Bi	3.10	3.47		1.12
Simple cubic				1.00

pivotal resonance to the three interlayer neighbors, some electrons in hybridized orbitals such as the sp^3d^3 type would be required to account for the weak interlayer bonds and satisfy their distorted octahedral disposition.[4] Since the Group V elements have only five valence electrons, these orbitals would be incompletely filled and the electrons therein would have metallic properties, accounting for the metallic conductivity of crystalline As, Sb, and Bi. The rhombohedral structure of these elements can be regarded as a distortion of the simple cubic structure in which d_1 and d_2 (Table 5-13) would be equal. The better developed the interlayer resonance bonds are, the closer should the ratio d_2/d_1 approach to the value of unity found in the simple cubic structure. Table 5-13 shows how the interlayer bonds increase in strength on going from As to Bi, and this is in agreement with the general trend in the properties of the semiconducting elements in any one Group toward metallic character with increasing atomic number.

The metallic properties of these substances depend on the interlayer bonds since the valence states describing the intralayer bonding alone clearly satisfy the conditions for semiconducting bonds. The breaking of the interlayer bonds should therefore lead to semiconductivity, and this is indeed found to be so in the amorphous forms of As and Sb, which generally consist of small portions of the double-layer structure found in the rhombohedral form, which are randomly arranged with respect to each other. In this condition the interlayer resonance bonds cannot develop, so that the second-nearest-neighbor distance d_2 increases from 3.15 Å in crystalline As to 3.75 Å in the amorphous form, the corresponding values for Sb being 3.37 Å and 3.75 Å respectively. No corresponding amorphous form of Bi exists at room temperature and the only amorphous form which has been described has superconducting properties and is stable only at the lowest temperatures. Because of its high atomic number, it is doubtful whether Bi could be a semiconductor in any form (cf. Pb and Po in neighboring Groups).

Black phosphorus has a two-layer structure (Figure 5-21) somewhat resembling that of As, but here the interlayer P–P distances of 3.88 Å correspond to the van der Waals diameter of P, so that the semiconductivity of black P is well understood. Intralayer P–P distances are 2 at 2.17 Å and 1 at 2.21 Å.

Selenium, Tellurium, and Polonium. Crystalline hexagonal selenium and tellurium show semiconducting properties, but those of selenium, particularly, are abnormal. The heaviest element of this group, polonium, in its two crystalline forms has a coordination number of six and is a metallic conductor. In the normal crystalline forms of selenium and

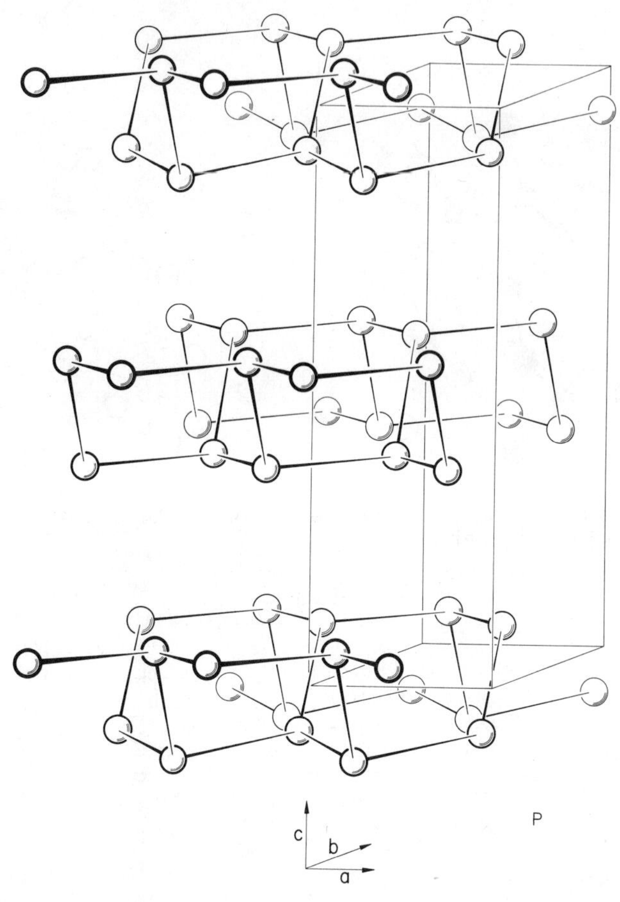

Figure 5-21. Pictorial view of the structure of black phosphorus ($oC8$).

tellurium with trigonal symmetry, the atoms form spiral chains running parallel to the *c* axis of the crystal, so that each has two nearest neighbors (Figure 5-22). As shown in Table 5-14, the distances to the four second-nearest-neighbors are definitely shorter than the van der Waals diameters,

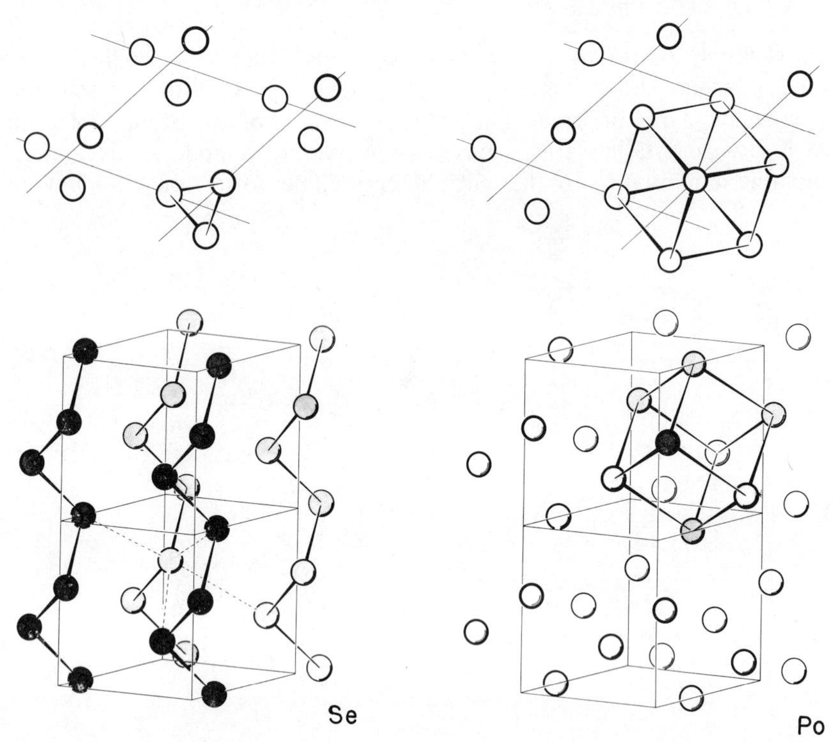

Figure 5-22. Pictorial views of the structure of Se or Te (*hP*3) compared with the simple cubic structure of α-Po (*cP*1).

Table 5-14

Element	Nearest-Neighbor Distance d_1 (Å)	Second-Nearest-Neighbor Distance d_2 (Å)	ratio d_2/d_1	van der Waals Diameter (Å)	Amorphous Form d_1 (Å)	d_2 (Å)
Se	2.32	3.46	1.49	4.0	2.35	3.7
Te	2.86	3.46	1.31	4.4		
α–Po (simple cubic)			1.00			

so that some interchain valence bonding is indicated, and although it is not so well developed as shown by the ratios of d_2 to d_1, the situation is similar to that already described for the layer structures of crystalline arsenic and antimony. When the intrachain bonding is satisfied, each atom attains a filled valence subshell through electron sharing, and there are no vacant orbitals to permit pivotal resonance of the bonds to the second-nearest-neighbors. In contrast to the metallic behavior of As and Sb, interchain bonding that arises from electrons in hybridized sp^3d^2 orbitals would not destroy semiconductivity in Se and Te, in agreement with their observed properties, since they each have 6 valence electrons. The anomalous physical properties of Se compared to Te result from the interchain bonds and the large energy band gap in Se (about 2.3 eV) compared to the small gap in Te (about 0.35 eV). For example, it is a general rule for the melting points of elements with the same crystal structure to decrease with increasing atomic number, yet Se melts at about 220°C and Te melts at about 450°C! The effect of thermal vibrations on heating Se or Te is to destroy the weak interchain bonding, but with the small energy band gap of 0.35 eV in Te, electrons are readily thermally excited into the conduction band so that the number of holes in the valence band increases with increasing temperature. These vacant orbitals in the valence shell of Te permit pivotal resonance of the electrons forming intrachain p bonds to the second-nearest-neighbors, so that an additional source of interchain bonding grows as the temperature rises. Se with a large energy band gap of 2.3 eV has virtually no thermal excitation to the conduction band by 220°C and, with the interchain bonds disrupted by thermal vibrations, it melts at an anomalously low temperature compared to Te. Both substances melt without rupture of intrachain bonds, so that the liquid at the melting point consists of a random arrangement of Se–Se or Te–Te chains. However, on further heating the Te chains soon break down to individual Te atoms, whereas at a much higher temperature the Se chains are still intact. Regarding intrachain bonds therefore, Se is much more thermally stable than Te, as expected on general principles, and the anomalously low melting point of crystalline solid Se only results from a combination of the weak interchain bonds and the large energy band gap.

An exactly parallel situation is found in respect of the expansion coefficients. Se with lower atomic number is expected to have smaller thermal expansion coefficients at room temperature than Te, but the expansion coefficient of Se in the basal plane (normal to the length of the chains) is notably larger than that of Te[9] (Table 5-15), and this results from the same factors as cause the anomalously low melting point.

Table 5-15 Thermal Expansion (Data from Straumanis[9])

Element	a in kX		Δa	c in kX		Δc
	18°C	25°C	18–25°	18°C	25°C	18–25°
Se	4.3544_8	4.3567_8	$+0.0023_0$	4.9496_2	4.9490_0	-0.0006_2
Te	4.4466_9	4.4475_9	$+0.0009_0$	5.9149_4	5.9148_7	-0.0000_7

***Phases with the Sb_2Se_3* (oP20) *Structure*.** The structure of the compounds Sb_2S_3, Sb_2Se_3, and Bi_2S_3 is illustrated in Figure 8-59, p. 426. The atoms are grouped together in columns running in the [001] direction, the arrangement being that of distorted fragments of rocksalt-like structure. In Sb_2Se_3 the bond lengths within a column vary from 2.58 Å to 3.22 Å and the bond angles lie between 86° and 99°. The interactions between atoms in neighboring columns are weaker than those between atoms in a column, although several of the intercolumnar approaches are considerably closer than the distances expected for van der Waals contacts (Table 5-16), and there definitely appears to be some direct chemical bonding between atoms in neighboring columns. The strongest intercolumnar interactions are between anions and cations and therefore, according to the rules given on p. 234, the formation of intercolumnar bonds need not interfere with the semiconducting bonds formed in intracolumnar bonding, and indeed these substances do appear to have normal semiconducting properties. The closest anion–anion distance in the structure, 3.65 Å, giving a bond number of only 0.01, occurs between Se(2) and Se(3) atoms which lie in the same column.

Table 5-16 Sb_2Se_3 Structure: Intercolumnar Distances

Neighboring Atoms in Different Columns	Observed Interatomic Distance (Å)	Distance for van der Waals Bonds in (Å)	Calculated Bond Number
Sb(1)–Se(2)	3.26	4.2	0.09
Sb(2)–Se(2)	3.46	4.2	0.04
Sb(1)–Se(3)	3.74	4.2	0.01
Se(2)–Se(3)[a]	3.65	4.0	0.01

[a] Intracolumnar distance.

***Phases with the Bi_2STe_2* (hR5) *Structure*.** In five-layer compounds with the Bi_2STe_2 structure, interlayer bonding between anions is indicated by the interlayer distances. Substances such as Bi_2Te_3, Sb_2Te_3, and Bi_2Se_3 with this structure have the five-layer sequence, Te(2)-Bi-Te(1)-Bi-Te(2) shown in Figure 8-67, p. 435. The Te(2) atoms at the boundaries of the five-layer units have 3 nearest neighbors in the same layer and 3 more neighbors at a greater distance in the adjacent layers. Their environment is therefore similar to that of the atoms in crystalline arsenic or antimony. Nevertheless, overall the atoms occupy approximately the sites of a close packed cubic arrangement with stacking sequence ABC... along the *c* axis of the hexagonal cell, and so the structure has much more the features of a metallic structure than many other structures of M_2X_3 compounds with directed valence bonds. Compounds with this structure have very much smaller energy band gaps than compounds with the Sb_2Se_3 type structure.

The intralayer bonding results in filled valence subshells on the anions and bonding involving sp^3d^2 orbitals[10] on Bi and Te(1) atoms† has been proposed, which is most satisfactory since it makes full use of the available valence electrons in forming bonds and conforms with the overall metal-like aspect of the structural arrangement. However, it leaves formal positive charges on the outer Te(2) atoms, whereas in fact the outer layer Te(2)–Te(2) distances are distinctly shorter than van der Waals distances expected for neighboring atoms with filled valence subshells, as shown in Table 5-17. These close Te(2)–Te(2) approaches suggest a small amount of direct chemical bonding rather than ionic repulsions between the five-layer units. Secondly, interlayer Te(2)–Te(2) bonding would seem to be an important factor in Bi_2Te_3 and similar compounds adopting the approximately close-packed metal-like structure, rather than the other distorted M_2X_3 type structures referred to above. Thirdly, although it may be unwise to draw anything but the broadest conclusions from the calculation of bond numbers from observed interatomic

Table 5-17

Compound	Interlayer $X(2)$–$X(2)$ Distance (Å)	van der Waals $X(2)$–$X(2)$ Distance (Å)
Bi_2Te_3	3.57 (Lange[11]) Massive crystals	4.4
	3.73 (Semiletov[12]): Thin films	
Sb_2Te_3		4.4
Bi_2Se_3	3.31	4.0

†Note that Drabble and Goodman refer to the crystallographic atom Te(1) as Te(2) and to Te(2) as Te(1).

distances and Pauling's covalent radii, the data shown in Table 5-18 indicate that the M–X(1) and M–X(2) bonds certainly do not have the approximately unit bond strengths expected from Drabble and Goodman's bond scheme.[10] The low bond numbers suggest the presence of nonbonding orbitals and the possibility of pivotal resonance (higher orbitals available) in the intralayer bonding. Table 5-18 also compares the bond numbers expected from this bonding with those determined from experimental data. Fourthly, substitution of S or Se for Te(1) is said to strengthen the Bi–X(1) bonds conforming with the increase in the energy band gap, although experimentally it is observed that the bond number for Bi–Se bonds in Bi_2SeTe_2 is only 0.15 compared to 0.28 for Bi_2Te_3! Such facts show that the interpretation of structural features and physical properties in terms of a single set of chemical bonds may be impossible. Undoubtedly, some few electrons on the outer Te(2) atoms partake in sp^3d^2 bonding between the five-layer units and since these would have metallic properties (5 electrons in 6 orbitals), their presence could form the basis of many of the anomalous electrical properties of this group of compounds.[13]

It has been shown that insoluble impurities disrupt the regularity of the chain sequences in Se and also the interchain bonds, and similar effects occur in mixed solid solutions of Se and Te. Although the data are not clear, it appears that mixed chains are formed and the variation of the c parameter (along the chain length) roughly follows Vegard's law, but the a parameter (interchain distances) has a positive deviation from Vegard's law. Wiese and Muldawer[14] have found a similar behavior in solid solutions of Bi_2Te_3 and Bi_2Se_3. The a parameter (parallel to the layers) varies linearly with composition as does the c parameter (normal to the layers) for the first 33 at.% added selenium when substitution is in the central Te(1) position only. Thereafter from about 40 to 100% Se, when there is a

Table 5-18 Bond Numbers for Bi_2Te_3 and Bi_2Se_3

Bonds	Observed Bond Numbers Calculated from Pauling's Covalent Radii and Interatomic Distances		Bond Numbers for Nonbonding Orbital Model with Orbitals for Resonance	Bond Numbers from Drabble and Goodman Model
	Bi_2Te_3	Bi_2Se_3		
Bi–Te(1) (Se(1))	0.28	0.23	0.33	1.0
Bi–Te(2) (Se(2))	0.41	0.32	0.67	1.0

mixing of Se and Te atoms on the outer Te(2) sites of the five-layer units, a positive deviation from Vegard's law occurs which implies (in part at least†) an expansion of the interlayer distance, corresponding to a reduction in the strength of the interlayer bonds when the layers contain two components randomly dispersed.

The CdI_2 Structure. $SiTe_2$, SnS_2, and $SnSe_2$ have the three-layer CdI_2 type of structure (Figure 8-69, p. 439). The interlayer chalcogenide distances indicate little more than van der Waals contacts between the anions in the structure, corresponding to the normal semiconducting properties which are found.

Finally, phases with the I_2, SbSBr, and BiI_3 types of structure also fall within the class of semiconducting compounds which have been discussed, but they do not fall within the scope of this book.

METALS, SEMICONDUCTORS, AND IONICS: COMMON STRUCTURAL TYPES

Ionic substances frequently take the same structures as metallic phases, because in each case the forces holding the atoms together are essentially nondirectional and the arrangement of the atoms is governed by the geometrical principles set out by Laves.[15] However, coordination numbers in ionic structures are strictly limited by the number of anions that can be assembled about the cation without inducing strong anion–anion repulsions, whereas no such restrictions are found in the structures of metallic substances. Semiconductors also take these same structures, when they are based on a close packing of the anions, because the tetrahedral and octahedral interstices therein provide the correct symmetry for the directed bond orbitals that are formed. The relationship between short-range order and chemical bonding in semiconductors and nonionic insulators makes it interesting to examine changes in coordination and structure on going from directionally to nondirectionally bonded phases in a series of chemically similar solids, particularly if the examination can be based on parameters describing the directional properties of the bonds.

Dehlinger[16] has pointed out that the principal quantum number n of the outermost occupied shell of an atom can be used as a measure of the directional character of the bonds formed between the atom and other atoms of the same kind. As n increases, the atomic orbitals involved in the formation of bonds, and hence the bonds themselves, gradually lose

†The distance between Te(2) atoms in neighboring layers involves not only the c lattice parameter, but also the atomic parameter $z_{Te(2)}$, which was not determined.

their directional properties. Although these observations refer strictly to the bonds in elements, it is possible to extend the discussion to include compounds by introducing a suitably averaged principal quantum number, $\bar{n} = \Sigma_i a_i n_i / \Sigma_i a_i$, where n_i and a_i are the principal quantum number of the valence shell and the number per formula unit of atoms of the ith kind respectively.[17] The directional character of the bonds in compounds does not, however, depend on the value of $\bar{n}$ alone, but also on the electronegativity difference, Δx, of the component atoms. Thus, regardless of the value of $\bar{n}$, the bonds become more ionic and hence more nondirectional, as the value of Δx increases. In order to deal with compounds having more than two components it is convenient to define Δx as follows: $\Delta x = |\bar{x}_A - \bar{x}_C|$, where $\bar{x}_A$ and $\bar{x}_C$ are the arithmetic means of the electronegativities of the component anions and the cations, respectively. Representation of the ionic character of a compound by the parameter Δx (the selected values of Gordy and Thomas are given in Table 3-1, p. 70), is satisfactory for classification except in extreme cases, such as the compound ZnO, where the ratio of the cationic and anionic covalent radii (1.77) deviates considerably from unity. When the cation is considerably larger than the anion, the ionic character of the compound may be significantly less than that gauged from the value of Δx alone (see below). These observations suggest that similar phases of the type A_iX_j ($i, j =$ 1, 2, 3, . . . should be examined on diagrams where the axes represent the $\bar{n}$ and Δx values of the component atoms in the compounds, since they may then be expected to separate into groups having the same structure and/or coordination numbers. Furthermore, directionally bonded phases would be expected to lie in the region where $\bar{n}$ and Δx have small values (here semiconducting compounds are found), ionic phases to lie in the region of high Δx values, and metallic phases in the region of large $\bar{n}$ and small Δx values. These expectations are remarkably well borne out in the diagrams shown below. The 'compounds' selected for these diagrams are generally those in which normal valence rules appear to be obeyed through the formation of cation–anion bonds alone, although some of these are metallic conductors because the simple valence expectations are not borne out. If all simple metallic phases A_iX_j were included generally in the diagrams, they would be expected to cover them widely, except in the regions of low $\bar{n}$ and high Δx, and the simple demonstration of the dependence of structure on the directional character of the bonds that is apparent in Figures 5-23 to 5-32 would be lost. Although this approach is purely phenomenological, a theoretical study of the same problem by Phillips and co-workers, which is described below (p. 249), achieves complete separation of MN valence compounds according to coordination numbers in terms of parameters derived from the optical spectra of the MN crystals. The results of these studies can better be appreciated after a slight digression to discuss the factors which control the building of ionic structures.

The structures of simple ionic compounds are governed on the one hand by Laves' principles of dense packing, high coordination, and high symmetry because the bonds between the ions are nondirectional in character, and on the other hand by the most favorable balance of electrostatic forces between the ions, which may be expressed as the Madelung constant[18] for the structure. The higher the value of the Madelung constant for a given stoichiometry MX, $MX_2, \ldots$, the more favorable the structure for a highly ionic compound. Other things being equal, high coordination of the cations by the anions and overall dense packing lead to structures with the highest Madelung constants. Thus for MX_2 compounds, the 8–4 coordinated fluorite structure has a higher Madelung constant than the 6–3 cadmium iodide structure, and is therefore preferred for highly ionic substances. For any given coordination number, the preferred anion coordination polyhedra and the method of stacking them together by sharing corners, edges, or faces is generally such as to give the largest possible separation between the cations. Thus for example, the 6–6 NaCl structure is preferred to the 6–6 NiAs structure for ionic compounds, because the sharing of edges of the anion coordination polyhedra in the former gives a larger separation of the cations than the sharing of some faces of the polyhedra in the NiAs structure.

Although these general principles guide the building of structures of ionic compounds, the actual structure adopted by any particular substance is determined primarily by the relative sizes of cations and anions. The distribution of the larger anions about the cations must be such that the anions are not in mutual contact leaving the cations free to "rattle" within the anion coordination polyhedron, since anionic repulsions would then render the structure unstable. Therefore, for any given coordination of anions surrounding the cations, there is a limiting ratio of cationic to anionic sizes for structural stability and Table 5-19 lists these ratios for various coordination polyhedra.

The principles governing the formation of structures of simple ionic compounds can be briefly summarized as follows: (i) the ratio of cationic to anionic radii R_C/R_A, and the stoichiometric proportions of the components MX, $MX_2, \ldots$ determine the maximum number of anions in the coordination polyhedra about the cations; (ii) symmetry and packing considerations lead to choice of the most symmetrical coordination polyhedral possible; and (iii) stoichiometric composition and electrostatic interactions control the arrangement of coordination polyhedra (to share corners, edges, or faces) and, when the ionicity is high, these

Table 5-19 Lower Limiting Values of Radius Ratios for Stability of Ionic Structures

Coordination about Cation	Distribution of Anions	Structural Examples	Lower Limiting Radius Ratio R_C/R_A for Stability
8	Corners of cube	CsCl Fluorite, CaF_2	0.732
6	Corners of octahedron	Rocksalt, NaCl Rutile, TiO_2	0.414
4[a]	Corners of square		0.414
4[a]	Corners of tetrahedron	Sphalerite, ZnS Wurtzite, ZnS Quartz, SiO_2	0.225

[a]With such low coordination, ionic character is no longer pronounced.

are arranged to give a structure with the largest possible Madelung constant. Weakly ionic substances may adopt structures with somewhat lower Madelung constants. These conditions have the result that simple ionic compounds of the type MX, MX_2, . . . are accommodated in a relatively small number of different crystal structure types, whereas simple metallic phases having essentially no coordination or valence restrictions, can and do adopt a large variety of different structure types. For this reason there is no single simple system of classifying the structures of all metallic phases.

Diagrams of n *vs.* Δx

The distribution of structures of simple valence compounds on $\bar{n}$ *vs.* Δx diagrams is examined by indicating the structure type of a given compound by a symbol which is located on the diagram according to the $\bar{n}$ and Δx values for the component atoms in the compound.

MX Compounds. Figure 5-23 shows a $\bar{n}$ *vs.* Δx diagram for valence compounds MX, formed by nontransition metals in their highest valence state. These compounds have the rocksalt ($cF8$), sphalerite ($cF8$), or wurtzite ($hP4$) structures. Octahedrally coordinated structures are sharply separated from those with tetrahedral coordination, except for the compounds ZnO, GaN, InN, and CuCl with tetrahedral coordination which lie in the region occupied by rocksalt structures. In these four compounds the ratios of the cationic to anionic *covalent* radii are considerably greater than unity (Table 5-20) and it has been shown[19] from consideration of physical properties such as energy band gaps,

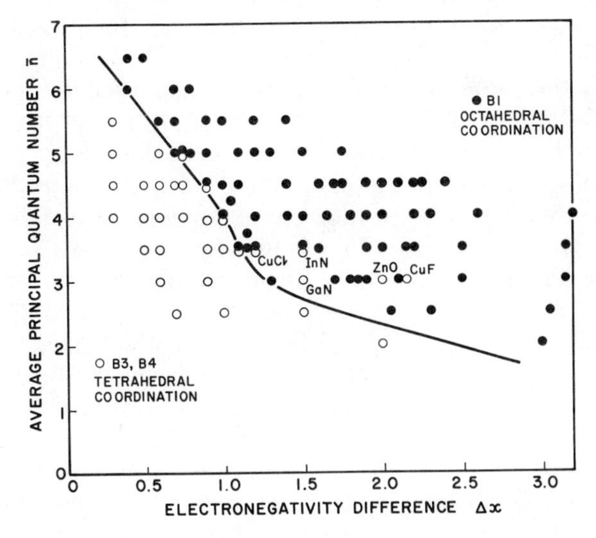

Figure 5-23. $\bar{n}$ *vs.* Δx diagram for nontransition metal *MX* normal valence compounds with the rocksalt, sphalerite, and wurtzite structures.[17]

Table 5-20

Compound	Covalent R_C/R_A	Δx
CuCl	1.36	1.2
ZnO	1.77	2.0
GaN	1.71	1.5
InN	1.95	1.5

microcleavage patterns, and interatomic distances, that when the cation is considerably larger than the anion (covalent sizes), the ionicity of the compound is less than that expected from consideration of the electronegativity difference of the component atoms. This effect can be explained by assuming a covalent model so that the result of an electronegativity difference between the components is to polarize the covalent bonds, shifting the electrons towards the "anions." When the anions are relatively small and the electronegativity difference is high, mutual repulsion of the electrons highly concentrated about the relatively small "anion," opposes polarization of the bonds, so that the extent of polarization—or the ionicity of the compound—is not as large as would be expected from consideration of the electronegativity difference alone. If this factor is taken into account by constructing, for example, a $\bar{n}$ *vs.* $\Delta x.R_A/R_C$ diagram for the MX compounds, these four compounds return to the region of directionally bonded phases and the separation of 4–4 and 6–6 structures on the diagram is practically complete.

The borderline between tetrahedrally and octahedrally coordinated phases (Figure 5-23) indicates that compounds may adopt the rocksalt structure, not merely because they are highly ionic, but rather because the bonding is nondirectional. Thus in the region of large $\bar{n}$ and relatively low Δx, the rocksalt structure is found for the phases with metallic character. In addition to the compounds shown in Figure 5-23, there are also some 12 compounds shown in Figure 5-24 which have the rocksalt structure or distorted forms thereof, and lie in the region for 4–4 coordinated directionally bonded phases. The reason for this is nevertheless clear, since they are compounds of the heavier Group IV cations in their *lower* valence state where they retain an unshared pair of *s* electrons forming resonating *p* bonds. In this valence state it is impossible for them to form the tetrahedrally disposed sp^3 hybrid bonds required in the sphalerite or wurtzite structures, and the rocksalt structure which they adopt is consistent with the symmetry required for resonating *p* bonds. Many of the interstitial carbides and nitrides of the transition metals

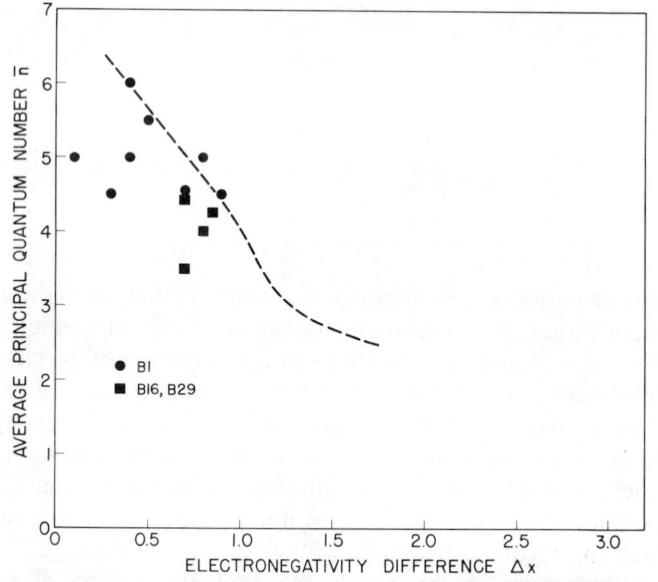

Figure 5-24. $\bar{n}$ *vs.* Δx diagram for MX compounds formed by cations with an unshared pair of *s* electrons:[17] ●, rocksalt structure; ■, GeS structure.

which have the rocksalt structure, also lie well within the region of directionally bonded phases on the $\bar{n}$ *vs*. Δx diagram. This is not surprising as they are not normal valence compounds, generally having variable nonstoichiometric compositions, their occurrence being governed by geometrical factors (see p. 52).

A theoretical study by Phillips and Van Vechten[20] of chemical bonding in *MN* valence compounds with four and six coordinated structures leads to the relationship $E_g^2 = E_h^2 + C^2$ where E_g is the energy gap between the bonding and antibonding states, E_h ($\propto a^{-2.5}$) is the part of the energy gap due to covalent bonding and C that part due to ionic change transfer, the values being derived from spectra of the *MN* crystals. The value of C for a compound *MN* amounts to a spectroscopically defined electronegativity difference, although the value is obtained only in terms of the

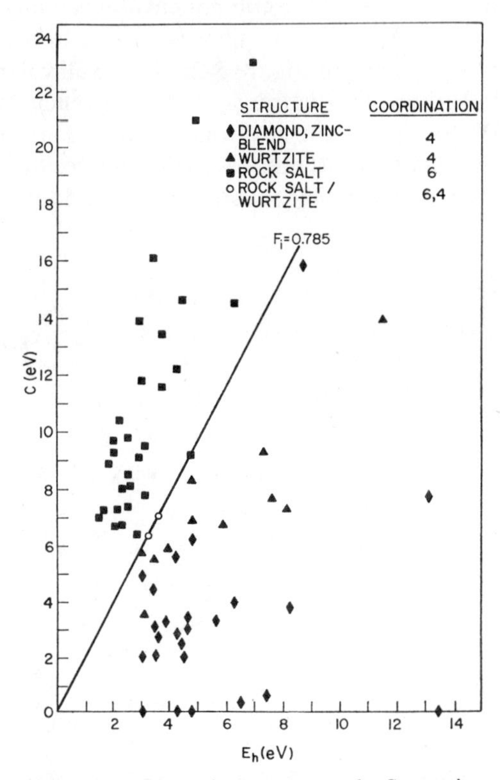

Figure 5-25. *MN* valence compounds: Separation of compounds with CN 6–6 from those with CN 4–4 on a diagram of the spectroscopically defined covalent and ionic energy band gaps E_h and C. (Phillips [20].)

compound MN, and not of the component atoms. A spectroscopic ionicity can then be defined as $f_i = C^2/E_g^2$, and a covalent fraction of the bond as $f_c = E_h^2/E_g^2$. Phillips has shown that there is a unique separation of MN valence compounds with tetrahedral and octahedral coordination on a diagram of C *vs.* E_h (Figure 5-25). The straight line which passes through $E_h = C = 0$ and corresponds to an ionicity value $f_i = 0.785 \pm 0.010$, separates compounds with the rocksalt structure from those with the sphalerite and wurtzite structures. Figure 5-25 demonstrates the success that can be achieved in predicting crystal structures when more rigorous criteria are applied than the phenomenological parameters used by Mooser and Pearson. However, these criteria have not yet been applied to more complex structural arrangements, and so the discussion continues in terms of the parameters $\bar{n}$ and Δx.

If the electronic character of the component atoms is taken into account, then a more detailed separation of phases can be achieved on a $\bar{n}$ *vs.* Δx diagram than that shown in Figure 5-23, where all valence compounds are examined together. In Figure 5-26, for instance, normal valence compounds, MX, formed by the A Group cations alone are shown. Now the regions of sphalerite, wurtzite, rocksalt, and cesium chloride structures are separated, and a similar separation is found for compounds of B

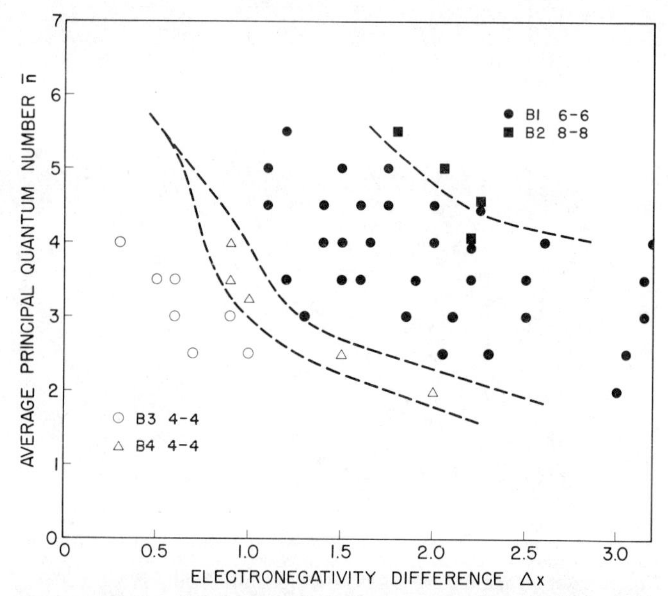

Figure 5-26. $\bar{n}$ vs. Δx diagram for nontransition metal MX compounds formed by A Group cations only:[17] ●, rocksalt; ■, CsCl; ○, sphalerite; △, wurtzite structures.

Group cations considered alone. The four structural types found along any line of constant $\bar{n}$ are, with increasing Δx value, in the order of increasing Madelung constants of the structures. The cesium chloride structure of normal valence compounds lies to the right of, and above the rocksalt structure (high Δx and $\bar{n}$), in accordance with the higher coordination of the ions and higher Madelung constant. Metallic compounds with the CsCl structure have not been included in Figure 5-23, since they are not normal valence compounds, but they locate in the region of high $\bar{n}$ and relatively low Δx, as shown in Figure 5-27, where all CsCl type compounds are plotted, together with the NaTl and ternary Heusler-alloy superstructures. Compounds with the CsCl structure, like those with the fluorite structure shown in Figure 5-35 and the CdI_2 structure (Figure 5-33), cover a very wide range of $\bar{n}$ and Δx values because the high coordination satisfies both metallic and ionic structures. Basically, the CsCl structure must be regarded as a metallic type because of the large number of metallic compounds which have the structure, but very little is known in detail about the electrical conductivity of many of these compounds. It seems probable that there must be a region of $\bar{n}$ and Δx values where semiconductors occur on passing from the metallic to ionic CsCl structure compounds, as in compounds with the fluorite structure. Indeed, the position of the phase CsAu which has semiconducting properties, calls for some comment. CsAu with $\bar{n} = 6$, $\Delta x = 1.5$, lies in a region of Figure 5-27 which would definitely indicate that the compound

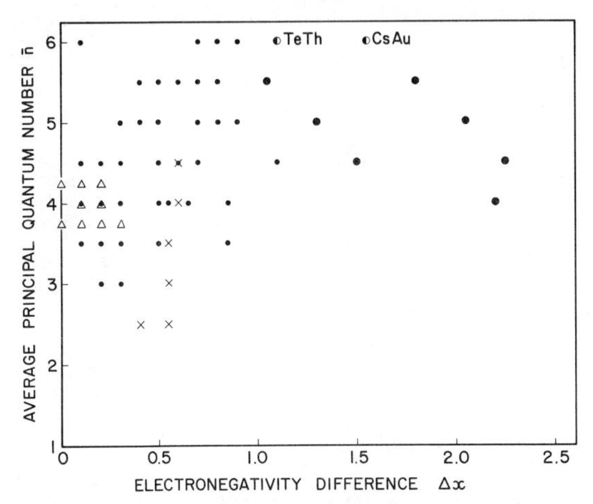

Figure 5-27. $\bar{n}$ *vs.* Δx diagram for all compounds (metallic, ●; semiconductor (?), ●; ionic, ●) with the CsCl type structure. Superstructures of the CsCl type: ×, NaTl; △, Heusler alloy structures. (Mooser and Pearson [17].)

did not have metallic properties. If its semiconducting properties had not already been discovered at the time the phase was first reported, a diagram such as Figure 5-27 would certainly indicate that the substance was probably not metallic, and that it should be examined for the occurrence of semiconductivity.

The transition metals are sometimes found in tetrahedral coordination in the inverse oxide spinels, but they normally prefer octahedral coordination, although the spherical symmetry of the half-filled *d* shell containing five unpaired electrons permits tetrahedral coordination in divalent Mn and trivalent Fe compounds. Thus tetrahedrally coordinated Fe^{III} is known in several compounds with the chalcopyrite structure, and there are metastable forms of MnS and MnSe which have the sphalerite and wurtzite structures. In general, however, the *MX* compounds of the transition metals adopt the hexagonal NiAs (*hP*4) structure, or the distorted MnP (*oP*8) form, when the $\bar{n}$ and Δx values are such as to suggest directional bonding. Figure 5-28 shows an $\bar{n}$ *vs.* Δx diagram of *MX* valence compounds of the transition metals. The line which separates compounds with the NiAs or MnP structures so effectively from the non-directionally bonded rocksalt structure is drawn in exactly the same position as the line in Figure 5-23. The distorted MnP structures lie at

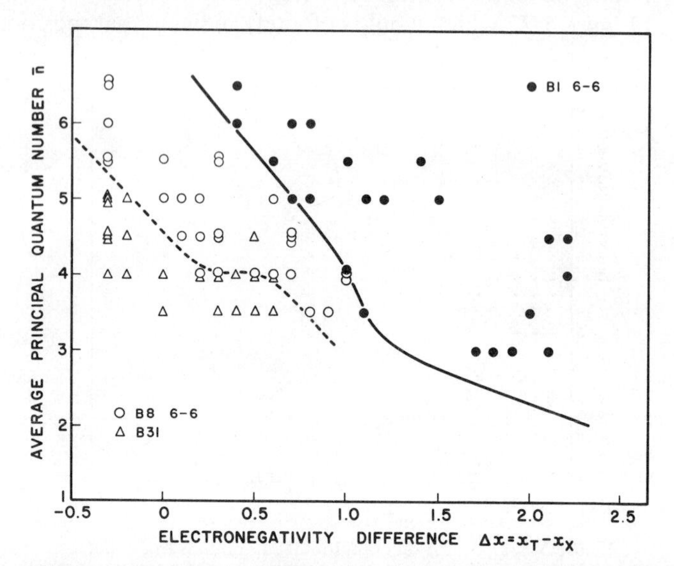

Figure 5-28. $\bar{n}$ *vs.* Δx diagram for transition metal *MX* normal valence compounds with the ●, rocksalt; ○, NiAs; △, MnP structures. (Mooser and Pearson [17].)

lower $\bar{n}$ and Δx values than the undistorted NiAs phases, which may be taken as an indication that the distortion arises from directional bonding.

There are several reasons, apart from avoiding tetrahedral coordination, why the transition metal compounds adopt the NiAs structure. In the first place when ionicity is relatively low and directional bonding may be important; the octahedral coordination of the transition metal atoms by the anions satisfies the requirements of their directional bonding in compounds such as FeS, whose axial ratios are close to the ideal value of 1.63. Secondly, the 8–6 coordination which arises because of the close approach of the transition metal atoms along the direction of the c axis of the crystal when the axial ratio, c/a, has low values, is favorably high for a metallic phase, and the stability of the structure is no doubt increased by bonding between the transition metal atoms along the c direction. Thirdly, the structure can conveniently minimize internal stresses, such as arise from antiferromagnetic ordering, without loss of crystal symmetry, merely by an adjustment of the axial ratio. Fourthly, the ionic radius ratio R_C/R_A for a T metal compound with a relatively high Δx value may be lower than the value of 0.414 required for the rocksalt structure to be stable. In this case it takes the NiAs structure with an expanded axial ratio.

Fairly ionic compounds with $\Delta x > 1$ prefer the rocksalt structure in which the octahedra of anions surrounding the cations only share edges, rather than the NiAs structure where the anion octahedra share faces normal to the c axis, bringing the cations closer together along [001] so that cationic repulsions may be important. For example the stable forms of MnS and MnSe with $\Delta x = 1.1$ and 1.0, respectively, have the rocksalt structure and the Ti and V chalcogenides (TiS, TiSe, VS, VSe, and perhaps CrS, CrSe) with similar $\bar{n}$ and Δx values would also be expected to have the rocksalt structure rather than the NiAs type. However, as Table 5-21 shows, the anions would be in contact in the rocksalt structure of these phases because the ionic R_C/R_A values are less than the limiting ratio of 0.414. The compounds are therefore forced to adopt the nickel arsenide structure. Nevertheless, in order to avoid strong cationic repulsions, the axial ratio of the NiAs structure in these compounds is increased considerably over the ideal value of 1.63 as shown in Table 5-21. The resulting structure is not very favorable for a compound with relatively high Δx value, as it is not in accord with Laves' principle of closest packing.

Only the six most common structures of normal valence MX compounds have been discussed above and, although there are a number of more rare types such as the PtS, CuS, TlF, CuO, PdS, and TlSe structures, these mostly lie in the region where directional bonding is strong

Table 5-21

Metal	Δx for MS	Δx for MSe	MO: Lattice Parameter (Å) for NaCl Structure	M^{2+} Ionic Radius[a] (Å)	Ionic R_M/R_S	Ionic R_M/R_{Se}	c/a for NiAs Structure MS	MSe
Ti	1.2^+	1.1^+	4.180	0.69	0.375	0.35	1.93	1.77
V	1.2^+	1.1^+	4.093	0.65	0.35	0.33	1.75	1.62
Cr	1.1	1.0	4.09[b]	0.62	0.34	0.32	1.64	1.63
Mn	1.1	1.0	4.444_5	0.82	0.44	0.414	NaCl structure[c]	

[a]Calculated with $R_{O^{2-}} = 1.40$ Å.
[b]CrO is probably unknown in the solid state. Estimated value.
[c]NaCl structure stable for $R_C/R_A > 0.414$.

in an $\bar{n}$ *vs.* Δx diagram. The special atomic arrangements found in these structures may therefore be attributed to the particular requirements of directional bonding, or to the ligand-field effects discussed by Dunitz and Orgel[21] on the ionic model.

MX_2 Structures with Anion Pairs. Both the calcium carbide and pyrite structures can be regarded as derived from the rocksalt structure, anion

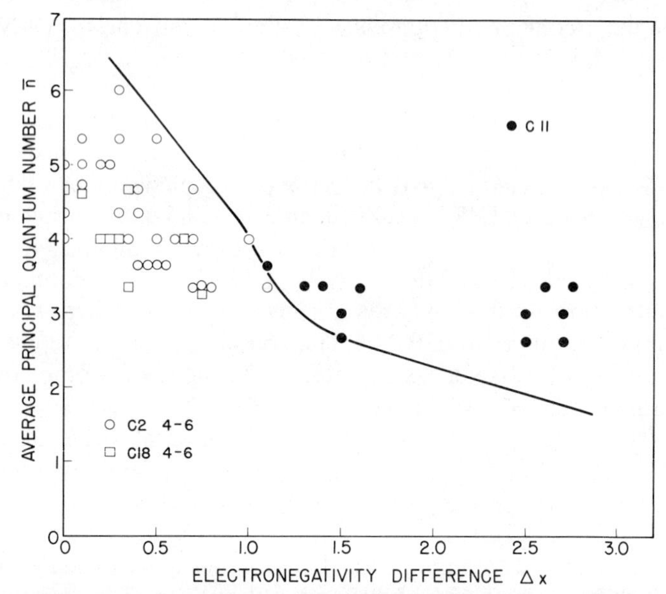

Figure 5-29. $\bar{n}$ *vs.* Δx diagram for MX_2 normal valence compounds containing anion pairs and having the calcium carbide, ●; pyrite, ○; marcasite, □ structures.[17]

pairs replacing the single anions of the rocksalt structure. Anion pairs also occur in the marcasite structure which is related to the pyrite structure as discussed on p. 412. In the ionic calcium carbide structure the bonds between the cations and anion pairs are largely nondirectional, whereas the bonds in the pyrite and marcasite structures are largely covalent and directional in character. It is interesting to find that these three structure types have a rather similar mutual relationship to that of the NiAs, MnP, and NaCl structures on the $\bar{n}$ *vs.* Δx diagram, as shown in Figure 5-29. The line separating the directionally and nondirectionally bonded phases in Figure 5-29 is identical with that drawn in Figure 5-23 or 5-27.

Anion-Rich Compounds MX_2, M_2X_3, MX_3. The change from open to more densely packed structures with increasing $\bar{n}$ and/or Δx values, which is apparent in the structures of MX phases with change from tetrahedral to octahedral coordination, is also found in anion-rich phases with compositions MX_2, MX_3, and M_2X_3. This is shown in Figures 5-30 to 5-32 where the structures lying to the right of the solid line all have high coordination numbers and, with the exception of the $PbCl_2$ and UCl_3 types, they are all based on a close-packed array of cations in whose

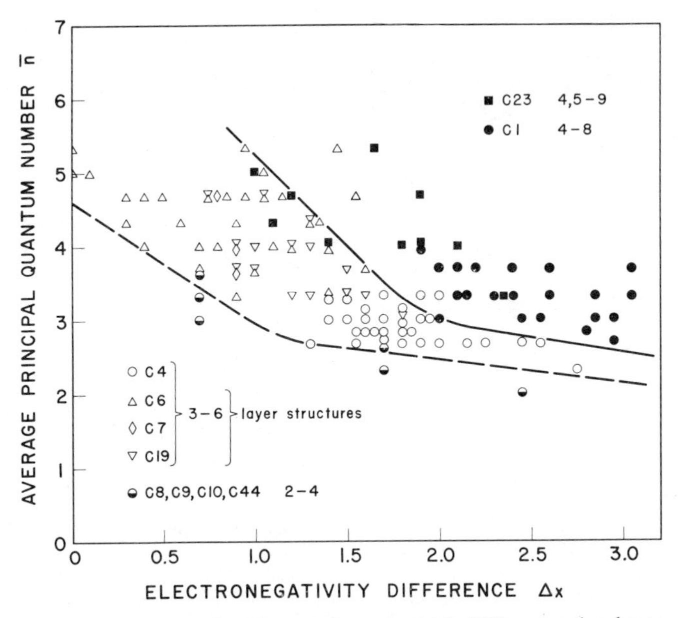

Figure 5-30. $\bar{n}$ *vs.* Δx diagram for anion-rich MX_2 normal valence compounds: ■, $PbCl_2$; ●, fluorite; ○, rutile; △, CdI_2; ◊, MoS_2; ▽, $CdCl_2$; ◓, GeS_2 and silica structures (Mooser and Pearson[17].)

tetrahedral and/or octahedral holes the anions lie. Because the number of anions exceeds that of the cations in phases with these structures, the degree of filling is rather high and densely packed structures are obtained which fulfil Laves' principles for the structures of substances with nondirectional bonding (Table 5-22). In contrast to this, most of the structures lying to the left of the solid line of Figures 5-30 to 5-32 are based on a close-packed array of *anions*, the smaller number of cations in these anion-rich compounds leading to a low degree of filling and open structures characteristic of phases with directional chemical bonds (Table 5-23). From the position of the structures of different anion-rich phases on the $\bar{n}$ *vs.* Δx diagrams, it follows that the directional character of the chemical bonds, or its absence, determines whether the

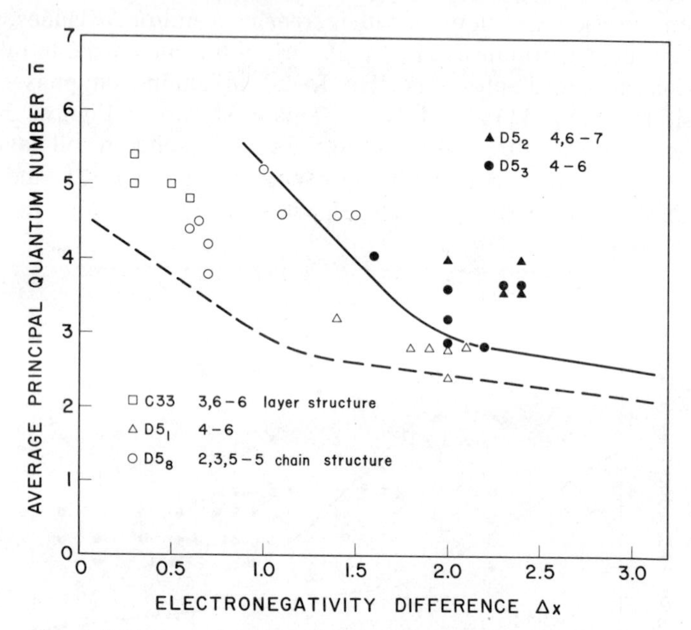

Figure 5-31. $\bar{n}$ *vs.* Δx diagram for anion-rich M_2X_3 normal valence compounds. ▲, La_2O_3; ●, Mn_2O_3; □, Bi_2STe_2; △, α-Al_2O_3; ○, Sb_2S_3 structures. (Mooser and Pearson [17].)

Table 5-22 Densely Packed Structures

Compound	MX_2		M_2X_3		MX_3	
Close-packing of cations	Cubic	Hex.	Cubic	Hex.	Cubic	Hex.
Structure type	CaF_2	—	Mn_2O_3	La_2O_3	BiF_3	LaF_3

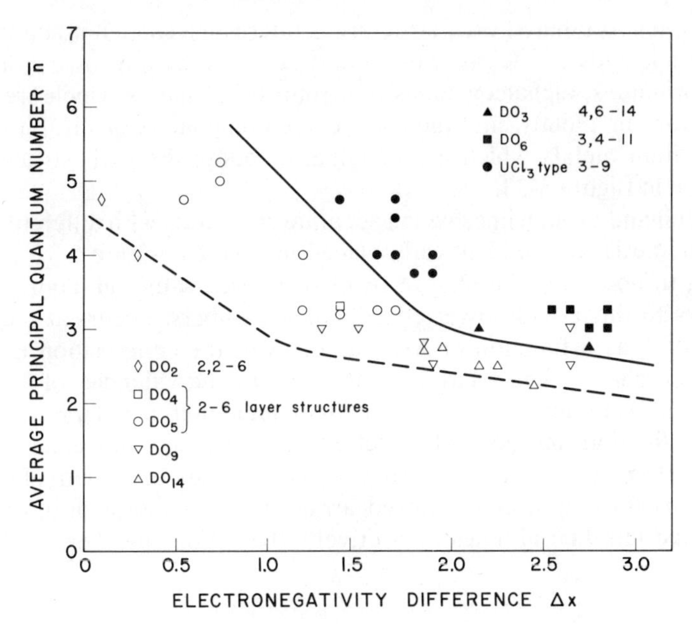

Figure 5-32. $\bar{n}$ *vs.* Δx diagram for anion-rich MX_3 normal valence compounds. ▲, BiF_3; ■, LaF_3; ●, UCl_3; ◊, $CoAs_3$; □, $CrCl_3$; ○, BiI_3; ▽, ReO_3; △, AlF_3 structures. (Mooser and Pearson[17].)

Table 5-23 Open Structures

Compound	MX_2		M_2X_3		MX_3	
Close-packing of anions	Cubic	Hex.	Cubic	Other	Cubic	Hex.
Structure type	$CdCl_2$	CdI_2	Al_2O_3	Bi_2Te_3	$CrCl_3$	BiI_3 AlF_3

anions or the cations lie on a close-packed sublattice. Dense packing required for nondirectionally bonded phases occurs only when the anions lie in the holes in a close-packed array of cations in anion-rich compounds.

Although in the more open structures the cations generally occupy the octahedral holes in the close-packed array of anions, the tetragonal structure of HgI_2 is an exception in which $\frac{1}{4}$ of the tetrahedral holes in the close-packed-cubic iodine sublattice are occupied by Hg. The 2–4 coordination to which this leads, is normally only met in the very open framework structures of SiO_2 and GeS_2, and the chain structure of SiS_2, which occur at the low $\bar{n}$ values to the left of the broken line in Figure 5-30.

The 3–6 coordinated CdI_2 structure is interesting since it is suitable for compounds with a wide range of properties. Thus, weakly ionic di-iodides and dibromides, dichalcogenides of Group IV elements which are semiconductors or metals, and dichalcogenide compounds of the Ni Group of transition metals which are metallic, all adopt the CdI_2 structure as indicated in Figure 5-33.

The full and broken lines which separate structures with different bonding characteristics are identically placed in Figures 5-30 to 5-32 and it is striking to note how the change from structures with high coordination numbers to those with lower coordination numbers, occurs at the same values of $\bar{n}$ as a function of Δx, regardless of the composition of these anion-rich phases. The relative n and x values and valencies of the component atoms in anion-rich valence compounds therefore largely specify the coordination and general structural arrangement. For example, the Hahn and spinel structures of MN_2X_4 compounds are relatively open, the anions forming a close-packed array in whose tetrahedral, or octahedral and tetrahedral holes respectively, the cations lie. Ternary anion-

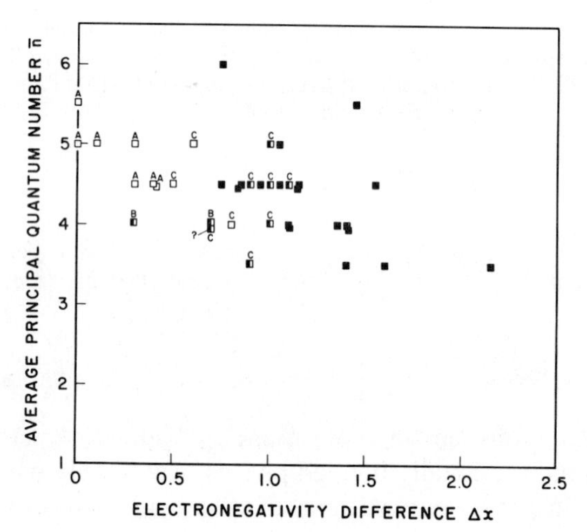

Figure 5-33. Compounds with the CdI_2 ($hP3$) structure shown on a $\bar{n}$ *vs.* Δx diagram: □, compounds known to be metallic; ◧, semiconductors; ■, compounds presumed to be ionic. Symbols marked *A*: S, Se, and Te compounds of Co and Ni Group transition metals. B: $SiTe_2$, SnS_2, $SnSe_2$. C: S, Se, and Te compounds of Groups IV and V transition metals. Unmarked: ionic compounds of OH^-, Cl^-, Br^-, and I^-. (Mooser and Pearson [17].)

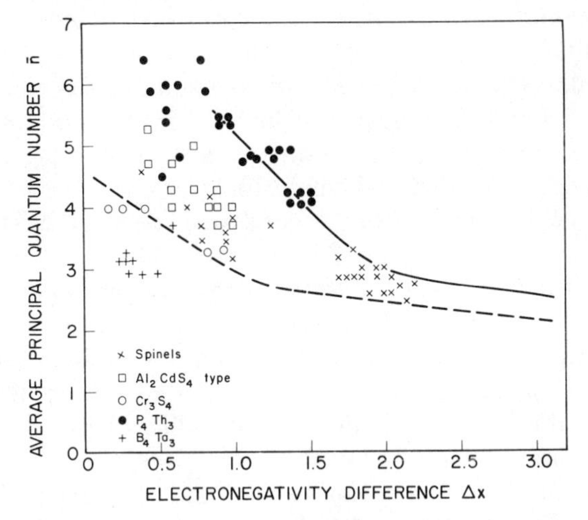

Figure 5-34. $\bar{n}$ *vs.* Δx diagram for some anion rich M_3X_4 and MN_2X_4 compounds. Structure types are indicated on the diagram.

rich chalcogenide and oxide phases with these structures are all found to have $\bar{n}$ and Δx values appropriately placing them between the full and broken lines of Figures 5-30 to 5-32 as shown in Figure 5-34. Also included in Figure 5-34 are various M_3X_4 phases. Those with the Th_3P_4 ($cI28$) structure in which the atoms have higher 8–6 coordination, generally lie above the spinels as expected, although the spread in their distribution increases at lower Δx values when the phases are certainly metallic. The few phases included with the Cr_3S_4 ($mC14$) structure which has 6–4 or 5 coordination comparable to that of the spinels, lie slightly below them. Phases with the Ta_3B_4 ($oI14$) structure which has the highest coordination of all, 16 or 12–9, lie well below the other phases, indicating that they do not belong to the same class of valence compounds as the other phases considered in Figure 5-34. Indeed, they are metallic interstitial phases, and it is doubtful whether boron should be included in the calculation of $\bar{n}$. If boron is ignored, then the phases lie well up in the region expected for metallic phases.

Cation-Rich Compounds, M_2X, M_3X_2, M_3X. In the structures of cation-rich compounds the anions can be regarded as forming close-packed arrays. As this is the characteristic of the directionally bonded anion-rich MX_2 phases, it might be imagined that the same condition indicates directional bonding in the cation-rich compounds also, but this is not necessarily so. Because the cations here outnumber the anions, the structures resulting from filling tetrahedral and octahedral centers in the anion sublattices, are densely packed and characterized by high coordination numbers, so that they are also suitable for compounds with nondirectional bonding. Thus in Figure 5-35 the antifluorite structure of the M_2X compounds (with the anions occupying the calcium, 0, 0, 0, sites) occurs over the whole range of $\bar{n}$ and Δx values. Although the structure of these M_2X compounds does not depend on $\bar{n}$ and Δx, their physical properties do, and accordingly metals, semiconductors, and ionic conductors are found in much the same regions as for compounds with the CsCl and CdI_2 types of structure discussed above. The cation-rich ternary compounds with an ordered antifluorite-like structure are expected to be semiconductors when the cations occupy the tetrahedral interstices of the anion sublattice, but metallic when they occupy half of the tetrahedral and the octahedral holes (see p. 204). Both types of ternary phase are included in Figure 5-35 and it is seen that the metallic phases are rather sharply separated from the semiconducting phases!

In general the most common structures adopted by the cation-rich valence compounds are the anti-types of the densely packed structures listed in Table 5-22. Thus the M_3X_2 compounds adopt the anti-La_2O_3, anti-Mn_2O_3 and closely-related Zn_3P_2 types, whereas the M_3X compounds favor the anti-BiF_3 or Na_3As (essentially the antitype of LaF_3)

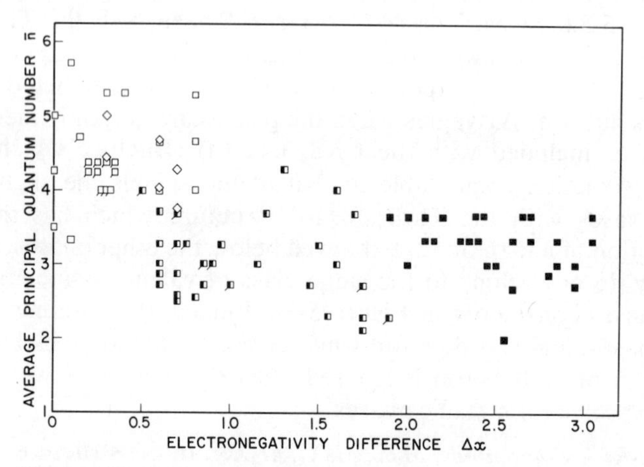

Figure 5-35. M_2X compounds with the anti-fluorite (and fluorite, MX_2) structure shown on a $\bar{n}$ *vs.* Δx diagram. Metallic phases: □, binary; □, ternary. Semiconductors: ◧, binary; ◧, ternary (M_2X). Ionic phases: ■, M_2X and MX_2. ◊, Rahlfs-type M_2X phases with partial occupation of octahedral interstices. (Mooser and Pearson[17].)

structures. These structure types also tend to spread over the whole range of possible $\bar{n}$ and Δx values because on the one hand they are densely packed, and on the other, their coordination configurations are well suited for directional bonding.

Summary

Consideration of the changes in the directional character of the bonds in a series of similar normal valence compounds can lead to a better understanding of the factors governing the choice of structure, than a discussion based on ionic structures and the degree of polarization, since it relates more naturally metals, covalent and ionic compounds. The change from directionally bonded covalent substances to those fulfilling Laves' geometrical principles of the best filling of space and the highest symmetry and coordination, follows naturally with increase of metallic or ionic character, when the discussion is founded initially on the directional properties of bonds as expressed in terms of $\bar{n}$ and Δx. Discussion of the distribution of structural types based initially on the ionic model is more limited in scope since the principal parameter, the ionic radius ratio, may lose its effectiveness when Laves' geometrical principles no longer hold. Although in the past there may have been a tendency to over-estimate the importance of ionic bonding in many solids of simple structure, the more recent developments of the ionic ligand field theory in accounting for structural features, suggest that ionic models cannot be ignored completely in considering the structures of many alloys. Finally, the work of Phillips and collaborators shows that the ionicity of a valence compound can be explicitly gauged from its optical properties in such a way that structure type can be accurately predicted—at least to the level of coordination number.

Ideally, structural distribution should be examined in three-dimensional space with $\bar{n}$, Δx, and R_M/R_X along the three coordinate axes. However, the least important of these in metals is the radius ratio, and Figures 5-23 to 5-35 indicate that this can be ignored without significant loss of definition, accepting the convenience of dealing with two dimensions only. It is only when R_M/R_X has extreme values, as in the case of ZnO discussed on p. 246, or Ta_3B_4 phases discussed on p. 259, that the approximation of considering only $\bar{n}$ and Δx becomes insufficient.

Although a sharp separation of metals and semiconductors is sometimes observed on $\bar{n}$ *vs.* Δx diagrams, this is not the general rule and the criteria discussed on p. 202 must be used to separate metals specifically from semiconductors. There is no sharp separation of semiconductors and ionic conductors; both are ideally insulators at the absolute zero of temperature and the above observations indicate that the electronegativity difference is not the only parameter controlling the occurrence of ionic or electronic conductivity in a nonmetal.

SPACE FILLING RELATIONSHIPS

In the previous section where the influence of the directional character of chemical bonds on the distribution of structure type was considered, geometrical characteristics, which might have been represented by radius ratio, R_C/R_A, were largely ignored. The space filling concept introduced by Laves[15] and extended by Parthé[22] provides a means of studying the influence of geometrical size effects on structures of ionic compounds although, since it is based on a hard sphere model of the atom, it is not suitable for metals. These are better examined with near-neighbor diagrams (pp. 52–68) based on a compressible atom model, which relate the structural dimensions of actual phases to the geometrical characteristics of the structures, rather than comparing the different degrees of space filling of different structures as the Laves model does. In the Laves–Parthé model the space filling factor, ψ, is defined as the ratio of the volumes of the atoms in the cell, considered as hard spheres of known radius, to the observed unit cell volume. Thus

$$\psi = \frac{4\pi/3 \,.\, \Sigma_{ij}\, nR_i^3 + mR_j^3 + \cdots}{U} \tag{6}$$

where n, m, ... are the numbers of atoms of types i, j, ... and radii R_i, R_j, ... in the unit cell of volume U. The variation of ψ for the ratios R_i/R_j, ... permitting atomic contacts i-i, i-j, j-j, ... can be calculated and compared for various structures.

For a binary system MN, the cell edge can be expressed as a function of the radii R_M and R_N for the special cases of M–M, M–N, and N–N contacts, thus enabling ψ to be given in terms of the ratio $\epsilon = R_M/R_N$; the variation of ψ with ϵ is represented on diagrams with log scales. For ionic compounds the occurrence of M–M or N–N contacts between like ions renders the structure unstable because of ionic repulsions, so that the space filling curves have a special significance for ionic structures. This significance escapes the structures of metals because there are no unique radii, R_M, R_N, associated with M–M, M–N, and N–N contacts Indeed, assuming appropriate R_M and R_N values for metals, it is often found that the ψ *vs.* ϵ points for particular metallic phases lie above the space filling curves for M–M, M–N, and N–N contacts (Figure 5-36), indicating a denser packing and emphasizing the lack of unique radii associated with M–M, etc. contacts. For example, the space

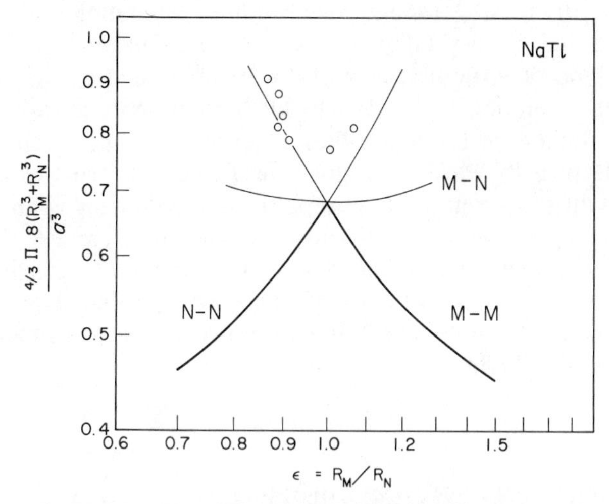

Figure 5-36. Space filling diagram for the NaTl structure: ○, phases with the NaTl structure, comparing the observed cell volume with the calculated volume.

filling value for the b.c. cubic ($A2$) structure (0.68) is distinctly less than that of the f.c. cubic ($A1$) structure (0.74), yet as shown on p. 140, there is generally less than 1% difference in atomic volume of metals in the $A1$ and $A2$ structures. For ionic compounds on the other hand, because ions have more or less fixed sizes, the ψ *vs.* ϵ points of actual compounds generally lie close to the space filling curve for M–N, contacts (Figure 5-37).

Providing that there are no variable atomic positional parameters or

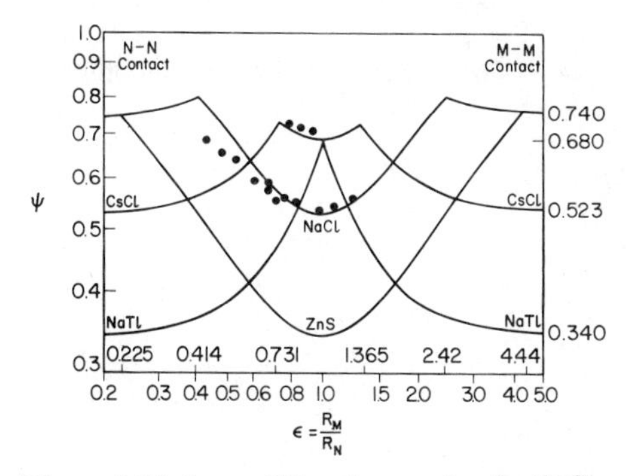

Figure 5-37. Space filling diagram for the CsCl, NaCl, and ZnS structures. (Parthé [22].) Points represent actual ionic phases with the NaCl and CsCl structures.

axial ratios, there is a unique space filling parameter, independent of atomic size, for the crystal structures of the elements. Thus for the f.c. cubic $A1$ type of structure, $\psi = (4 \times 4/3\pi R^3)/a^3$ and as the geometrical relationship assuming the atoms to be hard spheres gives $a = 2\sqrt{2}.R$, $\psi = 0.740$. Values of the space filling factor for various elemental structures so derived by Parthé are given in Table 5-24. In two component systems, with the same assumptions, three relationships can be derived for ψ which depend on the conditions for M–M, M–N, and N–N contacts. The relationship between ψ and ϵ for a particular structure is the envelope of these three ψ functions which gives the lowest ψ value at any value of ϵ.

Consider, for example, the NaCl structure where $\psi = [4 \times 4/3\pi(R_M^3 + R_N^3)]/a^3$. For M–M contacts $2\, R_M = a/\sqrt{2}$, therefore

$$\psi = \frac{16/3\pi\ (R_M^3 + R_N^3)}{16\sqrt{2}R_M^3} = \frac{\pi}{3\sqrt{2}} \cdot \frac{(\epsilon^3 + 1)}{\epsilon^3}; \tag{7}$$

for M–N contacts $R_M + R_N = a/2$, therefore

$$\psi = \frac{16/3\pi\ (R_M^3 + R_N^3)}{8(R_M + R_N)^3} = \frac{2\pi}{3} \cdot \frac{(\epsilon^3 + 1)}{(\epsilon + 1)^3}, \tag{8}$$

and for N–N contacts $a = 2\sqrt{2}\, .\, R_N$, therefore

$$\psi = (\pi/3\sqrt{2})(\epsilon^3 + 1). \tag{9}$$

Relationships (7) and (8) intersect at $\epsilon = 0.414$ and (8) and (9) at $\epsilon = 2.42$, giving the variation of ψ *vs.* ϵ shown in Figure 5-37, taken from Parthé. This figure also gives data for the ZnS and CsCl structures, and the inflection points in the ψ curves for each reflect the geometrical fact that anions come in contact at ϵ values of 0.225, 0.414, and 0.731, respectively, for the ZnS, NaCl, and CsCl structures, so that the structures are not expected

Table 5-24 Space Filling Values of Elemental Structures

Element	Structure Type	Space Filling Value	Element	Structure Type	Space Filling Value
Cu	$cF4$	0.740	Po	$cP1$	0.523
Mg	$hP2, c/a = 1.63$	0.740	Bi	$hR2, c/a = 2.60$	0.446
Zn	$hP2, c/a = 1.86$	0.650	Sb	$hR2, c/a = 2.62$	0.410
Pa	$tI2$	0.696	As	$hR2, c/a = 2.80$	0.385
In	$tI2$	0.686	Ga	$oC8$	0.391
W	$cI2$	0.680	Te	$hP3$	0.364
Hg	$hR1$	0.609	Diamond	$cF8$	0.340
Sn	$tI4$	0.535	Black P	$oC8$	0.285
α-U	$oC4$	0.534			

to be stable at ϵ values lower than these. This information is of course derived simply from consideration of the anion coordination polyhedra and recourse to space filling curves is quite unnecessary. Figure 5-37 also shows that above $\epsilon = 0.59$ the CsCl structure offers a higher space filling than the NaCl type, and it is interesting to note that under the influence of high pressure, NaCl structures of alkali halides with ϵ values less than 0.731 have been transformed to the CsCl type structure. Since they do not take account of relative free energies, space filling curves do nothing more than to indicate which structure may be regarded as more probable for a particular compound. Calculations by Parthé for the ionic case, suggest that the NaCl structure is energetically favored over the CsCl structure at all ϵ values, so that it should be stable throughout the region where ψ is controlled by M–N contacts. This is no doubt the reason why several alkali halides with $\epsilon > 0.731$ have the NaCl rather than the CsCl structure. Indeed, it has long been recognized that the radius ratio criteria only specify the values below which a particular ionic structure *cannot* be stable. When ϵ exceeds this critical value the structure type is possible, but whether it is adopted depends on relative electrostatic considerations appropriate to the particular compound. In the case of the NaTl structure which is a superstructure based on the CsCl arrangement, the ψ curves for M–M and N–N contacts intersect each other and that for M–N contacts at $\epsilon = 1$, so that this type of ordering would not be expected for ionic compounds, and indeed the only known compounds with the NaTl structure are metals.

Space filling curves for a binary compound have these general forms:

$$\text{for } M\text{–}M \text{ contacts} \qquad \psi = G_{M-M} \cdot \frac{(m\epsilon^3 + n)}{\epsilon^3} \tag{10}$$

$$\text{for } M\text{–}N \text{ contacts} \qquad \psi = G_{M-N} \cdot \frac{(m\epsilon^3 + n)}{(\epsilon + 1)^3} \tag{11}$$

$$\text{and for } N\text{–}N \text{ contacts} \qquad \psi = G_{N-N} \cdot (m\epsilon^3 + n) \tag{12}$$

where G_{M-M}, G_{M-N}, and G_{N-N} are geometrical factors depending on the particular structure and the other terms involve only the ratio $\epsilon = R_M/R_N$ and the numbers, m and n, of M and N atoms in the unit cell. From this it is apparent that compounds with the same composition have space filling curves differing only by a factor independent of ϵ, and so have the same shape for the three contacts, being severally shifted parallel to each other along the ψ axis on a diagram of ψ *vs.* ϵ on log scales. Equating (10) and (11) gives the point, ϵ_M, of intersection of the curves for M–M and M–N contacts. Similarly equating (11) and (12) gives ϵ_N, the intersection of curves for N–N and M–N contacts. ϵ_M–ϵ_N gives the range of ϵ values where M–N contacts occur for a particular structure, so that if $\epsilon_N > \epsilon_M$, there is no range of ϵ values where M–N contacts occur in the structure, and the M–M and N–N curves intersect at ϵ_x given by the solution of equations (10) and (12). If $\epsilon_N = \epsilon_M$, M–N contacts occur together with M–M and N–N contacts only at this particular radius ratio, as in the NaTl structure.

It is also apparent from equations (10) to (12) that the substructure due to each component fixes the values of ψ at $\epsilon = 0$ or $\epsilon = \infty$ (depending on whether the component is called M or N) and the ψ value is identical for all identical substructures and for elements with the same structure. Thus $\psi = 0.74$ (at $\epsilon = 0$ or ∞ appropriately) for Cu and for the f.c. cubic $A1$ substructures of Na, Cl, Zn, S, and Ca in NaCl, ZnS, and CaF_2, respectively; $\psi = 0.5023$ for the simple cubic substructure of Cs, Cl, and F_2 in CsCl and CaF_2, respectively, and for elemental Po, and $\psi = 0.34$ for the diamond cubic substructure of Na, Tl, and Mg in NaTl and Cu_2Mg, respectively, and for elemental diamond (Figure 5-37). Furthermore, when the substructures due to the two components are identical as are Na and Cl in NaCl, Zn and S in cubic ZnS, or Cs and Cl in CsCl, then the ψ curves are symmetrical about the value $\epsilon = 1$ (Figure 5-37). Structures of compounds with the same formulas which have the same nearest-neighbor coordination and interatomic distance ratios, d_{M-M}/d_{M-N} and d_{N-N}/d_{M-N} (i.e., polytypes), have identical space filling curves, even though the crystal systems may be different. Thus, cubic ZnS and the ideal wurtzite ZnS ($c/a = 1.633$) have the same space filling.

Derivation of space filling diagrams for complex structures with many interatomic distances is difficult and some suitable system of averaging is necessary. Figure 5-38 shows such curves for the CsCl, FeSi, FeB, CrB,

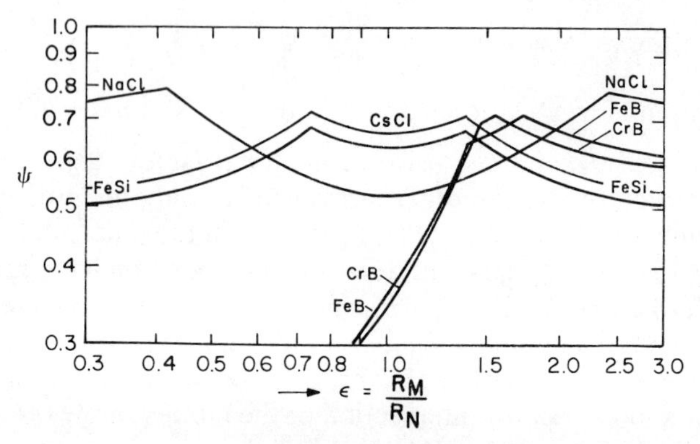

Figure 5-38. Space filling diagram for the CsCl, FeSi, FeB, CrB, and NaCl structures. (Parthé [22].)

and NaCl structures (taken from Parthé) common to MN compounds in which M–N interactions are important. Although most phases with these structures are metallic, they form an interesting series in which electron concentration increases in order from the CsCl or FeSi, to the FeB or CrB, to the NaCl structure. The approximate limits for which M–N contacts can occur can be read from the diagram:

	ϵ_N to ϵ_M
FeSi and CsCl	0.73–1.36
CrB	1.46–1.53
FeB	1.4–1.7
NaCl	Up to 2.4

Comparing these values with the covalent radius ratio of phases with the structures, it appears that these MN phases tend to choose the structure that provides the highest degree of space filling, provided that M–N contacts can be maintained at the particular radius ratio. Thus no phases with $\epsilon > 1.3$ are found to have the FeSi structure, and phases with the highest radius ratio take the NaCl structure. Indeed, Hägg observed that when R_X/R_M was less than about 0.59, simple interstitial structures of the NaCl or WC type occurred in preference to complex structures. This ratio, which is $1/\epsilon$, corresponds to an ϵ value of about 1.7 and coincides with ϵ_M for the FeB structure. Thus the reason for Hägg's rule is apparent, since it states that complex structures are to be expected at ϵ values smaller than 1.7. Furthermore, a condition for forming structures with M–N contacts at high ϵ values is obviously that the substructure due to the M component should have a high space filling value. Since the highest ψ values are obtained with the face centered cubic and ideal close packed hexagonal structures, it is not surprising that these frequently form the metal substructure in the Hägg interstitial compounds.

Although it may be possible to discuss the space filling in metallic structures in relation to that of ionic structures in this purely geometrical manner, it must not be forgotten that the geometrical space filling factor ψ for the structure, may have very little to do with the actual space filling achieved by a metallic phase which adopts the structure, and this is why geometrical considerations of space filling in the structures of metallic phases are unprofitable. The only practical criterion of relative space filling by a metallic phase is $\psi = (\Sigma_{ij}\, nV_i + mV_j + \cdots)/U$, the comparison of the unit cell volume, U, with the sums of the atomic volumes of the components nV_i, mV_j, ... , preferably determined from f.c. cubic or ideal hexagonal close-packed elemental structures. Such comparisons are essentially studies of Zen's law (see p. 174).

TRANSITION METAL COMPOUNDS

Crystal structure, ligand fields, spin-orbit coupling, and magnetic interactions are closely connected in semiconducting compounds containing transition metal atoms. The interactions depend on the numbers of *d* electrons and the degree of degeneracy of their orbitals, which in turn depend on the anion coordination polyhedra and their relative configurations, since these determine the extent of possible cation–anion–cation or cation–cation interactions. Much of the understanding of these compounds comes from a study of oxides, so these must be considered together with the transition metal chalcogenides and pnictides. One general principle seems to pervade the crystal chemistry of these compounds: The transition metal endeavors by one means or another to achieve a filled valence subshell of *d* electrons, the t_{2g}^6 subshell (octahedral coordination) being the most stable. There are many examples of this behavior; one will serve here: $IrSe_2$ with the marcasite or pyrite structure would have a d^7 configuration; instead it achieves a distortion of the marcasite structure in which Se–Se bonds occur only between alternate pairs of Se atoms (Figure 8-48, p. 415) with resulting d^6 configuration of the Ir atoms.

Both magnetic properties and the type of electrical conductivity are a source of information on interactions and *d* electron configuration occurring in these compounds, and therefore on expected distortions of crystal structure. Alternatively, observed structural distortions give information on probable *d* electron configurations and interactions. Octahedra which share corners permit only indirect cation–anion–cation interaction, the strength of this being favored by a C–A–C bond angle of 180°.[23] Octahedra which share edges or faces permit direct interaction between the cations centering the octahedra. Such interactions may lead to conducting wave-functions for the *d* electrons or to covalent pairing and structural distortion.[24] This occurs particularly when the *d* electron sharing removes a degeneracy in the lower-lying t_{2g} orbitals, and the *T* metal atoms are displaced from the centers of the surrounding octahedra, distinguishing the distortions from those of the Jahn–Teller and spin-orbit coupling type, where the *T* metal remains at the center of the distorted octahedra.

Whether or not the *d* electrons are localized has important consequences on the properties of transition metal compounds. Neglecting the case of delocalization that arises from overlap of partially filled *d* states with the valence band, the localized or collective state of the *d* electrons depends on the *T–T* distance, the spatial distribution of the *d* orbitals in the local crystal field, and particularly on the number of *d* electrons. As a result of the Madelung interaction, the spatial extent (radius) of any valence electron shell decreases rapidly as the number of electrons in it increases from zero to the number filling the shell. Thus, when the number of *d* electrons is small, they are likely to be collective and when it is large (say 6 to 10), they are likely to be localized. The number of *d* electrons on the transition metal can be determined either from an ionic or a covalent bond model, since both give the same result.

Ligand Field Effects: The Ionic Model

Electrostatic repulsion of the *d* electrons by the surrounding negative ions, causes splitting of the energies of the *d* levels. Important in both structural consequence and electron transport, are the number of *d* electrons and the energy of splitting of the *d* levels, Δ, because these determine the *d* electron distribution. The requirement of exchange forces that as many electrons as possible shall have parallel spins (Hund's rule) is in conflict with Δ, the separation of the degenerate *d* levels. A large value of the crystal field splitting means that the lowest over-all electron energy is obtained in spin-paired ("low-spin") configurations, whereby the lower degenerate levels are filled before the higher degenerate levels, whereas a small value of Δ means that both lower and higher levels are first filled in spin-parallel ("high-spin") configurations before spin pairing in the lower level occurs. Thus in the first place the electron configuration depends on the crystalline field; secondly, it depends on the atomic number (Period) of the transition metal atom, since a given crystalline field causes a larger splitting of the *d* levels of a third long Period transition metal, whose *d* electrons are further separated and screened from the nucleus, than of a first long Period transition metal. Thus low-spin configurations are expected in second and third long Period transition metals, whereas high-spin configurations may occur in first long Period *T* metals with weak crystalline fields. Ligands of ionic type O^{2-}, F^-, etc., in octahedral coordination generally give rise to high-spin weak splitting of the *d* electron levels in the first long transition Period, whereas "nonionic" dipolar ligands such as CN^-, NH_3 (not usually H_2O) give spin-paired strong-field splitting of the *d* levels. In the compounds of S, Se, Te, P, As, and Sb, although the ligand fields are weak, low-spin configurations frequently occur, and it appears that other factors such as spin-orbit coupling also influence the distribution of the *d* electrons.

The understanding of transition metal compounds has been considerably enhanced through the development of the electrostatic model or ligand-field treatment by the work of Griffith, Orgel, Dunitz, Nyholm, Ballhausen, Jörgensen, and others while Dunitz and Orgel[21] and Orgel,[25] particularly, have been responsible for setting out the principles and the results in simple terms, and reference should be made to these works for further information. Very briefly, the electrostatic field acting on a transition metal ion situated in an octahedral coordination polyhedron of anions causes a splitting of the energy of the d orbitals into two groups: a triplet t_{2g} and a doublet e_g. As indicated in Figure 5-39, showing the rocksalt structure, the t_{2g} orbitals (d_{xy}, d_{yz}, d_{zx}) point to the centers of the edges of the coordination octahedron where the electrostatic field is low. They are accordingly stabilized and have lower energy than the e_g orbitals ($d_{x^2} - d_{y^2}$ and d_{z^2}) which point directly to the anion ligands, and are thus energetically unfavored because of repulsion. In a tetrahedral electrostatic field due to four anions situated at the corners of a tetrahedron surrounding the transition metal ion, a similar splitting of the energies of the d orbitals occurs, but now the t_2 orbitals have higher energies than the e orbitals since they point to regions of slightly higher charge, as indicated in Figure 5-40.

The (approximate[26]) integral electron configurations for d electrons in octahedral and tetrahedral ligand fields are summarized in Tables 5-25 and 5-26.

Mechanisms of Structural Distortion

Jahn–Teller Effect. Within the framework of the ionic ligand field model, the Jahn–Teller[27] theorem may be stated as follows: The degen-

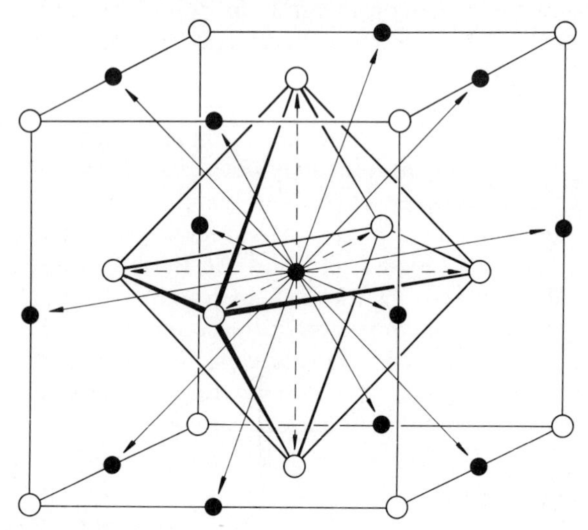

Figure 5-39. Rocksalt structure. Diagram indicating the directions of t_{2g} (full lines with arrows) and e_g (broken lines with arrows) orbitals of d electrons for a T metal atom in an octahedral anion ligand field.

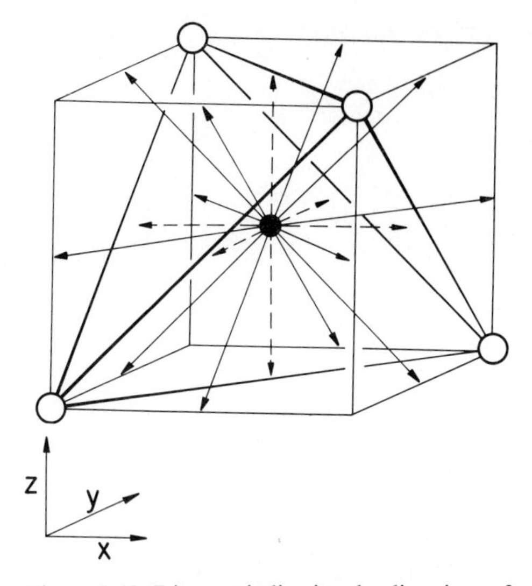

Figure 5-40. Diagram indicating the directions of t_2 (full lines with arrows) and e (broken lines with arrows) d electron orbitals of a T metal atom in a tetrahedral anion ligand field.

Table 5-25 Arrangement of *d* Electrons in Octahedral Ligand Fields

Number of *d* Electrons	High-Spin Weak-Field Arrangement					Low-Spin Strong-Field Arrangement				
	t			*e*		*t*			*e*	
1	↑			—		↑			—	
2	↑	↑		—		↑	↑		—	
3	↑	↑	↑	—		↑	↑	↑	—	
4	↑	↑	↑	↑		↑↓	↑	↑	—	
5	↑	↑	↑	↑	↑	↑↓	↑↓	↑	—	
6	↑↓	↑	↑	↑	↑	↑↓	↑↓	↑↓	—	
7	↑↓	↑↓	↑	↑	↑	↑↓	↑↓	↑↓	↑	
8	↑↓	↑↓	↑↓	↑	↑	↑↓	↑↓	↑↓	↑	↑
9	↑↓	↑↓	↑↓	↑↓	↑	↑↓	↑↓	↑↓	↑↓	↑

eracy of the ground state of a cation† may be removed with resulting crystal stabilization, by distortion of the anion coordination polyhedron to lower symmetry. In the first place therefore, the effect must be large enough to dominate other influential factors controlling the crystal struc-

†Providing that this is not a Kramer's doublet.

Table 5-26 Arrangement of *d* Electrons in Tetrahedral Ligand Fields

Number of *d* Electrons	High-Spin Weak-Field Arrangement		Low-Spin Strong-Field Arrangement	
	e	*t*	*e*	*t*
1	↑	—	↑	—
2	↑ ↑	—	↑ ↑	—
3	↑ ↑	↑	⇅ ↑	—
4	↑ ↑	↑ ↑	⇅ ⇅	—
5	↑ ↑	↑ ↑ ↑	⇅ ⇅	↑
6	⇅ ↑	↑ ↑ ↑	⇅ ⇅	↑ ↑
7	⇅ ⇅	↑ ↑ ↑	⇅ ⇅	↑ ↑ ↑
8	⇅ ⇅	⇅ ↑ ↑	⇅ ⇅	⇅ ↑ ↑
9	⇅ ⇅	⇅ ⇅ ↑	⇅ ⇅	⇅ ⇅ ↑

ture if it is to occur. Thus lattice vibrations (temperature) and spin-orbit effects must influence whether, and at what temperature, such distortions occur. Secondly, the degree of anion–cation interaction is important. In octahedral coordination, for instance, ground-state degeneracy occurs for d^1, d^2, d^4, d^5, d^6, d^7, and d^9 high and/or low spin configurations, but it is only in high spin configurations d^4 ($t_{2g}^3 e_g^1$) and d^9 ($t_{2g}^6 e_g^3$) that the distortions are well recognized (distortions are also expected to be found in the low-spin d^7, ($t_{2g}^6 e_g^1$) configuration), where the degeneracy occurs in e_g orbitals which interact strongly with the anions.

In tetrahedral coordination, Jahn–Teller distortions would be more probable in the high spin d^3, d^4, d^8, and d^9 configurations where the degeneracy concerns the t_2 levels which react most strongly with the anions, than in degeneracies which involve the *e* orbitals. However, since in tetrahedral coordination no *d* orbitals are directed immediately towards the anions as in the case of octahedral coordination, the tendency to Jahn–Teller type distortion is expected to be less, and secondly, for crystal distortion to occur, the effects must be large enough for one of the equiprobable *x*, *y*, or *z* axes to become unique in any given coordination polyhedron. It is presumably only when the distortions are of sufficient magnitude for interaction between neighboring ions, that a cooperative distortion of the crystal as a whole occurs. If the *z* axis is assumed to be unique, then it is seen from Table 5-27 that the d_{xy} orbital electrons repel the ligands more strongly than any electrons in d_{zx} and d_{yz} orbitals for d^3 and d^8 configurations, thus tending to give structural distortion with $c/a > 1$. For d^4 and d^9 configurations, it is the d_{zx} and d_{yz} orbitals which contain the most electrons, and hence repel the ligands most strongly,

Table 5-27

Configuration	Occupation of Orbitals by t_2 Electrons			
	d_{zx}	d_{yz}	d_{xy} (Unique)	
d^3	—	—	↑	$c/a > 1$
d^8	↑	↑	⇅	$c/a > 1$
d^4	↑	↑	—	$c/a < 1$
d^9	⇅	⇅	↑	$c/a < 1$

causing a tendency for structural distortion with $c/a < 1$. Figure 5-41 taken from Orgel,[25] shows the splitting of the energy levels expected for tetragonal distortion of structures containing d^8 Ni^{2+} and d^9 Cu^{2+} ions in tetrahedral coordination.

Jahn–Teller distortions are well recognized in octahedral T-metal coordination with d^4 or d^9 configuration. For example, the $(t_{2g})^6(e_g)^3$ configuration of the Cu^{2+} ion in tenorite CuO ($mC8$)† is orbitally degenerate with respect to the e_g orbitals. Suppose that two electrons occupy the d_{z^2} orbital and one is missing from the $d_{x^2-y^2}$ orbital pointing in the xy plane, the lessened nuclear screening in this plane then results in a stronger attraction of the anions by the copper nuclear charge than in the z direction. Hence the bond lengths in the x and y directions are shortened compared to those in the z direction, and the coordination octahedron distorts in the structure of tenorite. The greater nuclear screening in the z direction compared to the x and y directions leads to a stabilization and consequent lower energy of the d_{z^2} orbital relative to the $d_{x^2-y^2}$ orbital

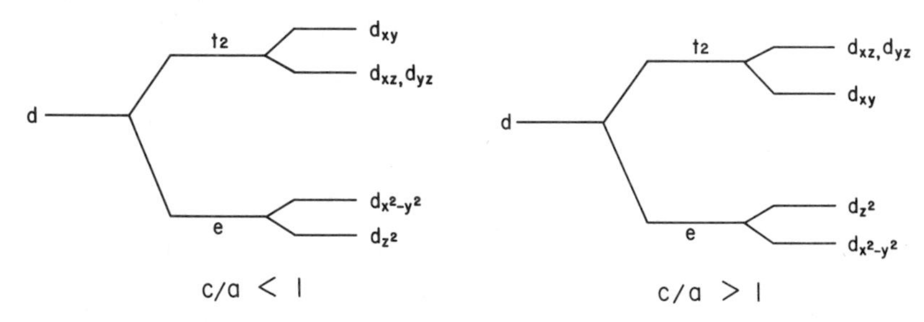

Figure 5-41. Splitting of the d orbital energy levels in a regular tetrahedral field, and then (right side of diagrams) in a tetrahedral field distorted by 'flattening' (left, $c/a < 1$) or "elongation" (right, $c/a > 1$) of the tetrahedron along an S_4 axis. (Orgel [25].)

†Dunitz and Orgel[21] believe that the square planar coordination of oxygen atoms about Cu in tenorite should be regarded as a limiting case of distortion of an octahedral environment to give 4 short and 2 long bonds, rather than as a consequence of dsp^2 covalent bonding.

concomitant with its full occupation. There is also a corresponding splitting of the t_{2g} orbitals as indicated in Figure 5-42. The converse situation in which the hole occurs in the d_{z^2} orbital should lead to a distortion with two short and four longer bonds. However, in recognized Jahn–Teller distortions of this type, four short and two long bonds are commonly found.[28]

An example of Jahn–Teller distortion in the high spin d^4 configuration is found in CrS (*mC*8), whose monoclinic structure giving 4 Cr–S = 2.43 Å and 2 Cr–S = 2.88 Å, is a distortion of the NiAs type structure in which the Cr atoms would be surrounded octahedrally by the six S atoms.

A characteristic feature of Jahn–Teller distortion of a transition metal compound is that the *T*-metal remains at the center of symmetry of the distorted anion coordination polyhedron. Such distortions are therefore quite distinct from those due to *T–T* covalent bonding in which generally the *T*-metal atom moves from the center of symmetry of the surrounding anion polyhedron (p. 276). Secondly, the shortening or lengthening of two of the bonds relative to the other four, which splits the degeneracy of the *d* levels, is also a characteristic feature of the Jahn–Teller distortion of octahedral coordination. The process is therefore quite different, say, to that of squashing or lengthening of the anion coordination octahedron in a compound with the nickel arsenide structure, when the axial ratio differs from the ideal value of 1.63. In this case the six ligands remain at equal distances from the central T-metal atom.

Jahn–Teller distortion is also found in the nonmetallic d^9 compounds CuF_2, $CuCl_2$, and $CuBr_2$, and high-spin d^4 compounds of Mn^{3+}, MnF_3, MMn_2O_4, and $HMnO_2$, but not in metallic compounds of Mn^{3+} such as MnAs, MnSb, and MnBi with the nickel arsenide structure. Evidently

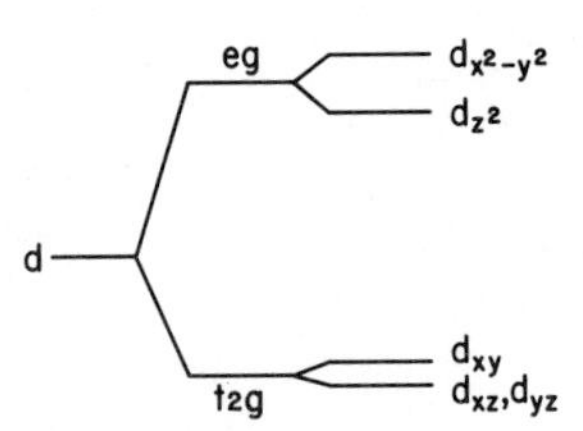

Figure 5-42. Splitting of the *d* orbital energy levels in a regular octahedral field, and then (right side of diagram) in a distorted octahedral field (extension along *z* axis, compression in *xy* plane). (Orgel [25].)

here the ligand fields are too weak, and spin-orbit coupling energies are comparable to those that could be derived by Jahn–Teller distortion removing the degeneracy of the e_g electrons.

The normal (AB_2X_4) and inverse ($B(A, B)X_4$) spinels provide numerous examples of Jahn–Teller distortion because the transition metal atoms can occupy either tetrahedrally coordinated positions when d^3, d^4, d^8, or d^9 configurations may lead to Jahn–Teller distortions, or octahedrally coordinated positions when d^4 or d^9 configurations may lead to structural distortions in the high-spin arrangement. Ligand field effects can influence whether a T-metal ion adopts octahedral or tetrahedral coordination, since the stabilization energy depends on the number and configuration of the d electrons, that for tetrahedral coordination being only about 40% of that for octahedral coordination. Derived octahedral and tetrahedral crystal field stabilization energies for various T-metal ions in oxygen environments, agree rather well with the observed distribution of M^{2+} ions in spinels of various types, as noted by Dunitz and Orgel[21] from whom Tables 5-28 and 5-29 are taken.

Spin-Orbit Distortions. Spin-orbit coupling may be important when the orbital angular momentum is not quenched by the ligand fields of the undistorted crystal. As Goodenough (1963)[24] points out, whereas Jahn-Teller distortions stabilize nondegenerate orbitals ordering the electron into a state without contribution to angular momentum, spin-orb coupling stabilizes a degenerate level. Jahn–Teller and spin-orbit coupling give distortions of opposite sign as indicated in Figure 5-43 taken from Goodenough, and hence there is a means of distinguishing between them

Table 5-28 Crystal-field Stabilization Energies (Kcal/mole) Estimated for Transition-Metal Oxides[21]

Ion	Octahedral Stabilization	Tetrahedral Stabilization	Excess Octahedral Stabilization
Mn^{2+}	0	0	0
Fe^{2+}	11.9	7.9	4.0
Co^{2+}	22.2	14.8	7.4
Ni^{2+}	29.2	8.6	20.6
Cu^{2+}	21.6	6.4	15.2
Ti^{3+}	20.9	14.0	6.9
V^{3+}	38.3	25.5	12.8
Cr^{3+}	53.7	16.0	37.7
Mn^{3+}	32.4	9.6	22.8
Fe^{3+}	0	0	0

Table 5-29 Distribution of Cations Between Octahedral and Tetrahedral Sites in Spinels[21]

	Values of δ^a						
	Mg^{2+}	Mn^{2+}	Fe^{2+}	Co^{2+}	Ni^{2+}	Cu^{2+}	Zn^{2+}
Aluminates	0	0	0	0	0.76		0
Chromites	0	0	0	0	0	0.1	0
Ferrites	0.9	0.2	1	1	1	1	0
Manganites	0	0	0.67(?)	0.67(?)	1	0	0
Cobaltites				0			0

$^a\delta = 0$ for normal spinels with M^{II} in tetrahedral sites
$\delta = 1$ for inverse spinels with M^{II} in octahedral sites
$\delta = 0.67$ for a random distribution of M^{II}.

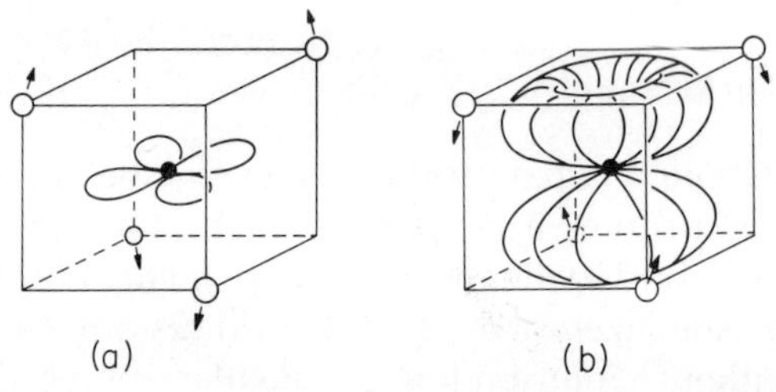

Figure 5-43. Distortions of a tetrahedral site due to the ordering of a single d electron. (a) Jahn–Teller ordering into a d_{xy} orbital. (b) Spin-orbit ordering into the $(d_{xy} \pm d_{zx})$ orbitals. (Goodenough (1963) [24].)

Thus the rhombohedral and tetragonal distortions in FeO and CoO below the Néel point, when the spins are colinear (atomic moments $\|[111]$ and $\sim \|[001]$, respectively), are of the spin-orbit coupling type.

Structural Distortion Arising from Covalent Bonds between Neighboring Cations. Cation–Cation interactions are possible when cation-occupied octahedra share edges since the t_{2g} orbitals are directed towards the centers of the edges, or when they share faces (hybridized d_{xy}, d_{yz}, d_{zx} directed through faces), but not when they share corners because of the intervening anions. In this case cation–anion–cation interactions are expected and these are strongest when the angle C–A–C is 180° and weakest when it is 90°. For high-spin states, when the number of d electrons, m, is equal to or less than 3, C–C interactions may be stronger than C–A–C interactions, but the latter dominate when $4 \leqslant m \leqslant 8$

with e_g levels degenerate. When $m = 4$, either Jahn–Teller distortion occurs, or the e_g orbitals appear to overlap the conduction or valence bands giving metallic conductivity (and ferromagnetism) as in Mn arsenide, antimonide, and bismuthide.

When the T–T distance exceeds the critical diameter, D_c, for collective electrons, C–C interactions are weak; when it is less than D_c, they are strong. Antiferromagnetic correlation between electrons on neighboring octahedral-site cations stabilizes the binding between them, since it permits the maximum charge in the region of overlap of the t_{2g} orbitals because of the antiparallel spins. This may be the situation with half or less filled t_{2g} orbitals, whereas ferromagnetic interaction is expected for interacting t_{2g} orbitals which are degenerate and more than half filled. When there are 5 or less t_{2g} electrons the degeneracy may not be maintained, and near-neighbor covalent bonding and loss of spin moment may occur. From considerations such as these, Goodenough (1960)[24] derived rules for the properties of compounds in which direct C–C interactions occur. Examples of structural transformations and distortions resulting from strong C–C interactions are taken from his works and from those of Hulliger and Mooser.[29] Interactions in oxides are discussed, both because in some cases they are metallic conductors, and because interactions occurring in dioxides resemble those in some dipnictides of the transition metals.

In the equiatomic transition metal oxides and nitrides with the NaCl structure ($cF8$) for which $m \leqslant 3$ and C–C interactions are expected to be of comparable strength to C–A–C interactions, transitions are recognized only in TiO, VO, and CrN. TiO transforms to a monoclinic structure below about 900°C and is metallic at all temperatures. VO transforms at $T_N = 114$°K on cooling, becoming an insulator-semiconductor. CrN becomes antiferromagnetic below 0°C, the structure distorting as indicated in Figure 5-44. Cation spins are arranged ferromagnetically in (110) planes and antiferromagnetically along [110] with the sequence $++--++--$. The structural distortion is such as to move cation planes with opposite spin closer and those with like spin further apart. Of the 12 like-neighbors of each cation, four come closer, four move away and four stay at the same distance. This distortion suggests covalent bonding via the d_{yz} and d_{zx} orbitals as indicated in Figure 5-44. The electron pairing is by no means complete, and in any case the remaining electrons in d_{xy} orbitals of d^3 CrN account for its metallic conductivity above and below the transformation.[24] In contrast, since d^3 VO is an insulator below T_N, it is suspected that all three t_{2g} orbitals in VO are used in covalent bonding giving effectively a filled t_{2g}^6 subshell.

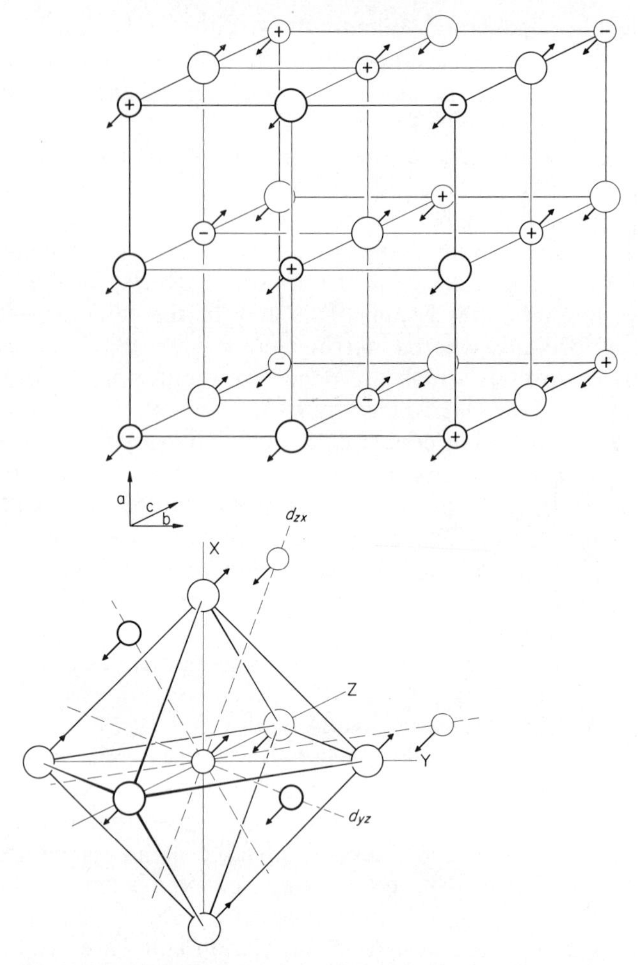

Figure 5-44. Rocksalt structure of CrN, indicating directions of spin-ordering and atomic displacements during transformation. (Goodenough (1960) [24].)

In the corundum structure (*hR*10) (Figure 8-80, p. 451) the cations occupy octahedral interstices in the close packed cubic array of anions in such a way that pairs of occupied octahedra share faces in the [001] direction of the hexagonal cell, and edges in the basal plane (Figure 5-45). Strong *C–C* interactions between pairs of cations are expected along [001], and *C–C* interactions can also occur in the basal plane. The angle for *C–A–C* interactions is about 135°. Formation of covalent bonds between pairs of cations along [001] can occur independently, causing

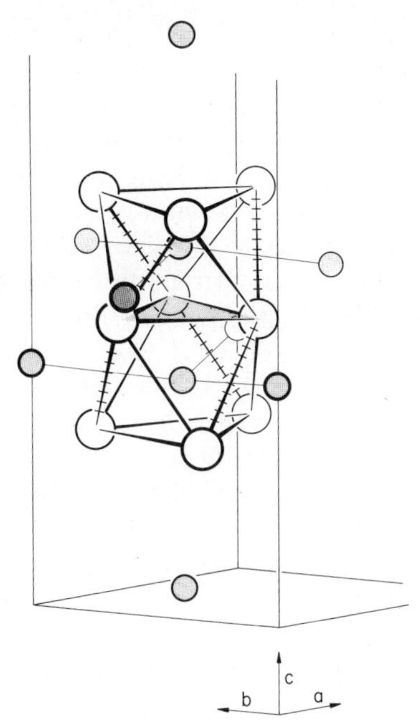

Figure 5-45. Portion of corundum structure showing basal plane and c axis T–T neighbors. Octahedra share a face parallel to (001), and edges marked by bars between basal-plane neighbors.

a change of axial ratio rather than a change of crystal symmetry, so that the transformation is non-cooperative, whereas basal plane bonding would be expected to occur at an abrupt cooperative transition. Both Ti_2O_3 and V_2O_3 with the corundum structure exhibit metallic properties at high temperature and semiconductor-insulator behavior at lower temperatures. The transformation in Ti_2O_3 which occurs in the neighborhood of 200°C, has the features of a non-cooperative transition. There is a gradual decrease of axial ratio with decreasing temperature, second-order type change of specific heat and a loss of magnetic moment, the low temperature phase having nearly temperature-independent magnetic susceptibility. This behavior is quite consistent with covalent pairing of the d^1 electrons of Ti_2O_3 between neighboring pairs of Ti atoms along the c axis, as also is the change of the low-temperature semiconductor phase back to a metallic conductor, by increasing the d electron concentration above one per cation through substitution. The extra d electrons enter a band running throughout the basal plane.

V_2O_3 with two d electrons should show two transitions: at high temperatures a non-cooperative one, as in Ti_2O_3, due to covalent bond formation between pairs of V atoms along [001], and at lower temperatures an abrupt transition due to the weak C–C interactions in the basal plane, the remaining d electron being paired in covalent bonds and causing crystal distortion and insulating properties. Goodenough[24] accounts well for the behavior of V_2O_3, since specific heat measurements indicate that a second-order transformation occurs between about 530° and 380°K on cooling. The conductivity is still metallic below the transformation because of the remaining d electrons in the bands running through the basal plane (weaker C–C interactions). A first-order transition to a monoclinic form occurs sharply at 150°K on cooling and 168°K on heating. The structure is said to be compatible with distortions expected for the d electron covalent bonding between V atoms in the basal plane, and the discontinuous decrease of electrical conductivity by many orders of magnitude and the almost complete loss of magnetic moment below the transformation, also point to the complete pairing of the d electrons in covalent cation–cation bonds.

In α-Fe_2O_3 with high spin d^5 configuration, the C–A–C interactions are definitely expected to be stronger than C–C interactions, so it is not surprising to find that it is a semiconductor-insulator with a high Néel temperature of 948°K. Cr_2O_3 is also a semiconductor-insulator with a lower Néel temperature of about 305°K resulting from weak antiferromagnetic C–A–C and C–C interactions, but it seems by no means certain why basal plane C–C interactions should apparently be so weak in this substance, since Cr has only three d electrons. The relevant C–C distances and the critical diameters, D_C, for these corundum compounds are given in Table 5-30. Because the C–C pairs along [001] are isolated, metallic interactions between them do not give overall

Table 5-30 Compounds with Corundum Structure

		C–C Distance (Å)		
Compound	d Configuration	Between Pairs along [001]	Between Atoms in Basal Plane	D_C for Localized d Band. Calc. by Goodenough
Ti_2O_3	t_{2g}^1	2.59	2.99	3.05
V_2O_3	t_{2g}^2	2.70	2.88	3.02
Cr_2O_3	t_{2g}^3	2.65	2.89	2.99
α-Fe_2O_3	$t_{2g}^3e_g^2$	2.89	2.97	2.93

metallic conductivity; there must be a conducting band formed between cations in the basal plane for this to occur, which means that for Ti_2O_3 $D_C > 2.99$, for V_2O_3 $D_C > 2.88$, but for Cr_2O_3 $D_C < 2.89$ Å and hence its insulating properties.

The rutile (*tP*6) and marcasite (*oP*6) structures are in many respects similar as noted below, and a similar distortion of both structures is found which sets the *T* atoms alternately close together and far apart along the direction of strings of anion octahedra which share edges (Figure 8-50, p. 417). The Ti atoms in rutile form a b.c. tetragonal sublattice ($c/a = 0.64$) which results in chains of nearest-neighbor Ti atoms running in the [001] direction. Stoichiometric TiO_2, having no *d* electrons, is an insulator; more interesting is d^1 VO_2 which is metallic above 340°K with a magnetic moment approaching that of one unpaired electron at high temperatures. Below 340°K it distorts monoclinically; there is a loss of magnetic moment and it becomes a semiconductor. The main feature of the structural distortion is the pairing of V atoms along the [001] direction, so that the V–V distances alternate between 2.65 and 3.12 Å, the behavior being consistent with covalent pairing of the d^1 electrons between neighboring V pairs.[30] The displacement of the V atoms from the center of symmetry of the anion coordination polyhedra clearly distinguishes the distortion from that of the Jahn–Teller type. d^2 MoO_2, WO_2, and d^3 TcO_2 and ReO_2 are isostructural with the monoclinic form of VO_2, the close *T–T* distances indicating the establishment of multiple bonds between pairs of *T* atoms (Table 5-31). d^2 CrO_2 and d^3 MnO_2, on the other hand, have the normal rutile structure, CrO_2 being metallic and MnO_2 a semiconductor. In the latter the *C–C* and weak antiferromagnetic *C–A–C* interactions are roughly comparable, leading to a more complex magnetic ordering than in rutile MnF_2 ($t_{2g}^3\, e_g^2$) where strong *C–A–C* interactions are dominant.

Table 5-31

	Alternating Distances (Å) along Cation Strings in the MoO_2 Structure ([001] direction of undistorted rutile structure)		Interatomic Distance in Pure Metal (Å)
VO_2	2.65	3.12	2.618
MoO_2	2.50	3.10	2.725
WO_2	2.49	3.08	2.741
TcO_2	(2.48)	(3.06)?	2.703 2.735
ReO_2	(2.49)	(3.08)?	2.741 2.760

The pnictides and chalcogenides with the marcasite structure are much less ionic than the oxides with the rutile structure. The essential difference between the two structures, apart from a change from tetragonal to orthorhombic symmetry, is a rotation of the anion octahedra surrounding the cations in marcasite, so that pairs of anions move towards each other and a bond occurs between them (see Figure 8-50, p. 417). The octahedra still share edges in (001) planes so that direct C–C interactions can occur along the strings of cations in the [001] direction.

TX_2 compounds with anion pairs X_2^{2-} and X_2^{4-} and d^6 configurations generally (though not invariably) adopt the pyrite structural arrangement, but those with low spin d^2, d^4, or d^5 configurations take the marcasite structure, or a distorted form thereof. In d^5 marcasites the cations lying in [001] strings are displaced towards each other in pairs, indicating the formation of covalent bonds between the unpaired t_{2g}^5 d electrons, whereby each cation completes a t_{2g}^6 subshell of increased stability. In agreement with this interpretation, the magnetic moment of the compounds formed by the second and third long Period T-metals is essentially zero. The moment found for compounds of the first long Period T-metals is intermediate between that of one and zero free spins, and the movement of the T atoms to form pairs is relatively less, indicating that the covalent pairing of the d electrons is incomplete. The T–T distances in several compounds with the $CoSb_2$ structure are compared with the T–T distances in the pure metals in Table 5-32. The compounds are all semiconductors and so, whether or not the pairing is complete, the longer T–T distances in the strings are greater than D_C. The T-atoms are clearly displaced from the centers of the anion octahedra (Figure 8-49), p. 416) so that the distortion cannot be of the Jahn–Teller type.

All TX_2 compounds with the t_{2g}^4 configuration (and also $CrSb_2$ with t_{2g}^2) take the marcasite structure, but it is severely compressed in the [001] direction, compared to the structures of compounds with t_{2g}^6 configurations. This results in a considerable shortening of the distances between

Table 5-32 *T–T* Metal Distances (in Å) in Some Compounds with the $CoSb_2$ Structure

Compound	Alternating T–T Distances in Chains ([001] of Marcasite)		T–T Distances in Pure Metal
$CoSb_2$	3.03	3.71	2.51
$RhSb_2$	3.04	3.90	2.69
$IrSb_2$	3.05	4.05	2.72
α-$RhBi_2$	3.24	3.97	2.69

T–T neighbors in the [001] direction as indicated in Table 5-33. The compression of the octahedra in [001] is indeed very strong, the shortening being up to 20% of the width expected from octahedra in d^6 compounds as Table 5-33 shows. The T atoms appear to be bonded together in the [001] chains, each T atom having two close neighbors. The T–T separations are very similar to those of the bonded pairs in the d^5 $CoSb_2$ compounds. An alternative account is that this is a Jahn–Teller type distortion with the destablized empty d_{xy} orbitals directed towards the T neighbors along [001] allowing their close approach.[29] However, no account of these structures is particularly satisfactory since it is now known that the T metal is not at the center of the anion coordination polyhedron, at least in $FeSb_2$ (also FeS_2, $FeTe_2$, and $CoTe_2$).[31] In the d^6 marcasites d_{xy}, d_{yz}, and d_{zx} orbitals are all filled so that there is no degeneracy and no Jahn–Teller distortion occurs. All of the d^2 and d^4 compounds with the compressed marcasite structure are semiconductors, and instead of a magnetic moment expected for two unpaired electrons, the moment is essentially zero, except perhaps for the $3d$ compounds, where the relative interatomic distances also indicate a somewhat incomplete pairing.

The compression of the d^4 marcasites along [001] also brings the anions into close contact in this direction. Indeed, in $FeSb_2$, for example, Sb–Sb distances along [001] are 3.20 Å. Since larger Sb–Sb distances of 3.37 Å between the double Sb layers in the elemental Sb structure cause metallic conductivity, the semiconductivity of these marcasites requires some explanation;[33] however, later studies call for still further evaluation.[34]

The NiAs (*hP*4) structure is complicated by the possibility of strong C–C interactions along the [001] direction where the anion octahedra share faces and in the basal plane where the octahedra share edges. However, the strongest C–C interactions would occur with any cations occupying the trigonal bipyramidal holes in the structure. In addition there are C–A–C interactions. The anion octahedra are frequently very much squashed along [001] (low c/a) and this strengthens the trigonal splittings of the d levels, which occur in addition to the octahedral splittings; NiAs phases with low axial ratios are invariably metallic. The ideal axial ratio 1.63 favors directed covalent bonds and any NiAs-type phases which are semiconductors have axial ratios close to this value. The most favorable Madelung constant for phases with ionic tendencies occurs at an axial ratio of 1.77,[35] and phases such as TiS which adopt the NiAs structure because the radius ratio, R_C/R_A, is too low to permit stability in the rocksalt structure, have axial ratios larger than the ideal value.

The only known Jahn–Teller distortion of a NiAs type structure occurs in d^4 CrS; the NiAs structure of other d^4 compounds such as MnAs,

Table 5-33 Comparison of Interatomic Distances (in Å) in Normal d_6 and Some "Compressed" d^6 and d^2 Marcasite Structures[32]

Compound	d Config.	Axial Ratios		T–$2T$ and X–$2X$ in [001]	T–$8T$	T–T Pure Metal	$2T$–X	$4T$–X	X–X
		b/a	c/a						
FeS_2	d^6	1.22	0.763	3.39	3.89_5	2.48	2.25	2.23	2.21
$NiAs_2$		1.218	0.744	(3.54)		2.49	2.34	2.40	2.45
$NiSb_2$		1.219	0.741	(3.84)			2.55	2.56	2.86
FeP_2	d^4	1.137	0.548	2.72	4.00	2.48	2.20	2.29	2.27
$FeAs_2$		1.129	0.544	2.88	4.25		2.33	2.44	2.41
$FeSb_2$		1.121	0.548	3.20			2.58	2.60	2.89
$RuAs_2$		1.139	0.547	2.97	4.37	{2.65 {2.71	2.52	2.45	2.36
$OsAs_2$		1.144	0.557	3.01	4.38	{2.68 {2.74	2.45	2.48	2.48
$CrSb_2$	d^2	1.140	0.543	3.27		2.50	2.75	2.69	2.84
Ortho-hexagonal $c/a = 1.63$		1.06	0.613						

MnSb, and MnSi is stabilized by spin-orbit coupling. Nevertheless, other distorted forms of the structure are known in the MnP, NiP, and FeS structures. The main distortion in the MnP structure is a movement of the T atoms from the centers of the distorted anion coordination polyhedra so that they attain two closer and two further T–T neighbors, rather than the 6 equidistant T–T neighbors in the NiAs structure. The T atoms thus form zigzag chains in the *bc* plane. In the NiP structure zigzag Ni chains are also formed and the Ni atoms move out of the center of the distorted anion octahedra, so that there is no question of either of these structural distortions being of the Jahn–Teller type.

REFERENCES

1. A. H. Wilson, 1931, *Proc. Roy. Soc.*, **A133**, 458.
2. H. Welker, 1952, *Z. Naturf.*, 7a, 744.
3. L. Pauling, 1960, *The Nature of the Chemical Bond*, 3rd ed, New York: Cornell University Press.
4. E. Mooser and W. B. Pearson, 1960, *Progress in Semiconductors*, Vol. 5, ed, Gibson, Kröger, and Burgess, New York: John Wiley and Sons Inc., p. 103.
5. E. Mooser, private communication.
6. E. Busmann, 1961, *Z. anorg. Chem.*, **313**, 90.
7. H. G. Grimm and A. Sommerfeld, 1926, *Z. Phys.*, **36**, 36.
8. E. Parthé, 1963, *Z. Kristallogr.*, **119**, 204.
9. M. Straumanis, 1940, *Z. Kristallogr.*, **A102**, 432.
10. J. R. Drabble and C. H. L. Goodman, 1958, *J. Phys. Chem. Solids*, **5**, 142.
11. P. W. Lange, 1939, *Naturwiss.*, **27**, 133.
12. S. A. Semiletov, 1954, *Trudy Inst. Krist. Akad. Nauk SSSR*, No. 10, 76.
13. E. Mooser and W. B. Pearson, 1958, *J. Phys. Chem. Solids*, **7**, 65; also *Idem*, 1956, *Can. J. Phys.*, **34**, 1369.
14. J. R. Wiese and L. Muldawer, 1960, *J. Phys. Chem. Solids*, **15**, 13.
15. F. Laves, 1956, *Theory of Alloy Phases*, Cleveland: American Society for Metals, p. 124.
16. U. Dehlinger, 1955, *Theoretische Metallkunde*, Berlin: J. Springer.
17. E. Mooser and W. B. Pearson, 1959, *Acta Cryst.*, **12**, 1015.
18. See, e.g., L. Pauling, 1960, *The Nature of the Chemical Bond*, 3rd ed, Ithaca: Cornell University Press, p. 508.
19. W. B. Pearson, 1962, *J. Phys. Chem. Solids*, **23**, 103.
20. See references in J. C. Phillips, 1970, *Rev. Mod. Phys.*, **42**, 317.
21. J. D. Dunitz and L. E. Orgel, 1960, *Adv. in Inorg. Chem. Radiochem.*, **2**, 1.
22. E. Parthé, 1961, *Z. Kristallogr.*, **115**, 52.
23. P. W. Anderson, 1950, *Phys. Rev.*, **79**, 350; 1959, *Ibid.*, **115**, 2.
24. See, e.g., J. B. Goodenough, 1960, *Phys. Rev.*, **117**, 1442; also 1963. *Magnetism and the Chemical Bond*, New York: Interscience.
25. L. E. Orgel, 1960, *An Introduction to Transition-Metal Chemistry. Ligand-Field Theory*, London, New York: Methuen-John Wiley.
26. See, e.g., J. H. van Santen and J. S. van Wieringen, 1952, *Rec. Trav. Pays-Bas*, **71**, 420.
27. H. A. Jahn and E. Teller, 1937, *Proc. Roy. Soc.*, **A161**, 220; see also L. E. Orgel, 1952, *J. Chem. Soc.*, 4756; J. D. Dunitz and L. E. Orgel, 1957, *Nature, Lond.*, **179**, 462.

28. U. Öpik and M. H. L. Pryce, 1957, *Proc. Roy. Soc.*, **A238**, 425; A. D. Liehr and C. J. Ballhausen, 1958, *Ann. Phys.*, **3**, 304.
29. F. Hulliger and E. Mooser, 1965, *Progress in Solid State Chemistry*, Oxford: Pergamon Press, p. 330.
30. B. Marrinder and A. Magnéli, 1957, *Acta Chem. Scand.*, **11**, 1635; also Goodenough.[24]
31. H. Holseth and A. Kjekshus, 1969, *Acta Chem. Scand.*, **23**, 3043; G. Brostigen and A. Kjekshus, 1970, *Ibid.*, **24**, 1925.
32. H. Holseth and A. Kjekshus, 1968, *Acta Chem. Scand.*, **22**, 3284.
33. W. B. Pearson, 1965. *Z. Kristallogr.*, **121**, 449.
34. G. Brostigen and A. Kjekshus, 1970, *Acta Chem. Scand.*, **24**, 2983; *Idem.*, 1970, *Ibid.*, **24**, 2993.
35. J. Zemann, 1958, *Acta Cryst.*, **11**, 55.

6

Metastable Phases, Interstitial Phases, and Martensitic Transformations

METASTABLE PHASES FORMED BETWEEN B GROUP ELEMENTS

Equilibrium diagrams between B elements of Groups II to V are (with the exception of the In–Bi system) characterized by the almost complete absence of extended solid solutions or intermediate phases, in striking contrast to the many intermediate phases formed between the Group I B elements and those of successive B Groups, or even to the compounds formed between the Group V and VI B elements. When, however, these alloys are very rapidly quenched at 10^7–10^8°C/sec by the splat-cooling method,[1] metastable intermediate phases are obtained, although it may be necessary to keep the metastable phases at liquid air temperatures to prevent decomposition to the stable regime.

Giessen[2] and co-workers who are responsible for most of the studies discussed here, find that the metastable phases generally have structures already recognized in stable phases. The main regions of metastable phases which they find are shown in Figure 6-1. A face centered cubic phase, α, occurring at electron concentrations of 2.2 to 2.4 e/a appears to result from a favorable electron concentration in this range, but the phase field also extends to the region of stable or metastable f.c. cubic solid solutions surrounding Al, and through stable phases about In to the f.c. cubic solid solutions surrounding Pb in the region of electron concentration from 3 to 4.2. Complex phases, ψ, of undetermined structure surround Ga, and f.c. tetragonal α' phases with axial ratio larger than unity

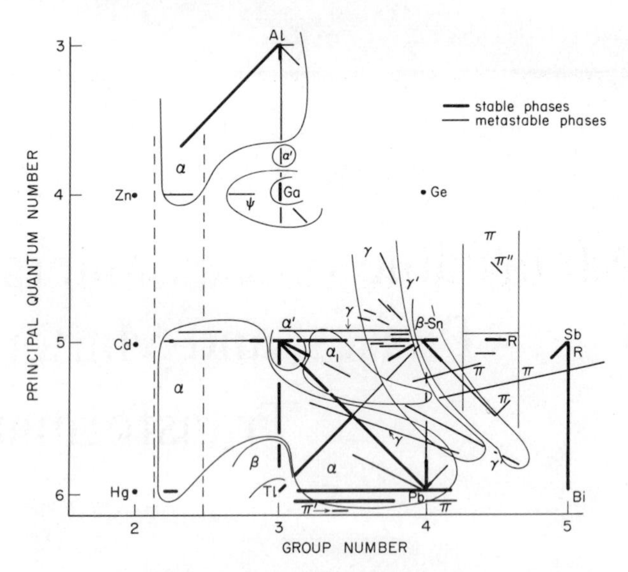

Figure 6-1. Diagram indicating metastable phases obtained by splat-cooling experiments. For details see text. Thin lines show the range of occurrence of metastable phases, thick lines of stable phases. (Giessen [2].)

(and increasing with electron concentration) surround In. The same metastable structure also occurs between Al and Ga. Bordering on the α' field and extending to higher electron concentrations, is another f.c. tetragonal field, α_1, with axial ratio less than unity and decreasing with increasing electron concentration. A body centered cubic field, β, lies between Tl and In, and in the In–Tl–Bi–Sn portion of the diagram β, or related Ni_2In or AlB_2 structures may be obtained. These are not shown in Figure 6-1. A simple hexagonal metastable phase field, γ, ($HgSn_{6-10}$ structure) occurs from an electron concentration of about 3.6 between the first and second long Periods, sweeping around with increasing average atomic number to an apparent electron concentration of about 4.4 between the second and third long Periods. A region of phases with the β-Sn structure runs parallel to the γ phase field at higher electron concentrations. Between these two phase fields, a metastable phase γ' is found whose structure is probably an intermediate distortion between the γ and β-Sn structures, which are related as shown in Figure 6-2. At higher electron concentrations (probably 4 to 5.2) metastable phases (π) with the simple cubic α-Po structure (or the rhombohedrally distorted β-Po structure) are obtained. Related to these phases are the rhombohedrally distorted simple cubic

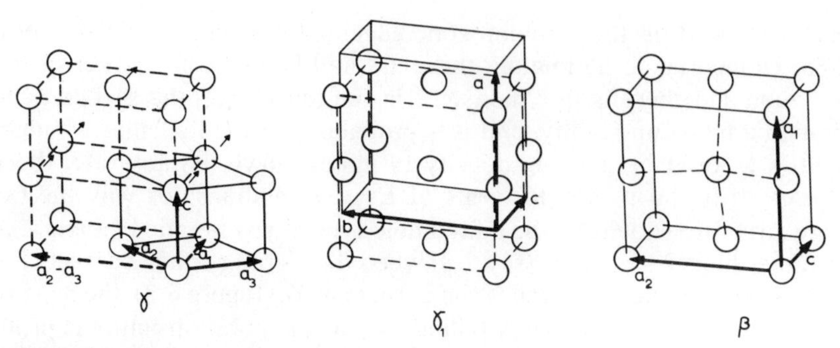

Figure 6-2. Diagram showing relationship between γ, γ', and β-Sn structures. See text. (Giessen [2].)

structures of Bi and Sb and the stable rhombohedrally distorted ordered structure of SnSb.

Such is a broad description of findings in this region, although details may provide some exceptions, and several other metastable phases of undetermined structure have also been observed on occasion. With the possible exception of the π phase, the characteristic of the regions of occurrence of these phases is that they generally extend to higher apparent electron concentrations as the average atomic number of the alloys increases. In alloys of metals of the third long Period, this may be a reflection of a lower effective valence state owing to the stability of the $6s^2$ electron pair which arises from the large separation of the $6s$ and $6p$ atomic levels (see p. 231). One of the more striking features of Figure 6-1 insofar as experimental observations have been made, is the apparent lack of connectivity of either metastable or stable solid solutions or intermediate phases between elements of the first and second long Periods.

In seeking to find reasons for the existence of these particular phases which are obtained metastably on rapid quenching, it is first to be noted that they are either extensions of existing stable phase regions, or are the simplest of structures (disordered) such as the atoms might readily adopt under the conditions of very rapid cooling, in preference to the equilibrium separation into two phases, with the attendant transport of atoms to give the compositions of the terminal solid solutions. The formation of these phases with sufficiently fast quenching rates rather than an amorphous mixture, implies that they have free energies at a level between those of the liquid and the stable phases, and therein electron concentration may play a role. For example in the f.c. cubic phase formed about an electron concentration of 2.3 e/a, the empty lattice model indicates that the Fermi surface, although overlapping the $\{111\}$ and $\{100\}$ faces of the first Brillouin zone into the second zone, has not yet reached point

U (Figure 6-3) on the common zone edge and overlapped into the third zone. Compared to aluminum, there are still holes in the corner of the first zone and there is no "monster" in the third zone; the second zone should be filling up readily, and it is probable that the f.c. cubic arrangement is favored by a rising density of states curve. At any rate, there appears to be no reason in terms of E *vs.* k relationships why the f.c. cubic structure should not be a favorable type at an e/a ratio of about 2.3. At rather higher e/a ratios (say 2.8) when the Fermi surface contacts the zone edge at U and approaches the corners at W (Figure 6-3), the density of states ($N(E)$) is presumably falling and the f.c. cubic structure is probably not a favored type. The face centered cubic structure also occurs about Al, between In and Cd and extensively in the region between In, Tl, and Pb at electron concentrations from 3 to 4 e/a.

The structure of the simple hexagonal disordered γ phase which is found in Sn alloys with electron concentrations from about 3.95 e/a to 3.6 e/a (although in two systems the phase extends to much lower electron concentrations), is related to that of β-Sn as shown in Figure 6-2. Taking the orthohexagonal cell for the γ phase and shifting one b.c. orthorhombic sublattice as shown, gives the β-Sn structure after slight dimensional changes ($a_\gamma \rightarrow c_{\text{Sn}}$, $a_\gamma \,.\, \sqrt{3} \rightarrow a_{\text{Sn}}$, $c_\gamma \rightarrow \frac{1}{2}a_{\text{Sn}}$). The orientation relationship is $[\bar{1}00]\gamma \parallel [001]_{\text{Sn}}$ and $[001]\gamma \parallel [100]_{\text{Sn}}$. Since the axial ratio of β-Sn increases with decreasing electron concentration, it is possible that the γ phase forms with equalization of the (220) and (211) spacings of β-Sn, which then merge into (10.1) planes of γ, as discussed

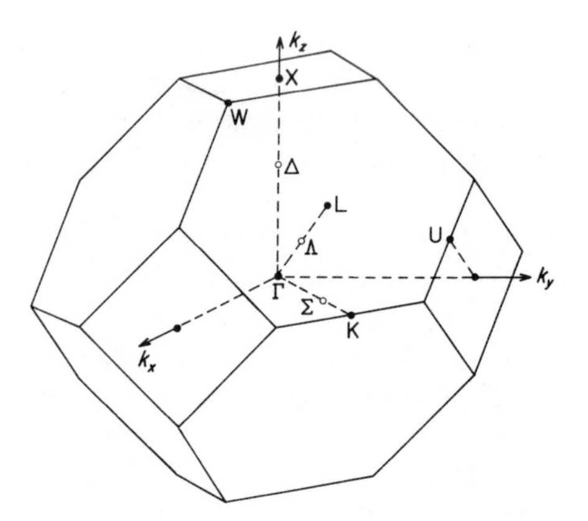

Figure 6-3. The Brillouin zone for the f.c. cubic lattice, showing naming of symmetry points.

by Kane and co-workers.[3] The γ phase is by far the most commonly formed metastable phase, some 14 examples being known. In addition to phases based on Sn, it is also formed on rapidly quenching isoelectronic InSb with added In to reduce the electron concentration,[4] and by Bi alloys. In Bi–Cd and Bi–In[5] alloys the γ phase extends to apparent electron concentrations considerably greater than 4.

Although determination of the exact composition of the γ phase is frequently difficult because it fails to occur as a single phase in the quenched metastable alloys, it is certain that the axial ratio of γ decreases with increasing electron concentration as shown in Figure 6-4 taken from Kane *et al.*[3] This was established from data on Sn alloys with Ag, Au, and Ga (single-phase γ alloys) and Al, Cd, Cu, and Zn. Data for equilibrium γ alloys[6] follow a similar curve, but appear to have slightly higher c/a values. The change of axial ratio with electron concentration may result from interaction of Fermi surface and Brillouin zone boundaries as has been discussed for other hexagonal phases (p. 116), and indeed the advantage of the hexagonal structure over cubic structures is that it can

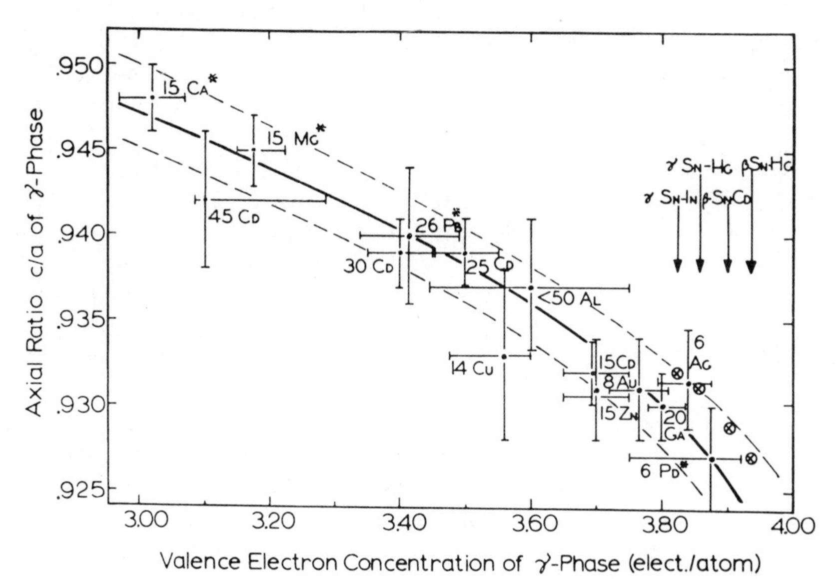

Figure 6-4. Axial ratios of γ phase structures of tin alloys as a function of electron concentration at $-190°C$. Numbers give the atomic percentage of the solute indicated in Sn–M alloys. Asterisk (*) indicates alloys located, not on the basis of composition, but in order to determine apparent valence electron concentration. ⊗ indicates points for stable alloys. (Kane *et al.* [3].)

readily adapt by change of axial ratio to internal stress resulting from electronic interactions of this type.

A metastable γ phase is found in Sn–Pb alloys, and its axial ratio (Figure 6-4) suggests that Pb exhibits a valency of 2, or slightly less in the alloy. Pd gives a metastable γ phase in rapidly quenched Sn–Pd alloys and the axial ratio (Figure 6-4) suggests that Pd also exhibits an effective valence electron concentration of 2. The axial ratio of the metastable Sn–Ca and Sn–Mg γ alloys (the latter being single phase) is not consistent with the normal valencies of the Group II elements, suggesting an apparent valency of -2! However, the axial ratio *vs.* apparent electronic concentration curve established for Sn alloys is not unique since the data for In–Bi alloys lie considerably above it, although perhaps the effective valence electron number for Bi is less than 5.

Studies of equilibrium alloys show that the axial ratio of alloys with the β-Sn type structure decreases as the electron concentration increases toward 4, and continues decreasing at electron concentrations above 4. In the metastable extensions of the Sn–Sb and Sn–Bi β-Sn solid solutions obtained by rapid quenching, the decreasing axial ratio is maintained.

The simple cubic π phase (α-Po structure) is found over a wide range of electron concentrations from at least 4 to 5.2 e/a. It occurs in Au–Sb, Au–Pd, Au–Ni, Au–Sn–Sb, Au–Bi,[7] Au–Te, Ag–Te,[8] and InSb–Sb alloys (Figure 6-5*a*). In the Au–Bi system a microcrystalline phase is obtained on splat cooling alloys with 30 to 45 at.% Au, which transforms at $-60°C$ (via π at $-130°C$) to a simple rhombohedral structure ($\alpha \sim 89°$). The phase is presumably disordered with the β-Po structure, rather than the rhombohedrally distorted NaCl structure of the stable SnSb phase. Another distorted form of the π phase (π'') is found in metastable Ge–Sb alloys. In equilibrium alloys of Sb or Bi, lowering the electron concentration appears to decrease the rhombohedral angle, α, rather than increasing it toward the undistorted value of 90°, so it is not envisaged that there is a continuous increase of α toward the 90° of the π phase. Indeed under high pressures, Sb and Bi change to the simple cubic type via a first order transition, and a wider analogy may develop between structures obtained in splat-cooling experiments and structural changes induced by the application of high pressures.

The β-Sn structure can also be regarded as a considerable distortion of the simple cubic structure of the π phases, so that the sequence of metastable or stable structures with decreasing electron concentration in the second long Period about Sn are all related by distortion one from the other, starting from the simple cubic π phase, through β-Sn and the γ' phase (3.95 e/a) to the undistorted simple hexagonal phase centered on electron concentrations of 3.9 to 3.6 e/a. On the other side of π which

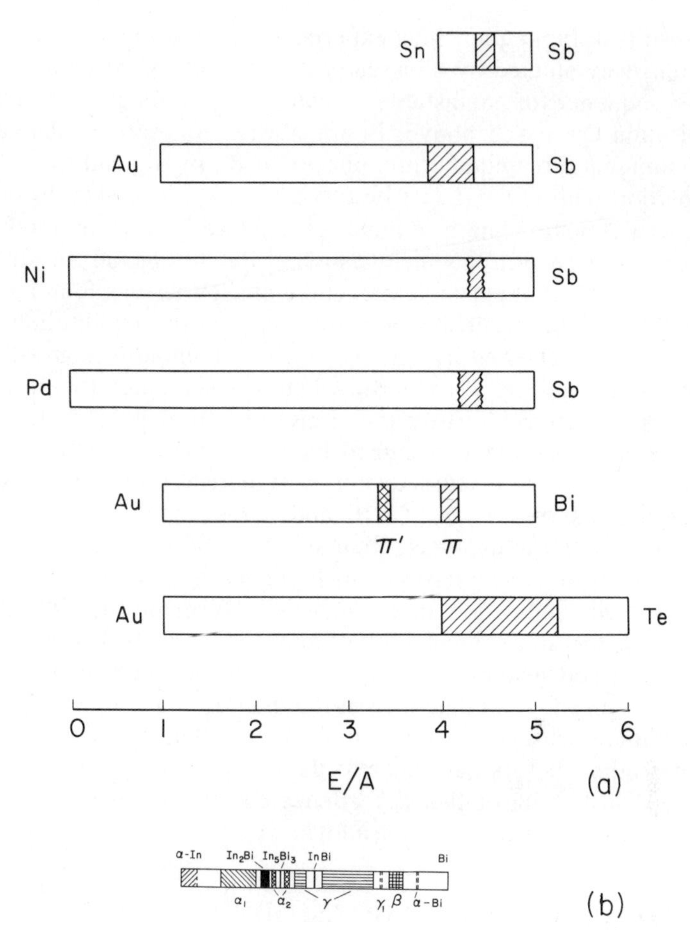

Figure 6-5. (*a*) Location of the simple cubic π phase in various alloy systems. (Giessen *et al.*[7].) (*b*) Diagram showing metastable and stable intermediate phases in the InBi system. (Giessen *et al.*[5].)

is centered about an electron concentration of 4.5 e/a, at an electron concentration of 5, a stable rhombohedral distortion of the simple cube is found in the Sb (As and Bi) structure. In the same Period about In, tetragonally distorted structures derived from a f.c. cube occur. From this it appears that the not-quite-stable structures about the second long Period, where changeover of bond type occurs, are f.c. cubic about $e/a = 2.3$, simple hexagonal about $e/a = 3.8$ and simple cubic about $e/a = 4.5$. In between these regions, either stable or metastable structures that can be regarded as distorted forms, are obtained.

The results of rapid quenching experiments on the In–Bi system which was extensively studied by Giessen *et al.*,[5] provide an example showing that the sequence of metastable structures as a function of electron concentration discussed above, is not always followed in detail. This system contains three equilibrium phases In_2Bi, In_5Bi_3, and InBi. The In solid solution with $c/a > 1$ can be retained to about 6 at.% Bi in splat-cooling experiments. The first metastable phase is α_1 from about 15 to 28 at.% Bi (Figure 6-5*b*), which also has the tetragonal In structure, but with $c/a < 1$. A metastable b.c. cubic phase (α_2) was found at about 35 and 40 at.% Bi, occurring on either side of the equilibrium In_5Bi_3 phase. This was followed by a range of the γ (simple hexagonal) metastable phase from 43 to 47 at.% Bi, which was separated by the equilibrium InBi phase from a further wide range of γ from about 53 to 72 at.% Bi. This covers the range occupied by the metastable simple cubic π phase in other alloys discussed above. A narrow range of metastable γ' phase occurs about 75 at.% Bi and a metastable phase with the tetragonal β-Sn structure was obtained from 78 to 83 at.% Bi at an apparent electron concentration much higher than expected. Finally, there was some indication of a metastable form of Bi with different lattice parameters at a composition of about 90 at.% Bi. If it is assumed that the observed general structural sequence of metastable alloys indicated in Figure 6-1 is influenced in the main by the effects of electron concentration, then it is also apparent from consideration of In–Bi alloys, either that other factors particular to the component atoms can control the phase distribution, or that the valence electron number assumed for Bi is considerably less than 5 in the alloys.

INTERSTITIAL ALLOYS OF TRANSITION METALS

Hägg[9] originally observed that compounds of the transition metals and the small nonmetal atoms H, B, C, and N generally take simple structures with the nonmetal occupying (octahedral) interstices in face centered or body centered cubic structures, or centering trigonal prisms formed by close packed layers in paired-layer stacking (WC or AlB_2 structures), when the radius ratio R_X/R_T is less than about 0.59; when it is greater, the compounds generally adopt complex structural arrangements. These observations have been remarkably well borne out in subsequent years, although Kiessling[10] noted that if B–B bonds occur in borides, simple metallic structures may still be formed when R_B/R_T exceeds 0.59. Thus for example, CrB_2 has the AlB_2 (*hP*3) structure even though $R_B/R_{Cr} = 0.69$.

As noted on p. 137, relative atomic sizes are rarely of particular importance to the stability of structures of metallic phases and compounds, but the Hägg interstitial phases provide one example where relative atomic size does control structure type. Normal valence relationships do not hold in the Hägg interstitial compounds and energy band considerations[11] suggest that some of the nonmetal atom *p* electrons are transferred to the transition metal atoms, accounting for the strong cohesion and very high melting points of many interstitial compounds. For this situation to arise there are definite restrictions on the distance between the transition metal atoms; if they are too far separated, the nonmetal does not empty electrons into the T metal energy bands; instead covalent shared-pair bonds are formed. The structure type changes, for example, from the NaCl to the CrB or FeB types where shared-pair covalent bonding becomes of increasing importance. The difference between interstitial and valence compounds is also illustrated by the interstitial transition metal hydrides and the covalent or ionic hydrides such as CaH_2 and LiH.

Binary transition metal interstitial compounds with structures derived from close packed T metal layers in c, hcc, hhcc, and h stacking arrangements ($1\underline{0}$, $3\underline{3}$, $3\underline{1}$, and $1\underline{1}$ in Ždanov–Beck notation) are recognized.[12] Different compositions are achieved by the presence of vacant sites in the layers of interstitial atoms in octahedral holes of these arrays. In the various arrangements of these that are observed, it appears to be an empirical rule that interstitial carbon atoms never sit one above the other in paired-layer stacking sequence.† Thus only one of the two octahedral holes above and below a close packed site of a hexagonally surrounded layer is occupied. This rule is satisfied by alternate layers of octahedral holes being vacant as in the CdI_2 (*hP*3) structure, or if sites in all layers of octahedral holes are occupied, then the occupied sites of one layer must be vacant sites in the layers above and below. For example, in the ϵ-Fe_2N structure, occupied octahedral holes in one layer form a 6^3 net and in the layers above and below, a 3^6 net whose nodes center the vacant centers of the hexagons of the 6^3 net. In the Co_2C (*oP*6) structure rectangular 4^4 nets of atoms occur in each octahedral hole layer, with the nodes of one net centering the rectangles of the nets above and below. For this reason interstitial compounds with hexagonal c.p. arrangements of T atom layers (h) cannot have more than 50% of the octahedral holes occupied by C. However, at least 6 different structural arrangements of vacancies for composition T_2X (with h stacking of T layers) have been recognized. Arrangements hcc, hhc_3 and hhcc of T

†The rule probably holds only for interstitial carbon, since two interstitial boride and one nitride phases have been reported to have the anti-NiAs (*hP*4) type structure.

metal layers cannot have more than 66.7, 66.7, and 75% occupation of the octahedral holes respectively, whereas all octahedral holes can be filled in a c.p. cubic T metal array. There does not appear to be any restriction on the occupation of the centers of neighboring trigonal prisms of T metal atoms by nonmetal atoms in interstitial phases, some of which take the WC ($hP2$) and AlB_2 ($hP3$) structures.

Binary interstitial phases also occur with lower interstitial content than the maximum allowed by the rule (e.g. Ni_3C ($hR8$), Ni_3N ($hP8$), β-$MoN_{0.4}$ ($tI \sim 11$), and α''-$Fe_{16}N_2$ ($tI18$)), and ranges of varying interstitial content are a common feature of the phases.

Many ternary interstitial phases such as $AlCr_2C$ ($hP8$), $MoUC_2$ ($oP16$), and Ti_3SiC_2 ($hP12$) are known where the nonmetal atom occupies octahedral holes in a close-packed array of metal atoms, and there are structures such as the Mn_5Si_3($hP8$) type wherein phases are often only stabilized by a small amount of nonmetal atoms in the octahedral holes of the atomic array. The perovskite structure of oxides such as $BaTiO_3$ may be largely ionic, but there is a metallic perovskite form which is stabilized in preference to the $AuCu_3$ type structure when an interstitial nonmetal atom such as carbon occupies the cell center, often only partially. Several examples of this type of structural stabilization have been discovered and the rôle of the nonmetal atom in random partial occupation of a siteset may be to increase the electron concentration favorably by emptying its p electrons into the energy band system of the other components, thus increasing the magnitude of both the enthalpy and entropy so that the free energy is lowered and the structure stabilized.

INTERSTITIAL STRUCTURES BASED ON B_{12} AND LARGER PSEUDOATOMS

A parallel to the interstitial structures of the transition metals is found in structures containing B_{12} icosahedra which behave as large spherical pseudoatoms with a diameter of about 5.1 Å. These have been found to pack together in close packed cubic (CCP), close packed hexagonal (HCP), or b.c. cubic {110} stacking of close packed layers (TCP), and to accept other atoms interstitially in the spaces between the icosahedra. In addition, another roughly spherical grouping of B atoms has been recognized by Matkovich and co-workers[13] on whose work these remarks are based. In this, a central icosahedron is surrounded icosahedrally by 12 other icosahedra. Taking half of the outer icosahedra (i.e., surrounding the central icosahedron by 12 pentagonal prisms) gives a nearly spherical group of 84 B atoms, which is a truncated icosahedron with 60 peripheral atoms forming 12 pentagonal and 20 hexagonal faces.

The B_{84} groups have a diameter twice that of the B_{12} groups and they are found to pack together in close packed arrangements similar to the B_{12} icosahedra, although they can accept much larger groupings in the interstices of their close packed arrays. Although the 12 outer icosahedra are each shared between neighboring inner B_{12} groups in these structures, resulting in B_{84} packing units, one structure is now known in which the complete unit of 13 icosahedra comprising 156 atoms is the packing unit.[14]

All known structures containing icosahedral B_{12} (or B_{84}, or B_{156}) groupings can be classified in terms of CCP, HCP, or TCP arrangements. In the notation used by Matkovich, the number preceding the packing symbol gives the ratio of the unit cell volume for a particular structure to that of the basic CCP, HCP, or TCP cells of B_{12} or B_{84} groups. v following the packing symbol indicates B_{84} rather than B_{12} groupings and u indicates B_{156} groupings. r, t, or o indicate the observed rhombohedral, tetragonal, or orthorhombic symmetry of the structures. The CCP and HCP basic cells are obvious; the former has an edge of 7.2 Å when built of B_{12} icosahedra ideally oriented and the latter has $a = 5.1$, $c = 8.3$ Å. The basic TCP body centered tetragonal cell is obtained by squashing a b.c. cube along c so that $c = \sqrt{2}/\sqrt{3} \, . \, a$. When built of B_{12} icosahedra, this cell has dimensions $a = 6.2$, $c = 5.1$ Å.

Several known structure types can be classified as having close packed a structure derived from the f.c. cubic arrangement of B_{12} pseudoatoms cubic arrangements of B_{12} or B_{84} groupings (Table 6-1). α-B ($hR12$) has by a slight rhombohedral distortion, the rhombohedral unit cell having $\frac{1}{4}$ of the volume of the basic cubic cell. This structure accepts elements C, Si, P, As, O, or S in the octahedral interstices, and up to three interstitial atoms can be accommodated per B_{12} group. The prototype for this group of interstitial compounds is the phase originally referred to as "B_4C" ($hR15$).

The B_{84} groups in β-B ($hR105$) are also arranged in a rhombohedrally

Table 6-1

Structure Type	Description Symbol
α-B ($hR12$) and "B_4C" types	$\frac{1}{4}$CCPr
β-B ($hR105$)	$\frac{1}{4}$CCPvr
α-AlB_{12}	$\frac{1}{2}$CCPvt
ReB_6	$\frac{1}{2}$CCPvt
tetragonal B	$\frac{1}{2}$CCPvt
YB_{66}	(NaCl)u
UB_{12}	B_{12} cubo-octahedra CCP

distorted CCP arrangement, with $\frac{1}{4}$ of the volume of a CCP array of B_{84} units. There are two B_{10} and a single 9-coordinated B atom per B_{84} which are accommodated in the interstices of the B_{84} array; it is probable that Al or Mg can substitute for the unique B atom.

Matkovich and co-workers suggest that the cell dimensions of the unsolved tetragonal α-AlB_{12} structure ($a = 10.16$, $c = 14.26$ Å) and isotypic BeB_6, are consistent with a close packed cubic arrangement of B_{84} groupings with some occupation of interstitial sites. Tetragonal boron ($a = 10.12$, $c = 14.14$ Å) has a cell resembling that of α-AlB_{12}, and it is also suggested that its structure is built of B_{84} units with interstitial boron atoms, although a description 4CCP*t* based on B_{12} icosahedra is also possible.

In the YB_{66} structure,[14] B_{156} groups in two orientations are packed together in a NaCl type arrangement, those at the midpoints of the cell edges being rotated 90° relative to those at the cell corners and face centers. This arrangement accounts for 1248 B atoms. Some 336 more B atoms and 24 Y atoms are located in large interstitial channels that run parallel to four-fold axes at $\frac{1}{4}, \frac{1}{4}, x$. The Y sites occur in pairs less than 2 A apart and only one is occupied at a time. The sites of the interstitial B atoms which form cages in the channels, are also partially occupied. Y has neighboring B atoms both from the cages and the icosahedral framework. Other phases presently referred to as HoB_{70}, TbB_{70}, YbB_{70}, GdB_{100}, PuB_{100} are probably isostructural with YB_{66}.

There is one other structure, the UB_{12} type (*cF*52), that can be described in the CCP classification, except that the B_{12} groups are cubo-octahedra instead of icosahedra. The cubo-octahedra are located about the sites of an *F* lattice forming a close packed cubic array. The space between them is made up of large truncated octahedra and truncated tetrahedra. The metal atoms are located at the centers of each of the truncated octahedral holes, the arrangement of the B_{12} pseudoatoms and U atoms being that of the NaCl structure.

The B_{12} icosahedra in C_4AlB_{24} (*oC*58) are in hexagonal close packing. The C atoms form bridges between icosahedra and Al occupies other interstitial space. The structure can be described as 2HCP*o*. The so-called γ-AlB_{12} structure has cell edges twice those of C_4AlB_{24} and it is built up of an orthorhombically distorted hexagonal close packing of B_{84} units, the description being 2HCP*vo*. This is also apparent from comparison of appropriate "hexagonal" cell dimensions with those α-AlB_{12}.

The tetragonal form of boron (*tP*50) can be described as a b.c. cubic {110} type stacking of close packed layers of B_{12} icosahedra with $\frac{1}{2}$ B per B_{12} unit located in tetrahedral interstices. The cell is described as 2TCP and its dimensions, $a = 8.74$, $c = 5.1$ Å, expressed in terms of the basic cell give an axial ratio $c_0/a_0 = 0.825$. This is very close to $\sqrt{2}/\sqrt{3} = 0.816$ required for equilateral triangles in the close packed layers of B_{12} icosahedra that lie on planes parallel to (110) of the basic body centered tetragonal cell. BeB_{12} (*tP*52) has a similar arrangement of B_{12} icosahedra with 4 Be per cell occupying tetrahedral interstices. The $Ca_2Al_3B_{48}$ (H.T.) structure also resembles that of tetragonal B although it is of lower symmetry. It is similarly constituted with the Ca and Al atoms accommodated interstitially. This high temperature phase decomposes to two related orthorhombic structures at low temperatures. The cell dimensions suggest that both of these can be described as 4TCP*o*.

MARTENSITIC TRANSFORMATIONS

The structures of phases produced by martensitic transformations have no special characteristics other than the possibility of achieving reasonable coherence with their parent structures on some suitable habit plane, and therefore they are not specifically discussed here. The only feature of relevant interest is how the competition between entropy and binding energy for the control of structural stability may be influenced by near-neighbor interactions and by the relative energies of electrons at the Fermi level, so that structural instabilities result which lead to diffusionless martensitic transformations at relatively low temperatures.

Martensitic and massive transformations result in change of crystal structure without change of composition, no appreciable diffusion being involved. Martensitic transformation causes macroscopic change of crystal shape, and the initial and final structures have a specific orientation relationship to each other and a coherent interface (habit plane) connecting the two structures, whereas in massive transformations growth is incoherent and parent and daughter structures do not necessarily have any specific orientation relative to each other. A thorough discussion of the characteristics of these transformations is given by Christian.[15]

Diffusionless martensitic transformations which involve shears and slight crystallographic distortions occur because of mechanical instability of the crystal lattice resulting from some internal stress, or from electronic instability resulting from the appearance of a distorted structure of lower free energy. Thus for example magnetic atoms may be stable in the f.c. cubic *A*1 structure at high temperatures because of the spin-disorder contribution to the entropy of the phase. At some lower temperature when the spin-disorder entropy term is decreased, exchange forces result in ordering of the magnetic spins, but in the f.c. cubic structure they cannot order so that up-spin atoms only have down-spin neighbors. The ordering may, however, be ferromagnetic (spin-parallel) in layers of atoms (say (001) planes) and antiferromagnetic (spin-opposed) between alternate planes of atoms along [001]. Thus the symmetry of the magnetic cell is lowered, but the induced uniaxial stress also results in transformation of the chemical cell from cubic to tetragonal, and if the temperature is sufficiently low, such transformations occur martensitically, as for example in some alloys of the Mn–Cu solid solution.[16] A somewhat similar situation occurs in AuMn with the CsCl structure, where the magnetic sheets below the Néel temperature are parallel to (010) or (001) planes depending on composition, and magnetic ordering is accompanied by martensitic transformation to a tetragonally distorted structure with c/a, respectively, greater or less than unity.[17]

Some intermetallic phases such as V_3Si with the β-W structure which is characterized by rows of T metal atoms running throughout the structure in three orthogonal directions, have also been found to undergo martensitic transformations at very low temperatures. Labbé and Friedel[18] have shown in a theoretical study how, as a result of the probable energy band structure, a tetragonal distortion of the structure can lower its energy. At a certain temperature, therefore, the β-W structure becomes electronically unstable relative to a tetragonally distorted structure and transformation occurs martensitically.

A second example of mechanical instability of the crystal structure causing martensitic transformations is found in β brasses discussed by Zener.[19] Such phases may be stabilized in the disordered β b.c. cubic structure at high temperatures by the entropy of vibration, but at lower temperatures the system becomes mechanically unstable with respect to a (110) $\langle 1\bar{1}0 \rangle$ shear because of interaction between the atom cores. Secondly, the entropy of positional disorder of the component atoms stabilizes the disordered structure at high temperatures, but at lower temperatures the core interactions result in an ordering of the structure because of the dominance of the binding energy contribution to the free energy as the entropy contribution declines. The temperatures for mechanical instability of the disordered phase, T_m, and for the ordering of the phase, T_{ord}, have no reason to be equal. Furthermore, the ordered phase, β', may itself become mechanically unstable at a sufficiently low temperature T'_m. The greater the difference of atomic core sizes, the greater the advantage of an ordered structure, and β'-AuMg with a difference of 0.71 Å in core radii, according to Zener, is stable up to a very high melting point.

If the free energy curves of the neighboring α and γ phases are such that β ceases to be stable with respect to decomposition into α and γ before ordered β' becomes stable on decreasing the temperature, and if T_m for β lies between these two temperatures, a sufficiently rapid quench to prevent $\beta \rightarrow \alpha + \gamma$ must result in martensitic transformation of β at T_m to a structure which is mechanically stable, before ordering to β' can occur at the lower temperature, T_{ord}. Behavior of this type can indeed be observed in the Cu–Al system on quenching β-Cu_3Al of appropriate composition. Low-temperature instability of the ordered β' phase is observed in AuZn, and in the β'-CuZn phase at compositions about 38 or 40 at. % Zn where it is one constituent of a two-phase alloy. However, no martensitic transformation occurs in equiatomic β'-CuZn. In equiatomic AgZn alloys β transforms to a phase with hexagonal structure, ζ at 280°C, but if it is quenched from above this transformation temperature, it is transformed to ordered β' which is metastable at the transformation temperature of about 270°C. No further transformation of β'-AgZn occurs on cooling it to the lowest temperatures.

REFERENCES

1. P. Duwez, R. H. Willens, and W. Klement, 1960, *J. Appl. Phys.*, **31**, 1136.
2. See, e.g., B. C. Giessen, 1969, *Advances in X-ray Analysis*, Vol. **12**, ed, Barrett, Newkirk, and Mallett, New York: Plenum Press, p. 23.
3. R. H. Kane, B. C. Giessen, and N. J. Grant, 1966, *Acta Met.*, **14**, 605.
4. B. C. Giessen, R. H. Kane, and N. J. Grant, 1965, *Nature, Lond.*, **207**, 854.
5. B. C. Giessen, M. Morris, and N. J. Grant, 1967, *Trans. Met. Soc., AIME*, **239**, 883.
6. G. V. Raynor and J. A. Lee, 1954, *Acta Met.*, **2**, 616.
7. B. G. Giessen, U. Wolff, and N. J. Grant, 1968, *Trans. Met. Soc., AIME*, **242**, 597.
8. H. L. Luo and W. Klement, 1962, *J. Chem. Phys.*, **36**, 1870.
9. G. Hägg, 1929, *Z. phys. Chem.*, **B6**, 221; 1931, *Ibid.*, **B12**, 33.
10. R. Kiessling, 1950, *Acta Chem. Scand.*, **4**, 209.
11. R. G. Lye, 1967, *Atomic and Electronic Structure of Metals*, Cleveland: American Society for Metals, p. 99.
12. See, e.g., K. Yvon and E. Parthé, 1969, *Acta Cryst.*, **B26**, 153.
13. V. I. Matkovich, R. F. Giese, and J. Economy, 1965, *Z. Kristallogr.*, **122**, 116.
14. S. M. Richards and J. S. Kasper, 1969, *Acta Cryst.*, **B25**, 237.
15. J. W. Christian, 1965, *Theory of Transformations in Metals and Alloys*. Oxford: Pergamon Press.
16. Z. S. Basinski and J. W. Christian, 1952, *J. Inst. Met.*, **80**, 659.
17. G. E. Bacon, 1962, *Proc. Phys. Soc., Lond.*, **79**, 938.
18. J. Labbé and J. Friedel, 1966, *J. de Phys.*, **27**, 153, 303, 708.
19. C. Zener, 1967, *Phase Stability in Metals and Alloys*, New York: McGraw Hill, p. 25.

7

Structures Based on the Close Packing of 3^6 Close Packed Nets, and Subdivisions Thereof

INTRODUCTION: SUMMARY OF BASIS OF STRUCTURAL CLASSIFICATION

The problem of organizing the structures of metals and alloys so that they can be discussed coherently has found only one practical solution, which is to consider them wherever possible as made up by the stacking of layers and atoms. Arrangements of layer networks of atoms can be classified and ordered with atomic coordination being adopted as the second parameter controlling the arrangement. This organization which covers 590 structure types of metals and alloys is indicated in a block diagram (Figure 7-1). Since there are only some 60 known structures remaining, and many of these are the structures of valence compounds which might have been included in the arrangement on one pretext or another, the assemblage must be considered as relatively successful, because therein one group of structures is naturally related to the next. However, it must be conceded that taking layer arrangements of atoms as the basic parameter for the organization is a sterile operation, since it cannot generally be related to the factors which control structural stability, except in the general case of close packed structures, and structures with inserted layers filling octahedral or tetrahedral holes of the close packed arrays. Secondly, in taking atomic coordination as the second parameter for gauging structural arrangement some structural groupings appear rather differently to those which are normally conceived. An example of this occurs where particular distinction is made

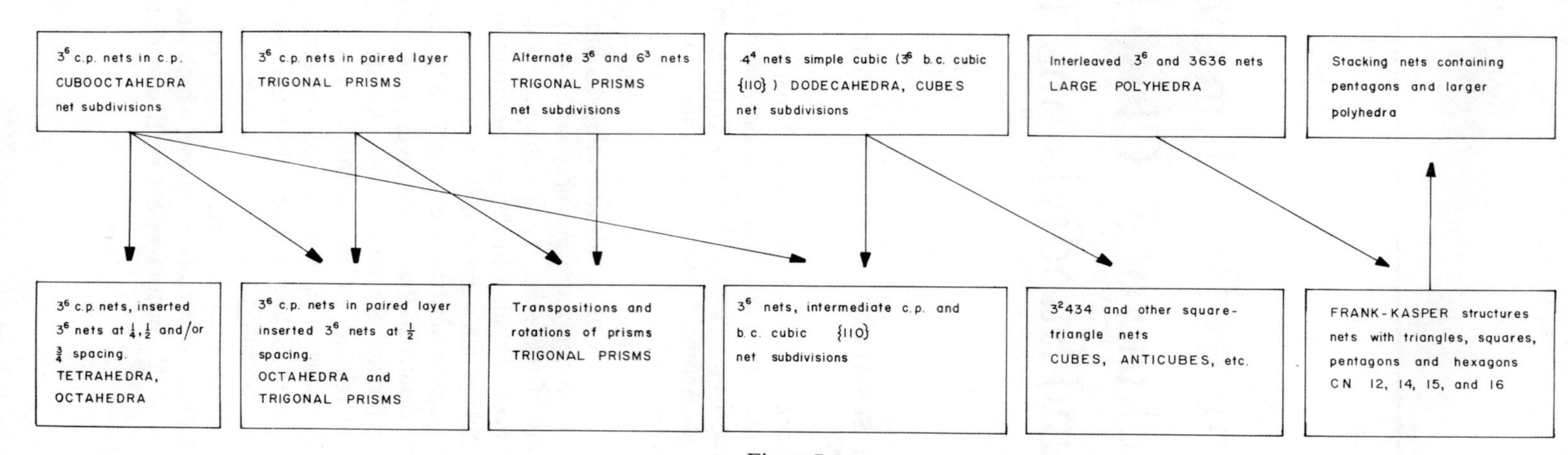

3⁶ c.p. nets in c.p.
CUBOOCTAHEDRA
net subdivisions
3⁶ c.p. nets in paired layer
TRIGONAL PRISMS
Alternate 3⁶ and 6³ nets
TRIGONAL PRISMS
net subdivisions
4⁴ nets simple cubic (3⁶ b.c. cubic
{110}) DODECAHEDRA, CUBES
net subdivisions
Interleaved 3⁶ and 3636 nets
LARGE POLYHEDRA
Stacking nets containing
pentagons and larger
polyhedra
3⁶ c.p. nets, inserted
3⁶ nets at 1/4, 1/2 and/or
3/4 spacing.
TETRAHEDRA,
OCTAHEDRA
3⁶ c.p. nets in paired layer
inserted 3⁶ nets at 1/2
spacing.
OCTAHEDRA and
TRIGONAL PRISMS
Transpositions and
rotations of prisms
TRIGONAL PRISMS
3⁶ nets, intermediate c.p. and
b.c. cubic {110}
net subdivisions
3²434 and other square-
triangle nets
CUBES, ANTICUBES, etc.
FRANK-KASPER structures
nets with triangles, squares,
pentagons and hexagons
CN 12, 14, 15, and 16

Figure 7-1

between paired layer stacking of close packed layers AA--- and close packed stacking of the layers ABC---, AB---, etc., because the former gives rise to trigonal prisms. No distinction is made between "anions" and "cations" in the classification of structures, although conventional thinking frequently attaches specific importance to the coordination about one or the other. The NiAs structure, for example, contains atoms which are octahedrally surrounded and those which are coordinated by trigonal prisms and it is so classified, whereas the CdI_2 structure only contains atoms that are surrounded octahedrally, and hence it occurs in a different grouping to the NiAs structure. Since no particular distinction attaches to whether "anions" or "cations" occupy a particular siteset, "types" and "antitypes" are not separately considered. Thus the fluorite structure is considered only as containing one set of equivalent points which are cubically surrounded, and one siteset whose points are tetrahedrally surrounded; it is not material to problems discussed in the following chapters whether, for example, oxygen in oxides with the fluorite structure occupies the four or eight coordinated sites.

THE DERIVATION OF ORDERED STRUCTURE TYPES FROM CLOSE PACKED AND FROM b.c. CUBIC ARRAYS OF ATOMS, AND THE DISTINCTION BETWEEN THEM

The Stacking of Triangular Nets on Various Sites

In order to classify families of ordered structures effectively, it is important first to compare those derived from close packed arrays of atoms (f.c. cubic or c.p. hexagonal) and those derived from a b.c. cubic array. The close packed layer (normal to $\langle 111 \rangle$ in f.c.c., or to [001] in h.c.p.) is a 3^6 net composed of equilateral triangles as shown in Figure 7-2. For such layers to be stacked in close packing, the nodes of one net must center alternate triangles of nets above and below. There are two equivalent possibilities B and C if the original net is at location A, as shown in Figure 7-2. In $\{110\}$ type planes the b.c. cubic structure also forms 3^6 nets of atoms with triangle angles approximately 55°, 55°, and 70° (Figure 7-2).† Each triangle thus has one unique corner and it is convenient to indicate this by chemical component by considering the ordered CsCl structure rather than the disordered b.c. cube, since in $\{110\}$ planes the triangles of the CsCl structure have two corners (55°) occupied by one component and one (70°) by the other. Indeed, it is seen

†In the $\langle 111 \rangle$ directions the b.c. cubic structure forms 3^6 nets with equilateral triangles. The structure can be described by a hexagonal cell that has axial ratio, $c/a = 1.225$ and layers spaced at $z = 0, \frac{1}{6}, \frac{1}{3}, \frac{1}{2}, \frac{2}{3}, \frac{5}{6}$ in ABCABC stacking sequence.

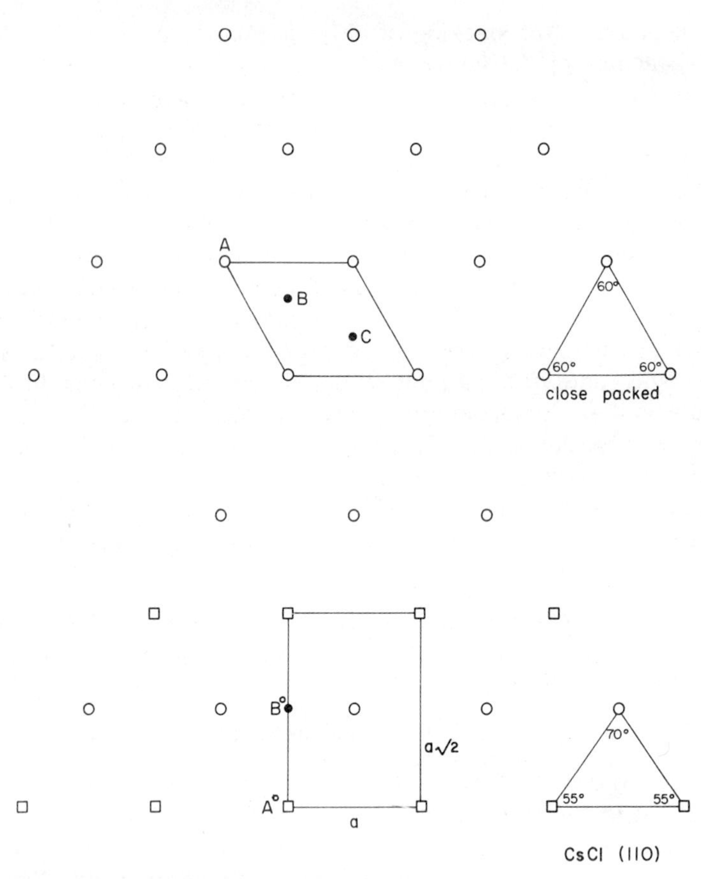

Figure 7-2. (Upper) Close packed plane of atoms showing three equivalent stacking sites A, B, and C. (Lower) CsCl (or b.c. cubic with one component atom) {110} plane of atoms, indicating triangle angles and two equivalent stacking sites A° and B°.

that the 3^6 net is made up of two interpenetrating 4^4 rectangular nets (Figure 7-2). In the b.c.c. or CsCl structures successive nets in [110] directions are stacked one above the other so that the unique corners of the triangles of one net, center the midpoint of the edge opposite the unique corner of the triangles in the nets above and below it, as shown in Figure 7-2, thus giving two stacking positions A° and B°. It can be seen that the two stacking positions B and C of the close packed stacking sequence have in the b.c. cubic or CsCl array, been reduced to one position B° midway between them on the triangle edge. Tetragonal distortion of the b.c. cubic or CsCl cell to *decrease* the axial ratio, c/a, changes the angles of the triangles in the nets parallel to the (110) plane toward 60° (Figure 7-3*a*) and when $c/a = \sqrt{2}/\sqrt{3} = 0.816$ the net triangles are equilateral. However, successive nets are still in b.c. cubic [110] stacking, A°B°, with a node of one net over the midpoint of a side of a triangle of the nets above and below. When, on the other hand, the b.c. cube is tetragonally distorted by an *increase* of axial ratio, the angles of the triangles of the 3^6 nets parallel to the (101) and (011) planes change toward 60° (Figure 7-3*b*), attaining this value when $c/a = \sqrt{2} = 1.41$. In this case the stacking of the planes also changes from b.c. cubic with a node of one net over an edge of a triangle of nets above and below, towards close packed stacking, ABC, with the nodes of one net in the center of triangles of the nets above and below, attaining this state exactly at $c/a = \sqrt{2}$. The (101) or (011) nets themselves then become exactly close packed, so that they are {111} planes of the f.c. cubes which are formed.

Such considerations are of some importance, for there are numerous ordered structures which strike a compromise between close packing and b.c. cubic packing with the equilateral triangles of the close packed 3^6 nets of atoms in the A°B° stacking sequence of the b.c. cubic array, rather than the ABC sequence of the close packed arrays. One such structure is the $MoSi_2$ (*tI*6) type with axial ratio, $c/a = 2.4$ (Figure 10-29, p. 591), where nearly equilateral-triangle 3^6 nets are stacked exactly in the b.c. cubic A°B° sequence rather than the close packed ABC sequences. In other structures such as the $TaPt_2$ or VAu_2 types (Figure 10-26, p. 588), both the net angles and the stacking positions are intermediate between the close packed and b.c. cubic arrangements.

In the b.c. cubic and CsCl type of arrangement the other two possible stacking positions C° and D° with the node of one net over either of the

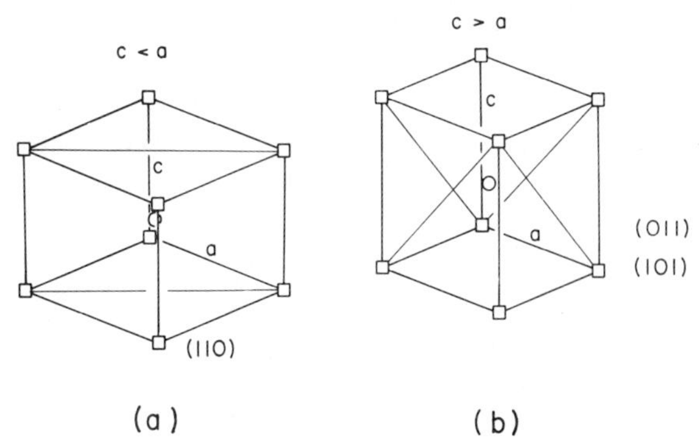

Figure 7-3. Distorted CsCl type cell. (*a*) $c < a$, indicating (110) plane: (*b*) $c > a$, indicating (011) and (101) planes.

other two sides of the triangles of the nets above and below, are equivalent to each other, but not to A° and B° which are mutually equivalent. When, however, the {110} nets of the b.c. cubic structure are distorted to equilateral triangles by decrease of c/a, so that the layers are themselves close packed nets, then all four stacking positions A°B°C°D° become equivalent. They are still not equivalent in the CsCl arrangement because the difference in components still maintains the two rectangular 4^4 subarrays. However, in the $TiSi_2$ ($oF24$) structure (Figure 10-27, p. 589) where the close packed 3^6 layers are subdivided into 6^3 nets occupied by Si and new larger 3^6 nets occupied by Ti, the four positions are equivalent, and the layers are stacked one above the other in the sequence A°B°C°D°.

The possible and observed subdivisions of 3^6 nets are important since they form the basis of families or polytypic structures obtained by stacking the close packed nets in various sequences. In the b.c. cubic array the 3^6 nets are already subdivided into two interpenetrating rectangular 4^4 nets by the geometry of the triangles. Further subdivisions of these rectangular nets are possible and occur in ordered derivatives such as the Heusler alloy or the $BiLi_3$ type structures. The close packed 3^6 nets can also be subdivided into two rectangular 4^4 arrays, each of which can be occupied by one component as in the AuCuI structure (Figure 7-6, p. 312), where the layers are then stacked in ABC sequence. Since the 6^3 array is the dual of the 3^6 array, and a 3^6 array gives one atom per planar hexagonal cell whereas the 6^3 array gives two, a 3^6 net can be subdivided into a 6^3 and a new larger 3^6 array for an ordered compound with MN_2 stoichiometry. With the N component occupying the two sites of the 6^3 array and the M component the single site of the new 3^6 array, superstructure ordering of layers of MN_2 stoichiometry is obtained as in the $MoPt_2$ structure which has ABC stacking of the layers.

Similarly a 3^6 net can be subdivided into a kagomé net, 3636, with three occupied sites per planar hexagonal cell and a new larger 3^6 net with one site per cell. With the N component occupying the three sites of the kagomé net and the M component the site of the 3^6 net, ordered layers with MN_3 stoichiometry are obtained such as occur in the family of polytypic superstructures with the $AuCu_3$ structure (ABC stacking sequence) as prototype. For MN_3 stoichiometry a 3^6 close packed net can also be subdivided into a rectangular 4^4 net with one site per rectangular cell which is occupied by the M component, and a $3636 + 3^26^2$ (1:2) hexagon-triangle net with three sites per rectangular cell which are occupied by the N component (Figure 7-21, p. 327). A family of polytypic superstructures based on various stackings of these nets is also known with the $TiAl_3$ structure and ABC stacking of the layers, as prototype. Further possibilities of ordered MN_3 structures arise through a combination of the triangular and rectangular subdivisions of 3^6 nets, introduced by stacking faults, as for instance in the $ZrAl_3$ structure.

More complex subdivisions of 3^6 nets to satisfy other stoichiometries such as MN_4, M_4N_7, M_2N_7, M_3N_8, etc. are well known. In all such structures where triangular layers occur, it is important to distinguish clearly the shape of the triangles and the nature of the layer stacking to see whether the structure is derived from close packed or b.c. cubic arrays, or is a compromise intermediate between the two.

Systematic discussion of subdivision of close packed nets and nomenclature for describing the subdivisions and stacking sequences can be found in a series of papers by Beck.[1]

CLOSE PACKED LAYERS OF ATOMS IN CLOSE PACKING: ELEMENTAL STRUCTURES

Four different elemental structures are known with the atoms in close packed arrays. Each atom is surrounded by a cubo-octahedron of 12 equally spaced neighbors in the cubic case, or by a twinned cubo-octahedron in the hexagonal structures. The cubo-octahedron has 6 square and 8 triangular faces (Figure 7-4).

Cu (*cF*4): The atoms are arranged in close packed layers parallel to $\{111\}$ planes of the cubic unit cell, the stacking sequence being ABC ($1\underline{0}$ in Ždanov–Beck notation). Parallel to $\{100\}$ planes the atoms are arranged on square 4^4 nets, alternate nets being displaced in $\langle 110 \rangle$ directions so that the nodes of one net lie over the centers of the squares of the nets above and below.

Mg (*hP*2): The atoms are arranged in close packed layers parallel to the basal (001) plane of the hexagonal cell, the stacking sequence being BC ($1\underline{1}$ in Ždanov–Beck notation). The ideal axial ratio, c/a of the hexagonal cell is 1.633 for close packed atoms. When the axial ratio differs from this value, the twinned cubo-octahedron about each atom is distorted and the distance to the 6 equatorial neighbors differs from that to the 6 neighbors located in planes above and below the central atom.

La (*hP*4): The dimensions of the hexagonal unit cell of La are $a = 3.770$, $c = 12.159$ Å, $Z = 4$. The atoms are arranged in close packed layers parallel to the (001) plane in a double hexagonal stacking sequence ABAC. The ideal value for the axial ratio is $c/2a = 1.633$. Half of the layers are in cubic stacking sequence and half in hexagonal stacking.

δ-*Sm* (*hR*3): The dimensions of the rhombohedral cell are $a = 8.996$ Å, $\alpha = 23°13'$, $Z = 3$. The corresponding hexagonal cell has $a = 3.621$,

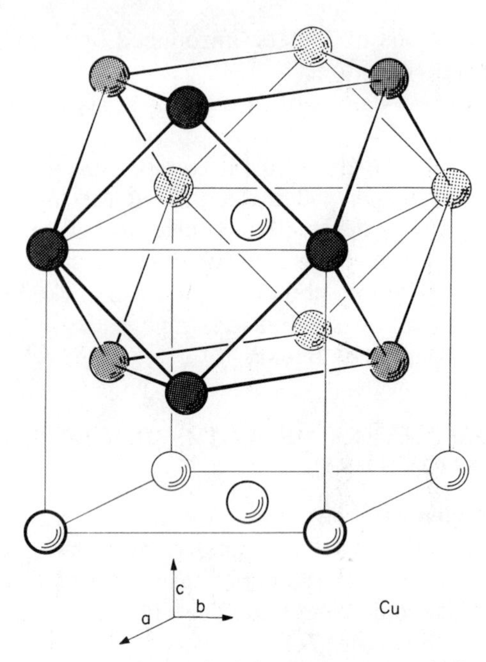

Figure 7-4. Unit cell of Cu ($cF4$) structure indicating cubo-octahedral coordination of the atoms.

$c = 26.25$ Å, $Z = 9$. The atoms in close packed layers parallel to (001) of the hexagonal cell, are stacked in the sequence ACACBCBAB ($2\underline{1}$ in Ždanov–Beck notation), so that one-third of the layers are surrounded cubically by neighboring layers and two-thirds are surrounded hexagonally. The structure is slightly distorted giving two interatomic distances of 3.59 and 3.62 Å.

Distortions of Close Packed Elemental Structures

Several elements and compounds have structures that can be considered as distortions of the elemental close packed structures:

In, Pa ($tI2$): With cell dimensions $a = 4.598$, $c = 4.947$ Å, $Z = 4$, the In structure can be regarded as a tetragonally distorted f.c. cubic cell with $c/a = 1.076$, whereas the Pa structure can better be regarded as a distorted form of the b.c. cubic structure (see p. 562).

Tb, Dy: The structures of several rare earth metals distort from hexagonal ($hP2$) to orthorhombic at low temperatures following magnetic transformations on cooling.

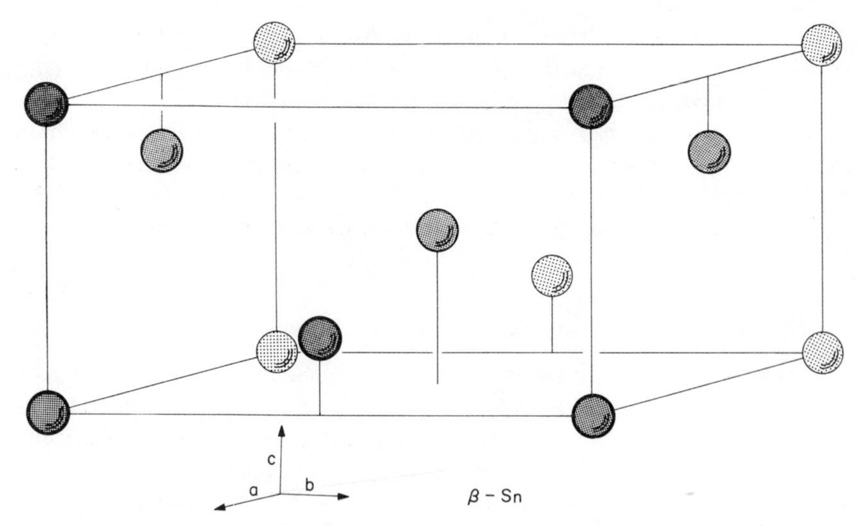

Figure 7-5. Unit cell of β-Sn ($tI4$) structure.

β-Sn ($tI4$): The β-Sn structure with cell dimensions $a = 5.831$, $c = 3.181$ Å, $Z = 4$, can be regarded as a very much distorted diamond type structure (Figure 7-5) which is intermediate between the diamond semiconductor structure of α-Sn and the f.c. cubic structure of Pb. Each Sn atom has four close neighbors, two more slightly farther apart, and four others at a considerably larger distance.

PbTh: PbTh has a b.c. or f.c. tetragonal cell. For the latter $a = 4.545$, $c = 5.644$ Å, $Z = 2$.[2] This tetragonal distortion of a f.c. cell may well be ordered, in which case the cell would likely be *I* centered and have $c' = 2c_0$.

β′-Cu_3Ti (H.T.) ($oC4$): β′-Cu_3Ti has cell dimensions $a = 2.572$, $b = 4.503$, $c = 4.313$ Å, $Z = 1$.[3] The atoms, which are randomly distributed, form close packed 3^6 triangular nets at $z = \pm\frac{1}{4}$ parallel to the (001) plane. These are stacked in close packing in AB stacking sequence, so that the structure is a slightly distorted derivative of the Mg structure.

SUPERSTRUCTURES OF CLOSE PACKED LAYERS STACKED IN CLOSE PACKING

MN: *M* and *N* on Alternate Close Packed Layers

CuPt I ($hR32$): CuPtI has rhombohedral cell dimensions $a = 7.59$ Å, $\alpha = 91°$.[4] The cell is a slightly distorted superstructure built up of a block of 8 f.c. cubic pseudocells. The atoms are arranged in alternate close packed layers of Cu and Pt in (111) planes of the cubic pseudocells, with a stacking repeat sequence $\underline{A}B\underline{C}A\underline{B}C$ (Cu underlined).

At a composition of Cu_3Pt_5, the "CuPt II" structure is found in which the Cu layers contain 25% Pt, the Cu atoms forming a kagomé net with the Pt atoms forming a large 3^6 net that centers the hexagons of the kagomé net. This arrangement can be described with a cell containing 8 atoms.

MN: *M* and *N* Each on a 4^4 Subnet in Each Close Packed Layer

Family of Polytypic Structures with* MN *Stoichiometry and Rectangular Arrangement of Components in Close Packed Layers. A close packed triangular array of atoms can be subdivided into two rectangular 4^4 arrays as shown in Figure 7-6. If one of these arrays is occupied by *M* atoms and the other by *N* atoms, close packed layers with *MN* stoichiometry are obtained. These can be stacked together in the usual close packed stacking sequences to give a family of polytypic structures, for which the AuCu I structure with ABC stacking of the layers can be regarded as the prototype, provided that the axial ratio of the tetragonal cell does not depart much from the ideal value of unity. Many phases with the AuCu I

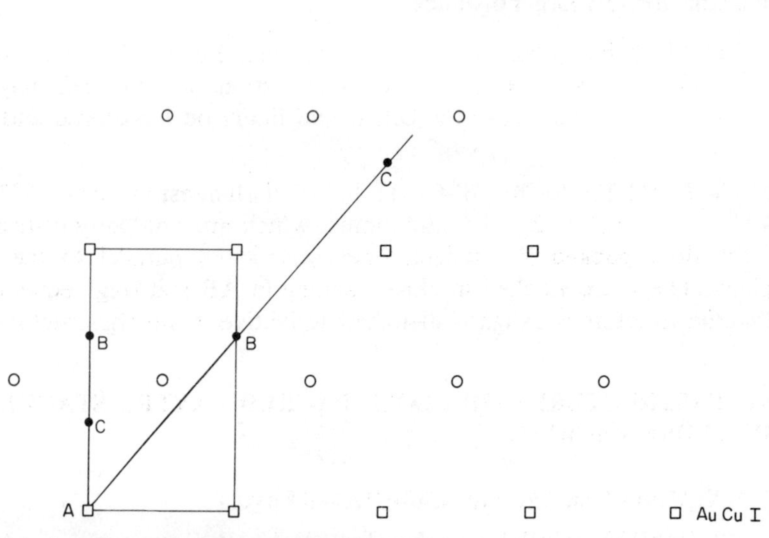

Figure 7-6. Atomic arrangement on close packed planes of the AuCu I (*tP*4) structure, showing the repeat cell for the arrangement and three equivalent stacking positions, A, B, C.

type structure, however, have considerably smaller axial ratios (say 0.80 to 0.85) and this represents an intermediate state between close packed layers in close packing and b.c. cubic {110} 3^6 layers in b.c. cubic ⟨110⟩ stacking. Indeed, at an axial ratio of $c/a = 0.707$, the transformation to exact b.c. cubic packing is complete.

In the AuCd structure (another member of the polytypic family) also, the angles of the 3^6 net triangles are found to be much more nearly those of a b.c. cubic {110} layer than of close packing, although the layer stacking positions are close packed rather than b.c. cubic ⟨110⟩. It must therefore be recognized that, although ideally the AuCu I structure is the prototype for a family of polytypic structures based on close packing, many phases with structures belonging to the family have packing which is intermediate between close packing and b.c. cubic packing. Although it might be presumed that the change towards b.c. cubic packing would be favored by component atoms of different size, there appears to be no very definite correlation between radius ratio of the component atoms and axial ratio of the unit cell for the AuCu I structure.

Four members of this family of polytypic structures have been recognized: AuCu I with ABC, AuCd with AB, α_1-RhTa with BABCAC and NbPb with ABABCBCAC stacking.

AuCuI (*tP*4): The AuCu I structure can also be described in terms of stacking alternate 4^4 layers of Au and Cu atoms in succession in the [001] direction. The axial ratio of the four-atom primitive cell is always less than unity, whereas in the two-atom primitive cell describing the structure, it is about 1.3. Although the atomic arrangement in the two-atom cell is similar to that of the CsCl structure, the cell is greatly elongated in the c direction, whereas in the true tetragonally distorted CsCl structures, such as those of the AuMn phases, c/a has a much smaller value of about 1.04.

Observed axial ratios for phases with the AuCu I structure generally vary from 1.00 to 0.80 (4 atom cell) with δ-CuTi having a value of 0.64 and AlTi and PtV having a value of 0.72, so that in many phases with the AuCu I structure the atoms occupy an intermediate position between close packing and b.c. cubic [110] packing.

AuCd (*oP*4): Typical cell dimensions for AuCd are $a = 4.767$, $b = 3.164$, $c = 4.855$ Å, $Z = 2$.[5] In planes parallel to (100) at heights $x = \frac{1}{4}$ and $\frac{3}{4}$ the atoms form 3^6 nets, but triangle angles approach more nearly those in 3^6 nets on {110} planes of a b.c. cubic array. The stacking of the layers nevertheless remains close packed. Both Au and Cd have 12 neighbors at distances up to 3.16 Å. The closest Au–Cd distance is 2.89 Å.

AuCd has the CsCl structure at high temperatures, and f.c. tetragonal[6] and rhombohedrally distorted CsCl structures[7] have also been described at low temperatures.

α_1-RhTa (*oP*12): The cell, with dimensions $a = 13.55$, $b = 2.822$, $c = 4.742$ Å, $Z = 6$,[8] contains six close packed layers stacked one over the other in a slightly distorted close packed stacking sequence, BABCAC ($3\underline{3}$ in Ždanov–Beck notation) along the *a* direction (Figure 7-7).

Each atom has 12 close neighbors. Close Rh–Ta distances range from 2.73 to 2.76 Å and close Rh–Rh and Ta–Ta distances from 2.82 to 2.86 Å. The phase has the composition $(Ta_{0.97}Rh_{0.21})Rh$.

NbRh (*mP*18): The dimensions of the monoclinic cell of the NbRh structure are $a = 2.806$, $b = 4.772$, $c = 20.250$ Å, $\alpha = 90.53°$, $Z = 9$,[8,9] in the setting with the *a* axis unique. The atoms are arranged in close packed 3^6 nets on nine planes approximately normal to the *c* axis of the crystal, each plane containing equal numbers of Nb and Rh atoms arranged in the rectangular repeat unit shown in Figure 7-8. These layers are stacked in close packed sequence ABABCBCAC ($2\underline{1}$ in Ždanov–Beck notation) and the structure is a member of the family of polytypic structures of which AuCuI is the prototype. The phase occurs in the range from 59 to 64 at.% Rh.

Two distorted structures based on the AuCuI arrangement are recognized:

HgNa (*oC*16): HgNa has cell dimensions $a = 7.19$, $b = 10.79$, $c = 5.21$ Å, $Z = 8$.[10] The Na and Hg atoms are arranged alternately in lines running

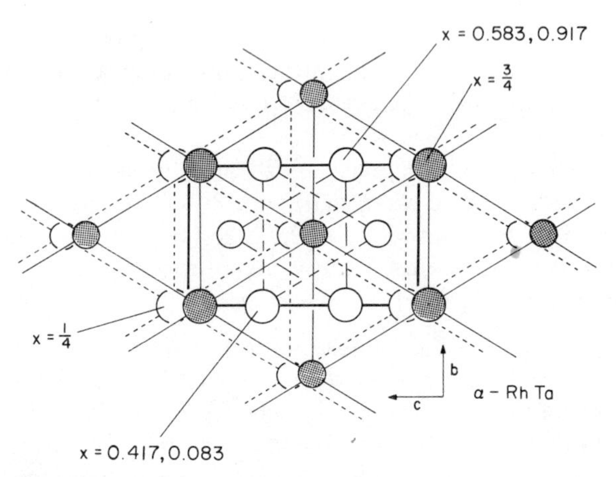

Figure 7-7. α_1-RhTa (*oP*12) structure projected down [100], indicating 3^6 nets of atoms at $x = 0.083$, 0.25, 0.417, 0.583, 0.75, 0.917. Rh small circles, Ta large.

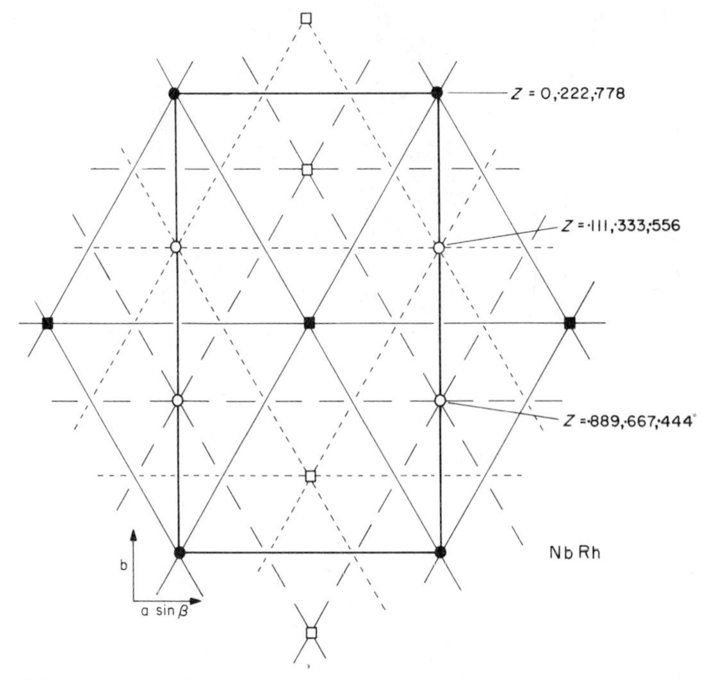

Figure 7-8. NbRh ($mP18$) structure projected down [001], indicating 3^6 nets of atoms at various heights: Nb, circles; Ta, squares.

in the a direction and forming distorted triangular nets parallel to the (001) plane at $z = \frac{1}{4}$ and $\frac{3}{4}$. These are stacked one above the other in AB sequence (Figure 7-9).

BiIn, PbO, FeTe ($tP4$): The BiIn structure ($a = 5.015$, $c = 4.781$ Å, $Z = 2$) can be regarded as a tetragonally distorted AuCuI structure in which c/a is decreased from unity to 0.95 and the Bi atoms are displaced from the Cu atom sites at $z = \frac{1}{2}$ to sites alternately at $z = 0.38$ and 0.62[11] (Figure 7-10a). In the PbO form of the structure the distortion has increased considerably so that the Pb atoms lie at heights of $z \sim 0.24$ and 0.76 and $c/a = 1.26$[12] (Figure 7-10b). In the antistructure of $FeTe_{\sim 0.9}$ (FeSe) the distortion of the unit cell has increased still further, c/a having a value 1.64 (1.47).[13] The Te (Se) atoms have heights $z = 0.285$ and 0.715 (0.26 and 0.74). The PbO and FeTe forms are therefore definitely three-layer structures with the O (Fe) atoms surrounded on either side by layers of Pb (Te) atoms. These groups of three layers are well separated from the neighboring groups of three layers.

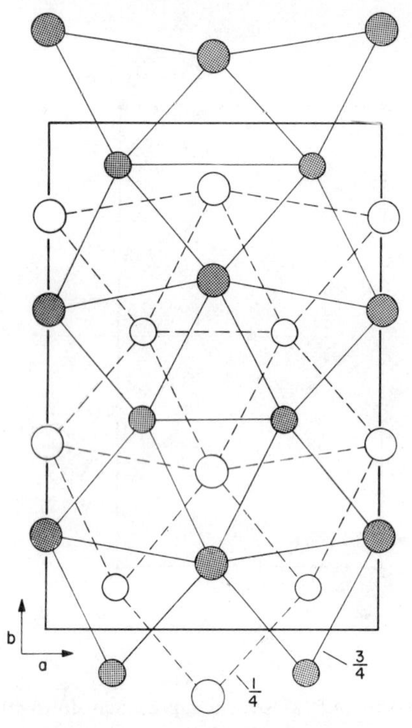

Figure 7-9. HgNa ($oC16$) structure projected down [001] indicating 3^6 nets of atoms at $z = \pm\frac{1}{4}$: Na, large circles: Hg, small circles.

Several superstructures of the AuCu I structures have been recognized:

AuCuII: In the AuCu II structure with cell dimensions $a = 3.97$, $b = 39.74$, $c = 3.70$ Å, $Z = 20$, 10 AuCu I type pseudocells are stacked in the [010] direction, there being a step-shift at the boundary of every fifth cell[14] as shown in Figure 3-20, p. 90. The arrangement in the close packed layers is shown in Figure 7-11.

Cu_2AuPd, $CuAu_2Pd$ ($tP4$): Both Cu_2AuPd and $CuAu_2Pd$ are ordered structures developed from AuCu I. In the former, Cu occupies 0, $\frac{1}{2}$, $\frac{1}{2}$, and $\frac{1}{2}$, 0, $\frac{1}{2}$, Au 0, 0, 0 and Pd $\frac{1}{2}$, $\frac{1}{2}$, 0 positions.[15] In $CuAu_2Pd$ the Cu and Au positions are interchanged. The axial ratio c/a, is significantly less in the former than in the latter. $AuCuPd_2$ is also known, but it has the AuCu I structure without further ordering. The ordered pattern

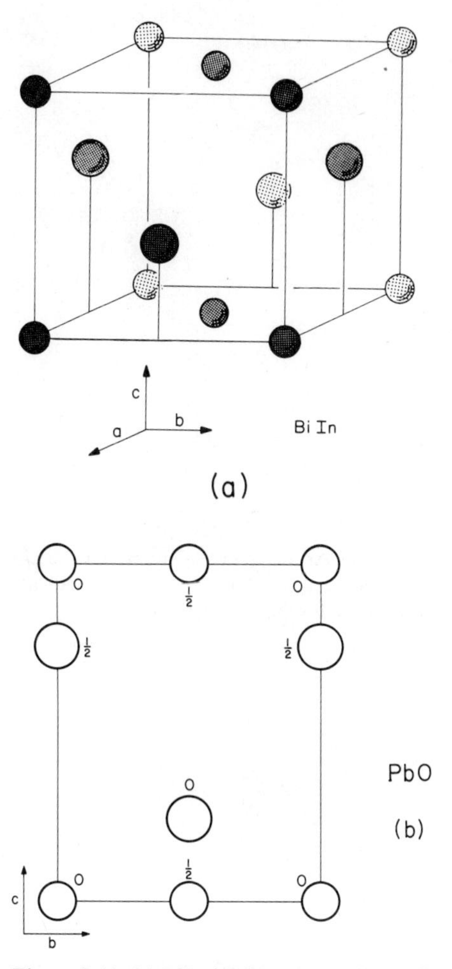

Figure 7-10. (*a*) BiIn (*tP*4) structure. In small circles, Bi large. (*b*) PbO (*tP*4) structure projected down [100]. Oxygen, small circles; Pb, large.

for the c.p. layers parallel to (111) in the Cu_2AuPd and $CuAu_2Pd$ structures is shown in Figure 7-12.

$Mn_{11}Pd_{21}$ (*tP*32): The phase has a range of homogeneity from 56.25 to 65.63 at.% Pd; at 66 at.% Pd it has cell dimensions $a = 2\sqrt{2}a_0 = 8.06$, $c = 2c_0 = 7.34$ Å where a_0 and c_0 refer to the AuCu I type structure of the partially disordered phase.[16] The superstructure results from secondary ordering of the Pd atoms in excess of 50 at.%. As the composition

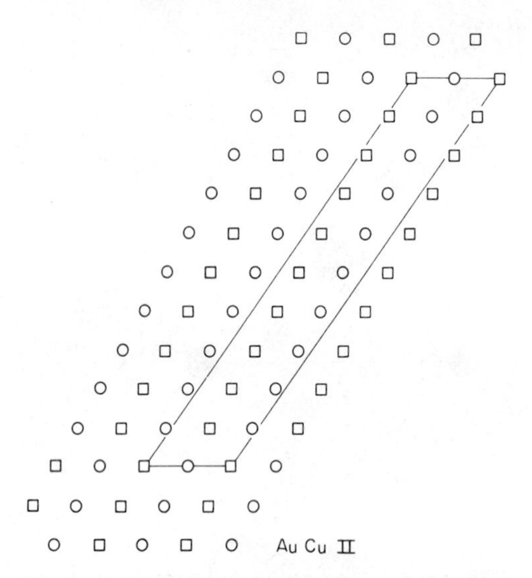

Figure 7-11. Arrangement of atoms on the close packed planes of the AuCu II structure, indicating the repeat cell for the arrangement.

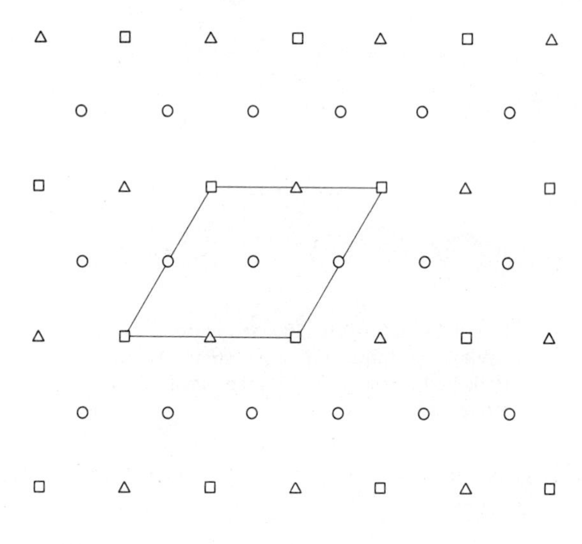

Figure 7-12. Arrangement of the atoms and their repeat cell on the close packed planes of the Cu_2AuPd and $CuAu_2Pd$ ($tP4$) structures.

increases over the range of homogeneity from $Mn_{14}Pd_{18}$ to $Mn_{11}Pd_{21}$, Mn and Pd atoms are located randomly, on positions 0, 0, $\frac{1}{2}$ and 0, $\frac{1}{2}$, 0; $\frac{1}{2}$, 0, 0, but at $Mn_{11}Pd_{21}$ these sites are completely filled with Pd. A surprising feature of the structure which contains alternate layers of Pd and Mn + Pd atoms (Figure 7-13), is that the Pd content is not the same in each of the mixed layers; one contains 2 Pd to 6 Mn, the other 3 Pd to 5 Mn.

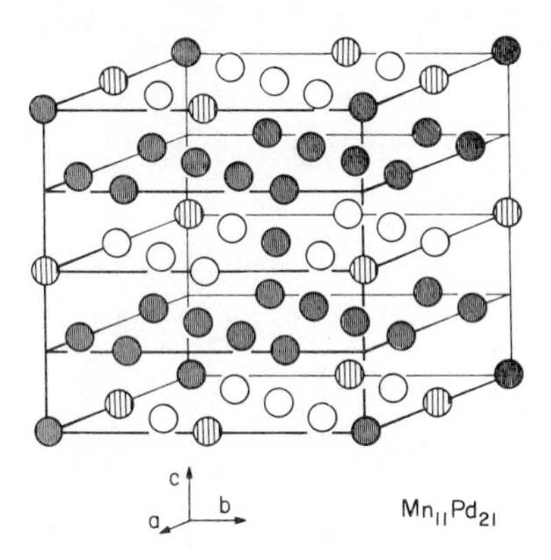

Figure 7-13. $Mn_{11}Pd_{21}$ structure:[16] ○, Mn; ●, Pd; ◍, sites occupied randomly by Pd and Mn in the concentration range from 56.25 to 65.63 at.% Pd; these are filled with Pd at 65.63% Pd corresponding to $Mn_{11}Pd_{21}$.

Two filled up AuCu I type structures have been described:

FeNiN (*tP*3): The FeNiN structure with cell dimensions $a = 2.830$, $c = 3.713$ Å, $Z = 1$, is a filled-up AuCu I structure type.[17] The Fe and Ni atoms are in AuCu I type arrangement and the nitrogen atoms are introduced into the basal planes so as to center the squares of Fe atoms.

Co_2Mn_2C (*tP*5): The Co and Mn atoms in Co_2Mn_2C ($a = c = 3.79$ Å) have the AuCu I structural arrangement with the C atom being located in the Co layer at the cell center, so that it centers every other square of the 4^4 Co network.[18] The structure requires confirmation by neutron diffraction.

MN_2: MN_2 Stoichiometry in Each Close Packed Layer. M on 3^6 Subnet, N on 6^6 Subnet

$MoPt_2$ (*oI*6): $MoPt_2$ with cell dimensions $a = 2.748$, $b = 8.238$, $c = 3.915$ Å, $Z = 2$,[8,19] is a close packed superstructure based on a f.c. cubic subcell

in which distorted close packed triangular layers lying parallel to the (101) plane of the superstructure, are stacked in close packed sequence ABC. Stoichiometry is satisfied in each of the layers which have the atomic arrangement shown in Figure 7-14, the Pt atoms forming a 6^3 net with

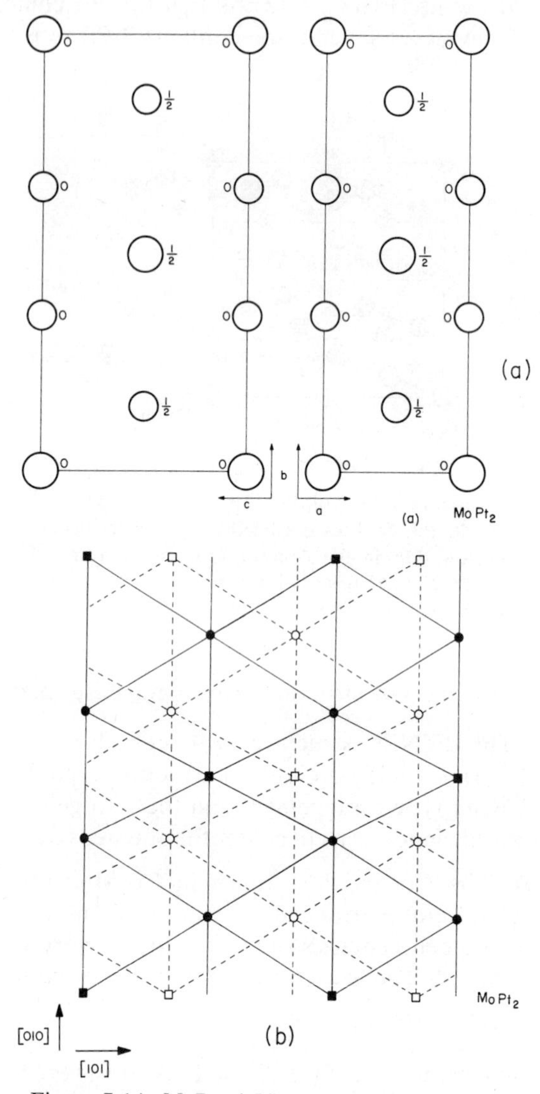

Figure 7-14. $MoPt_2$ (*cI*6) structure. (*a*) Projections down [100] and [001]. Large circles are Mo atoms. (*b*) Close packed layers of atoms parallel to the (101) plane. Squares are Mo.

the hexagons centered by Mo atoms which form a 3^6 net. Since the two components are of similar size, it appears that the close packed layers can be stacked in close packing, rather than in b.c. cubic {110} stacking adopted in the $TiSi_2$ or $CrSi_2$ structures where the component atoms have more disparate sizes.

Mo is surrounded by 12 neighbors in a polyhedron with 16 three and 2 four-sided faces; four neighbors have surface coordination six and the rest four. Thus Mo has 2 Mo at 2.75 Å, 8 Pt at 2.68 Å and 2 at 2.91 Å. Pt is also surrounded by 12 neighbors in a polyhedron with 12 three-sided faces and 4 four-sided faces; four neighbors have surface coordination 5 and the rest 4. Pt has 5 Mo neighbors, 1 Pt at the very close distance of 2.42 Å, 2 at 2.75 Å, and 4 at 2.93 Å.

MN_3: MN_3 Stoichiometry in Each Close Packed Layer. M on 3^6 Subnet, N on 3636 Subnet

Polytypic Structure Families with **MN_3** ***Stoichiometry and Triangular Arrangement of*** **M** ***Atoms.*** A close packed triangular array of atoms can be subdivided symmetrically into four subnets which are themselves each triangular 3^6 arrays. If one of these arrays is occupied by the M atoms and three of them are occupied by the N atoms, the close packed layer has MN_3 stoichiometry with the N atoms forming a 3636 kagomé net and the M atoms a larger 3^6 net as shown in Figure 7-15. These close packed

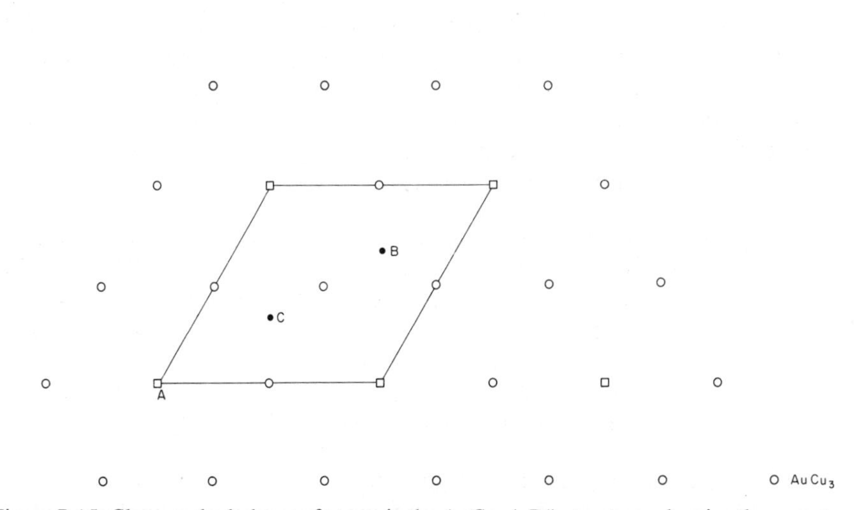

Figure 7-15. Close packed planes of atoms in the $AuCu_3$ ($cP4$) structure, showing the repeat cell and three equivalent stacking positions.

layers can be stacked one above the other in the normal close packed positions A, B, or C so that the M atoms have only N atoms as nearest neighbors in the layers above and below; at stoichiometry MN_3 there never need be $M-M$ inter- or intralayer nearest neighbors. With this condition satisfied, the stacking of MN_3 close packed layers in various close packed sequences generates a family of polytypically related structures for which the $AuCu_3$ structure with cubic ABC stacking sequence is the prototype. This structural arrangement has been referred to by Beck as the T mesh.

Ten recognized polytypes are described in the order of decreasing percentage of cubically surrounded layers.

$AuCu_3$ (*cP*4): In $AuCu_3$ the close packed layers, which lie normal to [111], are stacked in the sequence ABC, or in the Ždanov–Beck notation $1\underline{0},T$ so that all layers are surrounded cubically. Au has 12 Cu at 2.65 Å which are arranged in a cubo-octahedron and Cu has 4 Au and 8 Cu at the same distance and in the same arrangement.

The near-neighbor diagram (p. 53) shows the influence of the geometrical factor in phases with the $AuCu_3$ structure, consistent with the close packed arrangement of the atoms as a whole (Figure 3-6, p. 61). On the near-neighbor diagram the phases occur over a wide range of radius ratios and are distributed between lines for 12–4 M–N and 8 N–N contacts. The atomic volume of phases with the $AuCu_3$ structure is generally less than the appropriate sum of the elemental atomic volumes.

$Ti(Pt_{0.89}Ni_{0.11})_3$ (*hP*28): With cell dimensions $a = 5.491$, $c = 16.67$ Å, $Z = 7$,[20] $Ti(Pt_{0.89}Ni_{0.11})_3$ is a close packed superstructure with layer stacking sequence ACABCAB or in the Ždanov–Beck notation, $5\underline{2},T$. Two of the layers are hexagonally surrounded and five cubically surrounded giving 71.4% cubic stacking, the sequence being chcccch. Each atom has CN12 and there are no Ti–Ti contacts.

VCo_3 (*hP*24): The dimensions of the VCo_3 cell which contains 6 close packed layers of atoms are $a = 5.032$, $c = 12.27$ Å, $Z = 6$.[21] The stacking sequence is ACBABC, or in the Ždanov–Beck notation $3\underline{3},T$; 66.7% of the layers are cubically surrounded, the sequence being hcchcc. $PuAl_3$ is a slightly distorted form of this structure type.

All Co atoms have 4 V + 8 Co neighbors and V have 12 close Co neighbors. There are no V–V contacts either within or between layers.

$PuAl_3$ (*hP*24): The unit cell of the $PuAl_3$ structure, with dimensions $a = 6.10$, $c = 14.47$ Å, $Z = 6$,[22] contains six slightly nonplanar close packed layers of Pu and Al atoms stacked along [001] in the sequence

BABCAC, or in Ždanov–Beck notation $3\underline{3}, T$. The structure is a slightly distorted form of the CoV_3 structure.

$HoAl_3$ (*hR*20): $HoAl_3$ has a corresponding hexagonal cell with dimensions $a = 6.052$, $c = 35.93$ Å, $Z = 15$,[23] that contains 15 close packed layers stacked in the sequence is ABACABCBABCACBC (hchcc) corresponding to 60% of cubically surrounded layers. In Ždanov–Beck notation the stacking sequence is $3\underline{2}, T$.

$TiNi_3$ (*hP*16): In $TiNi_3$ with cell dimensions $a = 5.101$, $c = 8.3067$ Å, $Z = 4$,[24] the layers are stacked in double hexagonal close packed sequence ABAC in the [001] direction, so that 50% of the layers are surrounded cubically. In Ždanov–Beck notation, the stacking is described as $2\underline{2}, T$.

Both Ti(1) and (2) have 12 Ni neighbors at 2.55 Å, and Ni(1) and (2) have 4 Ti + 8 Ni neighbors each.

$PuGa_3$ (*hR*16): $PuGa_3$ with corresponding hexagonal cell dimensions $a = 6.178$, $c = 28.03$ Å, $Z = 12$,[25] has the stacking sequence CACABCBCABAB, or in Ždanov–Beck notation $3\underline{1}, T$. 50% of the layers are surrounded cubically, $(cchh)_3$. $(Ba_{0.95}Ca_{0.05})Pb_3$[26] has the same structure.

γ-$Ta(Pd_{0.67}Rh_{0.33})_3$ (*hP*40): With cell dimensions $a = 5.520$, $c = 22.43$ Å, $Z = 10$,[27] the structure contains 10 close packed layers normal to the *c* axis which are stacked in the sequence CBABCBCACB, or in Ždanov–Beck notation, $2\underline{2}1\underline{2}2\underline{1}, T$. 40% of the layers are in cubic surroundings, $(hhchc)_2$.

$BaPb_3$ (*hR*9): The dimensions of the corresponding hexagonal cell of $BaPb_3$ are $a = 7.287$, $c = 25.77$ Å, $Z = 9$.[28] The hexagonal cell contains 9 close packed layers in the stacking sequence ACACBCBAB, or $2\underline{1}$, *T* in the Ždanov–Beck notation. 33% of the layers are in cubic surroundings, $(chh)_3$.

$Ba(Pb_{0.8}, Tl_{0.2})_3$ (*hP*56): With cell dimensions $a = 7.342$, $c = 39.45$ Å, $Z = 14$,[29] the structure contains 14 close packed layers normal to the *c* axis. 28.6% of the layers are in cubic surroundings, hhhchhc. The structure was determined from X-ray powder data.

$SnNi_3$ (*hP*8): $SnNi_3$ has cell dimensions $a = 5.286$, $c = 4.243$ Å, $Z = 2$. The Sn and Ni atoms together form close packed layers at $z = \pm\frac{1}{4}$[30] which are arranged in hexagonal AB stacking ($1\underline{1}, T$ in Ždanov–Beck notation) with Ni occupying a kagomé subnet and Sn a larger 3^6 net. The structure is the hexagonally stacked polytype of the $AuCu_3$ structure. It is therefore a superstructure of a c.p. hexagonal structure (*hP*2) (Figure 7-16) in the same way that $AuCu_3$ is of the close packed cubic structure

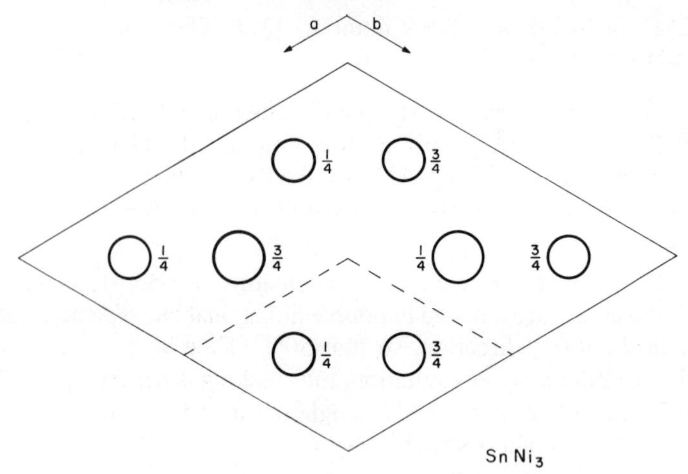

Figure 7-16. Plan of the $SnNi_3$ (*hP*8) structure down [001], showing the Mg type pseudocell. Sn large circles.

(*cF*4). Ni has 8 Ni at 2.615 and 2.64 Å and 4 Sn at 2.61 and 2.64 Å. Sn has 12 Ni neighbors.

The near-neighbor diagram for the $SnNi_3$ structure shows the influence of the geometrical factor in controlling the structural dimensions, since the phases are located between lines for 12–4 M–N and 8 N–N contacts and are found over a wide range of radius ratios of the component atoms, extending between about 0.75 and 1.3. Such behavior is to be expected for a superstructure based on a close packed array of atoms.

Several distorted and filled-up structures based on the $AuCu_3$ type are known:

$CuTi_3$ ($SrPb_3$) (tP4): The $CuTi_3$ structure with $a = 4.158$, $c = 3.594$ Å, $c/a = 0.86$, $Z = 1$,[31] is a tetragonal distortion of the $AuCu_3$ structure. About a dozen phases have been reported with this type of structure.

SiU_3 (tI16): In the SiU_3 structure U(2) atoms at $z = 0$ and $\frac{1}{2}$ form distorted square 1.4^4 nets. U(1) and Si atoms at $z = \frac{1}{4}$ and $\frac{3}{4}$ severally form $\frac{1}{2}.4^4$ nets centering the squares formed by the U(2) atoms (Figure 7-17). The cell dimensions, $a = 6.029$, $c = 8.696$ Å, $Z = 4$,[32] indicate that the SiU_3 structure is a distorted form of the $AuCu_3$ structure.

$BaTiO_3$ (cP5): The perovskite structure can be regarded as a filled-up derivative of the $AuCu_3$ structure. Ba occupies the cell corners, oxygen the face centers, and the Ti the body center (Figure 7-18). Ti is thus surrounded by 6 O at 1.90 Å; Ba has 12 O at 2.69 Å and O has 4 Ba and 2 Ti neighbors.

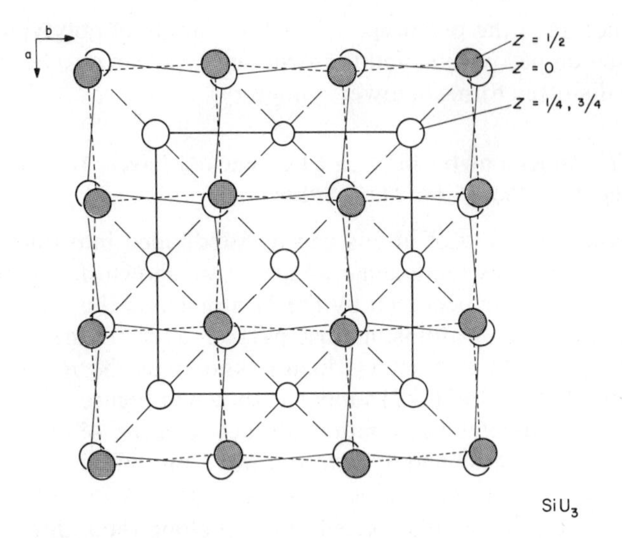

Figure 7-17. Plan of the SiU_3 (*tI*16) structure down [001]. U large circles.

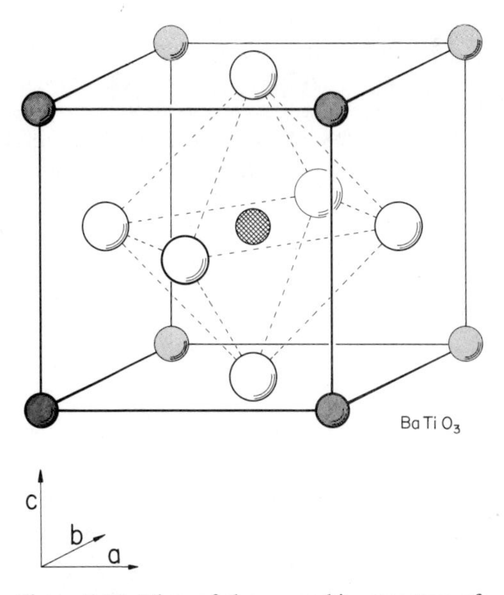

Figure 7-18. View of the perovskite structure of $BaTiO_3$ (*cP*5). Ba shaded atoms at cell corners; Ti atom at body center; oxygen large atoms.

The structure is the prototype of a whole family of polytypic arrangements based on perovskite, and the structure type can also be recognized in various distorted forms of lower symmetry.

MN_3: MN_3 Stoichiometry in Each Close Packed Layer. M on Rectangular 4^4 Subnet, N on $3636+3^26^2$ (1 : 2) Subnet

A triangular 3^6 array of atoms can be subdivided into four subarrays which are themselves rectangular 4^4 arrays as shown in Figure 7-19. If one of these arrays is occupied by the M atoms and the remaining three are occupied by the N atoms, a close packed layer of atoms is again obtained with MN_3 stoichiometry which is known as the R mesh. The N atoms form a $3636+3^26^2$ (1 : 2) subnet as shown in Figure 7-19.

Figure 7-20 indicates the relationship between the 4^4 arrangement of M atoms of the R mesh and 3^6 arrays of the T mesh found in $AuCu_3$ and polytypes. The former is obtained from the latter by introducing a step-shift of $\frac{1}{2}a$ at lines of atoms spaced every b along the b direction of the orthogonal cell. Introduction of the step-shift to generate the 4^4 nets of M atoms, creates two sets of sites for the M atoms. The coordinates of these for the different stacking positions referred to the rectangular repeat cell (Figure 7-21) are:

A	$0, 0$	A^+	$\frac{1}{2}, 0$
B	$0, \frac{1}{3}$	B^+	$\frac{1}{2}, \frac{1}{3}$
C	$0, \frac{2}{3}$	C^+	$\frac{1}{2}, \frac{2}{3}$

The condition that M atoms shall not be nearest neighbors when the layers are stacked one above the other is that the layers occupy primed

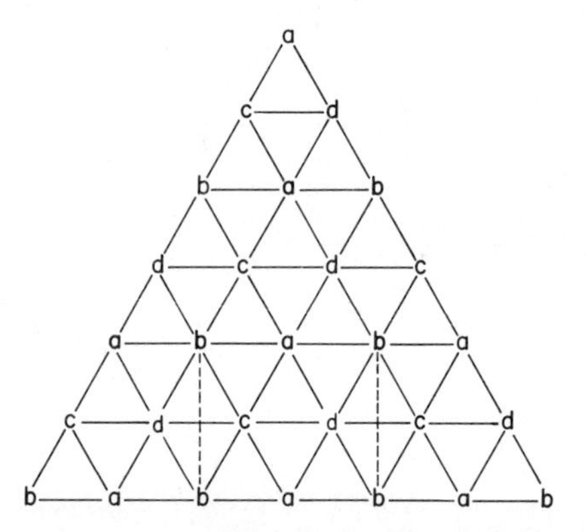

Figure 7-19. Rectangular subdivision of 3^6 net into four 4^4 subnets.

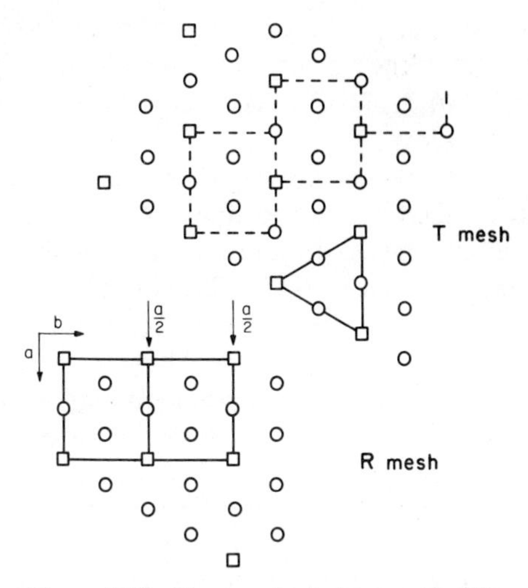

Figure 7-20. Close packed planes with MN_3 stoichiometry, showing relationship of triangular (T) and rectangular (R) arrangements of the M component atoms, and the displacements required to go from T to R.

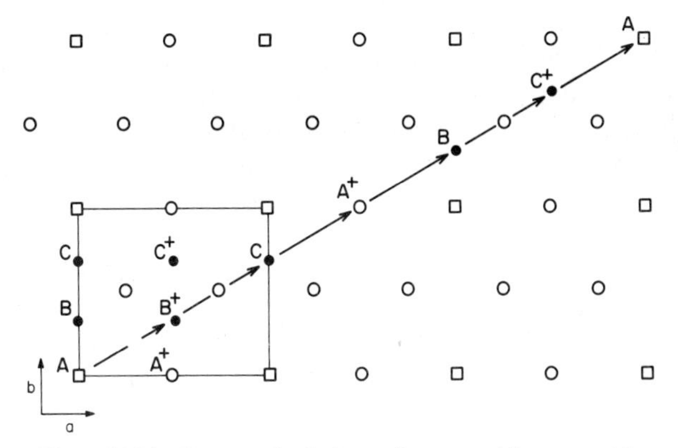

Figure 7-21. Close packed plane of atoms with composition MN_3 in the rectangular arrangement, showing the repeat cell and six stacking positions.

(with a $^+$) and unprimed sites alternately.[33] If the shift between layers is always in the same direction, it requires six layers to return to the original stacking position, the sequence being $AB^+CA^+BC^+$ as shown in Figure 7-21. This sequence of step-shifts generates the $TiAl_3$ (*tI*8) structure which can be regarded as the prototype for a family of polytypic structures resulting from other stacking sequences which satisfy the above conditions for the MN_3 R mesh. Other known structures which are polytypic include the $TiAl_3$, $TiCu_3$, $Mn_{26.5}Au_{73.5}$, $NbPt_3$, $Mn_{23}Au_{77}$, $NbPd_3$, and $Mn_{28}Au_{72}$ type structures described below.

The stacking sequence in polytypic structures involving MN_3 R meshes can be described along the lines of the Ždanov[34] notation (cf. Beck).[35] Displacements in two directions are possible, e.g. $A \rightarrow B^+$ called $+$ and $A \rightarrow C^+$ called $-$ (Figure 7-21). The stacking sequence is described by listing the number of successive displacements in the same direction of sequential layers throughout the unit cell. Thus, for example, the sequence $AB^+CA^+CB^+$ which involves three displacements in the positive direction followed by three in the negative direction, is completely described as $3\underline{3}$,R. The minimum sequence of displacements that repeats itself, describes the stacking sequence in the primitive cell.

The relationship between the MN_3 R and β-W ($cP8$) structures should be noted. In the former, coplanar $3636+3^26^2$ (1:2) nets of N atoms and 4^4 nets of M atoms form close packed layers which are stacked in close packing, whereas in the latter $3636+3^26^2$ (1:2) nets of N atoms, alternately rotated 90° in their plane relative to each other, are separated by 4^4 nets of M atoms equally spaced between them.

$TiAl_3$ ($tI8$): The $TiAl_3$ unit cell with dimensions $a = 3.848$, $c = 8.596$ Å, $Z = 2$,[36] is composed of two distorted $AuCu_3$ type subcells stacked one above the other along [001] with a step-shift at the interface between them. It is therefore a M = 1 structure in the terminology of Sato. The superstructure is built up of close packed MN_3 layers which lie parallel to (112) planes in the $TiAl_3$ cell (Figure 7-22), the stacking sequence being $AB^+CA^+BC^+$ (Figure 7-21), or in Ždanov–Beck notation $1\underline{0}$, R for the primitive cell. There are no Ti–Ti nearest neighbors in the structure.

Phases with the $TiAl_3$ structure (MN_3) divide into two groups: one with axial ratios $c/a \sim 2.05$ and the other with $c/a \sim 2.25$. The variation of c with a is quite different for phases in the two groups as shown in Figure 7-23, and this difference in behavior is also reflected in the near-neighbor diagram (Figure 7-23). Phases with $c/a \sim 2.05$ follow the lines for 8–4 $N(1)$–$N(2)$ and 4 $N(2)$–$N(2)$ contacts closely, showing that the N–N contacts control the structural dimensions, and in those phases with radius ratios greater than unity, the M–N contacts are compressed somewhat. Phases with $c/a \sim 2.25$, which include $TiAl_3$, are distributed close to the intersection of lines for 8–4 $N(1)$–$N(2)$ and 8–4 M–$N(2)$ contacts and generally between them, suggesting that now a geometrical or coordination factor controls the structural dimensions. The increased axial ratio in the second group of phases results from the failure of the a parameter to

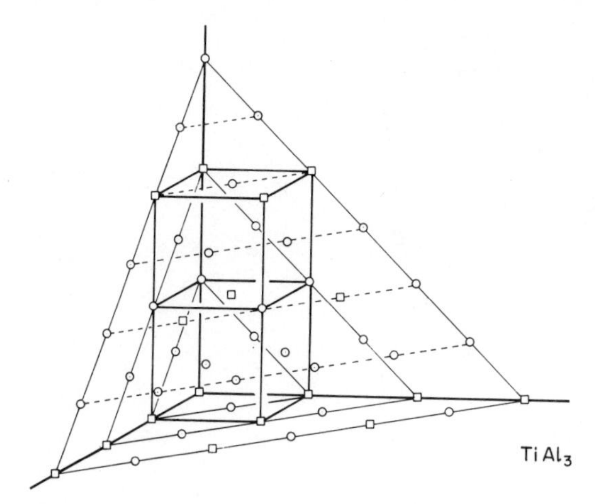

Figure 7-22. Diagram of the $TiAl_3$ ($tI8$) structure indicating the close packed planes of atoms.

match the increase in the c parameter (Figure 7-23), so that the 4–4 M–$N(1)$ contacts which lie in the basal plane of the structure are particularly compressed. This is reflected in the distortion of the close packed 3^6 nets in $TiAl_3$ and other phases with $c/a \sim 2.25$. The ratio of the edges of the rectangles formed by the M component in the (112) planes has an ideal value of $2/\sqrt{3} = 1.155$ for close packed layers. For $TiAl_3$ with $c/a = 2.234$ the ratio has the value of 1.030, whereas for the less distorted VNi_3 with $c/a = 2.036$, it is closer to the ideal value, being 1.136.

β-TiCu$_3$ (*oP*8): β-$TiCu_3$ has unit cell dimensions $a = 5.162$, $b = 4.347$, $c = 4.531$ Å, $Z = 2$.[3] The atoms are arranged in close packed layers parallel to the (010) plane (Figure 7-24). They are stacked in the sequence AB^+ in the notation of p. 326; the Ždanov–Beck description is $1\underline{1},R$. The ratio of the sides of the Ti rectangle, 1.139, is close to the ideal value of 1.155. Ti has 12 Cu neighbors at 2.58, 2.61, and 2.67 Å. Each Cu has 4 Ti and 8 Cu neighbors. There are no Ti–Ti nearest neighbors.

β-NbPt$_3$ (*mP*48): The axes of the β-$NbPt_3$ cell ($a = 5.54$, $b = 4.87$, $c = 27.33$ Å, $\beta = 90.5°$, $Z = 12$)[8,19] are very nearly orthogonal. Twelve close packed layers of atoms parallel to the (001) plane are stacked one above the other in the sequence $BC^+AB^+AB^+CA^+CA^+BC^+$ in the notation of p. 326 or $3\underline{1}, R$ in the Ždanov–Beck notation. There are no Nb–Nb nearest neighbors either within or between the layers. The value of $a/b = 1.137$ is slightly less than the ideal value of $2/\sqrt{3} = 1.155$ required for close packing in the layers.

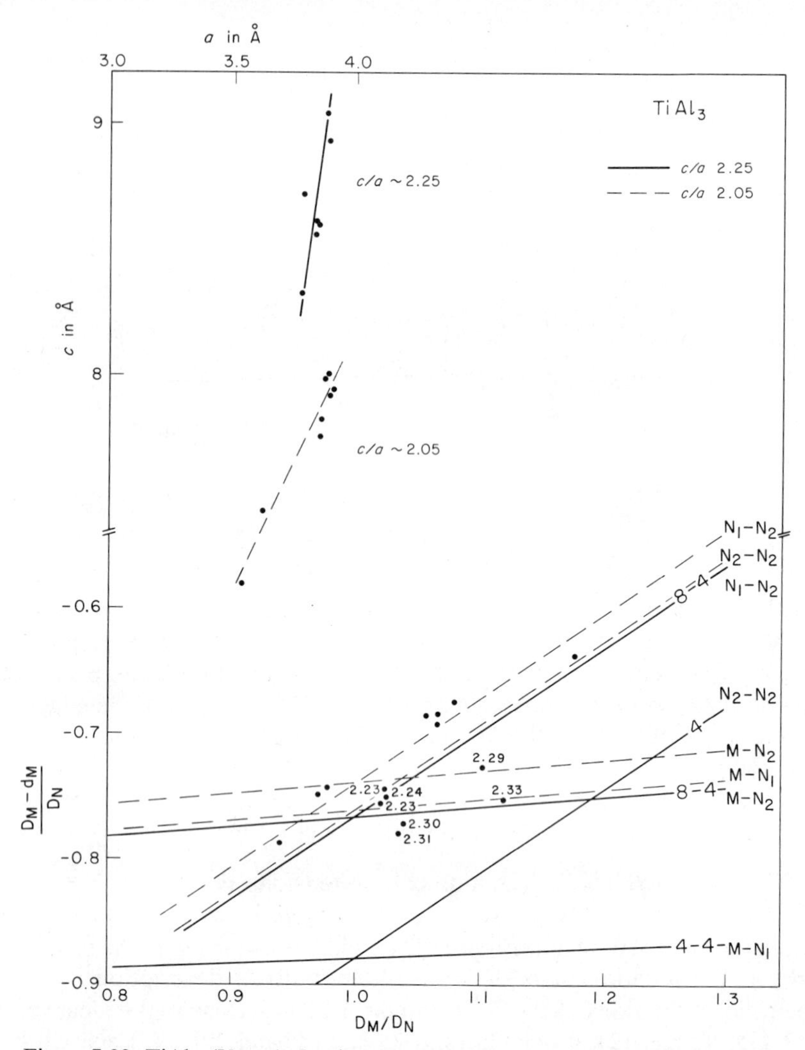

Figure 7-23. $TiAl_3$. (Upper): Lattice parameters c *vs.* a of phases with the $TiAl_3$ (MN_3) structure. (Lower) Near-neighbor diagram for phases with the $TiAl_3$ structure calculated for $c/a = 2.25$ full line, $c/a = 2.05$ broken line. Numbers on lines indicate numbers of N–N, M–N, etc. contacts. Small numbers give axial ratios, c/a, of phases with the $TiAl_3$ structure which have values about 2.25.

β-NbPd₃ (*oP*24): The β-$NbPd_3$ cell with dimensions $a = 5.486$, $b = 4.845$, $c = 13.60$ Å, $Z = 6$,[8] has close packed planes of atoms parallel to (001) at heights $\frac{1}{12}$, $\frac{1}{4}$, $\frac{5}{12}$, $\frac{7}{12}$, $\frac{3}{4}$, $\frac{11}{12}$. The layer stacking sequence is $C^+AC^+BA^+B$; in terms of the symbols given on p. 326, the Ždanov–Beck

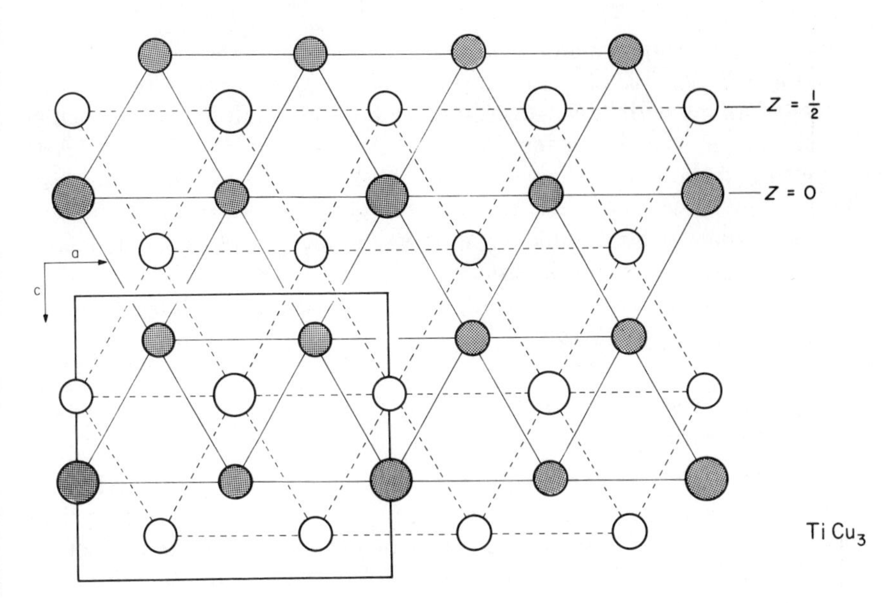

Figure 7-24. View of the $TiCu_3$ (*oP*8) structure down [010]. Large atoms Ti.

description is 33̲, *R*. There are no Nb–Nb nearest neighbors either within or between the layers. The ratio of $a/b = 1.132$ is close to the ideal value of $2/\sqrt{3} = 1.155$ for close packed layers.

$Mn_{23}Au_{77}$: Alloys about this composition have mainly the $5H^S$ type structure (see p. 99 for significance of symbols) formed by stacking close packed layers with stoichiometry approximately MN_3 and the *M* atoms in rectangular 4^4 array.[33,37] The layer stacking sequence in terms of the description given on p. 326 is $AB^+CA^+BA^+BC^+AB^+$ (in Ždanov–Beck notation it is 41̲, *R*). The dimensions and space group of the orthorhombic cell have not been given, but there are no Mn–Mn nearest neighbors either within or between the layers.

$Mn_{28}Au_{72}$: Alloys about this composition mainly have the two $6H^S$ structures[33] which are formed by stacking together close packed layers of approximate stoichiometry MN_3 and the *M* atoms arranged in rectangular 4^4 array. Both structures are six-layer stacking types and have orthorhombic symmetry, but details of space groups and cell dimensions have not been given. In the notation of p. 326 the stacking sequences are $AB^+CA^+CB^+$ for $6H_1^S$ and $AB^+CB^+CB^+$ for $6H_2^S$; in Ždanov–Beck notation the stacking sequences are respectively 33̲,*R* and 21̲12̲,*R*. Nearest-neighbor Mn–Mn approaches in or between layers are avoided, except for those that occur because the alloy contains 28 at. % Mn.

$Mn_{26.5}Au_{73.5}$: Alloys about this composition have the $3R^S$ and $6H^S$ structures.[33] The $3R^S$ structure results from the stacking of close packed layers of atoms with MN_3 stoichiometry and the M atoms in rectangular 4^4 array, in the sequence $AB^+CB^+CA^+CA^+BA^+BC^+BC^+AC^+AB^+$ according to the notation given on p. 326. In the Ždanov–Beck notation the stacking sequence is $2\underline{1},R$. The structure with the 18 layer stacking sequence, has orthorhombic symmetry, but details of cell dimensions and space group are not given. There are no nearest-neighbor Mn–Mn approaches in or between layers, except for those resulting from the Mn content exceeding 25 at. %.

MN_3: MN_3 Stoichiometry in Each Close Packed Layer. M on 3^34^2 Subnet, N on $3636 + 3^26^2$ Subnet

$ZrAl_3$ ($tI16$): The $ZrAl_3$ structure is a superstructure based on close packing. With cell dimensions $a = 4.014$, $c = 17.32$ Å, $Z = 4$, the height of the superstructure cell in the c direction corresponds to four f.c. cubic pseudocells (Figure 7-25).[36] Alternate layers of atoms in the c direction

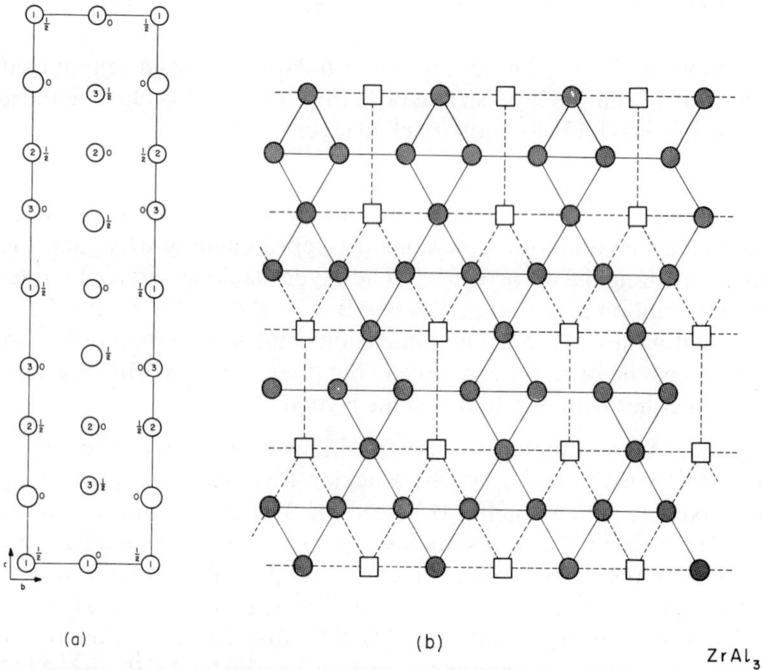

Figure 7-25. (*a*) View of $ZrAl_3$ ($tI16$) structure down [100]. Large atoms Zr. (*b*) Arrangement of Zr atoms in triangles and rectangles (TR) on close packed plane in $ZrAl_3$ structure.

are composed of either Al, or Al and Zr atoms in equal numbers, the Al(3) and Zr atoms in the mixed layers being displaced somewhat in the *c* direction from the exact close packed sites.

Each close packed layer has $ZrAl_3$ stoichiometry with the Zr atoms being arranged in a combination of the *R* and *T* arrangements (p. 326) as shown in Figure 7-25. This results from a displacement fault in alternate lines of Zr and Al atoms which transforms the *T* into the *R* arrangement and *vice versa*. Figure 7-26 shows a repeat cell in the close packed layers and indicates the positions of the twelve-fold repeat sequence in the layer stacking. The structure can alternatively be described by considering the arrangement of 4^4 nets along the *c* axis of the unit cell. Starting from the first mixed layer, there is a stacking fault at the second mixed layer, so

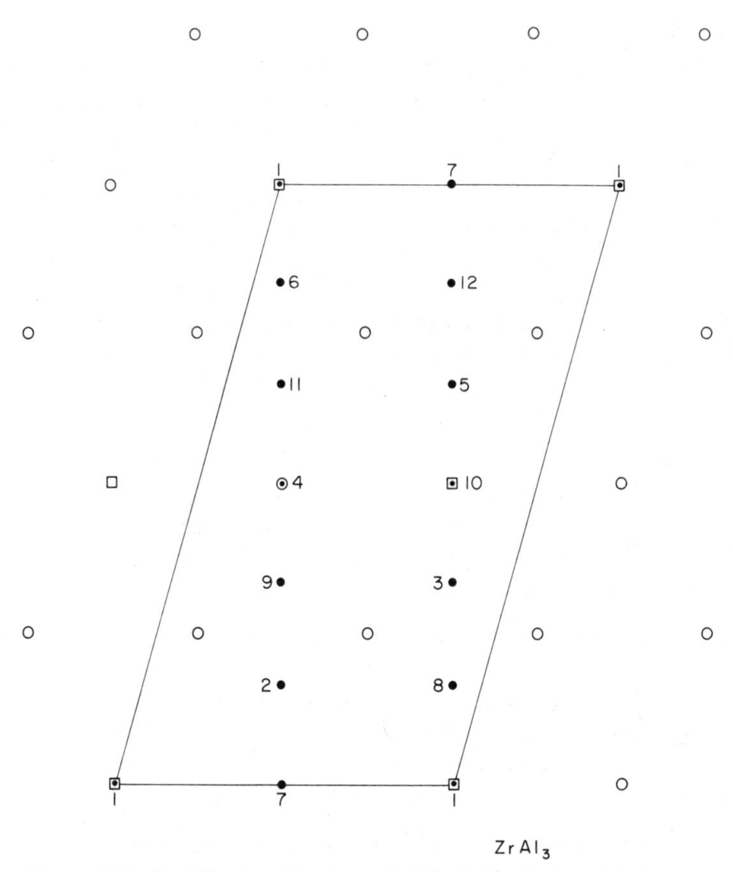

Figure 7-26. Further view of a close packed plane of atoms in the $ZrAl_3$ structure showing the repeat cell and 12 different stacking positions.

that compared to the AuCu structure, the Zr (Au) and Al (Cu) sites are interchanged. The third mixed layer maintains the same arrangement and a stacking fault occurs in the fourth mixed layer which returns the Zr (Au) atoms to their original site over the cell corners (0, 0, 0). This arrangement is maintained in the first mixed layer of the next cell.

The axial ratio, $c/4a$, in the $ZrAl_3$ structure has a value of 1.079. The Au_3Zn[H] phase also has this structure.

The $ZrAl_3$ structure is the prototype for a family of polytypic structures which arise through various stacking sequences, although in order to attain the maximum separation of the M atoms in successive layers, the stacking position number can only change by unity in a positive or negative sense on going from one layer to the next. The structures of several polytypes are described below.

$CdAu_3$ and alloys with 25–30 *at.%Cd*: These alloys have close packed MN_3 planes arranged in TR type nets. The stacking sequence (neglecting atomic ordering) was found to change with increasing Cd content as follows: ABC (25%), ABAC, $2\underline{2}$ (25.5%), ABCACB, $3\underline{3}$, or ABABAC, $2\underline{1}1\underline{2}$, (25 to 27%) and ABCBCACAB (27 to 30%).[38] The latter, having a 36 layer repeat when ordering is considered, is the $Cd_{10}In_{15}Au_{75}$ type structure described on p. 335. Two further complex structures found in the system at 29 to 35 at.%Cd are reported on p. 353.

The $CdAu_3$ structure[39] is a slight variant of the $ZrAl_3$ type, the displacements of the Cd atoms being in the opposite sense to those of Zr (i.e., in the bands of triangular arrangement they move toward each other rather than away from each other).

$Mg_{24}Au_{76}$ (*oC*64): The $Mg_{24}Au_{76}$ cell ideally contains 48 Au and 16 Mg atoms, the dimensions being $a = 5.747$, $b = 19.95$, $c = 9.437$ Å.[40] The structure has close packed $ZrAl_3$ type (TR) layers (Figure 7-26) lying parallel to the (001) plane and stacked successively on sites 1, 12, 1, 2 along the c direction of the $Mg_{24}Au_{76}$ cell. This is a 4H sequence, ABAC, or $2\underline{2}$ in Ždanov–Beck notation.

$Mg_{24}Au_{76}$ (*oC*160): Ideally the cell contains 40 Mg and 120 Au atoms, the dimensions being $a = 5.740$, $b = 19.83$, $c = 23.59$ Å in setting $Cm2m$.[40] The atoms are arranged in planes parallel to the (001) plane. The structure is a polytype of the $ZrAl_3$ structure with close packed (TR) layers stacked in the 10H sequence 1, 2, 3, 2, 3, 4, 3, 2, 3, 2 (Figure 7-26) along the [001] direction; $2\underline{1}2\underline{2}1\underline{2}$ in Ždanov–Beck notation.

$MgAu_{\sim 3}$: The $7R^s$ structure has been observed in ordered alloys about the composition $MgAu_3$.[41] The structure has close packed $ZrAl_3$-type (TR) layers (Figure 7-25, p. 332) stacked in an 84 layer sequence. The sub-

structure disregarding the ordering of the atoms has a 21 layer repeat. Two possible 7R arrangements are consistent with the experimental data and the sequence of the first 7 layers of these two arrangements fixes that throughout the 84 layers (see Figure 7-26):

(i) 1, 2, 3, 2, 1, 2, 3, ...
(ii) 1, 12, 1, 2, 1, 2, 3, ...

The order of preference of these Mg–Au, $ZrAl_3$-structure polytypes with increasing Mg content between 22 and 30 at. % is $1R^s$ or $ZrAl_3$, 4H, "$Mg_{23}Au_{77}$" structure, $7R^s$, $10H^s$.‡

$Cd_{12.5}In_{12.5}Au_{75}$: $Cd_{12.5}In_{12.5}Au_{75}$ has a pseudo-orthorhombic subcell with dimensions $a = 2.916$, $b = 5.037$, $c = 21.59$ Å, $\alpha = 89.73°$.[42] Assuming Cd and In to occupy the *M* component sites randomly, the structure is a polytype of the $ZrAl_3$ structure with close packed (*TR*) layers stacked in a 9 layer repeat sequence (neglecting atomic ordering), ABABCBCAC or $2\underline{1}$ in Ždanov–Beck notation. The superstructure has $a = 2a_0$, $b = 4b_0$, $c = 4c_0$ and a 36 layer repeat sequence, 1, 2, 1, 2, 3, 2, 3, 4, 3, . . ., 11, 12, 1, 12 expressed in terms of the layer repeat positions shown in Figure 7-26.

$Cd_{13.5}In_8Au_{78.5}$: $Cd_{13.5}In_8Au_{78.5}$, with hexagonal cell dimensions approximately $a = 2.91$, $c = 14.37$ Å, is a polytype of the $ZrAl_3$ structure with close packed (*TR*) layers stacked in the 6 layer sequence ABABAC neglecting ordering.[42] The ordered structure also has a 6 layer repeat sequence 1,2,1,2,1,12 in terms of the layer repeat positions shown in Figure 7-26. Cd and In occupy the *M* component sites randomly.

MN_3: Structures with M on Various Square-Triangle Subnets, MN_3 Stoichiometry in Each Layer

$SnCu_3$ (oC80): $SnCu_3$ with orthorhombic subcell $a_0 = 2.77$, $b_0 = 4.78$, $c_0 = 4.34$ Å, is made up of ordered close packed layers parallel to the (001) plane (Figure 7-27) which are described by Beck as TR_4. These are stacked along [001] in two positions AB, so that the repeat sequence in Ždanov–Beck notation is $1\underline{1}$. The *C*-face centered orthorhombic superstructures has $a = 2a_0$, $b = 10b_0$, $c = c_0$.[43] $Cd_{2.5}In_{22.5}Au_{75}$ has a similar structure.

$Cd_{10}In_{15}Au_{75}$: $Cd_{10}In_{15}Au_{75}$ with subcell dimensions $a_0 = 2.915$, $b_0 = 5.064$, $c_0 = 21.58$ Å, $\alpha = 89.75°$, has a pseudo-orthorhombic structure based on the stacking of close packed MN_3 layers of the type shown in

‡For notation see p. 99.

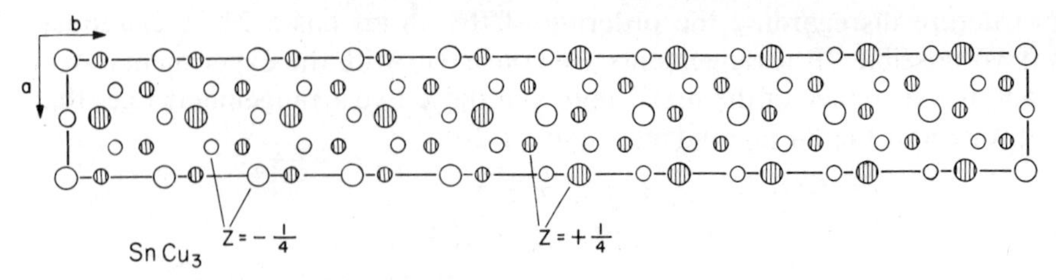

Figure 7-27. Plan of the $SnCu_3$ ($oC80$) structure down [001]. Sn atoms large.

Figure 7-28, according to X-ray powder investigations.[42] These are referred to as TR_2 by Beck, being a combination of triangular and rectangular arrangements of the M component (random arrangement of Cd and In). Neglecting the atomic ordering, the 9 layer stacking sequence of the subcell is ABABCBCAC or 2$\underline{1}$ in Ždanov–Beck notation. The superstructure has $a = 2a_0$, $b = 6b_0$, $c = 3c_0$ and a 27 layer repeat sequence, 1, 2, 1, 2, 3, 2, 3, . . . , 7, 8, 9, 8, 9, 1, 9 expressed in terms of the layer repeat positions shown in Figure 7-28.

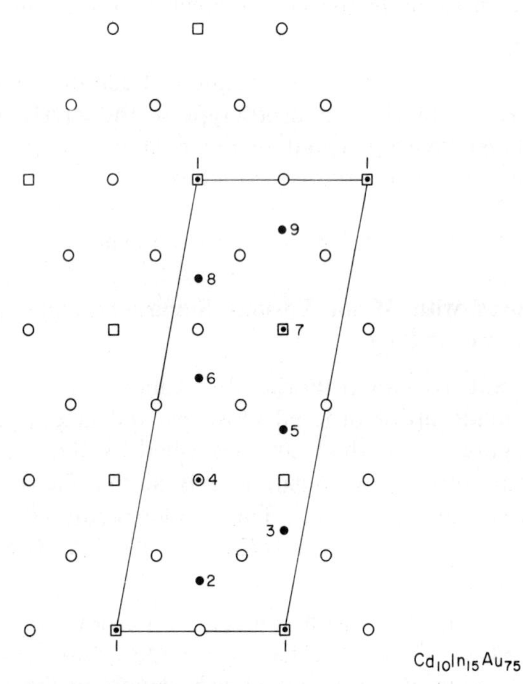

Figure 7-28. Arrangement of atoms in close packed planes of the $Cd_{10}In_{15}Au_{75}$ structure, showing the repeat cell and 9 different stacking positions.

$Cd_{3.3}In_{21.7}Au_{75}$: $Cd_{3.5}In_{21.5}Au_{75}$ with subcell dimensions $a_0 = 2.926$, $b_0 = \sqrt{3} . a$, $c_0 = 19.13$ Å[42] is a polytype of the $Cd_{10}In_{15}Au_{75}$ structure having MN_3 close packed TR_2 layers in an 8 layer repeat sequence for the subcell (i.e., neglecting the atomic ordering), ABABACAC, or $2\underline{1}1\underline{2}1\underline{1}$ in Ždanov–Beck notation according to X-ray powder investigation. The orthorhombic superstructure has $a = 2a_0$, $b = 6b_0$, $c = c_0$ and contains 192 atoms. The repeat sequence of the layers is still 8 with stacking sites 1, 2, 1, 2, 1, 9, 1, 9 expressed in terms of the layer repeat positions shown in Figure 7-28. Cd and In occupy the M sites randomly.

MISCELLANEOUS SUPERSTRUCTURES BASED ON CLOSE PACKING

$ZrGa_2$ ($oC12$): $ZrGa_2$ with cell dimensions $a = 12.894$, $b = 3.994$, $c = 4.123$ Å, $Z = 4$,[44] is a slightly distorted superstructure of a close packed array of atoms. In the a direction the cell corresponds to stacking three f.c. cubic pseudocells. Slightly distorted close packed planes of atoms normal to [111] of the pseudocells have the unusual distribution of Ga and Zr atoms shown in Figure 7-29. The 3×2 repeat parallelogram for the ordering sequence gives 9 different stacking positions, A′B′C′A″B″C″-ABC. The three independent Ga atoms and Zr each have 12 coordination. Ga–Zr distances range from 2.77 to 3.06 Å. Close Ga–Ga distances range from 2.76 to 3.07 Å and the Zr–Zr distances are 3.28 Å. The structure type probably needs further confirmation.

$HfGa_2$ ($tI24$): The $HfGa_2$ structure with cell dimensions $a = 4.046$, $c = 25.446$ Å, $Z = 8$,[44] contains 6 f.c. cubic pseudocells stacked along the c axis (Figure 7-30a), according to X-ray powder data. Proceeding along the c axis of the crystal, a layer of Ga(1) atoms is followed by two mixed layers of Ga(2) and Hf atoms which are not quite coplanar; then the sequence repeats. In the superstructure close packed layers of atoms parallel to {111} planes of the cubic pseudocell are stacked in close packing with the 18 layer repeat sequence indicated in Figure 7-30b. The close packed layers are not quite planar.

$MoNi_4$ ($tI10$): Mo (at cell corners) and Ni atoms form a $\frac{1}{4}.4^4$ net at $z = 0$ and also at $z = \frac{1}{2}$ (Mo at cell body center), the net at $z = \frac{1}{2}$ being displaced so that its atoms lie over the centers of the squares of the net at $z = 0$ and *vice versa*[45] (Figure 7-31a). The sequence of atoms along lines of the net is MoNiNiNiNi . . . . The unit cell dimensions, $a = 5.727$, $c = 3.566$ Å, $Z = 2$, indicate that the superstructure is based on a f.c. cubic, rather than a b.c. cubic pseudocell. The atoms form close packed layers perpendicular to the {111} planes of the f.c. cubic pseudocell

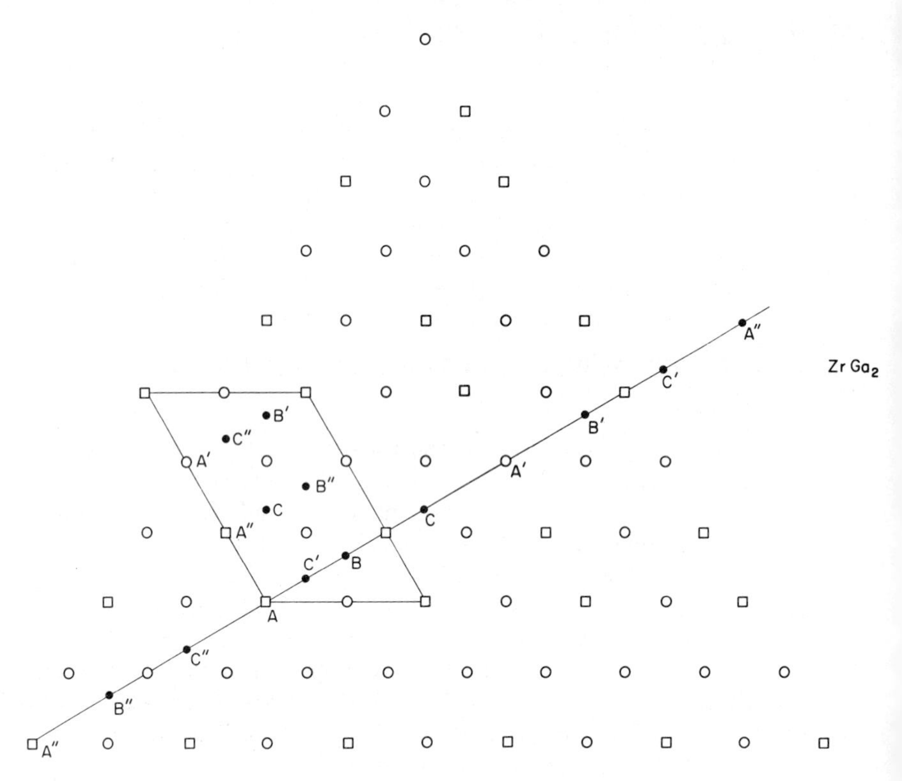

Figure 7-29. Arrangement of atoms on close packed planes in the $ZrGa_2$ (*oC*12) structure showing a repeat cell and the 9 stacking positions.

which are stacked in the 15 layer close packed repeat sequence indicated in Figure 7-31*b*. The structure needs further confirmation from single-crystal data.

$ZnAu_4$ (*oP*96): $ZnAu_4$ has a two-dimensional antiphase domain structure with $M_1 = 2.9$, $M_3 = 2.4$, according to single-crystal X-ray data. Atom positions have been determined for an ideal structure with $M_1 = 3$, $M_3 = 2$ and $a = 6a_0$, $b = b_0$, $c = 4c_0$, where $a_0 = 4.036$, $b_0 = 4.025$, and $c_0 = 4.061$ Å refer to the orthorhombically distorted f.c. cubic pseudocell.[46]

$ZrAu_4$ (*oP*20): $ZrAu_4$ with cell dimensions $a = 5.006$, $b = 4.845$, $c = 14.294$ Å, $Z = 4$ is a superstructure based on close packing.[47] The atoms form close packed layers parallel to the (010) plane at $y = \frac{1}{4}$ and $\frac{3}{4}$ and these are stacked in the [010] direction in AB sequence. The Zr atoms are arranged in the close packed layers in isolated bands of triangles running in the *a* direction, and the Zr atoms of one layer avoid those of the layers

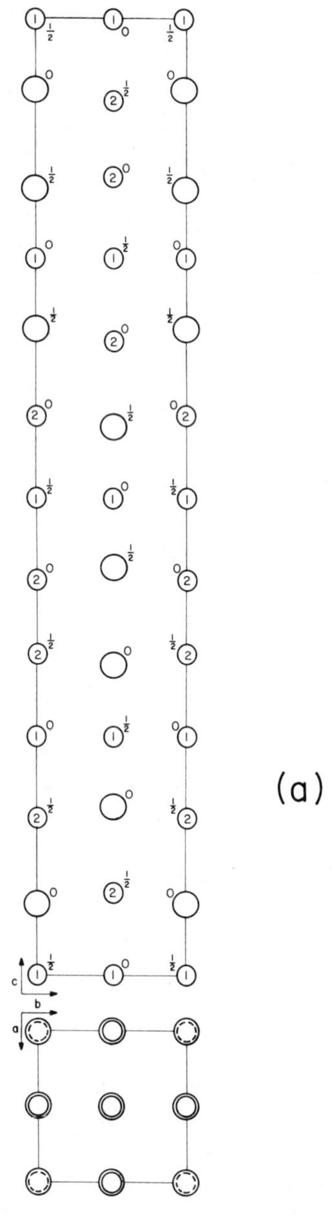

Figure 7-30. (*a*) Views of the $HfGa_2$ (*tI*24) structure along [100] and [001].

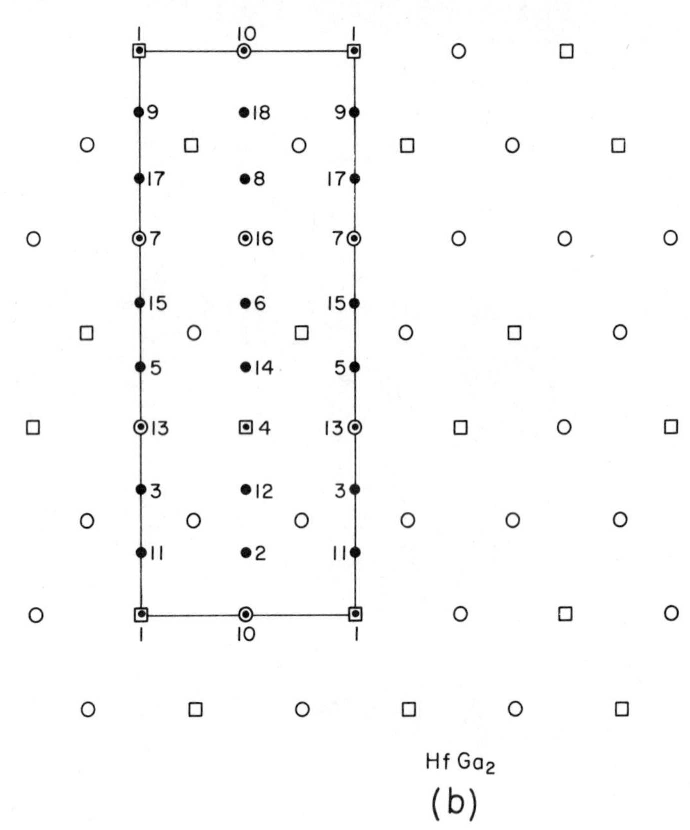

Figure 7-30 (*b*). Arrangement of atoms on close packed planes in the $HfGa_2$ structure showing a repeat cell and the 18 stacking positions.

above and below (Figure 7-32), so that there are no Zr–Zr contacts. The arrangement gives Zr 12 Au neighbors in a twinned cubo-octahedron at distances of 2.86 to 2.94 Å. The four independent Au atoms similarly have 12 neighbors. The structure, determined only from X-ray powder data, requires confirmation.

WAl_5 (*hP*12): The unit cell has dimensions $a = 4.902$, $c = 8.857$ Å, $Z = 2$.[48] The atoms form close packed layers at $z = 0$, $\frac{1}{4}$, $\frac{1}{2}$, and $\frac{3}{4}$ which are stacked in close packed sequence. The layers at $z = 0$ and $\frac{1}{2}$ correspond to the *T* arrangement found in the $AuCu_3$ structure and polytypes, with Al(1) and (2) atoms forming a kagomé net and the W atoms occupying a larger 3^6 net. These nets are stacked in two positions, 1 and 3, as indicated in Figure 7-33. At $z = \frac{1}{4}$ and $\frac{3}{4}$, Al(3) atoms alone form a 3^6 net which is stacked in two other positions, 2 and 4, as shown in Figure 7-33. This

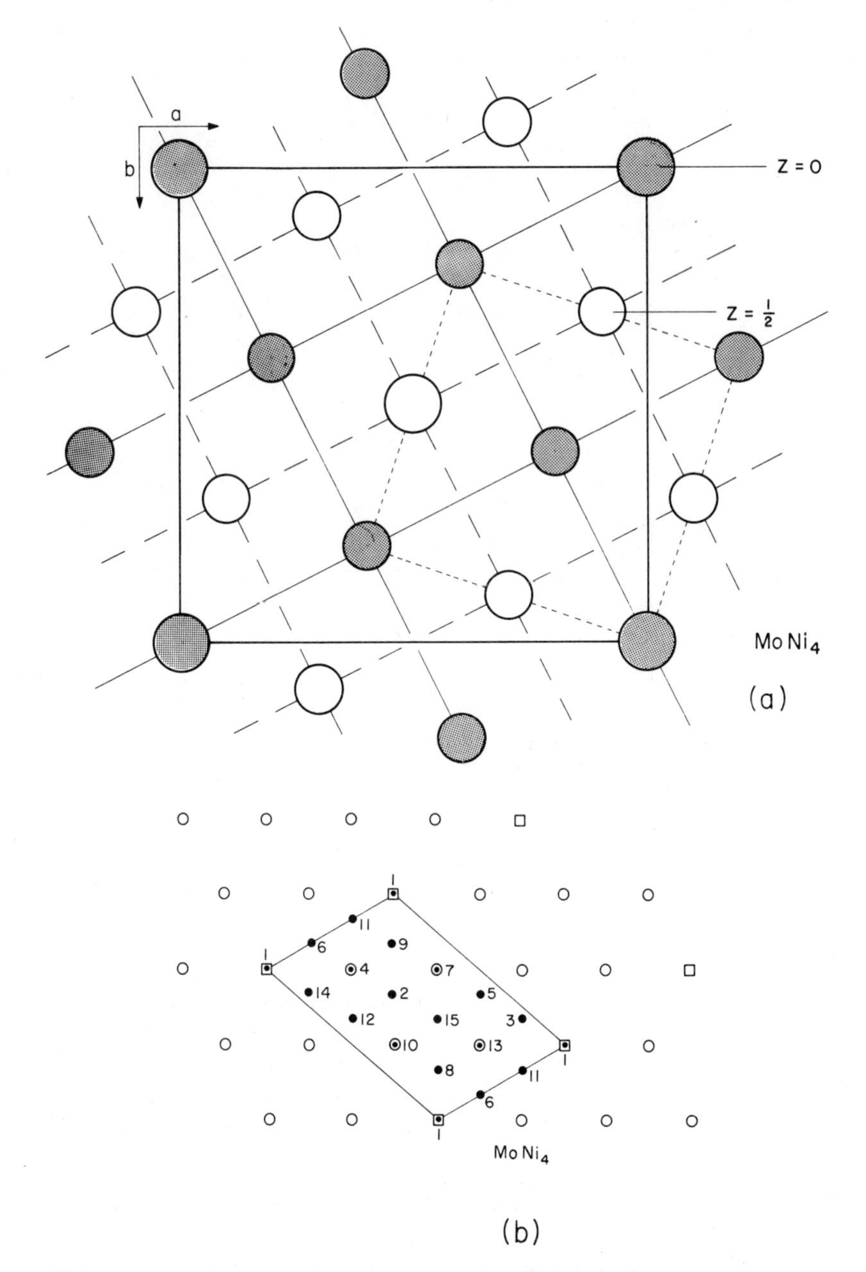

Figure 7-31. (*a*) Arrangement of atoms in the $MoNi_4$ (*tI*10) structure viewed down [001]. A f.c. cubic subcell is indicated. (*b*) Arrangement of atoms on close packed planes in the $MoNi_4$ structure showing a repeat cell and the 15 stacking positions.

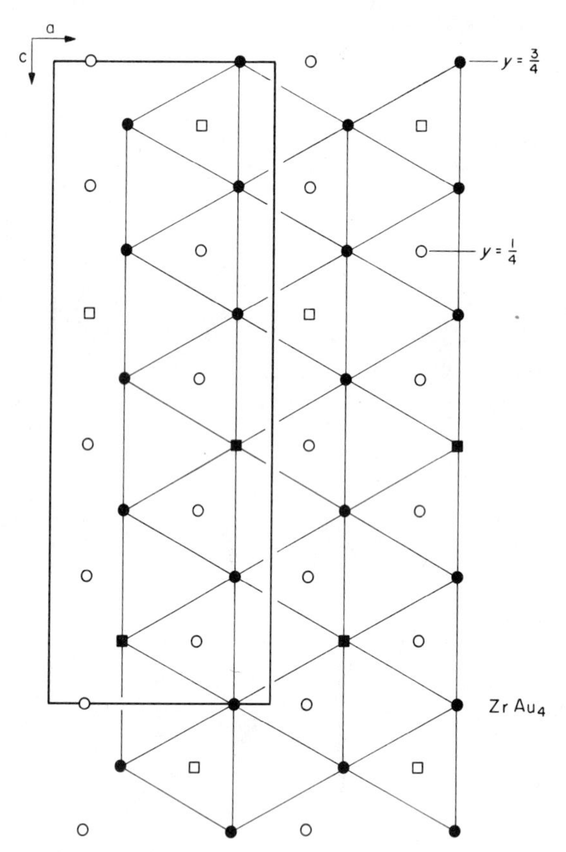

Figure 7-32. Arrangement of the atoms in the $ZrAu_4$ (*oP*20) structure on close packed planes parallel to the (001) plane. Squares indicate Zr.

arrangement separates the W atoms from each other as far as possible. All atoms are surrounded by cubo-octahedra, Al(1) and Al(2) by 9 Al (2.75 and 2.83 Å) and 3 W (2.83 Å), Al(3) by 10 Al (2.75 and 2.83 Å) and 2 W (2.75 Å), and W by 12 Al at 2.75 and 2.83 Å.

$MoAl_5$ (L.T.) is said to be a rhombohedral stacking variant of the WAl_5 structure.[49]

$GeCa_7$ (*cF*32): The large cubic cell ($a = 9.45$ Å, $Z = 4$) corresponds to 8 f.c. cubic cells and Ge occupies the corners and face centers of the large cell.[50] $GeCa_7$ is therefore an ordered superstructure of a close packed cubic arrangement. The close packed planes parallel to (111) are either all Ca or Ge+Ca in the ratio 1:3, with Ca occupying a kagomé net

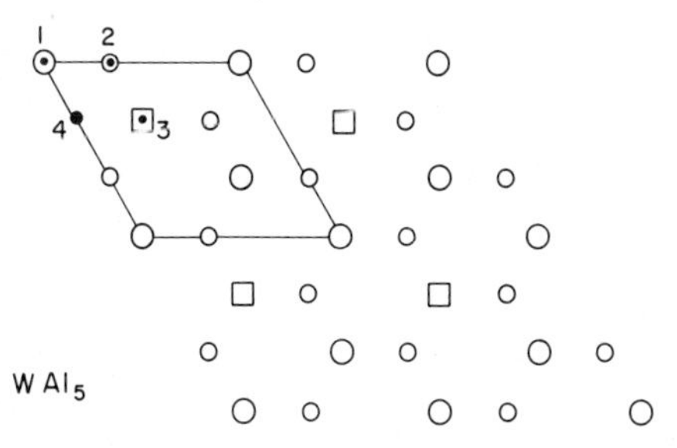

Figure 7-33. Atoms on two close packed planes of the WAl_5 (*hP*12) structure. Large symbols (squares W) indicate nets of W and Al which are stacked in positions 1 and 3, and small symbols indicate nets of Al atoms which are stacked alternately in positions 2 and 4.

and Ge a large 3^6 net which centers the hexagons of the kagomé net. The structure is thus similar to Cu_3Pt_5 (p. 312) except that in the mixed layers Cu occupies the kagomé net and Pt the 3^6 net. The close packed layers have a sixfold repeat sequence. All interatomic distances are 3.34 Å; Ca(1) at $\frac{1}{2}, \frac{1}{2}, \frac{1}{2}$ has 12 Ca(2) neighbors, Ca(2) has 10 Ca and 2 Ge, and Ge has 12 Ca(2) neighbors. $PtMg_7$ has the same structure.

$NbNi_8$ (*tF*36): $NbNi_8$ has cell dimensions $a = 10.8$ ($\sim 3a_0$), $c = 3.6$ Å, $Z = 4$.[51] The phase is a close packed superstructure, the close packed layers having a triangular ordering of the Nb atoms and a nine layer stacking repeat sequence according to electron diffraction data of single crystals. In the close packed layers at least two Ni atoms separate the Nb atoms. Figure 7-34 shows the unit cell.

$TiPt_8$ (*tI*18): The cell dimensions, $a = 8.312$, $c = 3.897$ Å, $Z = 2$, show $TiPt_8$ to be a superstructure of a close packed array of atoms.[52] The cell has a volume equal to that of 4.5 f.c. cubic pseudocells (Figure 7-35). The close packed layers of atoms lie parallel to planes of the type $(\bar{3}01)$ and $(0\bar{3}1)$. The structure requires confirmation by single-crystal methods.

Ti_2Ga_3 (*tP*10): The Ti_2Ga_3 structure ($a = 6.284$, $c = 4.010$ Å, $Z = 2$)[44] is formed by a $\frac{3}{4}.4^4$ net of Ga(1) and Ga(3) atoms at $z = 0$ which is centered at $z = \frac{1}{2}$ by a 4^4 net containing Ga(2) and Ti atoms in the proportion 1 to 4 (Figure 7-36*a*). Thus one-fifth of the Ga cubes formed by the $\frac{3}{4}.4^4$ nets at $z = 0$ and 1 is centered by Ga and four-fifths are centered by Ti, and the Ti atoms are slightly displaced toward the edges of the cubes formed by 4 Ga(3) atoms. The cell dimensions indicate a f.c. cubic rather

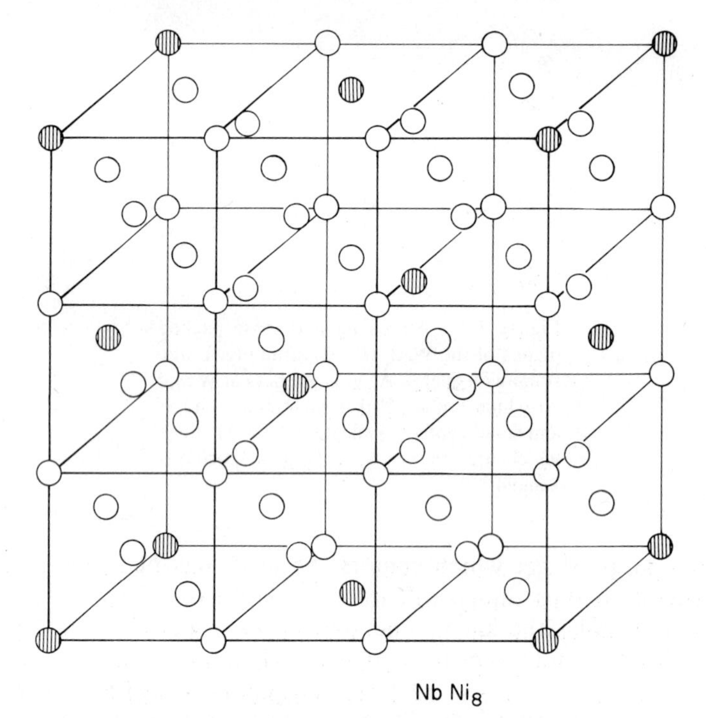

Figure 7-34. Diagram of the $NbNi_8$ ($tF36$) superstructure.

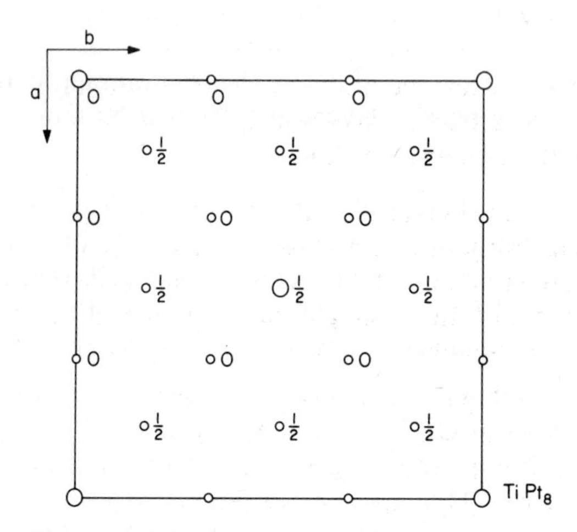

Figure 7-35. Diagram of the $TiPt_8$ ($tI18$) superstructure. Ti, large circles.

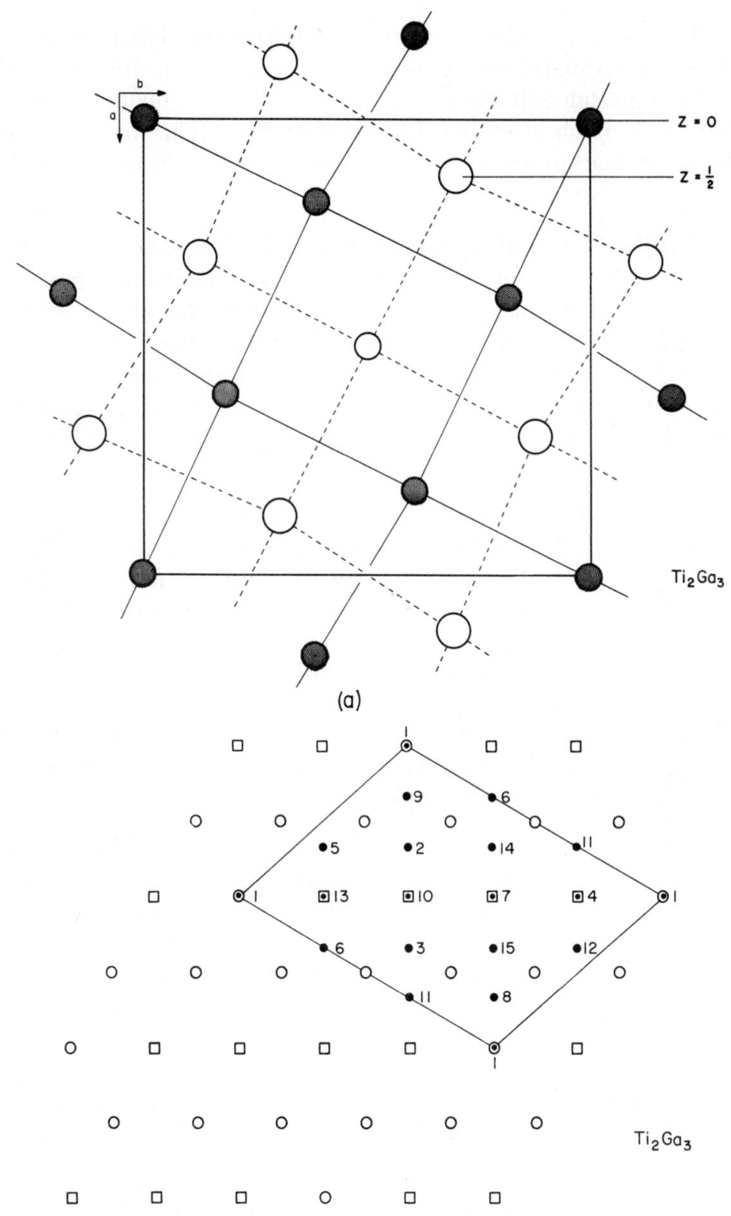

Figure 7-36. (*a*) Atomic arrangement in the Ti_2Ga_3 (*tP*10) structure viewed down [001]. A f.c. cubic pseudocell is indicated. (*b*) Arrangement of atoms on close packed planes in the Ti_2Ga_3 structure showing a repeat cell and the 15 stacking positions.

than a b.c. cubic pseudocell for the superstructure. The arrangement of atoms in approximately close packed layers parallel to the {111} planes of the cubic pseudocell and with a 15 layer repeat sequence, is shown in Figure 7-36*b*. Each atom has 12 near neighbors. The structure requires further confirmation since it was determined only from X-ray powder data.

Mn_2Au_5 (mC14): The dimensions of the Mn_2Au_5 cell, $a = 9.188$, $b = 3.954$, $c = 6.479$ Å, $\beta = 97.56°$, $Z = 2$,[53] indicate that it is a superstructure of a f.c. cubic rather than a b.c. cubic pseudocell. The atoms are arranged in ordered fashion on 4^4 nets parallel to the *ac* plane of the structure (Figure 7-37*a*) and in 3^6 nets on the slightly distorted close packed planes parallel to the {111} planes of the pseudocell. The atomic arrangement on the close packed planes and the stacking repeat sequence of 21 layers are indicated in Figure 7-37*b*.

Ga_3Pt_5 (oC16): Ga_3Pt_5 with cell dimensions $a = 8.031$, $b = 7.440$, $c = 3.948$ Å, $Z = 2$, appears to be a superstructure based on a close packed array of atoms,[54] although the structure requires verifying by single-crystal data. The unit cell corresponds to a block of 4 distorted f.c. cubic pseudocells. The atomic arrangement in close packed layers normal to a [111] direction of the pseudocell is shown in Figure 7-38. The repeat unit for the ordered distribution of Ga and Pt in these layers is a 4×2 parallelogram, and there is a threefold stacking sequence ABC. With these arrangements all atoms have 12 close neighbors at distances less than 3 Å. The closest Ga–Pt approach is 2.59 Å.

Co_5Ge_7 (tI24): The Co_5Ge_7 cell with dimensions $a = 7.64$, $c = 5.81$ Å, $Z = 2$,[55] corresponds to 8 very squashed f.c. cubic pseudocells, with displacement of the atoms from the ideal positions. Taking the cells over the base center as indicated in Figure 7-39, Co occupies the cell corners and alternate base centers (*ab* plane). Ge occupies the other base centers of these pseudocells and all base centers of the remainder, together with the other face centers (*ac* and *bc* planes), except that in alternate cells along *c*, these face centering atoms are absent. The structure is therefore a very distorted defect superstructure based on a f.c. cubic arrangement; however, it requires confirmation since it was only determined from X-ray powder data.

Nb_5Ga_{13} (oC36): Nb_5Ga_{13} with cell dimensions $a = b = 3.78$, $c = 40.10$ Å, $Z = 2$,[56] is a distorted superstructure based on close packing. The structure contains 9 f.c. cubic pseudocells stacked in the *c* direction. The cells are elongated somewhat in this direction and the atoms are displaced somewhat from their ideal sites. Figure 7-40 shows the ideal arrangement

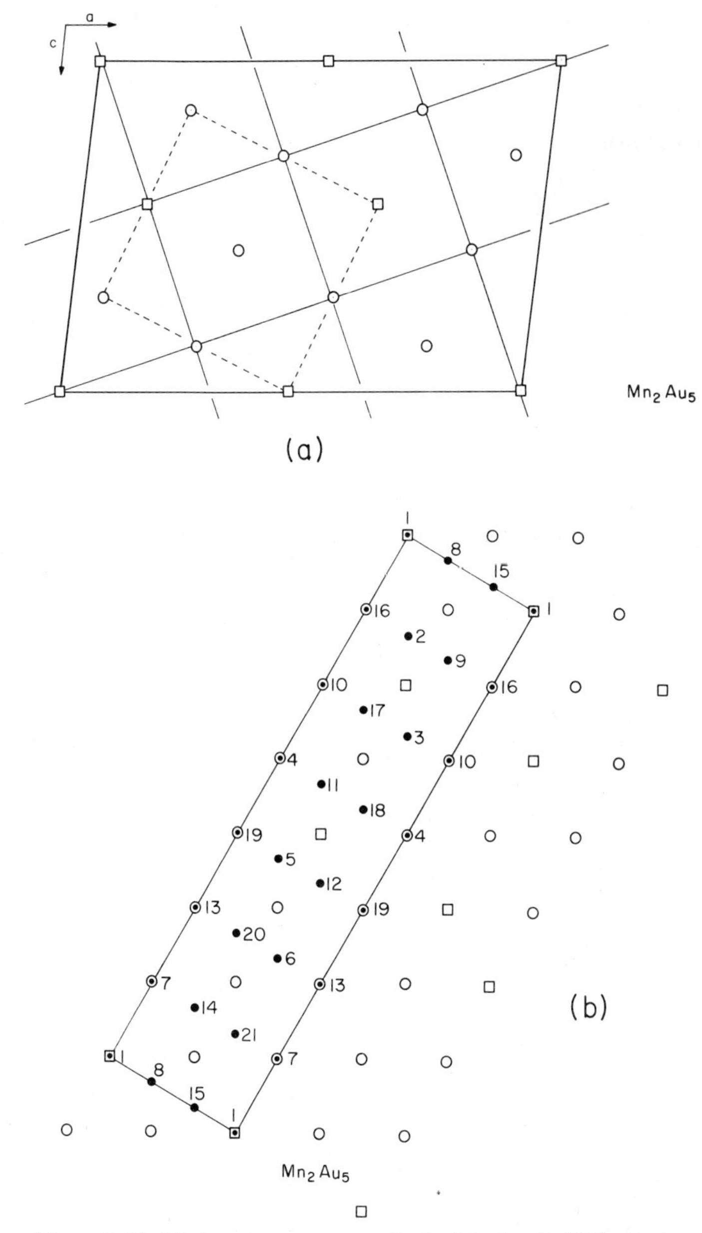

Figure 7-37. (*a*) Atomic arrangement in the Mn_2Au_5 (*mC*14) structure viewed down [010]. (*b*) Arrangement of atoms on close packed planes in the Mn_2Au_5 structure showing a repeat cell and the 21 stacking positions.

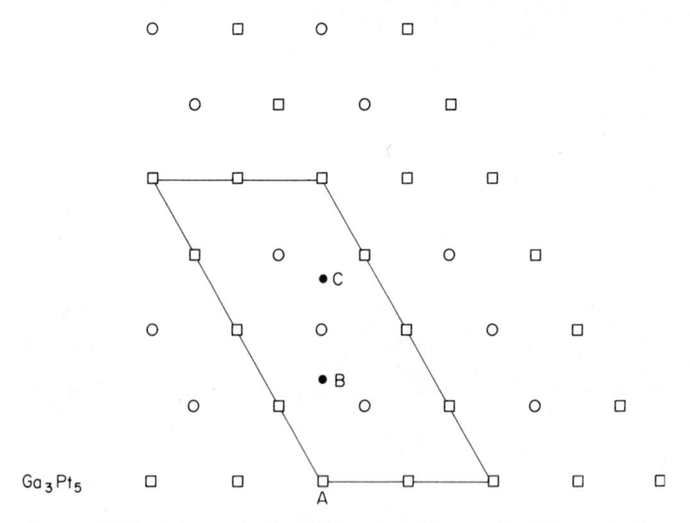

Figure 7-38. Arrangement of atoms on close packed planes in the Ga_3Pt_5 (*oC*16) structure showing a repeat cell and the 3 stacking positions.

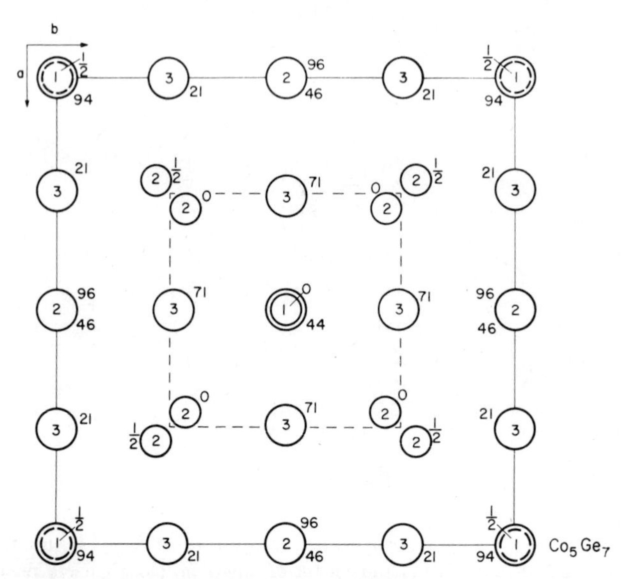

Figure 7-39. View of the structure of Co_5Ge_7 (*tI*24) down [001]. A f.c. cubic pseudocell is indicated.

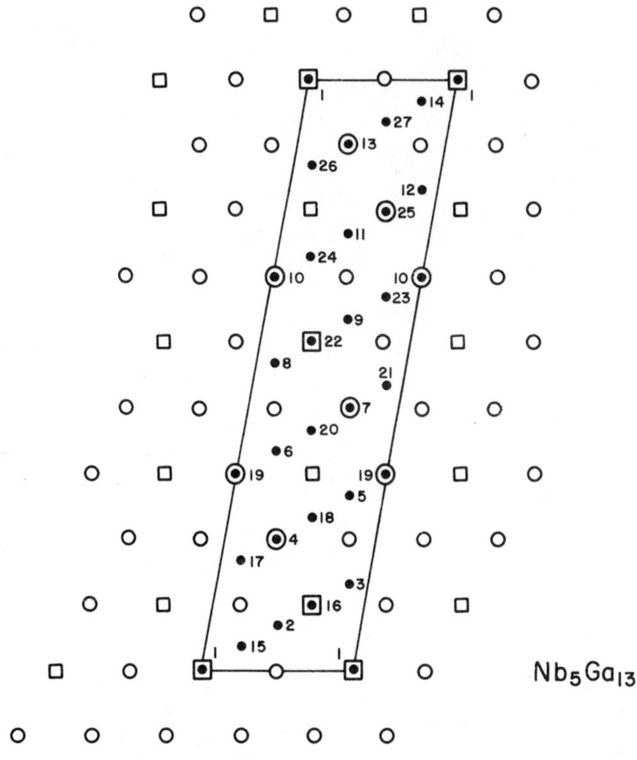

Figure 7-40. Arrangement of atoms on close packed planes in the Nb_5Ga_{13} ($oC36$) structure showing a repeat cell and the 27 stacking positions.

of the atoms in the close packed planes ((111) of pseudocell) with the repeat cell for the arrangement, and the 27-fold repeat sequence for the stacking of the layers. The structure requires further confirmation.

Mo_3Al_8 ($mC22$): In the Mo_3Al_8 structure ($a = 9.208$, $b = 3.638$, $c = 10.065$ Å, $\beta = 100°47'$, $Z = 2$)[57,58] the atoms are arranged in distorted $\frac{1}{6}.4^4$ nets parallel to the ac plane at heights $y = 0$ and $\frac{1}{2}$, with the atoms of one net centering the squares of nets above and below (Figure 7-41*a*). On these planes the Mo atoms are arranged in zigzag lines running in the c direction and generally avoid each other. The cell dimensions show that Mo_3Al_8 is a superstructure of a distorted f.c. cubic pseudocell, the unit cell volume corresponding to $5\frac{1}{2}$ cubic subcells. The arrangement of the atoms in the close packed layers normal to {111} of the f.c. cubic pseudocell is shown in Figure 7-41*b*, together with the stacking positions for the 33 layer repeat sequence.

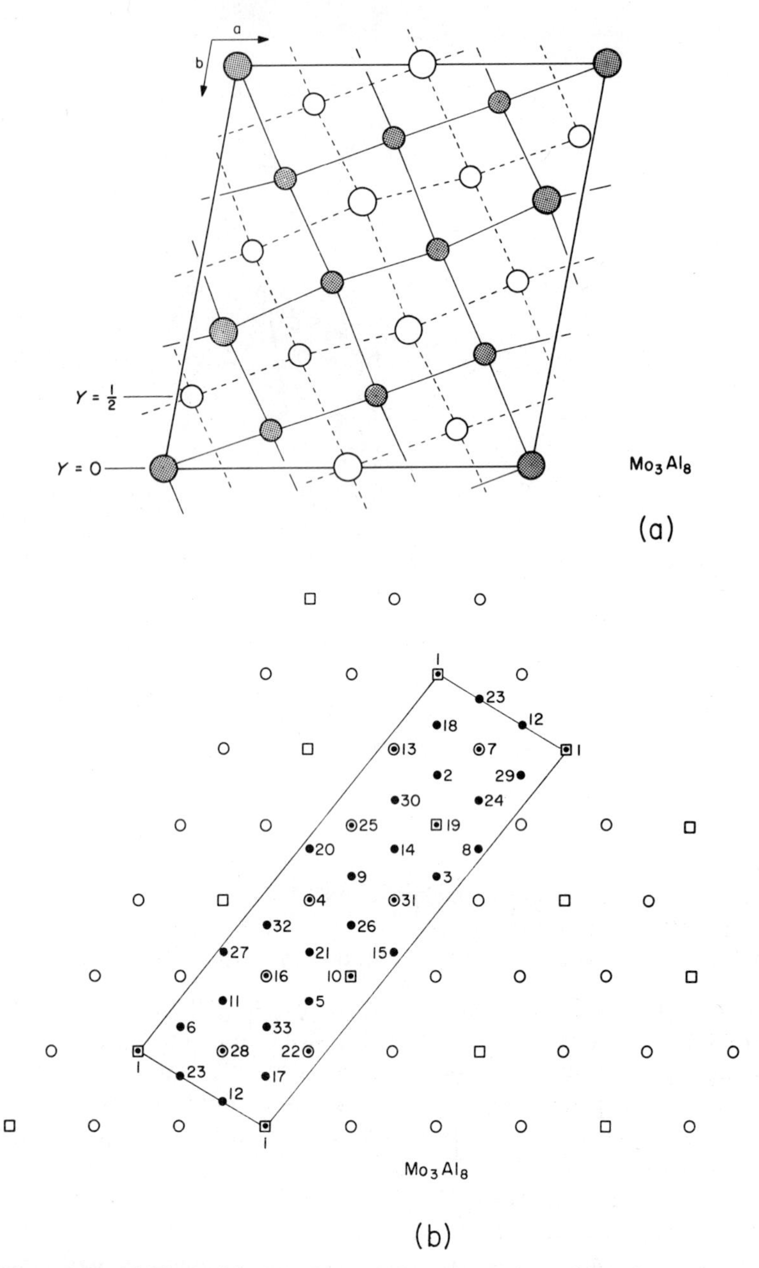

Figure 7-41. (*a*) View of the Mo_3Al_8 (*mC*22) structure down [001]. (*b*) Arrangement of atoms on close packed planes in the Mo_3Al_8 structure showing a repeat cell and the 33 stacking positions.

δ′-Sb_3Cu_{10} (hP26): The unit cell dimensions of δ′-Sb_3Cu_{10} are $a = 9.920$, $c = 4.319$ Å, $Z = 2$.[59] The atoms form slightly distorted close packed layers at $z = \frac{1}{4}$ and $\frac{3}{4}$ (Figure 7-42) which are stacked in an AB sequence so that the structure is a superstructure based on close packing. Sb_3Cu_{10} stoichiometry is satisfied in each layer and the Sb atoms are arranged in triangles joined at the corners giving a 3636 mesh.

Both Cu(1), with 9 Cu (2.60 and 2.71 Å) and 3 Sb (2.72 Å), and Cu(2) with 8 Cu (2.56 to 2.97 Å) and 4 Sb (2.71 to 2.86 Å) neighbors, are surrounded by polyhedra that have only 4 connected surface atoms. Cu(3) and Cu(4) are surrounded respectively by 8 Cu + 4 Sb and by 9 Cu and 3 Sb. Sb has 12 Cu neighbors at 2.65 to 2.86 Å.

$Au_{82}Mg_{26}$ ($Au_{77}Mg_{23}$) (hP108): $Au_{82}Mg_{26}$ has cell dimensions $a = 14.93$, $c = 9.441$ Å, $Z = 1$,[60] which are related to a hexagonal pseudocell, a_0, c_0, as follows: $a = 3\sqrt{3} \cdot a_0$, $c = 2c_0$. The atoms are arranged in close packed layers which (neglecting the ordering within the layers) are stacked in close packed 4H sequence ABAC at heights $z = 0$, $\frac{1}{4}$, $\frac{1}{2}$, and $\frac{3}{4}$.

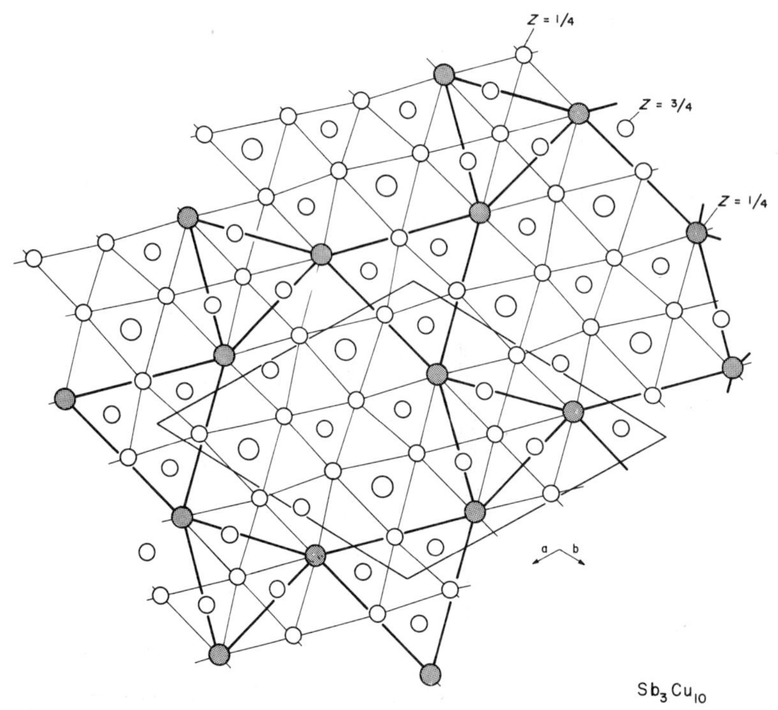

Figure 7-42. View of the δ′-Sb_3Cu_{10} (*hP*26) structure down [001] showing the close packed planes of atoms.

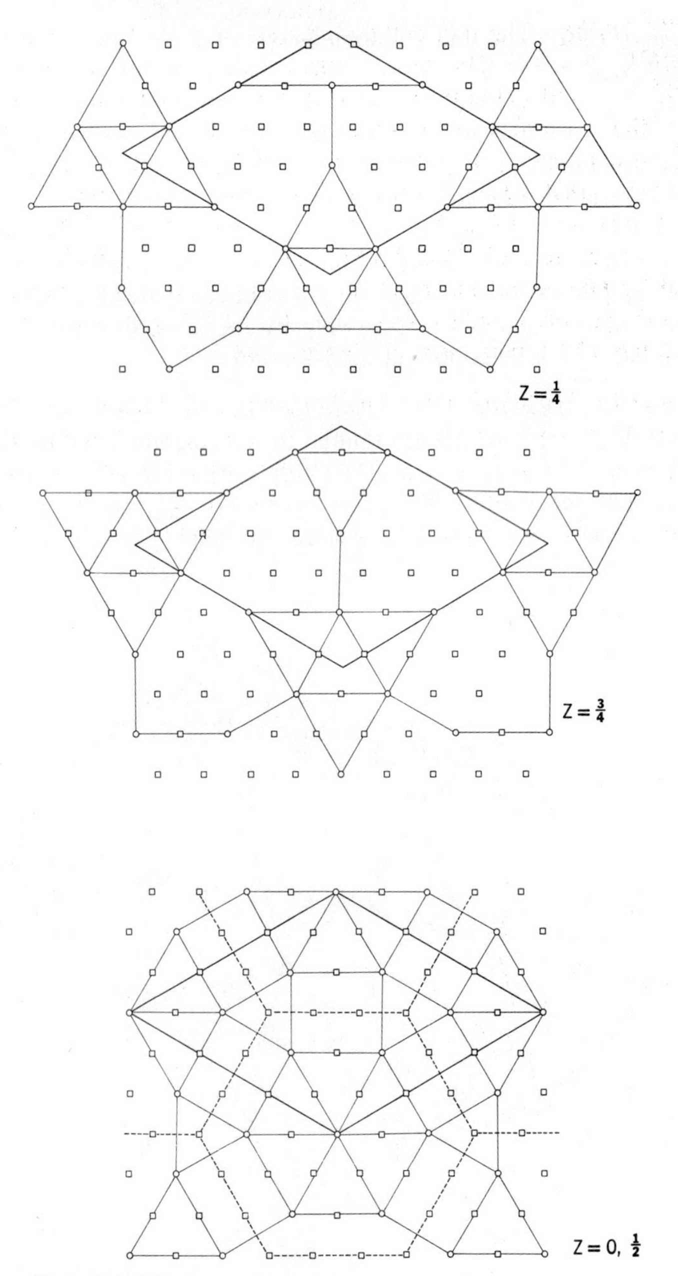

Figure 7-43. Atomic arrangements in various close packed planes of the $Mg_{23}Au_{77}$ (*hP*108) structure. Mg, circles. Antiphase boundaries are indicated in the lower figure.

Stoichiometry is satisfied in each layer, the Mg atoms being arranged in a mixture of triangles and rectangles at $z = 0$ and $\frac{1}{2}$, and in a mixture of triangles and six-sided polygons at $z = \frac{1}{4}$ and $\frac{3}{4}$ (Figure 7-43).

All atoms have 12 neighbors which are generally arranged in cubo-octahedra with 8 three-sided and 6 four-sided faces. Interatomic distances range from 2.76 to 3.10 Å.

$Au_{42}Cd_{12}$: A two-layer 2H polytype of the $Au_{77}Mg_{23}$ structure with $a = 9a_0$, $c = c_0$, where a_0 and c_0 refer to the hexagonal pseudocell, has been found in the Au–Cd system in the composition range about 29 at.% Cd.[61] The probable atomic arrangement in the close packed planes at $z = \frac{1}{4}$ and $\frac{3}{4}$, including sites for excess Cd, is shown in Figure 7-44 in respect of a smaller unit cell with $a' = 3\sqrt{3} \cdot a_0$. This was derived from electron diffraction data from single crystals.

$Au_{72}Cd_{26}$: A similar 2H superstructure with $a = 7a_0$, $c = c_0$ has been found at compositions richer in Cd, from 29 to 35 at.%.[61] The arrangement of the atoms in close packed planes at $z = \frac{1}{4}$ and $\frac{3}{4}$, as derived from electron diffraction data of single crystals, is shown in Figure 7-45.

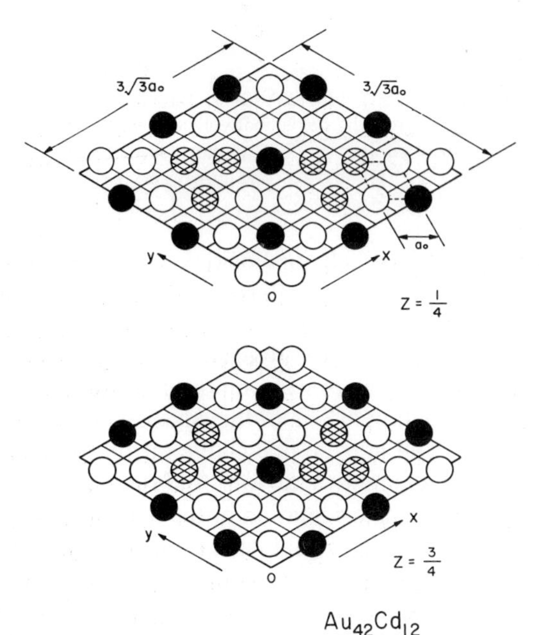

Figure 7-44. Structure of $Au_{42}Cd_{12}$ showing the atomic arrangement in the hexagonal cell: Cd, black circles; Au, open circles; hatched circles, sites where Cd replaces Au. (Hirabayashi and co-workers [61].)

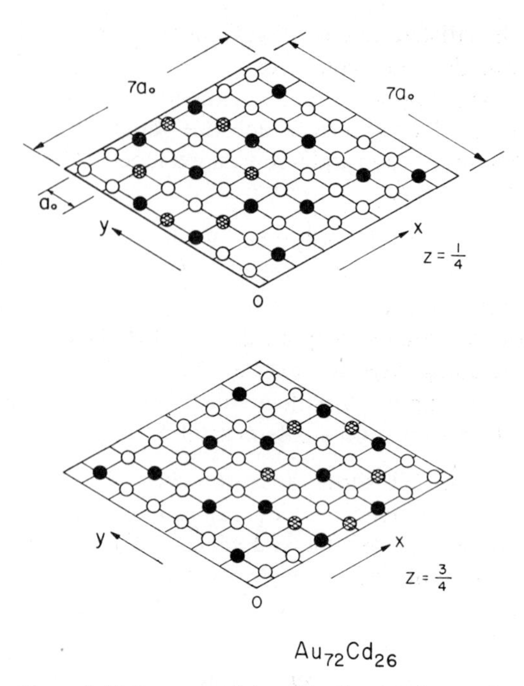

Figure 7-45. Structure of $Au_{72}Cd_{26}$ showing the atomic arrangement in the hexagonal cell: Cd, black circles; Au, open circles; hatched circles, sites where Cd replaces Au. (Hirabayashi and co-workers [61].)

Zr_2Al_3 (oF40): Zr_2Al_3 has cell dimensions $a = 9.601$, $b = 13.906$, $c = 5.574$ Å, $Z = 8$. According to the description of Renouf and Beevers[62] resulting from a single-crystal study, the Al(2) and Zr atoms form layers parallel to the (010) plane which give very rumpled triangular nets as shown in Figure 7-46*a*. The nets are composed of alternate lines of Zr and Al atoms along the direction of one of the edges, so that they do not have three-fold symmetry. The nets lie one above the other in the [010] direction on sites of close packing with the four-fold stacking sequence 1, 2, 3, 4 shown in Figure 7-46*b*. The orientation of the nets on sites 2 and 4 differs by 60° from that on sites 1 and 3. Between these nets there lie large planar triangular nets of Al(1) atoms, whose nodes are situated over triangles of the other nets, but displaced from the centers towards the corners occupied by Zr atoms. The structure can be regarded as a filled-up, *very* distorted superstructure based on close packing.

The Zr atoms have 8 Al neighbors up to 3.0 Å; the nearest Zr neighbors being 2×3.41 and 2×3.42 Å. Al(1) has 2 Al(2) at 2.67 Å and 2 at 2.71 Å.

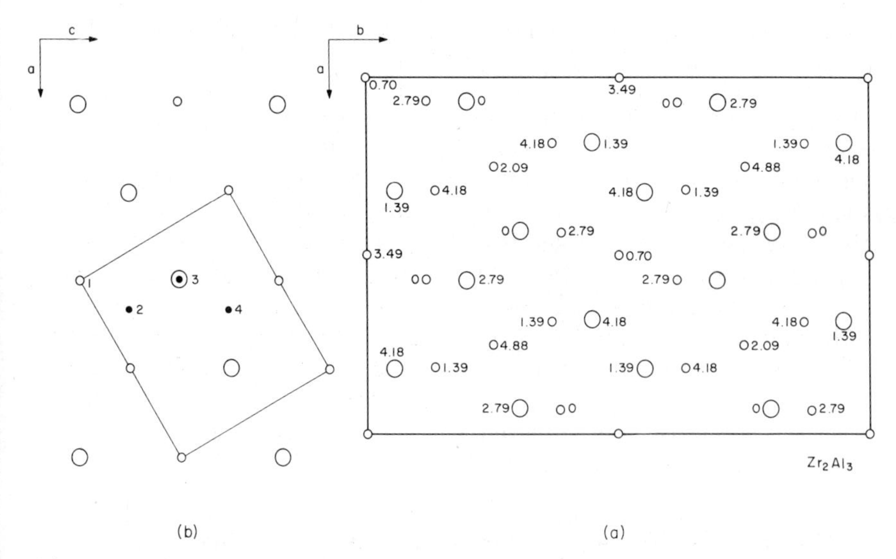

Figure 7-46. (*a*) Atomic arrangements in the Zr_2Al_3 (*oF*40) structure. Zr large circles. (*b*) Idealized close packed nets of atoms in the Zr_2Al_3 structure indicating a repeat cell and the four stacking positions.

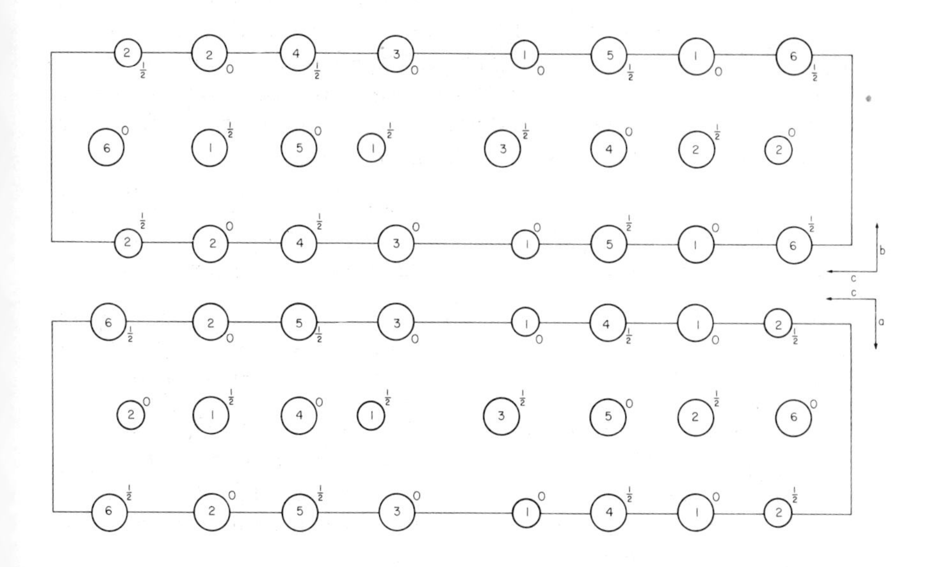

Figure 7-47. Plans of the $AsMn_3$ (*oP*16) structure down [100] and [010].

It also has 4 close Zr neighbors, 2×2.83 and 2×2.89 Å. Al(2) has 4 close Al neighbors and 6 Zr neighbors.

$AsMn_3$ (*oP*16): $AsMn_3$ which has cell dimensions $a = 3.788$, $b = 3.788$, $c = 16.29$ Å, $Z = 4$,[63] is made up of layers of somewhat distorted f.c. cubic arrangement, one and a half pseudocells thick, which lie parallel to the (001) plane. The unit cell contains two such layers in the [001] direction. The layers are separated from each other by the removal of a whole plane of atoms (parallel to (001)) from the f.c. cubic arrangement (Figure 7-47); otherwise the structure would have been a superstructure based on close packing. Starting from the cell origin and proceeding along the [001] direction, the first and sixth layers are absent. The second, fifth, seventh, and tenth layers are composed of As and Mn atoms alternately on interpenetrating 4^4 arrays, and the remaining layers (third, fourth, eighth, and ninth) are composed only of Mn atoms.

CLOSED PACKED LAYERS OF ATOMS NOT STACKED IN CLOSE PACKING

Polytypic Family of *MN* Compounds with Four-Layer Structures

Gallium monochalcogenides form four-layer structures, Se–Ga–Ga–Se, with the atoms arranged in triangular networks stacked in sequences of the type C$\underline{BB}$C (Ga underlined), so that Ga is in tetrahedral coordination. Distances between the chalcogenide atoms of different four-layer units correspond to van der Waals bonds. The three polytypic structures described below result from the different manner in which the four-layer units are stacked one above the other (Figure 7-48).

GaSe (*hR*4): The dimensions of the corresponding hexagonal cell of GaSe are $a = 3.747$, $c = 23.91$ Å, $Z = 6$.[64] The four-layer Se–Ga–Ga–Se units are stacked one above the other in the sequence ABC (Figure 7-48).

GaS (*hP*8): GaS has cell dimensions $a = 3.585$, $c = 15.50$ Å, $Z = 4$,[65] The unit cell contains two four-layer units along [001] with stacking sequence (Ga underlined) C$\underline{BB}$CB$\underline{CC}$B as indicated in Figure 7-48.

GaSe (*hP*8): With cell dimensions $a = 3.743$, $c = 15.92$ Å, $Z = 4$,[64] this form of GaSe has the structural arrangement illustrated in Figure 5-15, p. 226, the stacking sequence being A$\underline{BB}$AB$\underline{CC}$B (Ga underlined).[66] It is probable, however, that these three structural arrangements require further confirmation.

GaTe (*mC*24): The GaTe structure with $a = 17.37$, $b = 4.074$, $c = 10.44$ Å, $\beta = 104°12'$, $Z = 12$,[67] is a complex layer structure containing bands of

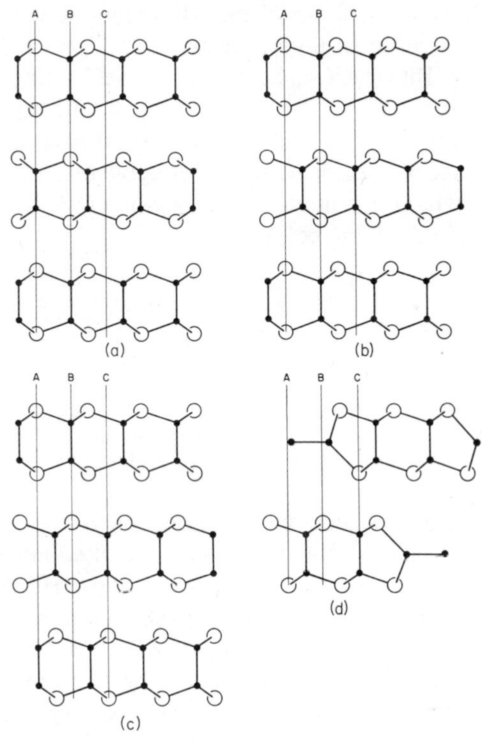

Figure 7-48. Diagram of stacking arrangements in (*a*) GaS (*hP*8), (*b*) GaSe (*hP*8), (*c*) GaSe (*hR*4),[66] and (*d*) GaTe (*mC*24).

atoms travelling parallel to the $(20\bar{1})$ plane. The bands are made up of Te–Ga–Ga–Te arrangements similar to those forming the four-layer units of the GaSe structures. However in GaTe, the Ga–Ga bonds are oriented not only perpendicular to the layer plane, but are also rotated 90° so that they lie in the plane of the bands of atoms (Figure 5-16 p. 228). Each Ga atom has 1 Ga (2.37 or 2.40 Å) and 3 Te neighbors (2.62–2.69 Å). Each Te has 3 Ga neighbors. Te–Te distances between the bands of atoms (3.93–4.26 Å) are approaching the van der Waals diameter of Te. SiAs has cell dimensions $a = 15.98$, $b = 3.668$, $c = 9.529$ Å, $\beta = 106.0°$, $Z = 12$,[68] and a structure similar or isotypic to GaTe.

Various Other Structures Containing Atoms Arranged in Triangular Close Packed Layers

Bi_2Se_2 (*hP*12): Bi_2Se_2 has cell dimensions $a = 4.18$, $c = 22.8$ Å, $Z = 3$.[69] The unit cell contains 12 layers of atoms in triangular nets parallel to

(001). The structure is made up of two Bi_2Se_3 five-layer units which are interleaved with a double Bi layer, the order of layers being Se(1), Bi(1), Se(2), Bi(2), Se(3), Bi(3), Bi(3), Se(3), Bi(2), Se(2), Bi(1), Se(1). Sb_2Te_2 has the same structure.

Studies of single crystals in the region of solid solution about BiSe as well as of the phase Bi_2Se_3, lead to the conclusion that various ordered structures occur which are made up of five-layer Bi_2Se_3 type units and Bi_2 units as shown in the table below.

		No. of Layers	Layer Sequence
Bi_2Se_3	$a = 4.13, c = 28.7$ Å	15	$(Bi_2Se_3)_3$
Bi_8Se_9	$a = 4.16, c = 97.4$ Å	51	$\{(Bi_2Se_3)_3Bi_2\}_3$
Bi_2Se_2	$a = 4.18, c = 22.9$ Å	12	$(Bi_2Se_3)_2Bi_2$
Bi_8Se_7		45	$Bi_2Se_3 . Bi_2 . (Bi_2Se_3)_2 . Bi_2 . Bi_2Se_3$ $Bi_2 . Bi_2Se_3 . Bi_2 . (Bi_2Se_3)_2 . Bi_2$
Bi_4Se_3	$a = 4.27, c = 39.9$ Å	21	$(Bi_2Se_3 . Bi_2)_3$

HoD_3 (*hP*24): The HoD_3 cell has dimensions $a = 6.308$, $c = 6.560$ Å, $Z = 6$.[70] At $z = \pm\frac{1}{4}$ the Ho atoms form close packed layers in very open hexagonal stacking AB. Some of the triangles of the Ho net are filled with H(1) at the same height. Others have H(2) slightly above or below at $z = \pm\frac{1}{6}$ or $z = \pm\frac{1}{3}$, and the remaining triangles have H(3) both above and below their centers at $z = 0.10$ and 0.40, or $z = 0.60$ and 0.90. This arrangement surrounds Ho by 9 D atoms as shown in Figure 7-49.

K_2S_2 (*Na_2O_2*) (*hP*12): The dimensions of the K_2S_2 cell are $a = 8.49$, $c = 5.84$ Å, $Z = 3$.[71] The K atoms at $z = 0$ and $\frac{1}{2}$ form somewhat distorted

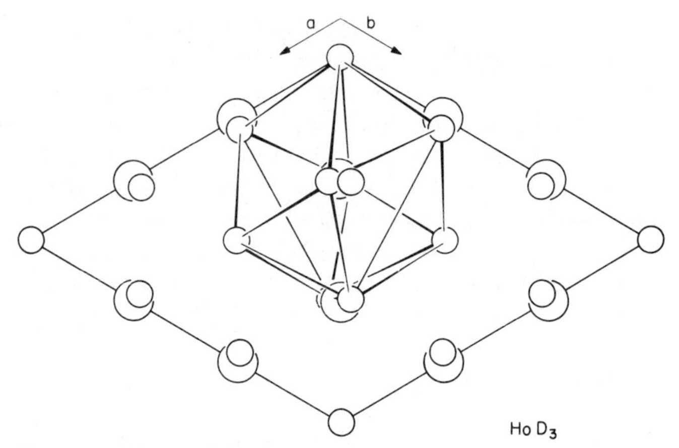

Figure 7-49. View of the HoD_3 (*hP*24) structure down [001], showing the coordination of Ho by 9 D atoms.

triangular nets stacked in the sequence BC, but too close for close packing. The S atoms are distributed on a larger 6^3 net (site a) at $z = \pm 0.180$ and on a larger 3^6 net at $z = \pm 0.320$ (site A) so that they are off-center of the distorted K octahedra. The S atoms each have 3 + 3 K neighbors (3.16 and 3.31 Å for S(1); 3.16 and 3.27 Å for S(2)) and 1 close S neighbor at 2.10 Å. K(1) has 2 + 4 S(1) and (2) at 3.16 Å; K(2) has 2 + 4 S(1) and (2) at 3.31 Å and 3.27 Å, respectively.

REFERENCES

1. P. A. Beck, 1967, *Z. Kristallogr.*, **124**, 101; 1968. *Acta Cryst.*, **B24**, 1477; 1969, In *Advances in X-ray Analysis*, Vol. **12**, ed, Barrett, Newkirk, and Mallett, New York: Plenum Press, p. 1.
2. A. Brown, 1961, *Acta Cryst.*, **14**, 856.
3. N. Karlsson, 1951, *J. Inst. Met.*, **79**, 391.
4. J. O. Linde, 1937, *Ann. Phys. Lpz.*, (5), **30**, 151.
5. A. Ölander, 1932, *Z. Kristallogr.*, **A83**, 145; L. C. Chang and T. A. Read, 1951, *J. Metals*, **3**, 47; L. C. Chang, 1951, *Acta Cryst.*, **4**, 320.
6. L. C. Chang and T. A. Read, 1951, *J. Metals*, **3**, 47.
7. K. Schubert *et al.*, 1957, *Naturwiss.*, **44**, 229.
8. B. C. Giessen and N. J. Grant, 1964, *Acta Cryst.*, **17**, 615.
9. D. L. Ritter, B. C. Giessen and N. J. Grant, 1964, *Trans. Met. Soc., AIME*, **230**, 1250.
10. J. W. Nielsen and N. C. Baenziger, 1954, *Acta Cryst.*, **7**, 277.
11. W. P. Binnie, 1956, *Acta Cryst.*, **9**, 686.
12. A. Byström, 1945, *Ark. Kemi, Min. Geol.*, **20A**, No. 11.
13. F. Grønvold, H. Haraldsen, and J. Vihovde, 1954, *Acta Chem. Scand.*, **8**, 1927.
14. C. H. Johnansson and J. O. Linde, 1936, *Ann. Phys. Lpz.*, (5), **25**, 1.
15. A. Nagasawa, 1955, *J. Phys. Soc. Japan*, **21**, 955.
16. A. Kjekshus, R. Møllerud, A. F. Andresen, and W. B. Pearson, 1967, *Phil. Mag.*, **16**, 1063.
17. R. J. Arnott and A. Wold, 1960, *J. Phys. Chem. Solids*, **15**, 152.
18. A. H. Holtzman and G. P. Conrad, 1959, *J. Appl. Phys.*, **30**, 103S.
19. B. C. Giessen and N. J. Grant, 1965, *J. Less-Common Metals*, **8**, 114.
20. A. K. Sinha, 1969, *Acta Cryst.*, **B25**, 996.
21. S. Saito, 1959, *Acta Cryst.*, **12**, 500.
22. A. C. Larson, D. T. Cromer, and C. K. Stambaugh, 1957, *Acta Cryst.*, **10**, 443.
23. J. H. N. van Vucht and K. H. J. Buschow, 1965, *J. Less-Common Metals*, **10**, 98; I. I. Zaluckij and P. I. Kripjakevi̇c, *Izv. Akad. Nauk, SSSR, Neorgan. Mater.* **2**, 264 [*Inorg. Mater.*, **2**, 226].
24. F. Laves and H. J. Wallbaum, 1939, *Z. Kristallogr.*, **A101**, 78.
25. A. C. Larson, D. T. Cromer, and R. B. Roof, 1965, *Acta Cryst.*, **18**, 294.
26. J. H. N. van Vucht, 1965, *J. Less-Common Metals*, **11**, 308.
27. B. C. Giessen and N. J. Grant, 1965, *Acta Cryst.*, **18**, 1080.
28. D. E. Sands, D. H. Wood, and W. J. Ramsey, 1964, *Acta Cryst.*, **17**, 986.
29. E. E. Havinga and J. H. N. van Vucht, 1970, *Acta Cryst.*, **B26**, 653.
30. P. Rahlfs, 1937, *Metallwirt.*, **16**, 343.

31. N. Karlsson, 1951, *J. Inst. Met.*, **79**, 391; E. Zintl and S. Neumayr, 1933, *Z. Elektrochem.*, **39**, 86.
32. W. H. Zachariasen, 1948, *Acta Cryst.*, **1**, 265; 1949, *Ibid.*, **2**, 94.
33. H. Sato, R. S. Toth, and G. Honjo, 1967, *J. Phys. Chem. Solids*, **28**, 137.
34. G. S. Ždanov, 1945, *Dokl. Akad. Nauk, SSSR*, **48**, 39.
35. P. A. Beck, 1967, *Z. Kristallogr.*, **124**, 101.
36. G. Brauer, 1938, *Naturwiss.*, **26**, 710; 1939. *Z. anorg. Chem.*, **242**, 1.
37. H. Sato, R. S. Toth, and G. Honjo, 1966, *J. Phys. Chem. Solids*, **27**, 413.
38. M. Hirabayashi, N. Ino, and K. Hiraga, 1967, *J. Phys. Soc. Japan*, **22**, 1509.
39. H. Iwasaki, M. Hirabayashi, and S. Ogawa, 1965, *J. Phys. Soc. Japan*, **20**, 89.
40. K. Burkhardt and K. Schubert, 1965, *Z. Metallk.*, **56**, 864.
41. H. Sato and R. S. Toth, 1968, *J. Phys. Chem. Solids*, **29**, 2015.
42. J. Wegst and K. Schubert, 1958, *Z. Metallk.*, **49**, 533.
43. K. Schubert, R. Kieffer, M. Wilkens, and R. Haufler, 1955, *Z. Metallk.*, **46**, 692.
44. M. Pötzschke and K. Schubert, 1962, *Z. Metallk.*, **53**, 474.
45. D. Harker, 1944, *J. Chem. Phys.*, **12**, 315.
46. H. Iwasaki, 1962, *J. Phys. Soc. Japan*, **17**, 1621.
47. E. Stolz and K. Schubert, 1962, *Z. Metallk.*, **53**, 433.
48. J. Adam and J. B. Rich, 1955, *Acta Cryst.*, **8**, 349.
49. K. Schubert *et al.*, 1969, *Naturwiss.*, **47**, 303.
50. O. Helleis, H. Kandler, E. Leicht, W. Quiring, and E. Wölfel, 1963, *Z. anorg. Chem.*, **320**, 86.
51. W. E. Quist, C. J. van der Wekken, R. Taggart, and D. H. Polonis, 1969, *Trans. Met. Soc., AIME*, **245**, 345.
52. P. Pietrokowsky, 1965, *Nature, Lond.*, **206**, 291.
53. S. G. Humble, 1964, *Acta Cryst.*, **17**, 1485; J. H. Smith and P. Wells, 1969, *J. of Phys. C, Solid State Phys.*, **2**, 356.
54. S. Bhan and K. Schubert, 1960, *Z. Metallk.*, **51**, 327.
55. E. Stolz and K. Schubert, 1962, *Chem. der Erde*, **22**, 709.
56. K. Schubert *et al.*, 1963, *Naturwiss.*, **50**, 41; H. G. Meissner and K. Schubert, 1965, *Z. Metallk.*, **56**, 475.
57. M. Pötzschke and K. Schubert, 1962, *Z. Metallk.*, **53**, 548.
58. J. B. Forsyth and G. Gran, 1962, *Acta Cryst.*, **15**, 100.
59. E. Günzel and K. Schubert, 1958, *Z. Metallk.*, **49**, 124.
60. K. Burkhardt, K. Schubert, R. S. Toth, and H. Sato, 1968, *Acta Cryst.*, **B24**, 137.
61. M. Hirabayashi, S. Yamaguchi, K. Hiraga, N. Ino, H. Sato, and R. S. Toth, 1969, *J. Phys. Chem. Solids*, **31**, 77.
62. T. J. Renouf and C. A. Beevers, 1961, *Acta Cryst.*, **14**, 469.
63. H. Nowotny, R. Funk, and J. Pesl, 1951, *Mh. Chem.*, **82**, 513.
64. K. Schubert, E. Dörre, and M. Kluge, 1955, *Z. Metallk.*, **46**, 216.
65. H. Hahn and G. Frank, 1955, *Z. anorg.* Chem., **278**, 340.
66. Z. S. Basinski, D. B. Dove, and E. Mooser, 1961, *Helv. Phys. Acta*, **34**, 373.
67. J. H. Bryden, private communication.
68. T. Wadsten, 1965, *Acta Chem. Scand.*, **19**, 1232.
69. M. M. Stasova, 1967, *Ž. Strukt. Khim.*, **8**, 655 [*J. Struct. Chem.*, **8**, 584]; 1965, *Izv. Akad. Nauk SSSR*, Neorg. Mater., **1**, 2134 [*Inorganic Materials*, **1**, 1930].
70. M. Mansmann and W. E. Wallace, 1964, *J. d. Physique*, **25**, 454.
71. H. Föppl, E. Busmann, and F. K. Frorath, 1962, *Z. anorg. Chem.*, **314**, 12.

8

Structures Derived by Filling Tetrahedral, Octahedral, and Other Holes in Close Packed Arrays of Atoms

Having discussed structures derived from a single array of close packed atoms, the next systematic grouping to consider is that in which interpenetrating close packed arrays occur with atoms of one located at the centers of the octahedral or tetrahedral holes of the other. Layers centering octahedral holes are at half of the spacing of the original close packed layers and those centering tetrahedral holes are at one and/or three quarters of the spacing. The layers centering octahedral holes in a close packed hexagonal array of atoms do not themselves form a close packed array as in the other cases, but a simple hexagonal array. This difference between hexagonal and cubic close packing complicates the classification according to coordination, of structures with mixed hexagonal and cubic stacking of the primary (generally anion) layers, because occupation of octahedral holes in cubically surrounded layers results also in octahedral surrounding of the anions by the cations, whereas in hexagonally surrounded layers it results in trigonal prismatic surrounding of the anions. Therefore a family of structures based on a close packed anionic array of atoms with cations in the octahedral holes thereof may have either octahedral coordination of all atoms, or mixed octahedral and trigonal prismatic coordination of the anions, depending on the particular stacking sequence of the anion layers. The octahedra formed by a close packed cubic array of atoms share edges only, whereas those formed by a close packed hexagonal array share faces parallel to the planes of the close packed layers. Tetrahedra formed by atoms in close packed cubic arrays share only edges, whereas in close packed hexagonal arrays in a direction normal to that of the close packed layers, the tetrahedra share alternately faces and corners, which results in the centers of pairs of tetrahedral holes along the [001] direction of the hexagonal cell being too close to be occupied simultaneously.

When all of the tetrahedral and octahedral holes in a close packed cubic array of atoms are occupied, giving four interpenetrating f.c. cubic arrays of atoms, the distribution of the atoms in space is identical to that in a block of eight b.c. cubic cells. Structure types such as the $BiLi_3$, NaTl, and the Heusler alloys built up of four interpenetrating f.c. cubic arrays of atoms, are more appropriately considered as superstructures derived from the b.c. cubic structure (pp. 572–578), rather than filled up derivatives of the f.c. cubic structure.

CLOSE PACKED LAYERS OF ATOMS WITH INSERTED LAYERS: TETRAHEDRAL COORDINATION

MN: *M* on 3^6 Nets. *N* on 3^6 Nets Inserted at $\frac{3}{4}$ Spacing Giving Tetrahedral Coordination

Polytypic Structures of* MN *Compounds with Atoms Surrounded Tetrahedrally. If layers of close packed atoms stacked in close packed sequence are penetrated by identical layers of atoms on the same stacking sites and $\frac{3}{4}$ of the interlayer distance away, then the atoms of the first array surround those of the second array tetrahedrally and *vice versa*.

The cubic prototype for a family of polytypically related structures formed on this principle is sphalerite (ZnS) whose stacking is $\underline{A}A\underline{B}B\underline{C}C$ (layers of atoms on one sublattice are underlined). The same geometrical conditions are preserved when the layers are in hexagonal stacking sequences as in wurtzite (ZnS) $B\underline{B}C\underline{C}$ and in mixed cubic and hexagonal sequences. Very many polytypes of ZnS itself have been recognized and characterized. Some of these and the layer sequence are listed in Table 8-1. The largest number of layers in regular sequence have, however, been found in the polytypes of SiC, examples of which are listed in Table 8-2 taken from the review of Schaffer.[1] One sequence of frequent occurrence is the $(3_{2n+1}2)_3$ series which gives rhombohedral polytypes 15R, 33R, . . ., $3[3(2n+1)+2]$R where $n = 0, 1, 2, \ldots$.

Diamond (*cF*8): The diamond structure is a three-dimensional adamantine network in which every atom is surrounded tetrahedrally by four

Table 8-1 Some Examples of Observed ZnS Polytypes

Polytype	Stacking Sequence	Polytype	Stacking Sequence
2H	11	24H	16, 4, 2, 2
3R	1		33242253
4H	22		33422433
6H	33		9564
8H	44	24R	$(53)_3$
10H	82	26H	17,4,2,3
	55	28H	9559
14H	5423	36R	$(6222)_3$
	77	48R	$(97)_3$
16H	88		$(10, 6)_3$
	14, 2		$(12, 4)_3$
	5335		$(7423)_3$
	332233		$(8422)_3$
18R	$(42)_3$		$(433222)_3$
20H	522362		$(13, 3)_3$
	533423	60R	$(18, 2)_3$
24H	7557		$(11, 432)_3$
	8943		$(522353)_3$
	15,9	72R	$(6, 11, 5, 2)_3$
			$(9546)_3$

neighbors (Figure 8-1). The 8 atoms in the unit cell form two interpenetrating f.c. cubic networks each of which is occupied by a different component atom in the derived sphalerite structure.

Hexagonal Diamond (*hP*4): A hexagonal form of diamond can be prepared at pressures over 130 kb and temperatures over 1000°C. The atoms

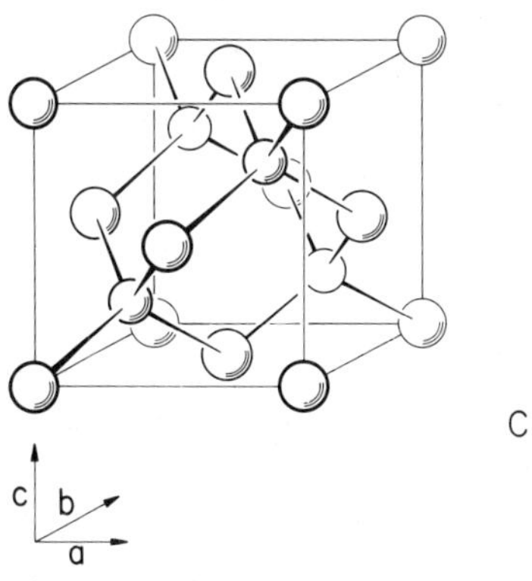

Figure 8-1. The cubic diamond structure.

Table 8-2 Summary of Known SiC Polytypes: from P.T.B. Shaffer[1]

No. of Layers and Symmetry of Unit Cell	Unit-Cell Dimensions			Layer Sequence
	Hexagonal[(1)] c_0(Å)	Rhombohedral a_0(Å)	α	
2H	5.028			11
2H	5.048			
3C	4.349			
4H	10.053			22
6H	15.079			33
7H	17.637			
8H	20.1064			44
9H	22.676			
10H	25.133			3223
14H	35.182[(3)]			$(22)_2 33$
15H	37.794			
15R	37.70	12.69	13°54½′	$(32)_3$
16H	40.208[(3)]			323323
18H	45.234[(3)]			$(22)_3 33$
19H	47.753			$22(23)_3$
21R	52.78	17.683	9°58′	$(34)_3$
24H	60.312[(3)]			
24R	60.47			
27R	67.859	22.689	7°46′	$(2223)_3$
33H	82.929[(3)]			
33R	82.94	27.704	6°21′, 30″	$(3332)_3$
36H	90.65			$(33)_2 32(33)_2 34$
39H	98.007[(3)]			$(33)_2 32(33)_2(32)_2$
39R	98.007[(3)]			$(3334)_3$
48H	120.624[(3)]			
48R	120.94			
51R(a)	128.178	42.763	4°7′	$[(33)_2 32]_3$
51R(b)	128.163[(3)]			$[(22)_3 23]_3$
54H	135.702[(3)]			
55H	138.58			
57R	143.52			$[(33)_2 34]_3$
57R	143.62			
58H	146.14			
60R	151.18			
66H	165.858[(3)]			
69R	173.85			$[(33)_3 32]_3$
72R	180.936[(3)]			
75R	188.497	62.857	2°48′	$[(32)_3(23)_2]_3$
78H	196.014[(3)]			
80H	201.57			
81R	204.09			
84R	211.117	70.395	2°30′	$[(33)_3(32)_2]_3$
87R	218.657	72.865	2°25′	$[(33)_4 32]_3$

Table 8-2 (*continued*)

No. of Layers and Symmetry of Unit Cell	Unit-Cell Dimensions Hexagonal[1] c_0(Å)	Rhombohedral a_0(Å)	α	Layer Sequence
90R	226.6			$[(23)_4 3322]_3$
93R	234.32			
105R	264.56			
105R	264.54			$[(33)_5 32]_3$
111R	279.68			$[(33)_5 34]_3$
120R	301.56[3]			
123R	309.91			
126R	316.638[3]			
141H	355.26			
141R	354.333	118.124	1°30′	$[(33)_7 32]_3$
144R	362.82			
147R	370.38			
150R	377.94			
153R	385.50			
159R	400.62			
168R	422.184[3]			$[(23)_{10} 33]_3$
174R	436.7			$[(33)_3 6(33)_5 4]_3$
192R	482.496[3]			
216R	544.23			
231R	582.03			
249R	627.38			
270R	680.29 (may be 267, 270, or 273R)			
	If 267R			$[(23)_{17} 22]_3$
	If 273R			$[(23)_{17} 33]_3$
339R	854.14			
354R	891.94			
393R	987.609	329.208	0°32′	$[(33)_{21} 32]_3$
400H	1005.2[3] (may be 1200R)			
417R	1050.7	350.4	30.4′	
453R	1141.4	380.6	28.8′	
513R	1292.6			
595R	1491.72[3]			
636R	1602.5			
1200R	3015.6[3] (may be 400H)			

[1] $a_0 = 3.073$ Å for all hexagonal polytypes.
[3] Calculated: Zx 2.513 Å.

form a tetrahedral network in hexagonal stacking AABB which is the degenerate structure of wurtzite. With dimensions $a = 2.52$, $c = 4.12$ Å, the cell has the ideal axial ratio of 1.63.[2]

ZnS (*cF*8): In the sphalerite structure both Zn and S are tetrahedrally coordinated by 4 atoms of the other kind with Zn–S = 2.35 Å (Figure 8-2), and each forms close packed layers (3^6 nets) normal to [111], the stacking sequence being $\underline{A}A\underline{B}B\underline{C}C$ (Zn underlined). As well as being the prototype for a family of polytypic structures, sphalerite is also the pseudocell for superstructures and defect structures with tetrahedrally coordinated atoms.

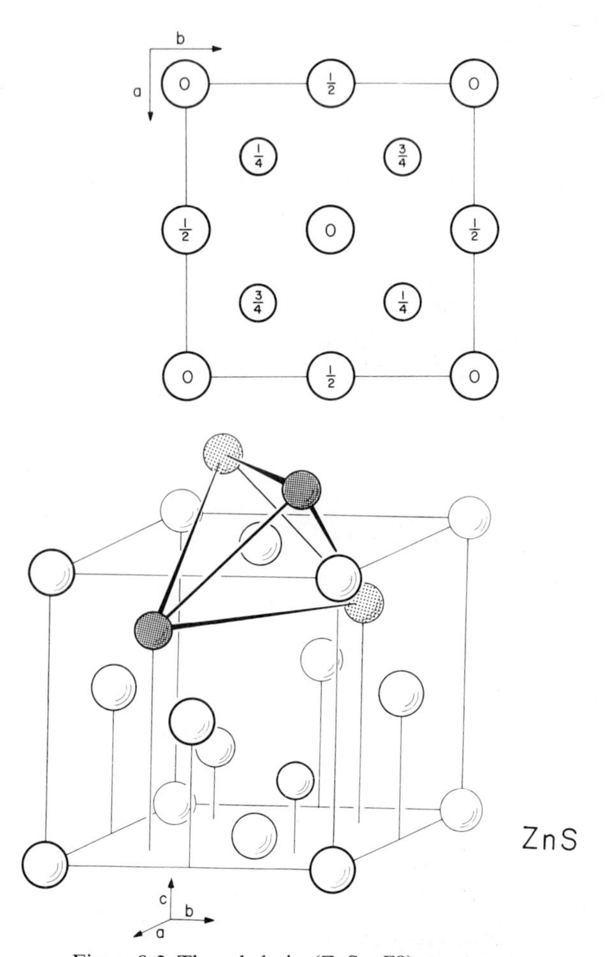

Figure 8-2. The sphalerite (ZnS, *cF*8) structure.

It is obvious that the tetrahedrally disposed M–N bonds control the structural dimensions in compounds with the sphalerite structure, and this is borne out by the near-nighbor diagram since all compounds lie about the line for M–N contacts (Figure 3-7, p. 62). The M–N bond factor controls the structural dimensions regardless of the wide range of radius ratios and electronegativity differences of the component atoms forming sphalerite-type structures.

ZnS (*hP*4): The dimensions of the wurtzite unit cell are $a = 3.823$, $c = 6.261$ Å, $Z = 2$. The atoms are in layers stacked along the [001] direction in the sequence $B\underline{B}C\underline{C}$ (Zn underlined) each atom being surrounded by a tetrahedron of the other (Figure 8-3). The structure can also be regarded as made up of rumpled layers of Zn and S atoms held together by Zn–S bonds running in the c direction of the crystal. The structure belongs to the polytypic family of tetrahedrally coordinated

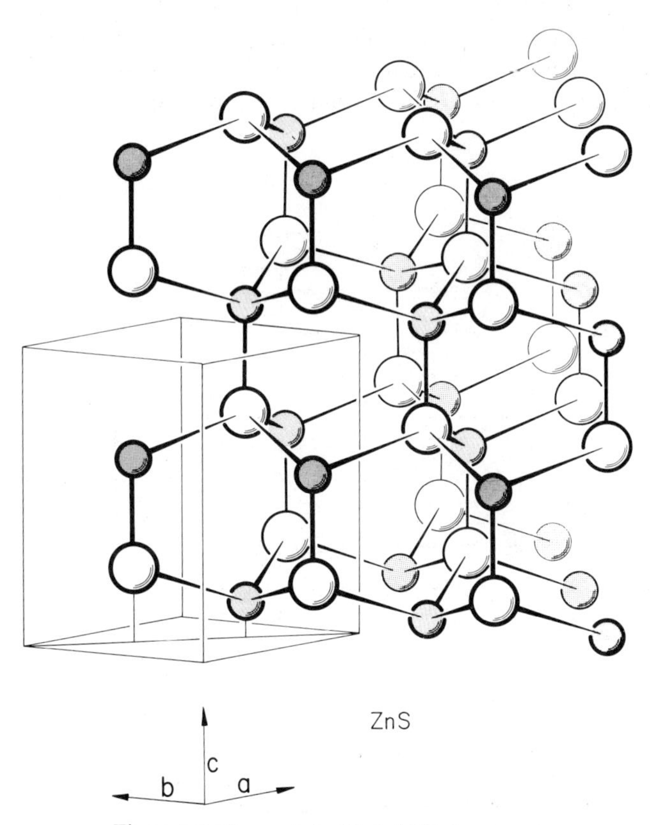

Figure 8-3. The wurtzite (ZnS, *hP*4) structure.

MN structures of which sphalerite is the prototype, and it is also an important basic type for the derivation of ordered and defect superstructures.

Some 20 phases are known to have the wurtzite structure; a characteristic of these is the constancy of the axial ratio which (with one possible exception) only varies about the ideal value of 1.633 from 1.595 to 1.658. Such behavior contrasts markedly with the wide variation of axial ratio found in the NiAs structure, and it is to be attributed to the rigidity of the tetrahedral sp^3 chemical bonds.

Superstructures and Defect Superstructures Based on Sphalerite

CuAsS (*oP*12): Lautite which has cell dimensions $a = 11.35$, $b = 5.456$, $c = 3.749$ Å, $Z = 4$,[3] is a superstructure of sphalerite, each atom having tetrahedral coordination. Cu has 1 As (2.42 Å) and 3 Cu neighbors (2.24–2.34 Å); As has 1 Cu, 1 S (2.25 Å), and 2 As neighbors (2.49 Å) giving zigzag As chains along [001], and S has 1 As and 3 Cu neighbors. Figure 8-4 shows the relationship of the CuAsS structure to sphalerite. The ordered dispositions of Cu, As, and S on the Zn and S sites is such as to give a repeat block of $3 \times 3 \times 1$ sphalerite type pseudocells.

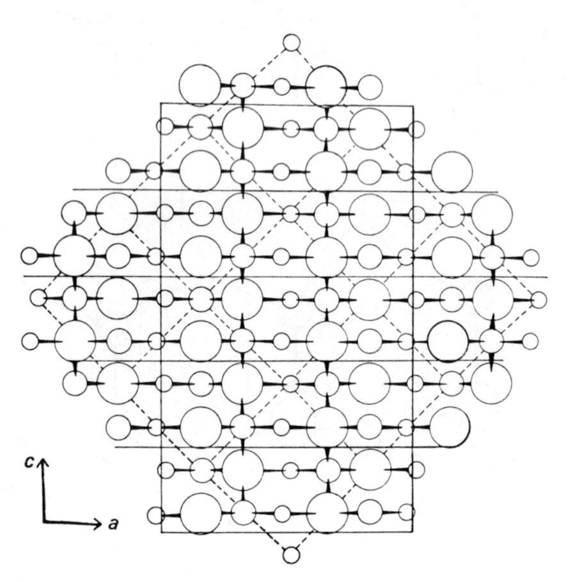

Figure 8-4. Arrangement of the atoms in the lautite CuAsS structure viewed down [010], indicating the relationship of the structure to that of sphalerite. (Craig and Stephenson [3].)

$CuFeS_2$ (*tI*16): The chalcopyrite structure ($a = 5.25$, $c = 10.32$ Å, $Z = 4$)[4] is a superstructure of sphalerite in which the two cations assume ordered positions. This ordering results in slight displacements of the anions from the exact sites of close packing. The superstructure cell corresponds to two sphalerite type cells stacked in the *c* direction (Figure 8-5), and it is observed that the axial ratio, $c/2a$, can be either greater or less than unity for phases with this structure. Chalcopyrite ordering does not occur in ternary phases with cations having the same valency.

Cu_2SiS_3 (*m*-28): With cell dimensions $a = 11.51$, $b = 12.04$, $c = 6.03$ Å, $Z = 4.66$,[5] Cu_2SiS_3 and the Se and Te compounds have a monoclinically distorted superstructure of the sphalerite type with 28 atoms per cell. An orthorhombic form of Cu_2SiS_3 is also known. The structure requires full characterization.

Cu_3AsS_4 (*tI*16): Luzonite with cell dimensions $a = 5.332$, $c = 10.57$ Å is a superstructure of sphalerite[6] with two pseudocells stacked along *c*. The atomic arrangement is shown in Figure 8-6. The closest Cu–S distance in the structure is 2.30 Å and the closest (As, Sb)–S distance is 2.265 Å.

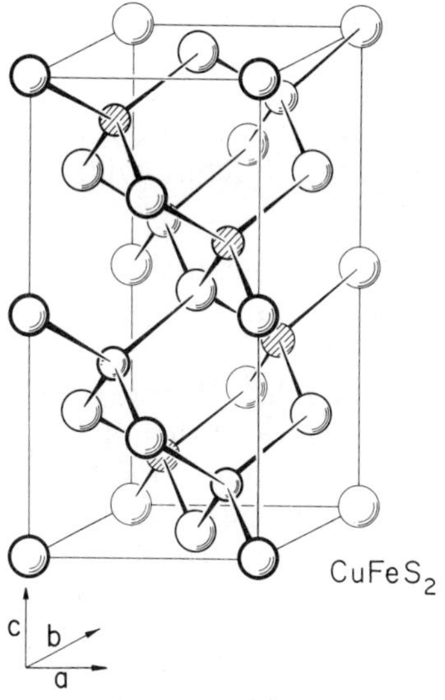

Figure 8-5. Diagram of the structure of chalcopyrite, $CuFeS_2$ (*tI*16). The origin of this idealized drawing has been transposed by $\frac{3}{4}$, $\frac{3}{4}$, $\frac{7}{8}$ from that normally used so as to put S atoms at the cell corners.

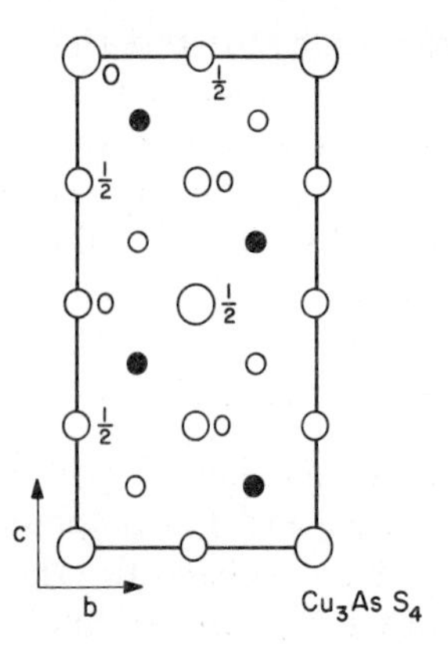

Figure 8-6. The arrangement of the atoms in luzonite, Cu_3AsS_4 (*tI*16), viewed along [100].[6]

Cu_2FeSnS_4 (*tI*16): Cu_2FeSnS_4 with cell dimensions, $a = 5.47$, $c = 10.75$ Å, $Z = 2$, is a superstructure of sphalerite[7] in which the Cu, Fe, and Sn atoms have ordered sites, and the S atoms are displaced slightly from the exact close packed sites. A number of phases are now known to have this structure, but accurate atomic positions must be determined.

CFe_4 (*cP*5): CFe_4 is said from an electron diffraction study of films, to have a defect sphalerite type structure with carbon occupying $\frac{1}{4}$ of the Zn sites in ordered fashion.[8]

$CdAl_2S_4$ (*TI*14): The $CdAl_2S_4$ structure ($a = 5.56$, $c = 10.32$ Å, $Z = 2$) is a superstructure of sphalerite with one quarter of the cation sites vacant in an ordered fashion.[9] The superstructure corresponds to two sphalerite type cells stacked in the c direction (Figure 8-7). The S atoms which have 2 Al + 1 Cd neighbors, are displaced slightly from the exact close packed sites, but their positions still require to be accurately determined. The axial ratio, $c/2a$, of phases with this structure varies from considerably below unity ($c/2a = 0.91$ for $CdAl_2S_4$) to slightly above.

$CdIn_2Se_4$ (*tP*7): The $CdIn_2Se_4$ structure ($a = 5.82$, $c = 5.82$ Å, $Z = 1$) is said to be an ordered defect derivative of the sphalerite structure[9] (Figure 8-8), but the work requires confirmation by single-crystal methods.

A-$CdIn_2Se_{4.15}$ (*oP*186): A superstructure of this composition is said to have orthorhombic cell dimensions $a = b = c = 17.44$ Å.[10] It is made up of 27 fundamental $CdIn_2Se_4$ cells with 30 metal atoms missing, so that some tetrahedra about Se (at heights between $z = \frac{1}{3}$ and $\frac{2}{3}$ have two

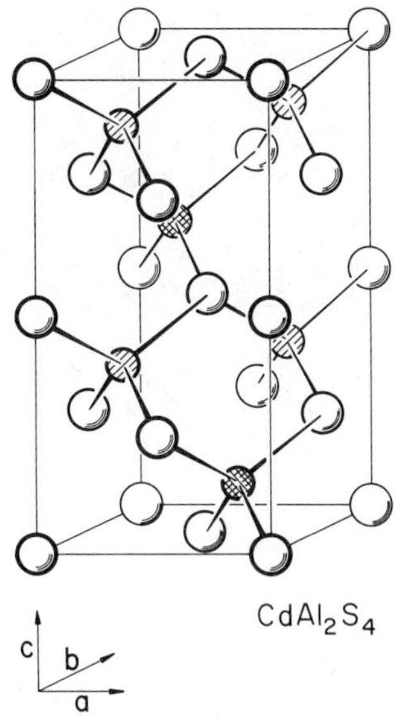

Figure 8-7. Diagram of the structure of $CdAl_2S_4$ ($tI14$). The origin of this idealized drawing has been displaced by $\frac{3}{4}$, $\frac{3}{4}$, $\frac{7}{8}$ from that normally used so as to put S atoms at the cell corners.

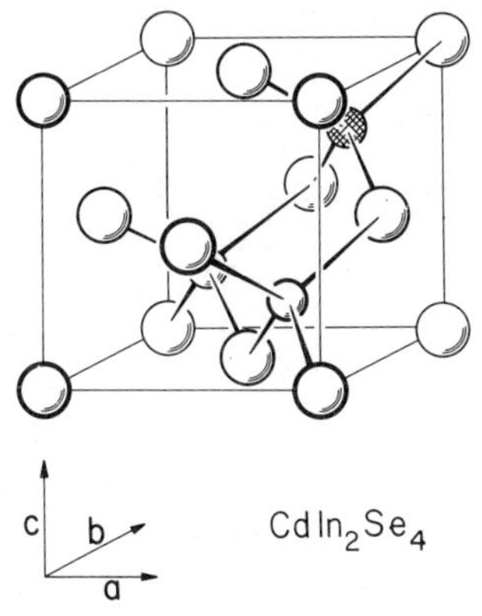

Figure 8-8. Diagram of the structure of $CdIn_2Se_4$ ($tP7$) with the origin transposed by $\frac{3}{4}$, $\frac{3}{4}$, $\frac{3}{4}$ to place Se atoms at the cell corners.

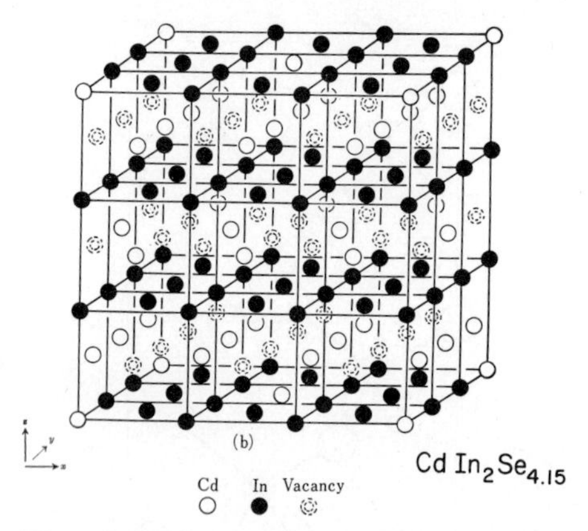

Figure 8-9. Diagram of the $CdIn_2Se_{4.15}$ ($oP186$) superstructure.[10]

corner atoms missing (Figure 8-9). The Se atoms in "tetrahedra" with one corner vacant, shift 0.20 to 0.40 Å toward the vacant corner; those with two vacant corners in the surrounding "tetrahedra" shift 0.15 Å toward the midpoint of the side joining two vacant corners. Another superstructure of $CdIn_2Se_4$ has also been reported.

Superstructures and Defect Superstructures Based on Wurtzite

$BeSiN_2$ ($oP16$): with cell dimensions $a = 4.977$ ($\sqrt{3}a_0$), $b = 5.747$ ($2a_0$), $c = 4.674$ Å (c_0), $Z = 4$.[11] The $BeSiN_2$ structure is a superstructure of wurtzite, corresponding to the chalcopyrite superstructure of sphalerite. N is surrounded tetrahedrally by 2 Be (1.755 Å) and 2 Si (1.755 Å) and both Be and Si are similarly surrounded by N.

$CuSbS_2$ ($oP16$): The wolfsbergite structure with cell dimensions $a = 6.120$, $b = 3.792$, $c = 14.485$ Å, $Z = 4$,[12] contains slabs of atoms in a wurtzite type arrangement which lie parallel to the (001) plane. These slabs terminate when the outside planes parallel to (001) are occupied by Sb atoms as shown in Figure 8-10. Thus Sb has only three close S neighbors (2.44 and 2×2.58 Å). Cu has 4 S neighbors at 2.26, 2.30, and 2×2.33 Å. The inner S(2) has 2 Cu and 2 Sb neighbors and the outer S(1) atoms have 2 Cu and only one close Sb neighbor. The structure requires redetermining although the original findings were obtained on single crystals.

Cu_3AsS_4 ($oP16$): According to an early structure determination which requires confirming, enargite with cell dimensions, $a = 7.43$, $b = 6.46$, $c = 6.18$ Å, $Z = 2$,[13] is a superstructure based on wurtzite (Figure 8-11*a*).

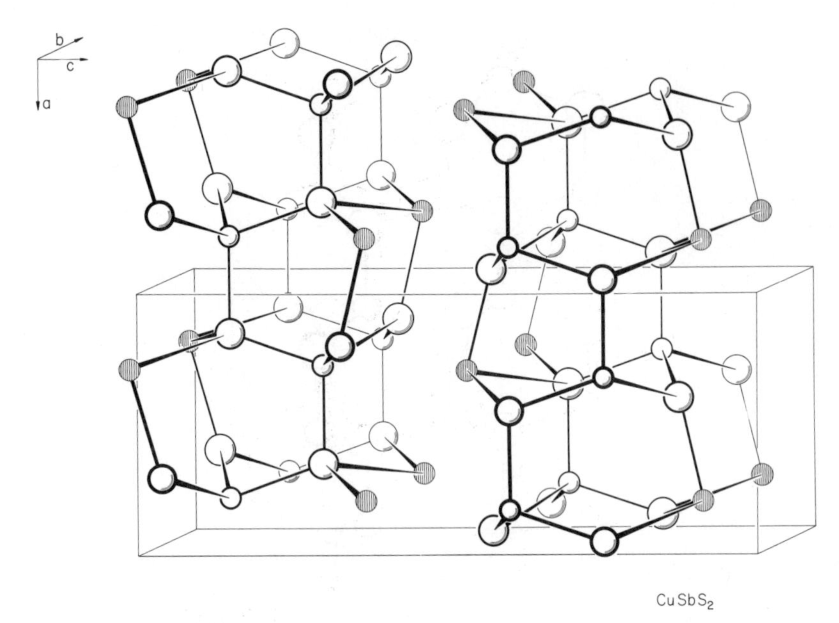

Figure 8-10. Diagram of the structure of wolfsbergite, $CuSbS_2$ ($oP16$): large circles, S; hatched circles, Sb.

Planes of sulfur atoms and of Cu and As atoms parallel to the (010) plane, are stacked in the sequence AB along the [010] direction. In the mixed close packed layers the Cu atoms form a $3636+3^26^2$ (1:2) net with the As atoms in a 4^4 rectangular net occupying the centers of the hexagons of the Cu net (Figure 8-11*b*). All atoms in the structure have four neighbors disposed tetrahedrally, although the atoms are slightly displaced from the ideal positions of the wurtzite pseudocell. Cu–S distances are 2.31 to 2.33 Å. As–S distances are 2.21 to 2.23 Å.

In_2Se_3 (H.T.) ($hP30$): Numerous structural forms of In_2Se_3 have been reported. This one, found by electron diffraction examination of annealed thin films, was originally called β. The structures of two rhombohedral forms $hR5$, called α and β are described on pp. 402 and 434.

The In_2Se_3 cell which has dimensions $a=7.11$, $c=19.30$ Å, $Z=6$ contains six close packed Se layers in hexagonal BC stacking.[14] The In atoms form hexagonal 6^3 nets which lie just under the Se layers and on the same stacking sites as the next Se layer below, so that the atoms occupy tetrahedral holes in the Se array, although they are slightly displaced from the pole of the tetrahedron toward the Se layer above. There are three In sites that can be left vacant relative to the two stacking positions

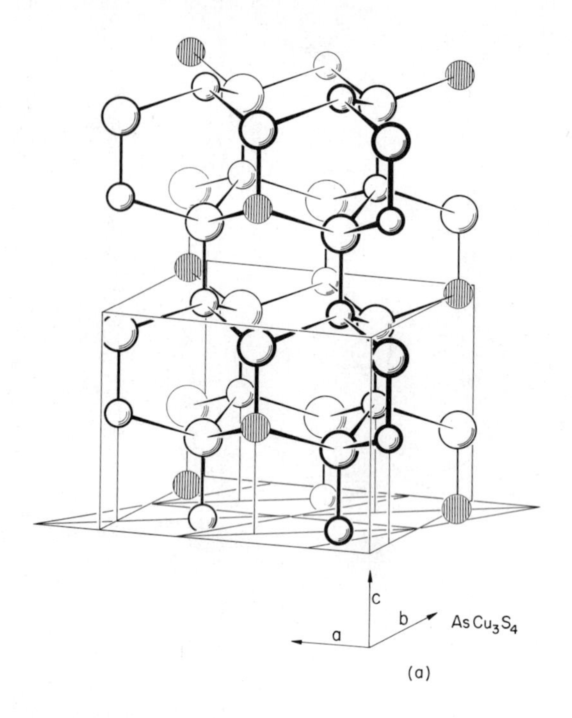

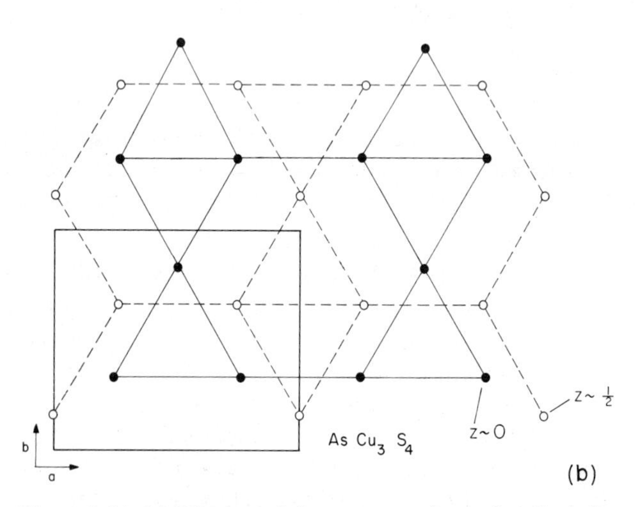

Figure 8-11. (*a*) Diagram of the structure of enargite, Cu_3AsS_4 (*oP*16), indicating the hexagonal pseudocells. (*b*) Arrangement of the Cu atoms on the closed packed planes of the enargite structure.

of the Se layers (Figure 8-12) so that the superstructure, which is of the wurtzite type, must contain 6 layers of Se atoms.

This arrangement results in In having 3 Se at 2.43 Å and one at 2.66 Å. Se(1) and (2) have 2 In at 2.43 Å and one at 2.66 Å and Se(3), that lies over the vacant In site, has only 2 In at 2.43 Å.

α-Ga_2S_3 (*mC*20): α-Ga_2S_3 with cell dimensions $a = 11.14$, $b = 6.41$, $c = 7.04$ Å, $\beta = 121.2°$, $Z = 4$, is a distorted defect superstructure of wurtzite.[15] Close packed layers of S atoms lie parallel to the *ab* plane and are stacked in hexagonal AB close packing in a direction normal to this plane. Hexagonal 6^3 nets of Ga atoms are stacked on the same sites at approximately $\frac{3}{4}$ of the way to the next S layer (Figure 8-13).

Ga(1) and (2) are surrounded by distorted tetrahedra of 4 S atoms at distances from 2.19 to 2.37 Å. S(1) has 2 Ga neighbors and S(2) and (3) have three each.

β-$ZnAl_2S_4$ (H.T.): The high temperature form of β-$ZnAl_2S_4$ has a complex structure made up of three pseudocells, A,B,C, with dimensions $a = 12.83$, $b = 7.50$, $c = 6.10$ Å, $Z = 4$, which are stacked along *c* in the sequence ACBCA.[16] The S atoms are in hexagonal close packing and on the average are surrounded tetrahedrally by Zn, 2Al, and a vacant site. The structure can be regarded as a defect superstructure of wurtzite.

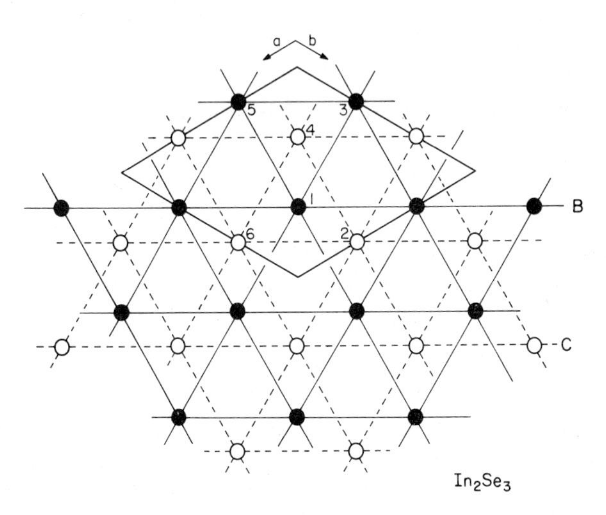

Figure 8-12. Part of the In_2Se_3 (*hP*30) structure showing the 3^6 nets of Se in BC stacking parallel to the (001) plane. The locations of the centers of the hexagons of the In 6^3 nets are numbered successively from 1 to 6 on proceeding up the unit cell from the origin.

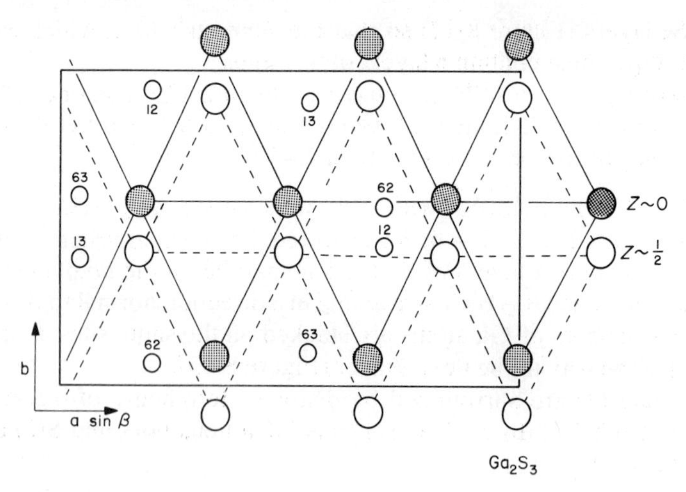

Figure 8-13. Projection of the structure of α-Ga_2S_3 ($mC20$) down [001]. S atoms large circles.

Cu_4SiS_4 ($oC144$), *$Cu_6Si_2S_7$* ($mC60$): In Cu_4SiS_4 with cell dimensions $a = 12.42$, $b = 15.21$, $c = 13.20$ Å, $Z = 16$,[17] and $Cu_6Si_2S_7$ with cell dimensions $a = 16.23$, $b = 6.32$, $c = 9.61$ Å, $\beta = 92.08°$, $Z = 4$,[17] the S atoms are said to be in c.p. hexagonal AB array, and the structures appear to be filled up superstructures of wurtzite.

Miscellaneous Structures Containing Tetrahedra

Isolated tetrahedra

NaGe ($mP32$): The dimensions of the monoclinic unit cell of NaGe are $a = 12.33$, $b = 6.70$, $c = 11.42$ Å, $\beta = 119.9°$, $Z = 16$.[18] The structure contains tetrahedra of Ge atoms with the Na atoms arranged in the intervening spaces (Figure 8-14). Ge thus has 3 Ge close neighbors at 2.54 to 2.59 Å. The closest Ge–Na approach is 2.94 Å.

NaSi ($mC32$): NaSi has cell dimensions $a = 12.19$, $b = 6.55$, $c = 11.18$ Å, $\beta = 119.0°$, $Z = 16$.[18] The structure contains regular tetrahedra of Si atoms between which the Na atoms are situated. The structure can be compared with that of NaGe. The unit cells are similar, but the arrangement of the tetrahedra differs somewhat (Figure 8-15).

The closest Na–Na distance is 3.22 Å and the closest Si–Na distance is 2.92 Å. Si–Si distances within the tetrahedra are 2.40 to 2.49 Å.

GeK ($cP64$): GeK has a large unit cell, $a = 12.78$ Å, $Z = 32$.[19] The Ge atoms form tetrahedra, Ge(1) with 3 Ge(1) neighbors at 2.57 Å and Ge(2)

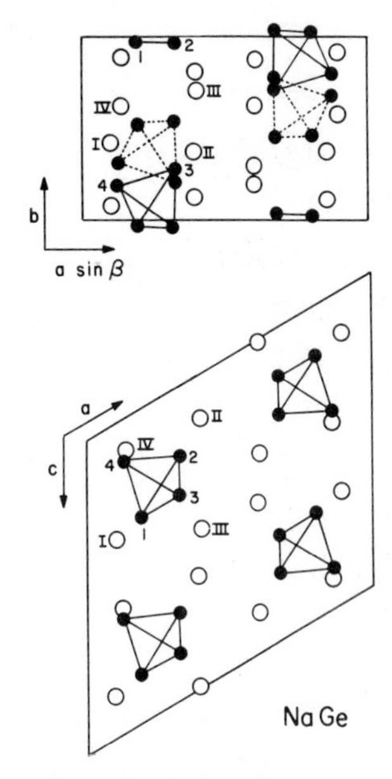

Figure 8-14. Atomic arrangement in the NaGe (*mP*32) structure. (Witte and Schnering[18].)

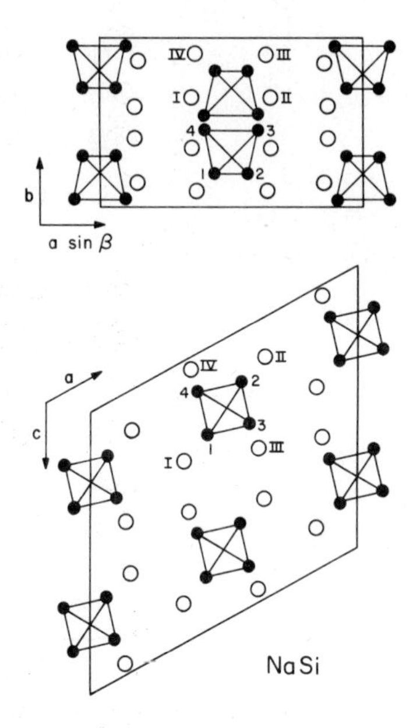

Figure 8-15. Atomic arrangement in the NaSi (*mC*32) structure. (Witte and Schnering[18].)

with 3 Ge(2) at 2×2.56 and 2.58 Å. The 8 tetrahedra are centered at 0, 0, 0; $\frac{1}{2}, \frac{1}{2}, \frac{1}{2} \pm (\frac{1}{4}, \frac{1}{2}, 0;\ 0, \frac{1}{4}, \frac{1}{2};\ \frac{1}{2}, 0, \frac{1}{4})$ which are the locations of the atoms in the β-W structure. The K atoms are distributed between the tetrahedra so that each Ge has 6 K neighbors at 3.37 to 3.63 Å. See Figure 5-5, p. 217.

NaPb (*tI*64): NaPb has cell dimensions $a = 10.58$, $c = 17.746$ Å, $Z = 32$.[20] The Pb atoms are arranged in tetrahedra centered at 0, 0, 0; $\frac{1}{2}, \frac{1}{2}, 0 + (0, \frac{3}{4}, \frac{1}{8};\ 0, \frac{3}{4}, \frac{5}{8};\ 0, \frac{1}{4}, \frac{3}{8};\ 0, \frac{1}{4}, \frac{7}{8})$ as shown in Figure 8-16. Pb–Pb distances along the tetrahedra edges are 3.146 and 3.162 Å. The Na atoms surround the Pb tetrahedra so that each Na is surrounded approximately tetrahedrally by four Pb_4 groups. The closest Na–Pb distance is 3.36 Å.

Na_2Tl (*oC*48): Na_2Tl has cell dimensions $a = 13.94$, $b = 8.880$, $c = 11.69$ Å, $Z = 16$.[21] The structure contains isolated tetrahedra of Tl atoms (distances 3.18 to 3.30 Å). Three Tl and 9 Na atoms surround each Tl in an irregular icosahedron, The Na–Tl distances being 3.13 to 3.62 Å for Tl(1) and 3.23 to 4.20 Å for Tl(2).

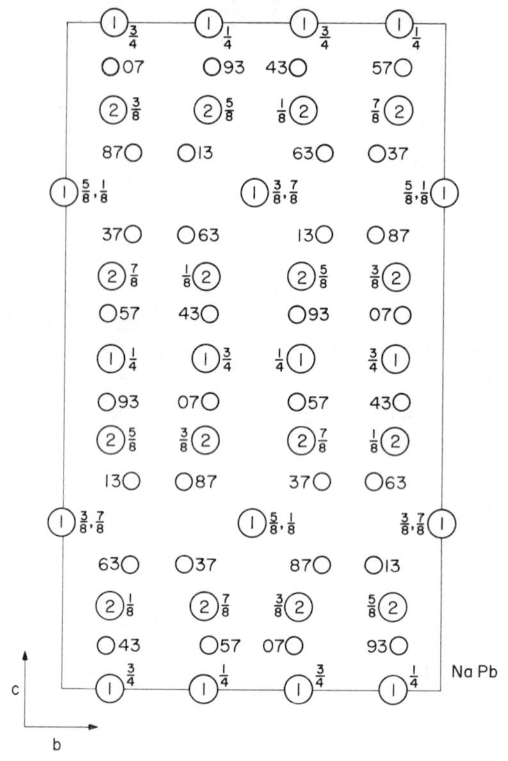

Figure 8-16. Structure of NaPb (*tI*64) viewed along [100]: Pb, small circles.

Connected and Interpenetrating Tetrahedra

GeS_2 (*oF*72): GeS_2 has cell dimensions $a = 11.68$, $b = 22.39$, $c = 6.87$ Å, $Z = 24$.[22] Chains of four-connected Ge atoms (centering S tetrahedra) and two-connected S atoms wander throughout the structure as shown in Figure 8-17. The structure is not derived from close packing.

GeS_2 II (H.P.) (*tI*12): GeS_2 II prepared under high pressure, has cell dimensions $a = 5.480$, $c = 9.143$ Å, $Z = 4$.[23] The structure contains distorted tetrahedra of S surrounding Ge, which share corners (Figure 8-18). SiS_2 II has the same structure.

ZnP_2 (*tP*24): The red tetragonal form of ZnP_2 has cell dimensions, $a = 5.08$, $c = 18.59$ Å, $Z = 8$.[24] The structure contains spiral chains of P atoms lying parallel to the (001) plane (Figure 5-9, p. 221). Both P(1) and P(2) are surrounded tetrahedrally by 2 P (2.16, 2.22 Å) and 2 Zn atoms

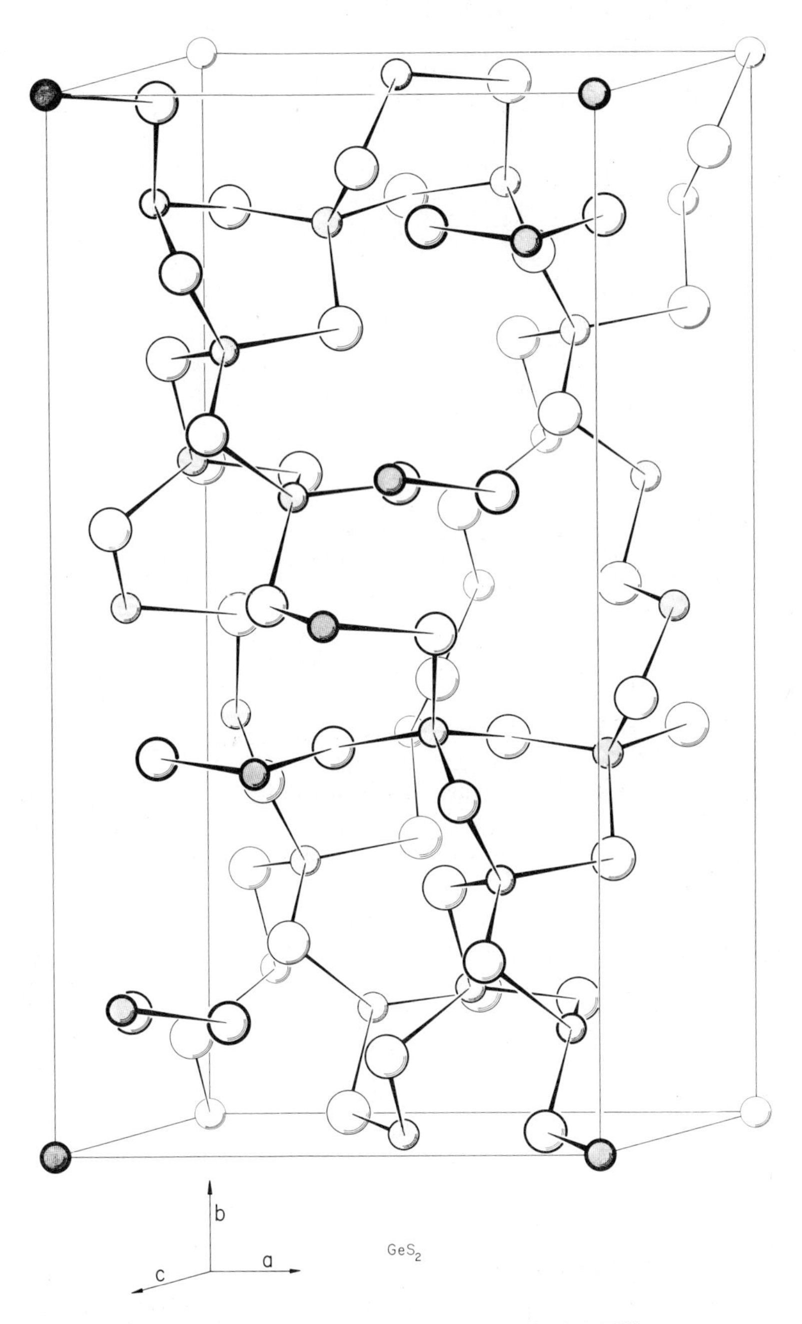

Figure 8-17. Diagram of the structure of GeS_2 ($oF72$).

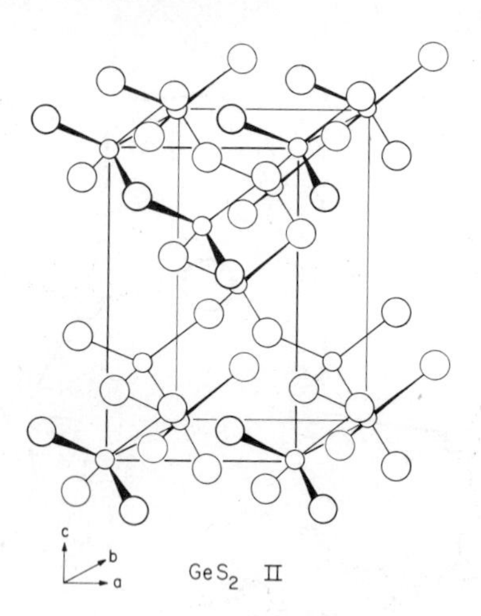

Figure 8-18. Diagram showing the atomic arrangement in GeS_2 II (*tI*12), the structure prepared under high pressure: Ge, small circles.

(2.36 to 2.43 Å) and Zn has 4 P neighbors; there are no Zn–Zn bonds as in the black monoclinic form.

ZnP$_2$ (*mP*24): The black monoclinic form of ZnP_2 has cell dimensions $a = 8.85$, $b = 7.29$, $c = 7.56$ Å, $\beta = 102.3°$, $Z = 8$.[25] The structure contains chains of P atoms which are cross linked by P(1)–P(1) bonds, and there are pairs of Zn(2) atoms (Figure 5-18, p. 228). There are large void spaces in the structure. All atoms are approximately tetrahedrally coordinated, Zn(1) having 4 P neighbors, Zn(2) having 1 Zn at 2.44 Å and 3 P neighbors, P(1) has 1 Zn and 3 P neighbors and P(2), (3), and (4) each have 2 Zn and 2 P neighbors.

FeKS$_2$ (*mC*16): $FeKS_2$ with cell dimensions $a = 7.05$, $b = 11.28$, $c = 5.40$ Å, $\beta = 112.5°$, $Z = 4$,[26] contains chains of S tetrahedra surrounding Fe,

```
  S       S
   \     / \     /
     Fe       Fe
   /     \ /     \
  S       S
```

which run throughout the structure in the *c* direction. The K atoms lie above or below Fe in the *b* direction separating the chains from each other (Figure 8-19).

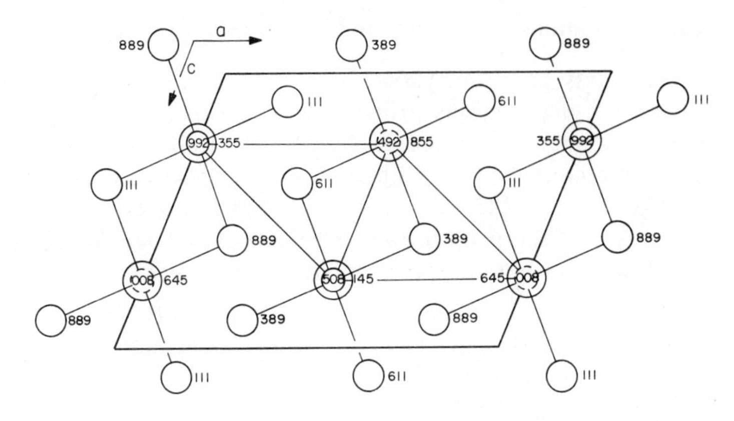

Figure 8-19. Atomic arrangement in the $FeKS_2$ (*mC*16) structure viewed down [010].

Fe has 2 S neighbors at 2.20 Å and two at 2.28 Å; it also has 2 Fe at 2.71 Å. The closest K–S and K–Fe distances are 3.31 and 3.85 Å respectively. S has 2 Fe neighbors.

TlSe (*tI*16) and *Si III* (*cI*16): The TiSe structure contains

```
      Se
 \   /  \   /
  Tl      Tl
 /   \  /   \
      Se
```

chains similar to the chains in the $FeKS_2$ structure, and the Si III structure is built up of very distorted interpenetrating tetrahedra. Both of these structures are, however, made up of 3^2434 nets of atoms and they are therefore described on pp. 605 and 632 together with other structures having this characteristic.

Ge III (*tP*12): Ge III ($a = 5.93$, $c = 6.98$ Å, $Z = 12$)[27] is formed at pressures over 120 kb and its structure is retained when the pressure is removed. Ge(1) has four Ge(2) neighbors that are almost equidistant, but the arrangements are considerably distorted from tetrahedral. The atoms form layers at $z \approx 0, \frac{1}{4}, \frac{1}{2}$, and $\frac{3}{4}$, with Ge(2) arranged in large, approximately square, $\frac{1}{2}.4^4$ nets and Ge(1) atoms being located somewhat off center of alternate distorted Ge(2) squares. The structure was determined only approximately.

Structures Containing Tetrahedra and Triangles or Pyramids

InS (*oP*8): InS has cell dimensions $a = 4.443$, $b = 10.64$, $c = 3.940$ Å, $Z = 4$.[28] The structure contains strings of In–S bonds running throughout in three-dimensional array (Figure 5-17, p. 228). In–In pairs also occur, a feature which recalls the Ga–Ga pairs found in structures of the gallium monochalcogenides (p. 223). Each S forms a pyramid with 1 In at 2.56 and 2 at 2.58 Å, whereas each In has 3 S neighbors and one In neighbor at 2.80 Å which are distributed approximately tetrahedrally.

CuS (*hP*12): CuS has cell dimensions $a = 3.794$, $c = 16.332$ Å, $Z = 6$.[29] The S atoms form six triangular nets normal to [001] which are stacked in the mixed close packed and pair-stacking sequences ABAACA ($c/3a = 1.44$). Cu(2) are situated in tetrahedral holes of this array, and Cu(1) lie in the S(1) arrays on B and C sites so that they center alternate triangles formed by S(1). The overall stacking sequence is therefore (Cu underlined) $A\underline{B}[B\underline{C}]\underline{B}AA\underline{C}[C\underline{B}]\underline{C}A$ (Figure 8-20). Although the existence of this structure has been confirmed at a later date, new work is required to establish the atomic positions accurately.

S(1) has three equatorial Cu(1) neighbors at 2.17 Å and two polar Cu(2) neighbors at 2.33 Å, being at the center of a trigonal bipyramid. S(2) has 1 S at 2.05 Å and 3 Cu(2) at 2.29 Å, being surrounded tetrahedrally (Figure 8-20). Cu(1) has 3 Si(1) neighbors and Cu(2) has 1 S(1) and 3 S(2) neighbors.

$GeAs_2$ (*oP*24): $GeAs_2$ with cell dimensions $a = 14.76$, $b = 10.16$, $c = 3.728$ Å, $Z = 8$,[30] contains chains of As(2) and (3) atoms bonded together and running in the [001] direction (Figure 5-4, p. 215). Each of these atoms is bonded to one Ge of the Ge and As(1) and (4) framework, where Ge is surrounded tetrahedrally by 4 As and the As atoms are each bonded to 3 Ge atoms.

β-Si_3N_4 (*hP*14): The β-Si_3N_4 structure ($a = 7.608$, $c = 2.9107$ Å, $Z = 2$)[31] resembles that of phenacite. The atoms all lie in layers at $z = \frac{1}{4}$ and $\frac{3}{4}$, Si forming a distorted kagomé net and N a triangular net. Some of the right trigonal prisms formed by Si are base centered by N(1). N(1) thus has 3 Si neighbors at 1.75 Å; N(2) also has 3 Si neighbors (2×1.72, 1×1.75 Å) and Si has four N neighbors. These connections result in a three-dimensional bond network running throughout the structure as shown in Figure 8-21.

α-Si_3N_4 (*hP*28): With cell dimensions $a = 7.748$, $c = 5.617$ Å, $Z = 4$,[32] α-Si_3N_4 is a layer structure with layers normal to the c axis of the cell. At $z = 0$, N(3) atoms form a large triangular net with sites fully occupied and

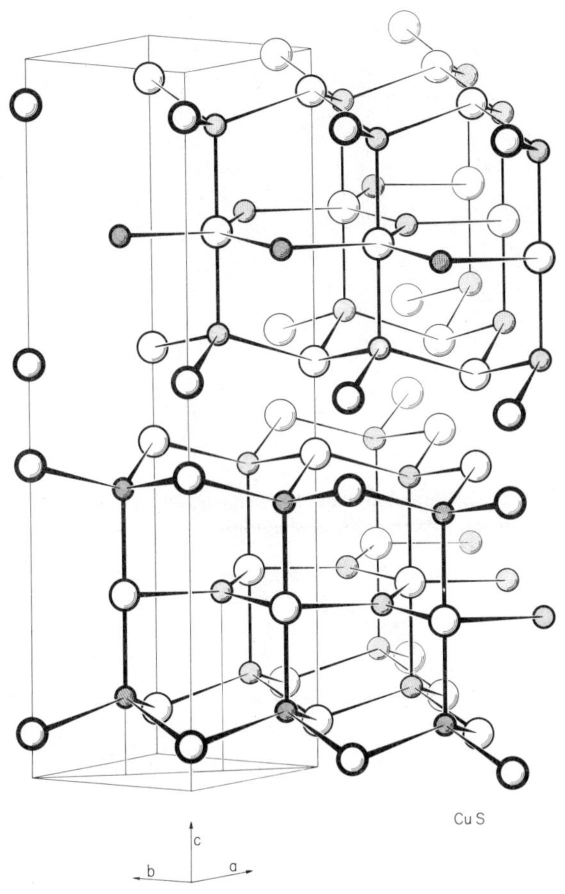

Figure 8-20. Diagram showing the atomic arrangement in the CuS ($hP12$) structure: S atoms, large circles.

N(1) and (3) together form a smaller triangular net with many sites vacant. Triangles of three Si(2) atoms surround the N(1) atoms (Figure 8-22). At $z = \frac{1}{2}$, N(1) and N(3) atoms occupy the same sites as at $z = 0$, but the orientation of the Si(2) triangles about N(1) is changed. At $z = \frac{1}{4}$, N(2) and N(4) atoms, respectively, play the same roles as N(1) and (3) at $z = 0$, N(2) being surrounded by a triangle of Si(1). At $z = \frac{3}{4}$, N(4) atoms have the same positions, but N(2) moves from the C to the B stacking site with its surrounding triangle of Si(1) atoms. Referring to the small triangular nets, the layers are so arranged that the vacant sites at one level ($z = 0$, $\frac{1}{2}$) are the filled sites at the other levels ($z = \frac{1}{4}, \frac{3}{4}$) and *vice versa*.

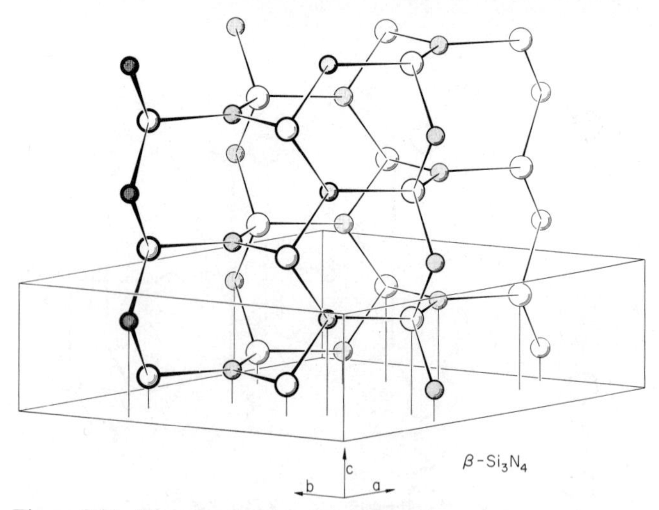

Figure 8-21. Diagram showing the arrangement of the atoms in the β-Si_3N_4 (*hP*14) structure: Si, large circles.

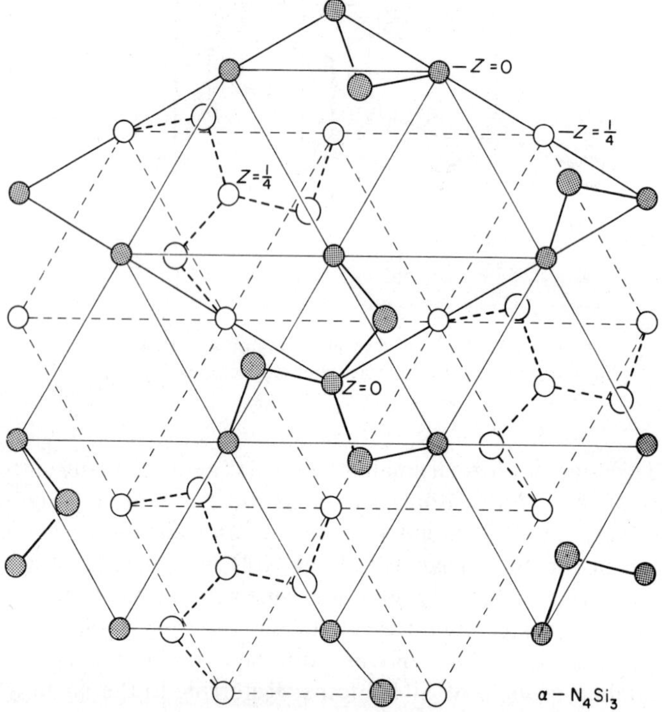

Figure 8-22. Diagram showing the arrangement of the atoms in the α-Si_3N_4 (*hP*28) structure: Si, large circles.

Thus Si atoms are surrounded tetrahedrally by four N atoms; N(1) and (2) are surrounded by 3 Si in a plane, and N(3) and (4) also have 3 Si neighbors, but they are not in the plane of the N atoms.

MN_2: M on 3^6 Nets in ABC Stacking. N on 3^6 Nets at $\frac{1}{4}$ and $\frac{3}{4}$ Spacing. Structures with Cubes and Tetrahedra

Structures with All Tetrahedral Holes Occupied in a Close Packed Array of Atoms. All of the tetrahedral holes in close packed layers of atoms in cubic ABC stacking sequence, can be filled by inserting two more close packed layers of atoms so that they are each $\frac{3}{4}$ of the interlayer distance from the layer whose stacking site they assume (e.g., $\underline{A}B_{1/4}A_{3/4}\underline{B}C_{1/4}$-$B_{3/4}\underline{C}A_{1/4}C_{3/4}$). The structure so generated is the fluorite CaF_2 type. There is no family of polytypic structures based on the fluorite arrangement because in hexagonally surrounded close packed layers, the two sets of tetrahedral holes which share a common tetrahedron face, are too close together to be simultaneously occupied by atoms. Nevertheless, a number of distorted structures and defect superstructures based on a hexagonal fluorite arrangement are known.

CaF_2 ($cF12$): In the fluorite structure the F atoms are located on the centers of both sets of tetrahedral interstices in the close packed cubic array of Ca atoms. Thus the structure comprises three interpenetrating f.c. cubic arrays of atoms. In planes normal to the [111] direction each component forms close packed layers and there are twice as many F layers as Ca, the overall stacking sequence along [111] being $\underline{A}BA\underline{B}$-$CB\underline{C}AC$ (Ca underlined) with F layers once removed from a Ca layer having the same stacking position.

In the antifluorite structure of Mg_2Sn ($a = 6.759$ Å) Sn is surrounded by a cube of 8 Mg at 2.93 Å and Mg is surrounded tetrahedrally by 4 Sn; it also has 6 Mg neighbors at 3.38 Å (Figure 8-23). The near-neighbor diagram (p. 53) for metallic and semiconducting phases (not oxides) with the fluorite MN_2 structure and for the ordered ternary AgAsMg† structure indicates the importance of M–N bonds in controlling the structural dimensions, and shows that at radius ratio values (D_M/D_N) less than unity, the 6 N–N contacts become increasingly important.

Compared to the sums of the volumes of the elements, the unit cell volume is expanded for the MMg_2 semiconductors and for the MAl_2 phases, and contracted slightly for the metallic phases with the fluorite structure.

†The diagram is constructed for the average diameter of the two N components on the F sites.

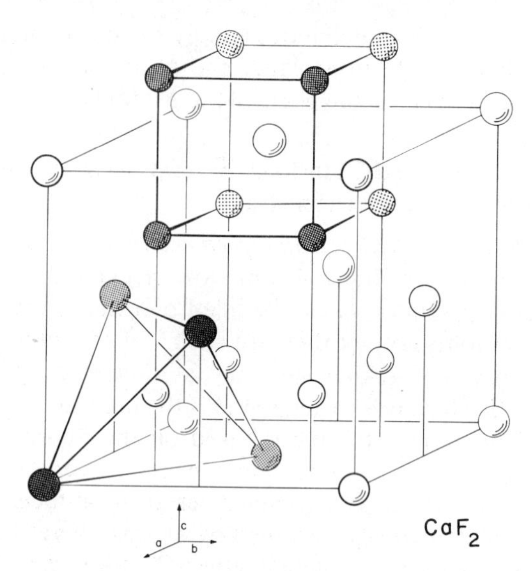

Figure 8-23. The atomic arrangement in the fluorite, CaF_2 ($cF12$) structure.

Distorted CaF_2 Type Structures and Superstructures

ThH_2 ($tI6$): The unit cell dimensions, $a = 4.10$, $c = 5.03$ Å, $Z = 2$,[33] suggest that the ThH_2 structure should be regarded as a squashed derivative of a f.c. cube, rather than an expanded derivative of a b.c. cube of Th atoms. The structure, therefore, appears as a tetragonally distorted ($c/a = 0.87$) fluorite structure with the hydrogen atoms occupying all tetrahedral holes in a f.c. cubic array of Th atoms.

AgAsMg ($cF12$): The AgAsMg structure is an ordered derivative of the fluorite structure with Ag and Mg severally ordered on the F atom sites so that they each occupy one set of tetrahedral holes in the f.c. cubic array of As atoms.[34] Thus each of the three components forms a f.c. cubic array. Phases such as AsAgMg or SbAgMg with the Group V atom on the Ca site are semiconductors, whereas phases such as CuSbMg with Cu on the Ca sites‡ are metallic.

$AlLi_3N_2$ ($cI96$): $AlLi_3N_2$ is a distorted antifluorite type superstructure with a cell ($a = 9.48$ Å, $Z = 16$) that corresponds to 8 fluorite type pseudo-

‡Or, changing the origin, with Sb on the Ca sites and Cu and Mg occupying one set of tetrahedral holes and the octahedral holes.

cells[35] (Figure 8-24). The N atoms occupy positions corresponding to Ca of fluorite, having 8 Li and Al neighbors at the corners of distorted cubes (N(1) to 6 Li at 2.15 Å and 2 Al at 1.89 Å; N(2) to 6 Li at 2.15–2.16 Å and 2 Al at 1.885 Å). Al is tetrahedrally surrounded by 4 N at 1.89 Å and Li has 4 N neighbors. Li–Li distances are 2.22 and 2.25 Å, Li–Al are 2.57 and 2.61 Å.

Li_7MnN_4 (*cP*96): The Li_7MnN_4 structure is a superstructure based on the antifluorite arrangement.[36] The large unit cell, $a = 9.571$ Å, $Z = 8$, corresponds to a block of 8 fluorite type cells in which the Mn atoms are ordered on two sitesets, 0,0,0; $\frac{1}{2}$ $\frac{1}{2}$,$\frac{1}{2}$ and $\pm$($\frac{1}{4}$,$\frac{1}{2}$,0; 0,$\frac{1}{4}$,$\frac{1}{2}$; $\frac{1}{2}$,0,$\frac{1}{4}$).

The N atoms have 7 Li + 1 Mn neighbors. N(2) is surrounded by a regular cube, all neighbors being at 2.07 Å. N(1) is irregularly surrounded, Li neighbors being at 2.02 to 2.24 Å and Mn at 1.91 Å. Mn have 6 Li at 2.39 Å and 4 N at 1.91 Å (Mn(1)) or 2.07 Å (Mn(2)). Li(1) to (5) all have 4 N and 6 Li or Mn neighbors. Li(1) has 2 Mn neighbors, Li(4) and (5) have one, and Li(2) and (3) have no Mn neighbors.

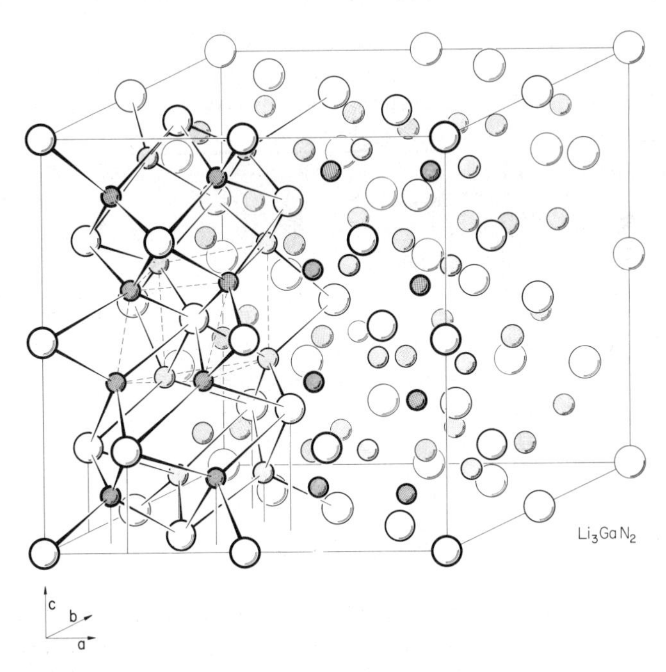

Figure 8-24. Diagram of the superstructure of $GaLi_3N_2$ (*cI*96). The origin has been transposed to put N atoms (large spheres) at the cell corners.

Defect Superstructures of CaF_2

β-"NiTe" (*hR*4): Electron diffraction studies of thin films reveal a phase with rhombohedral structure (hexagonal cell, $a = 3.88$, $c = 20.2$ Å, $Z = 6$)[37] which probably does not exist in massive alloys. The atoms are arranged in triangular layers normal to the c axis of the hexagonal cell. Although the composition of the phase may be Ni_3Te_2 and there is only partial ordering of the atoms in the layers, an ideal structure based on the composition NiTe and full ordering is described. The Te atoms are in close packed sequence CABCAB ($c/a = 1.73$) and the Ni atoms occupy both sets of tetrahedral holes between alternate pairs of Te atom layers giving four-layer structural units Te–Ni–Ni–Te stacked in sequences of the type $\mathrm{C}\underline{\mathrm{A}}\underline{\mathrm{C}}\mathrm{A}$ (Ni underlined). Each Ni has three Ni neighbors in its neighboring layer in addition to the 4 Te neighbors. The Te atoms have 3 Ni neighbors. The four-layer units are identical to the group of four layers, La–O(2)–O(2)–La, found in the La_2O_3 structure, where they are separated from other four-layer groups by triangular layers of O(1) atoms, and also to the C–Al–Al–C arrangements found in aluminum carbonitrides.

CuTe (*oP*4): In the CuTe structure with cell dimensions $a = 3.10$, $b = 4.02$, $c = 6.86$ Å, $Z = 2$,[38] the Te atoms form an approximately body centered orthorhombic cell with the Cu atoms occupying the four approximately tetrahedral holes located below the body center; those above it being empty (Figure 8-25). This distribution displaces the central Te atom upwards from the body center. Each Te has four Cu neighbors on the same side of it, either above or below (2×2.63, 2×2.72 Å). The closest Cu–Cu distance is 2.64 Å.

SiS_2 (*oI*12): The SiS_2 structure with cell dimensions $a = 9.57$, $b = 5.61$, $c = 5.54$ Å, $Z = 4$,[39] can be recognized as a distorted f.c. cubic array of S atoms with Si occupying two of the tetrahedral holes per f.c. cubic pseudocell, so that isolated

```
      S     S
     / \   / \
   Si   Si    Si
     \ /   \ /
      S     S
```

chains run throughout the structure in the [001] direction. This is shown in Figure 8-26 where the origin has been transposed and the array of sulfur atoms has been adjusted to the ideal f.c. cubic sites. The actual structure contains two distorted pseudocubic cells of S atoms along the a axis. Si has 4 S at 2.14 Å and S has 2 Si neighbors. The structure requires further confirmation.

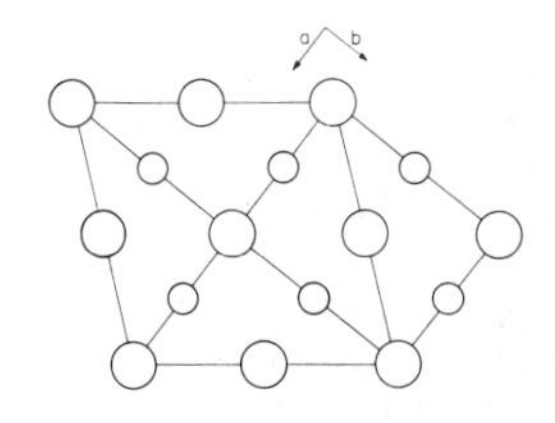

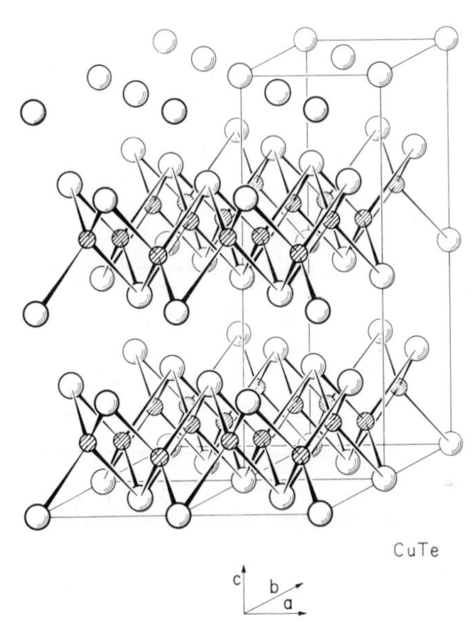

Figure 8-25. Idealized drawing of the CuTe (*oP*4) structure to show the three-layer arrangement of Cu and Te atoms. The relationship to the normal unit cell of CuTe is indicated at the top of the figure.

γ-ZrH$_{\sim 0.5}$ (*tP*6): $ZrH_{0.5}$ has unit cell dimensions $a = 4.586$, $c = 4.948$ Å, $Z = 4$.[40] The Zr atoms form a tetragonally distorted f.c. cubic array ($c/a = 1.08$) and the hydrogen atoms partially occupy the centers of 4 tetrahedral interstices arranged in a (110) type plane (Figure 8-27). Zr is surrounded by up to 4 H in a plane and H is surrounded by 4 Zr in a distorted tetrahedron. Atomic positions were determined by X-ray and neutron diffraction.

P_2Zn_3 (*tP*40): P_2Zn_3 with cell dimensions $a = 8.113$, $c = 11.47$ Å, $Z = 8$, is a superstructure of the fluorite type with two of the eight Zn sites surrounding P vacant, causing displacement of the atoms from their ideal

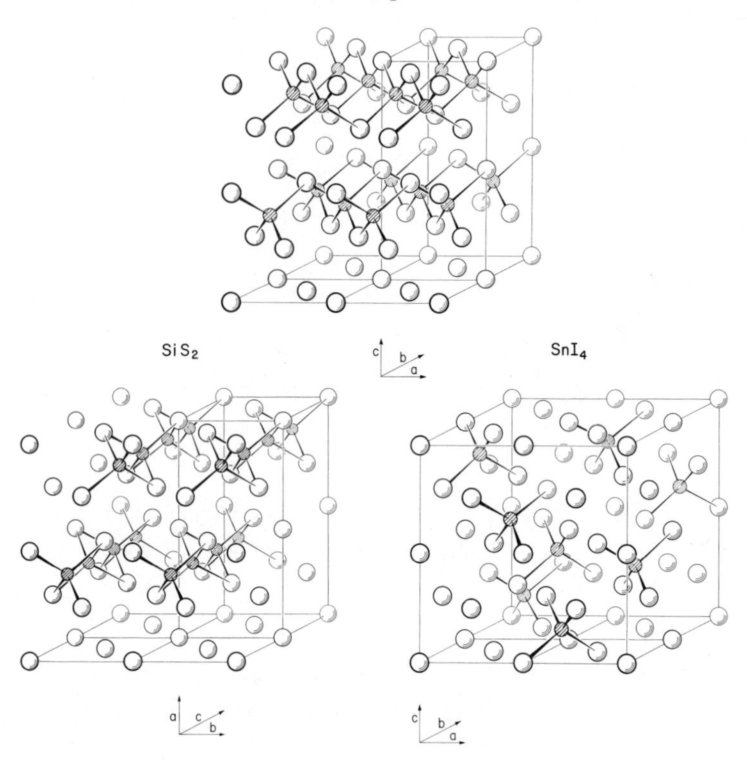

Figure 8-26. Diagram comparing the atomic arrangements in the HgI_2 ($tP12$), SiS_2 ($oI12$) and SnI_4 ($cP40$) structures. I or S atoms have been placed at the cell corners.

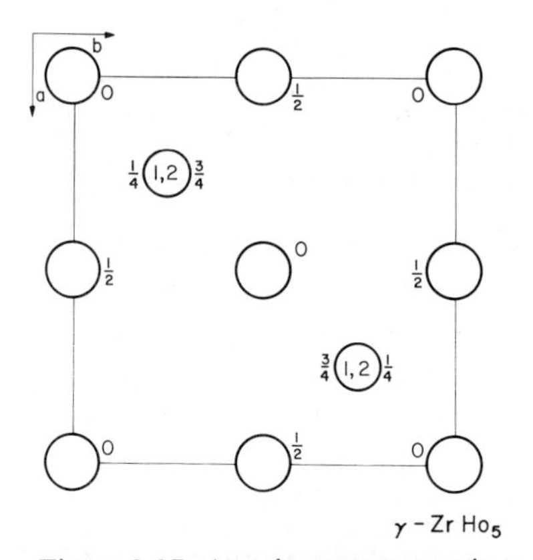

Figure 8-27. Atomic arrangement in the γ-$ZrH_{\sim 0.5}$ ($tP6$) structure viewed down [001]: Zr, large circles.

positions.[41] Although the structure was determined by single crystal methods, it does not yet appear to have been confirmed, and a different structure has been reported for some other substances such as As_2Cd_3, which were said[41] to have the P_2Zn_3 structure. The Zn atoms have 4 P neighbors distributed tetrahedrally at distances from 2.28 to 2.77 Å (Figure 8-28).

As_2Cd_3 (*tI*160): As_2Cd_3 has cell dimensions $a = 12.67$, $c = 25.48$ Å, $Z = 32$.[42] The structure is a defect superstructure of fluorite. The As atoms are in slightly distorted cubic close packing and the Cd atoms are located in the tetrahedral holes thereof. They are displaced from the ideal sites toward the vacant Cd sites which are ordered as indicated in Figure 8-29. As–Cd tetrahedra distances vary from 2.51 to 3.20 Å.

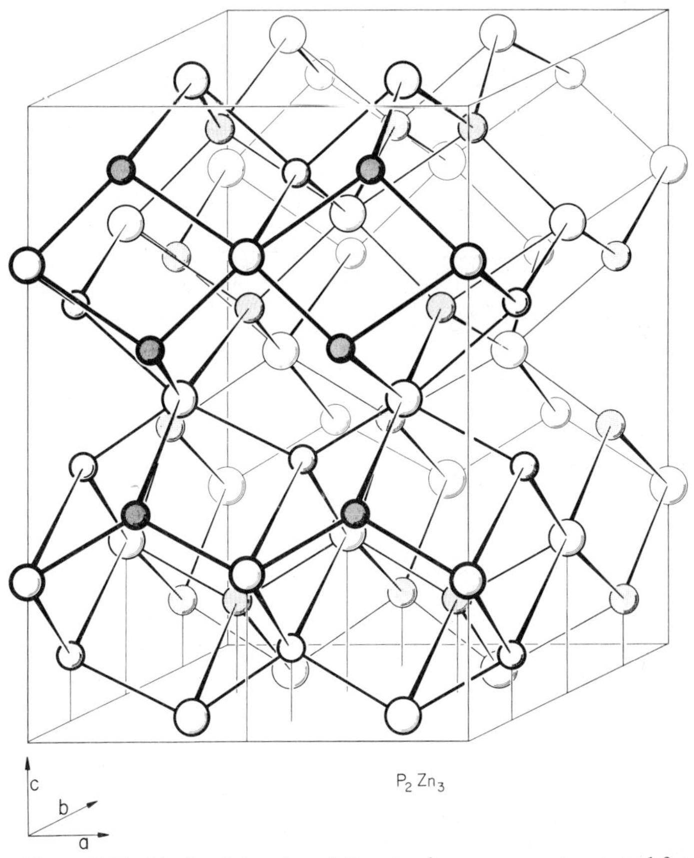

Figure 8-28. Idealized drawing of the atomic arrangement reported for P_2Zn_3 (*tP*40): P, large circles.

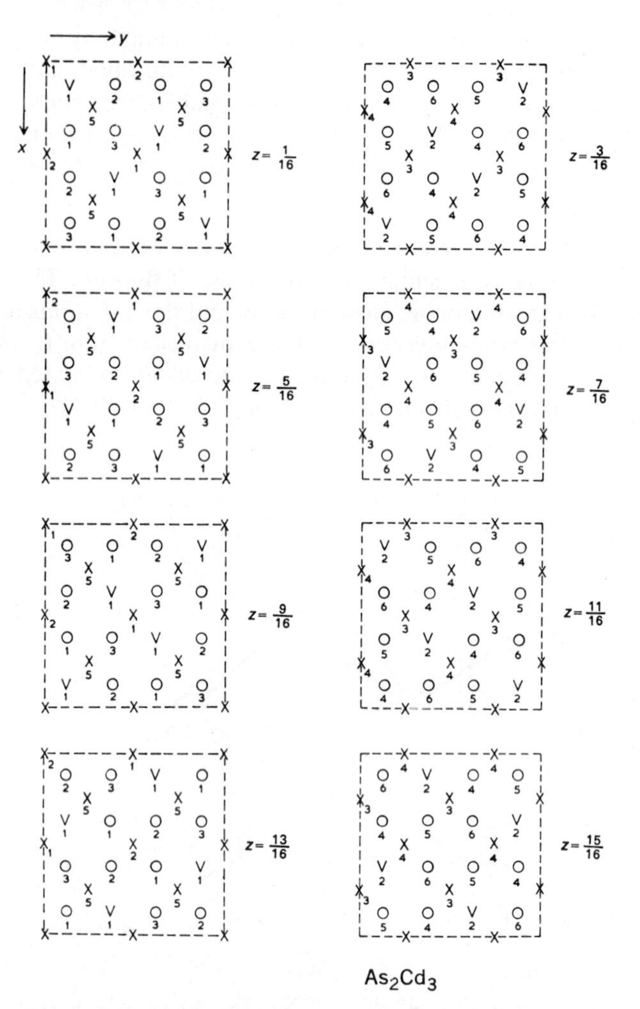

Figure 8-29. Idealized atomic arrangement on different planes in the As_2Cd_3 (*tI*160) structure. Circles mark Cd at the heights indicated, V indicating vacant Cd sites. Crosses mark As positions $z = 1/16$ below the planes indicated. Numbers indicate Cd sitesets 1 to 6 and As sitesets 1 to 5. (Steigmann and Goodyear[42].)

Mn_2O_3 (*cI*80): Numerous oxides have the Mn_2O_3 structure and several nitrides and phosphides have the antistructure. It is a distorted defect fluorite structure with a cell ($a = 9.408$ Å, $Z = 16$) that corresponds to eight fluorite type pseudocells.[43] One quarter of the oxygen sites are vacant so that Mn is surrounded by 6 O in two different arrangements.

Mn(1) has 6 equidistant O neighbors at 2.02 Å, whereas the 6 O surrounding Mn(2) are at 2.00, 2.01, and 2.03 Å. O is surrounded tetrahedrally by 1 Mn(1) and 3 Mn(2). In the nitrides and phosphides with the antistructure the P or N atoms have 6 metal neighbors and attain filled valence subshells.

Cu_3Se_2 (tP10): Cu_3Se_2 has cell dimensions $a = 6.406$, $c = 4.279$ Å, $Z = 2$.[44] Cu(1) atoms form a $\frac{1}{2}.4^4$ net at $z = 0$. The structure is a distorted defect fluorite arrangement which is considerably squashed along [001]. Cu(1) atoms form a defect f.c. tetragonal array in which the f.c. sites at $z = \frac{1}{2}$ are vacant. Within the body of the cell, Cu(2) and Se form distorted tetrahedra, but instead of these two forming a cube squashed along [001], the Se tetrahedron is rotated slightly and the Cu(2) tetrahedron is rotated considerably in the (001) plane from the fluorite sites lying on the cell body diagonals (Figure 8-30). Cu(1) is surrounded by a distorted tetrahedron of 4 Cu(2) at 2.63 Å and by a distorted tetrahedron of 4 Se at 2.49 Å. Cu(2) has 4 Se neighbors at 2.43 and 2.37 Å, 2 Cu(1), and one Cu(2) neighbor at 2.66 Å. Se has 2 Cu(1) and 4 Cu(2) neighbors.

VCu_3S_4 (cP8): In the sulvanite structure V occupies the cell corners, Cu the midpoints of the cell edges and S occupies essentially the same sites as in the sphalerite structure, although the atoms are slightly displaced along the body diagonals towards the cell corners ($x = 0.237$).[45] Vanadium is thus octahedrally surrounded by 6 Cu at 2.69 Å and tetrahedrally by 4 S at 2.19 Å; Cu is also surrounded tetrahedrally by 4 S at 2.28 Å and it has 2 V neighbors disposed linearly. S has 3 Cu at three corners of a tetra-

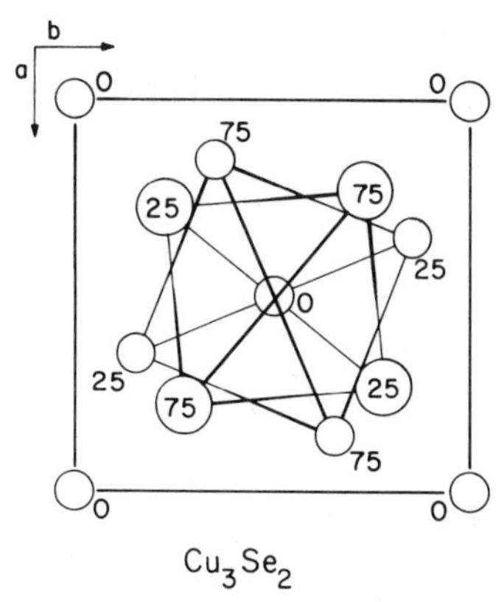

Figure 8-30. The atomic arrangement in the Cu_3Se_2 (*tP*10) structure viewed down [001]: Se, large circles.

hedron but its fourth neighbor, V, is located in the opposite direction from the vacant fourth corner of the tetrahedron (Figure 8-31). Neglecting the distortion of the S atoms from a c.p. cubic array, the structure is an ordered defect structure derived from fluorite with Cu in $\frac{3}{4}$ of one set of tetrahedral holes and V in $\frac{1}{4}$ of the other set.

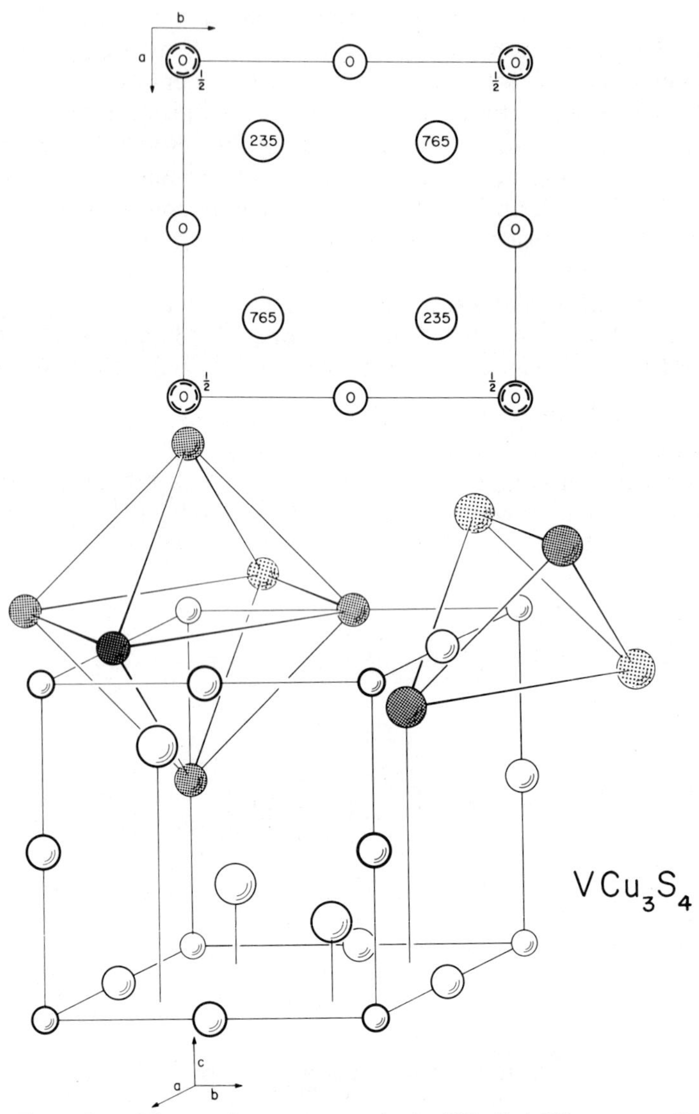

Figure 8-31. The atomic arrangement in the VCu_3S_4 ($cP8$) structure. V atoms small circles: S, atoms large circles.

$Sb_4Cu_{12}S_{13}$ (*cI*58): The S(2) atoms form a distorted defect c.p. cubic array in tetrahedrite, whose large cell ($a = 10.391$ Å) corresponds to 8 f.c. cubic pseudocells of S(2) with 8 sites vacant.[46] The S(1) atoms at the supercell corners and body center are approximately in the center of tetrahedral holes of the S(2) array, except that the 4 tetrahedrally disposed sites surrounding the S(1) atoms are those that are left vacant in the S(2) array.

The Sb atoms are located approximately at the centers of tetrahedral holes of the S(2) array in sites where one S(2) atom is missing, so that Sb has 3 S(2) neighbors at 2.45 Å. Cu(1) also sit approximately at the centers of tetrahedral holes in the S(2) array, but in those sites which are surrounded by 4 S(2), disposed approximately tetrahedrally at 2.34 Å. Cu(2) atoms are located off from the centers of tetrahedral holes that have the two neighboring S(2) sites vacant on the side where the S(1) atom sits. They are thus surrounded by 1 S(1) at 2.23 Å and 2 S(2) at 2.27 Å, all four atoms being located in the same plane. Thus S(1) atoms are surrounded by a regular octahedron of Cu(2). S(2) atoms have two Cu(1), one Cu(2), and one Sb neighbor arranged in a somewhat distorted tetrahedron.

The structure must be regarded as a considerably distorted defect superstructure of fluorite rather than sphalerite, since the S(1) atom is not in the same set of tetrahedral holes as Sb and Cu. Formulating fluorite as (Ca)(F)(F) the corresponding formulation for tetrahedrite is $(S(2)_{12}\square_4)$-$(Sb_4Cu_{12})(S(1)\square_{15})$, where □ represents a vacant site.

Defect Superstructures and Filled Up Derivatives of the Nonexistent Hexagonal Polytype of CaF_2

$CuFe_2S_3$ (*oP*24): In the cubanite structure[47] ($a = 6.233$, $b = 11.117$, $c = 6.46$ Å, $Z = 4$) the S atoms form close packed triangular layers parallel to the (100) plane. These are stacked in the hexagonal AB sequence along [100] ($c/a \sim 1.8$). The Fe and Cu atoms are located in almost undistorted tetrahedral holes in this array, half of the possible holes being filled (Figure 8-32).

However, the holes that are filled do not all belong to the same fourfold set, so that cubanite is not a superstructure of wurtzite, but of a defect hexagonal close packed analogue of the fluorite structure. As a result of a systematic changeover from one set to the other in the structure, each tetrahedron surrounding Fe shares one common edge with another tetrahedron, bringing the Fe atoms relatively close together in pairs.

$AsNa_3$ (*hP*8): $AsNa_3$ has cell dimensions $a = 5.098$, $c = 9.000$ Å, $Z = 2$.[48] Triangular layers of Na atoms are stacked in the sequence CACBAB, although they are much too close to be considered close packed. Na(1) on sites A at $z = \pm\frac{1}{4}$ form trigonal prisms, and two Na(2) atoms

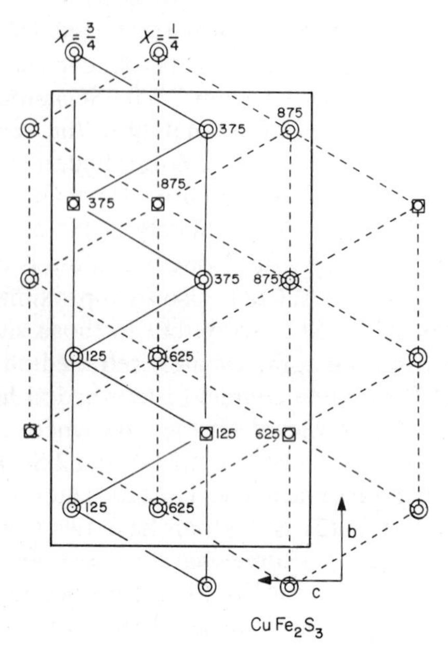

Figure 8-32. The cubanite, $CuFe_2S_3$ (*oP*24), structure viewed down [100]: ○ S; ◯, Fe; □, Cu.

(at $z = \pm 0.583, \pm 0.083$) and one As (at $z = \pm \frac{1}{4}$) are almost equally spaced along c on sites B and C (Figure 8-33). The As atoms are thus in hexagonal close packing, BC, and the Na(2) atoms lie approximately at the centers of both tetrahedral interstices in the c.p. hexagonal array of As atoms. As has 3 coplanar Na(1) neighbors at 2.94 Å, 6 Na(2) at 3.305 Å surrounding it in a trigonal prism, and two other Na(2) above and below it at 3.00 Å. Na(1) also has 11 neighbors similar to those of As. Na(2) has 4 As, 6 Na at 3.30 Å, and one Na(2) at 3.01 Å.

Compared to phases with the NiAs structure which have widely varying axial ratios, the twenty known phases with this structure are remarkable for the constancy of the axial ratio which has a value of 1.79 ± 0.03, with one possible exception. Such behavior is reminiscent of the wurtzite phases where the constancey of the axial ratio results from the tetrahedrally disposed array of chemical bonds. Herein lies an explanation of the dimensional behavior of phases with the $AsNa_3$ structure, since the arrangement of the As and Na(2) atoms is similar to that of the S and Zn atoms in wurtzite, except that Na occupies both sets of tetrahedral holes, whereas Zn occupies only one in the anion array. Pivotally resonating bonds around two sets of Na(2) tetrahedra about As can occur as in the

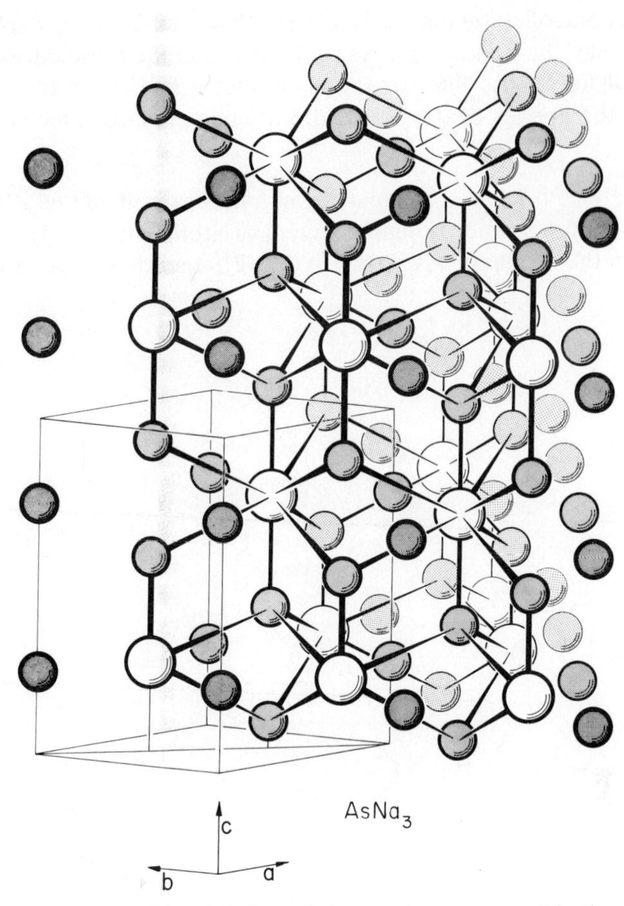

Figure 8-33. Pictorial view of the atomic arrangement in the $AsNa_3$ (*hP*8) structure.

$BiLi_3$ structure, where 2 sets of Li(2) tetrahedra form a cube about Bi. These tetrahedrally disposed bonds account for the relative constancy of the axial ratio as in the wurtzite structure. The role of the Na(1) atoms is to provide the extra electron required for the Na(2)–As bonding, and they can be regarded as occurring as Na^+ ions in much the same way as the Li(1) atoms occur as Li^+ in Li_3Bi. Thus the Na(1)–As distance of 2.94 Å is not excessively short. Phases with both types of structure are semiconductors, as the Group V atom attains a filled valence subshell in each case.

This explanation is confirmed by the near-neighbor diagram (Figure 3-8, p. 63), which shows beyond doubt that the dimensional behavior of the phases is controlled by the six Na(2) neighbors surrounding each As atom (that is to say the bonds from As to the bases of the tetrahedra), since the phases all lie on the line for these contacts. The other two Na(2)–As bonds to the apexes of the tetrahedra are compressed somewhat (Figure 3-8).

Pb_2Li_7 (hP9): Pb_2Li_7 has dimensions $a = 4.751$, $c = 8.589$ Å, $Z = 1$.[49] The unit cell contains nine triangular layers of atoms along [001]. These are stacked in the sequence AC$\underline{\text{B}}$ACBA$\underline{\text{C}}$B (Pb underlined), but they are too

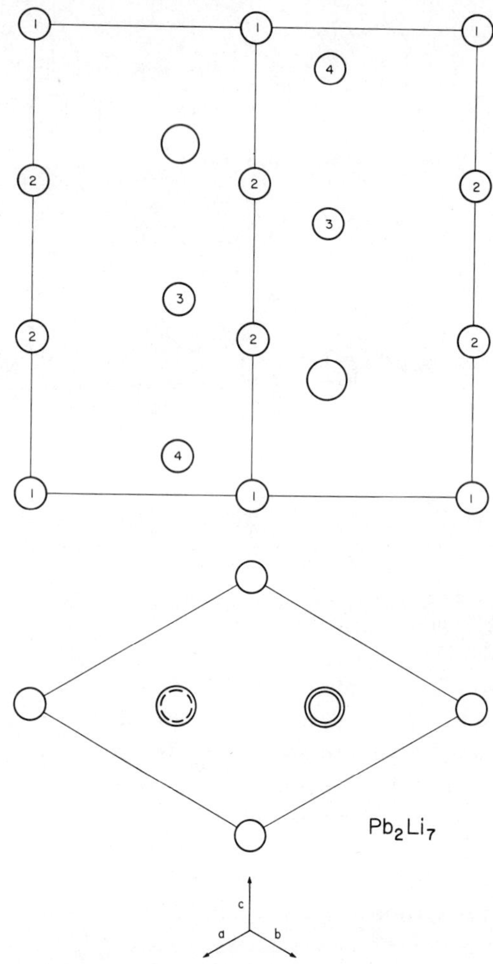

Figure 8-34. Projections of the structure of Pb_2Li_7 (*hP*9) down [110] and [001]: Pb, large circles.

extended in the basal plane and too compressed along c to be considered close packed. The structure can be regarded as a filled up and slightly distorted derivative of the $AsNa_3$ structure. Pb, Li(3), and Li(4) atoms are equally spaced along [001] on the B and C stacking sites, and the Li(2) atoms in stacking site A, which in the $AsNa_3$ structure would be coplanar with Pb, are displaced by $c/12$ toward the cell center. An additional layer of Li(1) atoms is introduced on A stacking sites at $z = 0$ (Figure 8-34). In keeping with the similarity to the $AsNa_3$ strucutre, c/a has a value of 1.807.

Pb has 11 Li at distances from 2.835 to 3.095 Å and 3 more at 3.48 Å. Li(2), (3), and (4) have 3 or 4 close Pb. The closest Li–Li distances are 2.83 Å. The structure requires confirmation from single-crystal data.

TETRAHEDRAL AND OCTAHEDRAL COORDINATION

Structures Based on Stacking of 3^6 Close Packed Nets and Both Tetrahedral and Octahedral Coordination

La_2O_3, $CaAl_2Si_2$ (hP5): La_2O_3 has cell dimensions $a = 3.938$, $c = 6.132$ Å, $Z = 1$.[50] The structure is isopuntal with the Ni_2Al_3 structure; the relative spacings of the layers differ. All atoms form close packed triangular networks parallel to (001) and the La layers are stacked in close packing BC sequence ($c/a = 1.56$) as shown in Figure 8-35. O(2) atoms fill both sets of tetrahedral holes between the La layers about the center of the cell, and O(1) atoms fill octahedral holes between the La layers at the outside of the cell ($z = 0$). The overall stacking sequence is thus A$\underline{B}$CB$\underline{C}$ (La underlined). Each La has 4 close O(2) neighbors (2.42 Å) and 3 O(1) neighbors at a larger distance (2.69 Å). O(1) has 6 La neighbors and O(2) has 4 La neighbors and 3 O(2) at 2.78 Å.

Several oxides and nitrides have this structure and three Group V–magnesium compounds have the antistructure (Figure 8-36). Ternary ordered La_2O_3 phases are also known in compounds such as $CaAl_2Si_2$ and $CaAl_2Ge_2$.

Ni_2Al_3 (hP5): The Ni_2Al_3 unit cell has dimensions $a = 4.036$, $c = 4.901$ Å, $Z = 1$.[51] The structure is isopuntal with La_2O_3; the relative spacing of the layers is quite different. There are three triangular layers of Al per cell parallel to the (001) plane. These are in a cubic stacking sequence ACB, but are much too close together to be considered close packed. In fact, alternate Al layers are at close packed spacings ($4c/3a = 1.62$) so that the intermediate Al layers occupy octahedral holes in this array. The Ni atoms are placed on the B and C sites so as to be equidistant

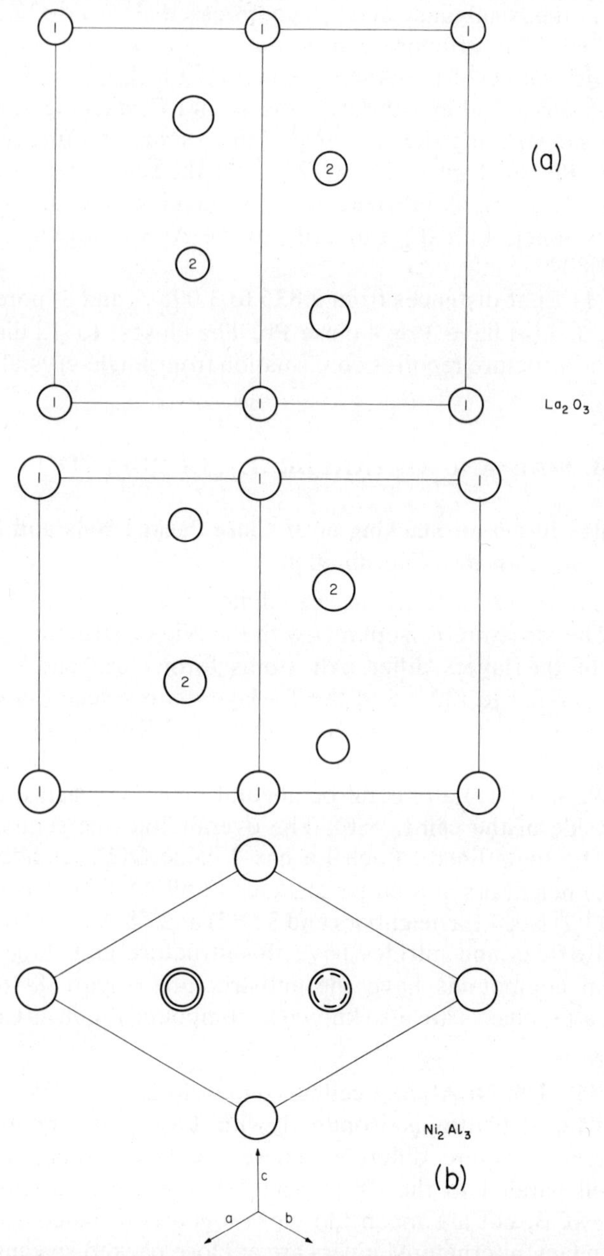

Figure 8-35. (*a*) Projection of the La_2O_3 (*hP*5) structure down [110]: La, large circles. (*b*) Projections of the Ni_2Al_3 (*hP*5) structure down [001] and [110]: Al, large circles.

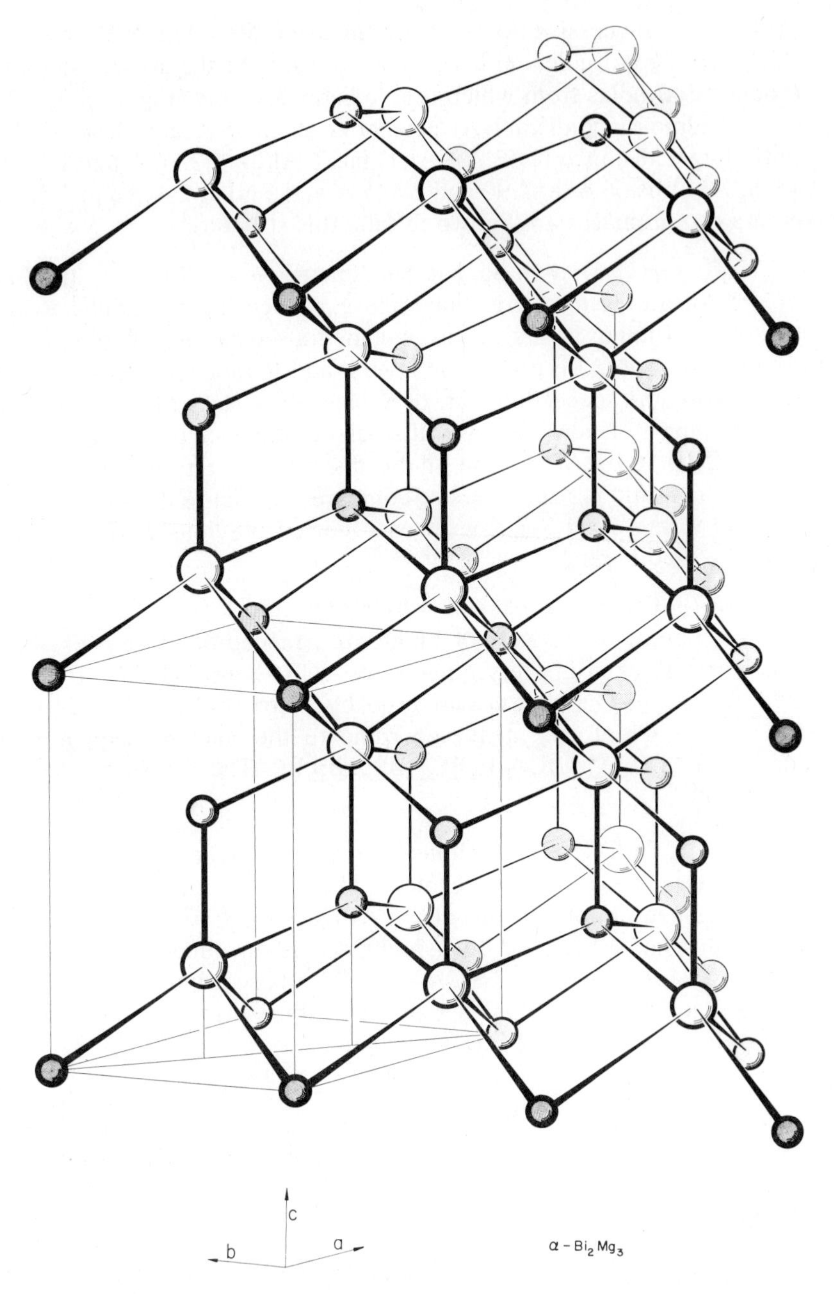

Figure 8-36. Pictorial view of the atomic arrangement in α-Bi_2Mg_3 with the anti-La_2O_3 structure.

($c/2$) between successive Al atoms on the same sites. The Ni atoms thus occupy tetrahedral holes, either in this Al array, or the array occupying the octahedral holes from which it is indistinguishable (Figure 8-35). Ni has $3+2$ almost equidistant Al neighbors at 2.445 Å and 3 more at a slightly larger distance (2.53 Å). Al(1) has 6 Al(2) at 2.90 Å and 6 Ni at 2.44 Å. Al(2) has 6 Al at 2.90 and 2.745 Å and 5 Ni at 2.445 and 2.53 Å. About a dozen phases are known to take this structure.

α-In_2Se_3 (L.T.) (*hR*5): α-In_2Se_3 has cell dimensions $a = 4.05$, $c = 28.77$ Å, $Z = 3$.[52] All atoms form triangular close packed layers parallel to the (001) plane. Indium layers occupy octahedral holes (but off-center) between one pair of close packed Se layers, and tetrahedral holes between the next pair; the succeeding pair of Se layers are close together without an intervening In layer. The overall stacking sequence is (In underlined) $A\underline{C}CB\underline{A}C\underline{B}BA\underline{C}B\underline{A}AC\underline{B}$. The In–Se distance for tetrahedral holes is 2.69 Å. The α and β structures are compared in Figure 8-37.

Several forms of In_2Se_3 have been reported previously, but this one appears to be different.

Th_3N_4 (*hR*7): The hexagonal cell, with dimensions $a = 3.871$, $c = 27.385$ Å, $Z = 3$,[53] contains 9 layers of Th atoms arranged on close packed triangular nets stacked in the sequence ACACBCBAB, which is found in the Sm structure. The N atoms are probably distributed in octahedral and tetrahedral holes of this array according to the stacking sequence (N underlined) $A\underline{B}C\underline{A}\underline{C}A\underline{B}C\underline{A}B\underline{C}\underline{B}C\underline{A}B\underline{C}A\underline{B}\underline{A}B\underline{C}$. These positions require confirmation.

Co_9S_8 (*cF*68): In Co_9S_8 the S atoms form a close packed cubic array and the Co atoms are located at the centers of half of the tetrahedral holes and $\frac{1}{8}$ of the octahedral holes in this array[54] (Figure 8-38). Co(1), surrounded octahedrally, has 6 S(2) at 2.39 Å, and Co(2) is surrounded tetrahedrally by 4 S (1 S(1) at 2.12 Å and 3 S(2) at 2.21 Å). It also has three close Co(2) neighbors at 2.52 Å. S(1) has 4 Co(2) neighbors disposed tetrahedrally and S(2) has 1 Co(1) and 4 Co(2) neighbors.

$CuGaO_2$ (hR_4): $CuGaO_2$ has the *Strukturbericht* $F5_1$, type of structure[55] with $x_O > 0.33$, leading to paired-layer stacking of the oxygens AABBCC (see $CrNaS_2$, p. 430 for comments). O is surrounded tetrahedrally by 3 Ga and 1 Cu; Cu is surrounded linearly by O, and Cr is surrounded octahedrally by O as shown in Figure 8-39.

$MgAl_2O_4$ (*cF*56): The spinel type structure of $ZnAl_2S_4$ has cell dimensions $a = 10.005$ Å, $Z = 8$.[56] The S atoms form a close packed cubic array which surrounds the Al atoms octahedrally and the Zn atoms tetrahedrally

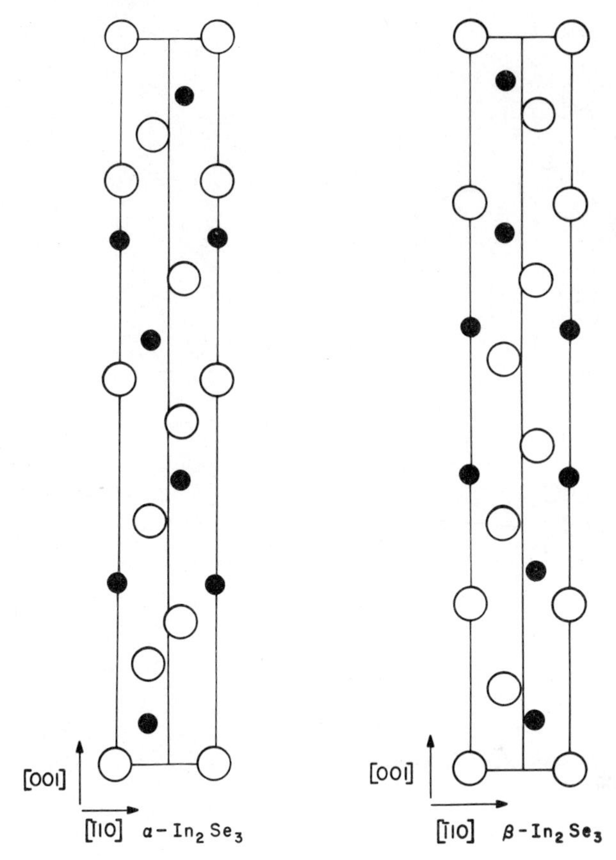

Figure 8-37. Projections of the structures of low temperature α-In_2Se_3 (*hR*5) and β-In_2Se_3 (*hR*5) structures down [110].

(Figure 8-40). One half of the octahedral sites are occupied by Al and $\frac{1}{8}$ of the tetrahedral sites by Zn, and the disposition is such as to separate Al and Zn by the greatest possible distance. Al has 6 S neighbors at 2.36 Å; Zn has 4 S at 2.43 Å, and S has 3 Al and 1 Zn neighbors. The Al and Zn atoms also form layers parallel to the close packed sulfur layers; there are no mixed layers, the layer sequence in the [111] direction being SAlSZnAlZnSAlSZnAlZnSAlSZnAlZn.

Spinels which have the divalent metal in the tetrahedral sites are known as *normal* and those which have it in octahedral sites are called *inverse*.

Al_2S_3 (*cF*53): The Al_2S_3 structure with $a = 9.93$ Å, $Z = 10.67$,[57] is a defect variation of the spinel structure with 5.33 Al(1) randomly in the 8 (*b*)

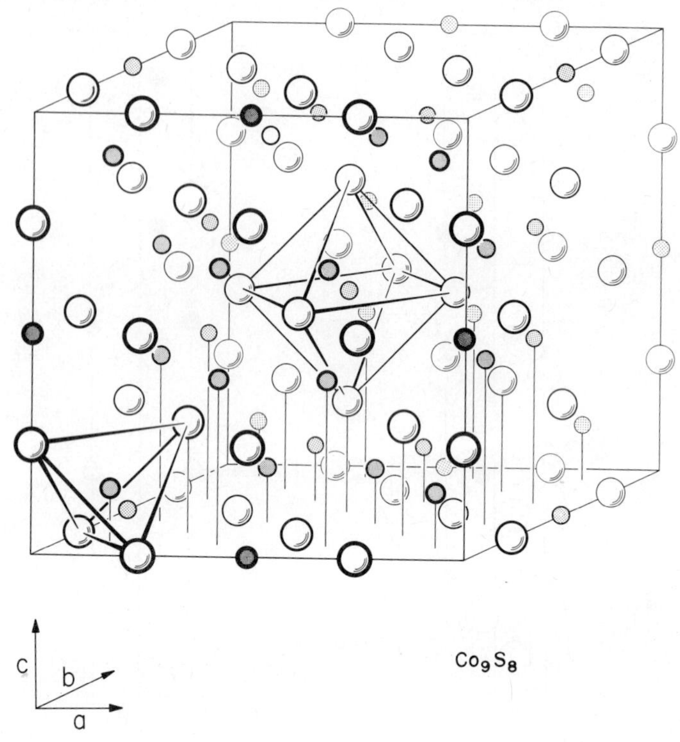

Figure 8-38. Pictorial view of the structure of Co_9S_8 (*cF*68): S, large atoms.

sites $\frac{1}{2}, \frac{1}{2}, \frac{1}{2}$; etc. and 16 Al(2) in 16 (*c*) $\frac{1}{8}, \frac{1}{8}, \frac{1}{8}$; etc. Three other modifications of Al_2S_3 have been recognized which have the wurtzite, corundum, or a hexagonal structure with $Z = 6$. In the form described here, As replaces some Al. The structure was examined by X-ray powder photographs only.

β-In₂S₃ (*tI*80): β-In_2S_3 with cell dimensions $a = 7.623$, $c = 32.36$ Å, $Z = 16$ has a cation deficient spinel type structure, with ordered cation defects.[58]

$SiMg_2O_4$ (*oP*28): The olivine structure[59] with cell dimensions $a = 10.20$, $b = 5.99$, $c = 4.76$ Å, $Z = 4$, is the hexagonal c.p. analogue of the spinel structure with O forming a c.p. hexagonal array in which Mg occupies half of the octahedral holes and Si one-eighth of the tetrahedral holes (Figure 8-41). Numerous sulfides and selenides of Groups II and IV metals take the olivine structure.

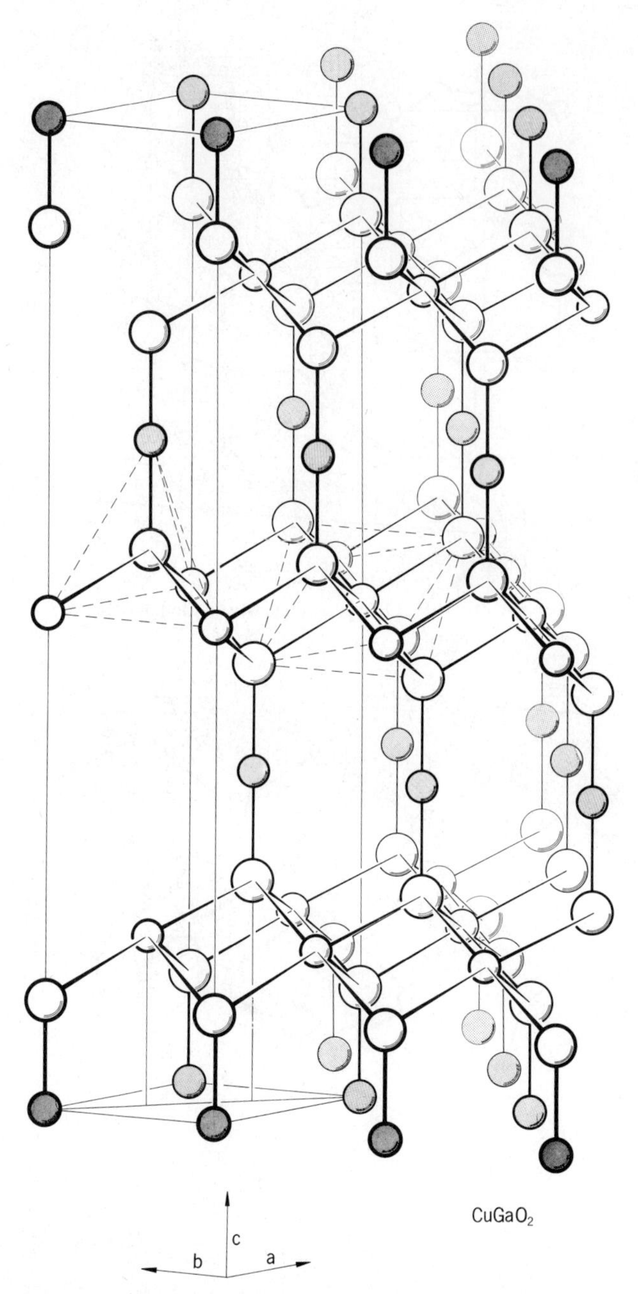

Figure 8-39. Atomic arrangement in the hexagonal cell of the $CuGaO_2$ ($hR4$) structure: large circles, oxygen; hatched circles, copper.

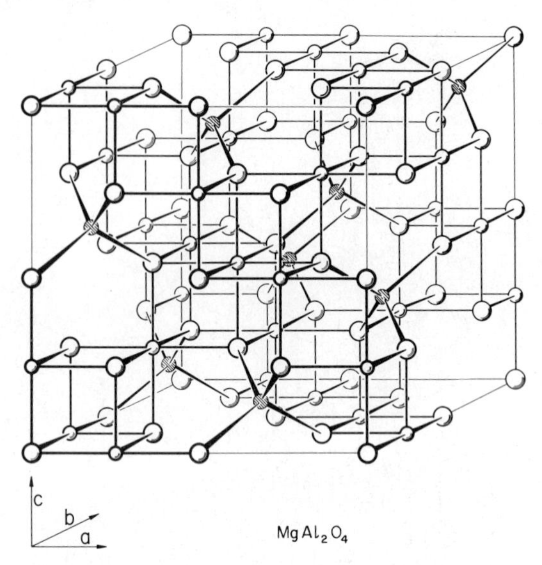

Figure 8-40. Pictorial view of the spinel, $MgAl_2O_4$ ($cF56$) structure: O large circles, Mg shaded circles.

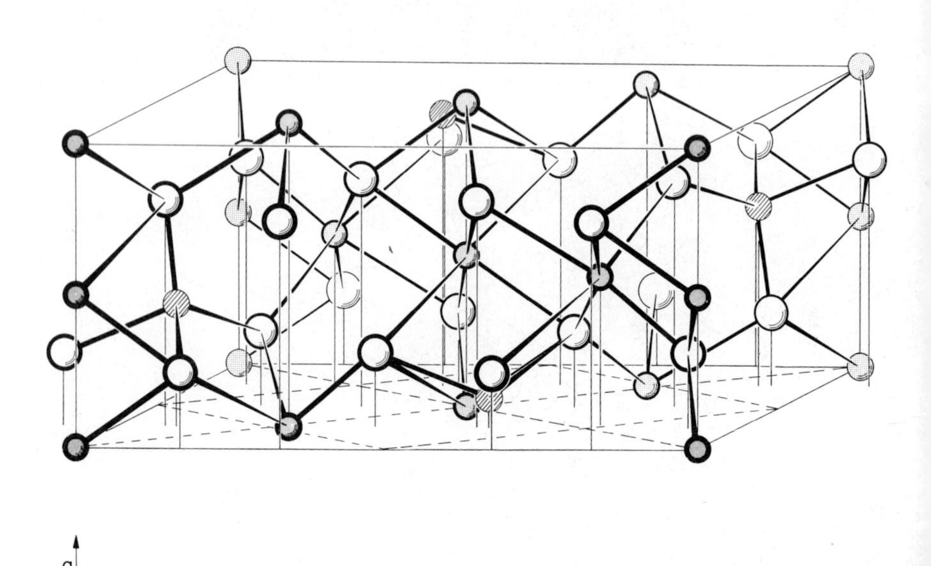

Figure 8-41. Pictorial view of the olivine, $SiMg_2O_4$ ($oP28$) type structure: O, large circles; Si, shaded circles.

$MgGa_2S_4$ ($mC84$): $MgGa_2S_4$ has cell dimensions $a = 12.74$, $b = 22.54$, $c = 6.43$ Å, $\beta = 108.8°$. $Z = 12$.[60] The S atoms form distorted close packed layers parallel to (100) at heights $x \sim \frac{1}{8}, \frac{3}{8}, \frac{5}{8}$, and $\frac{7}{8}$. The sequence of occupation of interstices between the layers is, starting from the cell origin: Mg ($x = 0$) in octahedral holes, Ga ($x \sim 0.3$) in tetrahedral holes, unoccupied holes and then Ga ($x \sim 0.7$) in tetrahedral holes. The arrangement changes in the center of the cell with Mg at $x = \frac{1}{2}$ as shown in Figure 8-42. The octahedra containing Mg share edges to form slabs running parallel to [001]. The tetrahedra of S atoms surrounding Ga are more distorted than the octahedra surrounding Mg. The S atoms are irregularly coordinated. Mg–S distances are 2.52 to 2.71 Å and Ga–S distances are from 2.18 to 2.35 Å.

In_2ZnS_4 ($hR7$): The dimensions of the hexagonal cell are $a = 3.85$, $c = 37.1$ Å, $Z = 3$.[61] The atoms form triangular nets parallel to the (001) plane. The S layers are stacked in close packed sequence A B C A C A B
V T* O T V T* O
C B C A B ($c/6a = 1.605$) and the In and Zn atoms occupy the tetra-
T V T* O T
hedral or octahedral holes therein as indicated, where T* and T are Zn and In, respectively, in tetrahedral holes. O is In in octahedral holes and V means that no holes between S layers are occupied.

$In_2Zn_3S_6$ ($hP11$): The semiconducting compound has cell dimensions $a = 3.85$, $c = 18.5$ Å, $Z = 1$.[62] The cell contains 6 close packed triangular layers of S atoms stacked in the sequence ABABAB. Between the first two layers, Zn and In atoms are disordered in tetrahedral holes (stacking position B). Between layers 2 and 3, Zn occurs in tetrahedral holes (stacking position A). In occurs between layers 3 and 4 in the octahedral

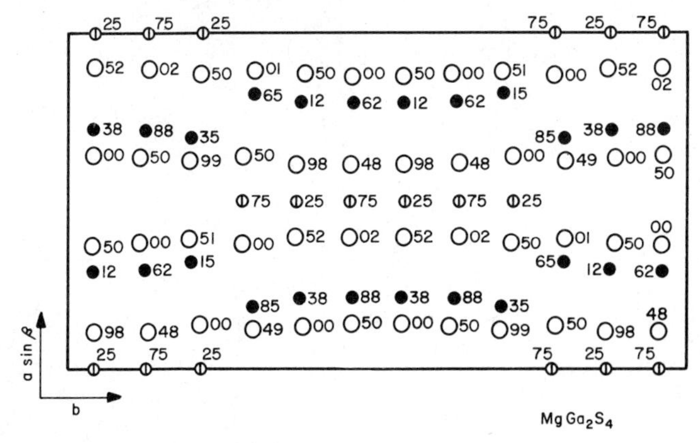

Figure 8-42. Projection of the $MgGa_2S_4$ ($mC84$) structure along [001]: ○, S; ⦶, Mg; ●, Ga.

holes (position C). Between layers 4 and 5, Zn is in tetrahedral holes (position B), and between layers 5 and 6 Zn and In occur together in tetrahedral holes (position A). No holes are occupied between layers 6 and 1 of the next cell. The structure is thus a mixture of wurtzite and rocksalt-like layer groups.

Aluminum carbonitrides. The structures of aluminum nitride (AlN), carbide (Al_4C_3) and carbonitrides are based on close packing of close packed layers of Al atoms with the carbon atoms in tetrahedral or octahedral holes and the nitrogen atoms in tetrahedral holes thereof, as indicated in Figure 8-43 taken from Jeffrey and Wu.[63] The observed pseudo-cell axial ratios, c/a, in the range from 1.25 to 1.45 indicate a closer than

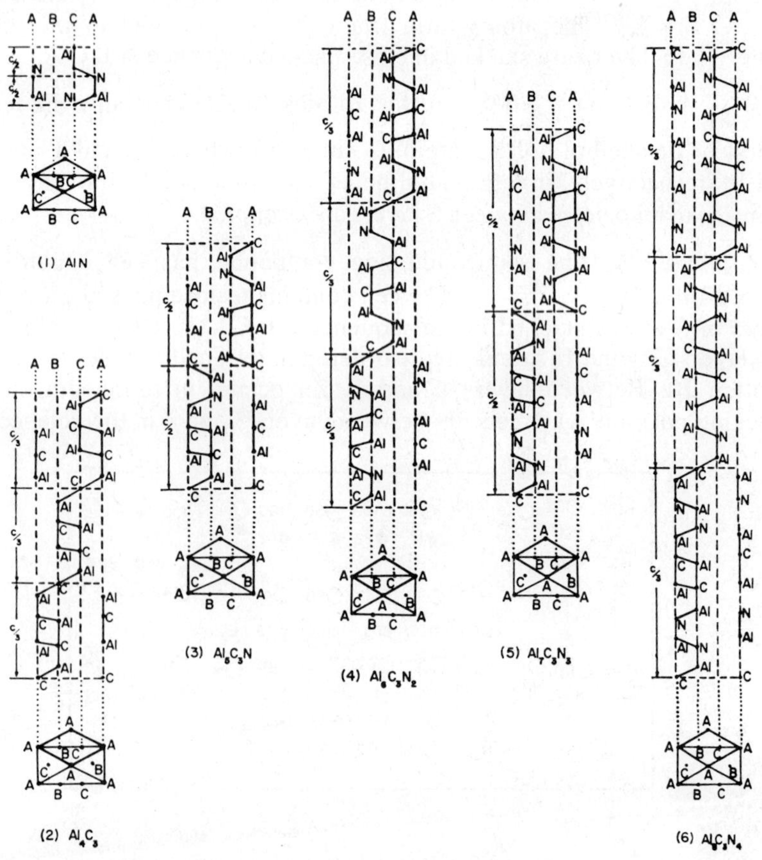

Figure 8-43. Diagrams of the stacking arrangement of the layers in AlN, Al_4C_3, and various aluminum carbonitrides. (Jeffrey and Wu [63].)

ideal packing of the Al layers. The arrangement of the layers is such that three building blocks can be recognized in the structures: (i) $(AlN)_n$ which is a wurtzite arrangement $\underline{A}A$, (ii) $(Al_2C_2)_n$ which consists of two adjacent close packed layers of Al with C in the tetrahedral interstices immediately above and below each Al atom, so that pairs of Al layers are enclosed between two C layers (Figure 8-43), and (iii) $(Al_2C)_n$ which consists of two close packed Al layers with an interleaved C layer in the third stacking position (e.g., $\underline{A}C\underline{B}$), so that C is surrounded octahedrally by Al. The (Al_2C_2) units in these structures are always surrounded by two further close packed Al layers on the same stacking sites as the C atoms, so that each C is surrounded tetrahedrally by 4 Al atoms, and the Al atoms of the pairs of Al layers are surrounded tetrahedrally by C. The separation of the two paired Al layers is about 10% less than the average spacing of the close packed Al layers.

Al_5C_3N (hP18): Al_5C_3N has cell dimensions $a = 3.281$, $c = 21.67$ Å, $Z = 2$.[63] The building unit arrangement along [001] in Al_5C_3N is $(Al_2C_2, AlN, Al_2C)_2$, as shown in Figure 8-43, the cell containing 10 close packed layers of Al atoms ($c/5a = 1.32$).

Al_4C_3 (hR7): The dimensions of the corresponding hexagonal cell of Al_4C_3 are $a = 3.330$, $c = 24.89$ Å, $Z = 3$.[63] The structure is made up of building units $(Al_2C_2, Al_2C)_3$ successively along [001] as shown in Figure 8-43. The stacking sequence is (with Al layers underlined) $A\underline{B}B\underline{A}\underline{B}A\underline{A}B\underline{C}C\underline{B}\underline{C}$-$B\underline{B}C\underline{A}A\underline{C}\underline{A}C\underline{C}$, the cell containing 12 close packed Al layers ($c/6a = 1.25$).

$Al_6C_3N_2$ (hR11): The dimensions of the corresponding hexagonal cell are $a = 3.248$, $c = 40.03$ Å, $Z = 3$.[63] The sequence of building blocks along [001] is $(AlN, Al_2C_2, AlN, Al_2C)_3$, as shown in Figure 8-43. The cell contains 18 close packed layers of Al atoms ($c/9a = 1.37$).

$Al_7C_3N_3$ (hP26): The dimensions of the hexagonal cell are $a = 3.226$, $c = 31.70$ Å, $Z = 2$.[63] The arrangement of building blocks along [001] is $(AlN, AlN, Al_2C_2, AlN, Al_2C)_2$ as shown in Figure 8-43. The 14 close packed layers of Al atoms are in the order BABABABCACACAC ($c/7a = 1.40$).

$Al_8C_3N_4$ (hR15): The dimensions of the corresponding hexagonal cell are $a = 3.211$, $c = 55.08$ Å, $Z = 3$.[63] The sequence of building blocks along [001] is $(AlN, AlN, Al_2C_2, AlN, AlN, Al_2C)_3$ as shown in Figure 8-43. The cell contains 24 close packed layers of Al atoms ($c/12a = 1.43$).

Various Structures Containing Tetrahedra and Octahedra

FeS_2 (*cP*12): In the pyrite structure the Fe atoms form a f.c. cubic array and the S atoms are arranged in pairs across the midpoints of the cell edges and the cell center[64] (Figure 8-44). The structure therefore can be regarded as a NaCl type in which the Cl atoms are replaced by pairs of S atoms aligned at an angle to ⟨100⟩ directions. Fe is octahedrally surrounded by 6 S at 2.26 Å, and S is tetrahedrally surrounded by 3 Fe and 1 S at 2.18 Å. The structure is compared with those of FeSi and $AuSn_2$ in Figure 11-7 p. 612, and with the ullmannite structure in Figure 8-45.

CoAsS (*oP*12): The cobaltite structure with cell dimensions $a = b = c =$ 5.582 Å, $Z = 4$ is essentially a pyrite type in which the As and S atoms have ordered positions in forming As–S pairs.[65] This is the low temperature form and also the structure of the mineral, although the ordering of As and S may not be perfect in all mineral samples. A high-temperature form has been recognized in which the As and S atoms are disordered on their sites.[66]

NiSSb (*cP*12): The ullmannite structure is an ordered ternary form of pyrite (Figure 8-45).[67] The Ni atoms form a slightly distorted f.c. cubic array and the S and Sb atoms form pairs similar to the S_2 pairs of pyrite, so that 3 S (at 2.35 Å) and 3 Sb (at 2.57 Å) surround Ni in a distorted octahedron. Both S and Sb have 3 Ni neighbors and one S–Sb neighbor (at 2.39 Å) distributed at the apices of a distorted tetrahedron.

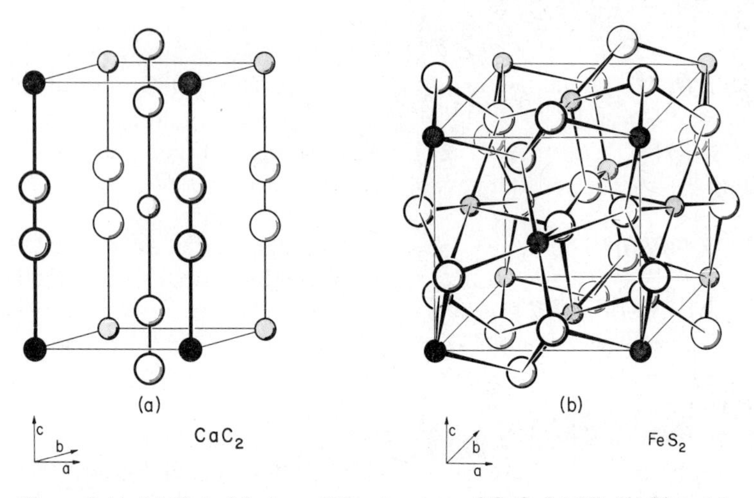

Figure 8-44. (*a*) Pictorial view of the structure of CaC_2 I (*tI*6). (*b*) Pictorial view of the structure of pyrite, FeS_2 (*cP*12).

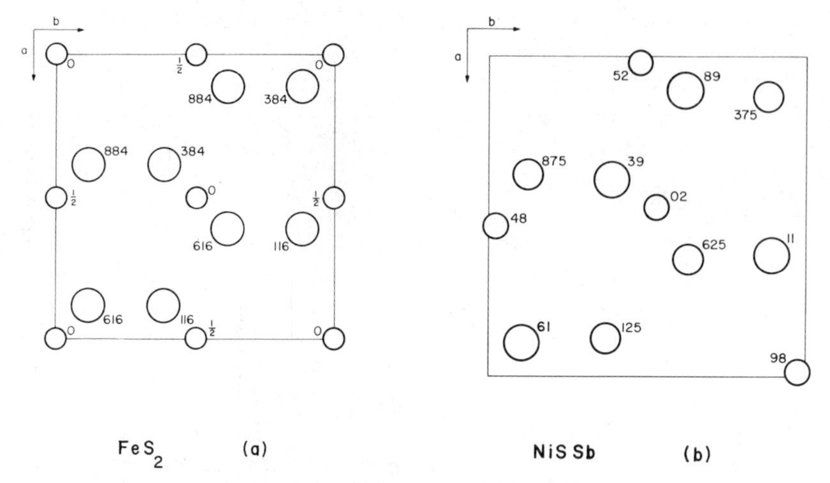

Figure 8-45. (*a*) Projection of the structure of pyrite, FeS_2 (*cP*12), down [001]. (*b*) Projection of the structure of ullmannite, NiSSb (*cP*12), down [001]: S or Sb largest circles, Fe or Ni smallest circles.

PdSe$_2$ (*oP*12): $PdSe_2$ with cell dimensions $a = 5.741$, $b = 5.866$, $c = 7.691$ Å, $Z = 4$,[68] is a distorted pyrite type structure in which, particularly, the c axis is greatly lengthened. This results in distortion of the octahedron of Se surrounding Pd, the bond distances being 2×2.438, 2×2.444 and 2×3.27 Å. The tetrahedral arrangement about Se (1 Se–Se contact at 2.36 Å and 3 Se–Pd contacts) is distorted similarly.

FeSi (*cP*8): In the FeSi structure,[69] the closest Fe–Si pairs (2.29 Å) lie parallel to the body diagonals of the unit cell and are centered about four f.c. cubic sites. Thus the structure can be regarded as a defect derivative of pyrite with Fe and Si corresponding to S_2 and the Fe sites of pyrite left vacant as indicated on Figure 11-7, p. 612. Fe has 4 neighboring Si atoms at 2.29 and 2.34 Å, and three other Si at 2.515 Å and 6 Fe at 2.75 Å. Si is surrounded by four Fe (1×2.29, 3×2.34 Å) and 3 other Fe at 2.515 Å, as well as 6 Si at 2.78 Å. The four closest neighbors are very distorted from a tetrahedral arrangement. The Fe and Si sites have equivalent surroundings and the same structure is obtained if Fe occupies Si sites and Si the Fe sites.

FeS$_2$ (*oP*6): Marcasite has cell dimensions $a = 4.445$, $b = 5.425$, $c = 3.388$ Å, $Z = 2$.[70] The iron atoms were said to occupy a b.c. orthorhombic cell with the S atoms distributed so as to surround Fe octahedrally (2×2.235, 4×2.255 Å) (Figure 8-50, p. 417). However, it has since been shown that the structure of $FeSb_2$, FeS_2, $FeTe_2$ and $CoTe_2$ lacks a mirror plane, and that the metal atoms are not located exactly at the center of the anion octahedra.[71] S has one close S neighbor (2.22 Å) and 3 Fe neighbors, the four atoms surrounding it approximately tetrahedrally.

As is apparent from the projection of the structure down [010] and from Figure 8-46, the structure can be derived from the NiAs type by removing in ordered manner half of the atoms from the Ni sites at $z = 0$ and $\frac{1}{2}$, and subsequent adjustment of the atom positions. Thus the structure can be compared to the Co_2N type, although the readjustment of atomic sites is more severe in marcasite. The relationship between the marcasite and pyrite forms of FeS_2 is a different stacking together of atoms in planes parallel to {100}. The marcasite and rutile structures contain similar features as is seen in Figure 8-50, p. 417. In the latter the O atoms are moved further apart so that there are no bonds between them, as between the paired S atoms in marcasite.

Binary compounds with the marcasite structure fall into two classes: The *marcasites* which have c/b values of about 0.615, and the *loellingites* with c/b values of about 0.485 (Table 8-3). The latter are compounds of Fe, Ru, and Os with P, As, and Sb, also $CrSb_2$. Since the decrease in axial ratio results essentially from contraction of c rather than change of a or b, the loellingites can be regarded as compressed in the [001] direction, bringing both the metal atoms and the nonmetals rather close

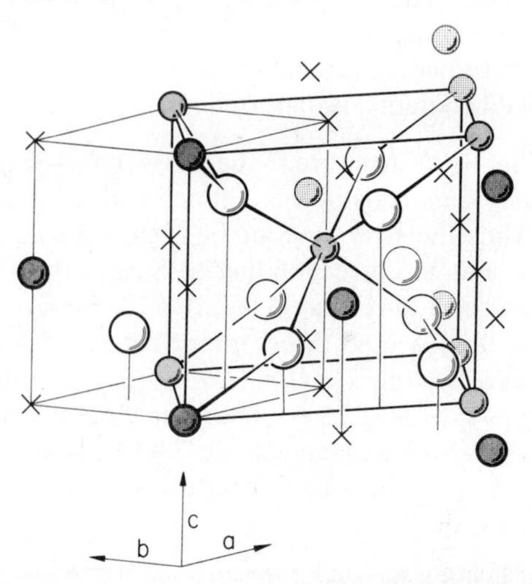

Figure 8-46. Drawing indicating the derivation of the marcasite, FeS_2 ($oP6$), structure from the NiAs structure. X marks atoms omitted in obtaining the marcasite structure after adjustment of the atomic positions.

Table 8-3

Phase	*c/b*	
As_2Fe	0.483	
As_2Ni		0.611
As_2Os	0.487	
As_2Ru	0.480	
$CoSb_{\sim 1.85}$	0.529	
$CoTe_2$ (L.T.)		0.616
$CrSb_2$	0.476	
FeP_2	0.482	
FeS_2		0.625
$FeSb_2$	0.489	
$FeSe_2$		0.620
$FeTe_2$		0.618
$NiSb_2$		0.608
OsP_2	0.495	
$OsSb_2$	0.480	
RuP_2	0.487	
$RuSb_2$	0.477	

together in lines along [001]. The octahedra surrounding the metal atoms share *edges* in planes normal to [001], whereas all other contacts between octahedra are at corners only, as in pyrite. The reason for this compression in the loellingites is uncertain, and it is also surprising that no binary compounds, with the possible exception of $CoSb_{\sim 1.85}$, have axial ratios intermediate between those of the marcasite and loellingite families, although ternary solid solutions have been prepared[72] in the (Fe, Co, Ni)As_2 series in which the axial ratio, c/b, varies continuously between the two extremes.

FeAsS (*aP*12): Arsenopyrite is a triclinically distorted form of marcasite[73] with the As and S atoms in ordered positions on the S sites of marcasite. Lines of As + S octahedra surrounding Fe and sharing edges, alternately As–As and S–S, run throughout the structure in the [101] direction (Figure 8-47). In this direction the Fe atoms have moved together in pairs (2.79 Å) across the S–S edge and away from each other (3.53 Å) across the As–As edge. This feature of the structure resembles $CoSb_2$ (*mP*12). $a = 5.74$, $b = 5.67$, $c = 5.78$ Å, $\alpha = 90°$, $\beta = 112.2°$, $\gamma = 90°$, $Z = 4$.

IrSe$_2$ (*oP*24): The $IrSe_2$ structure with cell dimensions $a = 20.94$, $b = 3.74$, $c = 5.93$ Å, $Z = 8$,[74] is an interesting development from the marcasite structure in which only half of the sulfur atoms form S_2 pairs

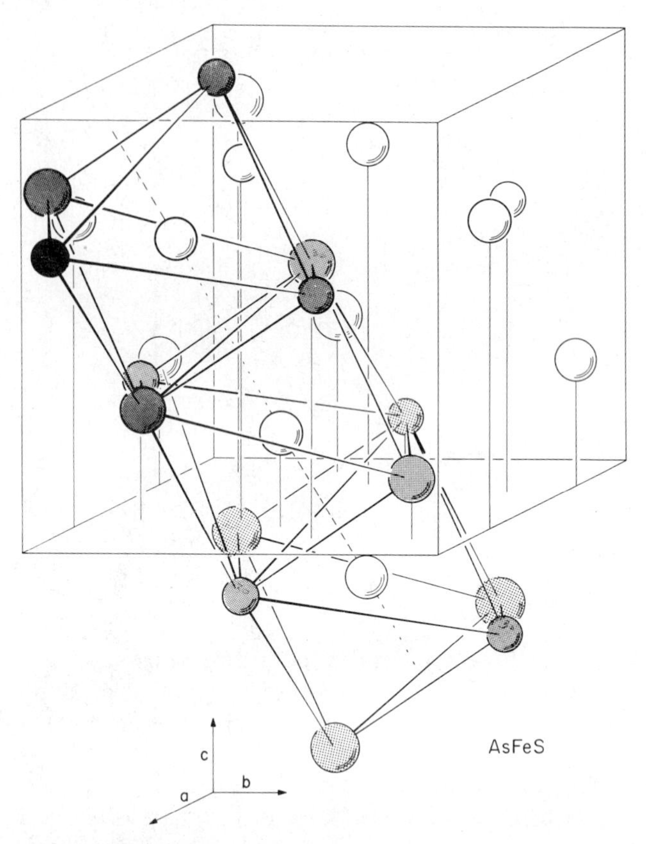

Figure 8-47. Pictorial view of the arsenopyrite, FeAsS (*oP*12) structure, showing lines of As + S octahedra about Fe which share edges: As, largest circles; S, smallest circles.

(Figure 8-48). This rearrangement permits Ir to have an average valency of 3, so that it attains a stable d^6 subshell.

$CoSb_2$ (*mP*12): The $CoSb_2$ structure which has cell dimensions $a = 6.52$, $b = 6.38$, $c = 6.55$ Å, $\beta = 118.2°$, $Z = 4$,[75] is a distortion of the marcasite structure in which the Co atoms move together in pairs in a direction corresponding to [001] of marcasite as indicated in Figure 8-49. In this direction which is normal to the shared edges of the octahedra of Sb about Co, the Co–Co spacings are alternately 3.02 and 3.69 Å. As in marcasite, the octahedra mutually share only two of their 12 edges; all other connections between octahedra are at corners. A further consequence of the close approach of pairs of Co atoms is the rather close approach of

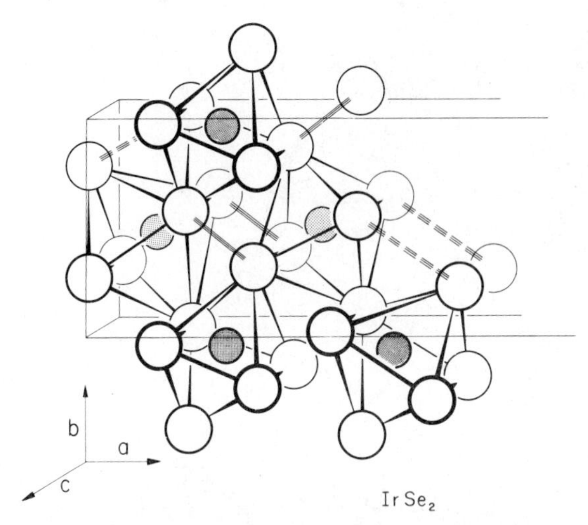

Figure 8-48. Pictorial view of a fragment of the $IrSe_2$ (*oP*24) structure indicating the atomic arrangement.[74]

an Sb(1) and Sb(2) atom at 3.18 Å in the same direction. Co–Sb distances vary from 2.48 to 2.66 Å in the surrounding octahedron of Sb atoms. Both Sb(1) and (2) atoms are surrounded tetrahedrally by 3 Co and one close Sb neighbor at 2.83 Å, in addition to the other close Sb–Sb contacts already referred to.

The form of the distortion of the $CoSb_2$ from the marcasite structure, is exactly analogous to that found in the low temperature VO_2 structure in its distortion from the rutile type (*vide infra*).

TiO_2 (*tP*6): In the rutile structure of TiO_2 ($a = 5.494$, $c = 2.962$ Å, $Z = 2$)[76] the Ti atoms are in a body centered arrangement, although the axial ratio ($c/a = 0.64$) is much distorted from a cube. The oxygen atoms lie on (110) and $(1\bar{1}0)$ planes and surround the Ti atoms octahedrally. The octahedra share edges in (001) planes; otherwise they share corners only (Figure 8-50). Thus the Ti atoms approach each other rather closely in the [001] direction. Each oxygen has three Ti neighbors. See also p. 281.

ϵ-Ti_2N has the antirutile type structure, with N surrounded octahedrally by Ti. Although the N–N distances of 3.04 Å along [001] are unlikely to be significant, it is noted that the axial ratio (0.61) is even smaller than in TiO_2.

VO_2 (L.T.) (*mP*12): The low temperature VO_2 structure stable below 340°K has the cell dimensions $a = 5.743$, $b = 4.517$, $c = 5.375$ Å,

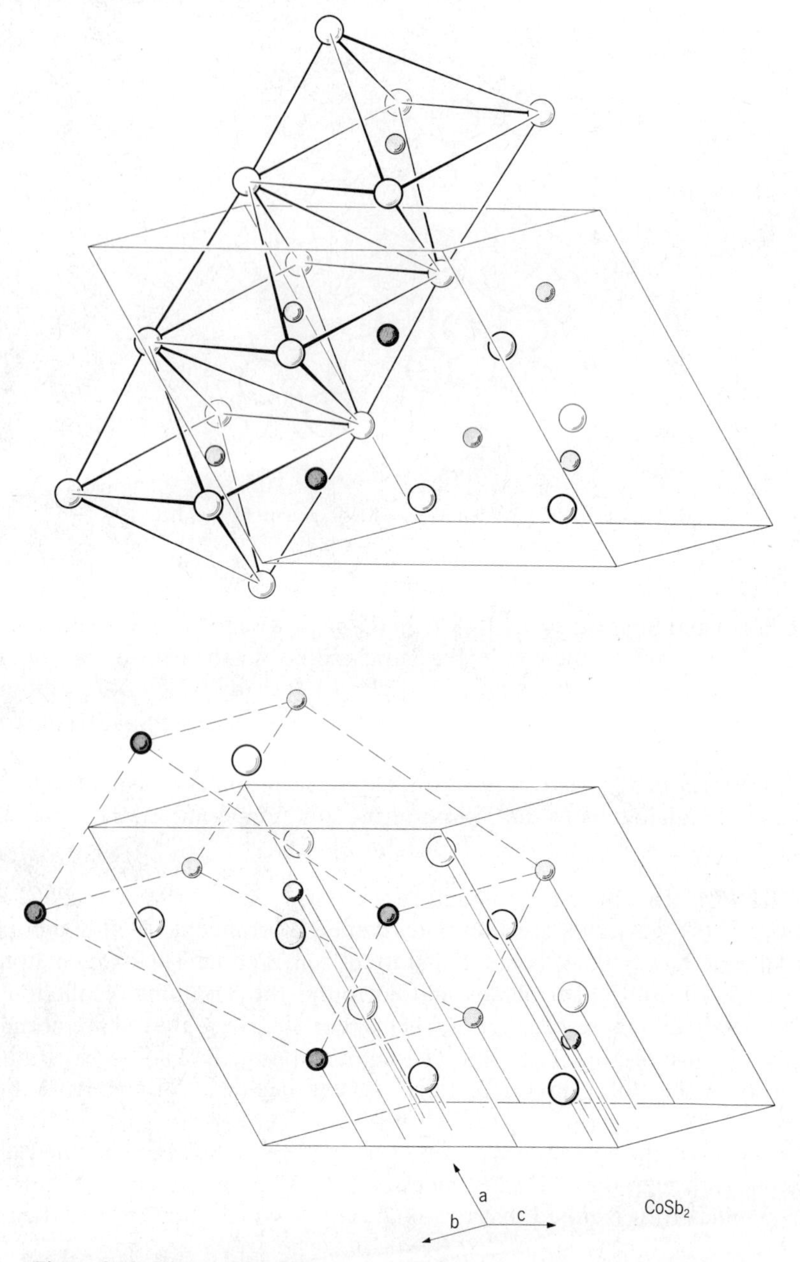

Figure 8-49. Views of the $CoSb_2$ ($mP12$) structure showing (upper) lines of octahedra of Sb surrounding Co and sharing edges, and (lower) the relationship of the $CoSb_2$ cell to that of a distorted marcasite arrangement.

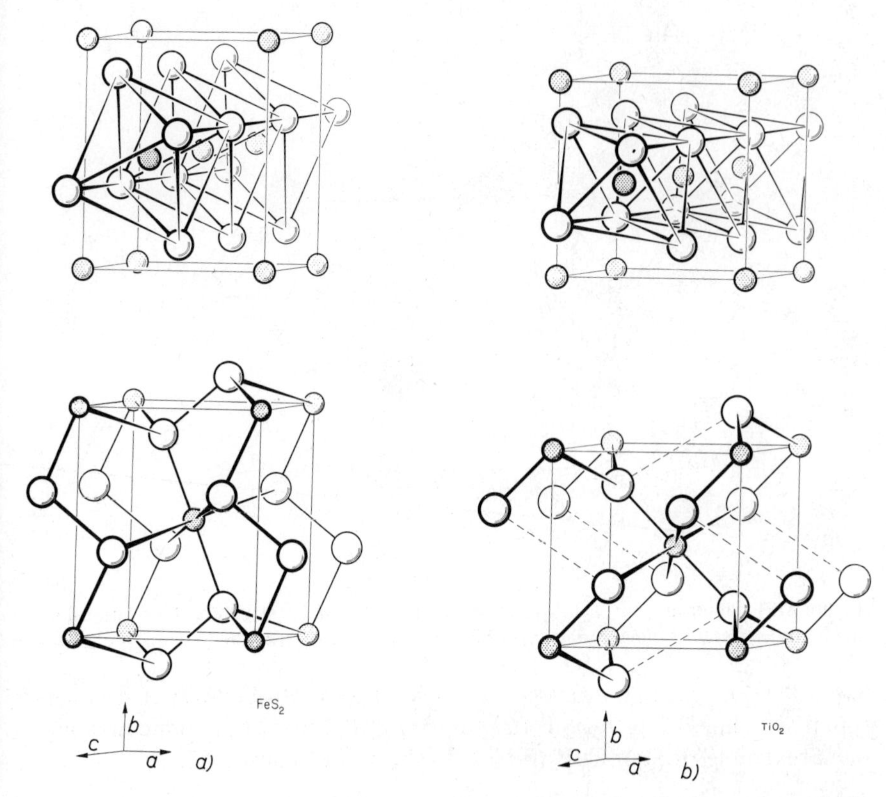

Figure 8-50. Pictorial views comparing (*a*) the marcasite, FeS_2 (*oP*6) structure, with (*b*) the rutile, TiO_2 (*tP*6) structure, showing in each case the lines of octahedra along [001] which share edges.

$\beta = 122.61°$, $Z = 4$.[77] It is a distorted rutile structure in which the V atoms move together in pairs (2.65 Å) in a direction corresponding to [001] of rutile. The nonpaired V–V distance is 3.12 Å. The analogy between the rutile and marcasite structure has been discussed on p. 412 and the monoclinic VO_2 structure is a distortion analogous to the $CoSb_2$ distortion of the marcasite structure. The V atoms in VO_2 are surrounded by a distorted octahedron of oxygen atoms at distances from 1.76 to 2.05 Å. These octahedra share edges which lie normal to the direction of the close V–V pairs (Figure 8-51). Both O(1) and (2) atoms have three close V neighbors.

CdP_4 (*mP*10): CdP_4 has cell dimensions $a = 5.27$, $b = 5.19$, $c = 7.66$ Å, $\beta = 80.32°$, $Z = 2$.[78] An array of bonds between the P atoms runs in three dimensions through the cell. P(1) has 2 P + 2 Cd neighbors and P(2)

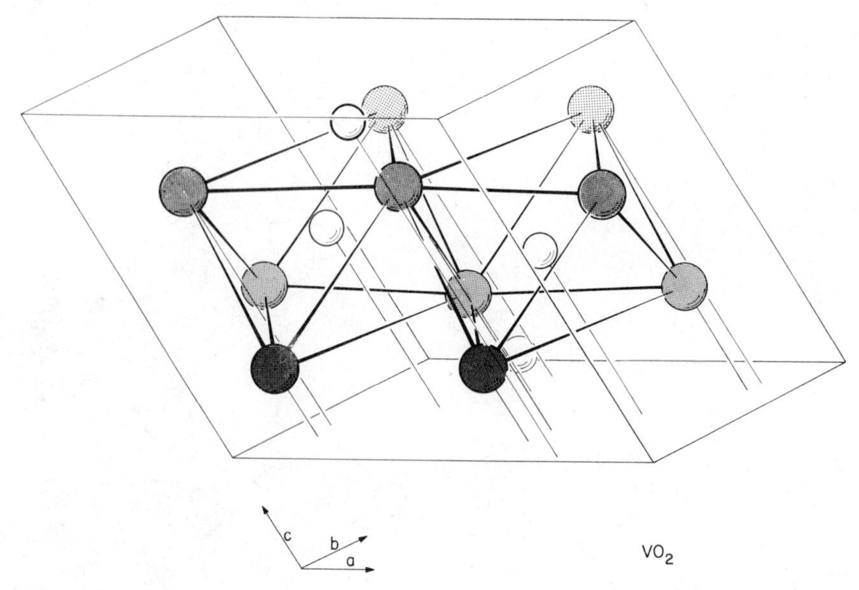

Figure 8-51. Pictorial view of the low temperature VO_2 (*mP*12) structure. Its relationship to the rutile structure, can be compared with that of the $CoSb_2$ structure to marcasite.

has 3 P + 1 Cd neighbors disposed more or less tetrahedrally. Cd is surrounded octahedrally by 6 P (Figure 5-14, p. 226). The atomic arrangement resembles that of CuP_2 (*mP*12) as shown in Figure 8-52.

CuP_2 (*mP*12): CuP_2 has cell dimensions $a = 5.802$, $b = 4.807$, $c = 7.525$ Å, $\beta = 112.68°$, $Z = 4$.[79] The structure is very similar to that of CdP_4 with Cu_2 (2.48 Å apart) replacing Cd as shown in Figure 8-52. Distorted P octahedra surround the Cu pairs and each Cu has 4 P (2.27 to 2.50 Å) and one Cu neighbor. P(1) has 3 Cu + 2 P (2.19 and 2.205 Å) neighbors, P(2) has 1 Cu and 3 P (2.19, 2.205, and 2.207 Å) neighbors.

Fe_3Th_7 (*hP*20): The dimensions of the unit cell of the Fe_3Th_7 structure are $a = 9.85$, $c = 6.15$ Å, $Z = 2$.[80] At $z \sim 0$ Th(3) and (1) form the network shown in Figure 8-53 and at $z = \frac{1}{2}$ they form a similar network reversed. At $z \sim \frac{1}{4}$ and $\frac{3}{4}$ Th(2) form triangles about the c edges of the cell giving octahedra which lie over the large voids in the networks at $z = 0$ and $\frac{1}{2}$. Fe atoms form larger triangles within the cell on the right side at $z \sim \frac{3}{4}$ and on the left at $z \sim \frac{1}{4}$. Th(1) and (3) atoms at $z \sim 0$ or $\frac{1}{2}$ form tetrahedra with two of these Fe atoms as indicated in Figure 8-53. The structure is somewhat related to those of Th_6Mn_{23} and $SrMg_4$ where an octahedron of large atoms is surrounded by a cage of other atoms. Also the structure as taken by Ru_7B_3 is similar to those of Mn_7C_3 and Fe_7C_3, as shown on p. 420.

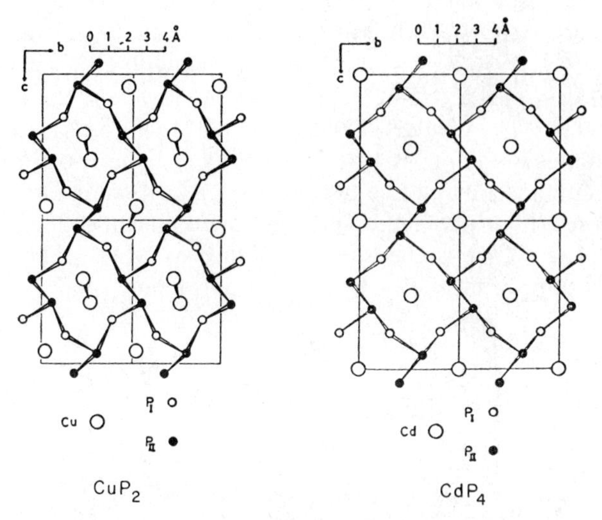

Figure 8-52. Comparison of the atomic arrangements in the CuP_2 ($mP12$) and CdP_4 ($mP10$) structures in projection down [100]. (Olofsson[79].)

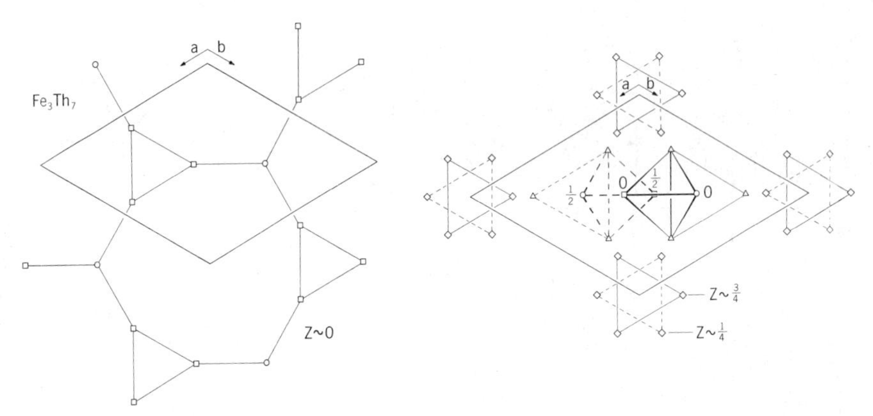

Figure 8-53. Atomic arrangements in the Fe_3Th_7 ($hP20$) structure viewed down [001]. (Left) ○, Th(1); □, Th(3). (Right) □ and ○, Th; △, Fe.

Fe has 6 Th atoms at 2.82 to 2.96 Å. Th(1) has 3 Fe at 2.96 Å and 12 Th at 3.57 to 3.87 Å. Th(2) has 3 Fe and 11 Th at 3.53 to 3.81 Å. Th(3) has 2 Fe and 11 Th at 3.42 to 3.87 Å. Each Th atom is surrounded by a convex polyhedron of 15 atoms, although several of the contacts are relatively long distances to Fe atoms. Although the structure was determined by single-crystal methods, it requires further refinement.

Mn_7C_3 (oP40): Cr_7C_3 (hP80): Mn_7C_3 which has cell dimensions $a = 4.546$, $b = 6.959$, $c = 11.976$ Å, $Z = 4$, is not isotypic with Cr_7C_3 which has cell dimensions $a = 14.01$, $c = 4.532$ Å, $Z = 8$.[81] However, the structures of Ru_7B_3 (Fe_3Th_7 type, Figure 8-53), Cr_7C_3 and Mn_7C_3 are related, the cells having volumes in the ratios 1:4:2 respectively. The Ru_7B_3 cell contains one metal octahedron and two tetrahedra, Cr_7C_3 4 octahedra and 8 tetrahedra, and Mn_7C_3 2 octahedra and 4 tetrahedra arranged as indicated in Figure 8-54. Fe_7C_3 has the same structure as Mn_7C_3. The structures appear to require confirmation by neutron diffraction.

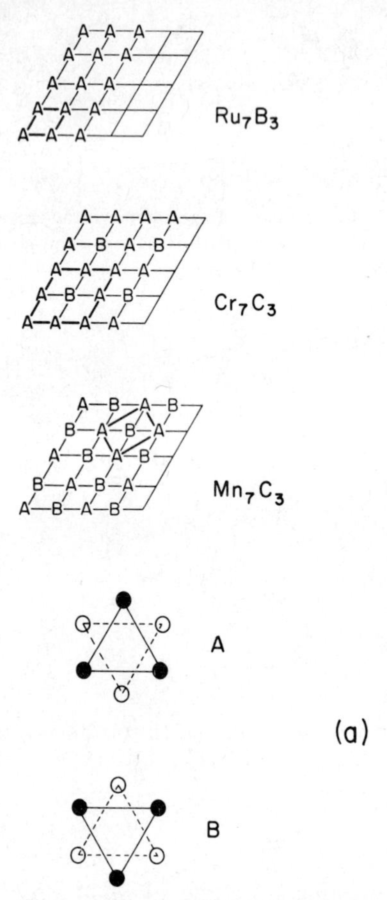

Figure 8-54. Comparison of the arrangements of metal atom octahedra in the Ru_7B_3 (Fe_3Th_7 (*hP*20) type), Cr_7C_3 (*hP*80), and Mn_7C_3 (*oP*40) type structures. (Bouchard *et al.* [81].)

$CuTaS_3$ (*oP*20): $CuTaS_3$ has cell dimensions $a = 9.49$, $b = 3.53$, $c = 11.82$ Å, $Z = 4$.[82] Ta atoms are surrounded irregularly by octahedra of S (2.21 to 2.73 Å). Pairs of these octahedra which are joined at an edge, form columns along *b*. Cu atoms are surrounded by tetrahedra of S (2.25 to 2.33 Å) which share corners to form strings running in the *b* direction. The tetrahedra also share an edge with the octahedra (Figure 8-55), and the short Cu–Ta distances (2.80 Å) resulting are accentuated by movements of the atoms toward each other from the centers of their polyhedra. The sulfurs have 3 or 4 Ta and Cu neighbors.

α-$Al_{13}Cr_4Si_4$ (*cF*84): The structure contains a continuous framework of linked tetrahedra which are centered by Si and composed of 1 Al(1) at 2.38 Å and 3 Cr at 2.41 Å.[83] Each Al(1) is apex to four tetrahedra and each Cr to three. Octahedra of Al(2) centered about 0, 0, $\frac{1}{2}$ and $\frac{1}{2}$, $\frac{1}{2}$, $\frac{1}{2}$ and octahedra of Al(3) centered about $\frac{1}{4}$, $\frac{1}{4}$, $\frac{3}{4}$ and $\frac{3}{4}$, $\frac{3}{4}$, $\frac{3}{4}$ fit within the framework of tetrahedra and are so related as to create a third distorted octahedron made up of Al(2) and Al(3). Al(2) has 4 Al(2) at 2.93 Å and 4 Al(3) at 2.89 Å as well as 2 close Cr at 2.46 Å and two Si at 2.81 Å. Al(3) has 4 Al(2) and 4 Al(3) at 2.94 Å as well as 2 Cr at 2.77 Å and 2 Si at 2.79 Å. Cr has 3 Si, 3 Cr, 3 Al(2), and 3 Al(3) neighbors arranged in a distorted icosahedron and Si has 7 Al and 3 Cr neighbors.

$U_3Si_2C_3$ (*oI*16): $U_3Si_2C_3$ with cell dimensions $a = 3.598$, $b = 3.535$, $c = 18.96$ Å, $Z = 2$ is pseudotetragonal.[84] The atoms form 1.4^4 or $\frac{1}{2}.4^4$ nets parallel to the (001) plane. C(1) is surrounded octahedrally by 4 U(1) and 2 Si. C(2) is surrounded tetrahedrally by 4 U(2) (Figure 8-56). Si is

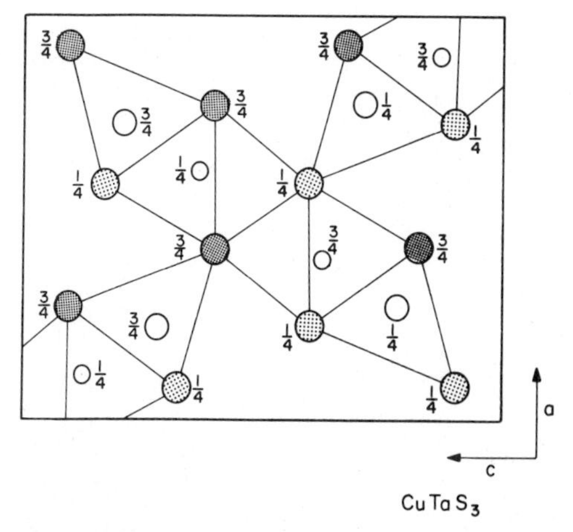

Figure 8-55. The $CuTaS_3$ (*oP*20) structure viewed down [010]: Cu, small circles; S, large circles.

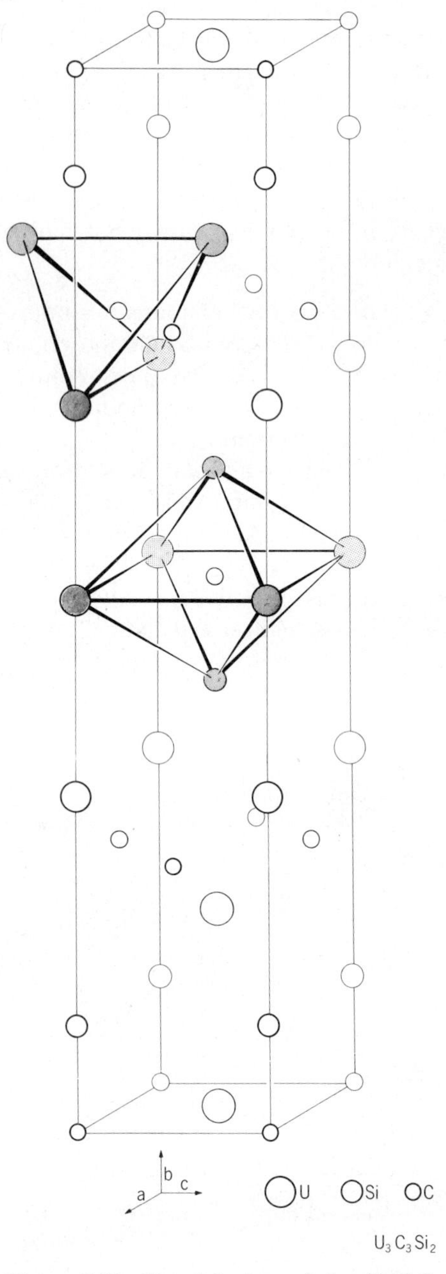

Figure 8-56. Pictorial view of the $U_3Si_2C_2$ (*oI*16) structure.

surrounded by a distorted cube of 4 U(1) and 4 U(2) with C(1) in one cube face center. Although the structure was determined by single-crystal methods, the carbon positions need to be checked by neutron diffraction.

OCTAHEDRAL COORDINATION

MN: *M* on 3^6 Nets in ABC Stacking: *N* on 3^6 Nets at $\frac{1}{2}$ Spacing. Octahedra Only.

An array of layers of close packed atoms stacked in a close packed cubic sequence ABC can be penetrated by an identical array of close packed layers half way between the original layers, giving a stack of equally spaced layers such that the atoms of the second array sit at the centers of the octahedral interstices of the first array and *vice versa*, as in the NaCl structure. When, however, the initial array of layers is in hexagonal stacking sequence, BC, as in the NiAs structure, the atoms of the second array placed at the centers of octahedral interstices of the first, no longer surround those of the first array octahedrally as in the NaCl structure, but form right trigonal prisms about them. When the first array has mixed cubic and hexagonal stacking sequences as for example in the TiAs structure, ABAC, then the atoms of the second array surround those of the first array either octahedrally or in trigonal prisms according to whether the local surrounding of layers is cubic or hexagonal. Since the near-neighbor coordination changes between these three structures, they are not members of a family of polytypic structures. Nevertheless, octahedral coordination alone is obtained when alternate layers of atoms centering octahedral holes in a c.p. hexagonal array of atoms are omitted, as in the CdI_2 structure, or when atoms in octahedral holes in one layer have vacant sites over them in the layers above and below, as in the ordered interstitial carbide structures.

NaCl (*cF*8): In the rocksalt structure both Na and Cl form interpenetrating f.c. cubic arrays and close packed triangular nets lying parallel to {111} planes. The Na and Cl layers are severally close packed in the sequence ABC and they are equally spaced in ⟨111⟩ directions on the appropriate sites so that each is situated at the center of the octahedral holes of the other array. The overall stacking sequence is thus $\underline{A}C\underline{B}A\underline{C}B$ (Na underlined).

The 6–6 octahedral coordination of the NaCl structure (Figure 8-57) makes the type suitable for all kinds of compounds: ionic substances, semiconductors and metals. A histogram of the number of known phases with the NaCl structure as a function of electron concentration (Figure

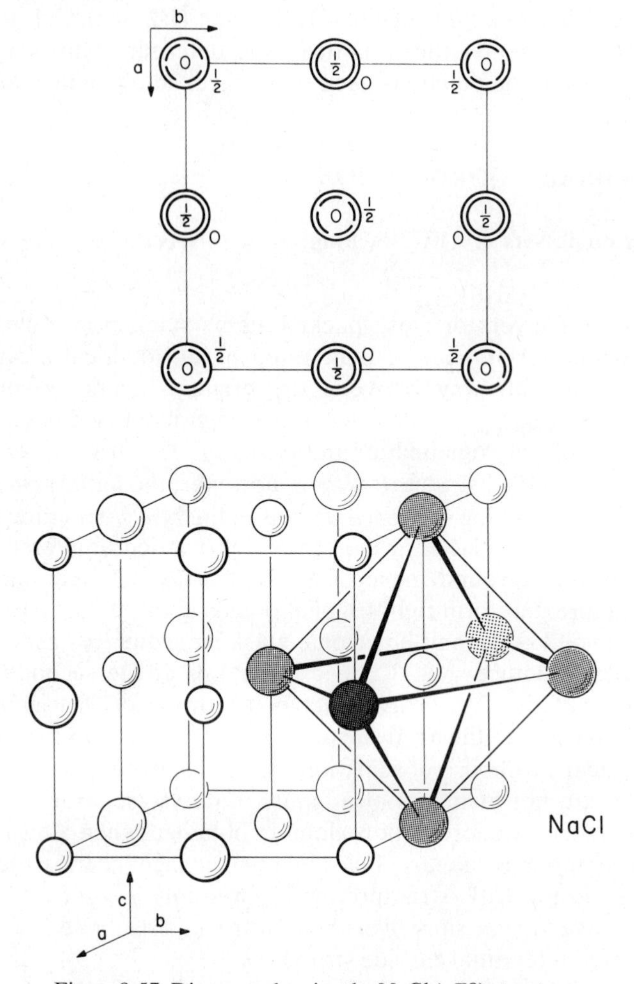

Figure 8-57. Diagrams showing the NaCl ($cF8$) structure.

3-32, p. 107), shows the importance of a high electron concentration (~ 4) for the occurrence of the structure, which is found at higher electron concentrations than the CsCl, FeB, and CrB phases. The importance of the *M–N* bond factor in non-ionic phases with the NaCl structure is demonstrated by the near-neighbor diagram (p. 53) where the phases clearly follow the line for *M–N* contacts, even though there is much scatter about it (Figure 3-11, p. 67).

Although the majority of phases with the NaCl structure form with a decrease in atomic volume compared to the averaged sums of the elemental volumes (Figure 10-10, p. 570), several phases appear to expand considerably during formation. Many of these are chalcogenides, and it is probable that the elemental volume assumed for S, Se, and Te is too small. Numerous interstitial compounds of the transition metals with C and N also appear to expand compared to the sums of the elemental volumes. The rocksalt structure is one of the main types adopted by Hägg interstitial compounds of the transition metals when the ratio of the metalloid radius to the metal radius is less than about 0.59. Indeed, Figure 10-11, p. 571, showing the occurrence of various MN structure types as a function of the volume ratio of the two component elements, emphasizes clearly the occurrence of the interstitial phases with the NaCl structure when the difference between the sizes of the component atoms is large. Apart from this, Figure 10-11 suggests that the NaCl structure occurs preferentially to the CrB and FeB types when the differences in atomic size of the component atoms is not too large. This is consistent with the fact that the NaCl structure is commutative, whereas the FeB and CrB types are not, and so might favor components of different size more readily.

Distorted Structures Based on NaCl, Including Structures Containing Fragments of NaCl-Like Arrangement

α-GeTe (*hR*8): α-GeTe with cell dimensions $a = 5.996$ Å, $\alpha = 88.18°$[85] is a rhombohedrally distorted rocksalt type structure with the atoms displaced from the exact NaCl sites of the distorted cell ($x = 0.237$ instead of $\frac{1}{4}$). The structure requires confirmation by single crystal methods.

GeS (*oP*8): The GeS structure with cell dimensions $a = 10.44$, $b = 3.647$, $c = 4.299$ Å, $Z = 4$,[86] can be regarded as a very distorted rocksalt structure as indicated in Figure 8-58. Ge has 6 S neighbors at distances from 2.475 to 3.01 Å and S similarly has 6 Ge neighbors. GeS and SnS were originally thought to have different structures, but it can be shown that they differ only very slightly in their atom positions.

Sb_2Se_3 (*oP*20): Sb_2Se_3 has cell dimensions $a = 11.77$, $b = 3.962$, $c = 11.62$ Å, $Z = 4$.[87] The structure is made up of columns of distorted fragments of rocksalt-like structure running in the b direction as indicated in Figure 8-59. The atoms are confined to planes at $y = \frac{1}{4}$ and $\frac{3}{4}$. The closest Se–Sb neighbors are for Se(1) 2.58 Å and 2×2.98 Å, for Se(2) 2×2.66 Å and for Se(3) 2.66 Å and 2×2.78 Å. Sb(1) and (2) each have 3 close Se neighbors and Sb(2) has in addition, 2 at 2.98 Å.

Sn_2S_3 (*oP*20): The Sn_2S_3 structure, with cell dimensions $a = 8.864$, $b = 14.02$, $c = 3.747$ Å, $Z = 4$,[88] is very similar to that of $KCdCl_3$. The

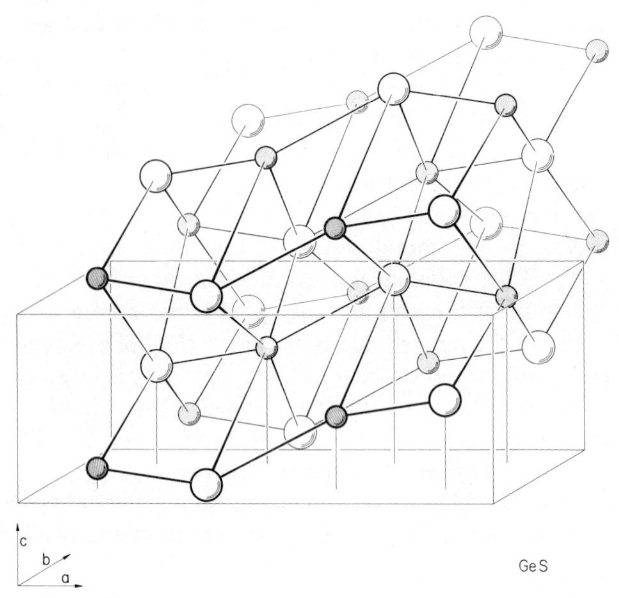

Figure 8-58. Pictorial view of the GeS (*oP*8) structure.

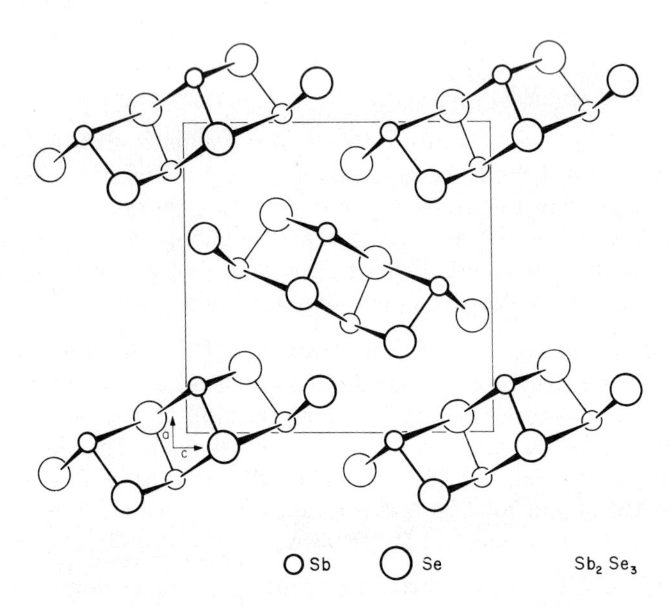

Figure 8-59. View of the Sb_2Se_3 (*oP*20) structure down [010].[87]

structure, like that of Sb_2Se_3 or Sb_2S_3, can be considered as containing fragments of rocksalt type structure running in columns in the [001] direction (Figure 8-60); however, the distortion of the fragments is not so severe as in the Sb_2Se_3 structure. The closest S–Sn neighbors are for S(1) 2.55 Å and 2 × 2.61 Å, for S(2) 2 × 2.54 Å and 2.74 Å and for S(3) 2.50 Å and 2 × 2.64 Å. Sn(1) is surrounded by an octahedron of 6 S atoms and Sn(2) at the end of the rocksalt-like fragments has 3 S neighbors.

Sc_2S_3 (oF80): Sc_2S_3 is an ordered cation-deficient superstructure based on the NaCl type. The cell dimensions, $a = 10.41$, $b = 7.39$, $c = 22.05$ Å, $Z = 16$ are related to $a_0 = 5.21$ Å of a NaCl type subcell as follows: $a = 2a_0$, $b = \sqrt{2}.a_0$, $c = 3\sqrt{2}.a_0$, so that the supercell has the volume of 12 NaCl-type subcells.[89]

Two ligands for each sulfur centered octahedron are missing so that each S has 4 Sc at distances from 2.57 to 2.61 Å. Each Sc is octahedrally surrounded by 6 S.

CaC_2 I (tI6): CaC_2 ($a = 3.88$, $c = 6.37$ Å, $Z = 2$)[90] and $MoSi_2$ have the same structure as regards space group and occupied sitesets. However, the axial ratios and atomic coordinates differ, resulting in different coordination of the atoms. For CaC_2, $c/a = 1.64$ compared to values of 2.5 to 3.5 for phases with the $MoSi_2$ structure. Furthermore, the larger value of the z parameter of the C atoms in the CaC_2 structure results in pairs of C

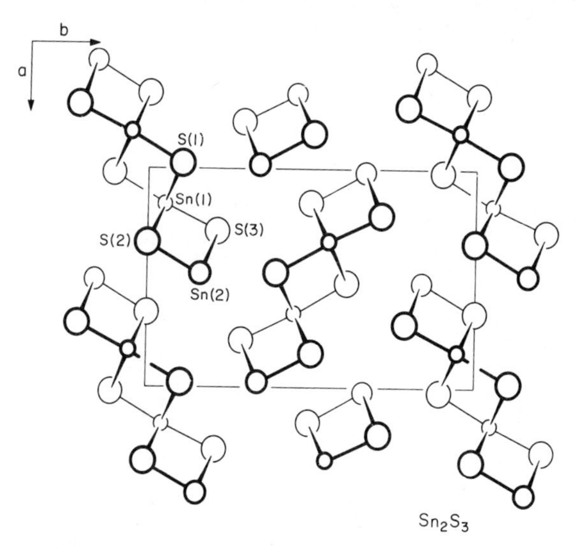

Figure 8-60. View of the Sn_2S_3 ($oP20$) structure down [001].[88]

atoms with their axis in the [001] direction (Figure 8-44*a*, p. 410). The carbon atom positions were determined by neutron diffraction.[91]

The structure is a tetragonally elongated derivative of the rocksalt structure with the pairs of C atoms replacing the Cl atoms, as is easily seen by considering the *F* cell.

ThC_2(mC12): In the ThC_2 structure ($a = 6.53$, $b = 4.24$, $c = 6.56$ Å, $\beta = 104°$, $Z = 4$)[92] the C atoms form pairs at 1.47 Å. The structure can best be regarded as a distortion of the CaC_2 type and therefore a derivative of the NaCl structure. A somewhat distorted b.c. tetragonal array of Th atoms can be recognized with the pairs of C atoms centered about the midpoints of the basal faces as in the CaC_2 structure (Figure 8-61). However, the axes of the C_2 pairs are rotated away from the *c* axis of the pseudocell like the S atom pairs in the marcasite structure. The carbon atom positions were determined by neutron diffraction.

In_6S_7(mP26): In_6S_7 has cell dimensions $a = 9.09$, $b = 3.89$, $c = 17.70$ Å, $\beta = 10.28°$, $Z = 2$.[93] The atoms, which occupy 13 sitesets, all lie in planes parallel to (010) at heights $y = \pm\frac{1}{4}$ (Figure 8-62). The S atoms form two fragments of slightly distorted close packed arrays with In atoms in octahedral holes thereof so that the structure is made up of fragments of rocksalt-like arrangement. One region of S close packing runs continuously throughout the structure across the middle of the unit cell as shown in Figure 8-62. The other region located at the ends of the cell

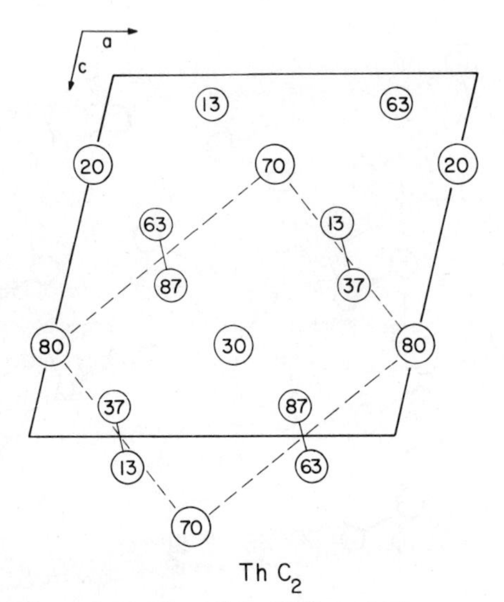

Figure 8-61. View of the ThC_2 (*mC*12) structure down [010]: large circles, Th.

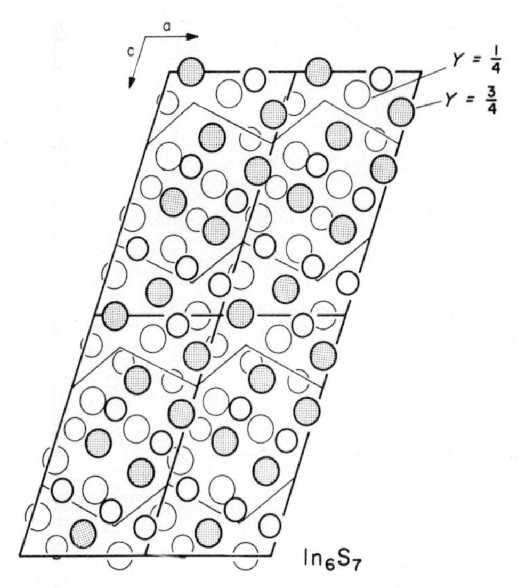

Figure 8-62. Projection of the In_6S_7 (*mP*26) structure down [010]: large circles, S.

with distortion in the region of the cell origin, is misoriented relative to the other by a rotation of 61.5° about an axis parallel to [010]. With the exception of In(5), the In atoms can be attributed to one or other S array. In(5) is slightly displaced from the center of an octahedral hole in each of the S arrays, and it has considerably larger average In–S distances than the other In atoms. In(5) thus has 6 S neighbors whereas the other In atoms only have 4 S neighbors because of the small extent of the fragments of close packed S arrays in the *c* direction.

$MoUC_2$ (*oP*16): $MoUC_2$ has cell dimensions $a = 5.625$, $b = 3.249$, $c = 10.980$ Å, $Z = 4$.[94] As in the $CrUC_2$ structure, the atoms are confined to planes parallel to (010) at heights $y = \frac{1}{4}$ and $\frac{3}{4}$, but here the similarity between the two structures ceases. In the $MoUC_2$ structure the atoms lie approximately in planes which are more or less parallel to (102) and (10$\bar{2}$). Figure 8-63 shows the atomic arrangement in one plane approximately parallel to (10$\bar{2}$). From this it can be seen that the second to sixth rows of atoms from the bottom form a portion of rocksalt-like arrangement with C occupying one site and U or Mo the other site. Then in the next rows this portion of structure has turned so that the axis that was perpendicular to (10$\bar{2}$) now lies in the plane running up the paper. The structure can therefore be regarded as made up of two differently oriented portions of rocksalt-like structure with Mo and U atoms occupying ordered positions on the Na sites and C occupying the Cl sites.

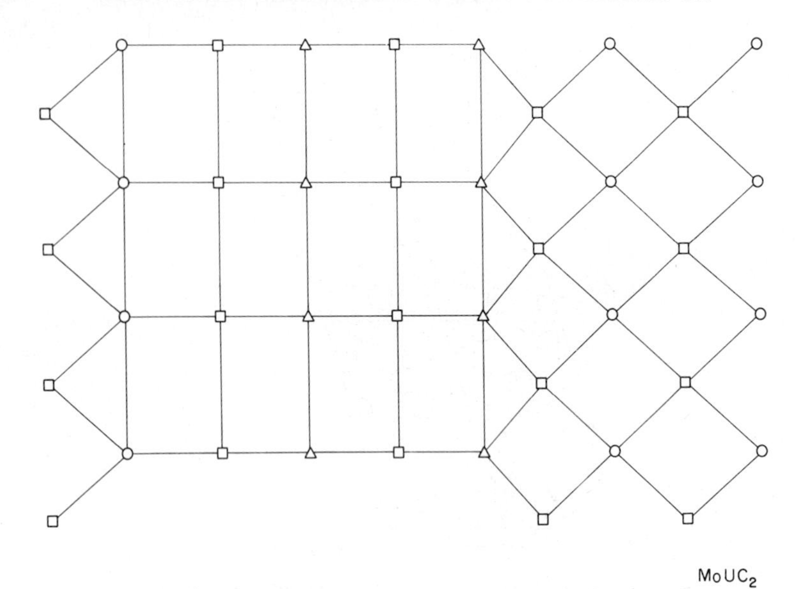

Figure 8-63. Atomic arrangement in one layer approximately parallel to a $(10\bar{2})$ plane in the $MoUC_2$ (*oP*16) structure: U, circle; C, square; Mo, triangle.

As a result C(1) is surrounded by a distorted octahedron of 2 Mo (2.11 and 2.20 Å) and 4 U atoms (2 × 2.395 Å, 2 × 2.51 Å). C(2) has similarly 3 Mo (2 × 2.16 Å, 2.20 Å) and 3 U (2.39 Å and 2 × 2.48 Å). Mo has 5 C neighbors and U has 7 at 2.395 to 2.51 Å. The structure, including C atom positions was determined by X-ray analysis of a single crystal.

Superstructures of NaCl

$CrNaS_2$ (*hR*4): The dimensions of the corresponding hexagonal cell are $a = 3.51$, $c = 19.57$ Å, $Z = 3$.[95] All atoms form 3^6 layers normal to [001] of the hexagonal cell. S layers are stacked in the sequence CABCAB with slightly uneven spacing between them. Cr and Na layers alternately separate the S layers, the metal atoms being surrounded octahedrally by S (Figures 8-64 and 8-65). The $CrNaS_2$ structure is therefore a superstructure based on the NaCl type with overall stacking sequence A$\underline{C}$B$\underline{A}$C$\underline{B}$-A$\underline{C}$B$\underline{A}$C$\underline{B}$ (S underlined). $AgBiSe_2$ has the same structure in the temperature range from 120 to 287°C.[96] The arrangement of the atoms is very similar to that in the low-temperature trigonal form described below.

The structure was the *Strukturbericht* $F5_1$ type which has caused some confusion in the literature. $NaCrS_2$ belongs to the $CsICl_2$ type with the atomic parameter $x_S < 0.33$ giving anion stacking sequences CABCAB.

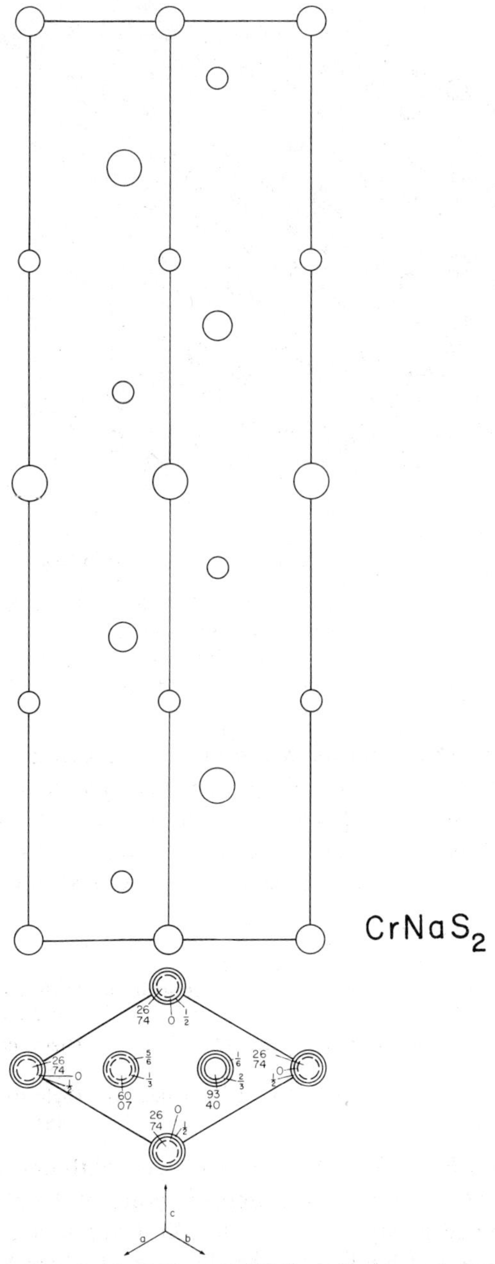

Figure 8-64. [110] and [001] projections of the hexagonal cell of the $CrNaS_2$ (*hR*4) structure: Na, largest circles; S, smallest circles.

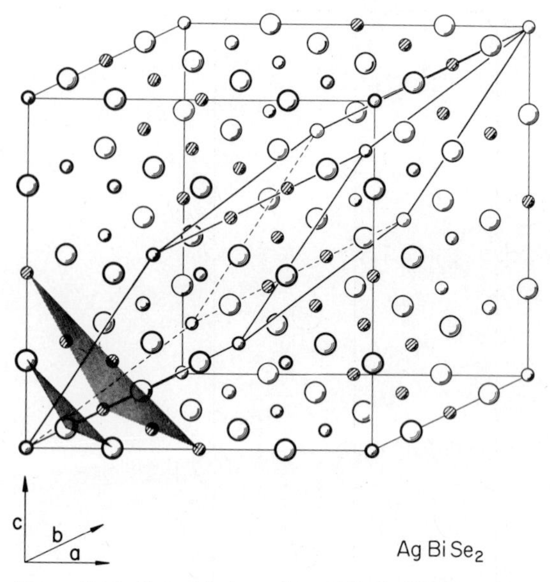

Figure 8-65. Pictorial view of the $CrNaS_2$ (*hR*4) structure of the $AgBiSe_2$ intermediate temperature phase, with the rhombohedral cell set in a cube, showing the close packed planes: large circles, Se; hatched circles, Bi.

When, however, $x_X > 0.33$ as in the $NaHF_2$ or $CuGaO_2$ $F5_1$ type (see p. 402), the anions take the paired-layer stacking sequence BBAACC rather than the close packed sequence. This causes the changes in atomic coordination, shown in the following table, which are also compared with $CuCrS_2$, that has a somewhat similar rhombohedral structure.

	$CrNaS_2$	$CrCuO_2$	$CrCuS_2$
	$x_S < 0.33$	$x_O > 0.33$	
S, O	Distorted octahedron (3 Cr + 3 Na)	Distorted tetrahedron (3 Cr + 1 Cu)	Distorted tetrahedron (3 Cr + 1 Cu)
Na, Cu	Distorted octahedron (S)	Linear (O)	Distorted tetrahedron (S)
Cr	Distorted octahedron (S)	Distorted octahedron (O)	Right-trigonal prism (S)

$AgBiSe_2$ (L.T.) (*hP*12): The $AgBiSe_2$ structure with unit cell dimensions $a = 4.18$, $c = 19.67$ Å, $Z = 3$, is a superstructure of the NaCl type.[96] The atoms are arranged in layers normal to [001], the unit cell containing 6 close packed layers of Se(1), (2), and (3) atoms ($c/3a = 1.57$) in stacking sequence CABCAB. Both the Ag and Bi atoms are located in octahedral holes of the Se array in the sequence (Ag underlined) $\underline{A}B\underline{C}A\underline{B}C$.

Ag(1) has 6 Se(2) at 2.82 Å and Ag(2) has 3 Se(1) at 2.83 Å and 3 Se(3) at 2.86 Å. Bi(1) has 6 Se(3) at 3.04 Å and Bi(2) has 3 Se(1) at 2.99 Å and 3 Se(2) at 2.98 Å. Se(1) has 3 Ag(2) at 2.83 Å, and 3 Bi(2) at 2.99 Å. Se(2) has 3 Ag(1) and 3 Bi(2) neighbors. Se(3) has 3 Ag(2) and 3 Bi(1) neighbors.

$YbAgS_2$ (*tI*16): $YbAgS_2$ has cell dimensions $a = 5.356$, $c = 11.803$ Å, $Z = 4$.[97] Yb is surrounded by a nearly regular octahedron of S, and Ag is surrounded by a very distorted octahedron of S (Ag–S = 2.71 to 3.66 Å). The structure is a distorted superstructure of rocksalt with the atoms displaced from their ideal sites; two rocksalt-type pseudocells are stacked along [001]. Ag and Yb are ordered on one set of sites and S occupies the other (Figure 8-66). The structure requires confirmation by single-crystal methods.

The monoclincially distorted form of this structure ($a_m = a_t + b_t$, $b_m = a_t - b_t$, $c_m = c_t$) adopted by rare earth compounds from Sm through to

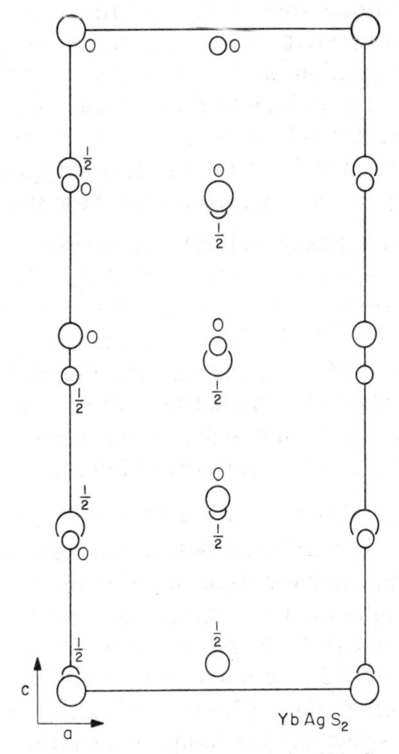

Figure 8-66. Projection of the $YAgS_2$ (*tI*16) structure down [010]: large circles, Yb; small circles, S.

Tm is a larger superstructure of NaCl. For $SmAgS_2$ $a = b = 7.78$ Å, $c =$ 12.32 Å, $\beta = 88.88°$, $Z = 8$.

Defect Superstructures Based on NaCl. Vacant *N* Component Layers

In_3Te_4, $SnSb_2Te_4$ (*hR*7): In_3Te_4 with hexagonal cell dimensions $a = 4.27$, $c = 40.9$ Å, $Z = 3$,[98] P_3Sn_4,[99] As_3Sn_4,[100] Bi_3Se_4,[101] and $SnSb_2Te_4$,[102] ($a =$ 4.31, $c = 41.7$ Å) have essentially the same structures. All atoms form triangular nets parallel to the (001) plane. The relative spacings of the layers differ slightly in the structures, but not enough to change the layer stacking sequence. The Te layers are stacked in the same close packed sequence CACABCBCABAB ($c/6a = 1.63$) as in Fe_3S_4, but the In atoms are arranged in the sequence AB□BCA□ABC□C (□ indicates an absent layer) with the unoccupied octahedral holes lying between pairs of Te layers (underlined) that are hexagonally surrounded, $A\underline{C}B\underline{A}\square\underline{C}B\underline{A}$ . . . . Thus, In always surrounds the Te layers octahedrally, in contrast to Fe_3S_4 where the Fe atoms always surround the S atoms in trigonal prisms.

The overall stacking sequence in $SnSb_2Te_4$, whose structure was examined by electron diffraction, is $A{*}C\underline{B}AC\underline{B}AC{*}B\underline{A}CB\underline{A}CB{*}A\underline{C}BA\underline{C}B$, with Sb underlined and Sn indicated by an asterisk. In P_3Sn_4 ($a = 3.968$, $c = 35.33$ Å) the mean Sn–P distance is 2.78 Å. The pairs of Sn layers which are not separated by P atoms, are closer together (Sn–Sn ~ 3.25 Å) than the other Sn layers. All coordination in these structures is octahedral.

$GeBi_2Te_4$(*hR*7): With hexagonal cell dimensions $a = 4.28$, $c = 39.2$ Å, $Z = 3$,[103] $GeBi_2Te_4$ has a seven-layer structure Te(1) BiTe(2)GeTe(2)-BiTe(1) with 21 layers parallel to (001) in the unit cell, according to electron diffraction studies of thin films. There are 12 close packed layers of Te in the cell with ABC stacking and the Ge and Bi atoms occupy $\frac{3}{4}$ of the layers of octahedral holes between them. $GeSb_2Te_4$[104] has similar cell and atomic parameters, although for some reason the relative arrangement of the components is thought to be different.

$Pb_2Bi_2Se_5$ (*hP*9): The structure ($a = 4.22$, $c = 17.42$ Å, $Z = 1$) contains 9 layers of triangular nets parallel to (001) which are stacked in the sequence ABCABCABC with components arranged in the sequence SePbSeBiSe-SeBiSePb.[105] The interatomic distances are Se–Se = 2.95 Å, Pb–Se = 3.02 Å, and Bi–Se = 3.11 Å. If the Se atoms in the nine-layer units are regarded as close packed layers and the Pb and Bi layers as occupying octahedral holes therein, then $c/2.5a = 1.65$ is close to the ideal value for close packing. The structure was found by electron diffraction from thin films.

β-In_2Se_3 (above 200°C) (*hR*5): β-In_2Se_3 has hexagonal cell parameters

$a = 4.05$, $c = 29.41$ Å at 250°C, $Z = 3$.[52] All atoms form triangular close packed layers parallel to (001). The Se layers are in close packed spacing with the octahedral holes between 2 out of 3 pairs of these layers occupied by In. Thus five-layer structural units are created. The overall stacking

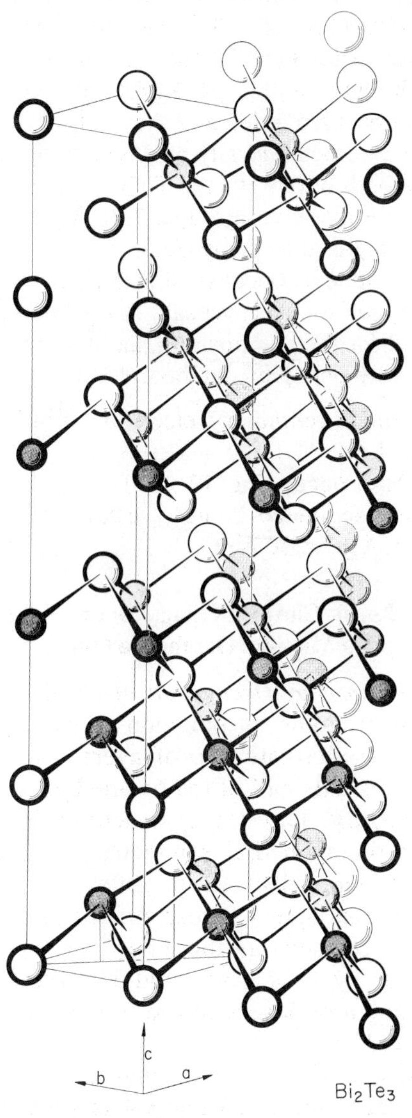

Figure 8-67. Pictorial view of the hexagonal cell of the Bi_2Te_3 (*hR*5) structure: Te, large circles.

arrangement is (In underlined) A$\underline{B}$CA$\underline{B}$C$\underline{A}$BC$\underline{A}$B$\underline{C}$AB$\underline{C}$. In–Se = 2.87 Å. Se–Se = 4.01 Å between the five-layer units. The α and β structures are compared in Figure 8-37, p. 403.

Bi_2STe_2 (Bi_2Te_3) (hR5): The dimensions of the hexagonal cell of Bi_2STe_2 are $a = 4.369$, $c = 30.42$ Å, $Z = 3$. The atoms form triangular close packed nets parallel to (001).[106] These are stacked in the sequence (Bi underlined, S with asterisk) A*$\underline{B}$CA$\underline{B}$C*$\underline{A}$BC$\underline{A}$B*$\underline{C}$AB$\underline{C}$. Thus the layers are all cubically surrounded and they form five-layer groups TeBi-SBiTe. Bi is surrounded octahedrally by 3 S at 3.06 Å and 3 Te at 3.13 Å; S is surrounded octahedrally by Bi, and Te has 3 Bi neighbors. The Te–Te distances between the five-layer groups, 3.70 Å, are closer than van der Waals distances. Compared to many other valence structures with M_2N_3 stoichiometry which appear, because of the vacant sites, to be fragments of other structural types such as rocksalt, Bi_2Te_3 appears to be a structure specifically designed for the composition, and it is not too bad a distortion from close packing with occupied octahedral holes (Figure 8-67).

Ge_2Te_3 (hP30): Rather preliminary electron diffraction results indicate that the Ge_2Te_3 cell ($a = 4.17$, $c = 53.0$ Å, $Z = 6$) contains 18 c.p. Te layers with Ge atoms occupying 12 of the layers of octahedral holes therein, the stacking sequence being (Ge underlined) AB$\underline{C}$A$\underline{B}$C$\underline{A}$B$\underline{C}$-ABC$\underline{A}$B$\underline{C}$A$\underline{B}$C$\underline{A}$BCA$\underline{B}$C$\underline{A}$B$\underline{C}$A$\underline{B}$C.[107]

MN_2: N on 3^6 Nets in Close Packing, M on 3^6 Nets at Half Spacing, Alternate Layers Being Missing. Octahedra Only

MN_2 *Structures Formed by Occupation of Octahedral Holes in a Close Packed Array of Atoms.* The introduction of close packed layers of M atoms halfway between alternate pairs of layers in an array of hexagonally close packed N atoms, fills half of the octahedral holes of the array and gives MN_2 stoichiometry. Three-layer structures formed by this process have hexagonal or rhombohedral symmetry and a family of polytypic structures can be developed by taking different close packed stacking sequences of the N atom layers. These structures thus differ from other ordered MN_2 structures such as the $TiSi_2$ type where stoichiometry is satisfied within the 3^6 layers of atoms, and from the α-MoS_2 type (p. 479) where pairs of neighboring layers of the N component are on the same stacking site.

Very many polytypes of the $CdCl_2$ structure formed by CdI_2 itself have been observed and characterized. Table 8-4 lists some of these.

$CdCl_2$ ($TaSe_2$) (hR3): The $CdCl_2$ structure of $TaSe_2$[108] has an equivalent hexagonal cell with $a = 3.437$, $c = 19.21$ Å, $Z = 3$. Ta layers are surrounded

Table 8-4 Some Observed Polytypes of CdI_2[109]

Polytype	Stacking Sequence (Cd layers underlined)
2H	$\mathrm{A\underline{C}B}$
4H	$\mathrm{A\underline{C}BC\underline{A}B}$
6H(a)	$\mathrm{A\underline{C}BC\underline{A}BA\underline{C}B}$
6H(b)	$\mathrm{A\underline{C}BC\underline{B}AC\underline{A}B}$
8H	$\mathrm{A\underline{C}BA\underline{C}BC\underline{A}BA\underline{C}B}$
10H	$\mathrm{A\underline{C}BC\underline{A}BA\underline{C}BC\underline{A}BA\underline{C}B}$
12H(a)	$\mathrm{A\underline{C}BC\underline{A}BA\underline{C}BC\underline{A}BA\underline{C}BA\underline{B}C}$
12H(b)	$\mathrm{A\underline{C}BA\underline{C}BC\underline{A}BC\underline{A}BA\underline{C}BA\underline{B}C}$
12H(c)	$\mathrm{A\underline{C}BC\underline{A}BA\underline{C}BC\underline{A}BA\underline{C}BA\underline{C}B}$
14H	$\mathrm{A\underline{C}BC\underline{A}BA\underline{C}BC\underline{A}BA\underline{C}BC\underline{A}BA\underline{C}B}$
16H	
18H	
20H	
24R	
24H(a,b,c,d)	
26H(a,b)	
28H(a,b)	
32H(a,b)	
40H(a,b,c)	
44H	
50H	
56H	
62H	
64H	

by two layers of Se giving three-layer units (Figure 8-68). These are stacked in the sequence ACB and there are three such units per cell in the c direction. The Se atoms surround Ta octahedrally and the interlayer Se–Se distances correspond to van der Waals bonding. The stacking sequence of the layers is $\mathrm{\underline{A}CA\underline{C}BC\underline{B}AB}$ (Ta layers underlined).

CdI_2 (*hP*3): The cell dimensions of PtS_2 with the CdI_2 type structure are $a = 3.543$, $c = 5.039$ Å, $Z = 1$.[110] The S atoms form a c.p. hexagonal array (stacking BC) with the Pt atoms occupying the octahedral holes (A sites) between alternate layers of S atoms (Figure 8-69). The structure is thus a three-layer type with van der Waals bond distances (or somewhat closer in some instances) between chalcogenide atoms of the outer layers. It can be regarded as derived from the NiAs structure by removal of alternate layers of metal atoms, and there are examples in which a continuous solid solution occurs from the NiAs to the CdI_2 type. In many cases, however, intervening superstructures of lower symmetry

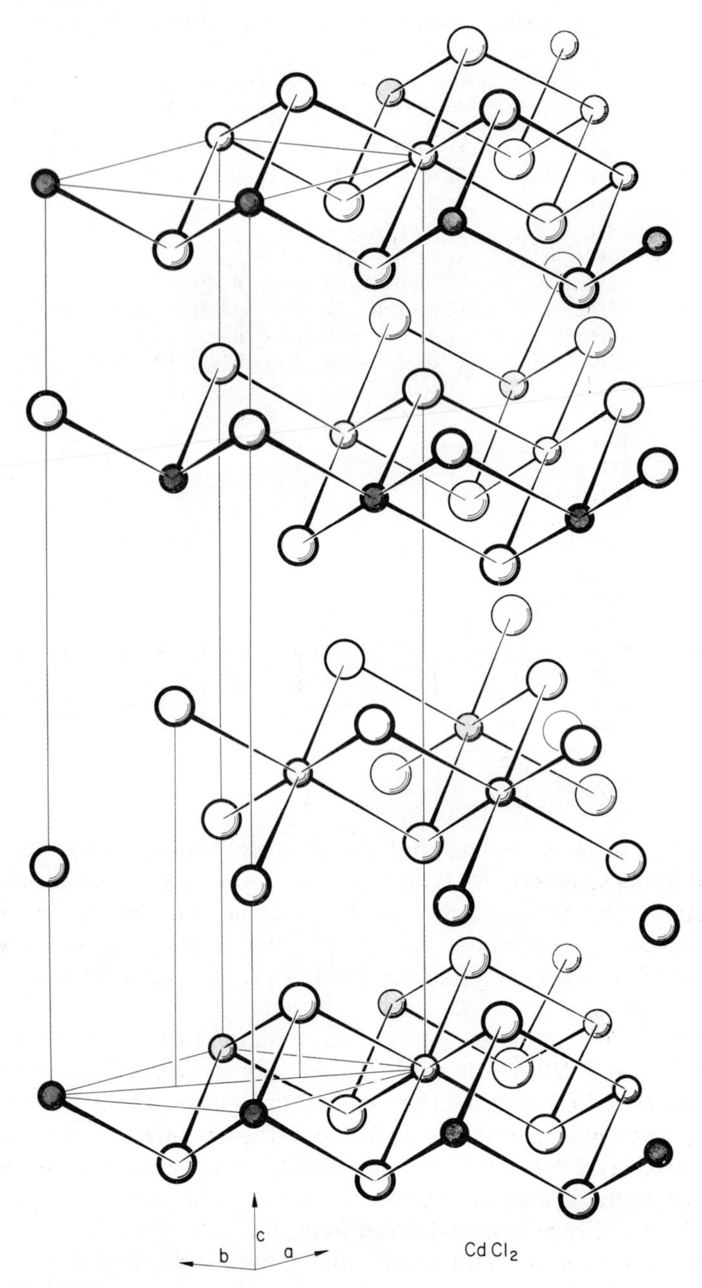

Figure 8-68. Pictorial view of the hexagonal cell of the $CdCl_2$ ($TaSe_2$, *hR*3) structure: Cl, large circles.

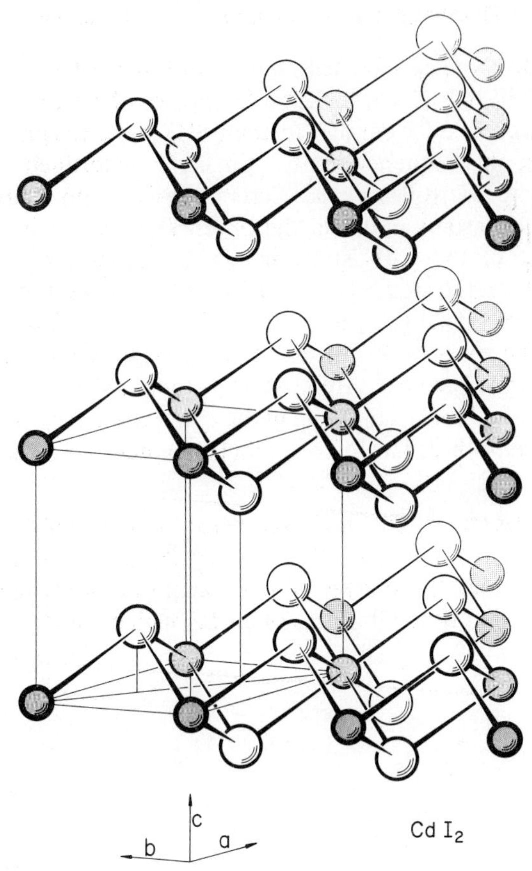

Figure 8-69. Pictorial view of the CdI_2 (PtS_2, *hP*3) structure: I, large circles.

occur due to the ordered removal of metal atoms from the alternate depleted layers.

The near-neighbor diagram (p. 53) indicates the importance of the 6–3 *M*–*N* contacts in controlling the cell dimensions of phases with the CdI_2 (MN_2) structure. The position of the line for *M*–*N* contacts changes with the axial ratio of the unit cell and, with one exception, the positions of phases on the diagram follow closely the position of the *M*–*N* contact line for the appropriate axial ratio as shown in Figure 3-9, p. 64. The axial ratio, c/a, of phases with this structure varies over wide limits from 1.82 to 1.27, showing considerable deviations from the ideal close packing of the *N* atoms (at $c/a = 1.633$).

Distorted and Related Forms of the CdI_2 Type Structure

$NbTe_2$ (*mC*18): $NbTe_2$ has cell dimensions $a = 19.39$, $b = 3.642$, $c = 9.375$ Å, $\beta = 134.58°$, $Z = 6$.[111] Rumpled planes of Te and Nb atoms form distorted triangular nets parallel to the (001) plane (Figure 8-70). These are stacked in the sequence A$\underline{B}$C (Nb layer underlined) in a direction perpendicular to the (001) plane. Thus three-layer units are formed and there are no Nb atoms between the outer Te layers. The structure is a distorted form of the CdI_2 structure. Nb(1) atoms are surrounded by relatively undistorted octahedra of Te (2.84–2.88 Å), but the octahedra about Nb(2) are considerably distorted (Nb(2)–6 Te at 2.69–2.91 Å). This results from the rows of Nb(2) atoms being displaced along *a* toward the Nb(1) rows. Thus the structure contains groups of three close Nb rows, Nb(2), Nb(1), and Nb(2) which run in the *b* direction. Nb(1) has 4 Nb(2) at 3.33 Å and each Nb(2) has 2 close Nb(1) neighbors. $TaTe_2$ has the same structure type.

$MoTe_2$ (H.T.) (*mP*12): The high temperature form of $MoTe_2$ has cell dimensions, $a = 6.33$, $b = 3.469$, $c = 13.86$ Å, $\beta = 93.92°$, $Z = 4$.[112] The atoms form rumpled three-layer units parallel to (001) (Figure 8-70). The layers are distorted triangular nets of atoms which are stacked perpendic-

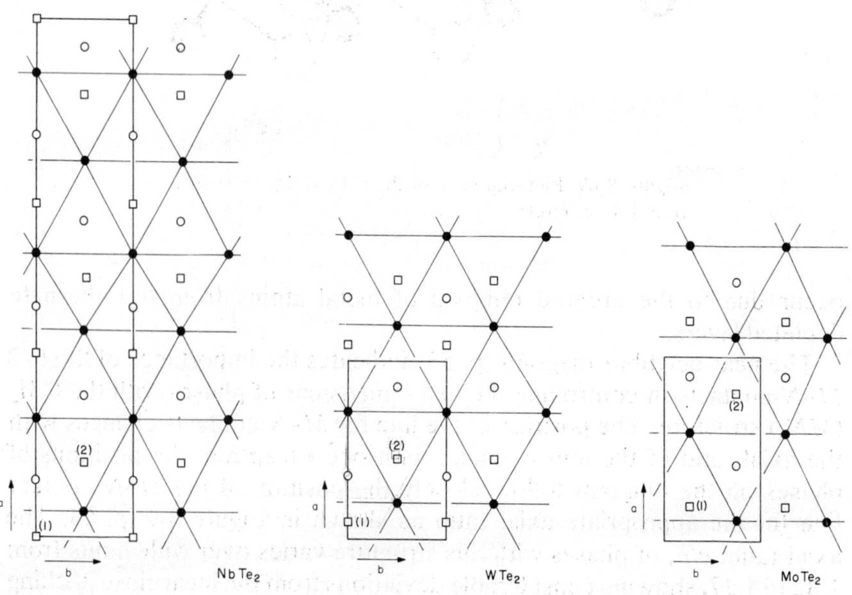

Figure 8-70. Atomic arrangements in the $NbTe_2$ (*mC*18), WTe_2 (*oP*12), and high temperature $MoTe_2$ (*mP*12) structures projected down [001]: Te, circles.

ular to the (001) plane in the sequence A$\underline{B}$C (Mo underlined). Mo is thus surrounded by distorted octahedra of Te atoms, but the Mo atoms are also displaced from the centers of the octahedra, because rows of Mo running in the *b* direction are moved along *a* toward each other in pairs. This gives each Mo two close neighbors at 2.90 Å resulting in the structure being a good electrical conductor in contrast to the low temperature form with the β-MoS_2 structure. The octahedral Mo–Te distances range from 2.70 to 2.82 Å, and the Te–Te distances between the three-layer groups are 3.86 to 3.90 Å.

The structure is a distorted form of the CdI_2 structure, which resembles closely the WTe_2 structure and also the $NbTe_2$ structure where three rows of metal atoms approach each other closely.

WTe_2 (*oP*12): WTe_2 has cell dimensions $a = 6.282$, $b = 3.496$, $c = 14.07$ Å, $Z = 4$.[112] Rumpled three-layer sequences of Te, W, Te atoms form distorted triangular nets parallel to the (001) plane (Figure 8-70). However, the rows of W atoms running in the *b* direction in these planes move toward each other in pairs along the *a* direction, so that the atoms are moved out of the centers of the distorted octahedra formed by the three-layer stacking in the sequence A$\underline{B}$C (W layer underlined). These movements give both W(1) and W(2) two close neighbors of the other kind at 2.86 Å and these metal–metal lines result in WTe_2 being a good electrical conductor. The W–Te octahedral distances range from 2.71 to 2.82 Å. Te–Te distances between the three-layer groups are 3.93 Å. The structure is a distorted form of the CdI_2 type.

$ReSe_2$ (*aP*12): $ReSe_2$ has cell dimensions $a = 6.727$, $b = 6.607$, $c = 6.720$ Å, $\alpha = 118.9°$, $\beta = 91.8°$, $\gamma = 104.9°$, $Z = 4$.[113] The Se atoms can be considered to form a distorted close packed hexagonal array parallel to (100) with alternate layers of octahedral holes occupied by Re. The structure is thus a distorted form of the CdI_2 structure. However, the Re atoms are displaced from the centers of their distorted octahedra towards one of the triangular faces, thus Re(1) obtains a very close Re(1) neighbor at 2.65 Å and 2 Re(2) neighbors at 2.84 Å. Similarly Re(2) has 2 Re(1) neighbors and one Re(2) at a slightly larger distance of 3.08 Å. The shortest Se–Se interlayer distance is 3.56 Å, although about 3.75 Å is the most common distance. The structure resembles those of $MoTe_2$ (*mP*12) and WTe_2 (*oP*12) (cf. Figure 8-71 with Figure 8-70).

$AuTe_2$ (*mC*6): The structure of calaverite with cell dimensions $a = 7.19$, $b = 4.408$, $c = 5.08$ Å, $\beta = 90°$, $Z = 2$,[114] is related to those of sylvanite and krennerite as shown in Figure 8-72. It is a distorted form of the CdI_2 type, each atom being surrounded octahedrally, Au by 2 Te at 2.67 Å and 4 Te at 2.98 Å, Te by 3 Au and 2 + 1 Te at 3.19 and 3.47 Å.

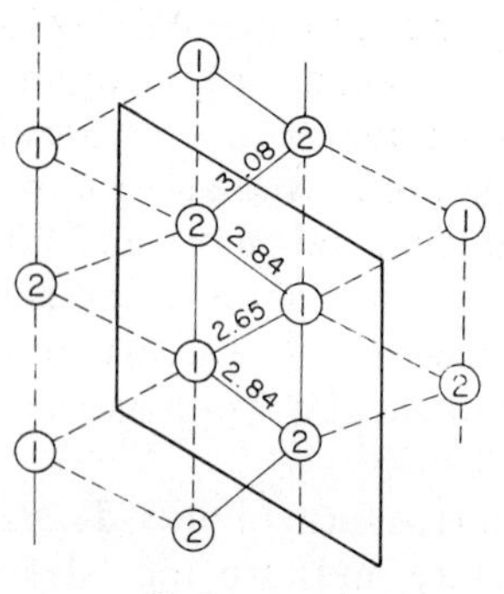

Figure 8-71. Arrangement of the Re atoms in the $ReSe_2$ (*aP*12) structure viewed down [010].

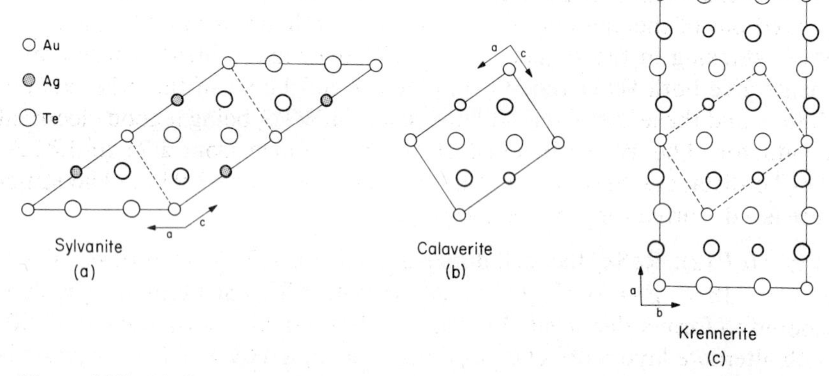

Figure 8-72. (*a*) Structure of sylvanite, $AgAuTe_4$ (*mP*12), viewed down [010]. (*b*) Structure of calaverite, $AuTe_2$ (*mC*6), viewed down [010]. (*c*) Structure of krennerite, $AuTe_2$ (*oP*24), projected down [001].[116]

$AuTe_2$ (*oP*24): The krennerite structure with cell dimensions $a = 16.54$, $b = 8.817$, $c = 4.458$ Å, $Z = 8$,[115] can be regarded as a distorted superstructure of calaverite (Figure 8-72), and therefore of the CdI_2 structure.

$AgAuTe_4$ (*mP*12): The unit cell of sylvanite, $AgAuTe_4$, with dimensions $a = 8.96$, $b = 4.49$, $c = 14.62$ Å, $\beta = 145°26'$, $Z = 2$,[116] is a superstructure of calaverite which results from the ordering of the Au and Ag atoms. The arrangement results in distorted octahedral coordination of Ag by Te (2×2.67, 2×2.95, 2×3.19 Å) and of Au by Te (2×2.69, 2×2.75,

2×3.25 Å). Te(1) has 2 Ag + 1 Au neighbors and Te(2) has 2 Au + 1 Ag neighbors which make a pyramid with Te. In addition Te(1) and (2) mutually have one close approach of 2.89 Å.

Tl_2S (*hR*27): Tl_2S has hexagonal cell dimensions $a = 12.22$, $c = 18.21$ Å, $Z = 27$. According to a diagram (but not the published coordinates),[117] the Tl atoms form slightly distorted close packed layers parallel to (001) (axial ratio 1.49) which are stacked in the hexagonal sequence AB, and the S atoms occupy the octahedral holes between every other pair of Tl layers. Tl_2S is therefore a distorted superstructure of the CdI_2 type.

H–*$AlCCr_2$* (*hP*8): *H*–$AlCCr_2$ has unit cell dimensions $a = 2.860$, $c = 12.82$ Å, $Z = 2$.[118] Al and Cr are in a somewhat squashed ($c/3a = 1.49$) hexagonal close packed array, B$\underline{C}$BC$\underline{B}$C (Al underlined). Between the pairs of Cr layers a layer of carbon atoms is inserted in the A stacking position at $z = 0$ and $\frac{1}{2}$, so that they are surrounded octahedrally by Cr at a distance of 1.985 Å. The structure is therefore a defect superstructure derived from the NiAs type with two out of every three layers of octahedral holes empty (Figure 8-73). Cr has 3 C neighbors, 3 Al at 2.67 Å and 3 Cr at 2.755 Å. Al is surrounded by a distorted cubo-octahedron of 6 Al at 2.86 Å and 6 Cr at 2.67 Å.

The structure belongs to the class of transition metal compounds that are stabilized by the presence of small nonmetal atoms in octahedral interstices. In some compounds with this structure the octahedral holes are probably incompletely filled with the interstitial atom. More than 35 carbides and nitrides are known to have the structure. The major component is always a transition metal atom, and the other noninterstitial component is generally from Group III or IV although three sulfides, $CSTi_2$, $CSZr_2$, and C_2SFeHf_2 (*sic*) are also known to have the structure. These have axial ratios of about 3.5 compared to a value of 4.4 ± 0.1 for most of the other phases.

β-Be_3N_2 (H.T.) (*hP*10): β-Be_3N_2 has cell dimensions $a = 2.8413$, $c = 9.693$ Å, $Z = 2$.[119] The Be atoms are in a very squashed close packed stacking sequence BABCAC along [001] ($c/3a = 1.15$); N(1) atoms in A sites at $z = 0$ and $\frac{1}{2}$ lie in distorted octahedral holes between Be(2) layers (N(1)–6 Be(2) = 1.79 Å). Apart from the difference in the stacking sequence of the Be atoms, the structure resembles that of $AlCCr_2$ (*hP*8) with every third layer of distorted octahedral sites occupied, and two extra N(2) atoms introduced in the Be layers at $z = \frac{1}{4}$ and $\frac{3}{4}$. N(2) are thus surrounded by 3 Be(1) (1.64 Å) and 2 Be(2) (1.70 Å) in a trigonal bipyramid. Be(1) has 3 N(2) neighbors and Be(2) has 1 N at 1.70 Å, 3 at 1.79 Å and 3 Be(2) at 2.19 Å.

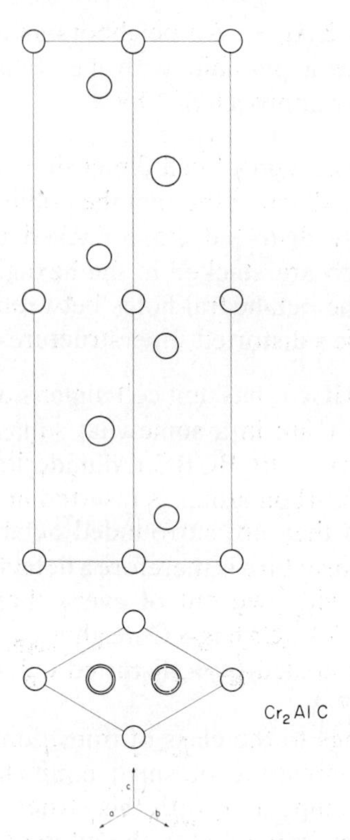

Figure 8-73. Structure of the *H* phase $AlCCr_2$ (*hP*8) projected down [001] and [110]: Al, largest circles; C, smallest circles.

Ti_3SiC_2 (*hP*12): Ti_3SiC_2 has cell dimensions $a = 3.068$, $c = 17.67$ Å, $Z = 2$.[120] The Ti and Si atoms form triangular close packed layers which are stacked in the close packed sequence, ABA̲BACA̲C (Si underlined), $c/4a$ having the value 1.44 Å. The carbon atoms occupy the centers of octahedral holes between pairs of Ti layers (Figure 8-74), two layers of octahedral holes thus being alternately filled or empty. The Ti–C distances are 2.135 Å and Ti–Si 2.696 Å. The carbon positions were determined by a difference-Fourier projection.

$CrCl_3$ (*hP*6): The $CrCl_3$ structure[121] ($a = 6.0$, $c = 17.3$ Å) is of potential interest for alloys. Cl forms a close packed array in cubic, ABC, stacking. Alternate layers of octahedral holes are completely empty and the other layers are $\frac{2}{3}$ filled by Cr atoms which form a 6^3 net (Figure 8-75).

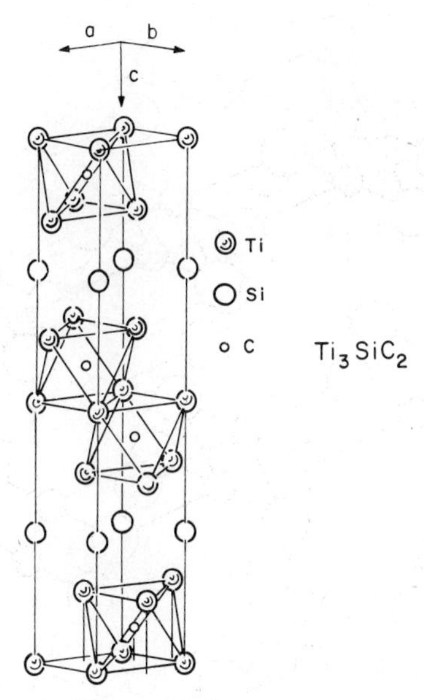

Figure 8-74. Pictorial view of the Ti_3SiC_2 (*hP*12) structure.

BiI_3 (*hR*8): The BiI_3 structure ($a = 7.50$, $c = 20.68$ Å, $Z = 6$ for the hexagonal cell)[122] is of potential interest for alloys since the I atoms form a close packed hexagonal array (axial ratio = 1.59), and the Bi atoms form 6^3 nets between alternate pairs of I layers filling $\frac{2}{3}$ of the octahedral holes between these layers, or $\frac{1}{3}$ of the octahedral holes in the whole structure (Figure 8-76). The structure is thus a three-layer type with distances between the I atoms of different three-layer I–Bi–I units equal to those of van der Waals bonds; it can be regarded as a defect CdI_2 type.

Defect Superstructures of NaCl or NiAs: Vacancies on All *M* Component Layers

ε-NFe_2 (*hP*9): ε-NFe_2 has cell dimensions $a = 4.79$, $c = 4.42$ Å, $Z = 3$ at the N-rich phase boundary. The N atoms occupy in ordered fashion octahedral holes in a close packed hexagonal array of Fe atoms. In one layer the N atoms form a triangular net and in the next graphite-like 6^3 net, such that occupied sites in one layer have vacant sites over them in the layers above and below. The structure has been determined by neutron diffraction for ε-Nb_2C[123] and for ε-W_2C.[124]

In the NNi_3 (*hP*8) structure which has a similar cell and arrangement of the Ni atoms, there is disagreement in the literature regarding which

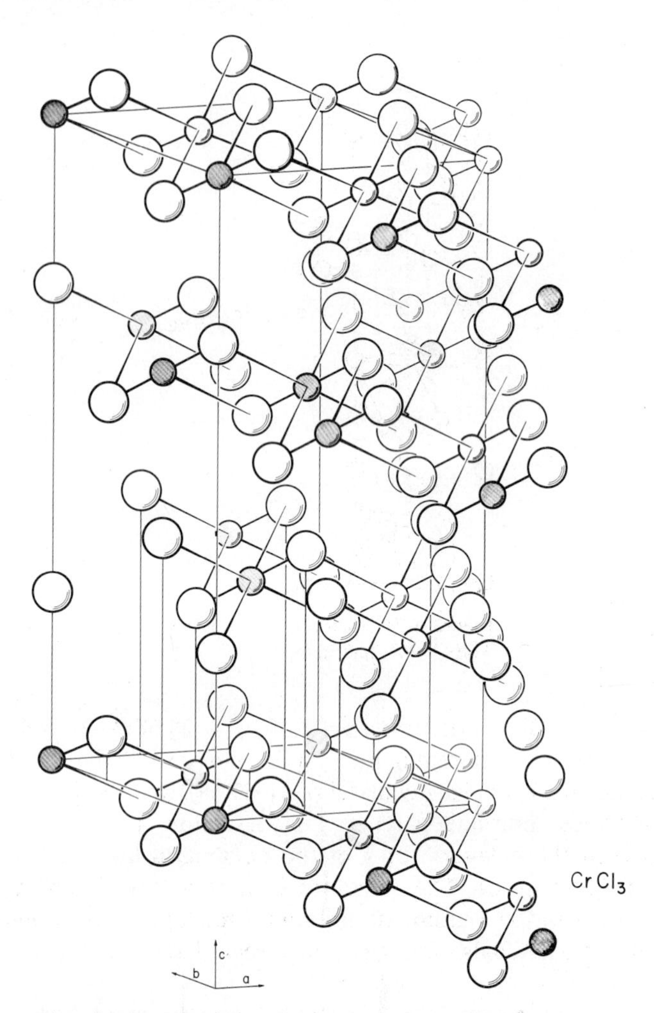

Figure 8-75. Pictorial view of the $CrCl_3$ (*hP*6) structure: Cl, large circles.

octahedral holes are indeed occupied.[125] In one description of the Ni_3N structure, one-third of the holes between *each* pair of metal atom planes are occupied by N, whereas in the other, the occupation only occurs between *alternate* planes of Ni atoms.

ζ-CNb$_2$ (*oP*12): ζ-CNb_2 has cell dimensions $a = 10.92$, $b = 3.09$, $c = 4.97$ Å, $Z = 4$.[123] The Nb atoms form a c.p. hexagonal array and the C atoms occupy octahedral holes in this array in ordered fashion. They are arranged in bands of triangles running in the *b* direction, such that

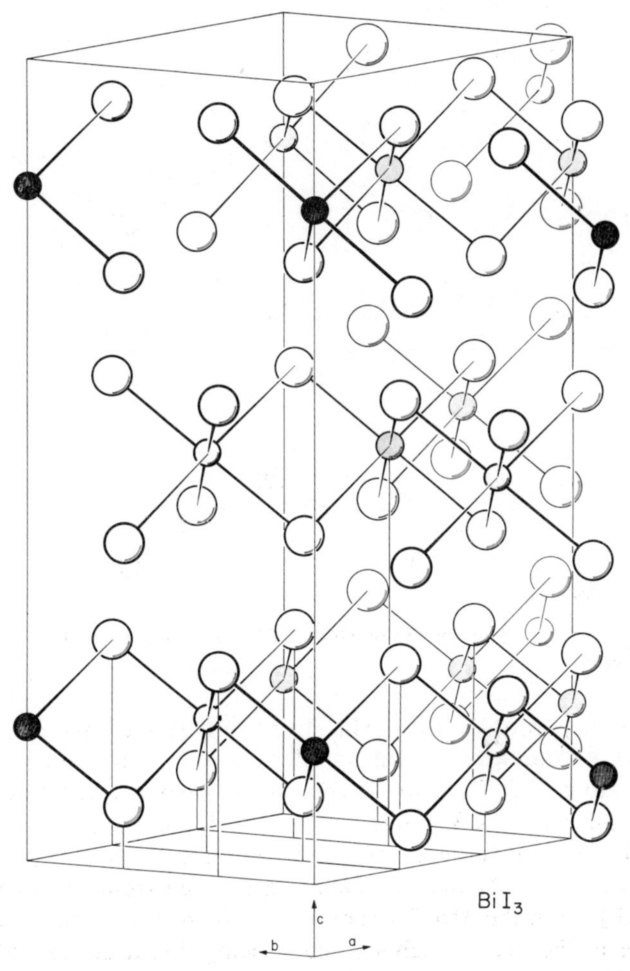

Figure 8-76. Pictorial view of the BiI_3 ($hR8$) structure: I, large circles.

bands at $z \sim 0$ lie over the vacant sites at $z \sim \frac{1}{2}$. The structure was examined by neutron diffraction.

ζ-NFe₂, α-CMo₂ ($oP12$): α-Mo_2C has cell dimensions $a = 4.724$, $b = 6.004$, $c = 5.199$ Å, $Z = 4$. The Mo atoms are in c.p. hexagonal array and the C atoms occupy octahedral holes in this array in ordered fashion, forming zigzag chains which run in the c direction (Figure 8-77). The chains in one layer lie over vacant sites in the layers above and below and they have a rectangular repeat pattern in (100) planes which is identical

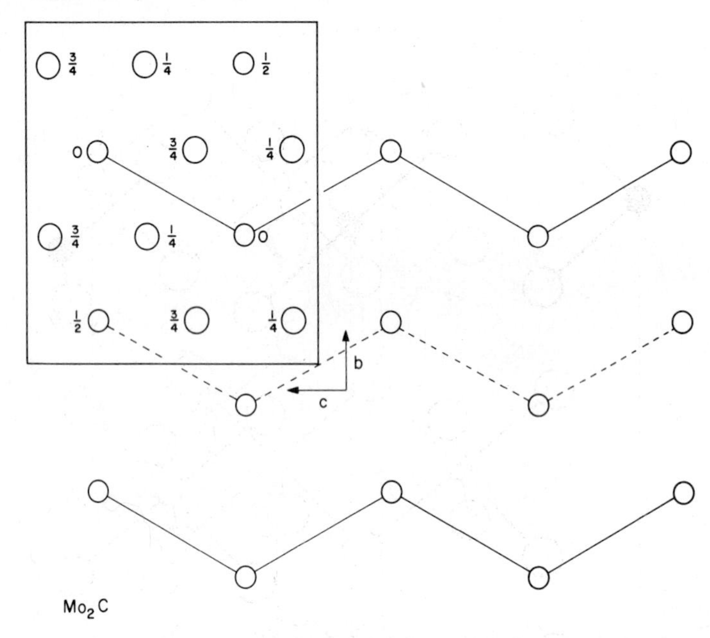

Figure 8-77. The structure of Mo_2C (*oP*12) projected down [100]: Mo, large circles.

with that in β-$MoN_{0.4}$ ($tI \sim 11$). The structures of α-Mo_2C[126] and ζ-V_2C[123] were examined by neutron diffraction. The structure can be compared with the MN_2 CdI_2 (*hP*3) type, on the one hand, where the N atoms are on exact close packed sites at $z = \frac{1}{4}$ and $\frac{3}{4}$, but the M atoms are all omitted from the plane with $z = \frac{1}{2}$, and with the Co_2N (*oP*6) type on the other hand. In Co_2N the M atoms are omitted from planes with both $y = 0$ and $\frac{1}{2}$, as in the Mo_2C structure ($x = 0$ and $\frac{1}{2}$), but the N atoms are no longer at the exact heights $\frac{1}{4}$ and $\frac{3}{4}$, being displaced up and down to $y = \pm 0.261$.

Each C atom in Mo_2C is surrounded octahedrally by 6 Mo at 2.10 Å and each Mo is surrounded by three carbon atoms which are almost coplanar with it. The Mo–Mo distance is 2.93Å. The atom positions were determined by neutron diffraction.

Co_2N (*oP*6): Co_2N has cell dimensions $a = 4.606$, $b = 4.344$, $c = 2.854$ Å, $Z = 2$.[127] The Co atoms are in close packed hexagonal arrangement, but half of the N atom sites, which occupy octahedral holes in this array, are vacant. Thus half of the nitrogen atoms in the basal plane and in the plane half way up the b axis are removed in ordered fashion leaving lines of N in the [001] direction, which have a rectangular repeat pattern (Figure

8-78). The structure can be compared with the marcasite type where the derivation from the NiAs type is the same (Figure 8-46, p. 412), but the atomic displacements are more severe. Co has 3 N neighbors at 1.88, and 2 at 1.94 Å which are almost coplanar with it, and 6 Co neighbors (2.63, 4 × 2.69, 2.78 Å). N is surrounded octahedrally by 6 Co (2 × 1.88, 4 × 1.94 Å). The structure type requires confirmation by a single-crystal study.

C_2Mo_3 (hP10): With cell dimensions $a = 3.01$, $c = 14.64$ Å, $Z = 2$,[128] the Mo atoms are arranged in a close packed stacking sequence 3$\underline{3}$ in Ždanov–Beck notation, the layer surroundings being hcc. Neutron diffraction studies have revealed that the carbon atoms are disordered in the octahedral holes of this array.[129]

β-$MoN_{0.4}$ (tI ~ 11): The β-$MoN_{0.4}$ structure with cell dimensions, $a = 4.200$, $c = 8.010$ Å,[130] has a slightly distorted f.c. cubic array of Mo atoms with N atoms inserted in certain of the octahedral interstices as shown in Figure 8-79, although even these are only partially occupied. The structure is thus a distorted defect NaCl type. The close packed (111) planes of the pseudocell are alternately occupied fully by Mo and partially by N. The partially occupied and vacant N sites run in zigzag chains in the close packed plane giving a rectangular repeat cell. A similar pattern is found in Mo_2C (*oP*12) (Figure 8-77). The structure appears to have been determined from X-ray powder photographs only.

NNi_4I (cP5): The Ni atoms form a f.c. cubic array and N is located at the body center of the cell,[131] the same positions being occupied as in the perovskite structure.

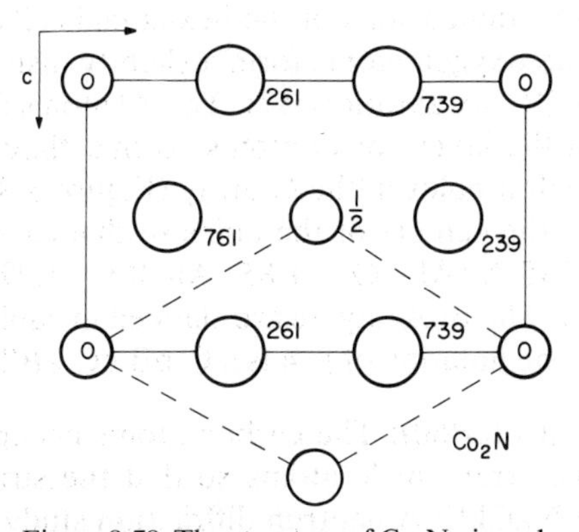

Figure 8-78. The structure of Co_2N viewed down [010], indicating a defective NiAs type pseudocell.

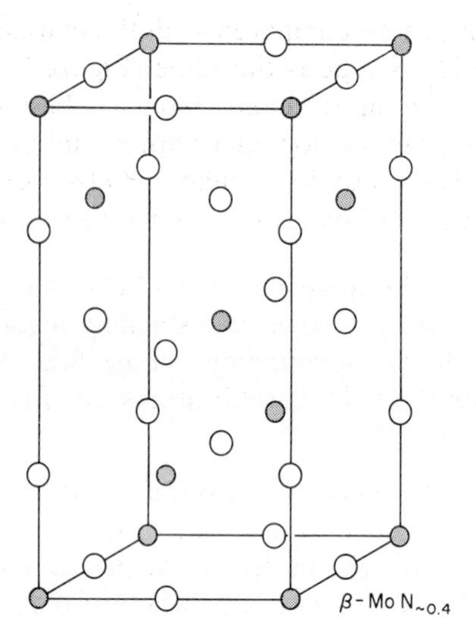

Figure 8-79. View of the β-$MoN_{0.4}$ ($tI \sim 11$) structure: Mo, open circles.

NNi₄ II: The tetragonal cell ($a = 3.72$, $c = 7.28$ Å, $Z = 2$) is made up of two f.c. cubic pseudocells of Ni atoms with N at 0, 0, $\frac{1}{4}$ and $\frac{1}{2}$, $\frac{1}{2}$, $\frac{3}{4}$, occupying $\frac{1}{4}$ of the octahedral holes.[131] The structure can therefore be regarded as a defect superstructure of NaCl. Both Ni_4N structures were determined by electron diffraction from polycrystalline foils. They require further confirmation.

α-Al_2O_3 ($hR10$): The dimensions of the hexagonal cell are $a = 4.759$, $c = 12.99$ Å, $Z = 6$. The oxygen atoms form a slightly distorted close packed hexagonal array, C′B′ (axial ratio 1.57). The Al atoms form very rumpled 6^3 arrays between the layers of O atoms so that they occupy $\frac{2}{3}$ of each layer of the octahedral holes in the O array (Figure 8-80); however, they are displaced from the centers of the holes so that each Al has one close Al neighbor at 2.735 Å (Al–3O = 1.85, Al–3O = 1.99 Å). Each oxygen has 4 Al neighbors. The 6^3 Al layers are stacked in cubic sequence, so the overall stacking in the notation of p. 4 is a C′bB′cC′aB′bC′cB′.[132]

V_6C_5 ($hP33$) and *V_8C_7* ($cP60$): The carbon atoms occupy ordered sites in close packed cubic arrays of V atoms so that the structures are defect superstructures of NaCl.[129] A neutron diffraction study has been made of V_8C_7.[133]

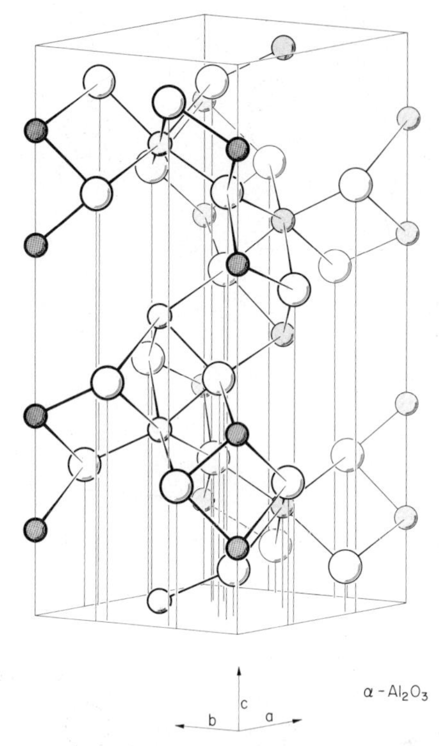

Figure 8-80. View of the α-Al_2O_3 ($hR10$) structure: oxygen, large circles.

α″-$Fe_{16}N_2$ ($tI18$): In the α″-$Fe_{16}N_2$ structure ($a = 5.720$, $c = 6.292$ Å, $Z = 1$)[134] the Fe atoms form a distorted f.c. cubic array with the N atoms located in octahedral interstices at alternate cell centers on proceeding along the [001] direction (Figure 8-81). Thus the structure is a distorted intermediate between the f.c. cubic ($cF4$) and NaCl ($cF8$) structures.

Derivatives of the NaCl Structure. Defects in Both *M* and *N* Layers

ReO_3 ($cP4$): In the ReO_3 structure the Re atoms occupy the cube corners and the O atoms the midpoints of the cell edges.[135] The structure is a defect NaCl type with the Re atoms at the face centers of the cell and the oxygen atom at the cell body center missing.

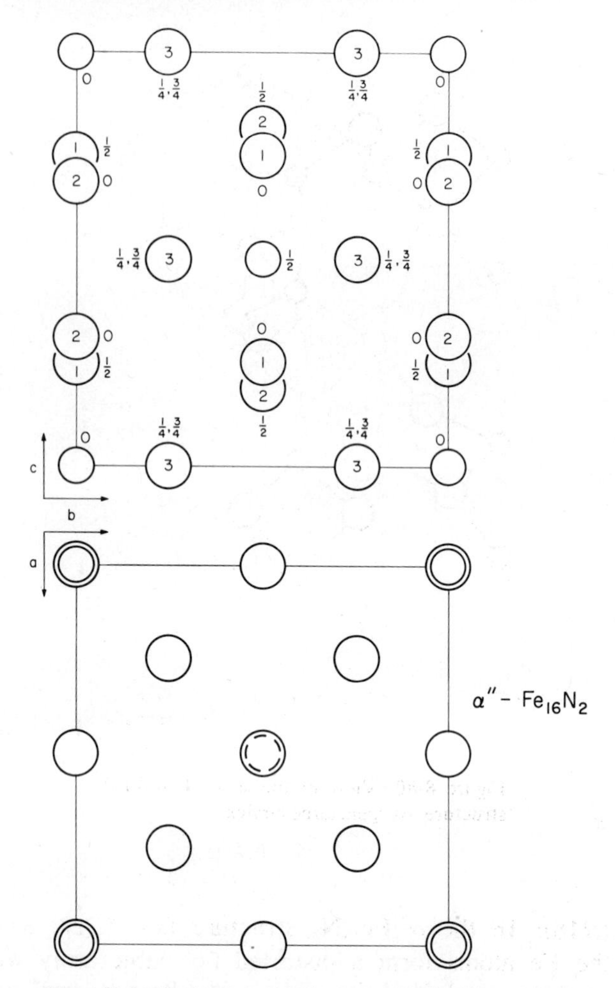

Figure 8-81. Projections of the α''-$Fe_{16}N_2$ ($tI18$) structure.

OCTAHEDRAL AND TRIGONAL PRISMATIC COORDINATION

MN: *M* on 3^6 Nets Stacked in Close Packing AB. *N* on 3^6 Nets at $\frac{1}{2}$ Spacing in Paired-Layer AA Stacking. Octahedra and Trigonal Prisms.

NiAs (*hP*4): The dimensions of the NiAs cell are $a = 3.619$, $c = 5.034$ Å, $Z = 2$. The As atoms form close packed layers in hexagonal stacking sequence BC and Ni atoms on sites A are located at the centers of all of the octahedral interstices in the As layers.[136] Ni thus has 6 As neighbors and As is surrounded by 6 Ni in a right trigonal prism (Figure 8-82). The As octahedra share faces normal to the *c* axis so that the Ni atoms are direct neighbors along the [001] direction. The structure is an important type because of many defect and filled up structures derived from it. When atoms are left out of the metal layers in ordered fashion, defect superstructures are obtained, and omitting all of the metal atoms in alternate layers, gives the CdI_2 type structure. On the other hand, atoms can be added to the structure in the same layers as the anions at the positions $\pm(\frac{2}{3}, \frac{1}{3}, \frac{1}{4})$ to give filled-up structures. Little is known at present about

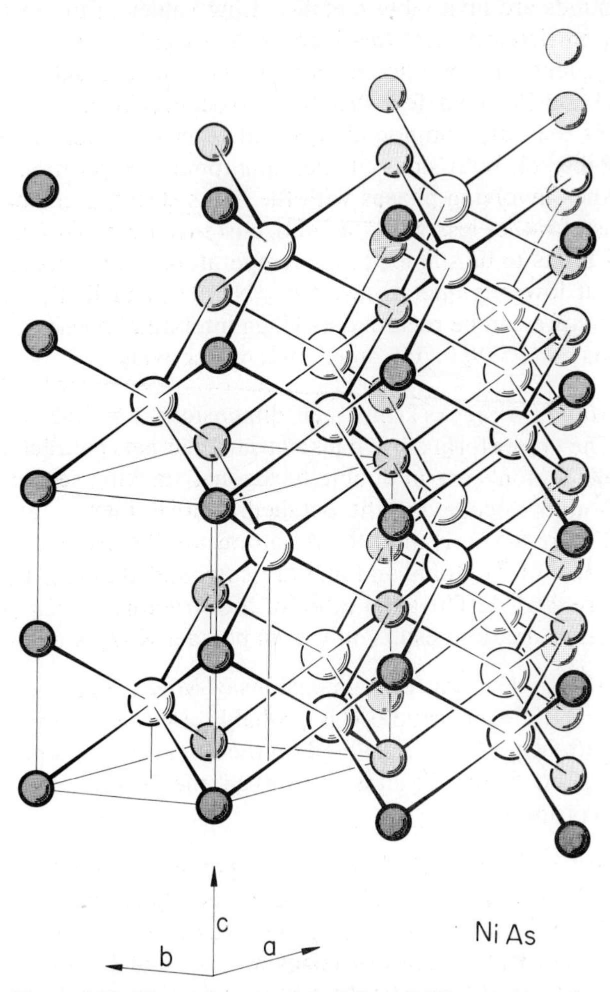

Figure 8-82. Pictorial view of the NiAs (*hP*4) structure: As, large circles.

possible partially filled-up superstructures, but when all of these positions are occupied, the Ni_2In structure type is obtained.

Some 55 binary phases are known to have the NiAs structure and, with the exception of AuSn, they all contain a transition metal atom. Axial ratios of phases with the NiAs structure vary from 1.93 in compounds such as TiS which, although metallic, have ionic tendencies (and are prevented from taking the NaCl structure because the ratio of cationic to anionic *ionic* radii is less than 0.414), through the ideal value of 1.63, where compounds may be semiconductors, to very small values (1.27) where the compounds are invariably metallic. Low values of the axial ratio are associated particularly with the filled-up NiAs phases. Generally speaking, the tendency is for filled-up structures to occur with anions from Groups III and IV and defect structures to occur with anions from Group VI. NiSb is the only compound reported to have a substantial range of solid solubility on both sides of the equiatomic composition. The more recent studies involving phases with the NiAs structure indicate that the structure is of rather less widespread occurrence than was first imagined, and that it tends to be stable at high temperatures and transform to other structures at lower temperatures. $Nb_{0.90}N$, PtB, and $RhB_{1.1}$ which have the antistructure can be regarded as Hägg interstitial phases although the radius ratios are 0.599, 0.707, and 0.792, respectively.

TiAs (γ'-*MoC*) (*hP*8): AsTi has cell dimensions $a = 3.65$, $c = 12.30$ Å, $Z = 4$.[137] The atoms form close packed triangular nets parallel to the (001) plane. The As atoms are in double hexagonal stacking sequence ACAB and the Ti atoms occupying the octahedral holes therein, are in BBCC stacking (Figure 8-83). Half of the As atoms are therefore surrounded by trigonal prisms of Ti atoms and half are surrounded octahedrally. As–Ti distances are 2.61 Å. The axial ratio for the structure is $c/2a = 1.72$. The structure determination results only from powder X-ray studies.

$MnTa_3N_4$ (*hP*8): The unit cell dimensions of $MnTa_3N_4$ are $a = 3.023$, $c = 10.49$ Å, $Z = 1$.[138] The structure is probably a ternary form of the AsTi structure with Mn + 1 Ta disordered on the As(1) sites, Ta occupying the As(2) sites, and N the Ti sites. However, the N sites still need to be determined properly.

W_2B_5 (*hP*14): The unit cell of the W_2B_5 phase has the dimensions $a = 2.984$, $c = 13.87$ Å, $Z = 2$.[139] The structure is a filled-up TiAs type, with W, B(1), and B(3) atoms occupying the sites of the TiAs structure. To these are added B(2) at the cell edges at $z = \frac{1}{4}$ and $\frac{3}{4}$, and B(4) atoms in B and C stacking positions slightly above and below the B(1) atoms on A sites (cell corners) at $z = 0$ and $\frac{1}{2}$. B(1) thus has 6 B(4) + 6 W neighbors

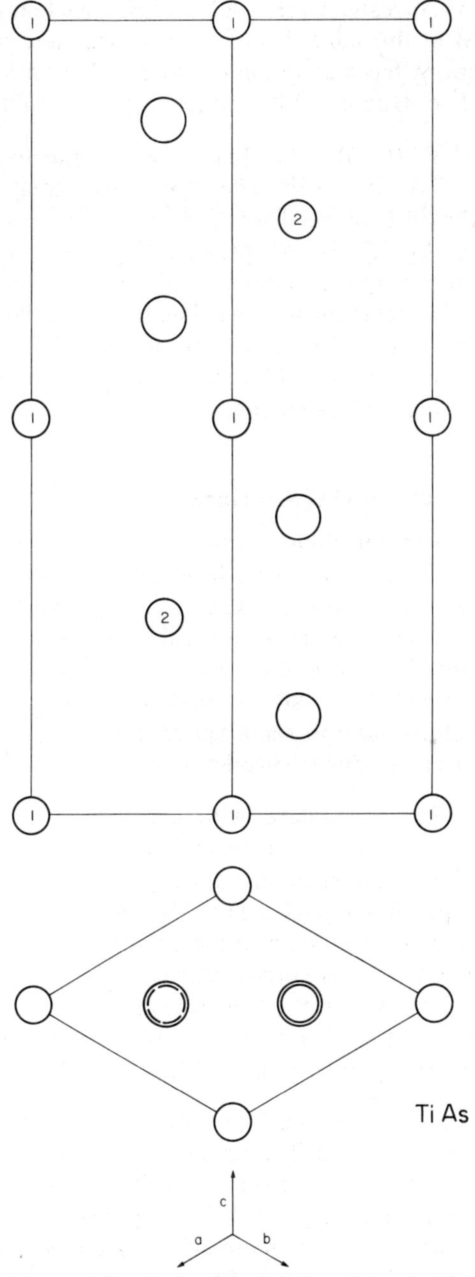

Figure 8-83. Projections of the TiAs (*hP*8) structure: As, large circles.

at 1.765 and 2.585 Å, respectively. B(2) and B(3) each have 6 W at 2.31 Å, and B(4) has 4 W at the same distance; they also have respectively 3, 3, and 6 B neighbors. W has a coordination shell of 20 comprised of 13 B and 7 W neighbors. The structure still requires to be accurately determined.

STi_{1-x} ($STi_{0.9}$) (H.T.) (*hR*6): The dimensions of the hexagonal cell are $a = 3.42$, $c = 26.5$ Å, $Z = 9$.[140] The atoms separately form triangular nets parallel to the (001) plane. These are stacked in a somewhat unevenly spaced sequence, $A\underline{B}C\underline{B}A\underline{B}C\underline{A}B\underline{A}C\underline{A}B\underline{C}A\underline{C}B\underline{C}$ (Ti layers underlined). Ti is always octahedrally surrounded by S, but S is either surrounded by a trigonal prism or an octahedron of Ti. The structure thus appears to be a mixture of the NaCl and NiAs types, but it requires confirmation from a good single-crystal study. The Ti deficiencies occur in the first and third layers of the pair-stacked Ti sequences, AAA, etc.[141]

Distortions of NiAs and TiAs Structures

CrS (*mC*8): CrS with cell dimensions $a = 3.826$, $b = 5.913$, $c = 6.089$ Å, $\beta = 101°36'$, $Z = 4$,[142] is a distorted form of the NiAs structure (Figure 8-84). The S atoms form close packed triangular layers parallel to the *ab* plane and Cr atoms occupy the octahedral holes between them, the octahedra being distorted because of the monoclinic angle. Cr has 2 S at 2.43, 2 at 2.47, and 2 at 2.88 Å. S is surrounded by 6 Cr in a distorted triangular prism. The structure has the same space group and sitesets as that of CuO, but the cell shapes differ considerably.

MnP (*oP*8): The MnP structure with cell dimensions $a = 5.258$, $b = 3.172$, $c = 5.918$ Å, $Z = 4$,[143] is a distortion of the NiAs type as is seen from the projection of the structure down [100], where it corresponds to the NiAs structure down [001]. The Mn atoms are displaced slightly from their ideal positions as indicated in Figure 8-85 which compares the MnP and NiAs cells. Mn is surrounded by a distorted octahedron of P atoms at 2.30 to 2.40 Å, and the Mn atoms form zigzag chains (2×2.695 Å) running in the [010] direction. Phosphorus is surrounded by 6 Mn in a distorted trigonal prism. Some 30 phases have the MnP structure.

Pt_6Si_5 (*mP*22): The atoms in the Pt_6Si_5 structure occupy 11 sitesets lying on planes at $y = \pm\frac{1}{4}$. The cell dimensions are $a = 15.46$, $b = 3.50$, $c = 6.17$ Å, $\beta = 86.6°$, $Z = 2$.[144] Chains of 6 Pt atoms at $y = \frac{1}{4}$ or $\frac{3}{4}$ have neighboring Si atoms in a very similar arrangement to that found in PtSi with the MnP (*oP*8) structure (Figure 8-86), as Gohle and Schubert point out. Pt(1) has 2 close Si at 2.35 Å, Pt(2) one at 2.37 Å and Pt(6) four at 2.22 to 2.27 Å. Other Pt–Si distances are generally in the range from 2.5

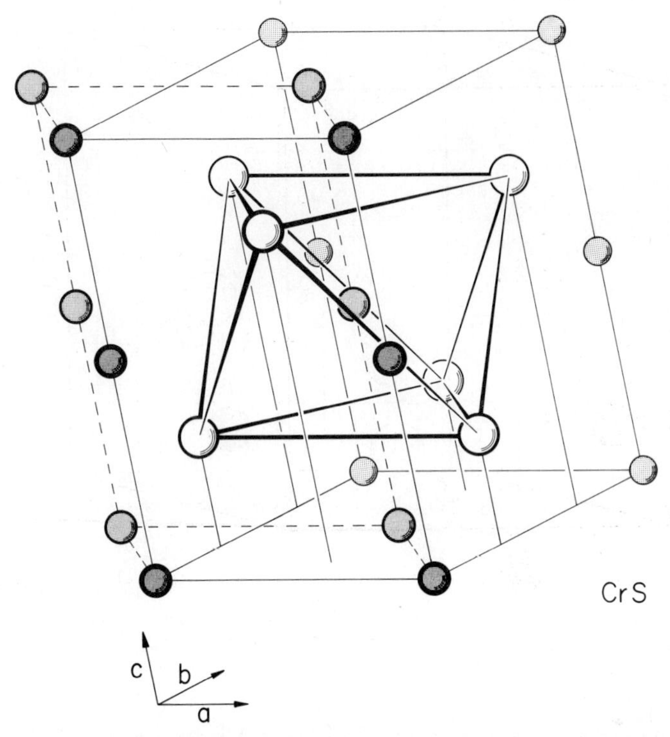

Figure 8-84. Pictorial view of the CrS (*mC*8) structure showing its relationship to a NiAs type pseudocell: S large circles.

to 2.7 Å. There are also some close Pt–Pt approaches of 2.72 and 2.81 Å. Si has either 6 or 7 Pt neighbors.

NiP (*oP*16): NiP has cell dimensions $a = 6.050$, $b = 4.881$, $c = 6.890$ Å, $Z = 8$.[145] The structure is a distortion of the NiAs type, differing from the MnP-type distortion as shown in Figure 8-87. A feature of the structure is P–P pairs at 2.43 Å; in addition Ni has two close Ni neighbors at 2.53 Å. The 6 Ni–P distances range from 2.23 to 2.34 Å with one at 2.95 Å.

α-FeS (L.T.) (*hP*24): α-FeS has unit cell dimensions $a = 5.968$, $c = 11.74$ Å, $Z = 12$.[146] In terms of the symbols given on p. 4, the arrangement of layers is a $A'\underline{B}'C'\underline{B}'A'\underline{B}'C'\underline{B}'$ (Fe layers underlined). This is a close packed stacking sequence which is identical with that of the NiAs structure and the α-FeS structure is a distortion superstructure of the NiAs type. Although the B′ and C′ layers in α-FeS are planar, the A′ sulfur layers are slightly nonplanar as indicated in Figure 8-88. Furthermore, although the Fe layers are planar, the atoms are slightly displaced from

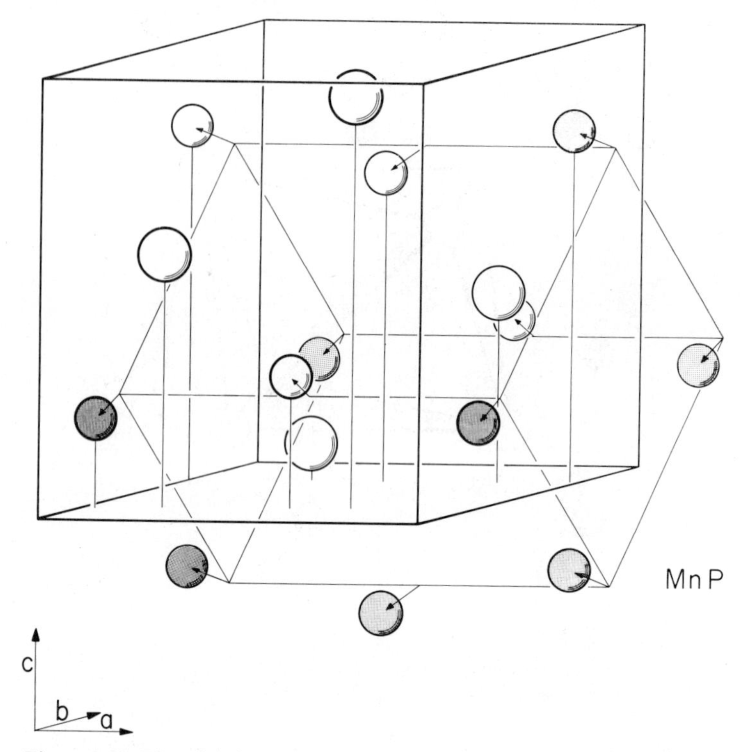

Figure 8-85. Pictorial view of the MnP (*oP*8) structure showing its relationship to the NiAs type structure.

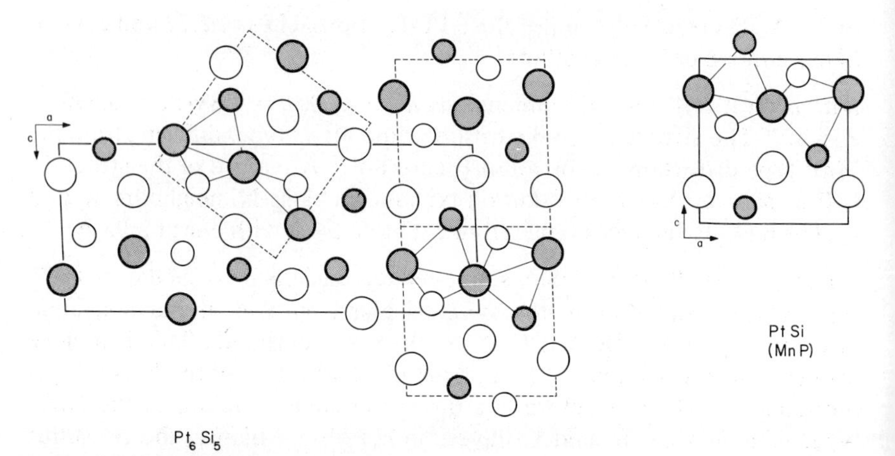

Figure 8-86. The structure of Pt_6Si_5 projected down [010], showing its relationship to the MnP type structure of PtSi.

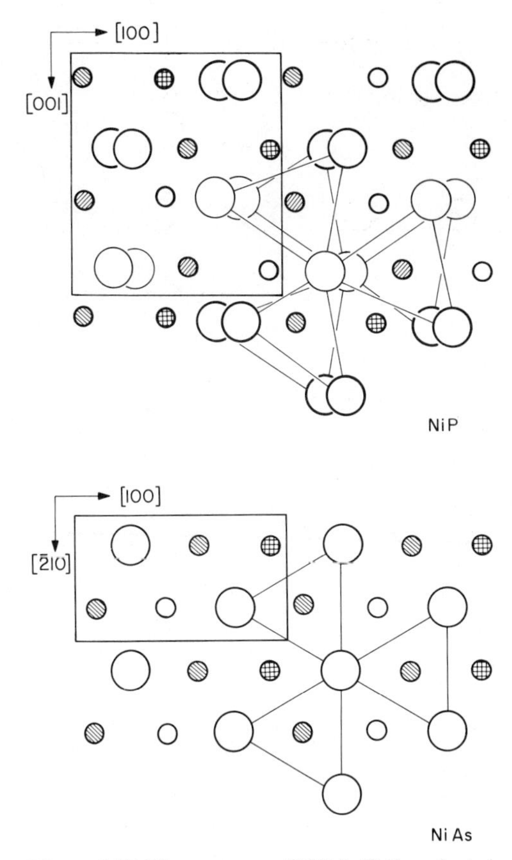

Figure 8-87. The structure of NiP ($oP16$) projected down [010], showing its relationship to the NiAs structure. (Larsson[145].)

the exact close packed sites within the layers. The distorted octahedra of S surrounding Fe have Fe–S distances varying from 2.35 to 2.64 Å. The closest Fe–Fe approach is 2.935 Å along c. S(1) has 6 Fe at 2.51 Å; S(2) has 6 Fe at 2.44 and 2.50 Å and S(3) has 6 Fe at 2.35, 3.45, and 2.64 Å.

β-Na_2S_2 (Li_2O_2) (hP8): β-Na_2S_2 has unit cell dimensions $a = 4.494$, $c = 10.228$ Å, $Z = 2$.[147] The atoms are arranged in layers on close packed stacking sites but the close packed layers are very much squashed together and extended in the basal plane ($c/2a = 1.14$). Otherwise the structure is the same as the AsTi type with Na(1) and (2) forming layers stacked ABAC and S stacked CCBB occupying distorted octahedral holes therein (S–Na 2.81 and 2.99 Å). Na(1) is surrounded by a distorted

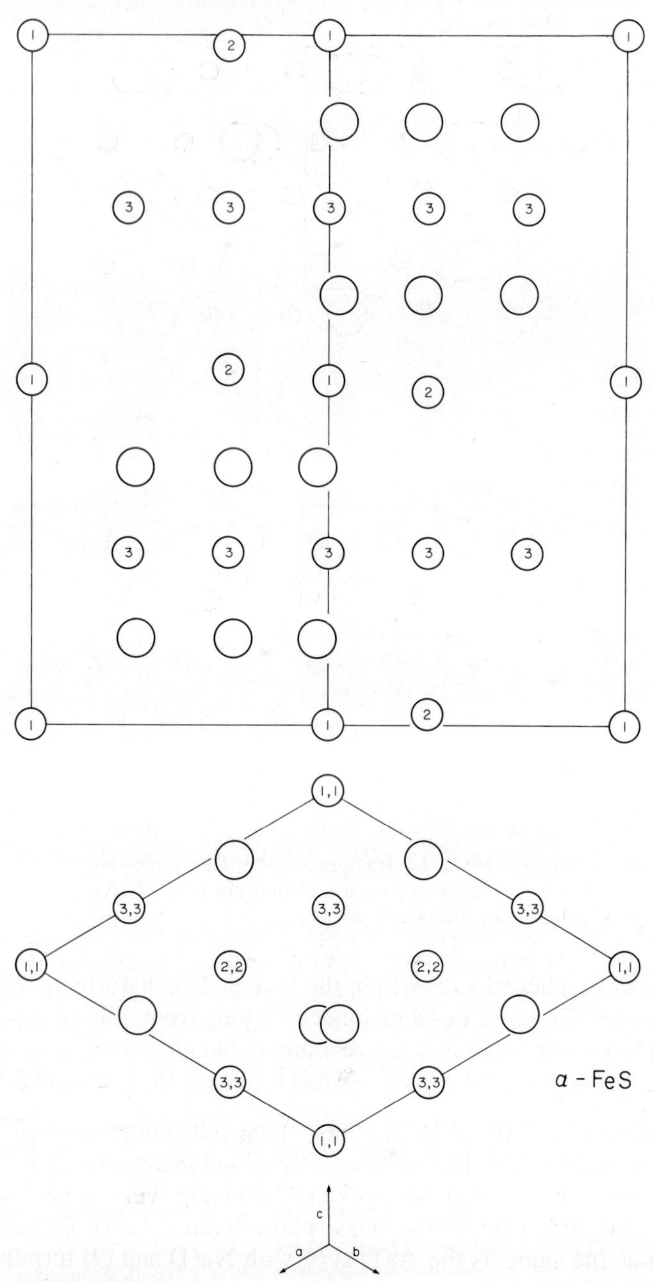

Figure 8-88. Projections of the structure of α-FeS down [110] and [001]: Fe, large circles.

octahedron of S (2.99 Å) and Na(2) is at the center of a trigonal prism of S (2.81 Å). Na–Na distances (6) are 3.64 Å.

Defect Structures Based on NiAs. Vacancies on Alternate Layers of the *N* Component

Cr_7S_8 (*hP*3.75): The unit cell has dimensions $a = 3.464$, $c = 5.763$ Å, $Z = \frac{1}{4}$.[142] The S atoms form layers in c.p. hexagonal stacking sequence, BC. Cr atoms are located in the octahedral holes at $z = 0$ and $\frac{1}{2}$. The deficiency of Cr is confined to the layer at $z = \frac{1}{2}$, although the distribution of Cr atoms and vacant sites is disordered therein.

Cr_7Se_8 (*mC*30): In setting $F2/m$, Cr_7Se_8 has cell dimensions $a = 12.67$, $b = 7.37$, $c = 11.98$ Å, $\beta = 90.95°$, $Z = 4$.[148] The structure is a slightly distorted defect superstructure of the NiAs type with omission of Cr atoms in alternate Cr layers on proceeding along the axis normal to the layers. There are 4 layers of Cr atoms in the cell along this direction (c) as indicated in Figure 8-89. The vacant Cr sites are identical to the occupied V sites in the defect layers of the V_5Se_8 structure and, allowing for change from occupied to unoccupied sites, the Cr_7Se_8 and V_5Se_8 structures are otherwise identical. The Se atoms have 5 or 6 Cr neighbors at

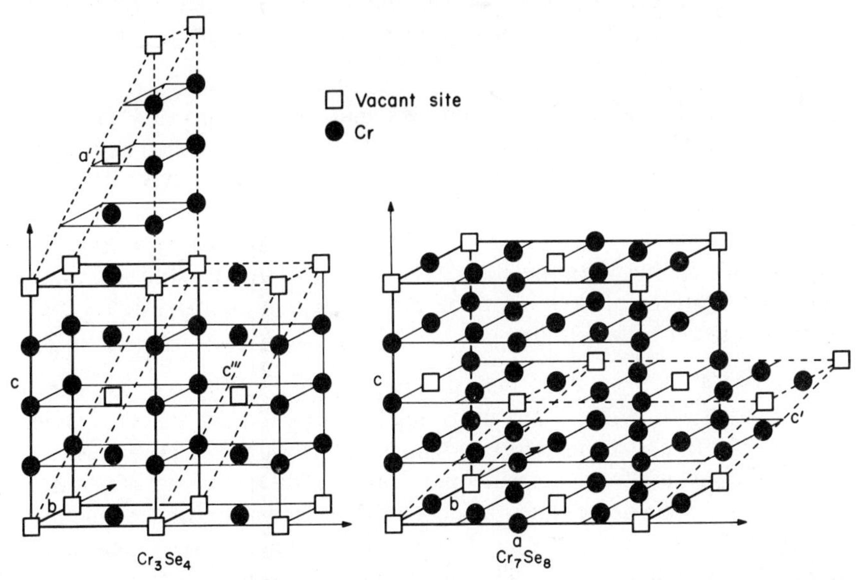

Figure 8-89. (Right) The arrangement of Cr atoms and vacant sites in the Cr_7Se_8 (*mC*30) structure in *F* setting (full lines) and in *C* setting (broken lines). Se atoms are omitted. (Left) The arrangement of the Cr atoms and vacant sites in the Cr_3Se_4 (*mC*14) structure in *I* setting. Broken lines show other settings of the cell. Se atoms are omitted.

2.50 to 2.82 Å and Cr are surrounded by octahedra of Se atoms. Closest Cr–Cr approaches are 2.995 Å and closest Se–Se distances are 3.445 Å. The structure requires confirmation by single-crystal methods.

Fe_7Se_8 (H.T.) (*hP*45): The high temperature Fe_7Se_8 structure is hexagonal, $a = 7.23$, $c = 17.65$ Å, $Z = 3$.[149] The ordered arrangement of vacant Fe positions in the orthohexagonal cell, $a = 12.53$, $b = 7.23$, $c = 17.65$ Å, $Z = 6$, is on sites 0,0,0; $\frac{1}{4}, \frac{1}{4}, \frac{1}{3}; \frac{1}{4}, \frac{3}{4}, \frac{2}{3} + (0,0,0; \frac{1}{2}, \frac{1}{2}, 0)$.

Fe_7Se_8 (L.T.) (*aP*120): The low temperature structure has a triclinic cell with $a = 12.53$, $b = 7.236$, $c = 23.54$ Å, $\alpha = 89.8°$, $\beta = 89.4°$ and $\gamma = 90.0°$, $Z = 8$.[149] The ordered vacant Fe sites are probably located at 0,0,0; $\frac{1}{4}, \frac{1}{4}, \frac{1}{4} + (0,0,0; \frac{1}{2}, \frac{1}{2}, 0)$.

Cr_5S_6 (*hP*22): Cr_5S_6 has unit cell dimensions $a = 5.982$, $c = 11.509$ Å, $Z = 2$.[142] The structure is identical to that of α-Cr_2S_3 except for the insertion of two more Cr atoms along the cell edges on the planes occupied by Cr(1) at $z = \pm\frac{1}{4}$. Thus sets of octahedral holes between the layers of S atoms are alternately completely filled and $\frac{2}{3}$ filled. The layers $\frac{2}{3}$ filled have Cr atoms on 6^3 nets stacked in two different positions so that the cell contains four layers of S atoms in the [001] direction. Cr atoms have 6 S neighbors disposed octahedrally and S have 5 Cr neighbors at 5 corners of a right-trigonal prism. The structure requires confirmation by single-crystal methods.

Fe_3S_4 (*hR*7): The dimensions of the hexagonal cell are $a = 3.47$, $c = 34.5$ Å, $Z = 3$.[150] All atoms form triangular nets parallel to (001). The S atoms are stacked in the close packed sequence CACABCBCABAB ($c/6a = 1.657$). The Fe atoms, arranged BBB□AAA□CCC□, occupy the octahedral holes in this array, every fourth layer of holes being vacant. The vacant holes lie between the pairs of S layers that are in cubic stacking so that the Fe atoms always surround S in a trigonal prism as in the NiAs structure, never octahedrally.

Cr_3S_4 (*mC*14): Cr_3S_4 with cell dimensions $a = 5.964$, $b = 3.428$, $c = 11.272$ Å, $\beta = 91.5°$, $Z = 2$,[142] in setting $I2/m$, is a defect distorted superstructure of the NiAs structure in which half of the Cr atoms are omitted in ordered fashion from alternate metal layers. Alternate lines of Cr atoms are omitted in the defect layers, and a different line is omitted in neighboring defect layers so that the cell has double the height of the NiAs subcell in the *c* direction (Figure 8-89). The layers of S atoms are close packed. Cr is surrounded octahedrally by 6 S (Cr(1); 4×2.47, 2×2.38 Å; Cr(2): 2.37–2.47 Å) and the Cr–Cr distance along *c* is 2.99 Å (Figure 8-90). S(1) has 5 Cr, and S(2) 4 Cr neighbors at the corners of a distorted defect

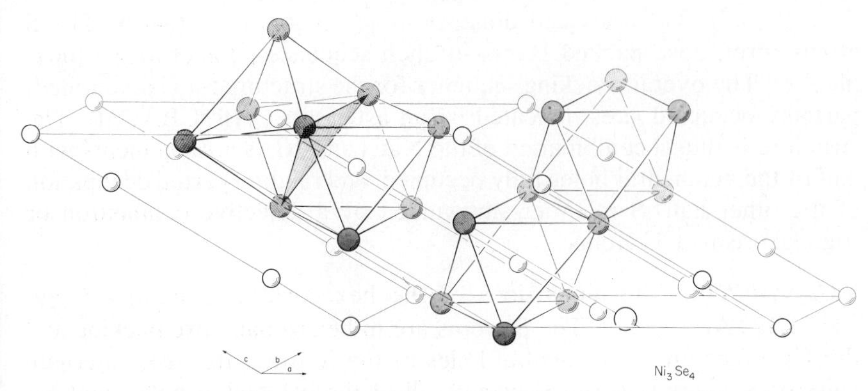

Figure 8-90. Pictorial view of the structure of Ni_3Se_4 (Cr_3S_4, *mC*14 type) in setting *C*2/*m*, showing connections between Se octahedra surrounding the Ni atoms: large circles, Se.

trigonal prism. The structure was determined from X-ray and neutron powder studies.

V_5Se_8 (mC26): In setting *F*2/*m*, V_5Se_8 has cell dimensions $a = 11.86$, $b = 6.96$, $c = 11.74$ Å, $\beta = 91°20'$, $Z = 4$.[151] The structure is a slightly distorted defect superstructure based on the NiAs type with vacant V atom sites in alternate V layers on proceeding up the axis normal to the layers. The close packed Se layers are in close packed stacking sequence AB. The V atoms in the defect layers form a large 3^6 net with atoms of one defect V net lying over the midpoint of a triangle edge of the V nets above and below, so that the *F* cell contains 4 layers of V atoms. The occupied V sites in the defect layers are identical with the unoccupied Cr sites in the defect layers of the Cr_7Se_8 structure (Figure 8-89, p. 461). The Se atoms have 3 or 4 close V atoms at 2.43 to 2.495 Å and V have 6 Se neighbors disposed octahedrally. Closest V–V distances are 2.935 Å and closest Se–Se approaches are 3.44 Å. The structure type requires confirmation by single-crystal studies.

Ti_5S_8 (hR6.4): The Ti_5S_8 structure with hexagonal cell dimensions $a = 3.418$, $c = 34.36$ A, $Z = \frac{3}{2}$, contains close packed layers of S parallel to the (001) plane with Ti occupying the octahedral holes therein.[152] The stacking sequence is, with Ti underlined and an asterisk indicating partially occupied Ti sites, $\underline{A}C\underline{A}^*B\underline{C}A\underline{B}^*C\underline{B}A\underline{B}^*C\underline{A}B\underline{C}^*A\underline{C}B\underline{C}^*A\underline{B}C$-$\underline{A}^*B$. The S atoms themselves are in hhcc arrangement. The structure thus combines the NaCl and NiAs arrangements, with half of the octahedral holes fully occupied and random partial occupation of the other half. Both the Ti_5S_8 and Ti_2S_3 structures require confirmation by single-crystal methods.

Ti_2S_3 (hP6.7): Ti_2S_3 has cell dimensions $a = 3.442$, $c = 11.44$ Å. The S atoms form close packed layers in chch sequence parallel to the (001) plane.[152] The overall stacking sequence for the structure is (Ti underlined; partially occupied sites indicated by an asterisk); $\underline{\mathrm{C}}\mathrm{A}\underline{\mathrm{B}}^*\mathrm{C}\underline{\mathrm{B}}\mathrm{A}\underline{\mathrm{C}}^*\mathrm{B}$. The structure is thus a combination of the NaCl and NiAs arrangements with half of the octahedral holes fully occupied, and random partial occupation of the other half. S is either surrounded by a defective octahedron or trigonal prism of Ti atoms.

α-Cr_2S_3 (hP20): The dimensions of the hexagonal cell are $a = 5.939$, $c = 11.192$ Å, $Z = 4$.[142] The S atoms are in hexagonal close packing and the Cr atoms are in octahedral holes of the S array. Between alternate pairs of S layers Cr(2) + (3) occupy all of the octahedral holes, and between the other pairs of S layers Cr(1) occupies one-third of the octahedral holes. There are 4 layers of S atoms per unit cell, two Cr(2) + (3) layers and two Cr(1) layers, the Cr(1) atoms being in BC stacking (Figure 8-91). The structure is thus an ordered defect superstructure of the NiAs type. Cr(1), (2) and (3) each have 6 S at 2.42 Å. Each S has 4 Cr neighbors at the corners of a defective trigonal prism. Both structures reported for Cr_2S_3 require to be confirmed by single-crystal methods.

β-Cr_2S_3 (hR10): β-Cr_2S_3 has hexagonal cell dimensions $a = 5.937$, $c = 16.698$ Å, $Z = 6$.[142] The S atoms form a c.p. hexagonal array stacked C′*B*′ (axial ratio 1.62). Layers of Cr atoms between the S atoms occupy alternately all and $\frac{1}{3}$ of the octahedral holes in the S array. The overall stacking sequence in terms of the symbols on p. 4 is (Cr underlined) $\underline{\mathrm{A}}'\mathrm{C}'\underline{\mathrm{B}}\mathrm{B}'$-$\underline{\mathrm{A}}'\mathrm{C}'\underline{\mathrm{A}}\mathrm{B}'\underline{\mathrm{A}}'\mathrm{C}'\underline{\mathrm{C}}\mathrm{B}'$, and the Cr atoms are distributed in cubic sequence in the layers of octahedral holes that are one-third filled. The S atoms are surrounded by a defective trigonal prism of Cr which has two corner sites vacant.

Sc_2Te_3 (hR6.7): The dimensions of the hexagonal cell are $a = 4.109$, $c = 40.59$ Å, $Z = 4$.[153] The Te atoms form triangular close packed layers (12 in the hexagonal cell) with the Sc atoms in octahedral holes thereof. The overall stacking sequence is (Sc underlined; asterisk indicates partially occupied sites) $\underline{\mathrm{A}}\mathrm{C}\underline{\mathrm{B}}^*\mathrm{A}\underline{\mathrm{B}}\mathrm{C}\underline{\mathrm{B}}^*\mathrm{A}\underline{\mathrm{C}}\mathrm{B}\underline{\mathrm{A}}^*\mathrm{C}\underline{\mathrm{A}}\mathrm{B}\underline{\mathrm{A}}^*\mathrm{C}\underline{\mathrm{B}}\mathrm{A}\underline{\mathrm{C}}^*\mathrm{B}\underline{\mathrm{C}}\mathrm{A}\underline{\mathrm{C}}^*\mathrm{B}$, so that Sc surrounds Te either in defective octahedra or in trigonal prisms and the structure is a mixture of the rocksalt and NiAs types. The structure determination is only approximate and requires confirmation.

Nb_2Se_3, Mo_2S_3 (mP10): Nb_2Se_3 has cell dimensions $a = 6.503$, $b = 3.434$, $c = 9.215$ Å, $\beta = 103.39°$, $Z = 2$.[154] The Se atoms form a distorted close packed array with chh layer surroundings. Octahedral holes between the adjacent h packed layers are filled by Nb(1) and half of the octahedral

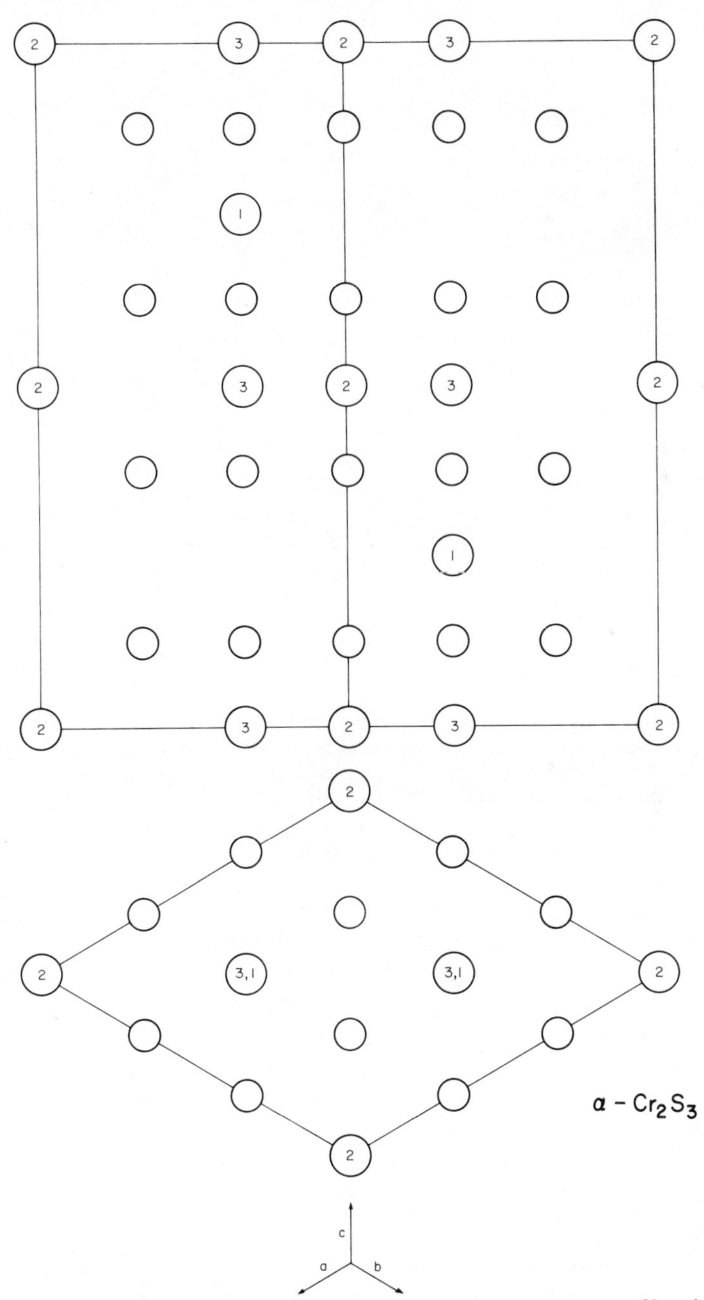

Figure 8-91. Projections of the α-Cr_2S_3 ($hP20$) structure down [110] and [001]: Cr, large circles.

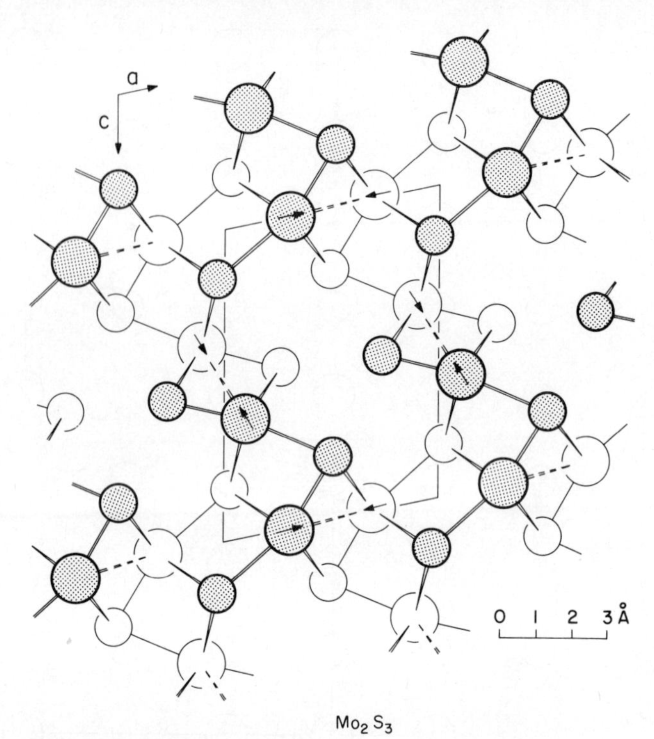

Figure 8-92. Atomic arrangement in the Mo_2S_3 ($mP10$) structure viewed down [010].[154] Mo large circles. Atoms represented by light circles are at a height $y = \frac{1}{4}$ and those by heavy circles are at $y = \frac{3}{4}$. Arrows on the metal atoms indicate movements of the metal atoms toward each other in the Nb_2Se_3 structure.

holes between the h and c packed layers are filled by Nb(2) atoms. The distortion in the structure brings the Nb atoms together in zigzag chains running the *b* direction; in those formed by Nb(1) the Nb–2Nb distances are short, 2.97 Å, whereas in those formed by Nb(2) they are longer, 3.13 Å (Figure 8-92). In the isotypic Mo_2S_3 structure the Mo–Mo distances in both Mo(1) and Mo(2) chains are more nearly equal and comparable to the Mo–Mo distances in the metal (Figure 8-92). Nb–Se octahedral distances lie between 2.50 and 2.89 Å.

Mixed Paired-Layer and Close Packed Stacking. Octahedra and Trigonal Prisms

$Nb_{1+x}Se_2$ ($x = 1.05$ to 1.10) ($hP6.2$): $Nb_{1+x}Se_2$ has cell dimensions $a = 3.454$, $c = 12.58$ Å, $Z = 2$.[155] The atoms form evenly spaced triangular nets parallel to the (001) plane, with Se in close packed and paired-layer

stacking and Nb occupying fully the centers of the trigonal prismatic holes and partially the centers of the octahedral holes in the Se array ($c/2a = 1.82$). The stacking sequence is (Nb underlined; asterisk denotes partially occupied layers); $\underline{A}C\underline{A}^*B\underline{C}B\underline{A}^*C$. $Ta_{1.10}Se_2$ has the same structure, and also $Nb_{1.4}S_2$ (H.T.) (*hP*7) and $Mo_{\sim 0.84}N$ (*hP* ~ 7.4), where the degree of filling of the partially occupied layers is higher. The $Nb_{1+x}Se_2$ structure described is probably only a pseudocell; at certain compositions further ordering occurs, and the structure requires determination by single-crystal methods.

$Nb_{\sim 1.2}S_2$ (*hR*3.2): The dimensions of the corresponding hexagonal cell of $Nb_{\sim 1.2}S_2$ are $a = 3.33$, $c = 17.80$ Å.[156] The S atoms are in close packed paired-layer stacking sequence BAACCB at approximately close packed spacing ($c/3a = 1.78$). The Nb atoms on filled sites occupy trigonal prismatic holes between the paired S layers and the partially filled Nb sites occupy octahedral holes between the pairs of S layers stacked on unlike sites. This and the following two structure types require confirmation by single-crystal studies.

$Ta_{1+y}S_2$ (*hR*5+): In $Ta_{1+y}S_2$ ($a = 3.315$, $c = 36.2$ Å, $Z = 6$)[157] the S atoms are in close packed and paired-layer stacking sequence CCAABBCCAABB at spacings somewhat extended from close packing ($c/6a = 1.91$). The Ta atoms occupy trigonal prismatic holes between the pairs of S layers stacked on the same sites (sequence ACCBBA). The excess Ta atoms enter octahedral holes between the other pairs of S layers.

TaS_2 (*hR*6): The dimensions of the corresponding hexagonal cell are $a = 3.335$, $c = 35.85$ Å, $Z = 6$.[157] All atoms form triangular nets parallel to the (001) plane. The S atoms are stacked in mixed close packing and paired-layer sequence ($c/6a = 1.79$). The Ta atom layers are spaced halfway between alternate pairs of S layers so that they occupy octahedral or trigonal prismatic holes in the array of S atoms. The overall stacking sequence is (Ta underlined) $B\underline{A}BA\underline{C}BA\underline{C}AC\underline{B}AC\underline{B}CB\underline{A}C$.

$TaSe_2$ (*hP*12): The unit cell dimensions are $a = 3.4575$, $c = 25.143$ Å, $Z = 4$. The atoms form close packed triangular nets parallel to the (001) plane. The Se layers are stacked in the paired-layer and hexagonal stacking sequence BCCBCBBC, and the Ta layers in A sites are inserted halfway between alternate pairs of Se layers so as to center the trigonal prismatic holes and half of the octahedral holes, the overall stacking sequence being $\underline{A}BC\underline{A}CB\underline{A}CB\underline{A}BC$ (Ta underlined). Thus the structure is made up of CdI_2 and MoS_2 type three-layer Se, Ta, Se units which are held together by van der Waals bonds.[158]

$Nb_{1-x}S_2$ (L.T.) (*hP*8): The unit cell has dimensions $a = 3.32$, $c = 12.92$ Å, $Z = 4$.[159] All atoms form close packed triangular nets parallel to (001), S being stacked in the sequence CCBB and Nb always in position A, so that it is surrounded either octahedrally or trigonal-prismatically by S. S is always surrounded by a trigonal prism of Nb.

Defect Structures Based on NiAs With Vacancies Occurring in All Layers of Octahedral Holes

β-V_2C (*hP*3): β-V_2C has cell dimensions $a = 2.88$, $c = 4.57$ Å, $Z = 1$ at 31 at. % C. In this structure which has been determined by neutron diffraction experiments,[124,160] the V atoms form a c.p. hexagonal array and the carbon atoms are randomly distributed in the octahedral holes.

Mo_2C (*oP*12) and Co_2N (*oP*6) structures already described on p. 447 can also be regarded as defect NiAs type structures.

ζ-V_4C_3 (*hR* ~ 20): The rhombohedral cell has dimensions $a = 9.428$ Å, $\alpha = 17°48'$ (hexagonal cell $a = 2.917$, $c = 27.83$ Å).[161] The vanadium atoms in close packed layers are arranged in a twelve-fold stacking sequence, $3\underline{1}$ in Ždanov–Beck notation, the layer surroundings being $(hhcc)_3$. About 8 carbon atoms occupy octahedral interstices between these layers, but it was not possible in the single crystal study carried out, to determine any ordering of the C atoms and vacant sites. With ordered vacancies the structure could be isotypic with Sn_4P_3 (p. 434).

Various Octahedra and Prisms

Cr_2VC_2, $Cr_3C_{1.6}N_{0.4}$, ~ Cr_7BC_4 (*oC*20): Cr_2VC_2 has cell dimensions $a = 2.87$, $b = 9.30$, $c = 6.99$ Å, $Z = 4$.[162] The atoms are arranged in layers parallel to the *bc* plane at heights $x = 0$ and $\frac{1}{2}$. The metal atoms form columns of isolated right triangular prisms along [100] which are centered by C(2) atoms (Figure 8-93). The corner atoms of the triangular prisms also form octahedra which are centered by the non-metal atoms (Figure 8-93). The octahedra form columns in the [100] direction by sharing edges in the (100) plane. These columns join at corners in the [001] direction and they are also joined to the columns of triangular prisms by sharing prism edges and corners. There are spaces in the structure between the filled octahedra and prisms.

Cr_7BC_4 has cell dimensions $a = 2.870$, $b = 9.260$, $c = 6.982$ Å, $Z = 4$.[163] The structure can be regarded as a filled up Re_3B type with B + C in the centers of the triangular prisms and C in octahedral holes. Although

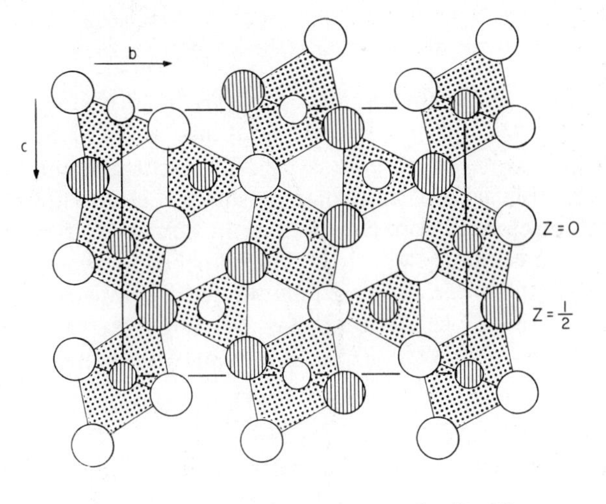

Figure 8-93. The Cr_2VC_2, $Cr_3(C,N)_2$, or Cr_7BC_4 type structure (*oC*20), projection down [100] showing triangular prisms and octahedra.

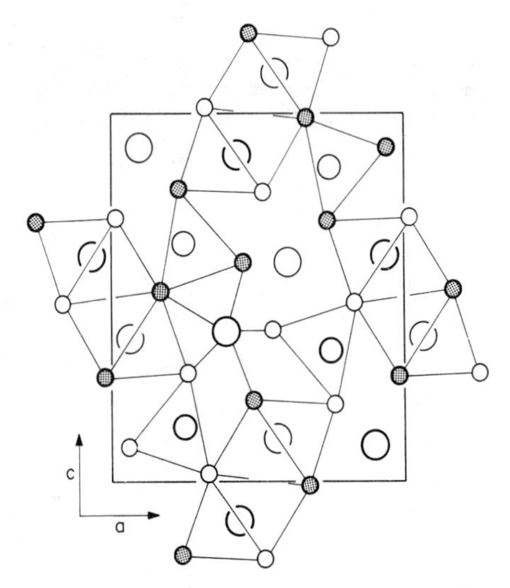

Figure 8-94. The structure of galenobismuthite, $PbBi_2S_4$ (*oP*28), projected down [010]: Bi atoms are the largest circles; S atoms are the smallest circles.

several of the structures were determined by single-crystal methods, the light atom positions require confirmation by neutron diffraction.

$PbBi_2S_4$ (oP28): Galenobismuthite ($a = 11.79$, $b = 4.10$, $c = 14.59$ Å, $Z = 4$)[164] contains lines of Bi and S atoms at $y = \frac{1}{4}$ and $\frac{3}{4}$, between which the Pb atoms lie (Figure 8-94). The arrangement is such as to surround Pb with a right triangular prism of S atoms (2.98 to 3.21 Å). Pb also has one close S atom out through the center of a rectangular prism face (2.85 Å), and another S atom much farther away (3.76 Å) through another rectangular face. Bi(1) is surrounded by a distorted octahedron of S atoms (2.63 to 3.12 Å). Bi(2) has 7 S neighbors at 2.78 to 3.10 Å. The structure of berthierite $FeSb_2S_4$ is somewhat similar, although not identical.

$Co_3W_9C_4$ (hP32): $Co_3W_9C_4$ has cell dimensions $a = 7.826$, $c = 7.826$ Å, $Z = 2$.[165] The W atoms form equilateral triangles about $\frac{1}{3}, \frac{2}{3}$ at $z = 0.075$ and

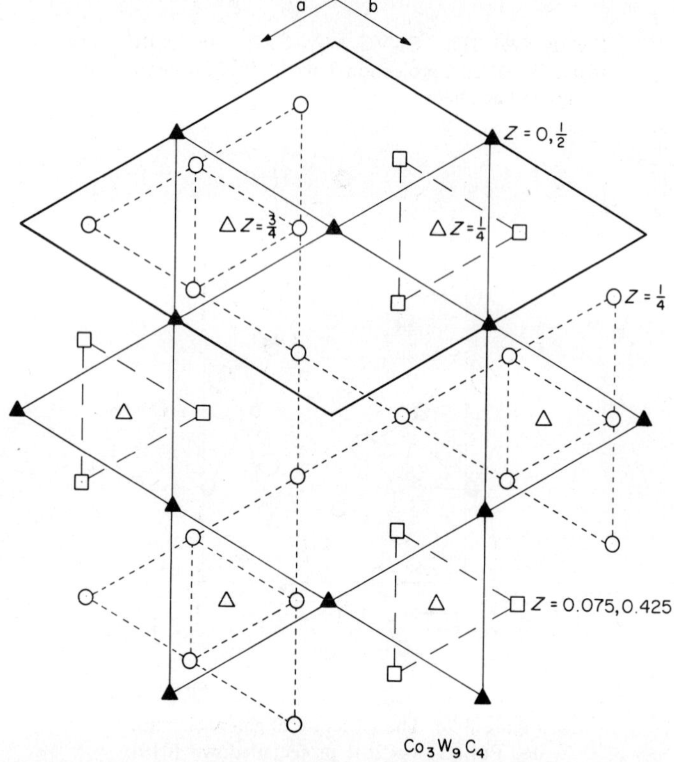

Figure 8-95. Atomic arrangement in the $Co_3W_9C_4$ (*hP*32) structure viewed down [001]: W, squares; (Co + W), circles; C, triangles.

0.425, and about $\frac{2}{3},\frac{1}{3}$ at $z=0.575$ and 0.925, giving trigonal prisms which are centered by C(1) at $z=\pm\frac{1}{4}$. Also lying at $z=\pm\frac{1}{4}$ is a partially filled 3^6 net of Co+W atoms that has voids through which the sides of the W trigonal prisms pass as indicated in Figure 8-95. The arrangement of these atoms is the same as that of the S nets in Th_7S_{12} (p. 539) or of Ho in Co_3Ho_4 (p. 543); however, the filling of the voids in the nets differs. Kagomé nets of C(2) atoms at $z=0$ and $\frac{1}{2}$ separate the W prisms, C(2) being surrounded octahedrally by 4 W and 2(Co+W)(2) atoms. The structure requires confirmation by neutron diffraction and single-crystal studies.

REFERENCES

1. P. T. B. Schaffer, 1969, *Acta Cryst.*, **B25**, 477.
2. F. P. Bundy and J. S. Kasper, 1967, *J. Chem. Phys.*, **46**, 3437.
3. D. C. Craig and N. C. Stephenson, 1965, *Acta Cryst.*, **19**, 543.
4. J. W. Boon, 1944, *Rec. Trav. Chim. Pays-Bas*, **63**, 69; L. Pauling and L. O. Brockway, 1932, *Z. Kristallogr.*, **A82**, 188.
5. H. Hahn, W. Klingen, P. Ness, and H. Schulze, 1966, *Naturwissenschaften*, **53**, 18.
6. F. Marumo and W. Nowacki, 1967, *Z. Kristallogr.*, **124**, 1.
7. *Strukturbericht*, **3**, 96.
8. Z. G. Pinsker and S. V. Kaverin, 1956, *Kristallografija*, **1**, 66.
9. H. Hahn, G. Frank, W. Klingler, A. D. Störger, and D. Störger, 1955, *Z. anorg. Chem.*, **279**, 241.
10. S. Kawano and I. Ueda, 1967, *Mem. Fac. Sci., Kyusyu Univ.*, Ser B, **3**, 127.
11. P. Eckerlin, 1967, *Z. anorg. Chem.*, **353**, 225.
12. W. Hofmann, 1933, *Z. Kristallogr.*, **A84**, 177.
13. *Strukturbericht*, **3**, 96.
14. S. A. Semiletov, 1960, *Kristallografija*, **5**, 704 [*Soviet Physics–Crystallography*, **5**, 673]; 1961. *Dokl. Akad. Nauk*, **137**, 584.
15. J. Goodyear and G. A. Steigmann, 1963, *Acta Cryst.*, **16**, 946.
16. G. A. Steigmann, 1967, *Acta Cryst.*, **23**, 12.
17. D. Thomas and G. Tridot, 1967, *C. R. Acad. Sci., Paris*, **264C**, 1385.
18. J. Witte and H. G. Schnering, 1964, *Z. anorg. Chem.*, **327**, 260.
19. E. Busmann, 1961, *Z. anorg. Chem.*, **313**, 90.
20. R. E. Marsh and D. P. Shoemaker, 1953, *Acta Cryst.*, **6**, 197.
21. D. A. Hansen and J. F. Smith, 1967, *Acta Cryst.*, **22**, 836.
22. W. H. Zachariasen, 1936, *J. Chem. Phys.*, **4**, 618.
23. C. T. Prewitt and H. S. Young, 1965, *Science*, **149**, 535.
24. J. G. White, 1965, *Acta Cryst.*, **18**, 217.
25. I. J. Hegyi, E. E. Loebner, E. W. Poor, and J. G. White, 1963, *J. Phys. Chem. Solids*, **24**, 333.
26. J. W. Boom and C. H. MacGillavry, 1942, *Rec. Trav. Chim. Pays-Bas*, **61**, 910.
27. F. P. Bundy and J. S. Kasper, 1963, *Science*, **193**, 340; J. S. Kasper and S. M. Richards, 1964, *Acta Cryst.*, **17**, 752.
28. K. Schubert, E. Dörre, and E. Günzel, 1954, *Naturwiss.*, **41**, 448.
29. N. Alsén, 1931, *Geol. Fören. Stockholm Förh.*, **53**, 111; I. Oftedal, 1932, *Z. Kristallogr.*, **A83**, 9.

30. J. H. Bryden, 1962, *Acta Cryst.*, **15**, 167.
31. D. Hardie and K. H. Jack, 1957, *Nature, Lond.*, **180**, 332.
32. S. N. Ruddlesden and P. Popper, 1958, *Acta Cryst.*, **11**, 465.
33. R. E. Rundle, C. G. Shull, and E. O. Wollan, 1952, *Acta Cryst.*, **5**, 22.
34. H. Nowotny and W. Sibert, 1941, *Z. Metallk.*, **33**, 391.
35. R. Juza and F. Hund, 1948, *Z. anorg. Chem.*, **257**, 13.
36. R. Juza, E. Anschütz, and H. Puff, 1959, *Angew. Chem.*, **71**, 161.
37. G. G. Dvorjankina and Z. G. Pinsker, 1963, *Kristallografija*, **8**, 556 [*Soviet Physics—Crystallography*, **8**, 448].
38. K. Anderko and K. Schubert, 1954, *Z. Metallk.*, **45**, 371, also R. V. Beranova and Z. G. Pinsker, 1964, *Kristallografija*, **9**, 104 [*Soviet Physics—Crystallography*, **9**, 83].
39. *Strukturbericht*, **3**, 37.
40. S. S. Sidhu, N. S. Satya Murthy, F. P. Campos, and D. D. Zauberis, 1963, "Nonstoichiometric Compounds," *Adv. Chem. Ser.*, **39**, 87.
41. M. V. Stackelberg and R. Paulus, 1935, *Z. phys. Chem.*, **B28**, 427.
42. G. A. Steigmann and J. Goodyear, 1968, *Acta Cryst.*, **B24**, 1062.
43. *Strukturbericht*, **2**, 38.
44. N. Morimoto and K. Koto, 1966, *Science*, **152**, 345.
45. F. J. Trojer, 1966, *Amer. Min.*, **51**, 890.
46. B. J. Wuensch, 1964, *Z. Kristallogr.*, **119**, 437.
47. M. J. Buerger, 1945, *J. Amer. Chem. Soc.*, **67**, 2056; *Amer. Min.*, **32**, 415.
48. G. Brauer and E. Zintl, 1937, *Z. phys. Chem.*, **B37**, 323.
49. A. Zalkin and W. J. Ramsey, 1956, *J. Phys. Chem.*, **60**, 234; A. Zalkin, W. J. Ramsey, and D. H. Templeton, *Ibid.* **60**, 1275.
50. W. C. Koehler and E. O. Wollan, 1953, *Acta Cryst.*, **6**, 741.
51. A. J. Bradley and A. Taylor, 1937, *Phil Mag.*, **23**, 1049.
52. K. Osamura, Y. Murakami, and Y. Tomiie, 1966, *J. Phys. Soc. Japan*, **21**, 1848.
53. R. Benz and W. H. Zachariasen, 1966, *Acta Cryst.*, **21**, 838.
54. S. Geller, 1962, *Acta Cryst.*, **15**, 1195; O. Knop and M. A. Ibrahim, 1961, *Canad. J. Chem.*, **39**, 297.
55. H. Hahn and C. de Lorent, 1955, *Z. anorg. Chem.*, **279**, 281.
56. *Strukturbericht*, **1**, 350; W. B. Pearson, 1967, *A Handbook of Lattice Spacings and Structures of Metals and Alloys*, Oxford: Pergamon, p. 69.
57. H. Schäfer, G. Schäfer, and Armin Weiss, 1963. *Z. anorg. Chem.*, **325**, 77.
58. G. A. Steigmann, H. H. Sutherland, and J. Goodyear, 1965, *Acta Cryst.*, **19**, 967.
59. *Strukturbericht*, **1**, 352.
60. C. Romers, B. A. Blaisse, and D. J. W. Ijdo, 1967, *Acta Cryst.*, **23**, 634.
61. F. Lappe, A. Niggli, R. Nitsche, and J. G. White, 1962, *Z. Kristallogr.*, **117**, 146.
62. F. G. Donika *et al.*, 1967, *Kristallografija*, **12**, 854 [*Soviet Physics—Crystallography*, **12**, 745].
63. G. A. Jeffrey and V. Y. Wu, 1963, *Acta Cryst.*, **16**, 559.
64. N. Elliott, 1960. *J. Chem. Phys.*, **33**, 903.
65. R. F. Giese and P. F. Kerr, 1965, *Amer. Min.*, **50**, 1002.
66. R. F. Giese, 1963, *Dissert. Abstr.*, **23**, 3406.
67. Y. Takeuchi, 1957, *Min. J. Japan*, **2**, 90.
68. F. Grønvold and E. Røst, 1962, *Acta Cryst.*, **15**, 11.
69. M. C. Farquhar, H. Lipson, and H. R. Weill, 1945, *J. Iron and Steel Inst.*, **152**, 457.
70. M. J. Buerger, 1937, *Z. Kristallogr.*, **A97**, 504.
71. H. Holseth and A. Kjekshus, 1969, *Acta Chem. Scand.*, **23**, 3043; G. Brostigen and A. Kjekshus, 1970, *Ibid.*, **24**, 1925.

72. E. H. Roseboom, 1963, *Amer. Min.*, **48**, 271.
73. N. Morimoto and L. A. Clark, 1961, *Amer. Min.*, **46**, 1448; M. Winterberger, 1962, *Bull. Soc. Fr. Minér. Crist.*, **85**, 107; M. J. Buerger, 1936, *Z. Kristallogr.*, **A95**, 83.
74. L. B. Barricelli, 1958, *Acta Cryst.*, **11**, 75.
75. G. S. Ždanov and R. N. Kuz'min. 1961, *Kristallografija*, **6**, 872 [*Soviet Physics–Crystallography*, **6**, 704].
76. D. T. Cromer and K. Herrington, 1955, *J. Amer. Chem. Soc.*, **77**, 4708.
77. G. Andersson, 1956, *Acta Chem. Scand.*, **10**, 623.
78. H. Krebs, K.-H. Müller, and G. Zurn, 1956, *Z. anorg. Chem.*, **285**, 15.
79. O. Olofsson, 1965, *Acta Chem. Scand.*, **19**, 229.
80. J. V. Florio, N. C. Baenziger, and R. E. Rundle, 1956, *Acta Cryst.*, **9**, 367.
81. J.-P. Bouchard and R. Fruchart, 1965, *Bull. Soc. Chim. Fr.*, p. 130; R. Fruchart, J.-P. Sénateur, J.-P. Bouchard, and A. Michel, 1965, *C. R. Acad. Sci. Paris*, **260**, 913.
82. C. Crevecoeur and C. Romers, 1964, *Kon. Ned. Akad. Wet. Proc. Sect. Sci.*, **B67**, 289.
83. K. Robinson, 1953, *Acta Cryst.*, **6**, 854.
84. P. L. Blum, P. Guinet, and G. Silvestre, 1965, *C. R. Acad. Sci. Paris*, **260**, 1911.
85. J. Goldak, C. S. Barrett, D. Innes, and W. Youdelis, 1966, *J. Chem. Phys.*, **44**, 3323.
86. W. H. Zachariasen, 1932, *Phys. Rev.*, **40**, 917; A. Okazaki, 1958, *J. Phys. Soc. Japan*, **13**, 1151; C. R. Vannewarf, A. Kelley, and R. J. Cashman, 1960, *Acta Cryst.*, **13**, 449.
87. N. W. Tideswell, F. H. Kruse, and D. J. McCullough, 1957, *Acta Cryst.*, **10**, 99.
88. D. Mootz and R. Kunzmann, 1962, *Acta Cryst.*, **15**, 913.
89. J. G. White and J. Dismukes, 1963, *Acta Cryst.*, Supplement, p. A25.
90. W. Borchert and M. Röder, 1959, *Z. anorg. Chem.*, **302**, 253.
91. M. Atoji, 1961, *J. Chem. Phys.*, **35**, 1950.
92. E. B. Hunt and R. E. Rundle, 1951, *J. Amer. Chem. Soc.*, **73**, 4777.
93. J. H. C. Hogg and W. J. Duffin, 1967, *Acta Cryst.*, **23**, 111.
94. D. T. Cromer, A. C. Larson, and R. B. Roof, 1964, *Acta Cryst.*, **17**, 272.
95. J. W. Boon and C. H. MacGillavry, 1942, *Rec. Trav. Chim. Pays-Bas*, **61**, 910.
96. S. Geller and J. H. Wernick, 1959, *Acta Cryst.*, **12**, 46.
97. R. Ballestracci, 1966, *C. R. Acad. Sci. Paris*, **262C**, 1253.
98. S. Geller, A. Jayaraman, and G. W. Hull, 1965, *J. Phys. Chem. Solids*, **26**, 353.
99. O. Olofsson, 1967, *Acta Chem. Scand.*, **21**, 1659.
100. G. Hägg and A. G. Hybinette, 1935, *Phil. Mag.*, **20**, 913.
101. S. A. Semiletov and Z. G. Pinsker, 1955, *Dokl. Akad. Nauk*, **100**, 1079.
102. A. G. Talybov, 1961, *Kristallografija*, **6**, 49 [*Soviet Physics–Crystallography*, **6**, 40].
103. K. A. Agaev and S. A. Semiletov, 1965, *Kristallografija*, **10**, 109 [*Soviet Physics–Crystallography*, **10**, 86].
104. K. A. Agaev and A. G. Talybov, 1966, *Kristallografija*, **11**, 454 [*Soviet Physics–Crystallography*, **11**, 400].
105. K. A. Agaev, A. G. Talybov, and S. A. Semiletov, 1966, *Kristallografija*, **11**, 736 [*Soviet Physics–Crystallography*, **11**, 630].
106. D. Harker, 1934, *Z. Kristallogr.*, **A89**, 175.
107. M. I. Chiragov and A. G. Talybov, 1965, *Kristallografija*, **10**, 409, [*Soviet Physics–Crystallography*, **10**, 331].
108. L. H. Brixner, 1962, *J. Inorg. Nucl. Chem.*, **24**, 257.
109. R. S. Mitchell, 1956, *Z. Kristallogr.*, **108**, 296.
110. S. Furuseth, K. Selte, and A. Kjekshus, 1965, *Acta Chem. Scand.*, **19**, 257.
111. B. E. Brown, 1966, *Acta Cryst.*, **20**, 264.
112. B. E. Brown, 1966, *Acta Cryst.*, **20**, 268.
113. N. W. Alcock and A. Kjekshus, 1965, *Acta Chem. Scand.*, **19**, 79.

114. G. Tunell and C. J. Ksanda, 1935, *J. Wash. Acad. Sci.*, **25**, 32.
115. G. Tunell and C. J. Ksanda, 1936, *J. Wash. Acad. Sci.*, **26**, 507.
116. G. Tunell and L. Pauling, 1952, *Acta Cryst.*, **5**, 375.
117. J. A. A. Ketelaar and E. W. Gorter, 1939, *Z. Kristallogr.*, **A101**, 367.
118. W. Jeitschko, H. Nowotny, and F. Benesovsky, 1963, *Mh. Chem.*, **94**, 672.
119. P. Eckerlin and A. Rabenau, 1960, *Z. anorg. Chem.*, **304**, 218.
120. W. Jeitschko and H. Nowotny, 1967, *Mh. Chem.*, **98**, 329.
121. *Strukturbericht*, **2**, 23.
122. *Strukturbericht*, **2**, 25.
123. K. Yvon, H. Nowotny, and R. Kieffer, 1967, *Mh. Chem.*, **98**, 34.
124. K. Yvon, H. Nowotny, and F. Benesovsky, 1968, *Mh. Chem.*, **99**, 726.
125. R. Juza and W. Sachsze, 1943, *Z. anorg. Chem.*, **251**, 201; K. H. Jack, 1950, *Acta Cryst.*, **3**, 392; N. Terao, 1958, *Naturwiss.*, **45**, 620; *Idem*, 1962, *J. Phys. Soc. Japan*, **17**, Suppl. B-II, 238.
126. E. Parthé and V. Sadagopan, 1963, *Acta Cryst.*, **16**, 202.
127. J. Clarke and K. H. Jack, 1951, *Chem. and Ind.*, **46**, 1004.
128. Erwin Rudy, Elisabeth Rudy and F. Benesovsky, 1962, *Planseeber, Pulvermet.*, **10**, 42.
129. E. Parthé and K. Yvon, 1970, *Acta Cryst.*, **B26**, 153.
130. D. A. Evans and K. H. Jack, 1957, *Acta Cryst.*, **10**, 833.
131. N. Terao, 1960, *J. Phys. Soc. Japan*, **15**, 227; 1962, *Ibid.*, **17** Suppl. B-II, 238.
132. *Strukturbericht*, **1**, 240.
133. A. W. Henfrey and B. E. F. Fender, 1970, *Acta Cryst.*, **B26**, 1882.
134. K. H. Jack, 1950, *Acta Cryst.*, **3**, 392.
135. *Strukturbericht*, **2**, 31.
136. *Strukturbericht*, **1**, 84.
137. K. Lukaszewicz and W. Trzebiatowski, 1954, *Bull. Acad. Polon. Sci.*, Cl III, **II**, 277; K. Bachmayer, H. Nowotny and A. Kohl, 1955, *Mh. Chem.*, **86**. 39.
138. N. Schönberg, 1954, *Acta Chem. Scand.*, **8**, 213.
139. R. Kiessling, 1947, *Acta Chem. Scand.*, **1**, 893.
140. H. Hahn and B. Harder, 1956, *Z. anorg. Chem.*, **288**, 241.
141. S. F. Bartram, 1958, *Dissert. Abstr.*, **19**, 1216.
142. F. Jellinek, 1957, *Acta Cryst.*, **10**, 620.
143. S. Rundqvist, 1962, *Acta Chem. Scand.*, **16**, 287.
144. R. Gohle and K. Schubert, 1964, *Z. Metallk.*, **55**, 503.
145. E. Larsson, 1965, *Ark. Kemi.*, **23**, 335.
146. F. Bertaut, 1954, *J. Phys. Radium*, **15**, 77s; 1956. *Bull. Soc. Fr. Minér. Crist.*, **79**, 276; A. F. Andresen, 1960, *Acta Chem. Scand.*, **14**, 919.
147. H. Föppl, E. Busmann, and F. -K. Frorath, 1963, *Z. anorg. Chem.*, **314**, 12.
148. M. Chevreton and F. Bertaut, 1961, *C. R. Acad. Sci. Paris*, **253**, 145.
149. A. Okazaki, 1961, *J. Phys. Soc. Japan*, **16**, 1162.
150. R. C. Erd, H. T. Evans, and D. H. Richter, 1957, *Amer. Min.*, **42**, 309.
151. S. Brunie and M. Chevreton, 1964, *C. R. Acad. Sci., Paris*, **258**, 5847.
152. E. Flink, G. A. Wiegers, and F. Jellinek, 1966, *Rec. Trav. Chim. Pays-Bas*, **85**, 869.
153. J. G. White and J. P. Dismukes, 1965, *Inorganic Chem.*, **4**, 1760.
154. F. Kadijk, R. Huisman, and F. Jellinek, 1968, *Acta Cryst.*, **B24**, 1102.
155. K. Selte and A. Kjekshus, 1964, *Acta Chem. Scand.*, **18**, 697.
156. F. Jellinek, G. Brauer, and H. Müller, 1960, *Nature, Lond.*, **185**, 376.
157. F. Jellinek, 1962, *J. Less-Common Metals*, **4**, 9.
158. B. E. Brown and D. J. Beerntsen, 1965, *Acta Cryst.*, **18**, 31.
159. See, e.g., F. Jellinek, 1963, *Ark. Kemi*, **20**, 447.

160. A. L. Bowman, T. C. Wallace, J. L. Yarnell, R. G. Wenzel, and E. K. Storms, 1965, *Acta Cryst.*, **19**, 6.
161. K. Yvon and E. Parthé, 1970, *Acta Cryst.*, **B26**, 149.
162. P. Ettmayer, G. Vinek, and H. Rassaerts, 1966, *Mh. Chem.*, **97**, 1258.
163. Ju. D. Kondrasev, 1966, *Kristallografija,* **11**, 559 [*Soviet Physics–Crystallography,* **11**, 492].
164. Y. Iitaka and W. Nowacki, 1962, *Acta Cryst.*, **15**, 69.
165. N. Schönberg, 1954, *Acta Met.*, **2**, 837.

9

Structure Types Dominated by Triangular Prismatic Arrangements

PAIRED-LAYER STACKING

Although many structures contain atoms in triangular prismatic coordination, there are two groups of structures in which it is the main feature of the atomic arrangement. One arises when close packed triangular layers are stacked not in close packing, but in the "paired-layer" sequence AA (or AACCBB) and have further close packed 3^6 layers inserted between all pairs of layers with the same stacking, the inserted layers being at half the layer spacing and stacked on a different site to the pair. Thus, for example, the WC structure type, A$\underline{B}$A$\underline{B}$ is obtained, or the α-MoS_3 structure, A$\underline{C}$AC$\underline{B}$CB$\underline{A}$B. In such structures the only coordination feature is trigonal prismatic. With hexagonal close packed stacking sequences of close packed layers and inserted layers at half spacing, or with paired-layer stacking sequences if the 3^6 layers inserted at half the layer spacing are inserted also between the nonpaired layers stacked on different sites, mixed octahedral and trigonal prismatic coordination is obtained. However, it was more convenient to consider structures having mixed octahedral and trigonal prismatic coordination with the derivatives of close packing (see p. 452). In all of the structures obtained by inserting triangular layers halfway between pair-stacked layers, only half of the trigonal prisms are centered.

The second case of structures dominated by trigonal prismatic coordination arises when 3^6 and 6^3 nets of atoms occupying the same stacking sequence are stacked alternately say, AaAa. Thus the AlB_2 structure is obtained which can be regarded as the prototype for this family of structures. In this arrangement the trigonal prisms are contiguous on all faces and all are centered. Other structure types are derived from this and the WC structure by transpositions to separate the prisms along various faces and by rotations of the separated or contiguous prisms or blocks of prisms. These can be discussed as on pp. 20–24 or as transposition structures after the method of Boller and Parthé.[1] One of the simpler structural types obtained from the AlB_2 structure by transpositions to separate slabs of prisms is the CrB structure. Further structural variants are obtained by altering or removing the atoms centering the trigonal prisms. Indeed, the WC structure discussed above is the simplest variant of this type where only half of the trigonal prisms are centered.

3^6 Nets in AA Stacking

γ-HgSn$_{6-10}$ (*hP*1): γ-$HgSn_{6-10}$ with cell dimensions $a = 3.21$, $c = 2.99$ Å, $Z = 1$ atom at 7.2 at. % Hg, has a disordered simple hexagonal structure.[2] This structure type is found in several stable phases and some 14 metastable phases.

InSb III (H.P.) (*oA*2): InSb transforms at about 30 kb pressure to an orthorhombic modification, $a = 2.92$, $b = 5.56$, $c = 3.06$ Å, $Z = 1$,[3] that is related to the β-Sn structure of InSb which is obtained by pressing to 30 kb and quenching to 77°K before release of pressure. The orthorhombic structure is obtained from the β-Sn structure by a simple shear process of shifting alternate (110) planes along [001] by $\frac{1}{4}$ in z. InSb III is made up of 3^6 nets parallel to (100) stacked in paired-layer sequence AA and it is therefore a distorted simple hexagonal structure with $c/a \sim 1$ (Figure 9-1). Interatomic distances are 2 × 2.92, 2 × 3.06, and 4 × 3.17 Å.

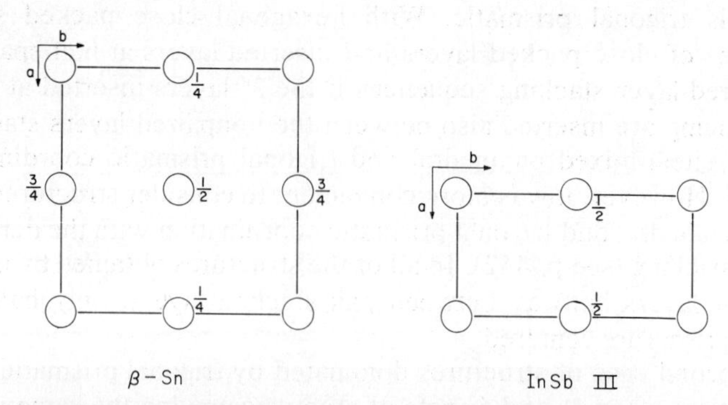

Figure 9-1. Comparison of InSb III (H.P.) structure with that of β-Sn.

MN: M on 3^6 Nets in AA Paired-Layer Stacking, *N* on 3^6 Net at $\frac{1}{2}$ Spacing. Trigonal Prisms Alternately Filled

WC, LiRh (*hP*2): The WC structure with cell dimensions $a = 2.906$, $c = 2.837$ Å, $Z = 1$, is one of the Hägg interstitial phases which are formed when the ratio of the nonmetal to metal radius is less than about 0.59. The carbon atom occupies alternate trigonal prisms in a simple hexagonal array of W atoms.[4] The prisms are contiguous on all faces giving a three-dimensional array with the prism axes lying in the *c* direction. The occupied prisms form columns running in the *c* direction. The axial ratio in compounds with this structure is close to unity.

LiRh has a structure which is formally similar, except that the unit cell dimensions, $a = 2.649$, $c = 4.357$ Å result in an axial ratio c/a of 1.64 instead of a value close to unity as found in the interstitial phases.[5] This presumably is the result of a phase adopting the structure under conditions where there are no restrictions on the relative sizes of the two components.

The WC arrangement ABA forms an important building block in hexagonal structures.

Polytypic Structures of MN_2 Compounds With Three-Layer Structures and Trigonal Prismatic Coordination

Rhombohedral α-MoS_2 with the S atoms in paired-layer stacking AA-CCBB and alternate trigonal prismatic holes between the pairs of similarly stacked layers occupied by Mo, is the prototype for a family of polytypic structures which depend both on the stacking sequence of the pairs of layers of the *N* component, and on the sequential distribution of the *M* components in the trigonal prismatic holes, since between any pairs of *N* component layers stacked AA, there are two stacking choices, B or C, for the *M* component. Such structures have hexagonal or rhombohedral symmetry.

α-MoS_2 (*hR*3): α-MoS_2 has a rhombohedral structure with dimensions $a = 3.163$, $c = 18.37$ Å, $Z = 3$ for the corresponding hexagonal cell.[6,7] The structure, which is composed of close packed 3^6 layers of atoms parallel to the (001) plane, is a three-layer type with Mo in trigonal prismatic coordination by the S atoms. There are only van der Waals bonds between S atoms of different three-layer MoS_2 units. The layer stacking sequence is ABACACBCB (Mo underlined), consistent with the rhombohedral symmetry of the unit cell.

β-MoS_2 (*hP*6): β-MoS_2 has cell dimensions $a = 3.160$, $c = 12.294$ Å, $Z = 2$.[8] The atoms form close packed triangular layers parallel to the

(001) plane. Mo atoms at $z = \pm\frac{1}{4}$ in BC stacking are surrounded on either side by S in CC and BB stacking, respectively, so that Mo centers alternate trigonal prisms formed by the S atoms (Figure 9-2), the overall stacking sequence being C$\underline{B}$CB$\underline{C}$B. The S,Mo,S layers form three-layer units which are separated from neighboring three-layer units by van der

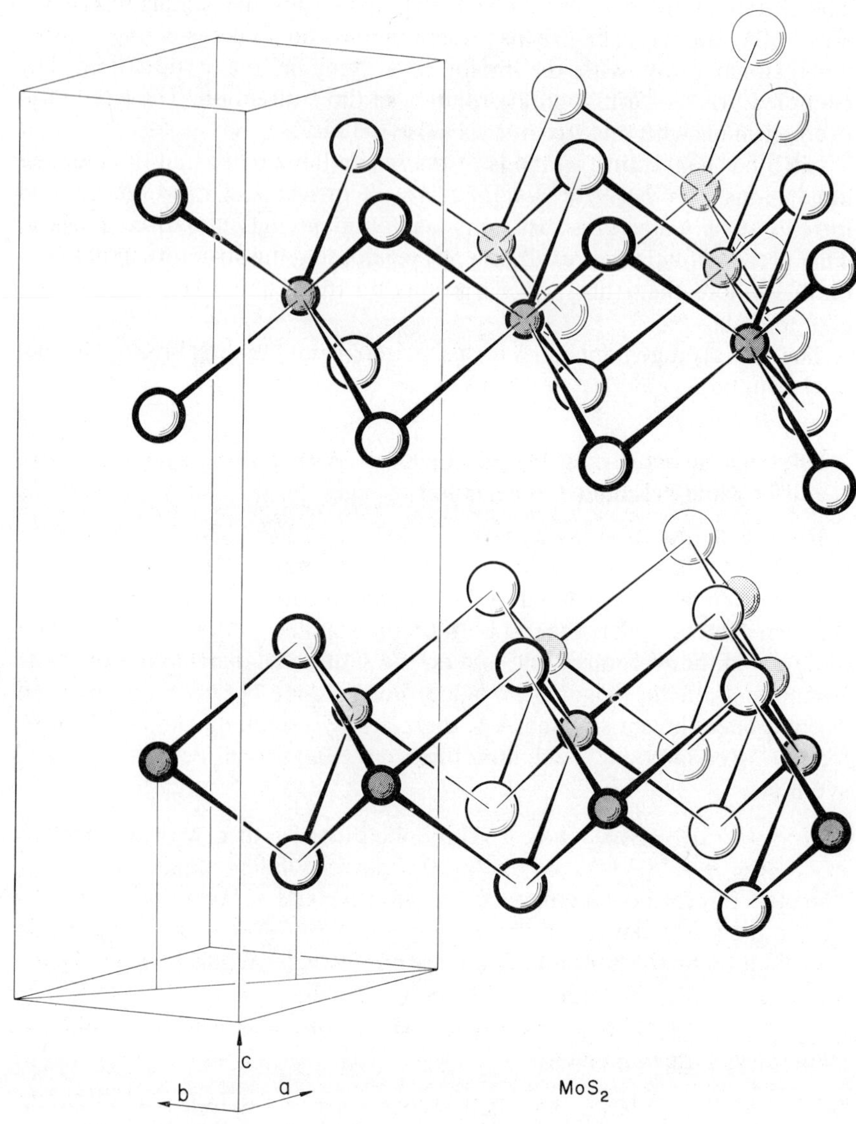

Figure 9-2. Pictorial view of the structure of β-MoS_2 ($hP6$): S, large circles.

Waals contacts between the S atoms. Mo has 6 S at 2.35 Å and 6 Mo at 3.16 Å. S has 3 Mo neighbors. S–S distances within the three-layer units are 2.98 Å(1) and 3.16 Å(6). Interlayer S–S distances are 3.66 Å.

NbS_2 (H.T.) (*hP*6): The unit cell dimensions of NbS_2 are $a = 3.31$, $c = 11.89$ Å, $Z = 2$.[6] The atoms form close packed triangular nets normal to the [001] direction. The S layers are in close packed and paired-layer stacking sequence BBCC, and the Nb atoms on A sites are inserted halfway between pairs of S layers with the same stacking position, giving the overall stacking sequence $\mathrm{B\underline{A}BC\underline{A}C}$ (Nb underlined). The Nb atoms center half of the trigonal prisms formed by S. Nb has 6 S at 2.42 Å and 6 Nb at 3.31 Å. S has 3 Nb. The S–S distance between the three-layer units is 3.53 Å. The structure requires confirmation by a single-crystal study.

Hf_2S (*hP*6)[9] has essentially the anti-NbS_2 type structure, except that there is bonding between all of the layers in the Hf_2S structure.

$NbSe_2$ (*hR*12): The unit cell has dimensions $a = 3.44$, $c = 25.24$ Å, $Z = 4$.[10] The atoms are arranged in close packed triangular nets normal to [001]. The 8 Se layers are in close packed and paired-layer stacking sequence, BCCAACCB. The Nb layers in stacking sequence AABA, are located halfway between the pairs of Se layers on similar stacking sites, so that they center half of the trigonal prisms formed by Se, the overall stacking sequence being $\mathrm{B\underline{A}BC\underline{A}CA\underline{B}AC\underline{A}C}$. The three-layer units are held together by van der Waals bonds.

ReB_2 (*hP*6), (*ReB_3*, *hP*8): ReB_2 has cell dimensions $a = 2.900$, $c = 7.478$ Å, $Z = 2$;[11] ReB_3 has essentially the same cell dimensions.[12] Close packed triangular nets of Re atoms at $z = \pm\frac{1}{4}$ are in hexagonal close packed stacking sequence, BC, but with the layer separation greatly increased ($c/a = 2.58$). In ReB_2 the boron atoms are inserted at $z = 0.048$ and 0.452 in C stacking and at $z = 0.548$ and 0.952 in B stacking. Re is thus surrounded by trigonal prisms of boron atoms, but unlike the similar arrangement in the β-MoS_2 structure, there is very little spacing between the boron atoms of one layer of prisms and those of the prisms above and below it. In ReB_3 further layers of boron atoms were said to be inserted at $z = 0$ and $\frac{1}{2}$ in the A stacking position, but this was later questioned by LaPlaca and Post.[11]

Re has 8 B neighbors (2×2.228, 6×2.255 Å) in ReB_2 and 6 Re at 2.90 Å. Each B has 3 B at 1.82 Å and 4 Re neighbors. In ReB_3 the coordination of Re is increased to 20 (8 B(1) at 2.245 Å, 6 B(2) at 2.51 Å and 6 Re at 2.90 Å). In addition to 3 B(1) and 4 Re, B(1) also has 3 B(2) neighbors at 1.715 Å. B(2) is surrounded by 6 B(1) and 6 Re. The structure requires further confirmation by single-crystal methods.

Derivatives of the WC Structure: by Translations (Along *c*), Rotations of 90°, and Filling Between Slabs of Triangular Prisms, etc.

NbAs (*tI*8): The NbAs structure with cell dimensions $a = 3.452$, $c = 11.68$ Å, $Z = 4$, is built up of a stack of 8 layers of Nb or As atoms in alternate sequence.[13] Each layer is a 4^4 net. The first at $z = 0$ is Nb at the cell corners; subsequent nets are displaced so that nodes lie either over the base center of the cell or at the centers of the cell edges (Figure 9-3).

The structure is made up of layers of triangular prisms of Nb surrounding As (or *vice versa*) which are parallel to the *ab* plane with the prism axes pointing either in the *a* or *b* direction. There are four such layers per cell and the axes of prisms in one layer are rotated 90° to those of the layers above and below. Next-but-one layers of prisms (those with axes in the same orientation) are transposed $(a+b)/2$ relative to each other (Figure 9-3). The As atoms only center alternate Nb prisms. The structure thus combines features of the WC and MoB structures, being similar to the $AgTlTe_2$ type (*tI*8), from which it differs in the relative distribution of the atoms centering the prisms.

$AgTlTe_2$ (*tI*8): The $AgTlTe_2$ structure with cell dimensions, $a = 3.92$, $c = 15.22$ Å, $Z = 2$, is formed by stacking eight 4^4 nets of atoms one above

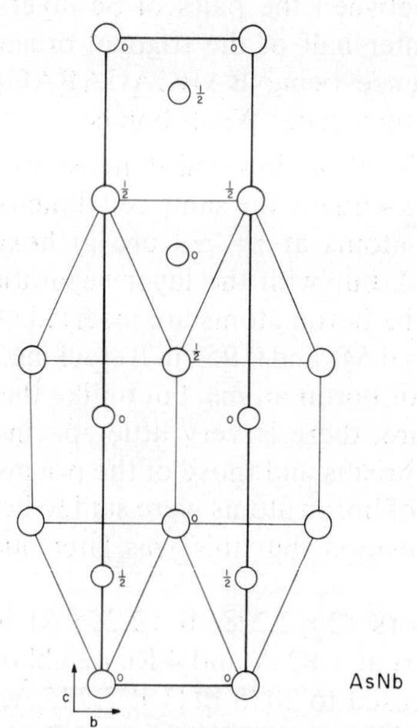

Figure 9-3. The NbAs (*tI*8) structure viewed down [100]. Nb, large circles.

the other in the c direction.[14] Starting with a layer of Tl atoms at the cell corners, the nodes of other nets either cover cell corners, midpoints of cell edges or the basal face center, the sequence of layers being Tl,Te,Ag, Te,Tl,Te,Ag,Te. This arrangement gives slabs of triangular prisms of Ag and Ti atoms parallel to the (001) plane. The slabs which are contiguous, have prism axes alternately along a and along b on proceeding up the c axis of the cell. Alternate prisms in lines through each of the slabs are centered by Te atoms (Figure 9-4), with the Te closer to the Ag than Tl prism edges. The structure is therefore a hybrid type combining the features of the WC and MoB structures. In addition to being a ternary phase, the structure of $AgTlTe_2$ differs from that of NbAs in the relative positions of the atoms which center alternate prisms. Ag has 4 close Te neighbors at 2.70 Å; Te has 2 close Ag neighbors, and 4 Tl neighbors at 3.39 Å; Tl has 8 Te neighbors. The structure was examined by electron diffraction of thin films.

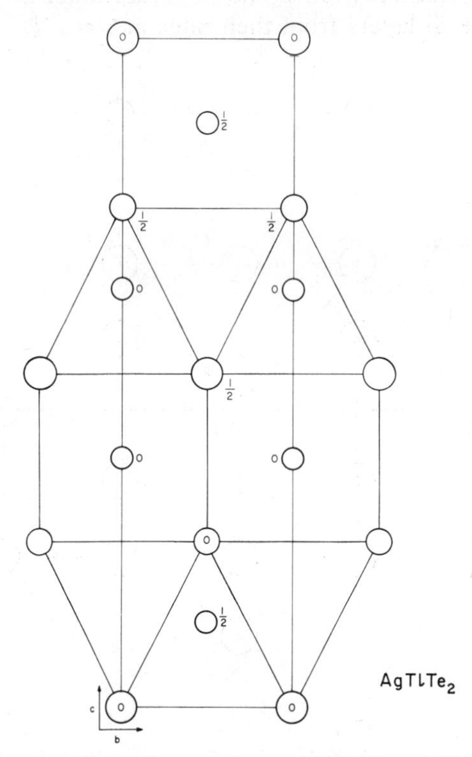

Fig. 9-4. The structure of $AgTlTe_2$ ($tI8$) viewed down [100]: Tl, largest circles; Te, smallest circles.

$BaAl_4$, $ThCu_2Si_2$ (*tI*10): The $BaAl_4$ structure with cell dimensions $a = 4.566$, $c = 11.25$ Å, $Z = 2$, is composed of 8 layers of 4^4 nets of Al or Ba atoms.[15] Starting from a Ba layer at $z = 0$ with atoms at the cell corners, subsequent layers have nodes of the 4^4 nets over cell corners, the mid-points of cell edges or the base center of the cell as shown in Figure 9-5.

The structure is a filled up form of the $AgTlTe_2$ (*tI*8) structure with Al(1) (Cu) and Ba (Th) forming layers of triangular prisms parallel to the *ab* plane with prism axes alternately along the *a* and *b* directions. The prisms are alternately centered by Al(2) (Si), and the added Al(1) (Cu) atoms at $\frac{1}{2}$, 0, $\frac{1}{4}$ and 0, $\frac{1}{2}$, $\frac{3}{4}$ center one of the rectangular faces of half of the prisms (Figure 9-5). The resemblance to the $AgTlTe_2$ structure is most striking in the ternary form of the $BaAl_4$ structure of phases such as $ThCu_2Si_2$.[16] The axial ratio $c/a = 2.40$ for the $ThCu_2Si_2$ structure suggests the $MoSi_2$ structure, and it can be seen from comparing Figure 9-5 with Figure 10-29, p. 591, that the $ThCu_2Si_2$ structure could be regarded as a distorted filled-up form of $MoSi_2$. The Cu layers added at $z = \pm\frac{1}{4}$ considerably displace the Si layers from their sites at $z = \frac{1}{6}, \frac{1}{3}, \ldots$ to $z = 0.121$, $0.379, \ldots$.

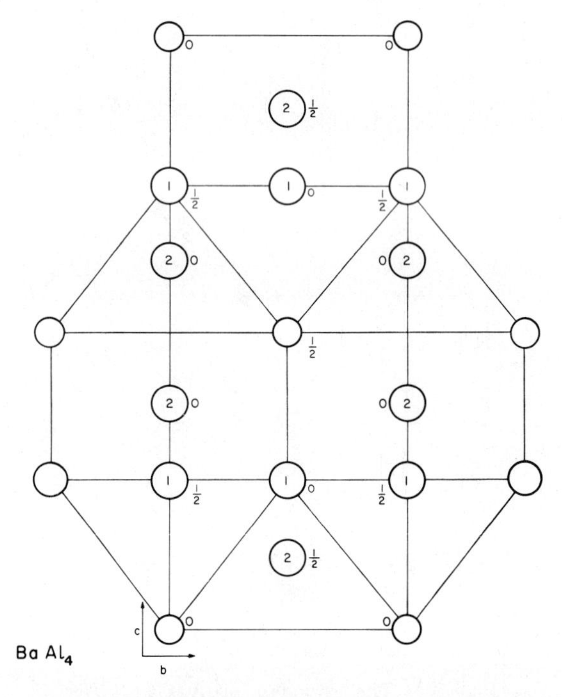

Figure. 9-5. The $BaAl_4$ (*tI*10) structure projected down [100]: Al, largest circles.

The arrangement gives Al(1) four close Al(2) neighbors at 2.69 Å and four Ba at 3.60 Å. Al(2) has four Al(1) neighbors, one Al(2) at 2.68 Å and four Ba at 3.48 Å. Ba has eight Al(1) and eight Al(2) neighbors.

Cu_2Sb, PbFCl, (ZrSiSe) (tP6): The Cu_2Sb structure[17] ($a = 4.00$, $c = 6.10$ Å, $c/a = 1.526$, $Z = 2$) can be regarded as a distortion of a f.c. cubic arrangement in which there is a displacement of $a/2$ or $b/2$ of the ordered atomic array after every two layers along [001], or in which every fourth plane of atoms (Cu) along the [001] direction is omitted.

In the PbFCl or ZrSiSe[18] form, Se occupies the Sb sites and Si occupies the Cu(1) sites at 0,0,0. The axial ratio in binary or ternary phases with this structure can vary over wide limits from ~ 1.5 to 2.3. In ZrSiSe with $c/a = 2.3$, the almost planar distribution of atoms at heights $z \sim \frac{1}{4}$ and $\frac{3}{4}$ is no longer apparent, and the structure has become one of two pairs of associated 4^4 layers of Zr and Se atoms separated by a layer of Si atoms. The structure now contains double slabs of triangular prisms of Se and Si atoms parallel to the (001) plane. The prism axes in one slab are along a and along b in the other. Alternate prisms in the slabs are either centered by Zr or base centered by Si on the common base joining the double slabs. The differences in the two forms of the structure are shown in Figure 9-6, but it is emphasized that the differences are not related to binary or ternary phases.

Binary phases with the structure separate into two groups: those with

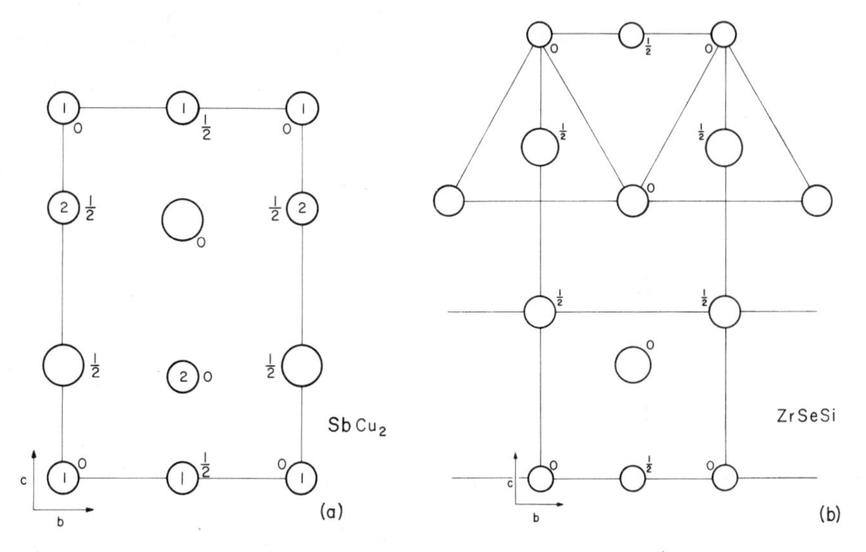

Figure 9-6. The $SbCu_2$ and ZrSeSi (*tP*6) structures projected down [100]: largest circles, Sb or Zr; smallest circles, Cu or Si.

axial ratios c/a, in the range from 1.5 to 1.75 and one set of atomic parameters, and those with axial ratios about 2.05 and another set of atomic parameters. In the former case it appears, according to the near-neighbor diagram (p. 53), that the 4–4 M–$N(2)$ contacts control the cell dimensions and that the 1–1 M–$N(2)$ and 4–4 M–$N(1)$ contacts are compressed (Figure 9-7). When the axial ratio is about 2.05, the structure appears to be influenced more strongly by the 4 $N(1)$–$N(1)$ and 4 $N(1)$–$N(2)$

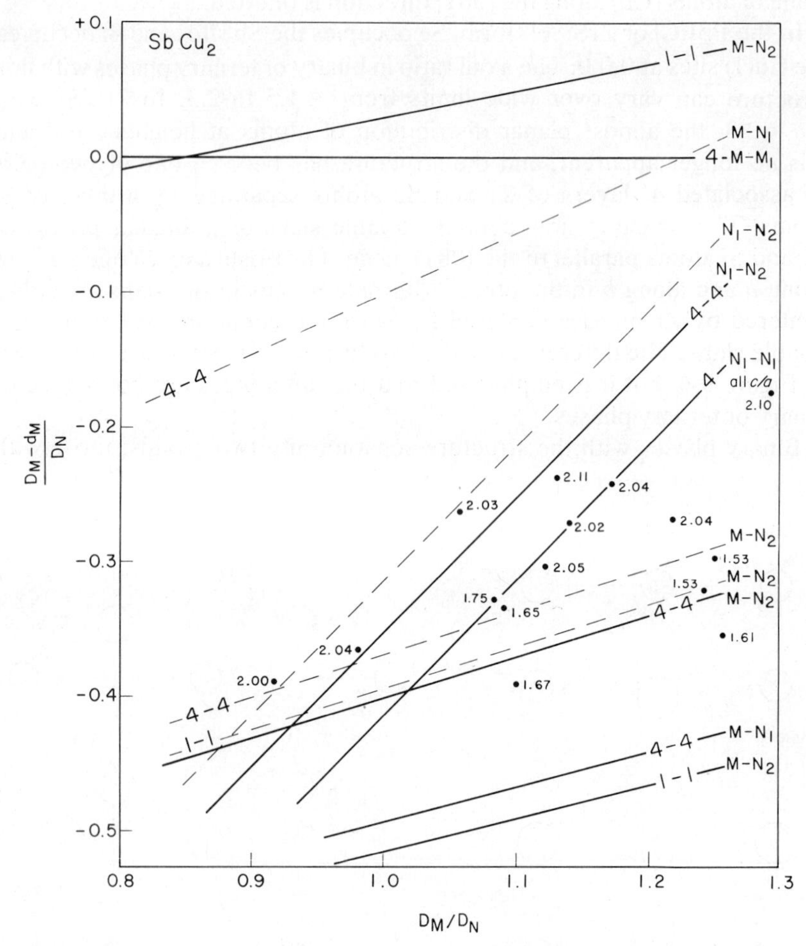

Figure 9-7. Near-neighbor diagram for phases with the $SbCu_2$ (MN_2) structure. Drawn for $c/a = 1.65$, $z_M = 0.73$, $z_N = 0.33$ full lines, and $c/a = 2.05$, $z_M = 0.63$, $z_N = 0.28$ broken lines. Numbers in large print indicate number of M–N, N–N, etc. neighbors. Numbers in small print indicate axial ratios, c/a, of phases with the $SbCu_2$ type structure.

contacts, and the location of the lines for 1–1 and 4–4 M–$N(2)$ contacts suggests an increased influence of the geometrical factor. No M–$N(1)$ contacts are made at this axial ratio. These differences correspond to a change of the metalloid atoms from the M component when $c/a \sim 1.6$ to the N component when $c/a \sim 2.05$.

$LaSe_2$ (tP24): $LaSe_2$, $CeSe_2$, and $NdSe_2$ form a superstructure of the Cu_2Sb type cell with $a = 2a_0$, $b = 2b_0$, $c = c_0$. $SmSe_2$ to β-$ErSe_2$ also form a superstructure of the Cu_2Sb type cell with $a = 2c_0$, $b = 4a_0$, $c = 3a_0$.[19]

α-$ErSe_{2-x}$ (L.T.) (*oC*132): The superstructure cell has dimensions $a = 16.22$, $b = 15.80$, $c = 11.88$ Å.[20] There is a Cu_2Sb type subcell, $a = 3.96$, $c = 8.10$ Å, and the superstructure arises from ordering the 12 vacant sites in α-$ErSe_{2-x}$.

$NdTe_3$ (*oC*16): $NdTe_3$ has cell dimensions $a = 4.35$, $b = 25.80$, $c = 4.35$ Å, $Z = 4$.[21] The structure is made up of stacked $NdTe_2$ (Cu_2Sb) type cells with an extra layer of Te inserted between cells which are alternately in mirror image orientation to each other about (001) planes (Figure 9-8). Nd is surrounded by 9 Te at 3.21 to 3.35 Å. Te(1) and (2) have each 4

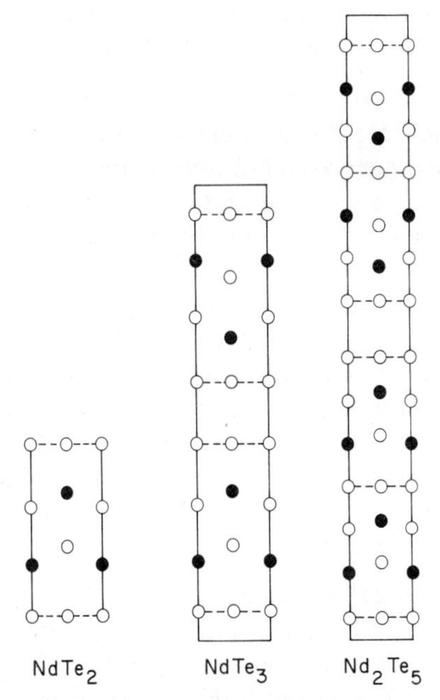

Figure 9-8. Comparison of the structures of $NdTe_2$ ($SbCu_2$ type), $NdTe_3$ (*oC*16), and Nd_2Te_5 (*oC*28): Te, open circles.

Te(1)–Te(2) neighbors equatorially at 3.08 Å, and 2 Nd at 3.35 Å. Te(3) has 5 Nd neighbors, and 8 Te at distances greater than 3.86 Å.

Nd_2Te_5 (*oC*28): The Nd_2Te_5 structure, with cell dimensions $a = 4.409$, $b = 44.1$, $c = 4.409$ Å, $Z = 4$, is derived from the Cu_2Sb type structure of $NdTe_2$.[22] The structure contains pairs of $NdTe_2$-type unit cells joined in the c direction. Two such units separated from each other in the [001] direction by an added layer of Te atoms, comprise the unit cell of Nd_2Te_5 as shown in Figure 9-8. The two cells in each pair are in mirror image orientation about (001) planes. The structure, determined only by powder methods, is related to the $NdTe_3$ type which is similarly constituted from single $NdTe_2$ type cells.

$Ni_{3+x}Te_2$ (*tP*5): $Ni_{3+x}Te_2$ ($a = 2.865$, $c = 6.62$ Å) has a defect Cu_2Sb type structure with the octahedral holes in the distorted close packed array of Te atoms only partly occupied randomly by Ni(2).[23] Below 140°C on the Ni-rich side, a monoclinic superstructure (*mP*10) occurs with ordered disposition of Ni(2) atoms ($a = 7.540$, $b = 3.799$, $c = 6.089$ Å, $\beta = 91.2°$ at $Ni_{3.3}Te_2$). On the tellurium-rich side there is a different ordering of vacant Ni(2) sites. The superstructure (*tP*20) ($a = 7.564$, $c = 6.062$ Å at $Ni_{2.86}Te_2$) is made up of four Cu_2Sb-type pseudocells. The structure requires confirmation by single-crystal methods.

α-$PdBi_2$ (L.T.) (*mC*12): In the α-$PdBi_2$ structure ($a = 12.74$, $b = 4.25$, $c = 5.665$ Å, $\beta = 102°35'$, $Z = 4$)[24] the atoms lie in layers at $y = 0$ and $\frac{1}{2}$. The Bi atoms are arranged so as to give strips of parallelograms running in the c direction. Strips of wide and narrow parallelograms alternately separate those of equal width, and the narrow parallelograms at $y = 0$ overlie the wide strips at $y = \frac{1}{2}$ and *vice versa* (Figure 9-9). These strips of parallelograms give rise to triangular prisms and the arrangement is such that the wide prisms are body centered by Bi, the narrow ones are uncentered and those of equal width are alternately body centered by Bi, or base centered by Pd with a second Pd atom near the center of the smallest rectangular face (Figure 9-9). Pd thus has 7 Bi neighbors at 2.77 to 3.185 Å, together with 2 Pd at 2.92 Å. Bi(1) has 4 Pd at 2.77 to 3.155 Å and Bi neighbors at 3.5 Å upwards. Bi(2) has 3 Pd at 2.84 and 2.95 Å and Bi neighbors at 3.32 Å upwards.

Ta_2P (*oP*36): Ta_2P has cell dimensions $a = 14.42$, $b = 11.55$, $c = 3.399$ Å, $Z = 12$.[25] Figure 9-10, showing a projection of the structure on (001), indicates that it is made up of triangular prisms of Ta atoms. Those with axes along [001] are centered by P; those with axes in the (001) plane are alternately centered by P atoms. There are no chains of connected P atoms running through the structure, in contrast to many boride structures.

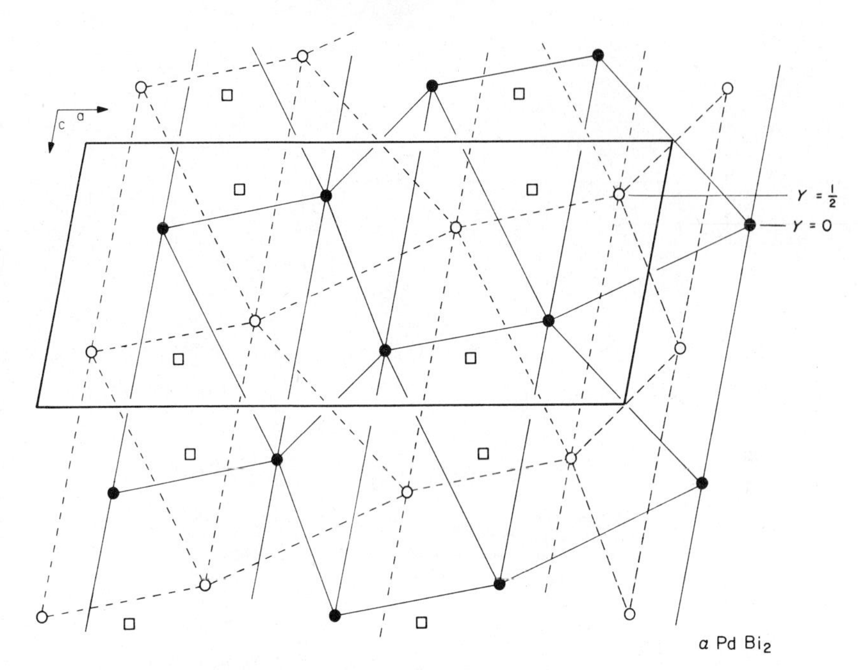

Figure 9-9. Atomic arrangement in the α-$PdBi_2$ (*mC*12) structure viewed down [010]: Pd, squares.

Ta(3) atoms occupy voids left between the triangular prisms, being surrounded by 8 Ta in a distorted cube. All P have 6 Ta prism neighbors; P(1) also has one out through a rectangular prism face, P(2) has two Ta neighbors out through rectangular prism faces and P(3) has two plus a third Ta at a greater distance. The shortest Ta–Ta distance in the structure is 2.87 Å. Ti_2S and Zr_2Se have the same structure.

Cr_3C_2 (*oP*20): Cr_3C_2 has cell dimensions $a = 5.53$, $b = 2.827$, $c = 11.48$ Å, $Z = 4$.[26] The atoms lie on planes parallel to (010) at $y = \frac{1}{4}$ and $\frac{3}{4}$, forming triangular prisms of Cr with axes lying either along [010] or in the (010) plane. All prisms with axes along [010] are centered by C, whereas only alternate prisms with their axes in the (010) plane are centered (Figure 9-11), C atom positions were determined by neutron diffraction. The structure was refined by Rundqvist and Runnsjö.[27]

Rh_4P_3 (*oP*28): Rh_4P_3 has cell dimensions $a = 11.66$, $b = 3.317$, $c = 9.994$ Å, $Z = 4$.[28] The atoms form layers at $y = \frac{1}{4}$ and $\frac{3}{4}$. The Rh atoms form layers of triangular prisms that extend in the (001) plane. These prisms have their axis along the [010] direction and they are alternately centered by P atoms. Other distorted prisms with axes in the *ac* plane connect

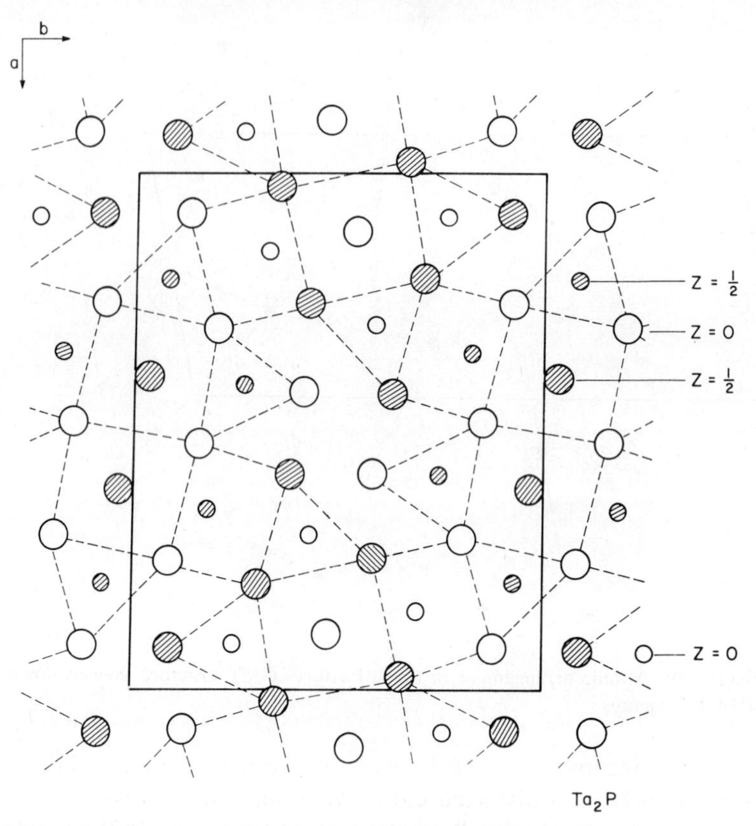

Figure 9-10. The Ta_2P (*oP*36) structure viewed down [001]: Ta, large circles.

these slabs of prisms together as shown in Figure 9-12. These prisms are alternately centered by P atoms. There are no P–P chains in the structure. In addition to the 6 Rh prism atoms surrounding P, P(1) and (2) have one Rh neighbor out through the center of a rectangular prism face. P(3) centering prisms with axes in the *ac* plane have no such neighbors. The closest Rh–Rh distance in the structure is 2.80 Å.

Mo_4P_3 (*oP*56): With cell dimensions $a = 12.43$, $b = 3.158$, $c = 20.44$ Å, $Z = 8$,[29] the Mo_4P_3 structure is composed of an array of triangular prisms of Mo. Most of those with their axis along *b* are centered by P (Figure 9-13), but only half of the prisms with their axis in the *ac* plane are centered. All P atoms have 7 Mo neighbors with the exception of P(2) which has 6. Mo–P distances average about 2.48 Å. There are no P–P bonds in the structure. The closest Mo–Mo approach is 2.84 Å. The structural architecture resembles that of Rh_4P_3 (Figure 9-12).

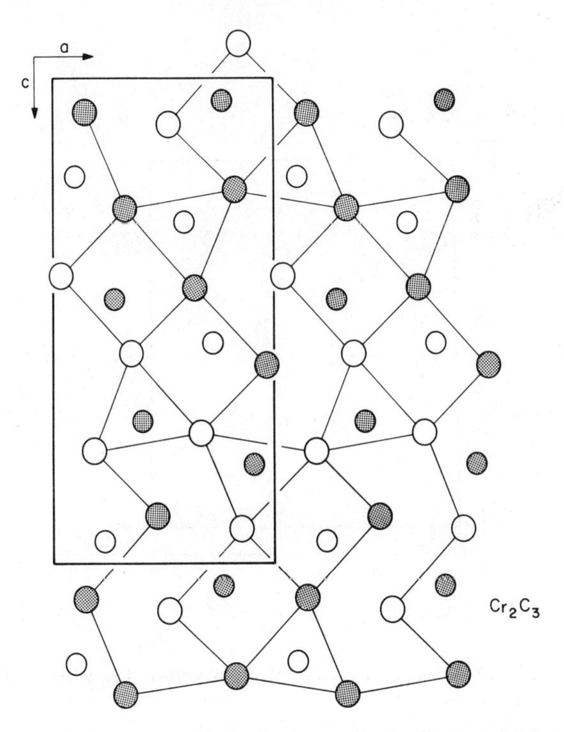

Figure 9-11. Atomic arrangement in the Cr_3C_2 (*oP*20) structure viewed down [010]: Cr, large circles.

Nb_7P_4 (*mC*44): Nb_7P_4 has cell dimensions $a = 14.950$, $b = 3.440$, $c = 13.848$ Å, $\beta = 104.74°$, $Z = 4$.[30] Atoms are confined to planes at $y = 0$ and $\frac{1}{2}$ where they form an irregular network of triangles and distorted squares and rectangles. The structure is made up of triangular prisms with axes either along [010] or in the (010) plane. Prisms with axes along [010] are centered by P. Those with axes in the (010) plane are alternately centered by P. Distorted cubes of Nb centered by Nb(1) or (2) are, looking down [010], located over the cell corners, base center, and midpoints of the cell base edges. The array of these polyhedra contains "free space" as indicated in Figure 9-14. The P atoms centering the Nb prisms also have 1, 2, or 3 more Nb neighbors out through the centers of the rectangular prism faces. There are no P–P contacts, the nearest approach being 3.44 Å. The closest Nb–Nb distance is 2.86 Å.

$Nb_{21}S_8$ (*tI*58): The $Nb_{21}S_8$ structure ($a = 16.794$, $c = 3.359$ Å, $Z = 2$)[31] contains layers of Ni and S atoms at $z = 0$ and $\frac{1}{2}$ which form the network illustrated in Figure 9-15. Both S(1) and S(2) have CN 7, being surrounded by a triangular prism of Ni atoms with a seventh Ni neighbor equatori-

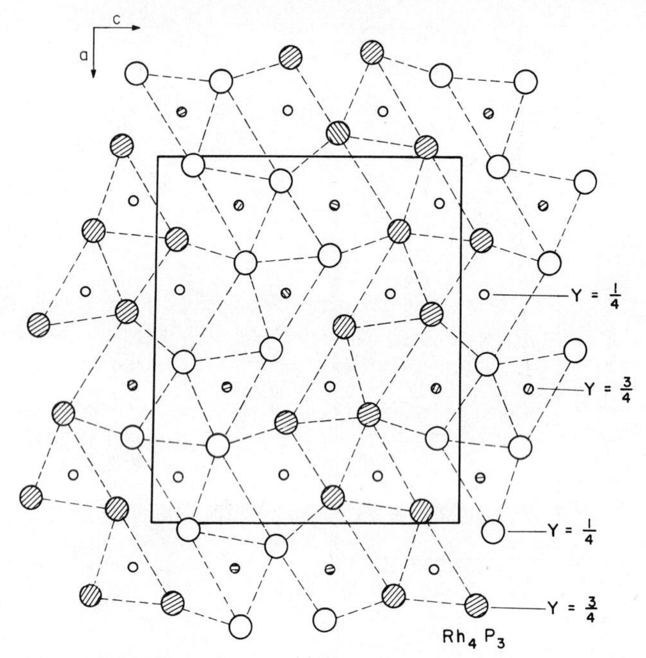

Figure 9-12. Projection of the Rh_4P_3 structure down [010]: Rh, large circles.

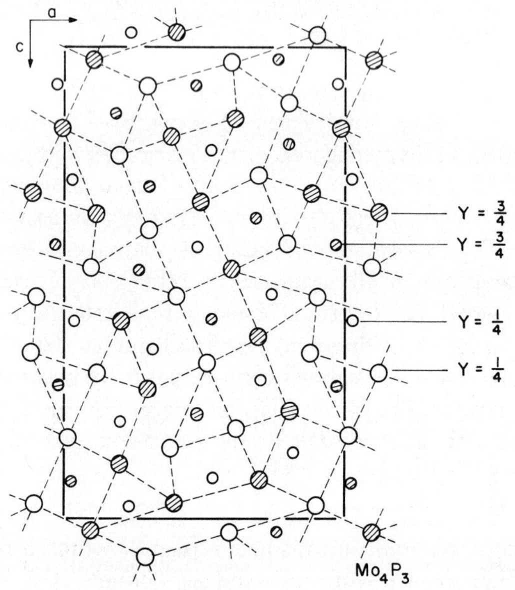

Figure 9-13. The Mo_4P_3 (*oP*56) structure viewed down [001]: Mo, large circles.

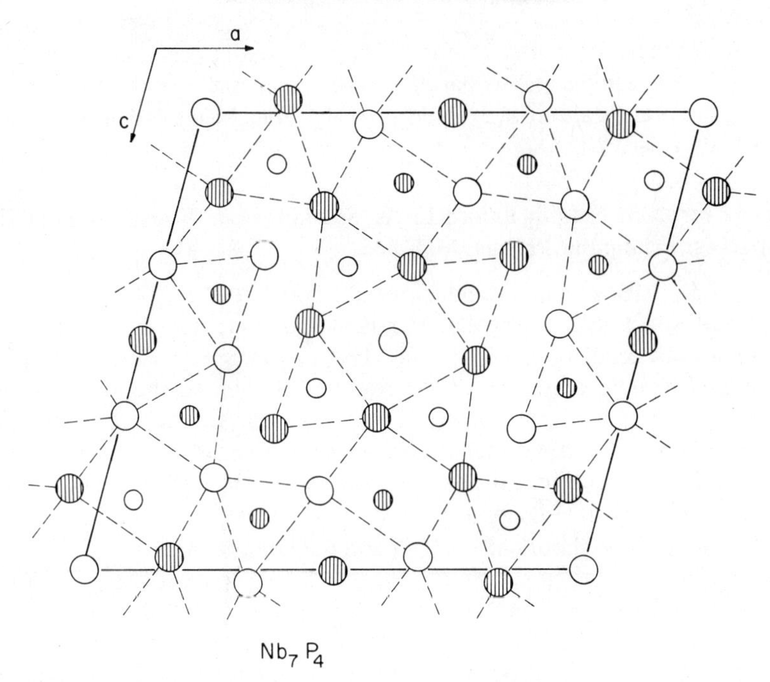

Figure 9-14. The Nb_7P_4 (*mC*44) structure viewed down [010]: Nb, large circles.

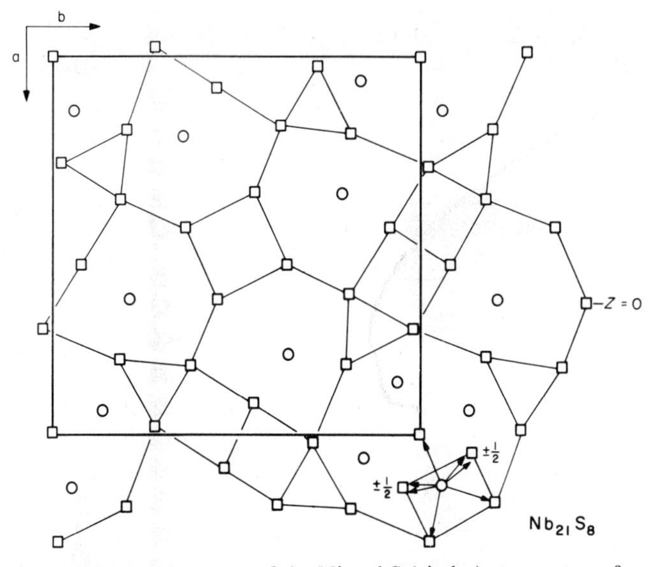

Figure 9-15. Arrangement of the Ni and S (circles) atoms at $z = 0$ in the $Nb_{21}S_8$ (*tI*58) structure. Also shown is a triangular prism of Ni surrounding S.

ally out through the center of one of the rectangular prism faces. The axis of prisms containing S(1) is parallel to the basal plane of the cell, and the axis of the prisms about S(2) is along [001]. The closest Nb approaches in the structure are 2.82 Å.

MN_2: M on 3^6 Nets in Paired-Layer Stacking AA. N on 6^3 Nets at Half Spacing. Triangular Prisms, All Filled.

AlB_2 ($hP3$): AlB_2 has cell dimensions $a = 3.006$, $c = 3.252$ Å, $Z = 1$. The Al atoms forming a triangular 3^6 net at $z = 0$ are in simple hexagonal stacking with axial ratio $c/a = 1.08$. The B atoms at $z = \frac{1}{2}$ which form a hexagonal 6^3 net, center all of the trigonal prisms, so the structure is a filled-up WC type (Figure 9-16). The trigonal prisms are contiguous on all faces giving a three-dimensional array of prisms with the prism axes lying in the c direction. Al has 12 B neighbors at 2.38 Å; also it is surrounded by 6Al at 3.01 Å and two polar Al at 3.25 Å giving CN 20. Each B has 3 B neighbors at 1.735 Å and 6 Al neighbors.

There are more than 75 binary phases known to have the AlB_2 (MN_2) structure and axial ratios vary over a wide range from 0.59 to 1.2. The radius ratios (R_M/R_N) of components forming the structure also vary over wide limits from about 0.94 to 1.8. N components are not confined

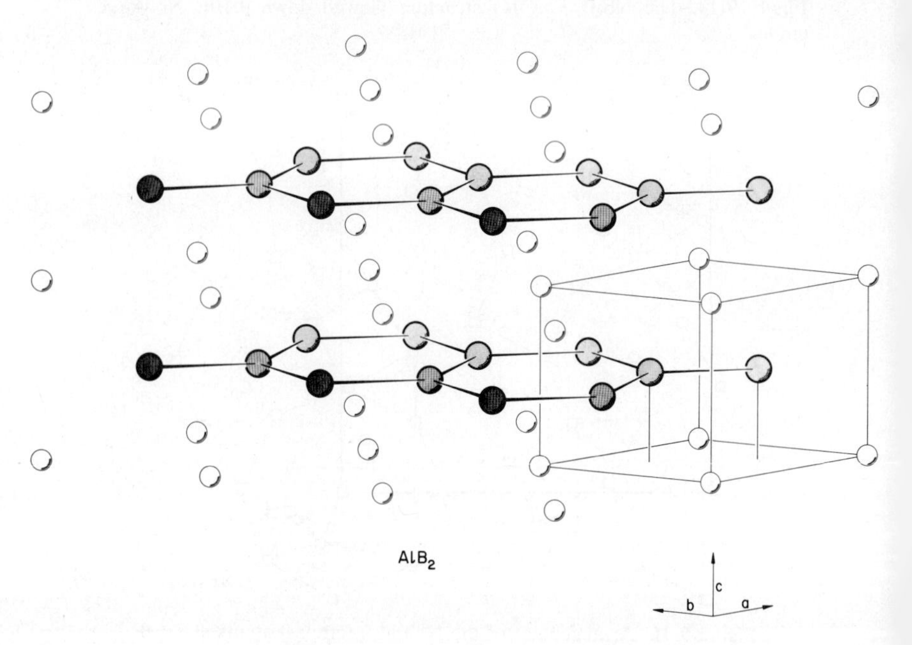

Figure 9-16. Pictorial view of the AlB_2 ($hP3$) structure.

to Group III and IV elements; compounds occur in which Ni and the noble metals form the N component. Many of the phases with the AlB_2 structure are deficient in the N component and formulas $MN_{\sim 1.5}$ are common.

The axial ratio of phases with this structure which varies between such wide limits, has to be taken into account in constructing a near-neighbor diagram (p. 53). The diagram shown in Figure 9-17 indicates three important axial ratios. At a c/a value of 1.075 the lines for M–M_{I},† M–N,† and N–N contacts intersect at $D_M/D_N = 1.732$; at $c/a = 0.866$ ($= \sqrt{3}/2$) the lines for M–M_{II},† M–N, and N–N contacts intersect at $D_M/D_N = 1.5$, and at a c/a value of 0.577 the lines for N–N_{I}† and N–N_{II}† contacts coincide. These relationships suggest that if a coordination factor alone controlled the occurrence of the structure, it would form for components with D_M/D_N values in the range from about 1.5 to 1.75 with unit cell axial ratios varying respectively from 0.86 to 1.08, instead of at the observed radius ratios from 0.94 to 1.8 and axial ratios respectively changing from 0.59 to 1.27.

The unit cells of phases whose N components come from Groups III and IV (and Ni) all have axial ratios of 0.95 and larger. They lie in a fairly narrow band following, but not parallel to the line for N–N_{I} contacts which itself is independent of the axial ratio and depends only on the radius ratio of the components. Within this band there does not appear to be any strong dependence of axial ratio on radius ratio although high axial ratios do tend to be found at high radius ratios (Figure 9-18). This band of phases straddles the lines for M–N contacts, but again there is no more than a general tendency for those phases with axial ratios of 1.0 ± 0.05 to lie about the M–N line for this ratio, those with ratios 1.1 ± 0.05 to lie about the M–N line for that ratio and similarly for axial ratios of about 1.2. This behavior is to be expected if structural dimensions, as well as being influenced by M–N contacts, depend to a large extent on the graphite-like net of N atoms, since the line for the 3 N–N_{I} contacts, which are somewhat compressed, is independent of axial ratio. Thus both the N–N bond factor and a geometrical factor give stability to phases formed from components with radius ratios from 1.1 to 1.8 and axial ratios larger than 0.95. The formation of 12–6 M–N and 3 N–N_{I} contacts gives an overall coordination of 12–9. In addition, a few of the phases make 6 M–M_{I} contacts.

When the axial ratio has a value of 0.90 and less, there is a quite different and more marked relationship between radius and axial ratios, as

†The distances $d_{M-M_{\text{I}}}$, etc. are as follows: $d_{M-M_{\text{I}}} = a$, $d_{M-M_{\text{II}}} = c$, $d_{N-N_{\text{I}}} = a/\sqrt{3}$, $d_{N-N_{\text{II}}} = c$, $d_{M-N} = \sqrt{(a^2/3)+(c^2/4)}$.

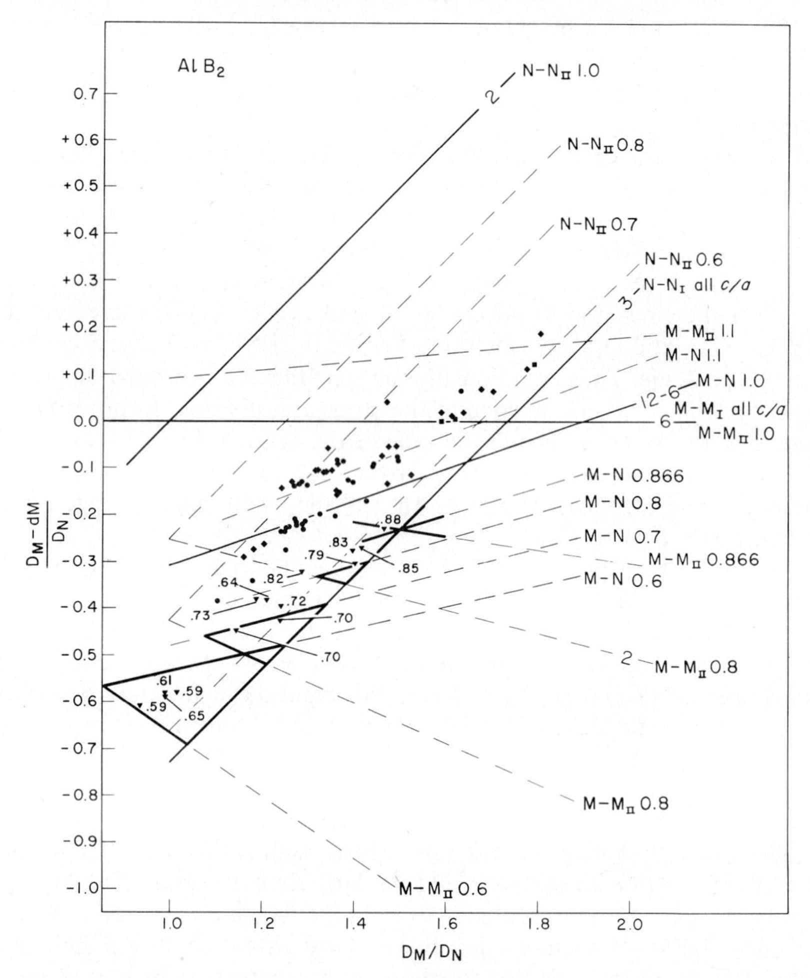

Figure 9-17. Near-neighbor diagram for phases with the AlB_2 structure (MN_2). Drawn for $c/a = 1.0$ full lines, and various other axial ratios (broken lines) indicated by the number following the symbols, M–N, etc. designating the contact. Numbers of the contact lines designate the number of neighbors involved in the contacts. Small numbers by points indicate the axial ratio, c/a, of phases with the AlB_2 structure: ▽, phases with $c/a < 0.88$; ○, ◊, □ phases with axial ratios in groups from 1.0 to 1.10, 1.10 to 1.20, and > 1.20.

shown in Figure 9-18. This results from the added influence of the M–M_{II} bond factor. These phases with axial ratios less than 0.9 show the influence of a coordination factor since the axial ratio varies with radius ratio in such a way that they are all located within, or very close to the triangle where (depending on axial ratio) the lines for 3 N–N_I, 12–6 M–N, and 2

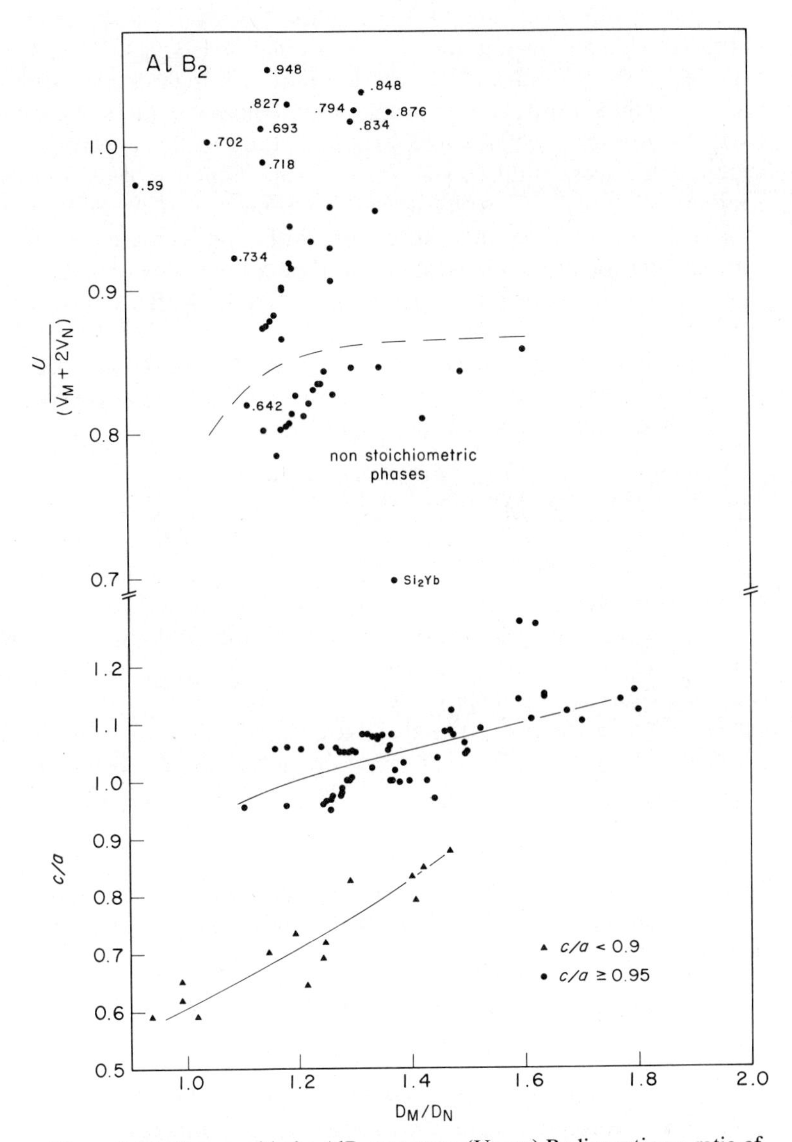

Figure 9-18. Phases with the AlB_2 structure. (Upper) Radius ratio *vs.* ratio of the unit cell volume to the volumes of the atoms in the cell, calculated from the elemental atomic volumes. Numbers by points indicate axial ratio, c/a, of phases. (Lower) Axial ratio, c/a, *vs.* radius ratio.

M–M_{II} contacts meet. These three lines intersect at an axial ratio of 0.866 (Figure 9-17). At $c/a = 0.6$ the line for 2 N–N_{II} contacts also lies within this triangle increasing the M–N coordination from 14–9 to 14–11. Thus, through change of axial ratio of the unit cell, high coordination is achieved at radius ratio values much lower than would be expected (1.5 to 1.75, *vide supra*). The most striking feature of this control of cell dimensions with decreasing D_M/D_N values is the influence of the two M–M_{II} contacts (Figure 9-17) which result from lines of M atoms running throughout the structure in the c direction. Although they are part of the general coordination or geometrical factor, their influence is such that they must be regarded as exerting a specific bond factor which controls the unit cell dimensions.

In agreement with the near-neighbor diagram, the observed structural compression, $U/(V_M + 2V_N)$, is less for phases with low axial ratios than for those with high axial ratios as shown in Figure 9-18.

Distortions of the AlB_2 Type Structure

ω-CrTi (*hP*3): The ω-CrTi structure is formed on quenching several c.p. hexagonal solid solutions of Ti, Zr, or Hf with other transition metal atoms; it is also found as a high pressure modification of Ti and Zr. It is not an ordered structure; the two components are randomly arranged. One-third of the atoms form a triangular net at $z = 0$ giving trigonal prisms and two-thirds form at rumpled 6^3 net at $z \sim \frac{1}{2}$ so that all of the trigonal prisms are centered; however, the atoms are slightly displaced alternately up and down from the centers of the trigonal prisms.†[32] Although the structure is approximately isotypic with AlB_2, the axial ratio of the unit cell, instead of being close to unity, is very much smaller, have a value of 0.62 ± 0.01 for the known phases. Indeed Figure 9-17 indicates that an axial ratio of about 0.6 is appropriate for phases with a random atomic arrangement (i.e. $D_M/D_N = 1.0$). The axial ratio of the δ-UZr_2 structure (0.62), which has the same atomic arrangement, indicates that it also is an ω phase, although it has not been generally so described.

The dimensions of the unit cell of ω-CrTi with 5% Cr are $a = 4.616$, $c = 2.827$ Å, $Z = 1$.

ϵ-PtZn$_{1.7}$ (*hP*2.7): The ϵ-$PtZn_{1.7}$ structure ($a = 4.111$, $c = 2.745$ Å)[34] approximately determined from X-ray powder data, appears to be a partially filled up ω-CrTi structure. Pt atoms form a simple hexagonal cell whose trigonal prisms are centered by partially occupied Zn sites slightly

†A b.c. cubic cell, $a = 9.80$ Å, $Z = 54$, $I\bar{4}3m$ has also been described for the phase as a result of single crystal investigation.[33]

displaced alternately up and down from the prism centers. Other Zn atoms partially occupy the centers of the rectangular faces of the prisms. The ratio of prism height to base (0.67) is close to that of the ω phases (0.62).

$CaIn_2$ (*hP*6): The cell dimensions are $a = 4.895$, $c = 7.750$ Å, $Z = 2$.[35] Ca atoms in A sites at $z = \pm\frac{1}{4}$ form trigonal prisms which are slightly off-centered alternately up and down by In atoms in B and C sites ($z = 0 \pm 0.045$ and $\frac{1}{2} \pm 0.045$). The In atoms thus form a rumpled 6^3 graphite-like layer with the Ca atoms over the centers of the hexagons (Figure 9-19). In has three close Ca neighbors on one side (along c) and three at normal

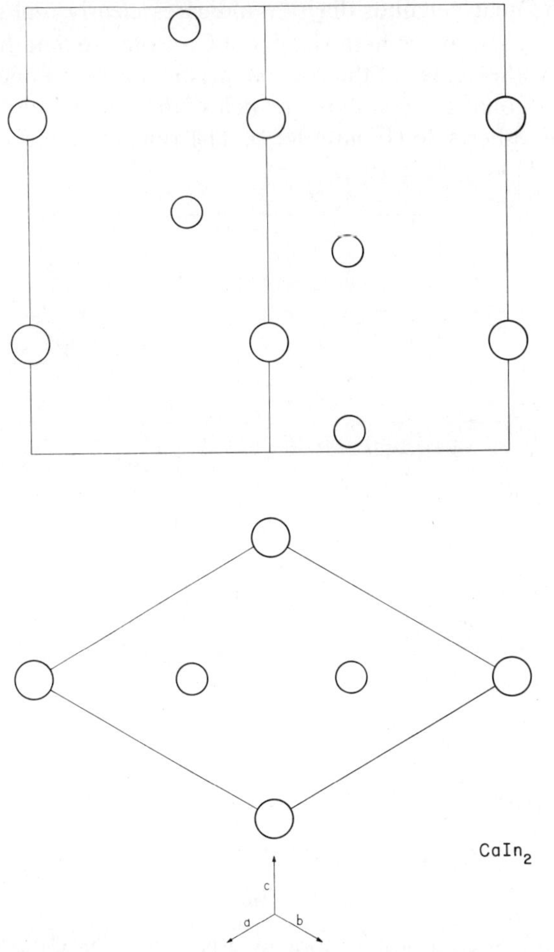

Figure 9-19. Projections of the $CaIn_2$ (*hP*6) structure down [110] and [001].

Ca–In distances on the other side. The direction of the displacement of In from the prism center alternates in neighboring prisms both parallel to the basal plane and along c (Figure 9-19). Ca has 6 In at 3.24 Å and 8 more at 3.635 and 3.875 Å. In has 3 In at 2.91, one at 3.18 Å, 3 Ca at 3.24 Å and three at 3.635 Å.

According to the near-neighbor diagram (p. 53) the atoms are compressed in the contacts 6–3 Ca–In (M–N_I) at 3.24 Å, 3 In–In at 2.91 Å and 2 Ca–Ca at 3.875 Å until the 6–3 Ca–In contacts (M–N_{II}) at 3.635 Å are formed, since all phases are distributed closely along the line for M–N_{II} contacts (Figure 9-20). Since the structure can be regarded as a distorted AlB_2 type, the near-neighbor diagram indicates clearly that the nature of the distortion is in drawing in the layers of Ca atoms so that they approach the In atom at the center of the trigonal prism closely on one side. In the neighboring trigonal prisms above or below they approach the In atoms closely on the other side (Figure 9-19). The remaining three In–Ca con-

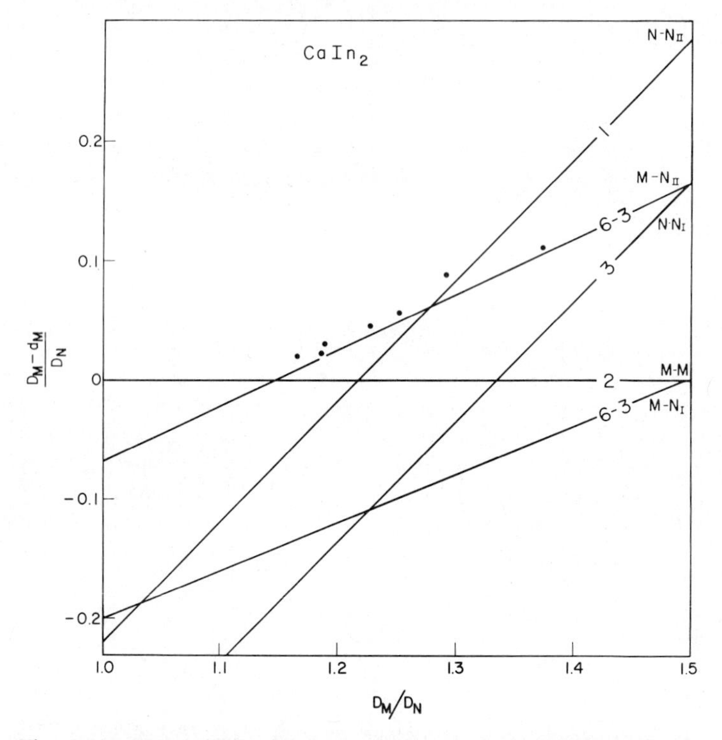

Fig. 9-20. Near-neighbor diagram for phases with the $CaIn_2$ (MN_2) structure. Calculated for $c/a = 1.59$ and $z_{In} = 0.455$. Numbers indicate numbers of neighbors for contacts N–N, M–N, etc.

tacts within any trigonal prism (M–N_{II} of the near-neighbor diagram) are at the normal Ca–In interatomic distances.

$CeCu_2$ (oI12): $CeCu_2$ has cell dimensions $a = 4.43$, $b = 7.05$, $c = 7.45$ Å, $Z = 4$.[36] The atoms form planar pentagon-triangle nets parallel to the (100) plane at $x = 0$ and $\frac{1}{2}$. Each pentagon is made up of 1 Ce and 4 Cu atoms. The Ce atoms of one layer lie over the pentagons of the layers above and below and are located away from the pentagon corner where the Ce atom is situated. Copper atoms lie over the triangles of nets above and below (Figure 9-21*a*).

Regarded down the [010] direction, the structure can be seen as a distortion of the AlB_2 type structure (here appearing in its orthohexagonal cell). The Ce atoms are moved from the corners of the AlB_2 type cell and the hexagons of the 6^3 net become non-planar moving away from the larger Ce atoms (Figure 9-21 *b*). Cu has three close Cu neighbors within the net (2×2.55 Å, 1×2.56 Å), one further Cu neighbor at 2.81 Å and 6 Ce neighbors at 3.025 to 3.21 Å. Ce is surrounded by 12 Cu at these distances.

K_5Hg_7 (oP48): The K_5Hg_7 structure ($a = 10.06$, $b = 19.45$, $c = 8.34$ Å, $Z = 4$) is shown in Figure 9-22 taken from the paper of Duwell and

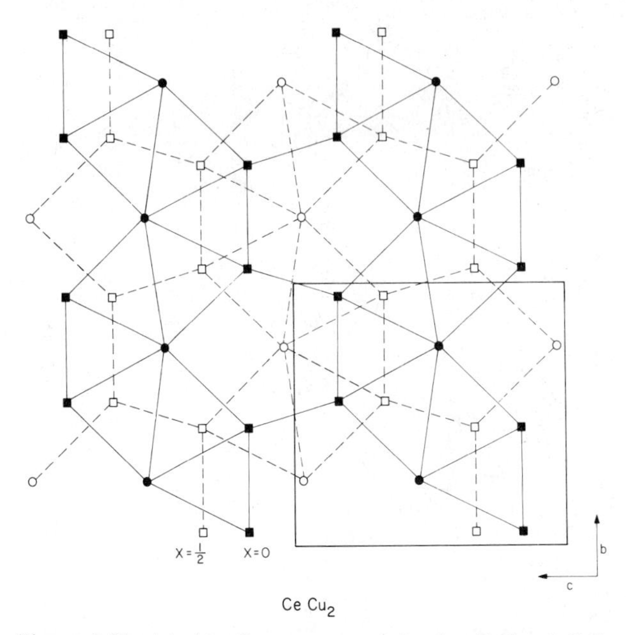

Figure 9-21. (*a*) Atomic arrangement in the $CeCu_2$ (*oI*12) structure viewed down [100]: Ce, circles.

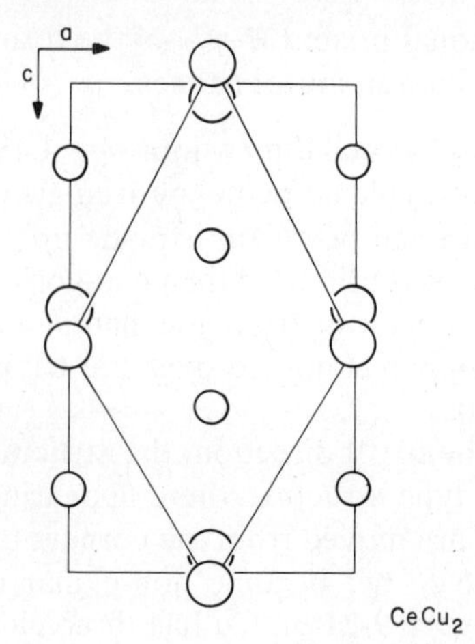

Figure 9-21. (*b*) $CeCu_2$ structure viewed down [010]: Ce, circles.

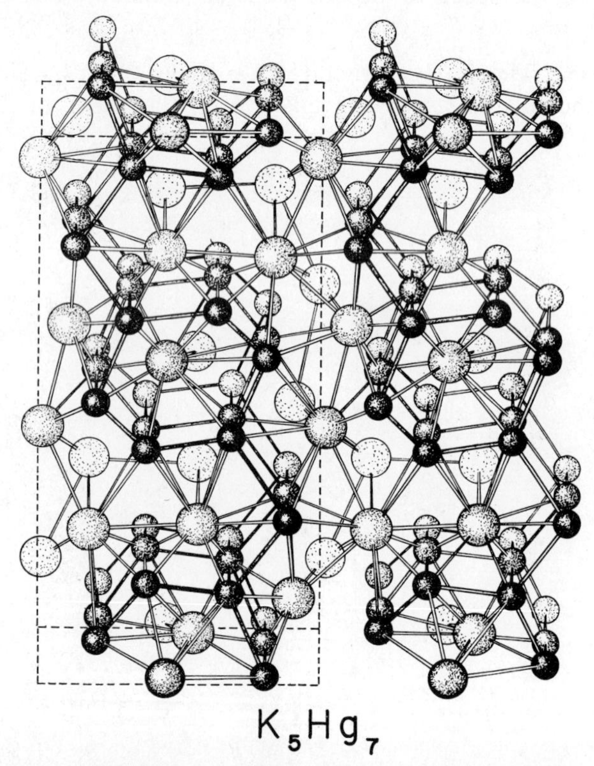

Figure 9-22. Pictorical view of the structure of K_5Hg_7 (*oP*48): K atoms, large circles. (Duwell and Baenziger[38].)

Baenziger.[37] The structure has features of substitutional distortion from the AlB_2 type.

WB_4 (hP20): The WB_4 structure has cell dimensions $a = 5.200$, $c = 6.340$ Å, $Z = 4$.[38] The probable structure determined from single crystal X-ray data, is a superstructure of AlB_2 in which $\frac{1}{3}$ of the W atoms in each of the metal atom layers at $z = \pm\frac{1}{4}$ are replaced by pairs of boron atoms above and below, i.e., at $z = 0.115$ and 0.385 and $z = 0.615$ and 0.885. The phase appears to contain additional boron.

Filled Up Derivatives of the AlB_2 Structure

Li_3N (hP4): Li_3N cell dimensions are $a = 3.665$, $c = 3.890$ Å, $Z = 1$.[39] One way of describing the structure is to consider that the triangular net of Li(1) atoms at $z = \frac{1}{2}$ form trigonal prisms which are centered by a 6^3 net of Li(2) at $z = 0$. The N atoms at $z = 0$ occupy the midpoints of the prism edges and center the hexagons of Li(2) atoms. Li(1) thus has 2 N at 1.94 Å and 6 Li at 3.665 Å. Li(2) has three very close Li neighbors at 2.12 Å and 3 N at the same distance. N has 2 Li(1) and 6 Li(2) neighbors.

TaZrNO (hP4): The unit cell dimensions are $a = 3.645$, $c = 3.881$ Å, $Z = 1$.[40] The Zr atoms at $z = 0$ form a simple hexagonal array and the Ta atoms at $z = \frac{1}{2}$ center alternate Zr trigonal prisms in columns. The other trigonal prisms are probably centered by nitrogen, and oxygen at $z = 0$ centers the base of the trigonal prisms which contain Ta. The axial ratio of the unit cell is slightly greater than unity (1.06). Ta and N form mixed graphite-like rings at $z = \frac{1}{2}$. Zr has 6 N at 2.86 Å and 3 O at 2.10 Å. Ta has 3 N at 2.10 Å and 2 O at 1.94 Å. N is surrounded by 6 Zr and 3 Ta, oxygen by 3 Zr and 2 Ta.

Structures Derived From AlB_2 Through Translations or 90° Rotations of Slabs of Prisms, With or Without Filling Between Slabs

CrB (oC8): The CrB structure has cell dimensions $a = 2.969$, $b = 7.858$, $c = 2.932$ Å, $Z = 4$.[41] It is made up of independent layers of triangular prisms of Cr atoms parallel to (010) with the prism axes running in the [100] direction (Figure 9-23). The prisms are centered by B atoms which form zigzag chains running in the [001] direction. Thus boron has 2 B neighbors at 1.74 Å and 6 Cr at 2.19 Å. Cr has 6 B neighbors at the same distance and one at 2.31 Å, as well as 10 Cr at 2.65 to 2.97 Å. The structure can be regarded as a development from the AlB_2 type in which the contiguous triangular prisms are separated into independent layers and then displaced relative to each other by $\boldsymbol{a}/2$, $\boldsymbol{c}/2$ referred to the axes of the CrB cell.

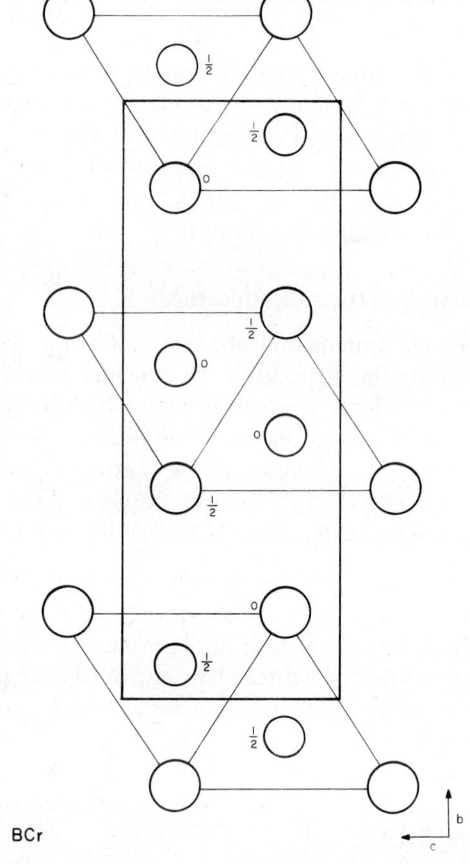

Figure 9-23. The CrB ($oC8$) structure viewed along [100]: Cr, large circles.

Several transition metal borides have the CrB structure as well as many rare earth compounds of Groups II, III, and IV elements and of the Co and Ni group transition metals. The structure was originally characterized for TlI (*Strukturbericht* type *B*33). Apart from halides and hydroxides with the CrB structure, the phases fall into two groups: group 1 phases are formed between alkaline or rare earth metals or transition metals (*M*) and B, Al, Ga, Si, Ge, Sn, or Pb (*N*). Group 2 phases are formed between alkaline earth, rare earth or Group III and IV transition metals (*M*), and Group VIII transition metals or the Cu Group metals (*N*). Schob and Parthé[42] point out that with the exception of three aluminides, the two groups separate on a diagram of the axial ratios b/a *vs.* a/c

(Figure 9-24). This separation corresponds to different ratios for the triangular prism edges formed by the larger M atoms. The abscissa of Figure 9-24, a/c, gives the ratio of the height to base of the prisms in the CrB structure, whence it can be seen that group 1 phases have prism edge ratios from 0.95 upwards, the prisms generally being elongated (ratio > 1). Group 2 compounds on the other hand have squashed prisms (ratio < 1), Similar observations are made for the FeB structure (see p. 520). In the MoB structure where prisms oriented 90° differently have contiguous square prism faces, the ratio must of course be unity. The different prism edge ratios for group 1 and 2 structures might suggest some fundamental difference in bonding or electronic arrangement. However, the near-neighbor diagram (p. 53) does not reveal any particular difference between the two groups, both are distributed along the lines for M–N contacts (Figure 9-25).

Rieger and Parthé[44] show that the distance between the N components along the chains remains nearly constant despite increase in size of the M component in compounds of Si, Ge, or Sn with Ca, Sr, and Ba. This results in an increase in the y parameter of the N component and in the chain angle as the size of the M component increases.

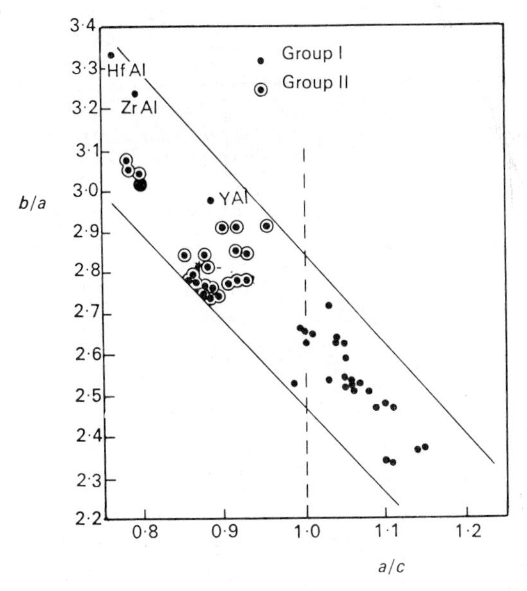

Figure 9-24. Axial ratios b/a *vs.* c/a for phases with the CrB structure: large circles are phases with the ratio of prism height to base < 1; filled circles those with ratio > 1. (Hohnke and Parthé [143].)

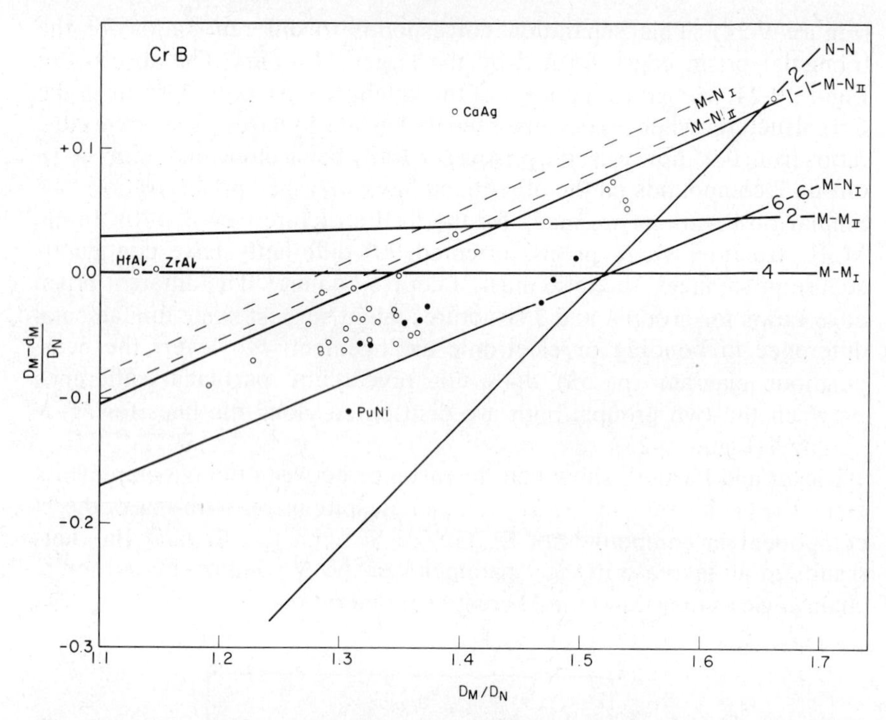

Figure 9-25. Near-neighbor diagram for phases with the CrB (MN) structure. Calculated for $y_M = 0.46$, $y_N = 0.440$, $b/a = 2.647$, $c/a = 0.9875$. Open circles: group 1 phases with prism height to base ratio >1. Filled circles: phases with ratio <1. Numbers indicate the number of neighbors involved in contacts N–N, M–N, etc.

Ta_3B_4 (oI14): The Ta_3B_4 structure with cell dimensions $a = 3.284$, $b = 13.98$, $c = 3.129$ Å, $Z = 2$,[45] can be regarded as a development from the AlB_2 (*hP*3) type structure. It contains double layers of Ta triangular prisms parallel to the (010) plane. The prisms are joined alternately face to face or apex to apex in the [001] direction, and all prisms are centered by chains of B atoms which run in the *c* direction (Figure 9-26). These double layers of Ta prisms can be derived from the AlB_2 type arrangement by transpositions $\boldsymbol{a}/2$, $\boldsymbol{b}/7$, and $\boldsymbol{c}/2$ referred to the axes of the Ta_3B_4 cell. The arrangement of Ta atoms on the outside of the double layers of prisms, relative to the Ta atoms of the next double layer, is the same as in CrB.

The structure type has been confirmed by a single-crystal study of Cr_3B_4;[46] there were no abnormally short B–B distances such as were reported for Ta_3B_4.

AlMoB (oC12): AlMoB has cell dimensions $a = 3.212$, $b = 13.985$, $c = 3.102$ Å, $Z = 4$.[47] The structure is made up of slabs of Mo triangular

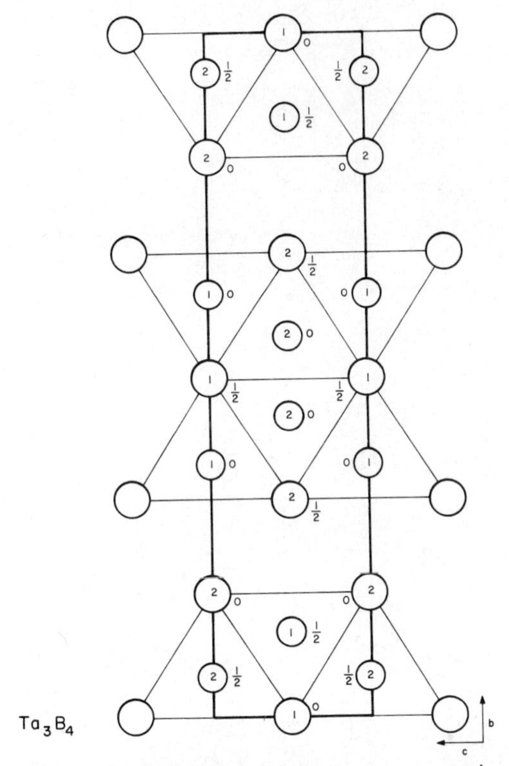

Figure 9-26. Diagram of the atomic arrangement in Ta_3B_4 ($oI14$) looking down [100]: Ta, large circles.

prisms with the prism axis along a, each prism being centered by B, so that zigzag B chains run in the c direction (Figure 9-27). Two 4^4 nets of Al atoms parallel to the (010) plane separate the slabs of Mo prisms. Mo has 6 Mo neighbors at 2.935 to 3.21 Å, 4 Al at 2.69 Å and 4+2 B at 2.36 Å; Al has 4 Al neighbors at 2.70 Å, 4 Mo at 2.69 Å and 1 B at 2.25 Å. Each B has 2 B at 1.83 Å, 1 Al at 2.25 and 6 Mo at 2.36 Å.

$ZrSi_2$ ($oC12$): The $ZrSi_2$ structure ($a = 3.721$, $b = 14.68$, $c = 3.683$ Å, $Z = 4$)[48] is a filled up CrB ($oC8$) type with slightly rumpled 4^4 nets of Si(1) atoms at $z = \frac{1}{4}$ and $\frac{3}{4}$ inserted between the slabs of triangular prisms of Zr atoms. These lie parallel to the (010) plane, with their axes running in the [100] direction (Figure 9-28). The Zr prisms are centered by Si(2) atoms (Si–Zr: 4 × 2.71 Å, 2 × 2.95 Å). Si(1) has 4 Si(1) neighbors at 2.62 Å and 4 Zr neighbors at 2.83 and 2.88 Å. Zr has 10 Si neighbors and 6 Zr at 3.52 to 3.72 Å.

Mo_2B_5 ($hR7$): The dimensions of the hexagonal cell are $a = 3.011$, $c =$

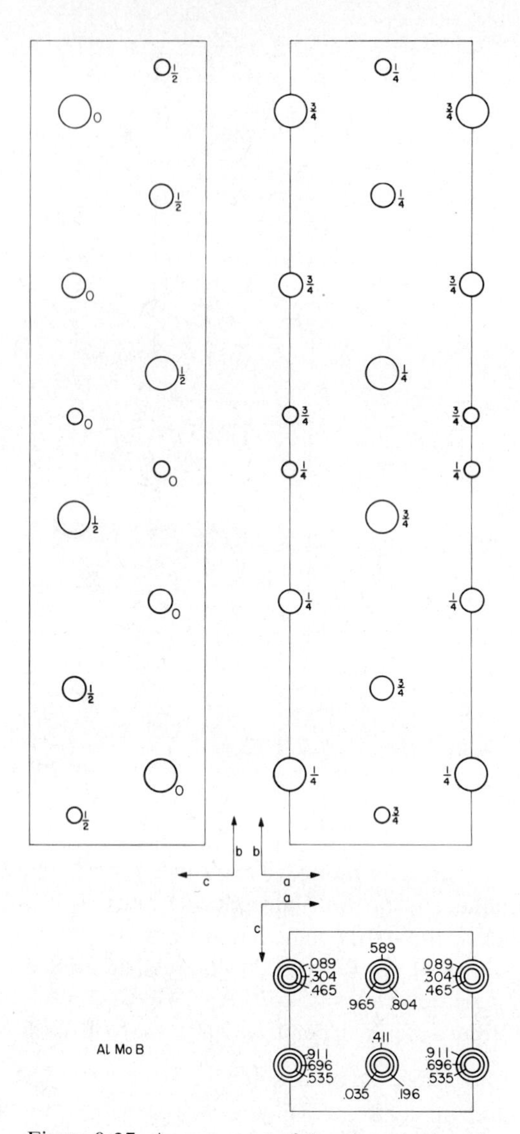

Figure 9-27. Arrangement of the atoms in the AlMoB ($oC12$) structure: Mo, largest circles; B, smallest circles.

20.93 Å, $Z = 3$.[49] The atoms form 3^6 and 6^3 nets parallel to (001). The stacking sequence in the notation of p. 4 is with Mo underlined: a$\underline{A}$(CBA)$\underline{C}$c$\underline{C}$(BAC)$\underline{B}$b$\underline{B}$(ACB)$\underline{A}$. The pair-stacked Mo layers are separated by a 6^3 net of boron so that Mo lies over the centers of the B hexa-

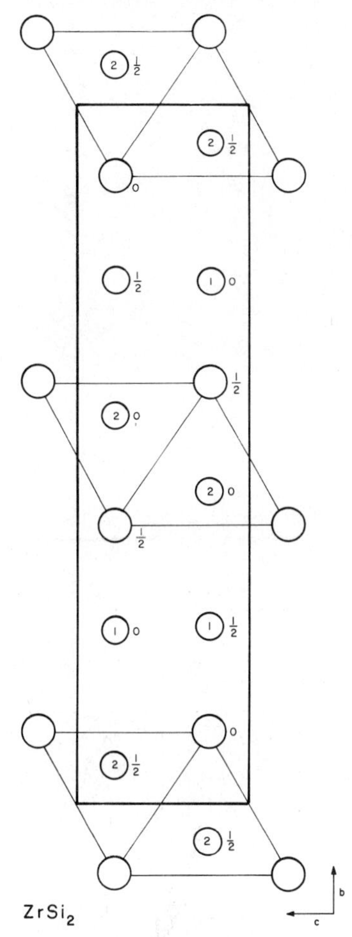

Figure 9-28. Atomic arrangement in the $ZrSi_2$ ($oC12$) structure viewed down [100]: Zr, large circles.

gons. The structure is therefore made up of layers of AlB_2 type cells separated by three 3^6 layers of B atoms packed in a close sequence (Figure 9-29) like the three 3^6 layers in the Laves phase unit γ(CAB). The structure requires confirmation by single-crystal methods.

UBC ($oC12$): The UBC structure with cell dimensions $a = 3.598$, $b = 11.97$, $c = 3.346$ Å, $Z = 4$,[50] is a filled-up CrB ($oC8$) type structure, with the cell somewhat elongated in the *b* direction. The U and B atoms occupy the same sites as Cr and B, and the C atoms in UBC are inserted so as to lie just outside the centers of the external rectangular prism faces of the

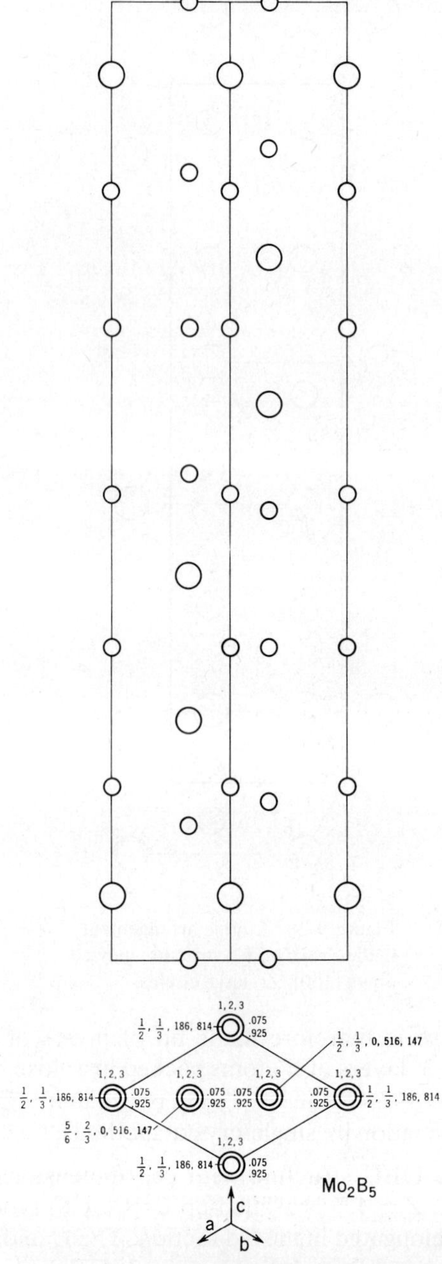

Figure 9-29. Atomic arrangement in the Mo_2B_5 ($hR7$) structure: Mo, large circles.

layers of triangular prisms, which lie parallel to the (010) plane (Figure 9-30). Thus each B atom centering the triangular prism of U atoms, in addition to having 6 U at 2.69 and 2.72 Å and 2 B neighbors at 1.87 Å, has one C neighbor at 1.64 Å. Each U has 6 B neighbors, 1 C at 2.35 Å and 4 C at 2.50 Å, as well as 4 U neighbors. Each carbon has 1 B and 5 U neighbors distributed in a distorted octahedron. Structural details require confirmation by neutron diffraction.

Mo_2BC (oC16): Mo_2BC has cell dimensions $a = 3.086$, $b = 17.35$, $c = 3.074$ Å, $Z = 4$.[51] The structure is a filled up CrB (*oC*8) type with double

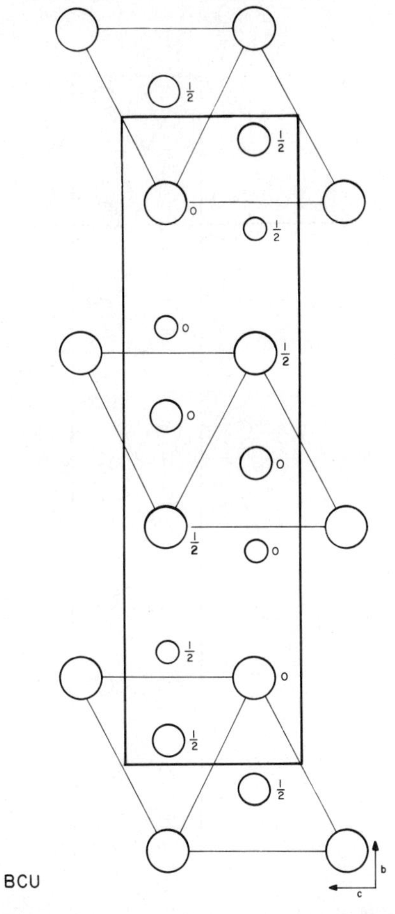

Figure 9-30. Atomic arrangement in the UBC (*oC*12) structure viewed down [100]: U, largest circles; C, smallest circles.

layers of slightly rumpled 4^4 nets of Mo(2) and C atoms inserted between the slabs of boron-centered triangular prisms of Mo(1) atoms, which lie parallel to the (010) plane (Figure 9-31). Each of the 4^4 nets is made up of Mo(2) and C atoms alternately, and the Mo atoms of one net lie over carbon atoms in the other 4^4 net so that each C is surrounded by 5 Mo(2) and *vice versa*. The C–Mo(2) distances are 1×2.11 Å and 4×2.17 Å.[52] In addition, C has one Mo(1) neighbor at 2.08 Å which completes the octahedron of Mo surrounding it. The B atoms are surrounded by a triangular prism of 6 Mo(1) atoms at 2.31 Å, and each B has 2 B neighbors at 1.79 Å,

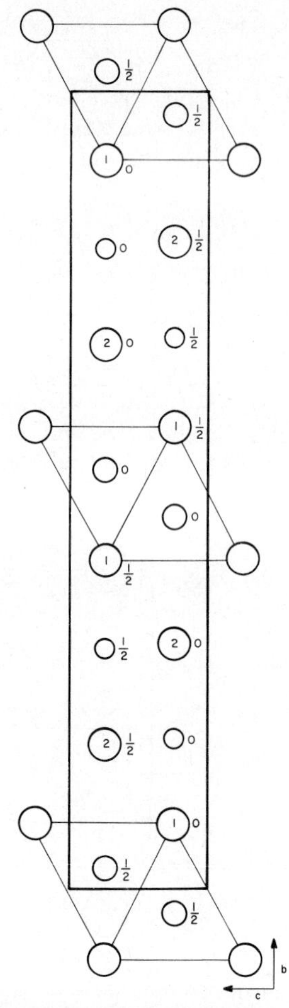

Fig. 9-31. Arrangement of the atoms in the $BCMo_2$ (*oC*16) structure: Mo, largest circles; C, smallest circles.

giving the zigzag chains of B atoms which run in the c direction. Mo(1) has 6 B and 1 C neighbors. It also has 10 Mo neighbors at 2.93 to 3.09 Å. Mo(2) has 5 C neighbors and 12 Mo at 2.94 to 3.10 Å. The B and C atom positions were determined accurately.[52]

$SmSb_2$ (oC24): $SmSb_2$ has cell dimensions $a = 6.171$, $b = 6.051$, $c = 17.89$ Å, $Z = 8$.[53] The structure can be considered as made up of slabs of triangular prisms of Sm atoms parallel to the (001) plane. These can be selected with their axes either along [100] or along [010]. The unit cell contains two slabs of prisms which are separated from each other. The two rectangular prism faces within the prism slabs contain Sb(2) atoms at $\frac{1}{4}$ and $\frac{3}{4}$ of the prism height and the third outer face is centered by Sm. Sb(1) atoms lie just outside the centers of the four edges of this face (Figure 9-32). Sm has 9 Sb neighbors at 3.13 to 3.48 Å; Sb(1) has 5 Sm neighbors and 1 Sb(1) at 2.72 Å. Sb(2) has 4 Sm and 4 Sb(2) at 3.03 and 3.09 Å. The closest Sm–Sm approach is 4.20 Å.

$NiZrH_{2.7}$ (oC ~ 19): $NiZrH_{2.7}$ has cell dimensions $a = 3.53$, $b = 10.48$, $c = 4.30$ Å, $Z = 4$.[54] The structure can be regarded as a filled up CrB (*oC*8) type. The Zr atoms form slabs of triangular prisms which lie parallel to the (010) plane and have their prism axis in the a direction. These are centered by Ni atoms and also base centered by H(1). Between the slabs of prisms there are inserted two rectangular 4^4 layers of H(2) atoms as indicated in Figure 9-33. Ni has 4 H neighbors at 1.77 and 1.79 Å

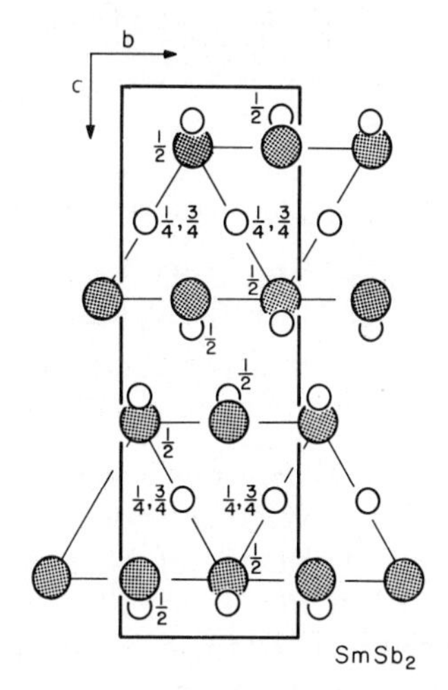

Figure 9-32. Atomic arrangement in $SmSb_2$ (*oC*24) viewed along [100]: large circles, Sm.

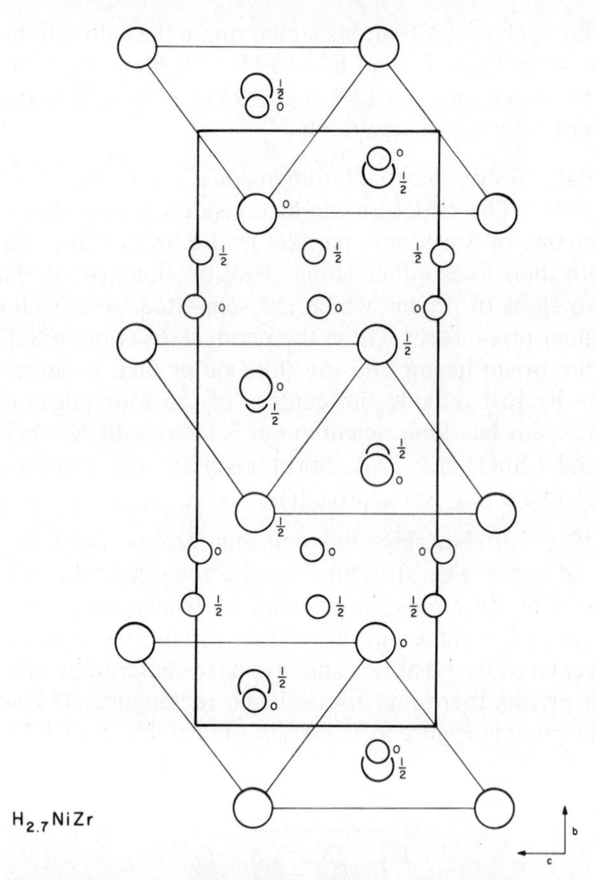

Figure 9-33. Arrangement of the atoms in $NiZrH_{2.7}$ ($oC \sim 19$) viewed down [100]: Zr, largest circles; H, smallest circles.

and two Ni at 2.60 Å. Zr has 3 H at 1.93 Å and 2 × 1.99 Å, 4 at 2.15 Å and 2 at 2.37 Å. H(1) has 2 Ni and 1 Zr neighbors and H(2) has one close Ni, one close Zr and two further Zr neighbors at 2.15 Å. The H atom positions were determined by powder neutron diffraction.

$NiMg_2$ (*hP*18): $NiMg_2$ has cell dimensions $a = 5.19$, $c = 13.22$ Å, $Z = 6$.[55] The Mg atoms have 9 equally spaced layers along c. They are separated into groups of three layers by planes of Ni atoms at $z = \frac{1}{6}$, $\frac{1}{2}$, and $\frac{5}{6}$, the Ni atoms being arranged in lines in the [010], [100], and [110] directions respectively. The groups of 3 Mg layers are arranged on very large 3^6, 6^3, and 3^6 nets in AaA type stacking (p. 4), so that the nodes of the 3^6 nets center the hexagons of the 6^3 nets. Each group of 3 Mg layers is oriented at 60° to neighboring groups of 3 layers. The structure resembles

that of $CuMg_2$ in which there are four groups of 3^6, 6^3, and 3^6 Mg layers. Geometrically it can be regarded as a derivative of the AlB_2 type, since slabs of large centered trigonal prisms of Mg are separated by displacement and rotation and layers of Ni atoms are inserted between the slabs.

$CuMg_2$ (oF48): $CuMg_2$ has cell dimensions $a = 5.284$, $b = 9.07$, $c = 18.25$ Å, $Z = 16$.[55] The atoms are arranged in layers parallel to the (001) plane. Planes of Cu atoms occur at $z \sim \frac{1}{8}, \frac{3}{8}, \frac{5}{8}$, and $\frac{7}{8}$ in which the atoms lie in lines in the [110], [$\bar{1}$10], [110], and [$\bar{1}$10] directions respectively. Between each of these layers there are three layers of Mg atoms arranged on very large 3^6, 6^3, and 3^6 nets so that the atoms of the 3^6 nets lie over the centers of the hexagons of the 6^3 nets. The four groups of three Mg layers are stacked on different sites, neighboring groups of three layers being displaced parallel to the (001) plane, relative to each other. The atomic arrangement resembles that in $NiMg_2$, where there are three groups of 3^6, 6^3, and 3^6 nets of Mg atoms. Geometrically, the structure can be regarded as a derivative of the AlB_2 structure, since slabs of filled triangular prisms of Mg are separated by displacement and Cu layers are inserted between the slabs.

$Th_{0.9}Ge_2$ (oC12): $Th_{0.9}Ge_2$ has cell dimensions $a = 16.64$, $b = 4.023$, $c = 4.160$ Å, $Z = 4$.[56] The Th atoms form 3^34^2 nets parallel to the (001) plane. The structure is a filled up variant of the CrB (*oC*8) structure which can be compared with the $ZrSi_2$ (*oC*12) type. In $Th_{0.9}Ge_2$ the Th atoms form slabs of triangular prisms which lie parallel to the (100) plane with their axes in the [001] direction. These prisms are centered by Ge(3) atoms (Ge–4 Th = 3.06 Å, Ge–2 Th = 3.39 Å), but the arrangement differs from that in the CrB and $ZrSi_2$ structures, since the Th atoms of different slabs of prisms lie directly over each other in projection on the basal plane of the structure (Figure 9-34). 4^4 nets of Ge(1) and (2) atoms are inserted between the slabs of Th prisms as in the $ZrSi_2$ structure. Each Th atom has 10 Ge neighbors at 3.06 to 3.39 Å, and 6 Th at 4.02 to 4.19 Å. Ge(1) has 4 Th neighbors at 3.07 Å and 4 Ge(2) at 2.89 Å. Ge(2) similarly has 4 Th at 3.12 Å and 4 Ge(1) neighbors. In addition to the 6 Th neighbors, each Ge(3) has 2 Ge(3) neighbors at 2.62 Å, giving chains of Ge(3) running in the [010] direction which are comparable to the boron chains of the CrB structure.

$AlMn_2B_2$ (oC10): $AlMn_2B_2$ has cell dimensions, $a = 2.92$, $b = 11.08$, $c = 2.89$ Å, $Z = 2$.[57] The Mn atoms form a 3^34^2 net parallel to the (001) plane at $z = \frac{1}{2}$. The cubes formed by these nets are centered by Al at $z = 0$ and the triangular prisms are centered by B, also at $z = 0$ (Figure

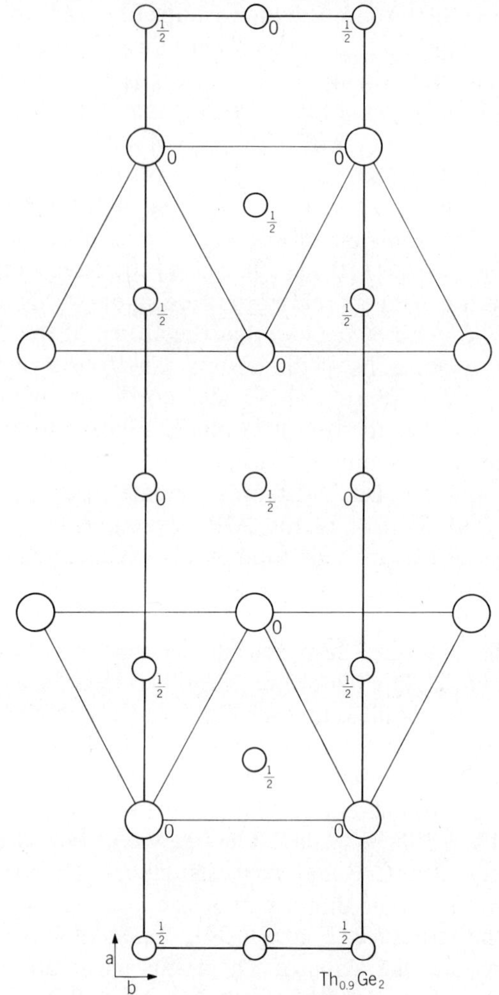

Figure 9-34. Atomic arrangement in the $Th_{0.9}Ge_2$ (*oC*12) structure viewed along [001]: Th, large circles.

9-35), so that zigzag chains of B atoms run in the [100] direction (B–2B = 1.72 Å). Mn has 6 Mn neighbors at 2.76 to 2.92 Å, 6 B at 2.17 Å, and 4 Al at 2.60 Å. Al has 8 Mn at 2.60 Å, 4 Al at 2.89 and 2.92 Å, and 2 B at 2.30 Å.

MoB (L.T.) (*tI* 16): The MoB structure ($a = 3.110$, $c = 16.95$ Å, $Z = 8$)[49,58] is built up of 4^4 nets of Mo or B atoms stacked in the *c* direction in the sequence of two Mo layers followed by two B layers. Nodes of the various

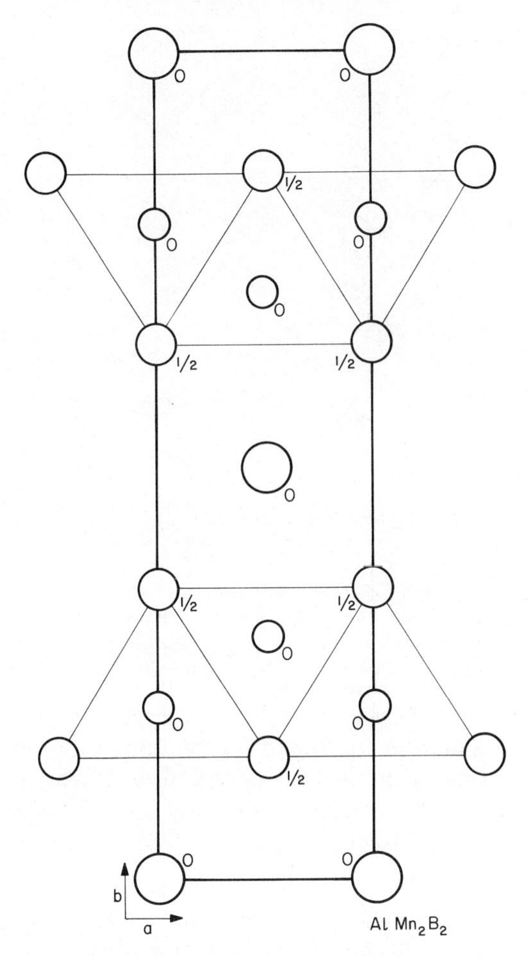

Figure 9-35. Arrangement of the atoms in the $AlMn_2B_2$ (*oC*10) structure viewed down [001]: Al, largest circles; B, smallest circles.

4^4 nets lie either over cell corners, midpoints of the basal edges or over the base center of the cell. The Mo atoms have 6 B neighbors at 2.23 or 2.345 Å and one B at 2.64 Å, together with 6 close Mo neighbors at 2.86 Å and 4 more at a rather larger distance (3.11 Å). Each B atom has two close B neighbors and 6 Mo neighbors which surround it at the corners of a triangular prism. This arrangement results in independent layers of triangular prisms of Mo atoms lying parallel to the basal plane of the cell, with the prism axes of successive layers lying alternately along the *a* and *b* directions (Figure 9-36). All prisms are centered by B atoms which

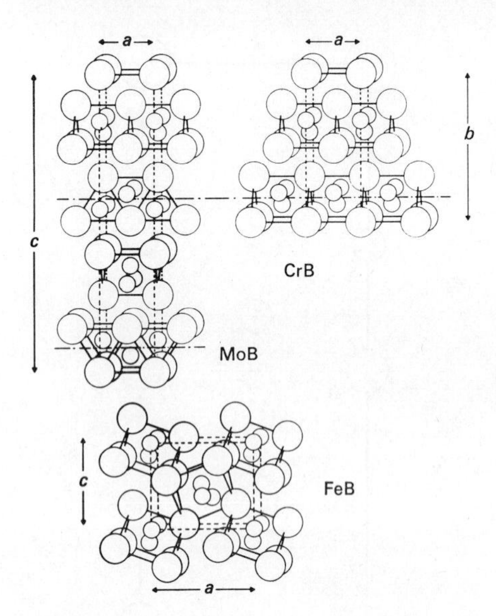

Figure 9-36. Pictorial view of the atomic arrangements in the low temperature MoB (*tI*16), CrB (*oC*8), and the FeB (*oP*8) structures. (Schob and Parthé [42].)

form chains alternately in the b and a directions on proceeding along c from one prism layer to the next. The structure which is a variant of the CrB type, requires to be confirmed by single-crystal methods, although GaZr was reported to be isotypic as the result of a single-crystal study.

α-ThSi$_2$ (H.T.) (*tI* 12): The α-$ThSi_2$ structure with cell dimensions $a = 4.135$, $c = 14.375$ Å, $Z = 4$,[59] is made up of a succession of layers of 4^4 nets of Si or Th atoms in the sequence Th (at $z = 0$) followed by two Si layers, which repeats four times along the c axis in the unit cell. The 4^4 nets have their nodes above the cell corners, centers of the cell edges, or above the center of the basal plane. The arrangement is such as to give contiguous layers of triangular prisms of Th atoms which are *all* centered by Si atoms, the prism axes in one layer being rotated 90° in the basal plane relative to those of the prism layers above and below it, as shown in Figure 9-37. The structure is therefore a filled up NbAs or $AgTlTe_2$ (*tI*8) type, differing in the centering of all of the prisms instead of only half of them. Each Si has 6 close Th neighbors at 3.16 Å; it also has three close Si neighbors at 2.39 and 2.40 Å which result from a three connected framework shown diagramatically in Figure 9-38. This framework is reminiscent of the planar graphite net, but its simultaneous existence in

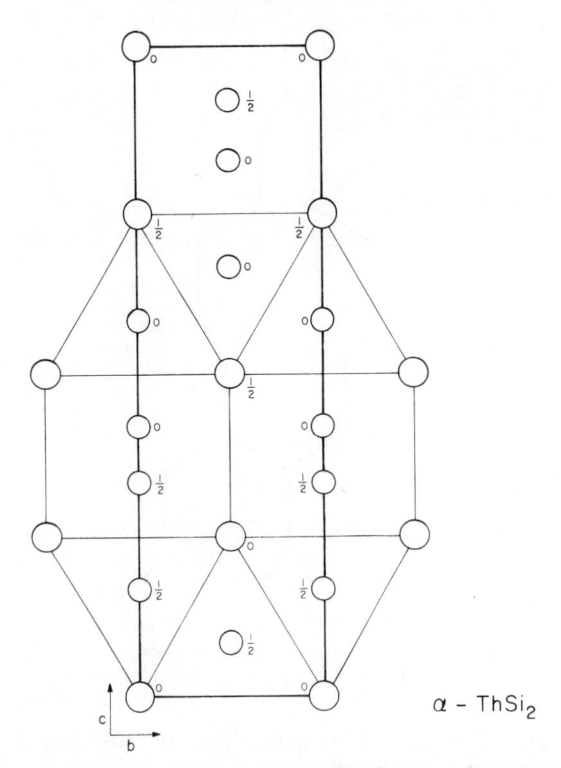

Figure 9-37. Atomic arrangements in the high temperature α-$ThSi_2$ ($tI12$) structure viewed down [100]: Th, large circles.

planes parallel to both (100) and (010), through interconnections, has the result that one corner of each hexagon is always missing. Th has 12 Si neighbors and 8 Th at 4.13 to 4.15 Å.

Some 25 binary phases are known to take the α-$ThSi_2$ structure (MN_2). The N component is always a Group III or IV atom, and its sites are incompletely occupied in many of the phases, deficiencies of 30% ($MN_{1.4}$) being common. The phases form from components with a relatively narrow range of radius ratios from 1.16 to 1.42, the range from 1.3 to 1.4 being most popular. Axial ratios for the unit cell vary from 3.10 to 3.75. The near-neighbor diagram indicates that the stability of phases with the structure is controlled by the chemical bond, rather than the coordination factor (p. 55). Cell dimensions are primarily controlled by the M–N contracts with the N–N contacts exerting a secondary influence.

Compared to the sum of the elemental volumes, the unit cell volume indicates a compression of some 5 to 15% for phases with the α-$ThSi_2$

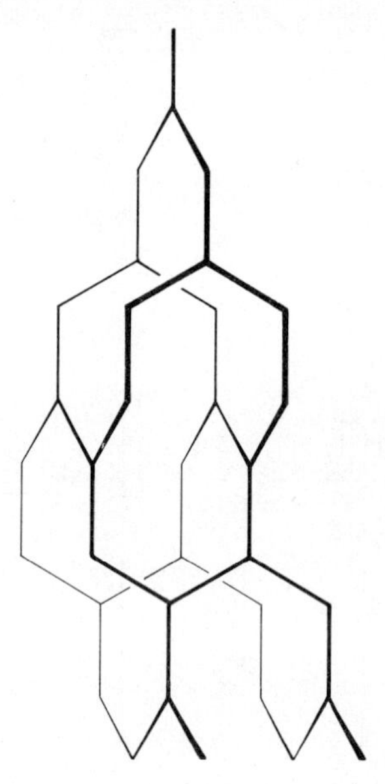

Figure 9-38. The silicon atom framework in the α-$ThSi_2$ structure.

structure (Figure 3-4, p. 57), reflecting the compression of the *N* atoms indicated by the near-neighbor diagram.

α-$GdSi_2$ ($GdSi_{1.4}$) (oI 12): α-$GdSi_2$ with cell dimensions $a = 4.09$, $b = 4.01$, $c = 13.44$ Å, $Z = 4$,[60] is a filled up form of the NbAs or $AgTlTe_2$ structures (*tI*8), which differs only in details and symmetry from the α-$ThSi_2$ structure (*tI*12). The Gd atoms form contiguous slabs of triangular prisms parallel to the (001) plane with their axes running in either the *a* or *b* direction, the axes of successive layers of prisms being rotated 90° in the (001) plane. All prisms are centered either by strings of Si(2) atoms running in the *b* direction, or by strings of Si(1) atoms which run in the *a* direction (Figure 9-39).

Gd has 12 Si neighbors at distances from 2.95 to 3.10 Å and 8 Gd neighbors at distances of 3.91 to 4.09 Å. Si(1) has 2 Si(1) at 2.265 Å, one Si(2) at 2.39 Å and 6 Gd(1) at 2×2.95 and 4×3.10 Å. Si(2) has 2 Si(2) at 2.234 Å, 1 Si(1) neighbor and 6 Gd neighbors (2×2.98 and 4×3.10 Å).

FeB (oP8): In the FeB structure ($a = 5.506$, $b = 2.952$, $c = 4.061$ Å, $Z = 4$)[61] the Fe atoms form columns of triangular prisms by each sharing two rectangular faces. These run in the [010] direction. Each prism shares

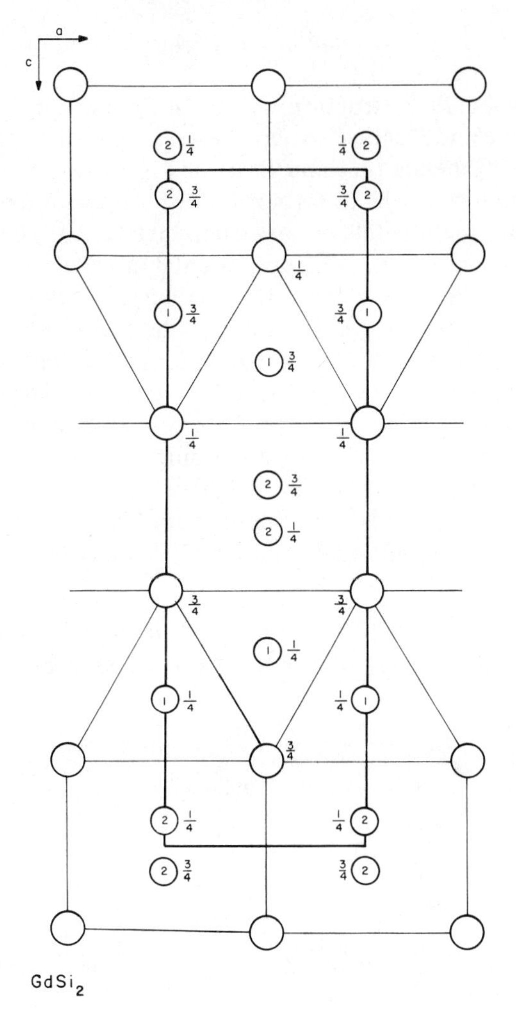

Figure 9-39. Atomic arrangement in the α-$GdSi_2$ ($oI12$) structure viewed down [010]: large circles, Gd.

one basal edge at each end and the apexes opposite these edges with 4 other columns of prisms as shown in Figure 9-36, p. 518. The columns of prisms are otherwise isolated and alternate rows of columns in the *a* direction are canted about + and − 20° to the direction of the *a* axis. The prisms are centered by B atoms which form chains running in the *b* direction. The structure is compared with the CrB ($oC8$) and MoB ($tI16$) types in Figure 9-36.

Each B is surrounded by 6 Fe prism neighbors at 2.145 to 2.17 Å, and

2 B at 1.80 Å. Fe has 6 B neighbors at 2.14 to 2.17 Å, one at 2.21 Å and 10 Fe neighbors at 2.63 to 2.95 Å.

Phases with the FeB structure can be separated into two groups as noted by Hohnke and Parthé:[43] group 1 phases are formed between transition or rare earth metals (*M*) and B, Si or Ge (*N*); group 2 phases are formed between rare earths or Group III or IV transition metals (*M*) and Group VIII transition metals or Cu Group metals (*N*). These two groups of phases are separated on a diagram of axial ratios a/b *vs.* c/b, although all together they lie in a narrow band (Figure 9-40). This separation corresponds to a difference in the ratio of the height to base of the triangular prisms formed by the *M* components. Group 1 phases have prism-edge ratios greater than 0.95 and generally greater than unity; group 2 phases have squashed prisms. Contours of constant prism-edge ratio are shown on Figure 9-40. This apparent difference between group 1 and 2 phases is not reflected in the near-neighbor diagram (Figure 9-41). The phases are distributed parallel to the lines for *M–N* contacts, which are generally somewhat compressed, and they lie mainly between these lines and those for *M–M* contacts.

NiY (*mP*8): NiY has a distorted FeB type structure which results particularly from the *x* parameters of the Ni and Y atoms departing from $\pm\frac{1}{4}$.[62]

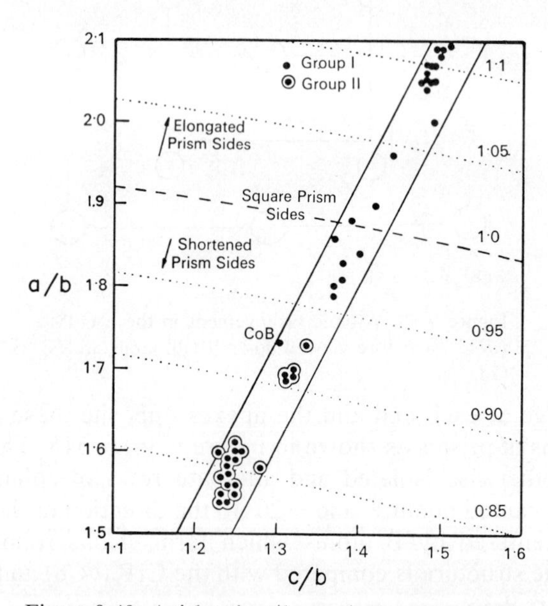

Figure 9-40. Axial ratio a/b *vs.* c/a for phases with the FeB (*oP*8) structure. (Hohnke and Parthé [43].)

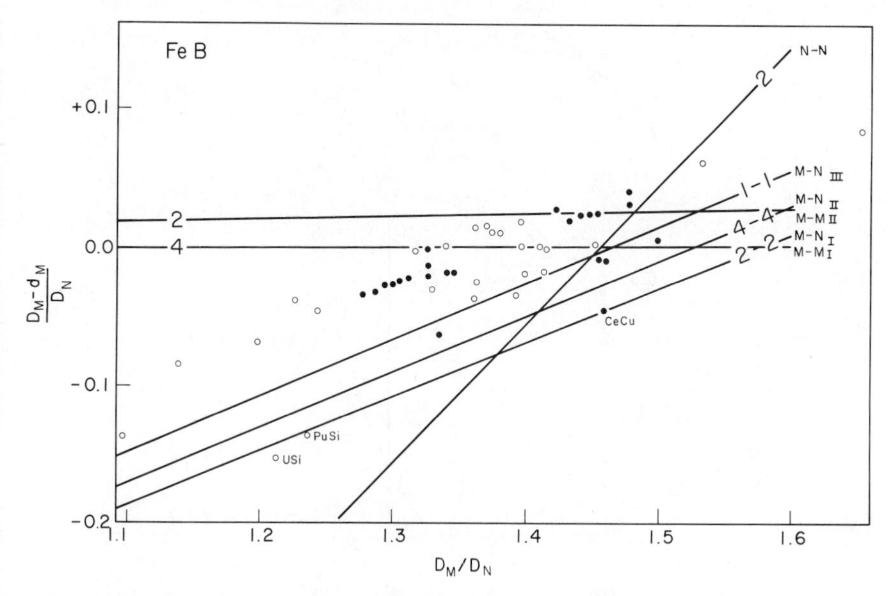

Figure 9-41. Near-neighbor diagram for phases with the FeB structure (MN). Drawn for $x_M = 0.180$, $z_M = 1/8$, $x_N = 0.031$, $z_N = 0.620$, $b/a = 0.5361$, $c/a = 0.7376$. Open circles: phases with ratio of prism height to base >1; filled circles: those with ratio <1. Numbers indicate number of neighbors involved in contacts N–N, M–N, etc.

Ni_4B_3 (oP28): The orthorhombic form of Ni_4B_3 has cell dimensions $a = 11.95$, $b = 2.981$, $c = 6.568$ Å, $Z = 4$.[63] The Ni atoms form an array of interconnected triangular prisms with axes either along [010] or in the (010) plane (Figure 9-42). The B(3) atoms centering prisms with axes along [010] are isolated; B(1) and (2) centering the other prisms form zig-zag chains running in the [010] direction (B–B, 1.73 or 1.89 Å), and they also have one Ni neighbor out through a rectangular prism face. In addition to the 6 prism Ni surrounding B(3), there are 3 other Ni neighbors out through the rectangular prism faces.

W_2CoB_2 (oI10): W_2CoB_2 with cell dimensions $a = 7.075$, $b = 4.561$, $c = 3.177$ Å, $Z = 2$,[64] is made up of triangular prisms of 4 W and 2 Co which are centered by B. The prisms form rows parallel to the (100) plane by sharing alternately along [010], the rectangular face made up of 4 W and the edge made up of 2 Co, and sharing triangular faces along [001] (Figure 9-43). Thus the B atoms occur in pairs at 1.82 Å; Co–B = 2.10, W–B = 2.34 and 2.50 Å.

Ti_3P (tP32): Ti_3P has cell dimensions $a = 9.959$, $c = 4.987$ Å, $Z = 8$.[65] The P atoms are surrounded by 6 Ti forming a distorted triangular prism

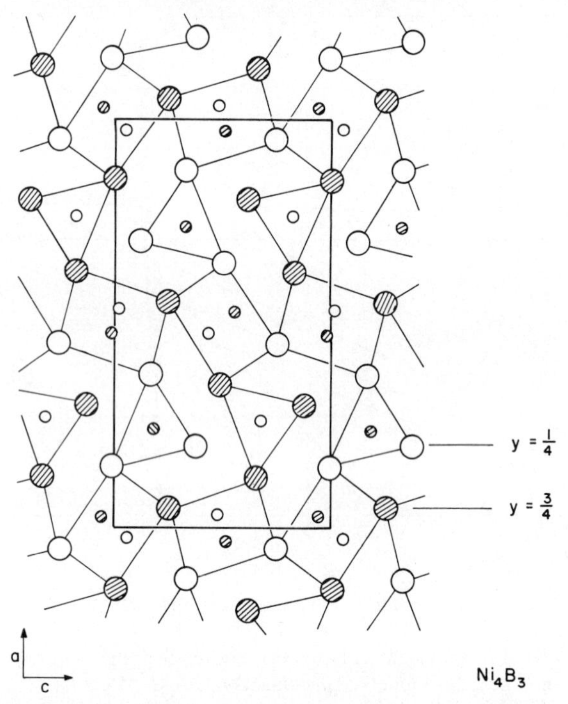

Figure 9-42. Atomic arrangement in the Ni_4B_3 (*oP*28) structure viewed down [010]: Ni, large circles.

(P–Ti 2.49 to 2.57 Å) and three more Ti neighbors out through the rectangular faces of the prism at distances 2.60, 2.61, and 2.63 Å. The Ti prisms are very distorted, heights being 3.11, 3.70, and 3.79 Å and basal edges 2.84 to 3.44 Å. The shortest Ti–Ti distance in the structure is 2.78 Å. Zr_3P and Nb_3P have the same structure.

MoP_2 (*oC*12): MoP_2 has cell dimensions $a = 3.145$, $b = 11.184$, $c = 4.984$ Å, $Z = 4$.[66] The Mo atoms are surrounded by triangular prisms of 2 P(1) and 4 P(2) atoms which share edges between P(2) atoms (Figure 9-44). Thus Mo has 2 P(1) neighbors at 2.52 Å and 4 P(2) at 2.47 and 2.51 Å. There is a further P(1) neighbor at 2.45 Å out through one of the rectangular faces of the prism. Nearest Mo–Mo neighbors are 2 at 3.145 Å and 2 at 3.25 Å. P(1) has 3 Mo neighbors and one P(2) at 2.17 Å. P(2) has the P(1) neighbor and 4 Mo neighbors. The structure needs confirmation by single-crystal methods.

$Ru_{11}B_8$ (*oP*38): $Ru_{11}B_8$ has cell dimensions $a = 11.61$, $b = 11.34$, $c = 2.836$ Å, $Z = 2$.[67] The structure is built up of interconnected triangular

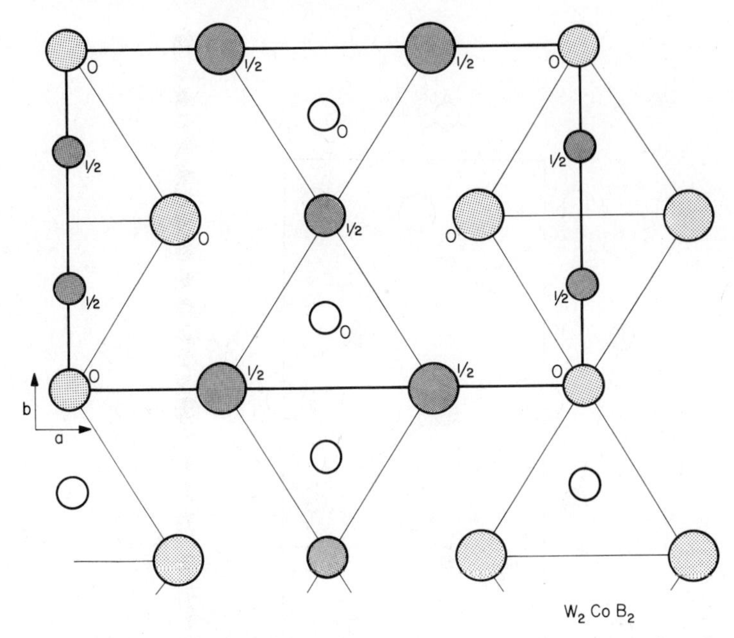

Figure 9-43. Atomic arrangement in the W_2CoB_2 ($oI10$) structure viewed down [001]: W, largest circles; B, smallest circles.

prisms of Ru atoms (Figure 9-45), and it contains boron chains of the type shown in the Figure, which run in the [001] direction. The B(1) atoms have no boron neighbors, but 3 Ru neighbors out through the rectangular faces of the prisms. B(3) has two such neighbors; B(2) has one and B(4) none. Ru–Ru distances average 2.79 Å (shortest 2.58 Å) and Ru–B distances 2.18 Å.

Re_3B, $CoPu_3$ ($oC16$): Re_3B has cell dimensions $a = 2.890$, $b = 9.313$, $c = 7.258$ Å, $Z = 4$.[68] The structure contains independent columns of triangular prisms of Re along a which is the direction of the prism axes. The prisms are centered by B as shown in Figure 9-46. The structure is thus a development from the CrB type with the prism layers further separated so as to give columns only. B is surrounded by 4 Re(2) and 2 Re(1) at 2.23 Å, arranged in a triangular prism. It also has 2 Re(2) at 2.53 Å and one Re(1) at 2.96 Å through the centers of the rectangular faces of the prisms. Re(2) has 3 B and 11 Re neighbors at 2.66 to 3.04 Å. Re(1) has 2 B neighbors and 10 Re at 2.75 to 3.04 Å.

$CoPu_3$ with cell dimensions $a = 3.475$, $b = 10.98$, $c = 9.220$ Å, $Z = 4$,[69] has the same structure as Re_3B. The Co–Pu distances in the triangular prism of 6 Pu surrounding Co are 4×2.69 and 2×2.80 Å. The Pu–Pu

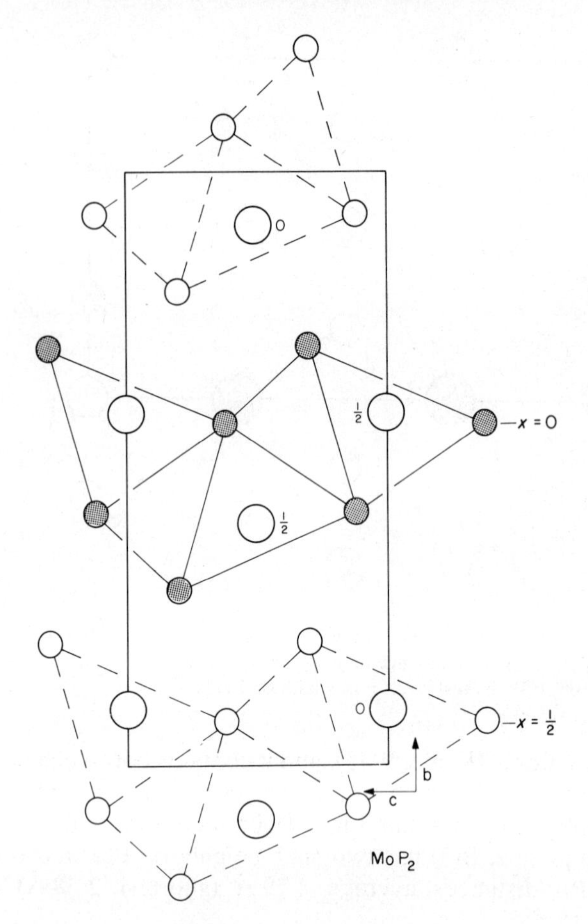

Figure 9-44. Arrangement of the atoms in the MoP_2 ($oC12$) structure viewed down [100]: Mo, large circles.

distances in the 10 Pu surrounding Pu(1) are 3.36 to 3.475 Å. Pu(2) on the other hand has one close Pu(2) neighbor at 3.08 Å.

$NbAs_2$ (mC 12): $NbAs_2$ has cell dimensions $a = 9.357$, $b = 3.382$, $c =$ 7.792 Å, $\beta = 119.46°$, $Z = 4$.[70] The As atoms form pairs of distorted triangular prisms which form columns running in the [010] direction (Figure 9-47). These prisms are centered by Nb (Nb–6As at 2.54 to 2.82 Å). Nb also has 2 As out through rectangular prism faces (2.66 and 2.74 Å) and one Nb at 3.01 Å through the shared prism face. As(1) has 5 Nb neighbors and As(2) three. As(2) has a close As neighbor at 2.45 Å and two other neighbors at 2.84 Å.

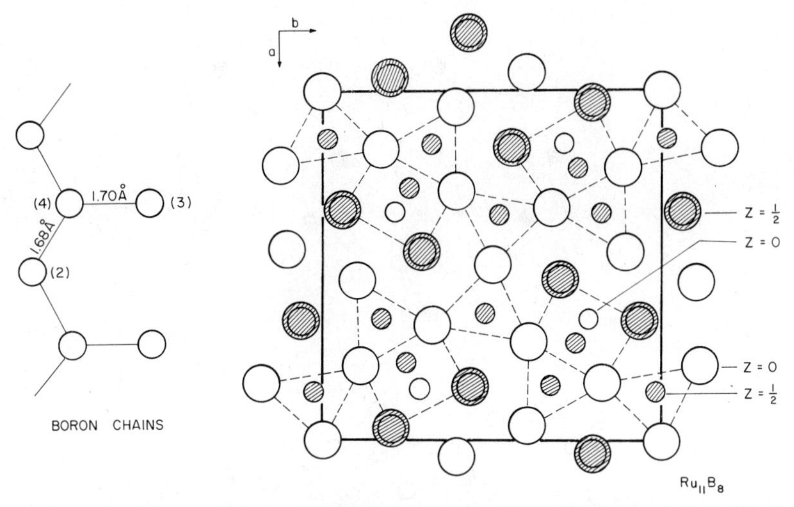

Figure 9-45. The $Ru_{11}B_8$ (*oP*38) structure viewed down [001] (Ru, large circles). The form of the boron chains is indicated on the left of the diagram.[67]

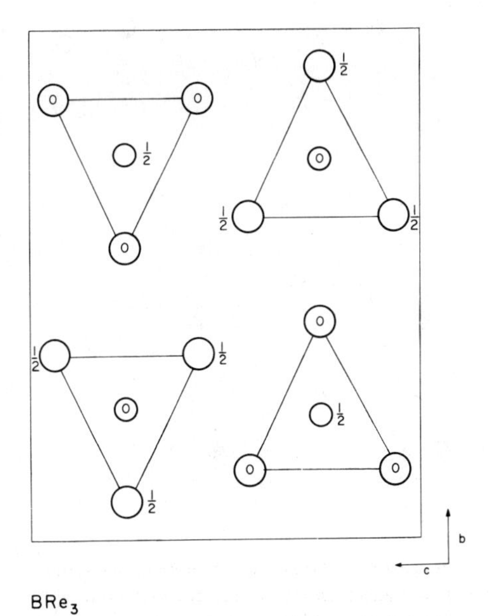

Figure 9-46. The Re_3B (*oC*16) structure viewed down [100]: Re, large circles.

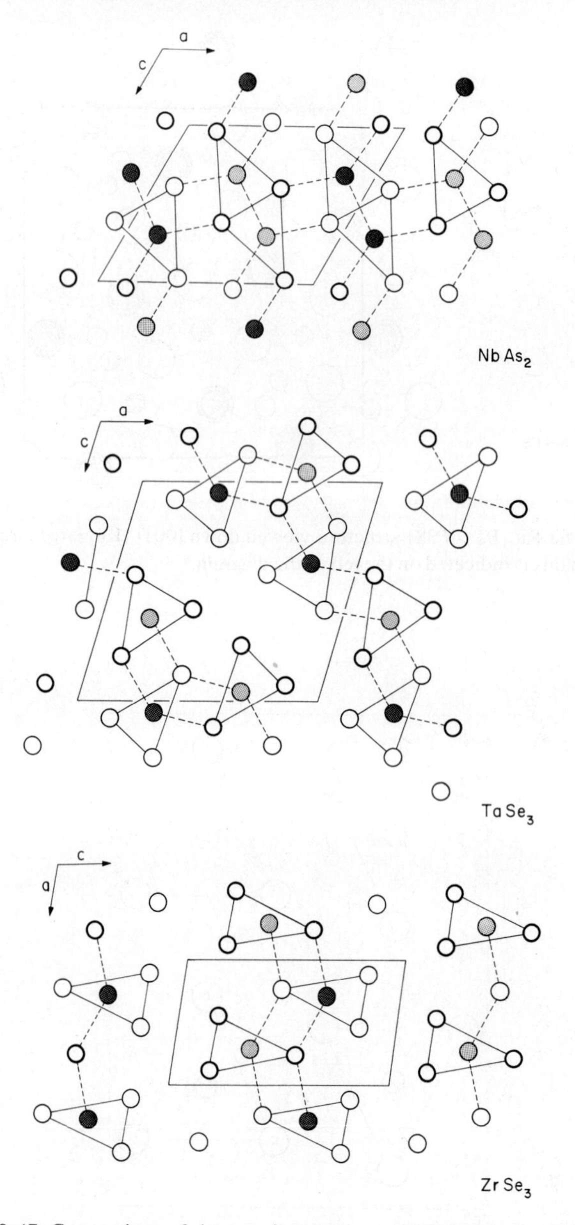

Fig. 9-47. Comparison of the atomic arrangements in the $NbAs_2$ (*mC*12), $TaSe_3$ (*mP*16), and $ZrSe_3$ (*mP*8) structures viewed down [010].[70] Shaded circles are *T* metal atoms. Thick lined or black atoms are at $y \sim 0$ ($NbAs_2$) and $y \sim \frac{3}{4}$ ($TaSe_3$ and $ZrSe_3$), the remainder are at $y \sim \frac{1}{2}$ ($NbAs_2$), and $y \sim \frac{1}{4}$ ($TaSe_3$ and $ZrSe_3$).

ZrSe$_3$ (*mP*8): The monoclinic cell ($a = 5.41$, $b = 3.77$, $c = 9.45$ Å, $\beta = 97.5°$)[71] contains columns of wedge-shaped triangular prisms of Se atoms that are centered by Zr (6 Se–Zr at 2.73 to 2.75 Å). The columns run in the [010] direction and are separated from each other (Figure 9-47), but the Se atoms of one prism also make contacts to Zr in the neighboring prisms through the centers of the two large rectangular prism faces (2.87 Å) as shown in Figure 9-48. Se(2) and (3) have 2 Zr neighbors each at 2.73 and 2.745 Å respectively, whereas Se(1) has 4 Zr neighbors at 2.73 to 2.89 Å. Zr thus has CN 8, and the columns of Se prisms are linked by bonds from Zr at the center of one prism, to Se of neighboring prisms to form slabs parallel to the (001) plane. These

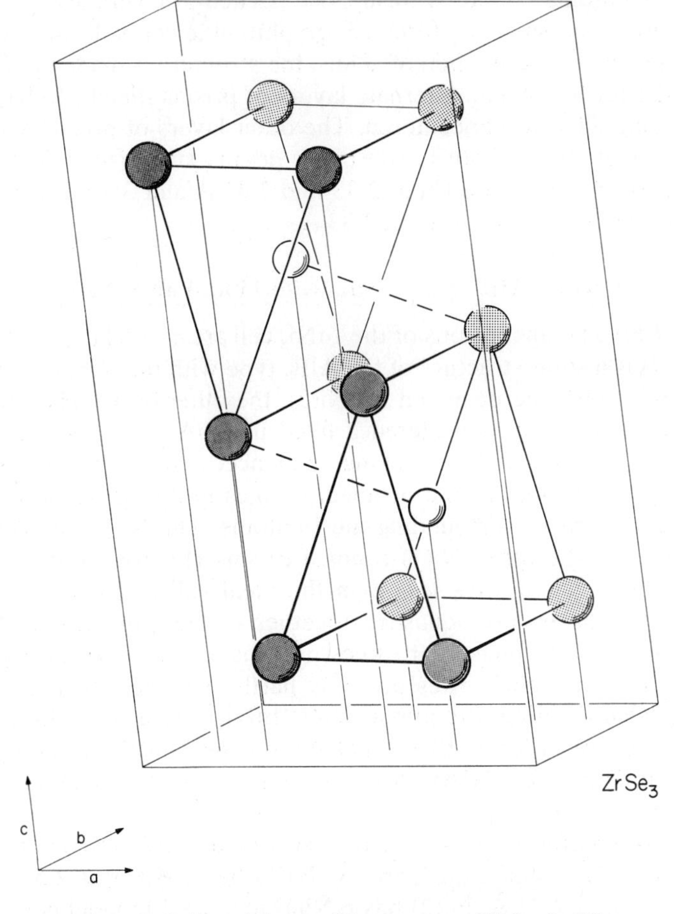

Figure 9-48. Pictorial view of the structure of $ZrSe_3$ (*mP*8).

slabs are well separated from neighboring slabs in the [001] direction, the closest Se–Se distance being 3.87 Å.

$TaSe_3$ (*mP*16): $TaSe_3$ has cell dimensions $a = 10.411$, $b = 3.494$, $c = 9.836$ Å, $\beta = 106.36°$, $Z = 4$.[72] The atoms form slightly rumpled layers at $y \sim \frac{1}{4}$ and $\frac{3}{4}$ (Figure 9-47). The Ta atoms are surrounded by distorted triangular prisms of Se with two more Se neighbors out through rectangular faces of the prisms (2.59 to 2.80 Å). The Se prisms are independent of each other in the (010) plane, forming columns in the [010] direction. The shortest Se–Se distance is 2.53 Å. The metal atoms are similarly coordinated in the $TaSe_3$, $ZrSe_3$, and $NbAs_2$ structures.

Cu_2Te (L.T.) (*hP*6): Cu_2Te has cell dimensions $a = 4.246$, $c = 7.289$ Å, $Z = 2$.[73] Te atoms on site A form close packed 3^6 layers at $z = \pm 0.160$, and Cu atoms on sites "a" form a 6^3 graphite-like net, with the Te atoms lying over the hexagons thereof. Thus, the structure is made up of layers of trigonal prisms of Te. Alternate layers of prisms (height to base ratio 0.666) along [001] are uncentered. The other layers of prisms each contain two Cu atoms at about $\frac{1}{4}$ and $\frac{3}{4}$ of the prism axis (prism height to base ratio 1.050). Cu has 1+3 Cu at 2.33 and 2.45 Å and 3 Te at 2.67 Å. Te has 6 Cu neighbors and one Te at 2.83 Å.

Superstructures of AlB_2: 6^3 Nets Subdivided into Two 3^6 Nets

$InNi_2$ (*hP*6): The dimensions of the $InNi_2$ cell are $a = 4.179$, $c = 5.131$ Å, $Z = 2$.[74] It is a superstructure of the AlB_2 type with the 6^3 net subdivided into two 3^6 nets, one occupied by Ni(2), the other by In. However, the structure is generally considered a filled up NiAs type. The In atoms have a c.p. hexagonal BC stacking sequence, and the Ni(1) atoms are located at the centers of the octahedral holes in this array, so that they form trigonal prisms surrounding the In atoms. The Ni(2) atoms occupy the centers of the other Ni(1) trigonal prisms (Figure 9-49). The axial ratio of the cell is always less than the ideal value of 1.63 so that the close packed layers are squashed together in the *c* direction. Over 30 binary and ternary phases are known to have the $InNi_2$ structure; in most of them the Ni(2) sites are only partly occupied. In the range of partially filled up NiAs structures (defect $InNi_2$), there has so far not been a recognition of ordering of the defects on the Ni(2) sites comparable to that found in defect NiAs structures in the range between the limiting NiAs and CdI_2 types.

The coordination of In is 11; it is surrounded by 6 Ni(1) at 2.73 Å, 3 Ni(2) at 2.41 Å and 2 at 2.565 Å. Ni(1) has 2 Ni(1) at 2.565 Å and 6 Ni(2)+6 In at 2.73 Å. Ni(2) has 6 Ni(1) and 3+2 In neighbors at 2.41 and 2.565 Å.

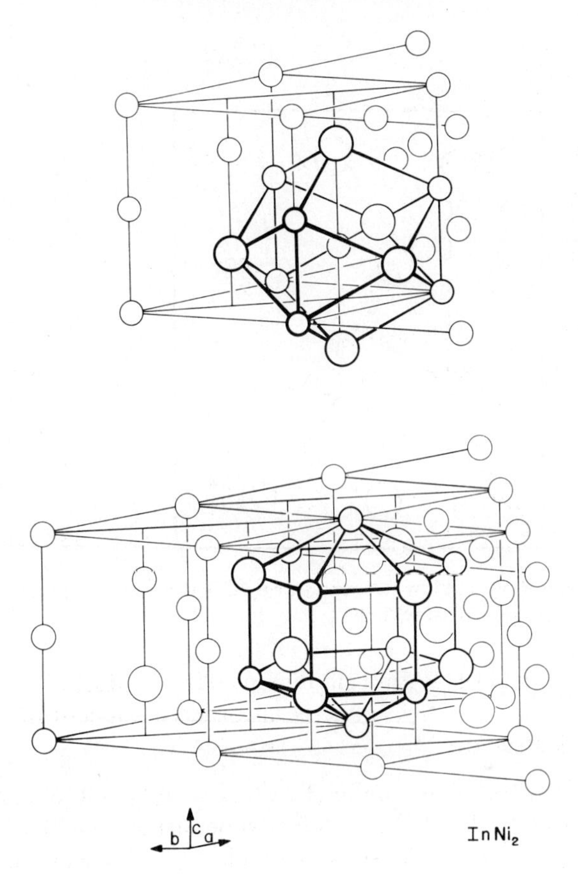

Figure 9-49. Views of the $InNi_2$ (*hP*6) structure showing coordination of the two types of Ni atoms (small circles).

Defect Superstructures of $InNi_2$

Sn_2Ni_3 (L.T.) (*oP*20): With cell dimensions $a = 7.11$, $b = 5.211$, $c = 8.23$ Å, $Z = 4$,[75] Sn_2Ni_3 is a superstructure derived from the high temperature "disordered" $InNi_2$ structure of the phase as indicated in Figure 9-50; $a = a_0 . \sqrt{3}$, $b = 2a_0$, $c = c_0$, where a_0, c_0 refer to a $InNi_2$-type pseudocell. The Ni(2) atoms have ordered sites and the other atoms are displaced slightly from their sites of the $InNi_2$ structure.

In_4Ni_7 (L.T.) (*mP*22): In_4Ni_7 which has cell dimensions $a = 7.40$, $b = 4.26$, $c = 10.46$ Å, $\beta = 90.1°$, $Z = 2$, is an ordered superstructure of $InNi_2$[75] but its structure and that of Ge_4Ni_7 require complete determination.

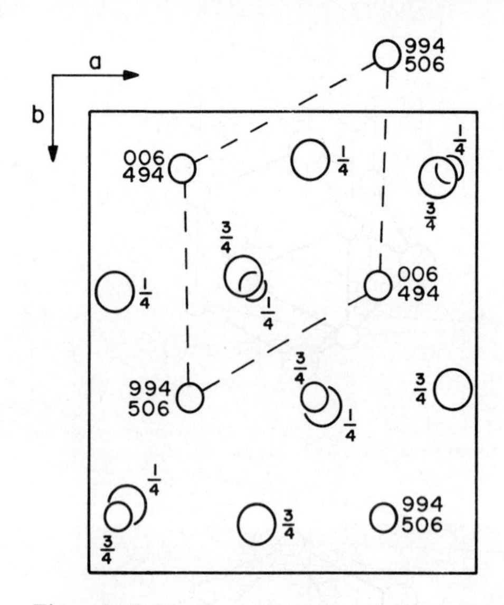

Figure 9-50. The low-temperature Sn_2Ni_3 ($oP20$) structure viewed along [001]: Sn, large circles.

Ge_4Ni_7 (L.T.) (*mC*44): With cell dimensions $a = 10.13$, $b = 7.80$, $c = 6.83$ Å, $\beta = 90.4°$, $Z = 4$, Ge_4Ni_7 is an ordered superstructure derived from the $InNi_2$ structure.[75]

η-$Ge_8(Fe,Ge)_{12.6}$ (*hP*20.6): η-$Ge_8(Fe,Ge)_{12.6}$ with cell dimensions $a = 7.984$, $c = 4.999$ Å, $Z = 1$,[76] is a defect superstructure of $InNi_2$ with $a \sim 2a_{InNi_2}$, $c \sim c_{InNi_2}$. The 2 Ni sitesets divide into 4 sitesets in the superstructure, and Fe vacant sites are confined to two of these sitesets, one of which has 0.56 Ge atoms, but the structure requires confirmation by single-crystal methods.

$SiCo_2$, $PbCl_2$, E-*TiNiSi* (*oP*12): $SiCo_2$ has cell dimensions $a = 4.918$, $b = 3.737$, $c = 7.109$ Å, $Z = 4$.[77] The structure is a distortion of the $InNi_2$ type as can be seen from the projection down [100]. Parallel to the (010) plane the Co and Si atoms form distorted square-triangle nets at $y = \frac{1}{4}$ and $\frac{3}{4}$ (Figure 9-51). The Si atoms of one net lie over the diamonds of the nets above and below, and the Co atoms lie over the distorted squares. This arrangement surrounds Si with 6 Co at the corners of a right triangular prism and gives it 4 more Co neighbors out through the three prism faces (two through one face, Figure 9-52), the distances being 2.32 to 2.57 Å. Co(1) has 5 Si at 2.32 to 2.62 Å and 8 Co at 2.52 to 2.68 Å. Co(2) also has 5 Si neighbors (2.34 to 2.57 Å) and 8 Co at 2.50 to 2.68 Å.

$SiCo_2$ is representative of a group of $PbCl_2$ structure phases having

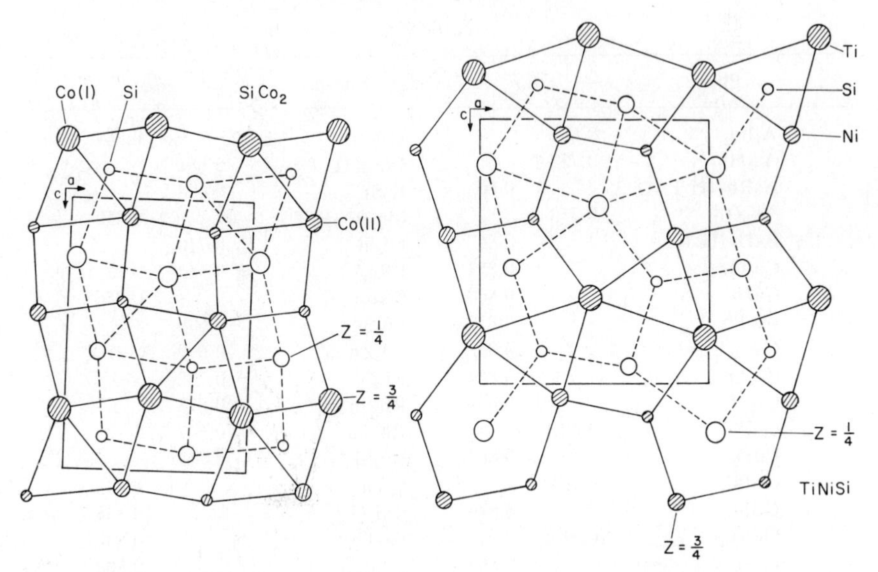

Figure 9-51. The $SiCo_2$ and *E*-TiNiSi (*oP*12) structures viewed down [010].

axial ratios *a*/*c* in the range from 0.67 to 0.73, whereas most other phases with the structure, including $PbCl_2$ itself, have *a*/*c* ratios in the range from 0.83 to 0.88 (see Table 9-1). Although the distortion from the $InNi_2$-type structure is larger in structures such as that of *E*-TiNiSi[78] with a large *a*/*c*

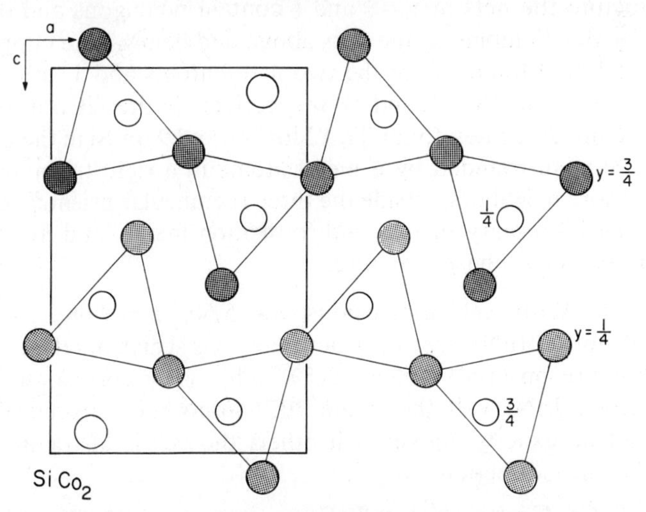

Figure 9-52. A further view of the $SiCo_2$ (*oP*12) structure showing the arrangement of triangular prisms of Co.

Table 9-1

Phase	a/c		Phase	a/c	
$AlPd_2$	0.696		H_2Yb		0.868
As_2Hf	0.757		$InPd_2$ (L.T.)	0.681	
$AsRh_2$ (H.T.)		0.805	Ir_2Si	0.694	
As_2Zr	0.753		Mg_2Pb (H.T.)		0.850
BaH_2 (H.T.)		0.867	Ni_2Si	0.710	
Ca_2Ge		0.853	PRe_2	0.552	
CaH_2		0.868	PRu_2		0.856
Ca_2Pb		0.837	Pd_2Sn	0.696	
Ca_2Si		0.852	Pd_2Zn	0.699	
Ca_2Sn		0.834	Rh_2Si	0.732	
Co_2P		0.854	Rh_2Sn	0.673	
Co_2Si	0.692		Rh_2Ta	0.667	
EuD_2		0.867	Ru_2Si	0.712	
$GaPd_2$	0.703		S_2Th		0.843
$GdSe_2$		0.876	β-S_2U		0.838
$GeRh_2$	0.719		Se_2Th		0.840
H_2Sr		0.867	β-Se_2U		0.808
			E-TiNiSi		0.876

value (in order to accommodate the larger Ti atom), there is no observable correlation between a/c value and radius ratio of the component atoms of the binary phases. Projections of the structure of $SiCo_2$ and of the ternary *E*-TiNiSi phase down [010] are compared in Figure 9-51. In the ternary phase structure the nets at $y = \frac{1}{4}$ and $\frac{3}{4}$ contain pentagons and these are centered by the Ti atoms of the nets above and below. The coordination which would be 14 and 11 for the two metal atoms and 11 for Si in the $InNi_2$ structure, and 13, 13, and 10 respectively in the $SiCo_2$ form of the $PbCl_2$ structure, becomes 15 for Ti, 12 for Ni, and 9 for Si in the *E*-TiNiSi structure. Si is surrounded by 6 metal atoms in a right triangular prism with three more neighbors outside the three rectangular prism faces.

More than 20 ternary phases with transition metals and Si or Ge are known to have the *E* phase structure.

UPt_2 (oC12): With cell dimensions $a = 5.60$, $b = 9.68$, $c = 4.12$ Å, UPt_2 is a distorted $InNi_2$ type of structure,[79] as is seen from the projection of the structure on (100) (Figure 9-53). This view corresponds to the orthohexagonal $InNi_2$ cell, the U and Pt(2) atoms being displaced so that they do not lie exactly above each other; the cell is also considerably elongated in the b direction.

Ge_3Rh_5 (oP16): Ge_3Rh_5 has cell dimensions $a = 5.42$, $b = 10.32$, $c = 3.96$ Å, $Z = 2$.[80] The structure is made up of layers of atoms parallel to

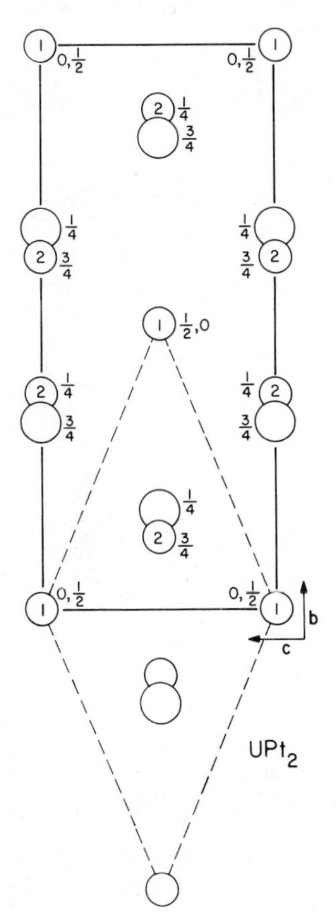

Figure 9-53. Projection of the UPt_2 ($oC12$) structure down [100] indicating the relationship of the structure to the NiAs type.

the (001) plane at $z = 0$ and $\frac{1}{2}$. The layer at $z = \frac{1}{2}$ (Figure 9-54), particularly, is reminiscent of the distorted square triangle layers found in the $SiCo_2$ ($oP12$) structure. This is perhaps not surprising as it has been pointed out that the Ge_3Rh_5 structure can be thought of as containing portions of distorted $InNi_2$ type arrangement. The layer at $z = \frac{1}{2}$ differs from that at $z = 0$ where alternate rows of atoms lie in straight lines in the a direction (Figure 9-54). The arrangement is such as to form an array of triangular prisms of Rh at $z = 0$ with axes along [001] and centered by Ge. These prisms share corners so as to build hexagonal prisms which contain a Ge atom surrounded by a rectangle of Rh at $z = \pm\frac{1}{2}$ (Figure

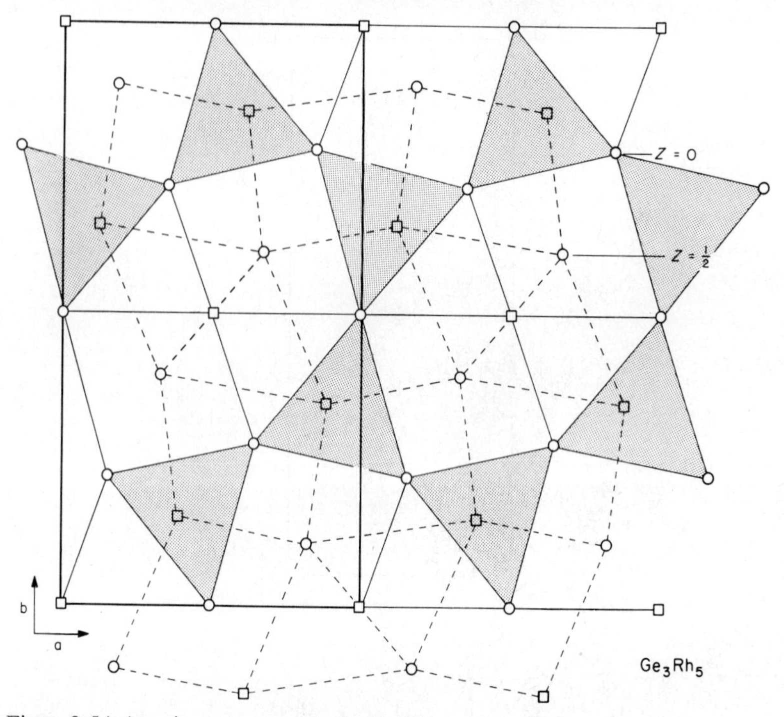

Figure 9-54. Atomic arrangement in the Ge_3Rh_5 ($oP16$) structure shown in projection down [001]: squares, Ge; the unit cell is outlined.

9-54). The atoms are surrounded by convex polyhedra of 13, 14, or 15 atoms. The closest Ge–Rh approaches are 2.415 Å and the closest Rh–Rh approaches are 2.78 Å.

MN: 3^6 Nets Stacked in Paired-Layer Sequence AA, Subdivided into $3^6 + 3636$. *N* on Larger 6^3 Nets

CoSn (*hP*6): CoSn has cell dimensions $a = 5.279$, $c = 4.258$ Å, $Z = 3$.[81] Sn(1) atoms at $z = 0$ form a simple hexagonal array ($c/a = 0.807$) and a kagomé net of Co atoms, also at $z = 0$, centers the edges of the Sn(1) triangles. At $z = \frac{1}{2}$, Sn(2) atoms form a 6^3 graphite-like array. The Sn(1) atoms lie over the centers of the hexagons of both the Sn(2) and Co arrays. In terms of the symbols given on p. 4, the structural arrangement is $[\underline{A}\alpha]\underline{a}$. The structure can be regarded as made up of layers of trigonal prisms of Co and Sn(1) (height to base ratio = 1.61), one quarter of which are centered by Sn(2).

Co has 4 Co + 2 Sn(1) at 2.64 Å and 4 Sn(2) at 2.62 Å in a characteristic coordination polyhedron (Figure 9-55). Sn(1) has 6 Co neighbors and

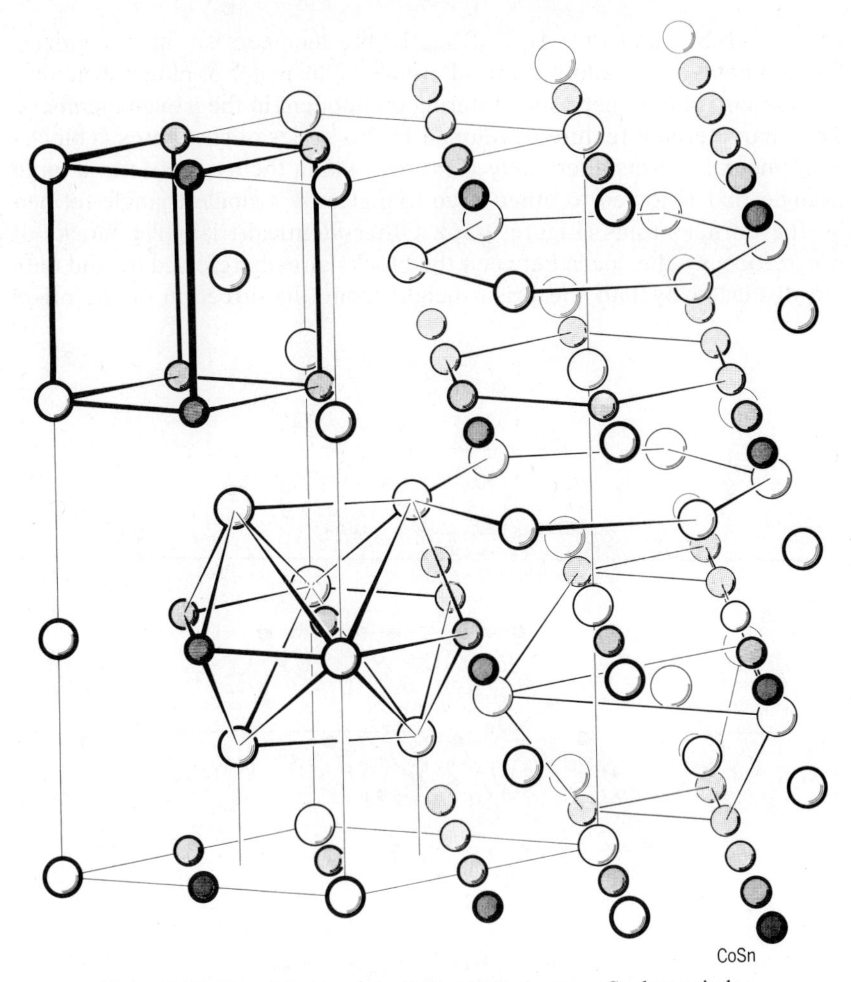

Figure 9-55. Pictorial view of the CoSn (*hP*6) structure: Sn, large circles.

Sn(2) has 6 Co and 3 Sn at 3.05 Å. The structure requires confirmation by single-crystal methods.

HOMOLOGOUS SERIES OF COMPOUNDS WHOSE STRUCTURES ARE MADE UP OF TRIANGULAR PRISMS

Several series of homologous compounds have been recognized, whose structures are built up of linked triangular prisms of atoms arranged in characteristic fashion. The prisms are generally centered.

The first series to be recognized was $M_{(n^2+n)}N_{(n^2-n+1)}$[82] whose first member ($n = 1$) is the AlB_2 structure. This is followed by ($n = 2$) P_3Fe_6,

($n = 3$) Th_7S_{12}, and ($n = 4$) $Si_{13}Rh_{20}$. In the members so far recognized, the nonmetal or metalloid atoms alternate as the major or minor structural component, being the prism-centering component in the n even members. The characteristic feature developed in this series of structures is blocks of triangular prisms alternately centered, which themselves form a large triangle that is joined to other large triangles by a single triangle formed by the corner atoms (Figure 9-56). Other identical triangular blocks of prisms occupy the space between the blocks already referred to, and they are displaced by half the prism height along the direction of the prism

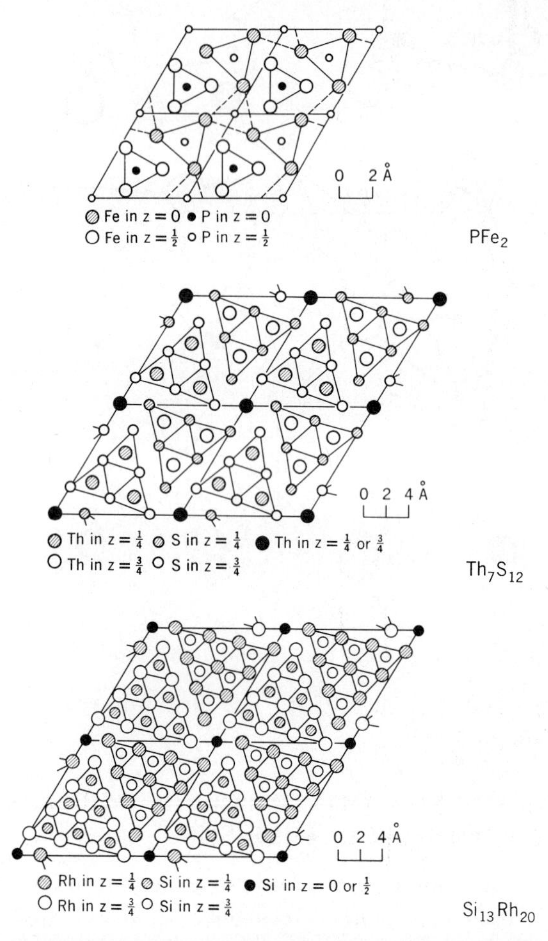

Figure 9-56. Atomic arrangements in the PFe_2 ($hP9$), Th_7S_{12} ($hP19$), and $Si_{13}Rh_{20}$ ($hP33$) structures, members of the series $M_{(n^2+n)}N_{(n^2-n+1)}$. (Engström[82].)

axes relative to the other prisms. The triangles formed by the corners of these blocks overlie the others antiprismatically forming an octahedron which is either centered or empty. The blocks of prisms are contiguous on their basal faces so that they form columns running in the direction of the prism axes.

The $M_{(n^2+5n)}N_{(n^2+5n)/2}$ series of structures[83] differs from that described above in that a block of centered prisms with prism apexes pointing outwards, is surrounded by a continuous network of centered prisms joined corner to corner with apexes alternately pointing inwards and outwards in pairs. This network is displaced by half of the prism height in the direction of the prism axes relative to other prism blocks. Since the prisms are contiguous base to base, they form columns in the direction of the prism axes. The first member of this series ($n = 1$) is Fe_6P_3, the second is $M_{14}P_7$, and the third $M_{24}P_{12}$ (Figure 9-57). In this series there are twice as many metal as phosphorus atoms in the formula, and the metal atoms can be of two kinds in ordered disposition.

$M_{(n^2+n)}N_{(n^2-n+1)}$ Series

Fe_2P, $InMg_2$ (*hP*9): The Fe_2P cell has dimensions $a = 5.865$, $c = 3.456$ Å, $Z = 3$.[84] The Fe atoms at $z = 0$ or $\frac{1}{2}$ form the pattern of triangular prisms shown in Figure 9-58. The P atoms at $z = 0$ or $\frac{1}{2}$ center these prisms. The Fe and P atoms together form a triangle-pentagon net at $z = 0$. This net is similar to that formed by the Pd atoms in the Th_3Pd_5 (*hP*8) structure. Fe(1) has four P neighbors at 2.22 and 2.29 Å and Fe(2) has 1 P at 2.38 Å and 4 at 2.48 Å. Fe(1) also has 2 Fe at 2.60 Å, 2 at 2.63 Å, and 4 at 2.71 Å. Fe(2) has 2 Fe at 2.63 Å, 4 at 2.71 Å, and 4 at 3.08 Å. The P atoms are surrounded by 9 Fe atoms, six prism neighbors, and three out through the centers of the rectangular faces of the prisms. Numerous ordered ternary phases have the Fe_2P structure in which the largest transition metal atom occupies the Fe(2) sites.

The atomic positions and c unit cell edge of the $InMg_2$ structure are practically identical with those of the Fe_2P structure, but the a cell edge is some 39% longer ($a = 8.27$, $c = 3.42$ Å). In(1) has 6 Mg neighbors at 2.68 Å. In(2) has 6 Mg at 3.03 Å and 4 at 3.16 Å. Both Mg atoms have 4 In neighbors.

Th_7S_{12} (*hP*19): The dimensions of the Th_7S_{12} unit cell are $a = 11.086$, $C = 4.010$ Å, $Z = 1$.[85] The structure is shown in plan in Figure 9-56. It contains blocks of four distorted triangular prisms of S atoms with bases at either $z \sim \frac{1}{4}$ or $\frac{3}{4}$. The three outer prisms are centered by Th(2) atoms and each Th atom has two other S neighbors out through rectangular prism faces. Th(1) atoms over the cell corners at $z = \pm\frac{1}{4}$ are

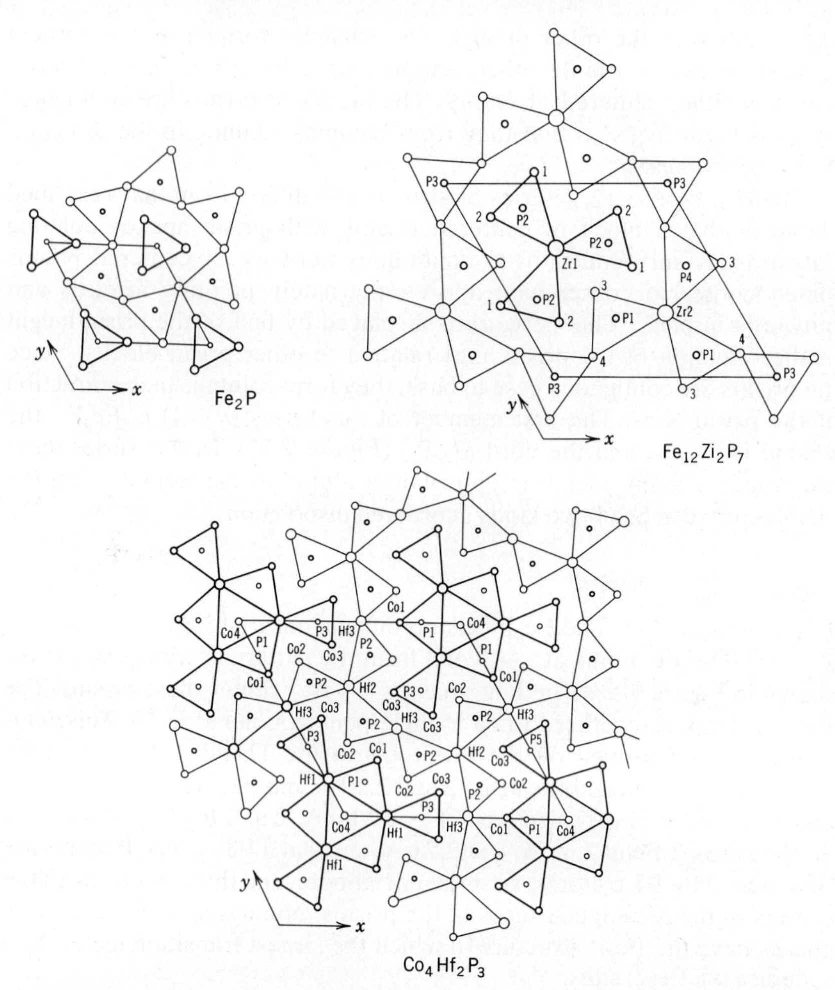

Figure 9-57. Atomic arrangements in the Fe_2P ($hP9$), $Fe_{12}Zr_2P_7$ ($hP21$), and $Co_4Hf_2P_3$ ($hP36$) structures, members of the series $M_{(n^2+5n)}N_{(n^2+5n)/2}$. For Fe_2P and $Fe_{12}Zr_2P_7$ thick lines mark atoms at $z = 0$, thin at $z = \frac{1}{2}$. For $Co_4Hf_2P_3$ thick lines mark atoms at $z = \frac{1}{2}$ and thin lines atoms at $z = 0$. The largest circles indicate Fe, Zr, and Hf atoms, respectively, and the smallest circles indicate phosphorus. (Ganglberger[83].)

surrounded octahedrally by S, but only half of these Th sites are occupied, randomly. The S atoms actually form distorted 3^6 nets at $z \sim \pm\frac{1}{4}$ with one-seventh of the sites vacant, and Th(1) and (2) similarly form distorted 3^6 nets with three-sevenths of the sites vacant. S(1) has 5 Th(2) neighbors at 2.95 to 2.96 Å. S(2) has 1 Th(1) neighbor at 2.605 Å and 3 Th(2) at 2.80 to 2.84 Å. Th(1) has 3 S at 2.605 Å, and Th(2) has 8 S neighbors.

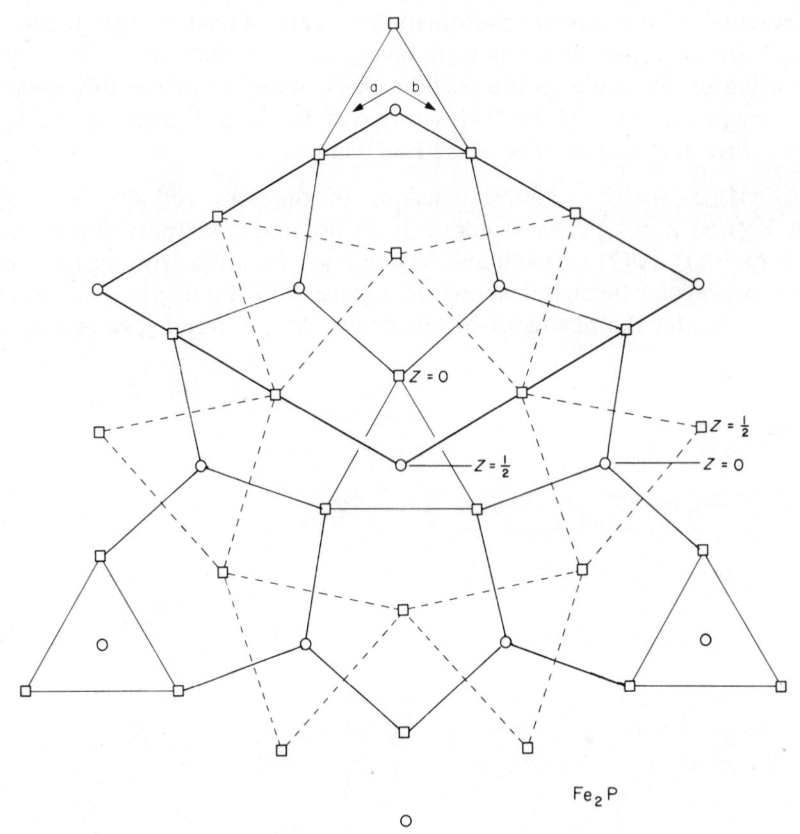

Figure 9-58. Atomic arrangement in the Fe_2P (*hP*9) structure: squares, Fe atoms.

$Rh_{20}Si_{13}$ (*hP*33): $Rh_{20}Si_{13}$ has dimensions $a = 11.851$, $c = 3.623$ Å, $Z = 1$.[82] The structure which is shown in plan down [001] in Figure 9-56, p. 538 contains blocks of 9 triangular prisms of Rh, six of which are centered by Si(1) and (2) atoms. Neighboring blocks are displaced by $c/2$ relative to each other along [001]. In addition to the 6 prism neighbors, Si(1) has 2 Rh neighbors out through rectangular prism faces, and Si(2) has one such neighbor. Rh–Si distances vary from 2.39 to 2.52 Å. Si(3) lies between the blocks of Rh prisms and is octahedrally surrounded by 6 Rh at 2.38 Å, only half of these sites being occupied by Si.

TiFeSi (*oI*36): The unit cell with dimensions $a = 6.997$, $b = 10.830$, $c = 6.287$ Å, $Z = 12$,[86] is a distortion superstructure of the Fe_2P structure in which the atoms are somewhat displaced from their ideal positions.

Nevertheless the atomic coordination is very similar to that found in Fe_2P. In the supercell atoms form layers at $x \sim 0$ and $\frac{1}{2}$ and $x = \pm\frac{1}{4}$. The ordering of the atoms in the pseudocell is similar to that in the ordered ternary phases with the Fe_2P structure with the large Ti atoms occupying sites corresponding to those of the Fe (2) atoms.

$Al_8FeMg_3Si_6$ ($hP18$): The dimensions of the unit cell are $a = 6.63$, $c = 7.94$ Å, $Z = 1$.[87] The structure is an ordered quaternary superstructure of Fe_2P. Al(2) at $z = 0$ and Mg at $z = \frac{1}{2}$ form the array of triangles and non-regular hexagons shown in Figure 9-59. Fe at $z = 0$ and Al(1) at $z = \frac{1}{2}$ center the hexagons of this array. At $z \sim \frac{1}{4}$ and $\frac{3}{4}$, Si and Al(3)

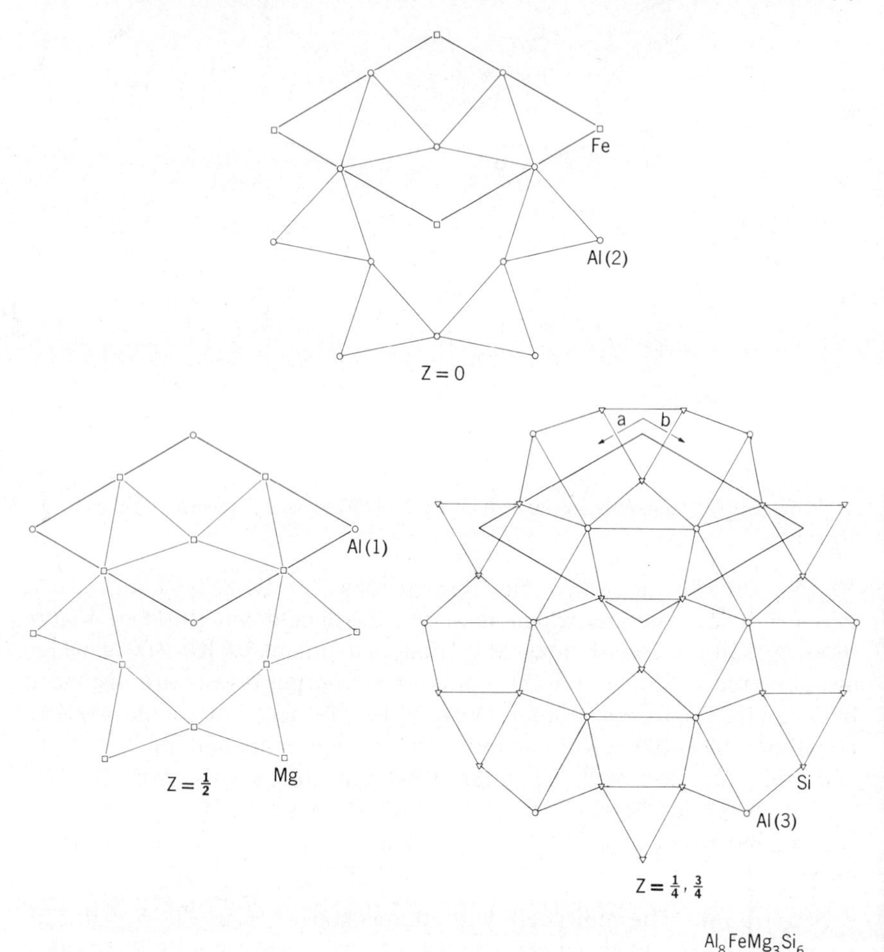

Figure 9-59. Layer arrangements of atoms in the $Al_8FeMg_3Si_6$ ($hP18$) structure.

form a slightly rumpled triangle-pentagon net (Figure 9-59) such that Fe and Al(1) center triangular prisms of Si, and Al(3) center distorted triangular prisms of Al(2) and Mg.

Al(1) is surrounded by 3 Mg at 2.94 Å and 6 Si at 2.76 Å; Al(2) has 4 Al(3) neighbors at 2.73 Å, 6 Si at 2.90 and 2.93 Å and 1 Fe at 2.67 Å; Al(3) has 3 Mg at 2.89 Å, 3 Al(2) neighbors and 3 Si at 2.53 Å. Fe is surrounded by 3 Al(2) and 6 Si at 2.42 Å; Mg has 1 Al(1), 4 Al(3) and 2 Si at 3.00 Å, as well as 4 Mg and 4 Si at 3.38 Å. Si has 1 Mg, 1 Al(1), 3 Al(2), 2 Al(3), 1 Fe neighbor, and 2 Si neighbors at 2.87 Å.

$M_{(n^2+5n)}N_{(n^2+5n)/2}$ Series

$Fe_{12}Zr_2P_7$ (hP21): $Fe_{12}Zr_2P_7$ has cell dimensions $a = 9.000$, $c = 3.592$ Å, $Z = 1$.[83] The atoms form layers at $z = 0$ and $\frac{1}{2}$ (Figure 9-57, p. 540). The P atoms center triangular prisms of 4 Fe and 2 Zr and have three more Fe neighbors out through the rectangular prism faces (Fe–P = 2.17 to 3.35 Å). Ignoring the difference between the Fe and Zr atoms, $Fe_{12}Zr_2P_7$ is the $n = 2$ member of this homologous series of structures.

$Co_4Hf_2P_3$ (hP36): $Co_4Hf_2P_3$ has cell dimensions $a = 12.056$, $c = 3.625$ Å, $Z = 4$.[83] P(1) and (2) are surrounded by triangular prisms of 4 Hf and 2 Co, and P(3) are surrounded similarly by 4 Co and 2 Hf. In addition, P(1) and (2) have 3 Co neighbors out through the rectangular faces of the prisms and the P(3) atoms have 2 Co and 1 Hf neighbors similarly disposed (Figure 9-57, p. 540). Ignoring the difference between the Co and Hf atoms, $Co_4Hf_2P_3$ is the $n = 3$ member of this homologous series of structures.

$M_{\{2(n-1)^2+2\}}N_{(n^2+n)}$ Series

$Co_{3.1}Ho_4$ (hP21): $Co_{3.1}Ho_4$ has cell dimensions $a = 11.4$, $c = 3.99$ Å, $Z = 3$.[88] In the structure, groups of 4 Co atoms at $z \sim 0.48$ and 0.98 are surrounded by a rumpled network of Ho atoms at the same heights (Figure 9-60). Partially filled Co(1) sites at $z = \frac{1}{4}$ and $\frac{3}{4}$ lie between the largest triangles in the Ho network so that they are surrounded octahedrally. The groups of 4 Co atoms are located about the B site in the cell at $z \sim 0.98$ and about the C site at $z \sim 0.48$. The structure contains very short Co(2)–Co(3) distances.

The Ho network is the same as the S network at $z = \frac{1}{4}$ or $\frac{3}{4}$ in Th_7S_{12}, and this suggests that the structure may belong to a homologous series similar to several others. Indeed, assuming the Co sites to be fully occupied, it is the $n = 3$ member of a series which can be described as $M_{\{2(n-1)^2+2\}}N_{(n^2+n)}$ where $n = 1, 2, 3, 4, \ldots$. The first member of the series

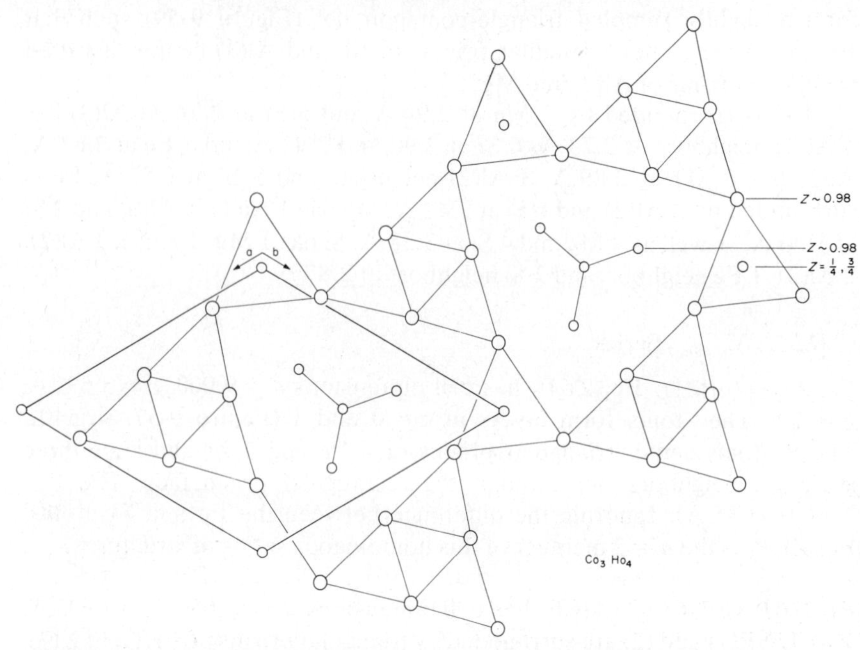

Figure 9-60. Atomic arrangement in the $Co_{3.1}Ho_4$ (*hP*21) structure viewed down [001]: Ho, large circles.

is the NiAs structure, the second does not appear to be known, although it has the stacking sequence A[Cγ] (notation of p. 4) which is the basic stacking unit on (111) planes of the perovskite and $PtHg_4$ structures.

Structures with Arrangements of Atoms in Layers Resembling Those Described Above

Nb_3Te_4 (*hP*14): Nb_3Te_4 has a cell with dimensions $a = 10.671$, $c = 3.6468$ Å, $Z = 2$.[89] All atoms are located on planes at $z = \frac{1}{4}$ and $\frac{3}{4}$. The structure is made up of blocks of four triangles of Nb and Te atoms in locations at $z = \frac{1}{4}$ or $\frac{3}{4}$ as shown in Figure 9-61. These form columns of triangular prisms, the center ones formed by Nb being centered by Te(1) atoms with Nb–Te distances of 2.88 Å. The structure contains large voids about the cell corners. Te(2) has 2 Nb neighbors in the same plane and one more in each of the planes above and below, but the arrangement is not tetrahedral. Each Nb atom has 6 Te neighbors, two in the same plane and four in planes above and below; the arrangement is a distorted octahedron. Nb is displaced from the center of these octahedra towards a face sharing 2 edges with other octahedra, the result being the

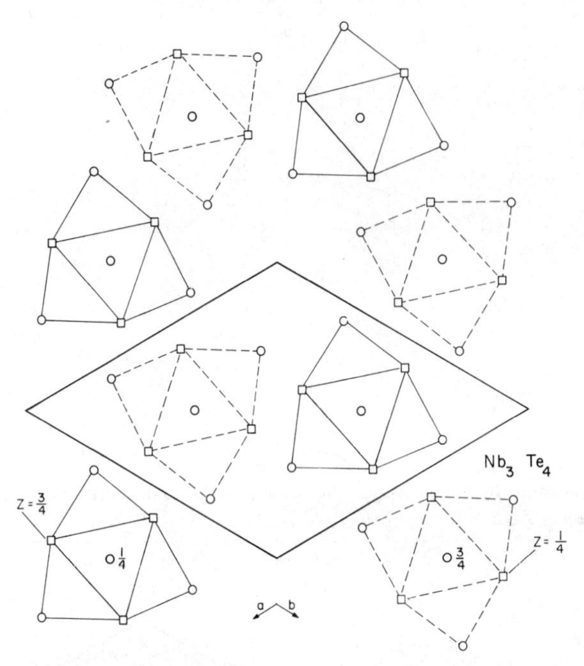

Figure 9-61. Atomic arrangement in the Nb_3Te_4 structure viewed down [001]: circles, Te.

formation of zigzag chains of Nb (2.97 Å) running in the [001] direction. In Nb_3S_4 with the same structure[90] the 2Nb–Nb distances are 2.88 Å.

$BaLi_4$ (*hP*30): $BaLi_4$ has cell dimensions $a = 11.026$, $c = 8.891$ Å, $Z = 6$.[91] The structure can be regarded as a layer structure, with distinctly nonplanar hexagon-triangle layers at $z \sim 0$ and $\frac{1}{2}$. These layers shown in Figure 9-62, amount to a mixture of empty and filled hexagons of Li(2) and (4) atoms, with the filled hexagons centered by Li(1) over the cell corners. At $z = \frac{1}{4}$ and $\frac{3}{4}$, Ba and Li(3) form blocks of four triangles arranged similarly to those of Nb_3Te_4, except that the Li atoms are closer together over the cell corners where they form triangles (Figure 9-62). The Ba atoms lie over the empty hexagons at $z = 0$ and $\frac{1}{2}$ and slightly off-center. The nonplanarity of the $z \sim 0$ and $\frac{1}{2}$ layers results from them bulging away from the Ba atoms on one side of the cell at $z = \frac{1}{4}$ and on the other side at $z = \frac{3}{4}$. This arrangement gives Ba 14 Li neighbors at 3.84 to 4.09 Å. Li(1) has a CN 12 polyhedron with surface coordination of 5 for each atom and 20 three-sided faces, the 12 Li neighbors being at 2.96 and 3.14 Å. Li(2) has 4 Li and 6 Ba neighbors and Li(3) and (4) each have 12 (Li + Ba) neighbors in polyhedra with 20 three-sided faces and surface coordination

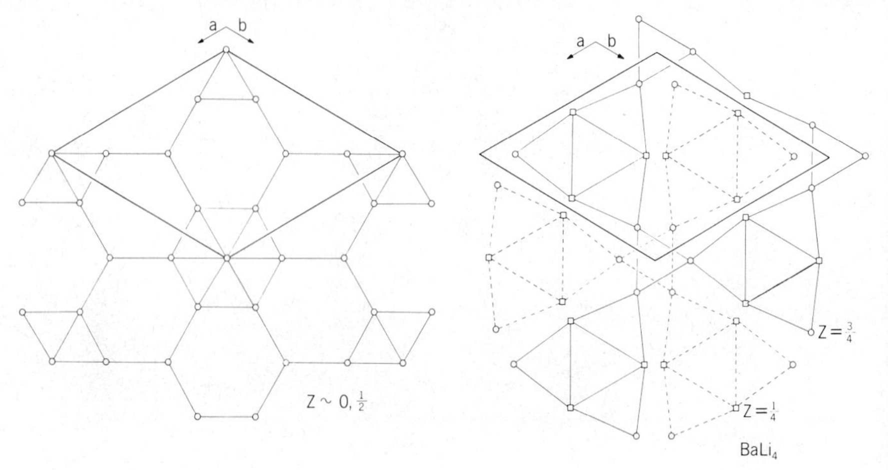

Figure 9-62. Atomic arrangement in the $BaLi_4$ (*hP*30) structure viewed down [001]: squares, Ba atoms.

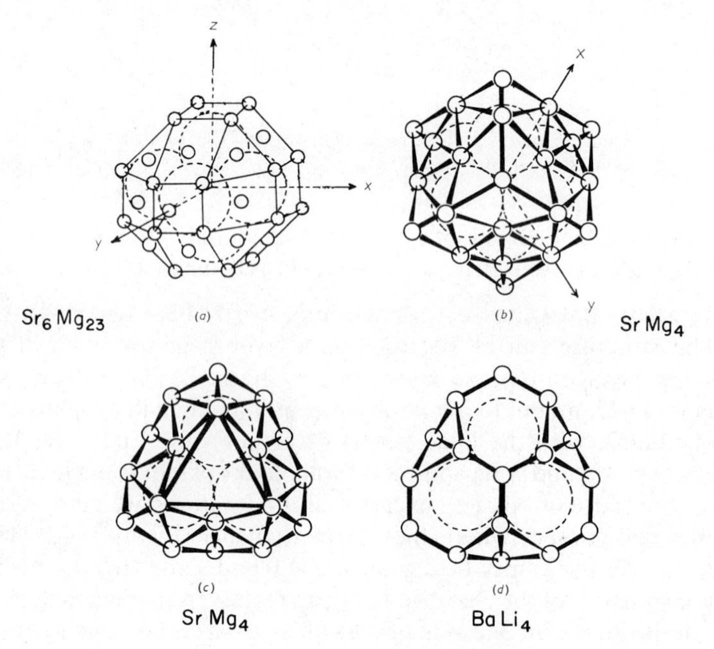

Figure 9-63. (*a*) 44 Mg atoms surrounding an octahedral cluster of 6 Sr atoms in the Sr_6Mg_{23} structure. A further 6 Mg atoms lie along the *x*, *y*, and *z* axes, making 50 in all. (*b*) 50 Mg atoms surrounding an octahedral cluster of 6 Sr atoms in the $SrMg_4$ structure. (*c*) 3 Sr atoms surrounded by 33 Mg in $SrMg_4$. (*d*) 3 Ba atoms surrounded by 29 Li atoms in $BaLi_4$.[91]

of 5 for each atom. The arrangement of the 29 Li atoms surrounding the triangles of Ba atoms is shown in Figure 9-63*d*.

Co_2Al_5 (hP28): Co_2Al_5 has cell dimensions $a = 7.672$, $c = 7.609$ Å, $Z = 4$. Layers of Al(2) and Co atoms at $z = \frac{1}{4}$ and $\frac{3}{4}$ form the hexagon-pentagon-triangle net shown in Figure 9-64. This has marked similarities to the layers of atoms formed at $z = \frac{1}{4}$ and $\frac{3}{4}$ in the Nb_3Te_4 and $BaLi_4$ structures. At $z \sim 0$ and $\frac{1}{2}$, Al(1) and (3) form rectangle-triangle nets which are considerably rumpled so as to avoid the three Al(2) atoms that occur at the corners of pentagons in the other layers at $z = \frac{1}{4}$ on one side of the cell and at $z = \frac{3}{4}$ on the other side. Thus the distortion of these layers resembles that found at $z \sim 0$ and $\frac{1}{2}$ in the $BaLi_4$ (*hP*30) structure. In fact, these two structure types are very similar. Compared to $BaLi_4$, Co_2Al_5[92] has one extra atom in each of the layers at $z \sim 0$ and $\frac{1}{2}$. The $3^6 + 3^2434$ layers occurring at $z = 0$ and $\frac{1}{2}$ are a common feature of several structure types.

β-V_4Al_{23} (hP54): With cell dimensions $a = 7.693$, $c = 17.04$ Å, $Z = 2$, β-V_4Al_{23} is a layer structure.[93] At $z = \frac{1}{4} \pm 0.134$ and $z = \frac{3}{4} + 0.134$ partially occupied triangular networks of Al(4) and Al(5) atoms occur, the unit

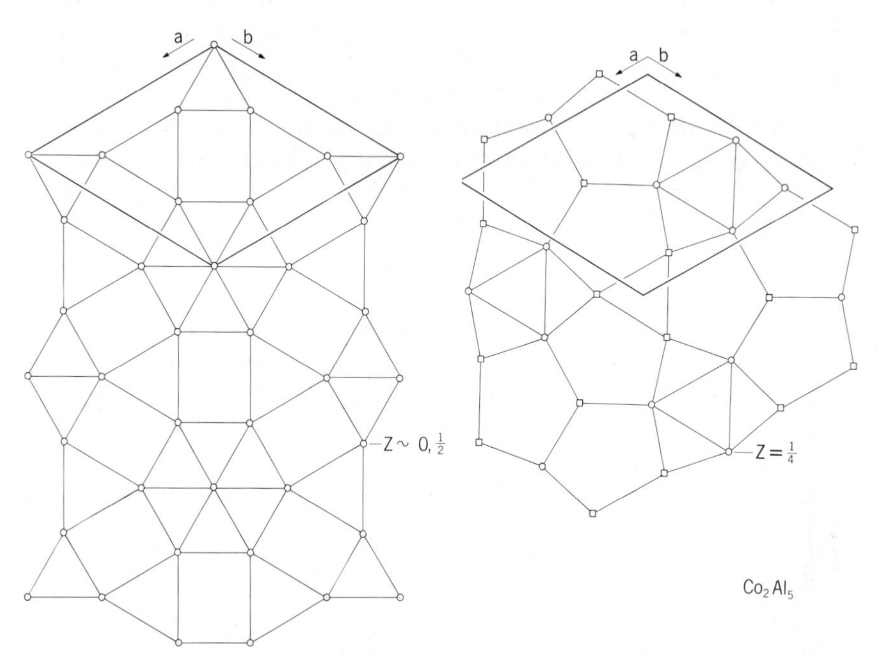

Figure 9-64. Layer arrangements of atoms in the Co_2Al_5 (*hP*28) structure: squares, Co atoms.

cell containing six occupied and three unoccupied sites. At $z=\frac{1}{4}\pm0.134$ the occupied sites occur on the left of the cell and the central triangle is cented by an Al(1) atom (Figure 9-65*a*), whereas at $z=\frac{3}{4}\pm0.134$ the occupied sites occur on the right of the cell (Figure 9-65*c*.) These pairs of layers are interleaved with a similar arrangement of V(2) and Al(2) atoms at $z=\frac{1}{4}$ (Figure 9-65*b*) and $\frac{3}{4}$. The V(2) atoms located on the left of the cell at $z=\frac{1}{4}$ center the triangular prisms formed by Al atoms at $z=\frac{1}{4}\pm0.134$, and similarly for the V(2) atoms on the right of the cell at $z=\frac{3}{4}$. At $z\sim0$ and $\frac{1}{2}$ the layers of Al(3) and V(1) atoms shown in Figure 9-65*d* occur. Similar layer arrangements are found in the Co_2Al_5 (*hP*28), β-Al_9Mn_3Si (*hP*26), and $U_{20}Si_{16}C_3$ (*hP*39) structures. These layers are rumpled so that the Al atoms on the left of the cell move respectively down and up away from the Al atoms in the layers at $z=0.116$ and 0.384, similarly for the Al(3) atoms on the right of the cell with respect to the Al atoms in the layers at $z=0.616$ and 0.884. The closest Al–V distance is 2.52 Å. All atoms in the structure have 12 neighbors at distances less than 3 Å.

α-Sn₅Ti₆ (*hP*22): In the α-Sn_5Ti_6 cell which has dimensions $a=9.22$, $c=5.69$ Å, $Z=2$,[94] the atoms are arranged in layers at $z=0$, $\frac{1}{4}$, $\frac{1}{2}$, and $\frac{3}{4}$. Ti(1) atoms form a kagomé network at $z=0$ and $\frac{1}{2}$. Pentagon-triangle layers at $z=\pm\frac{1}{4}$ which resemble those in the Co_2Al_5 structure, are made up of Ti(2) and Sn(1), (2), and (3) atoms as shown in Figure 9-66. Columns of fused pentagonal antiprisms formed by Sn and Ti at $z=\frac{1}{4}$ and $\frac{3}{4}$ are centered by Ti atoms on the kagomé nets at $z=0$ and $\frac{1}{2}$. Sn(1) has 3 Ti at 2.635 Å and 2 Sn at 2.845 Å; Sn(2) has 3 Ti at 2.69 Å and 6 at 3.03 Å; Sn(3) has 8 Ti at 2.80 to 3.00 Å. Ti(1) has 6 Sn at 2.80 and 3.02 Å, together with 2 Ti at 2.845 Å and 4 at 3.03 Å. Ti(2) has 6 Sn at various distances and 4 Ti at 3.03 Å.

PCu₃ (*hP*24): PCu_3 has unit cell dimensions $a=6.954$, $c=7.149$ Å, $Z=6$.[95] In the notation given on p. 4 the stacking sequence of the structure along [001] is $AC'B\underline{B}'CC'AB'B\underline{C}'CB'$ (with P layers underlined). The layers are essentially equally but very closely spaced, somewhat distorted in their planes and displaced from the exact sites described by the layer symbols. The structure can also be regarded as made up of rather rumpled layers of atoms about $z\sim0$, $\frac{1}{4}$, $\frac{1}{2}$, and $\frac{3}{4}$ which have the arrangements shown in Figure 9-67. In the layer at $z=\frac{3}{4}$ the triangles cover those of the $z=\frac{1}{4}$ layer antisymmetrically. These layers resemble those occuring in Fe_2P.

Cu(1) is surrounded by 6 P at 3.19 Å and 6 Cu at 2.50 Å. Cu(2) has 3 P at 2.25 Å and 9 Cu at 2.62 and 2.73 Å. Cu(3) has 3 P at 2.305 to 2.475 Å and 8 Cu at 2.50 to 2.73 Å. P has 8 Cu in pairs at 2.25, 2.305,

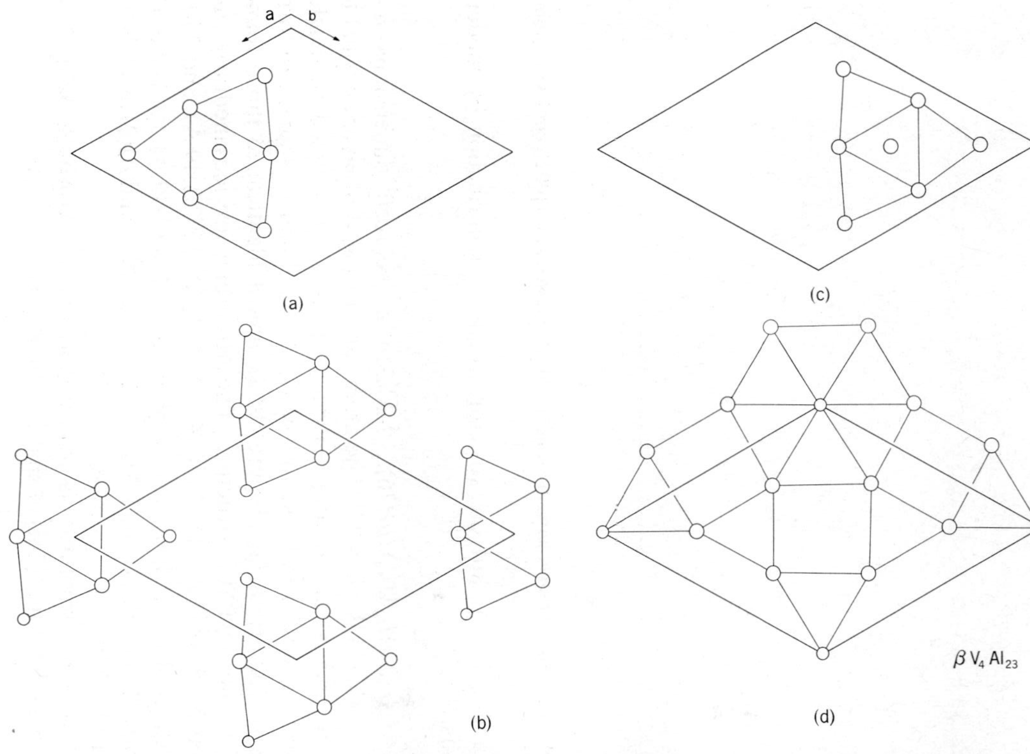

Figure 9-65. Layer arrangements of the atoms in the β-V_4Al_{23} (*hP*54) structure viewed down [001]: V, small circles.

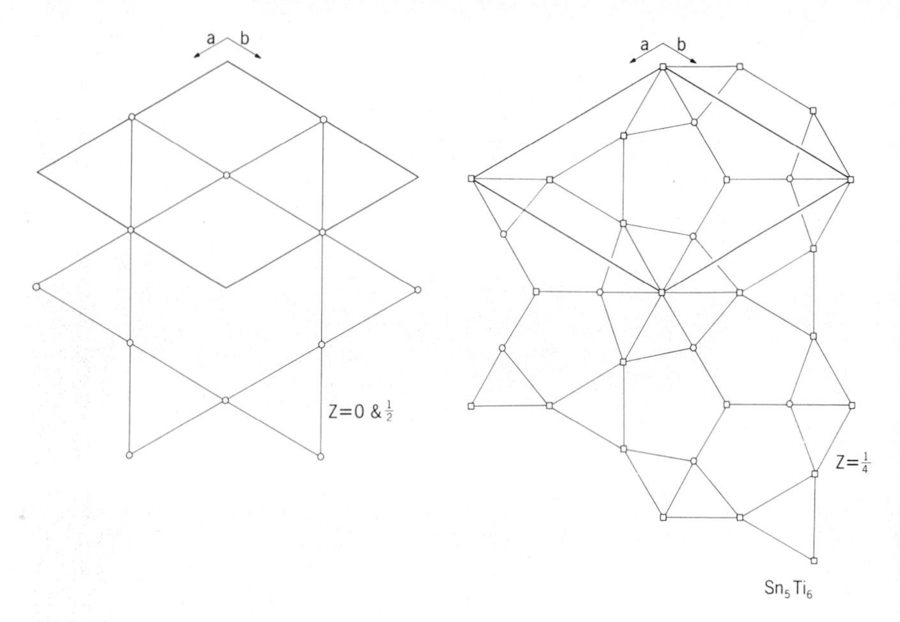

Figure 9-66. Atomic arrangements in α-Sn_5Ti_6 ($hP22$) viewed down [001]: circles, Ti atoms.

2.41, and 2.475 Å, but the structure requires confirmation by modern methods.

β-Al_9Mn_3Si (φ-$Al_{10}Mn_3$) (hP26): β-Al_9Mn_3Si has cell dimensions $a = 7.513$, $c = 7.745$ Å, $Z = 2$.[96] The atoms are arranged in layers normal to [001]. Si and Al atoms at $z \sim 0$ and $\frac{1}{2}$ form very rumpled rectangle-triangle nets similar to those in Co_2Al_5. At $z = \frac{1}{4}$ and $\frac{3}{4}$, Al and Mn form groups of four triangles (Figure 9-68) almost identical to those found at $z = \pm\frac{1}{4}$ in the $BaLi_4$ structure. The Mn atoms are icosahedrally surrounded by 8 Al (2×2.42, 2×2.67, 4×2.68 Å), 2 Si (2.49 Å), and 2 Mn (2.70 Å), as also are Si by 6 Al (2.66 Å) and 6 Mn (2.49 Å). Al(1) has 10 Al (2.805, 2.77, 2.97 Å) and 2 Mn (2.42 Å) neighbors and Al(2) has CN 13 (9 Al, 1 Si, 3 Mn).

The $Al_{10}Mn_3$ structure[97] is identical to that of Al_9Mn_3Si, with 2 Al occupying the Si sites (2(*a*) of space group $P6_3/mmc$).

$U_{20}Si_{16}C_3$ (hP39): $U_{20}Si_{16}C_3$ has cell dimensions $a = 10.385$, $c = 8.013$ Å, $Z = 1$.[98] The U atoms (1) to (4) form a square-triangle net at $z = 0$ and $\frac{1}{2}$ which resembles the nets found in the Co_2Al_5 ($hP28$) and Al_9Mn_3Si ($hP26$) structures. At $z = \pm 0.285$ U(5) atoms form a kagomé net, the nodes of which lie over the centers of the squares of the U nets at $z = 0$

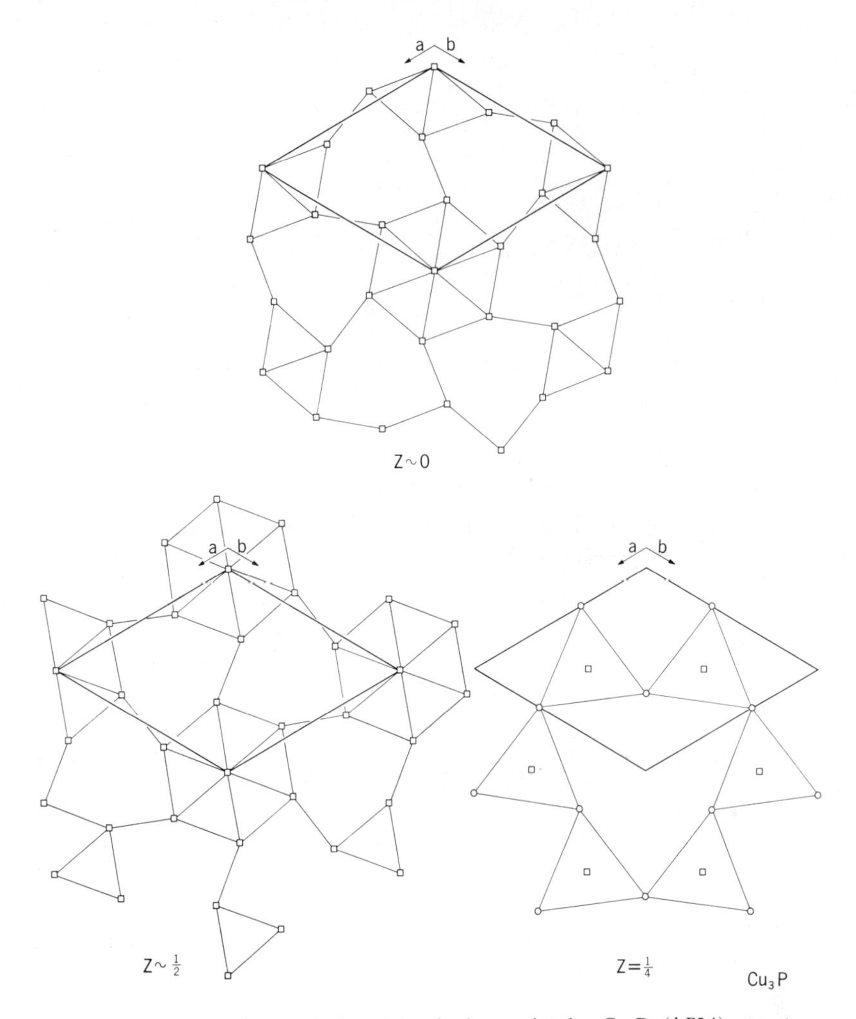

Figure 9-67. Arrangement of the atoms in layers in the Cu_3P (*hP*24) structure: squares, Cu atoms.

and $\frac{1}{2}$ and the Si nets at $z = \frac{1}{4}$ and $\frac{3}{4}$. Si atoms at $z = \pm\frac{1}{4}$ form a hexagon-square-triangle net (Figure 9-69), such that the atoms center the triangular prism of U atoms formed by the layers at $z = 0$ and $\frac{1}{2}$. The C atoms at $z = 0$ lie at the centers of nearly regular octahedra of U atoms. The structure is thus made up of triangular prisms of U centered by Si, and cubes of U off-centered by U atoms, with the C atoms centering half of the octahedra formed by this array of centered cubes of U atoms. The carbon atom positions need confirmation by neutron-diffraction methods.

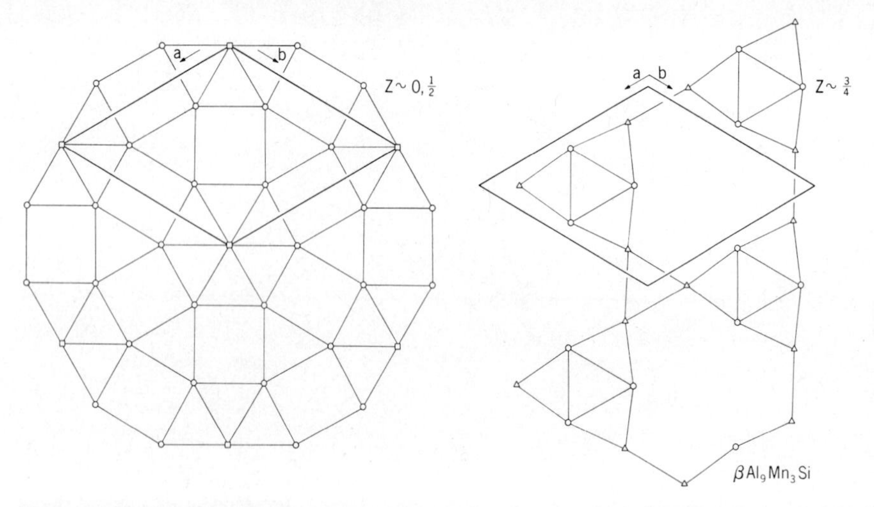

Figure 9-68. Layer arrangements of the atoms in the β-Al_9Mn_3Si (*hP*26) structure: Al atoms are circles; Mn, triangles; Si, squares.

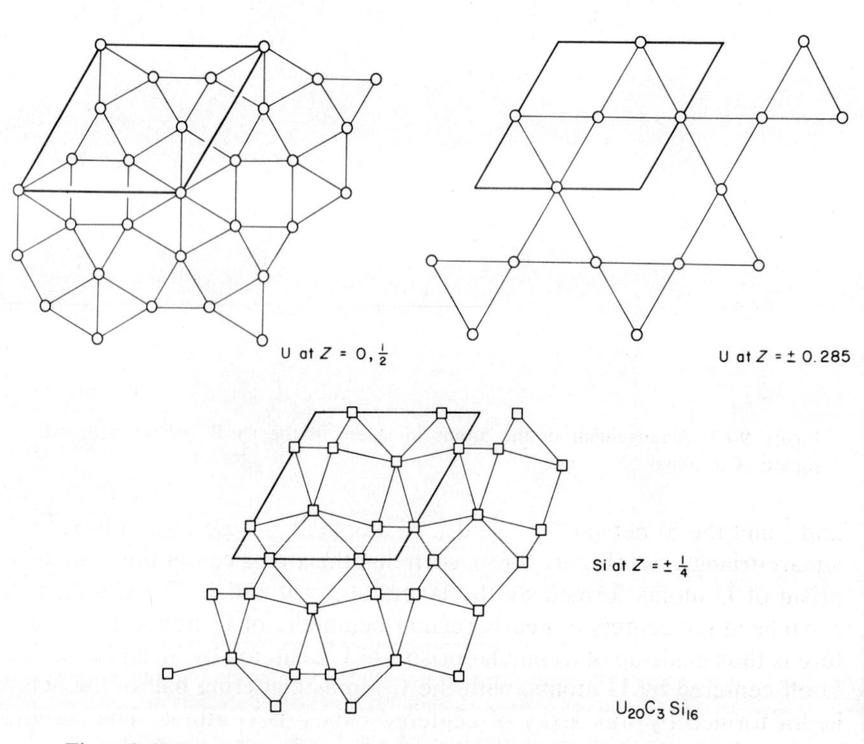

Figure 9-69. Layer arrangements of the atoms in the $U_{20}C_3Si_{16}$ (*hP*39) structure.

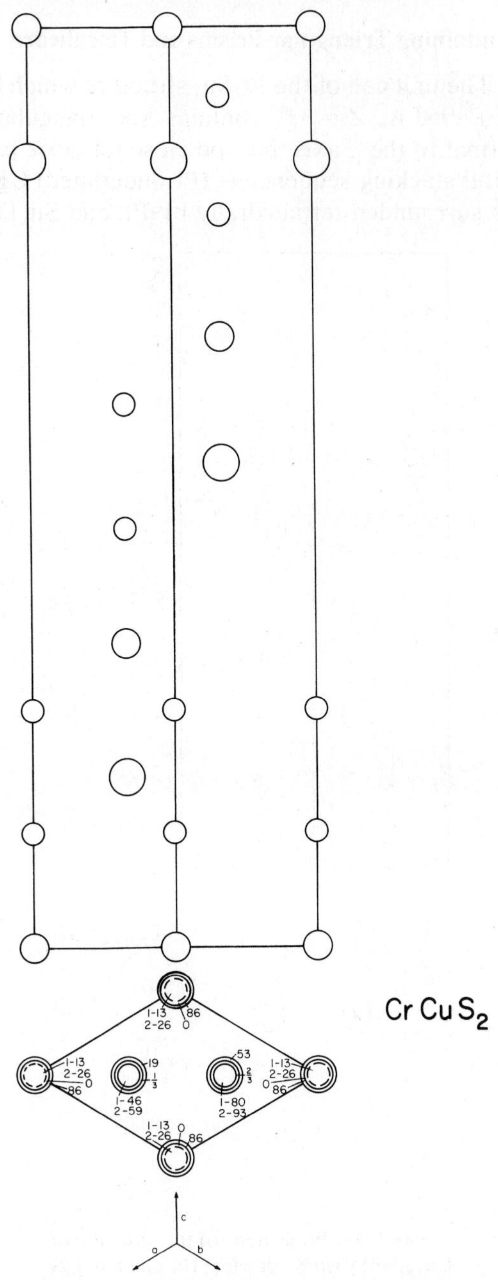

Figure 9-70. Projections of the structure of $CrCuS_2$ (*hR*4) down [110] and [001]: largest circles, Cr; smallest, S.

Structures Containing Triangular Prisms and Tetrahedra

Pt_2Sn_3 (*hP*10): The unit cell of the Pt_2Sn_3 structure which has dimensions $a = 4.334$, $c = 12.960$ Å, $Z = 3$,[99] contains six triangular layers of Sn atoms lying normal to the c axis, but too close for close packing ($c/3a = 1.00$). The overall stacking sequence is (Pt underlined) $\mathrm{C\underline{B}A\underline{B}CB\underline{C}A\underline{C}B}$ so that Sn(2) is surrounded tetrahedrally by Pt, and Sn(1) is surrounded

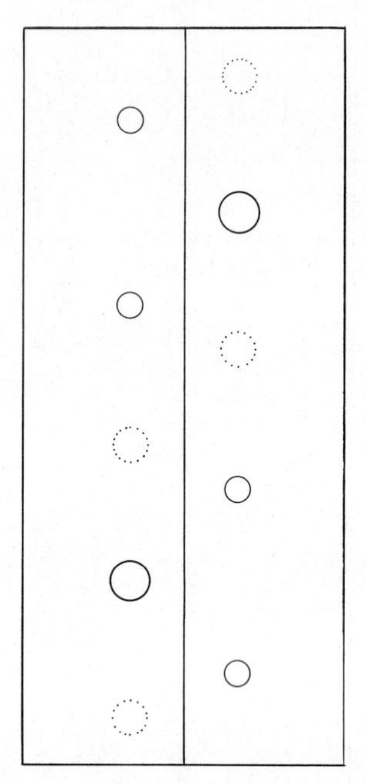

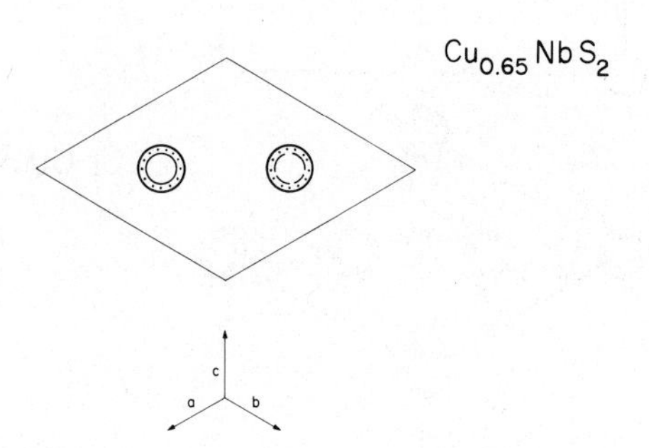

Figure 9-71. Projections of the structure of $Cu_{0.65}NbS_2$ (*hP*8) down [110] and [001]. S atoms are the smallest circles. Broken circles represent sites partially occupied by Cu.

by trigonal prisms of Pt, with a ratio of prism height to base of 0.62 as is found in the ω phase. $Au_4In_3Sn_3$ is said to be isotypic.[100]

Pt has 7 Sn neighbors at 2.675 to 2.86 Å and 1 Pt at 2.77 Å. Sn(1) has 6 Pt at 2.86 Å and Sn(2) has 3 Pt at 2.675 Å, 1 at 2.76 Å and 3 Sn(2) at 3.09 Å.

$CrCuS_2$ (hR4): The dimensions of the corresponding hexagonal cell are $a = 3.49$, $c = 18.87$ Å, $Z = 3$.[101] The structure contains groups of four equally spaced 3^6 layers of Cr, Cu, S(1), and S(2) on the same stacking site, and three of these groups are stacked in the sequence ACB, one above the other along [001] of the hexagonal cell. These groups of four layers interpenetrate so that Cr centers trigonal prisms of S atoms and Cu is surrounded tetrahedrally by S (Figure 9-70). S(1) is surrounded tetrahedrally by 3 Cr and 1 Cu and S(2) centers a trigonal prism of 3 Cr and 3 Cu. See also p. 432. The structure requires confirmation by single-crystal methods.

$Cu_{0.65}NbS_2$ (hP8): $Cu_{0.65}NbS_2$ has cell dimensions $a = 3.35$, $c = 13.13$ Å, $Z = 2$.[102] In this structure the Nb and S atoms are arranged as in the β-MoS_2 structure (*hP*6) with much the same layer separations so that S atoms surround Nb in a trigonal prism. The Cu atoms (1.3 Cu randomly on fourfold sites) are inserted in the tetrahedral holes between the other pairs of S layers (Figure 9-71), giving SCuCuS units similar to the CAlAlC units found in Al_4C_3 and aluminum carbonitrides (p. 408). The structure requires confirmation by single-crystal methods.

REFERENCES

1. H. Boller and E. Parthé, 1963, *Acta Cryst.*, **16** 1095.
2. K. Schubert *et al.*, 1954, *Z. Metallk.*, **45** 643; G. V. Raynor and J. A. Lee, 1954, *Acta Met.*, **2**, 616.
3. J. S. Kasper and H. Brandhorst, 1964, *J. Chem. Phys.*, **41**, 3768.
4. J. Leciejewicz, 1961, *Acta Cryst.*, **14**, 200; D. Meinhardt and O. Krisement, 1962, *Arch. Eisenhüttenw.*, **33**, 493; E. Parthé and V. Sadagopan, 1962, *Mh. Chem.*, **93**, 263.
5. S. S. Sidhu, K. D. Anderson, and D. D. Zauberis, 1965, *Acta Cryst.*, **18**, 906.
6. F. Jellinek, G. Brauer, and H. Müller, 1960, *Nature, Lond.*, **185**, 376.
7. S. A. Semiletov, 1961, *Kristallografija*, **6**, 536 [*Soviet Physics–Crystallography*, **6**, 428].
8. *Strukturbericht*, **1**, 164.
9. H. F. Franzen and J. Graham, 1966, *Z. Kristallogr.*, **123**, 133.
10. B E. Brown and D. J. Beerntsen, 1965, *Acta Cryst.*, **18**, 31; L. H. Brixner, 1962, *J. Inorg. Nucl. Chem.*, **24**, 257.
11. S. LaPlaca and B. Post, 1962, *Acta Cryst.*, **15**, 97.
12. B. Aronsson, E. Stenberg, and J. Åselius, 1960, *Acta Chem. Scand.*, **14**, 733.
13. H. Boller and E. Parthé, 1963, *Acta Cryst.*, **16**, 1095; S. Furuseth and A. Kjekshus,

1965, *Ibid.*, **18**, 320; G. S. Saini, L. D. Calvert, and J. B. Taylor, 1964, *Canad. J. Chem.*, **42**, 630.
14. R. M. Imamov and Z. G. Pinsker, 1964, *Kristallografija*, **9**, 743.
15. K. R. Andress and E. Albertini, 1935, *Z. Metallk.*, **27**, 126.
16. Z. Ban and M. Sikirica, 1965, *Acta Cryst.*, **18**, 594.
17. M. Erlander, G. Hägg, and A. Westgren, 1936–38, *Ark. Kemi, Min. Geol.* **12B**, No. 1.
18. A. J. K. Haneveld and F. Jellinek, 1964, *Rec. Trav. Chim. Pays–Bas*, **83**, 776.
19. R. Wang and H. Steinfink, 1967, *Inorg. Chem.*, **6**, 1685.
20. D. J. Haase, H. Steinfink, and E. J. Weiss, 1965, *Inorg. Chem.*, **4**, 538; **6**, 1685.
21. B. K. Norling and H. Steinfink, 1966, *Inorg. Chem.*, **5**, 1488.
22. M.-P. Pardo and J. Flahaut, 1967, *Bull. Soc. Chim. Fr.*, No. 10, 3658.
23. R. B. Kok, G. A. Wiegers, and F. Jellinek, 1965, *Rec. Trav. Chim. Pays-Bas*, **84**, 1585.
24. L. S. Zevjn, G. S. Ždanov, and N. N. Žuravlev, 1953, *Ž. Eksper. Teor. Fiz. SSSR*, **25**, 751; N. N. Žuravlev and G. S. Ždanov, 1953, *Ibid.*, **25**, 485.
25. A. Nylund, 1966, *Acta Chem. Scand.*, **20**, 2393.
26. D. Meinhardt and O. Krisement, 1962. *Arch. Eisenhüttenw.*, **33**, 493.
27. S. Rundqvist and G. Runnsjö, 1969, *Acta Chem. Scand.*, **23**, 1191.
28. S. Rundqvist and A. Hede, 1960, *Acta Chem. Scand.*, **14**, 893.
29. S. Rundqvist, 1965, *Acta Chem. Scand.*, **19**, 393.
30. S. Rundqvist, 1966, *Acta Chem. Scand.*, **20**, 2427.
31. H. F. Franzen, T. A. Beineke, and B. R. Conrad, 1958, *Acta Cryst.*, **B24**, 412.
32. Ju. A. Bagarjackij and G. I. Nosova, 1958, *Kristallografija*, **3**, 17 [*Soviet Physics–Crystallography*, **3**, 15]: J. M. Silcock, 1958, *Acta Met.*, **6**, 481.
33. A. E. Austin and J. R. Doig, 1957, *J. Metals, N.Y.*, **9**, 27.
34. H. Nowotny, E. Bauer, A. Stempfl, and H. Bittner, 1952, *Mh. Chem.*, **83**, 221.
35. A. Iandelli, 1964, *Z. anorg. Chem.*, **330**, 221.
36. A. C. Larson and D. T. Cromer, 1961, *Acta Cryst.*, **14**, 73.
37. E. J. Duwell and N. C. Baenziger, 1960, *Acta Cryst.*, **13**, 476.
38. P. A. Romans and M. P. Krug, 1966, *Acta Cryst.*, **20**, 313.
39. E. Zintl and G. Brauer, 1935, *Z. Elektrochem.*, **41**, 102.
40. N. Schönberg, 1954, *Acta Chem. Scand.*, **8**, 627.
41. R. Kiessling, 1949, *Acta Chem. Scand.*, **3**, 595.
42. O. Schob and E. Parthé, 1965, *Acta Cryst.*, **19**, 214.
43. D. Hohnke and E. Parthé, 1966, *Acta Cryst.*, **20**, 572.
44. W. Rieger and E. Parthé, 1967, *Acta Cryst.*, **22**, 919.
45. R. Kiessling, 1949, *Acta Chem. Scand.*, **3**, 603; 1950, *Ibid.*, **4**, 209.
46. M. Ellström, 1961, *Acta Chem. Scand.*, **15**, 1178.
47. W. Jeitschko, 1967, *Mh. Chem.*, **97**, 1472.
48. H. Schachner, H. Nowotny, and H. Kudielka, 1954, *Mh. Chem.*, **85**, 1140; P. A. Vaughan and A. Bracuti, 1955, *Abstr. Amer. Cryst. Assn.*, Summer Meeting, p. 8; A. J. J. Bracuti, 1958, *Dissert, Abstr.*, **19**, 1217.
49. R. Kiessling, 1947, *Acta Chem. Scand.*, **1**, 893.
50. L. Toth, H. Nowotny, F. Benesovsky, and E. Rudy, 1961, *Mh. Chem.*, **92**, 794.
51. W. Jeitschko, H. Nowotny, and F. Benesovsky, 1963, *Mh. Chem.*, **94**, 565.
52. G. S. Smith, A. G. Tharp, and Q. Johnson, 1969, *Acta Cryst.*, **B25**, 698.
53. R. Wang and H. Steinfink, 1967, *Inorg. Chem.*, **6**, 1685.
54. W. L. Korst, 1962, *J. Phys. Chem.*, **66**, 370; S. W. Peterson, V. N. Sadana and W. L. Korst, 1964, *J. de Physique*, **25**, 451.
55. K. Schubert and K. Anderko, 1951, *Z. Metallk.*, **42**, 321.
56. A. Brown, 1962, *Acta Cryst.*, **15**, 652; A. Brown and J. J. Norreys, 1963, *J. Less-Common Metals*, **5**, 302.

57. H. J. Becher, K. Krogmann, and E. Peisker, 1966, *Z. anorg. Chem.*, **344**, 140.
58. R. Kiessling, 1950, *Acta Chem. Scand.*, **4**, 209.
59. G. Brauer and A. Mitius, 1942, *Z. anorg. Chem.*, **249**, 325.
60. J. A. Perri, I. Binder, and B. Post, 1959, *J. Phys. Chem.*, **63**, 616; J. A. Perri, E. Banks, and B. Post, 1959, *Ibid.*, **63**, 2073.
61. T. Bjerström and H. Arnfelt, 1929, *Z. phys. Chem.*, **B4**, 469.
62. J. F. Smith and D. A. Hansen, 1965, *Acta Cryst.*, **18**, 60.
63. S. Rundqvist and S. Pramatus, 1967, *Acta Chem. Scand.*, **21**, 191.
64. W. Reiger, H. Nowotny, and F. Benesovsky, 1966, *Mh. Chem.*, **97**, 378.
65. T. Lundström and P.-O. Snell, 1967, *Acta Chem. Scand.*, **21**, 1343.
66. S. Rundqvist and T. Lundström, 1963, *Acta Chem. Scand.*, **17**, 37.
67. J. Åselius, 1960, *Acta Chem. Scand.*, **14**, 2169.
68. B. Aronsson, M. Bäckman, and S. Rundqvist, 1960, *Acta Chem. Scand.*, **14**, 1001.
69. A. C. Larson, D. T. Cromer, and R. B. Roof, 1963, *Acta Cryst.*, **16**, 835.
70. S. Furuseth and A. Kjekshus, 1965, *Acta Cryst.*, **18**, 320.
71. W. Krönert and K. Plieth, 1965, *Z. anorg. Chem.*, **336**, 207.
72. E. Bjerkelund and A. Kjekshus, 1965, *Acta Chem. Scand.*, **19**, 701.
73. K. Anderko and K. Schubert, 1954, *Z. Metallk.*, **45**, 371.
74. F. Laves and H. J. Wallbaum, 1941–42, *Z. angew. Min.*, **4**, 17.
75. P. Brand, 1967, *Z. anorg. Chem.*, **353**, 270.
76. K. Kanematsu, 1965, *J. Phys. Soc. Japan*, **20**, 36.
77. S. Geller, 1955, *Acta Cryst.*, **8**, 83.
78. C. B. Shoemaker and D. P. Shoemaker, 1965, *Acta Cryst.*, **18**, 900.
79. B. A. Hatt and G. I. Williams, 1959, *Acta Cryst.*, **12**, 655.
80. S. Geller, 1955, *Acta Cryst.*, **8**, 15.
81. O. Nial, 1938, *Z. anorg. Chem.*, **238**, 287.
82. I. Engström, 1965, *Acta Chem. Scand.*, **19**, 1924.
83. E. Ganglberger, 1968, *Mh. Chem.*, **99**, 559.
84. S. Rundqvist and F. Jellinek, 1959, *Acta Chem. Scand.*, **13**, 425.
85. W. H. Zachariasen, 1949, *Acta Cryst.*, **2**, 288.
86. W. Jeitschko, 1970, *Acta Cryst.*, **B26**, 815.
87. H. Perlitz and A. Westgren, 1942, *Ark. Kemi, Min. Geol.*, **B15**, No. 16.
88. R. Lemaire and J. Schweizer, 1967, *C. R. Acad. Sci. Paris*, **264B**, 642; R. Lemaire, J. Schweizer, and J. Yakinthos, 1969, *Acta Cryst.*, **B25**, 710.
89. K. Selte and A. Kjekshus, 1964, *Acta Cryst.*, **17**, 1568.
90. A. F. J. Ruysink, F. Kadijk, A. J. Wagner, and F. Jellinek, 1968, *Acta Cryst.*, **B24**, 1614.
91. F. E. Wang, F. A. Kanda, C. F. Miskell, and A. J. King, 1965, *Acta Cryst.*, **18**, 24.
92. J. B. Newkirk, P. J. Black, and A. Damjanovic, 1961, *Acta Cryst.*, **14**, 532.
93. J. F. Smith and A. E. Ray, 1957, *Acta Cryst.*, **10**, 169.
94. J. H. N. van Vucht, H. A. C. M. Bruning, H. C. Donkersloot, and A. H. Gomes de Mesquita, 1964, *Philips Res. Repts.*, **19**, 407.
95. *Strukturbericht*, **6**, 7.
96. K. Robinson, 1952, *Phil. Mag.*, **43**, 755.
97. M. A. Taylor, 1959, *Acta Cryst.*, **12**, 393.
98. P. L. Blum and G. Silvestre, 1966, *C. R. Acad. Sci., Paris*, **263B**, 709.
99. K. Schubert and H. Pfisterer, 1949, *Z. Metallk.*, **40**, 405.
100. K. Schubert, H. Breima, and R. Gohle, 1959, *Ibid.*, **50**, 146.
101. H. Hahn and C. de Lorent, 1957, *Z. anorg. Chem.*, **290**, 68.
102. K. Koerts, 1963, *Acta Cryst.*, **16**, 432.

10

Structures Based on Simple Cubic and Body Centered Cubic Packing

SIMPLE CUBIC, INCLUDING DISTORTED STRUCTURES

α-Po (*cP*1): α-Po has a simple cubic structure, the atoms occupying the corners of a cube of edge, $a = 3.345$ Å, at about 10°C[1] (Figure 5-22, p. 238). Several alloys such as Au–Te and Ag–Te, retained metastably by very rapid quenching, also have a simple cubic structure.

Hg, *β-Po* (*hR*1): Hg has a rhomodedral cell with $a = 3.005$ Å, $\alpha = 70°32'$, $Z = 1$ at 227°K[2] (Figure 10-1). The atoms are stacked in ABC sequence along the [111] direction, so that Hg is in octahedral coordination. The structure, like that of β-Po which has $\alpha = 98°$, is a distorted form of the simple cubic structure of α-Po.

Se (*hP*3): The unit cell dimensions are $a = 4.366$, $c = 4.959$ Å, $Z = 3$. The Se atoms form spiral chains with a pitch of 3 about the c edges of the hexagonal cell[3] (Figure 5-22, p. 238). Se has two close neighbors along the chains (2.33 Å) and four more at 3.47 Å making a very distorted octahedron. Indeed, the structure can be regarded as a very distorted simple cubic form.

As (*hR*2): As has a rhombohedral cell with $a = 4.1318$ Å, $\alpha = 54°8'$, $Z = 2$. The structure is a two-layer type (Figure 5-20, p. 235) with As–As distances between the double layers considerably longer than the distances within the layers.[4] Each As atom has three neighbors within the double layers at 2.51 Å and three in the next double layer at 3.15 Å.

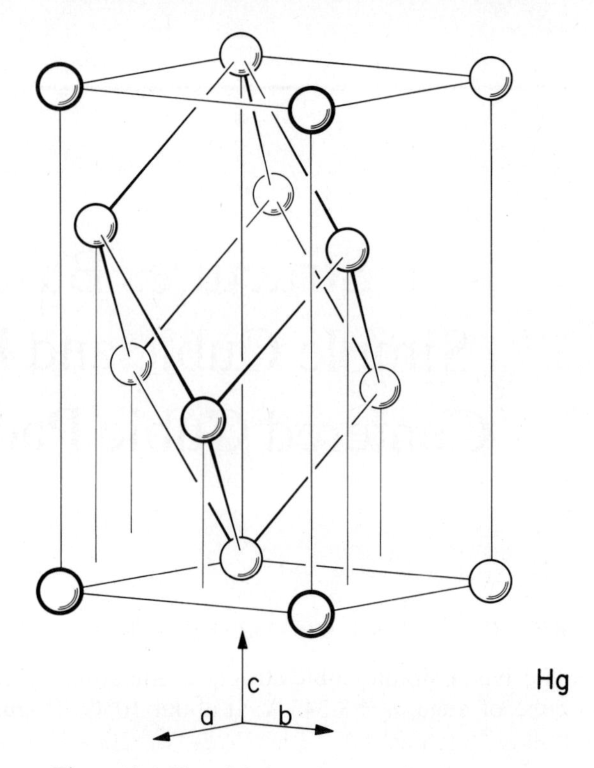

Figure 10-1. Pictorial view of the Hg (*hR*1) structure.

The structure is a distortion of a simple cube which can also be described by a large rhombohedral cell comparable to the cell of the rocksalt structure.

$CaSi_2$ (*hR*6): The dimensions of the hexagonal cell are $a = 3.685$, $c =$ 30.47 Å, $Z = 6$.[5] The Si atoms form rumpled double layers resembling those of the As structure (Figure 5-6, p. 218). Regarding these as a single rumpled 6^3 net, they are arranged in the sequence abc and are interleaved with 3^6 nets of Ca atoms giving an overall stacking sequence (Ca underlined): a$\underline{A}$b$\underline{C}$c$\underline{C}$a$\underline{B}$b$\underline{B}$c$\underline{A}$. Ca thus has 7 Si neighbors at 2.98 to 3.11 Å. Si(1) has 3 Si(1) at 2.51 Å and 4 Ca neighbors, whereas Si(2) has 3 Si(2) at 2.52 Å and 3 Ca at 2.98 Å. The structure can be regarded as a filled up As-like arrangement, but must be redetermined by modern single-crystal methods.

PtS (*tP*4): The PtS structure with cell dimensions $a = 3.470$, $c = 6.110$ Å, $Z = 2$,[6] contains a tetragonally distorted ($c/2a = 0.88$) simple cubic array of S atoms, the height of the unit cell in the c direction being equal to two of these cells. The Pt atoms center two opposite faces of these cells alternately in the (100) and (010) planes on proceeding along [001].

Pt and S each have 4 neighbors of the other type, those of Pt being planar and those of S forming a distorted tetrahedron (Figure 10-2).

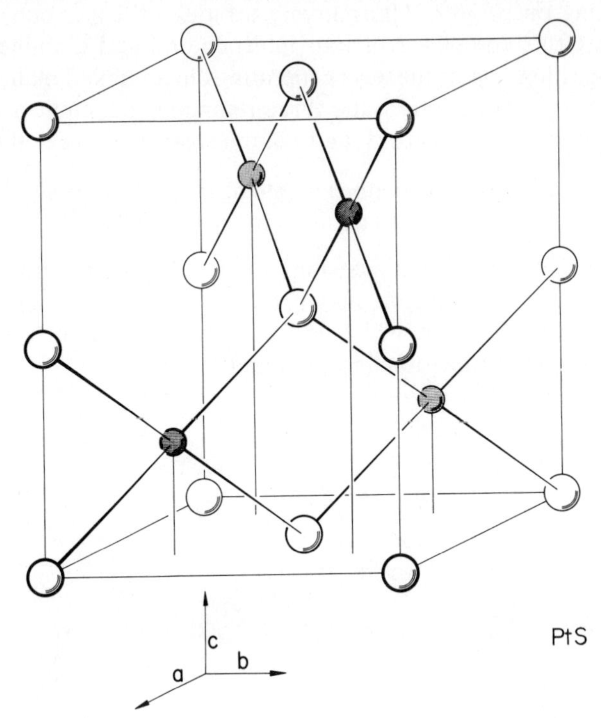

Figure 10-2. Pictorial view of the PtS ($tP4$) structure.

BaS_3 ($oP16$): BaS_3 has unit cell dimensions $a = 8.34$, $b = 9.66$, $c = 4.83$ Å, $Z = 4$.[7] The structure is composed of a block of 4 simple tetragonal pseudocells of Ba atoms ($b/2 = c$), each of which is centered on four faces by S(1) and (2) atoms. A further S atom (3) is set in the cells so that it has one S(1), one S(2), and one Ba neighbor.

STRUCTURES BASED ON B.C. CUBIC PACKING

Body Centered Cubic Stacking of b.c. Cubic {110} Type 3^6 Nets (or Close Packed 3^6 Nets and 3^6 Nets at $\frac{1}{4}$, $\frac{1}{2}$, and $\frac{3}{4}$ Spacing): Cubes and Dodecahedra

W ($cI2$): In the body centered wolfram structure one atom occupies the cube corners and the other the body center, giving each a cube of near neighbors at $a\sqrt{3}/2$ with 6 other neighbors disposed octahedrally at a. The 14 atoms form a dodecahedron with 12 four-sided faces (Figure 10-3).

The arrangement of atoms in triangular nets parallel to {110} planes in the b.c. cubic structure has already been described and compared with close packing on p. 306. Alternatively a block of 8 b.c. cubic cells can be considered as composed of four interpenetrating f.c. cubic arrays of atoms. Regarding one of these as generating close packed planes of atoms normal to the [111] direction, the three remaining f.c. cubic arrays introduce c.p. planes of atoms at $\frac{1}{4}$, $\frac{1}{2}$, and $\frac{3}{4}$ of the spacing between these layers.

Pa (*tI*2): The Pa structure with $a = 3.925$, $c = 3.238$ Å, $Z = 2$,[8] is a distorted form of the b.c. cubic structure with $c/a = 0.825$.

β-Np (*tP*4): The unit cell dimensions of the β-Np structure $a = 4.897$, $c = 3.388$ Å, $Z = 4$,[9] indicate that it is a slightly tetragonally distorted b.c. cubic arrangement in which the b.c. atoms at height $z = \frac{1}{2}$ are displaced alternately up and down to $z = \frac{3}{8}$ and $\frac{5}{8}$.

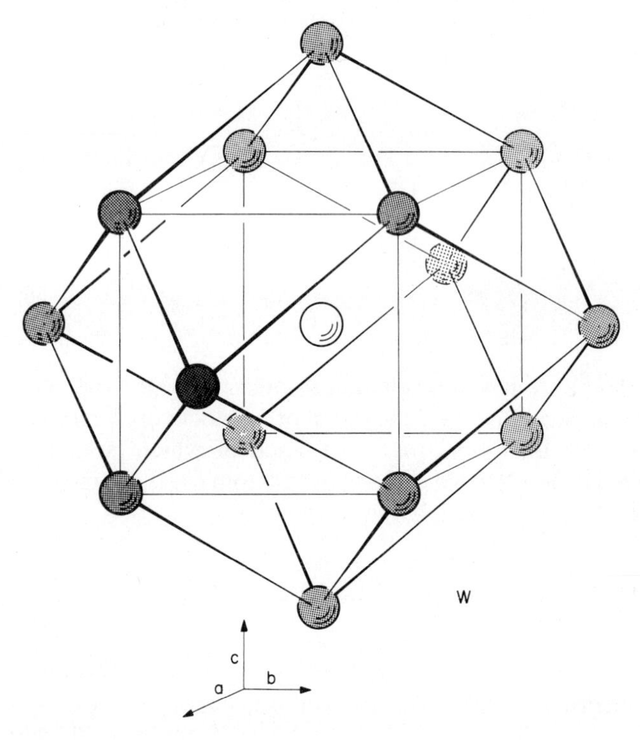

Figure 10-3. The b.c. cubic structure of W (*cI*2) showing the dodecahedral coordination.

FILLED UP DERIVATIVES OF THE B.C. CUBIC STRUCTURE

Martensite (*tI*2): Basically martensite is a tetragonal distortion of the b.c. cubic cell of Fe ($a = 2.85$, $c = 2.98$ Å at 1% C) with carbon randomly distributed in the octahedral holes at the midpoints of the *c* edges of the cell and of the basal faces of the cell parallel to (001). However, it has been shown[10] that the iron atoms have a range of structural parameter along the fourfold axis, the resulting displacement from the cell corners and body center is some 0.2 Å so as to enlarge the octahedral interstices containing carbon.

Cu_2O (*cP*6): In the Cu_2O structure Cu occupies the same sites ($\frac{1}{4}, \frac{1}{4}, \frac{1}{4}$) as S in the sphalerite structure and O occupies the cell corners and cell center.[11] The structure contains two entirely independent interpenetrating Cu–O–Cu networks with O surrounded tetrahedrally by 4 Cu at 1.845 Å and Cu having 2 O neighbors disposed linearly (Figure 10-4).

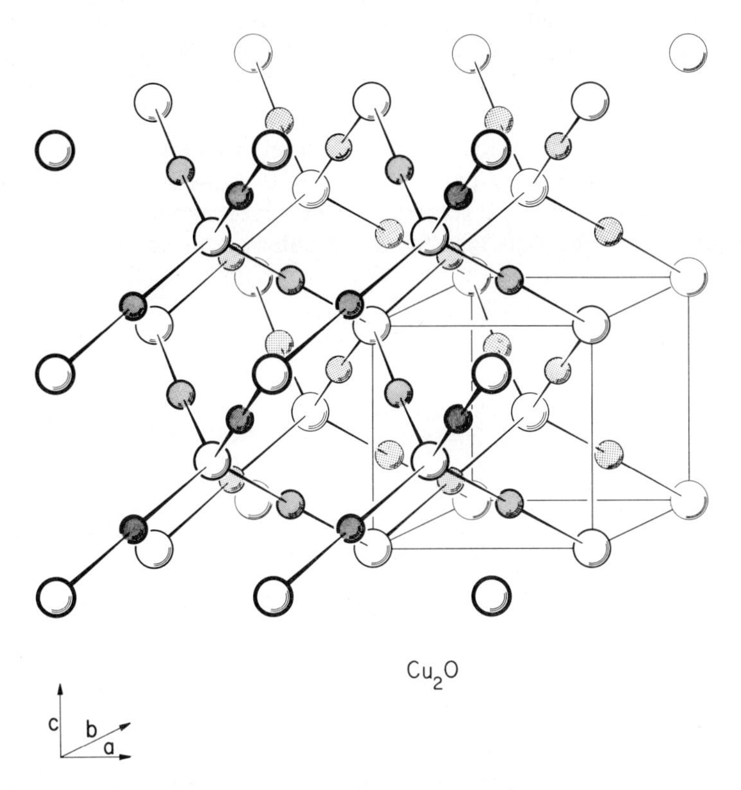

Figure 10-4. View of the Cu_2O (*cP*6) structure showing the independent interpenetrating Cu–O networks.

α-Ni_3S_2 (L.T.) (*hR*5): The hexagonal cell of Ni_3S_2 has dimensions $a =$ 5.730, $c = 6.964$ Å, $Z = 3$.[12] The rhombohedral cell is only slightly distorted from a cube and the S atoms form a slightly distorted b.c. cubic array. The Ni atoms are located in the centers of distorted tetrahedral holes therein (Figure 10-5). They are so arranged that they form spirals running along [001] in the hexagonal cell. The S atoms have 3 Ni neighbors at 2.27 Å and 3 at 2.285 Å, whereas Ni has 4 S neighbors, 2 Ni at 2.48 Å and 2 at 2.51 Å. The structure[13] requires confirmation by single-crystal methods.

Rh_2S_3 (*oP*20): Rh_2S_3 has cell dimensions $a = 8.462$, $b = 5.985$, $c = 6.138$ Å, $Z = 4$.[14] The structure contains pairs of distorted octahedra of S atoms surrounding Rh. The Rh atoms form slightly irregular 3^6 nets at $y \sim \frac{1}{4}$ and $\frac{3}{4}$ which correspond to (110) layers from a b.c. cubic array, since the average angles in the triangles are very approximately 55°, 55°, and 70°. Furthermore, these layers are in b.c. cubic A°B° stacking positions since the nodes of one net lie approximately over the midpoints of one triangle edge of the nets above and below. The S atoms surround Rh octahedrally, but the irregularity in the 3^6 nets of Rh atoms and the number of S atoms are such as to create pairs of distorted octahedra surrounding two Rh atoms and sharing a common face (Rh–Rh = 3.21 Å). The pairs of octahedra are parallel to the (010) plane, and the layer about $y \sim \frac{1}{4}$ is displaced by $c/2$ relative to the layer about $y \sim \frac{3}{4}$ as shown in Figure 10-6. The structure therefore differs in principle from that of

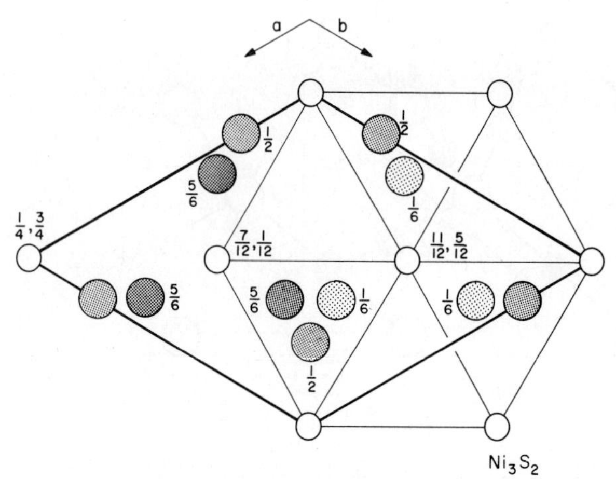

Figure 10-5. View of the hexagonal cell of the low temperature phase α-Ni_3S_2 (*hR*5) projected down [001], indicating the slightly distorted b.c. cubic arrangement of the S atoms (small circles).

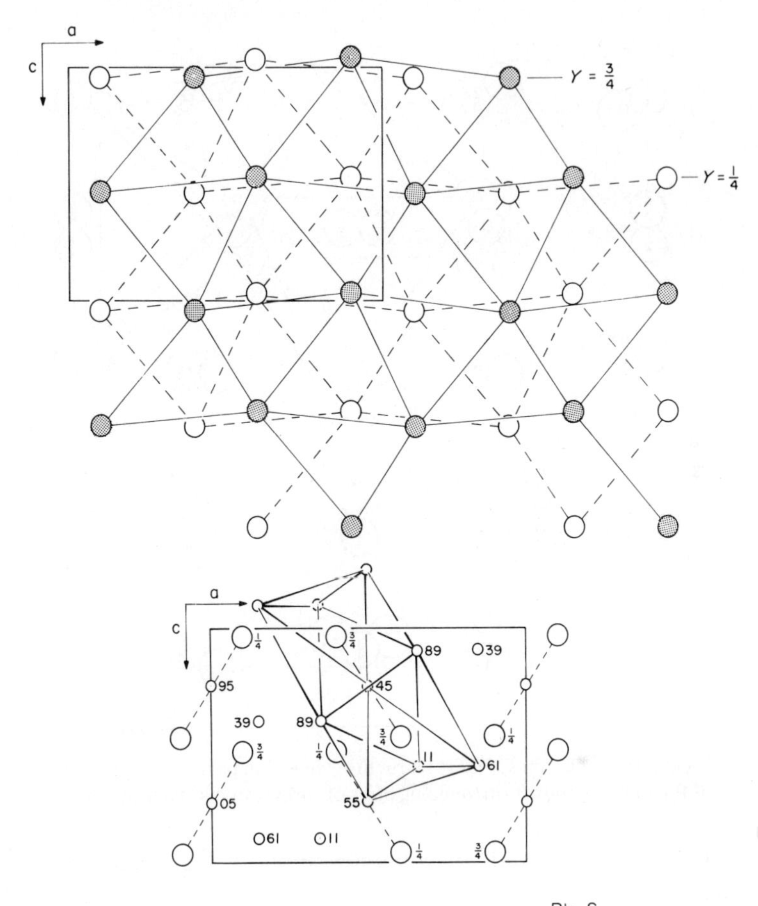

Figure 10-6. (Upper) Distorted 3^6 nets of Rh atoms in the Rh_2S_3 (*oP*20) structure viewed down [010]. (Lower) View showing the pairs of distorted S octahedra sharing a common face.

Sb_2S_3 which can be regarded as made up of fragments of rocksalt structure. The S atoms have 4 Rh neighbors at the corners of quite distorted tetrahedra at distances from 2.31 to 2.40 Å.

Pu_2C_3 (*cI*40): The Pu atoms occupy positions displaced 0.704 Å along ⟨111⟩ directions from the sites of a b.c. cubic array of atoms, the Pu_2C_3 cell being a stack of 8 b.c. cubic pseudocells.[15] Pu has 3 Pu at 3.35 Å, 2 at 3.52 Å and 6 at 3.70 Å, and 9 C neighbors (3 × 2.48, 3 × 2.51, 3 × 2.83 Å). The two octahedra surrounding the paired carbon atoms combine to give the polyhedron shown in Figure 10-7. The structure can be

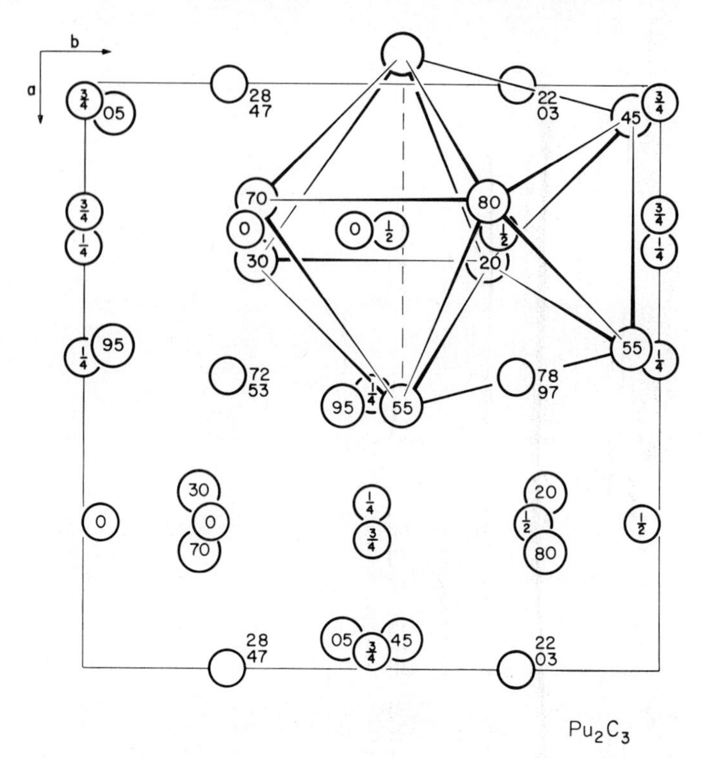

Figure 10-7. The Pu_2C_3 (*cI*40) structure viewed in plan. A polyhedron of Pu atoms is shown surrounding pairs of carbon atoms at a height of $\frac{1}{2}$.

regarded as a partly filled up superstructure of a distorted b.c. cubic structure.

PtHg$_4$ (*cI*10): In the $PtHg_4$ structure ($a = 6.186$ Å) the Pt atoms form a b.c. cubic array and the Hg atoms occupy the fluorine positions of the CaF_2 structure[16] (Figure 10-8). The structure is therefore built up of 4^4 layers of atoms. Pt is surrounded by a cube of Hg at 2.68 Å and Hg has 2 Pt disposed linearly and is also surrounded octahedrally by 6 Hg at 3.09 Å. The structure was determined by X-ray powder photographs only. It can be regarded as a filled up Cu_2O type, bearing the same relationship to Cu_2O as CaF_2 does to sphalerite, ZnS.

Pd$_4$*Se* (*tP*10): The S atoms in Pd_4Se ($a = 5.232$, $c = 5.647$ Å, $Z = 2$)[17] have a distorted b.c. cubic arrangement ($c/a = 1.08$), and Pd form tetrahedra about some of the octahedral holes in this array. However, the Pd–Pd distances within the tetrahedra (2.80, 2.80, and 3.10 Å) are not the closest Pd distances in the structure; in addition there are Pd–Pd

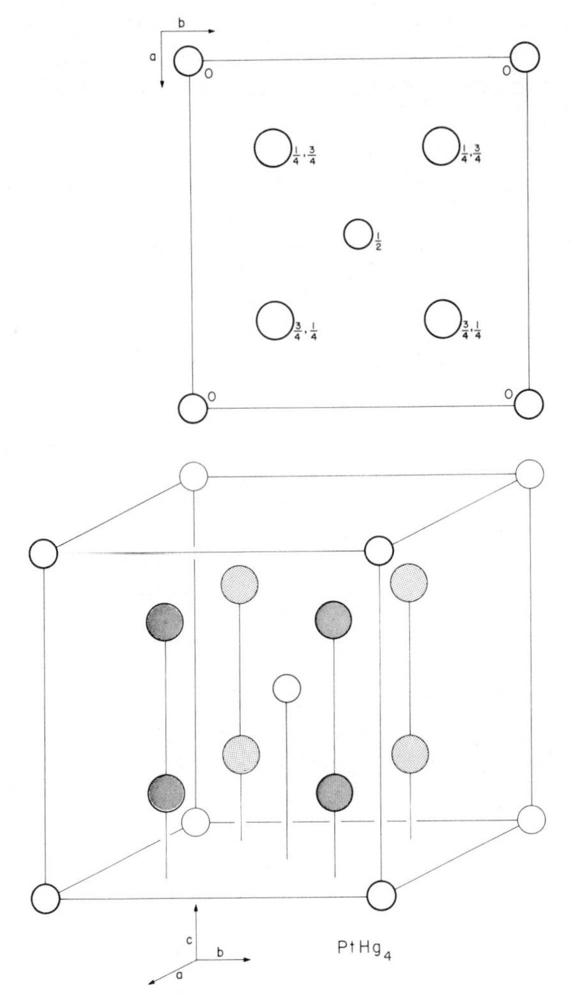

Figure 10-8. Views of the $PtHg_4$ ($cI10$) structure.

distances of 2.76, 2.84, 2.84, 2.93, 2.93, 3.115, and 3.115 Å and each Pd has two close Se neighbors. Se has 4 Pd at 2.46 and 4 at 2.49 Å.

$AuAg_3Te_2$ (L.T.) ($cI48$): The Te atoms are slightly displaced from the sites of a b.c. cubic array, with the unit cell corresponding to a stack of 8 b.c. cubic Te pseudocells.[18] One Au atom is located in each pseudocell on the body diagonals, $\frac{1}{4}$ of the distance between corners. They therefore have two linearly disposed Te neighbors at 2.535 Å. The Ag atoms occupy sites slightly displaced from the centers of the tetrahedral holes in the b.c. cubic array of atoms, filling three of the 12 holes available in each b.c. pseudocell. Each Ag is surrounded tetrahedrally by four Te neighbors (2 × 2.895 Å, 2 × 2.95 Å). In addition, Ag has 2 Ag and 2 Au at 3.09 Å. Au has 6 Ag at 3.09 Å and Te has 6 Ag arranged more or less octahedrally and 1 Au neighbor.

SUPERSTRUCTURES BASED ON B.C. CUBIC PACKING

MN: *M* on 4^4 Nets and *N* on 4^4 Nets

CsCl (*cP*2): In the CsCl structure one component occupies the cube corners and the other the body center, so that alternate layers of the two components occur along ⟨100⟩ directions. However, compared to the AuCu I structure, the separation of the atoms within the {100} planes is relatively larger and the spacing between the planes is relatively smaller corresponding to $c/a = 1$, compared to $c/a \sim 1.3$ for phases with the AuCu I structure. The Cs and Cl atoms together form triangular nets parallel to {110} planes with Cs occupying one of the rectangular 4^4 subnets resulting from the geometry of the triangles of the 3^6 net, and Cl occupying the other. These layers are stacked A°B° in the ⟨110⟩ directions (Figure 7-2, p. 306). Originally the CsCl structure was thought of as an ionic structure, but now that several hundred metallic phases have been found to have the structure, it must be regarded essentially as a metallic structure. Each atom has 8 neighbors of unlike kind (for AuZn the Au–Zn distance is 2.77 Å) and 6 of like kind (for AuZn, Au–Au, and Zn–Zn = 3.20 Å).

The Hume-Rothery β'-phases with the CsCl structure which are formed between Cu, Ag, and Au and the following B Group elements, have an upper limiting electron concentration of about 3.0 per primitive cell containing two atoms. However, the great majority of nontransition metal phases with the CsCl structure have electron concentrations of 4 or 5 per primitive cell (Figure 3-32, p. 107) and the near-neighbor diagram (p. 53) suggests that the CsCl structure occurs primarily because of *M–N* interactions which control the unit cell dimensions (Figure 10-9), although a general geometrical effect involving also *M–M* interactions, giving 14–8 coordination is also important. It is only in the alloys of the noble metals and the following Group II to IVB metals, that the influence of electron concentration in controlling the phase boundaries can be recognized. The CsCl structure is generally formed with a contraction of the unit cell volume relative to the sum of the elemental volumes of the component atoms as shown in Figure 10-10, which compares data for phases with the CsCl, FeSi, CrB, FeB, and NaCl structures. Figure 10-11

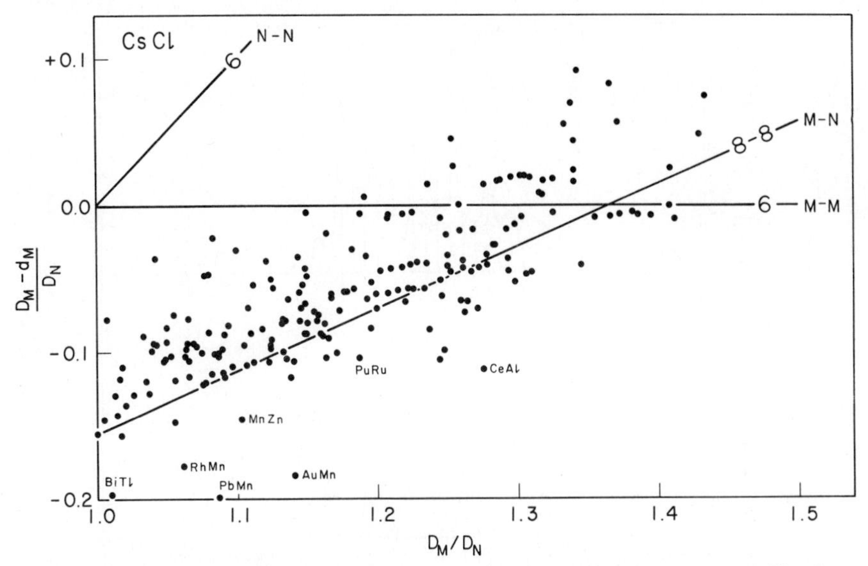

Figure 10-9. Near-neighbor diagram for phases with the CsCl (MN) structure. Numbers indicate the number of neighbors for the contacts M–N, M–M, etc.

compares the occurrence of phases with these structures, as a function of the volume ratio, V_M/V_N, of the component atoms. It is notable that there are very few phases with the CsCl structure that have volume ratios larger than about 2.55 which corresponds to the limiting radius ratio of 0.732 (or 1.366) for stability of ionic phases with the CsCl structure.

A block of 8 CsCl cells has the same site occupation as a f.c. cubic array of atoms in which all octahedral and tetrahedral interstices are occupied, giving four interpenetrating f.c. cubic arrays of atoms. Thus for example the CsCl structure has the same site occupation, but different ordering to the $BiLi_3$ structure, the Heusler alloy structure and also to the NaTl type structure (Figure 10-12), all of which have F cells.

Distorted CsCl Type Structures

α-IrV (*oC*8): α-IrV with cell dimensions $a = 5.791$, $b = 6.756$, $c = 2.976$ Å, $Z = 4$,[19] is a superstructure of the CsCl type arising from distortion of the atom positions and unit cell edges (Figure 10-13). The IrV cell corresponds to 8 distorted CsCl type cells. The atoms form distorted 3^6 nets on (110) planes with triangle angles similar to those of the CsCl structure. These nets are stacked one above the other in A°B° stacking with the node of one net lying over the midpoint of one of the triangle edges of the nets above and below. Ir has 4 V at 2.61 and 4 at 2.675 Å,

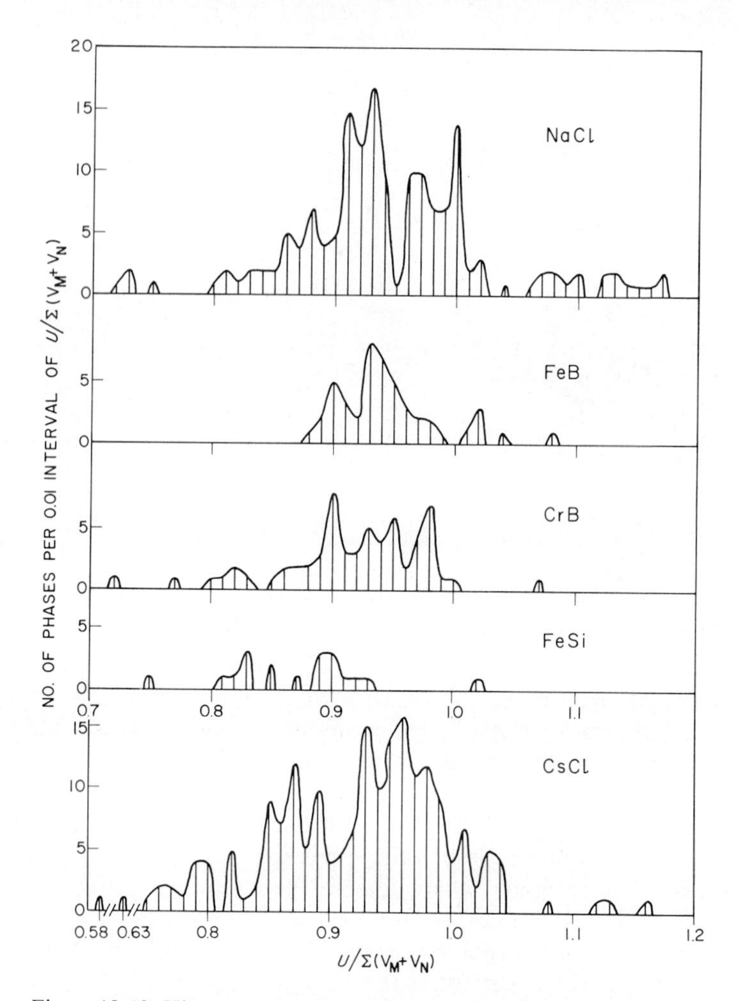

Figure 10-10. Histogram showing number of phases per 0.01 interval of $U/\Sigma(V_M+V_N)$ for phases with the CsCl, FeSi, CrB, FeB, and NaCl type structures. $U/\Sigma(V_M+V_N)$ represents the ratio of the unit cell volume to the sum of the atomic volumes (determined from the elemental structures) of the atoms contained therein.

together with 5 Ir at 2.80 to 2.97 Å. V has 8 Ir neighbors, one V at 2.55 and two at 2.80 Å. The structure, however, requires confirmation by single-crystal methods.

CoU (*cI*16): The CoU structure ($a = 6.356$ Å) can be regarded as a very distorted CsCl type with the unit cell[20] corresponding to a block of 8

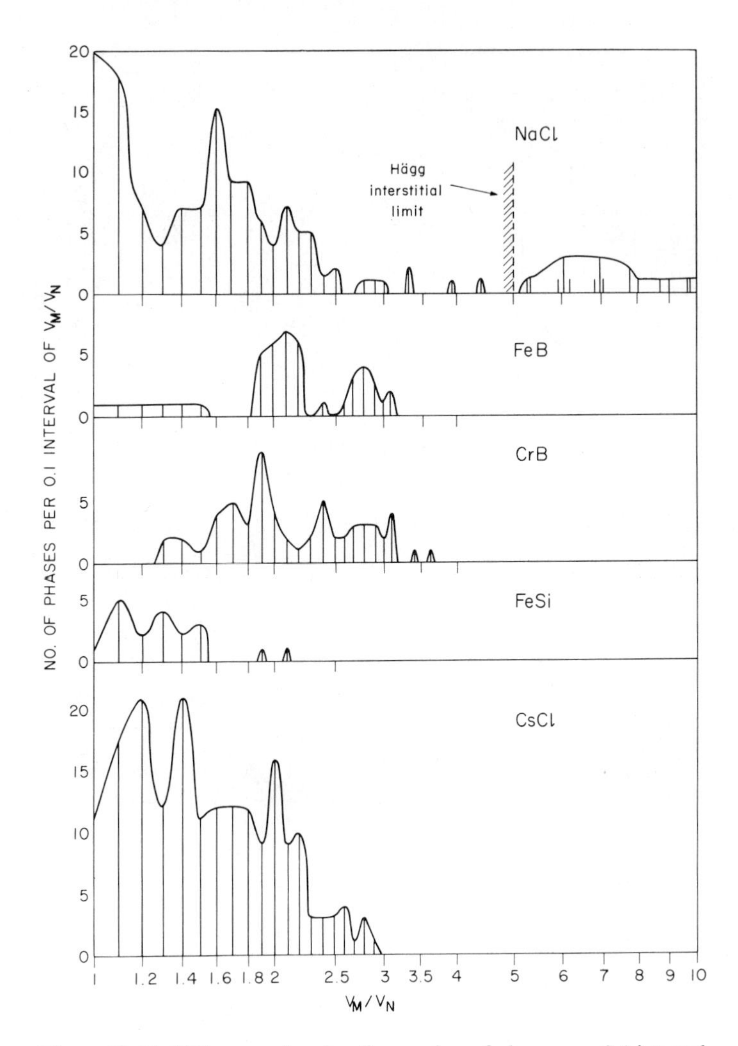

Figure 10-11. Histogram showing the number of phases per 0.1 interval of V_M/V_N, the atomic volume ratio of the components M and N, which take the CsCl, FeSi, CrB, FeB, and NaCl structures.

CsCl type cells (Figure 10-14). Thus each atom has 8 neighbors of the opposite kind and 6 of the same kind. Co–U distances are 2.65 Å (one) to 2.875 Å. Whereas Co has 3 Co at 2.68 Å, the other 3 Co are at 3.78 Å; similarly U has 3 U at 2.77 Å and 3 U at 3.65 Å. The structure requires confirmation by a single-crystal study.

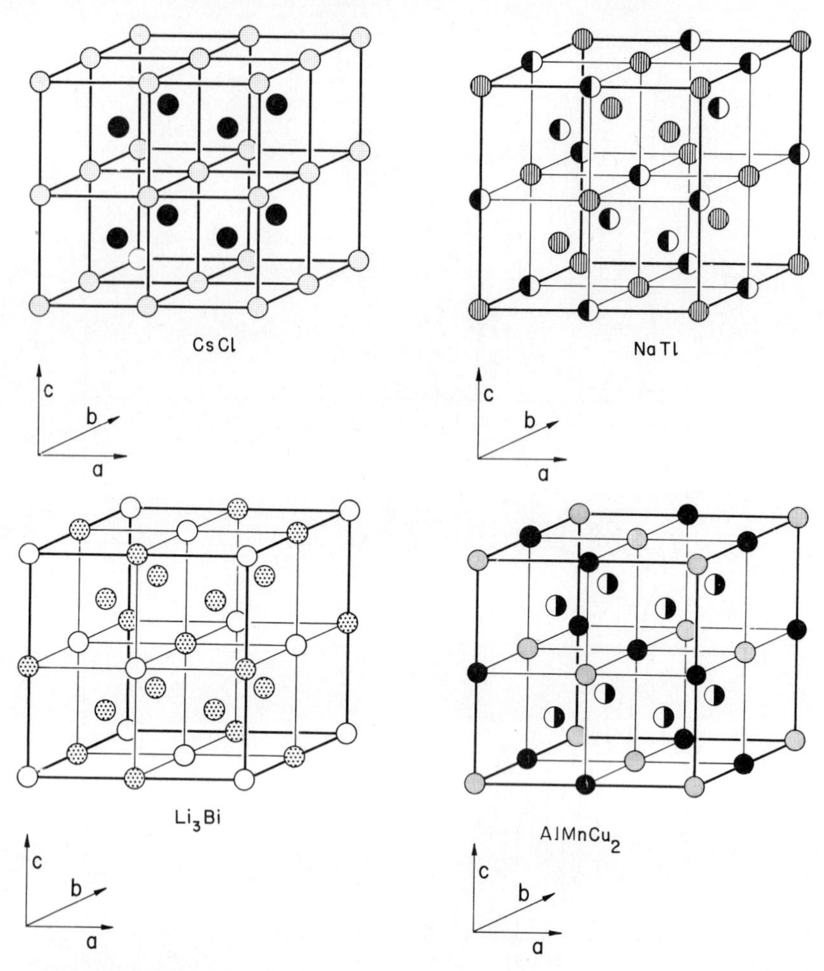

Figure 10-12. Diagram comparing atomic arrangements in the CsCl, NaTl, Li_3Bi, and $AlMnCu_2$ type structures.

Other Superstructures Based on b.c. Cubic Packing

γ-CuTi (*tP*4): The γ-CuTi structure with cell dimensions $a = 3.118$, $c = 5.921$ A, $Z = 2$,[21] can be regarded as an ordered derivative of the b.c. cubic structure involving two cells stacked in the [001] direction in which two 4^4 layers of Cu atoms [one displaced relative to the other by $(a+b)/2$] are followed by two 4^4 layers of Ti atoms (one similarly displaced relative to the other). This ordered arrangement of the b.c. cubic (*cI*2) structure in which pairs of ordered layers of the two components

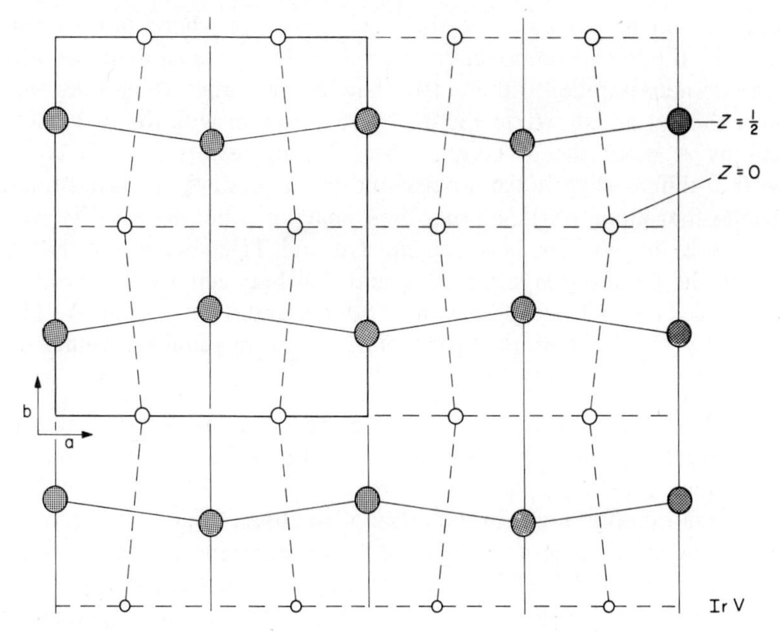

Figure 10-13. Atomic arrangement in the α-IrV ($oC8$) structure viewed down [001].

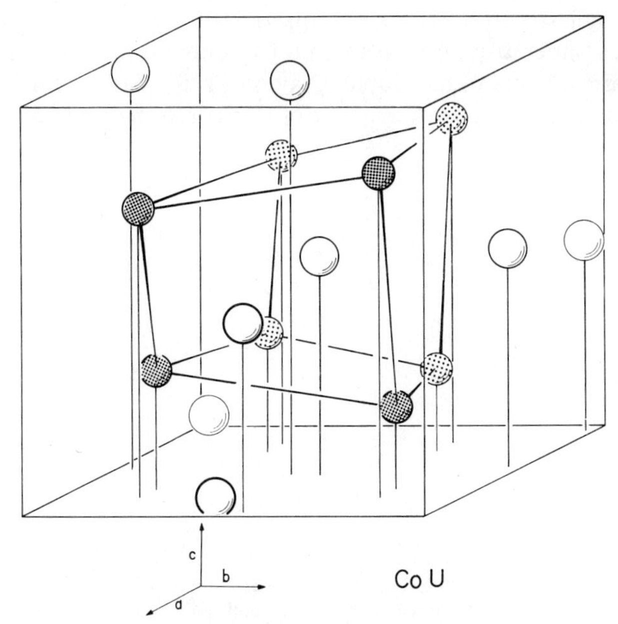

Figure 10-14. Pictorial view of the CoU ($cI16$) structure.

alternate, can be compared to the CsCl structure where single ordered layers of the two components alternate. The arrangement of atoms on the 3^6 nets parallel to the (110) plane of the structure and the repeat unit of the net are shown in Figure 10-15, together with the two stacking positions A°B° of the 3^6 layers. The ordering results in an axial ratio $c/2a$ that differs slightly from unity and the layers are displaced relative to each other along [001], so that their spacing is not necessarily $c/4$. In CuTi itself the spacing between the Cu and Ti layers is $c/4$, but that between the Cu atom layers is less, and that between the Ti atom layers is more. Cu has 5 Ti neighbors at ~ 2.66 Å and 4 Cu at 2.50 Å. Ti has 5 Cu and 4 Ti neighbors at 2.83 Å. The structure requires confirmation by single-crystal methods.

$BiLi_3$ ($cF16$): The $BiLi_3$ or BiF_3 structure is composed of four interpenetrating f.c. cubic arrays of atoms.[22] With Bi at the cell corners and face centers, Li(1) occupies the centers of the octahedral holes and Li(2) all of the tetrahedral holes in the array of Bi atoms (Figure 10-12, p. 572). Bi and Li severally form close packed 3^6 nets of atoms normal to the [111] direction of the cubic cell, but the spacing of the layers is only $\frac{1}{4}$ of that for close packing. The Bi layers are in cubic close packing sequence ABC with three equally spaced Li layers lying between each Bi layer. Since the space filling in this structure is identical to that in 8 cells of the b.c. cubic or CsCl structure, the most appropriate structural description of $BiLi_3$ is as a superstructure based on a b.c. cubic array. Bi and Li atoms together form 3^6 nets of b.c. cubic array on {110} planes of the supercell or b.c. pseudocell. The arrangement of these and the rectangular repeat

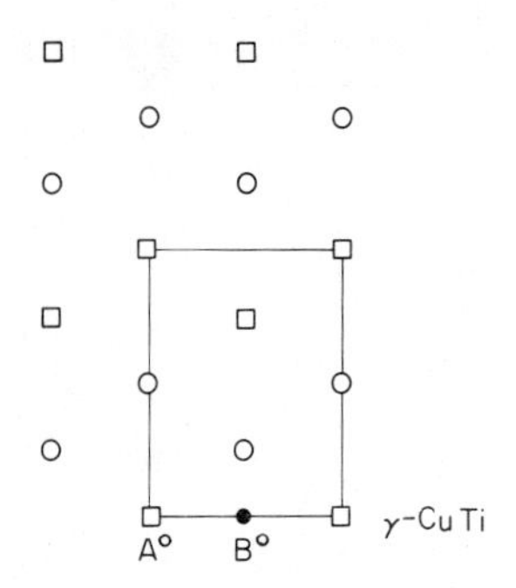

Figure 10-15. Arrangement of the Cu and Ti atoms on planes parallel to (110) in the γ-CuTi ($tP4$) structure, showing the repeat cell and the two stacking positions.

pattern are shown in Figure 10-16; they are stacked in b.c. cubic ⟨110⟩ sequence A°C° with the nodes of one net lying over the midpoints of one of the sides of the triangles of nets above and below.

All atoms have CN 14 in dodecahedra with 12 four-sided faces. Bi has 8 Li(2) neighbors at 2.91 Å and 6 Li(1) at 3.36 Å. Li(1) has 8 Li at 2.91 Å

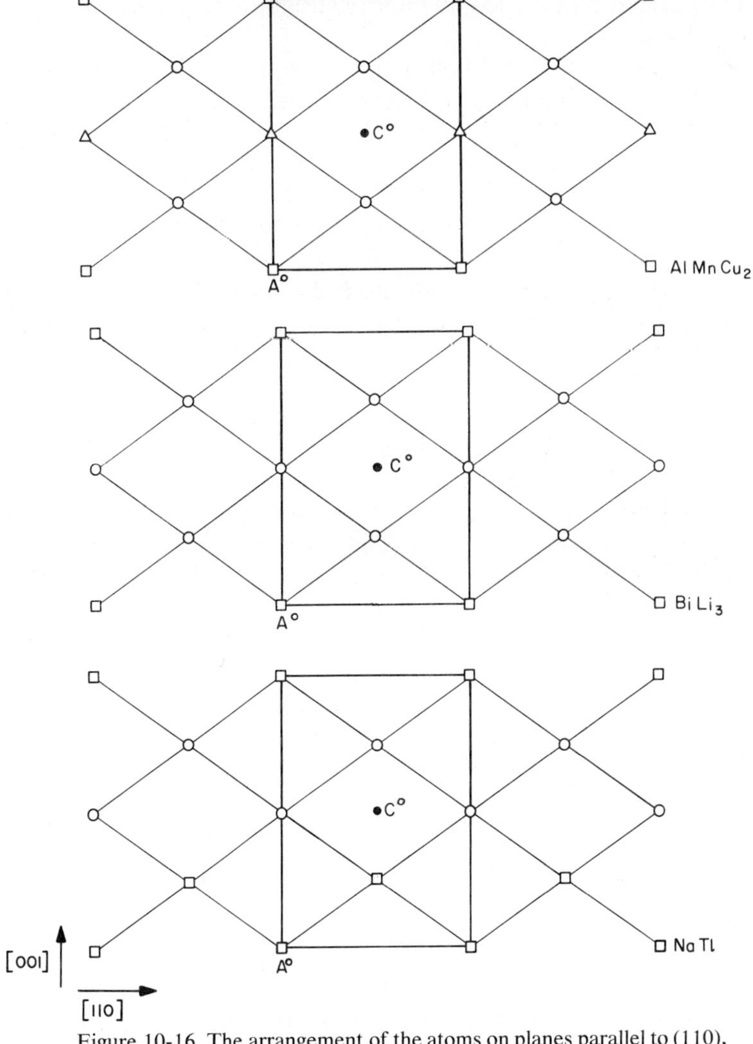

Figure 10-16. The arrangement of the atoms on planes parallel to (110), the rectangular repeat unit and stacking positions for the $AlMnCu_2$ (*cF*16), $BiLi_3$ (*cF*16), and NaTl (*cF*16) structures: □, Al, Bi, or Na; ○, Cu, Li, or Tl; △, Mn.

and 6 Bi neighbors. Li(2) has 4 Li(1) and 4 Bi, both at 2.91 Å, and 6 Li(2) at 3.36 Å.

Compounds such as $BiLi_3$ are semiconductors, which suggests that Li(1) is present as an ion that supplies the extra electron needed for the system of Bi–Li(2) covalent bonds giving Bi a filled valence subshell. This indicates that, from the point of view of physical properties, $BiLi_3$ is better considered as a filled up fluorite structure, than a superstructure of a b.c. cubic array.

An ordered form of the $BiLi_3$ structure containing two cationic components is recognized in Li_2MgSn, but according to Laves[23] both ordered (Sn: 0, 0, 0; . . . , Li(1): $\frac{1}{2}, \frac{1}{2}, \frac{1}{2}$; . . . , Li(2): $\frac{1}{4}, \frac{1}{4}, \frac{1}{4}$; . . . , Mg: $\frac{3}{4}, \frac{3}{4}, \frac{3}{4}$; . . .) and partially disordered (Sn: 0, 0, 0; . . . , Li(1): $\frac{1}{2}, \frac{1}{2}, \frac{1}{2}$; . . . , (Li(2) + Mg): $\frac{1}{4}, \frac{1}{4}, \frac{1}{4}$; . . . , $\frac{3}{4}, \frac{3}{4}, \frac{3}{4}$; . . .) forms occur simultaneously. Nevertheless the order-disordered situation is unusual because of the two different sitesets upon which cation disorder can occur. The unexpected manner in which one of the Li atoms rather than the lone Mg atom always occupies the octahedral holes, suggests that the compound bears a stronger similarity to the semiconductor $BiLi_3$, than to metallic phases with occupied tetrahedral and octahedral holes.

NaTl (*cF*16): The ordered arrangement of Na and Tl atoms in a cell containing 16 atoms results in the same space filling as that of 8 b.c. cubic or CsCl type cells. Although the structure can be regarded as a completely filled up f.c. cubic arrangement formed by four interpenetrating f.c. cubic arrays of atoms, in which each component occupies a diamond-like array of sites, the most appropriate structural description is as a superstructure based on a b.c. cubic array of atoms. This is similar to that described above for $BiLi_3$; the ordering of the Na and Tl atoms on the triangular nets and the rectangular repeat unit are shown in Figure 10-16. All atoms have CN 14, the dodecahedra having 12 four-sided faces. Na has 4 Na and 4 Tl neighbors at 3.24 Å and 6 more Tl at 3.74 Å. Tl has 4 Na and 4 Tl at 3.24 Å and 6 Na at 3.74 Å.[24] The structure does not yet appear to be confirmed by a single-crystal study.

The NaTl structure is formed by five binary alloys of Groups IA and III metals and two of Groups IA and IIB metals. Its stability was originally thought to result from ionic transfer from the Group I to III metal which was then believed to form a strong diamond-like array of bonds, although the later discovery that Group II metals also formed the structure compromised this belief. Furthermore the postulate of strong covalent bonds between like atoms that are negatively charged is also objectionable. The near-neighbor diagram (p. 53) based on covalent sizes for CN 8 shows that both M–M and M–N contacts are compressed until N–N contacts form in all phases with the NaTl (Figure 10-17). In this figure the smaller atom is taken as the N component and even in InLi where this is Li, N–N contacts are established; yet it can scarcely be argued that the In atoms are compressed to make a strong diamond network formed by the Li atoms! Indeed, it can be shown that phases with the NaTl structure are characteristically metallic phases.

The site occupation is identical in the CsCl and NaTl structures with interpenetrating cubes giving 8–8 nearest-neighbor coordination in each case. In the CsCl structure this coordination is achieved by 8–8 M–N neighbors which largely control the unit cell dimensions (Figure 10-9, p. 569), whereas in the NaTl structure it is obtained by 4 M–N, 4 M–M, and 4 N–N contacts. Therefore, if the compressibility of the M atoms is such as to allow M–N and N–N contacts to be established, 8–8 CN is achieved at a smaller average atomic volume relative to the volumes of the component atoms in the NaTl structure than in the CsCl structure. Other things being equal, a lower free energy may therefore be expected

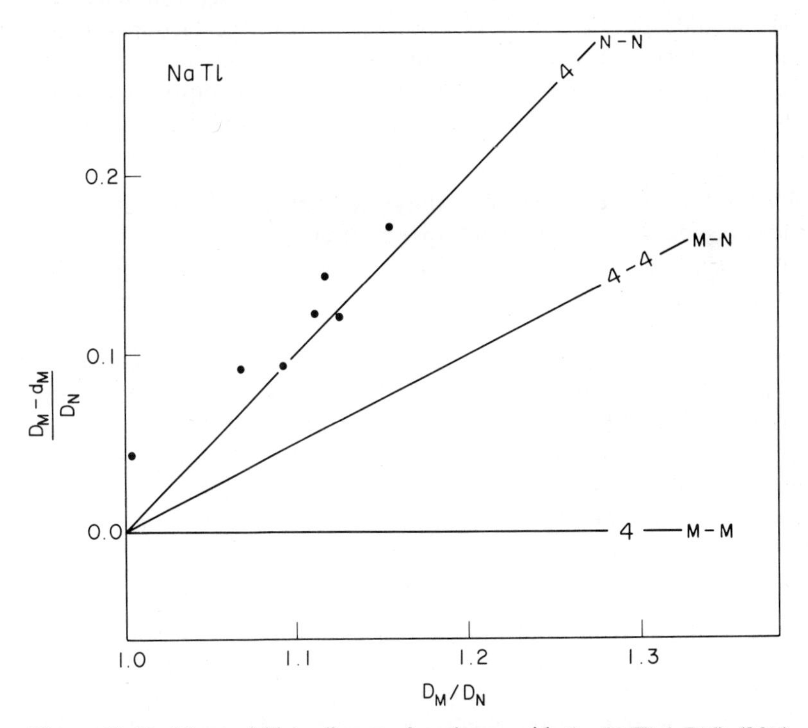

Figure 10-17. Near-neighbor diagram for phases with the NaTl ($cF16$) (MN) structure. M is taken as the larger of the two atoms. Numbers indicate the number of neighbors for contacts N–N, M–N, etc.

for a phase in the NaTl than the CsCl structure, provided that the M component is readily compressible and the radius ratio does not depart much from unity.

Thus, it is apparent that phases with the NaTl structure are neither "ionic," nor are they dominated by strong covalent diamond-type bonds, but they are characteristically metallic phases with 8–8 coordination like the metallic phases with the CsCl structure. A phase may have a lower free energy in the NaTl structure than in the CsCl structure when two favorable circumstances occur simultaneously: (i) the larger M component is highly compressible, and (ii) the axial ratio is not much larger than unity. These conditions are satisfied in phases with the alkali metals Li and Na as the M component and radius ratio values about 1.1. In the two phases with the NaTl structure where the alkali metal is the smaller N component, the axial ratio is closer to unity (1.004 and 1.067). LiTl with $R_{Tl}/R_{Li} = 1.10$ has the CsCl structure and it is apparent that the radius ratio is too large for the Tl atoms to be sufficiently compressed for both Li–Tl and Li–Li contacts to occur in the NaTl structure. Similar considerations would apply to phases of potassium with the Group III metals; even for the most favorable case of "KTl," the value of $R_K/R_{Tl} = 1.397$ is too large for Tl–Tl contacts to be formed by compression of the K atoms. Equiatomic phases of the alkali metals with the Group IV elements generally have complex structures.

AlMnCu$_2$ (*cF*16): The Heusler alloy structure is a filled up f.c. cubic structure, composed of four interpenetrating f.c. cubic arrays of atoms, such that the Al atoms are located at the cell corners and face centers with Mn in the octahedral holes of the Al array, and Cu in the tetrahedral holes[25] (Figure 10-12, p. 572). However, since the space filling is the same as in 8 b.c. cubic or CsCl cells, the structure is most appropriately considered as a superstructure of a b.c. cubic array of atoms. The description is similar to that of Li_3Bi above. The arrangement of the components on {110} planes, the rectangular repeat unit in the plane and the stacking sequence of the planes are indicated in Figure 10-16. All atoms have CN 14 in a dodecahedra with 12 four-sided faces. Al has 6 Mn at 2.97 Å and 8 Cu at 2.58 Å. Mn has 6 Al and 8 Cu at 2.58 Å. Cu has 4 Al, 4 Mn, and 6 Cu neighbors at 2.97 Å.

VTl$_3$*S*$_4$ (*cI*16): The V atoms occupy the cube corners and body center, and the Tl atoms occupy the face centers and midpoints of the cell edges.[26] The S atoms lie on the body diagonals, as shown in Figure 10-18, so that they surround V tetrahedrally at 2.28 Å. Tl also has 4 S at 3.07 Å in a very distorted tetrahedral arrangement. Each S has 1 V and 3 Tl neighbors. The structure thus corresponds to a superstructure based on 8

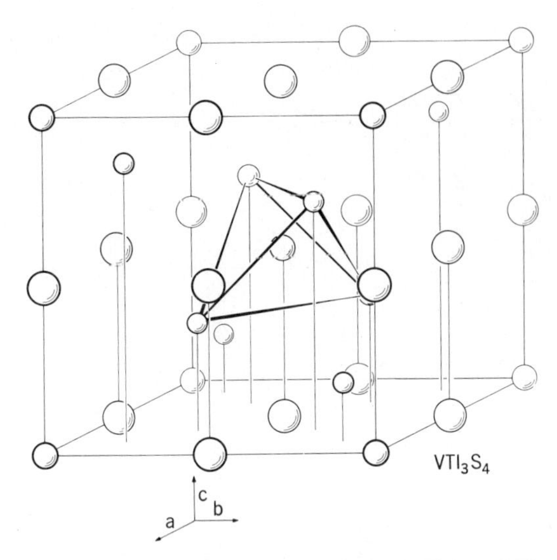

Figure 10-18. Pictorial view of the VTl_3S_4 ($cI16$) structure.

b.c. cubic cells (c.f. Li_3Bi) except that the S atoms with $x = 0.175$ instead of 0.250, are displaced from the pseudocell body centers. The structure requires confirmation by single-crystal methods.

Sb_2Tl_7 ($cI54$): Sb_2Tl_7 is said to have a filled up Fe_3Zn_{10} type structure with two additional Tl atoms occupying the vacant sites at the cell body center and corners, and Sb occupying the Fe positions of Fe_3Zn_{10}.[27] However, these results require confirmation by a single-crystal study. The structure is a superstructure built up of 27 b.c. cubic pseudocells. Each atom has 14 neighbors.

Pb_3Li_8 ($mC22$): The Pb_3Li_8 structure has cell dimensions $a = 8.240$, $b = 4.757$, $c = 11.03$ Å, $\beta = 104°25'$, $Z = 2$.[28] It is a superstructure based on the b.c. cubic structure. The atoms form 3^6 nets with triangle angles corresponding to those of the b.c. cubic structure on $\{110\}$ planes, and they are stacked in b.c. cubic sequence with the nodes of one net lying over the midpoints of the sides of the triangles of the nets above and below. The Pb atoms are in ordered array in each 3^6 layer, with the repeat unit for ordering equal to the ac base of the unit cell and two stacking positions, A°B° as indicated in Figure 10-19. Stoichiometry is satisfied in each 3^6 layer. The Pb atoms are of two types, Pb(2) forming pairs at 2.91 Å. All atoms have CN 14, the coordination polyhedra being dodecahedra with 12 four-sided faces. The closest Pb–Li and Li–Li distances are 2.91 Å. The structure can be compared to that of Mo_3Al_8 (p. 349) where

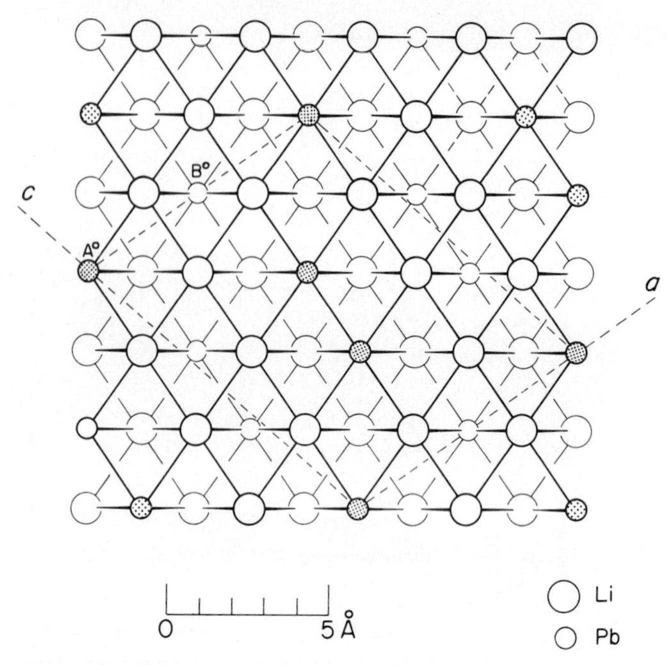

Figure 10-19. Atomic arrangement in the Pb_3Li_8 (*mC*22) structure viewed along [010].

the atoms form close packed 3^6 nets which are stacked in close packed sequence.

V_4Zn_5 (*tI*18): The V_4Zn_5 structure is made up of slightly distorted 1.4^4 nets of V + Zn atoms at $z = 0$ and $z = \frac{1}{2}$, the net at $z = \frac{1}{2}$ being displaced so as to center the squares of the net at $z = 0$. The unit cell contains three rows of the net in the *a* and *b* directions (Figure 10.20*a*).

The cell dimensions, $a = 8.910$, $c = 3.224$ Å, $Z = 2$,[29] indicate that the superstructure is based on an expanded b.c. cubic pseudocell with $c_0/a_0 \sim 1.08$, rather than a considerably more squashed f.c. cubic pseudocell. Indeed, parallel to the (101) plane of the pseudocell where the atoms form 3^6 nets, the angles are more nearly those of b.c. cubic {110} layers than of close packing (60°), and the stacking of successive layers also is not much displaced from b.c. cubic [110] stacking with the nodes of one net over the midpoints of edges of the triangles of nets above and below. Thus V_4Zn_5 is a superstructure of a b.c. cubic cell with the atoms somewhat displaced, and with overall slight distortion toward close packing. The arrangement of atoms in the 3^6 nets approximately parallel to (031) planes of the supercell is shown in idealized form in Figure 10-20*b*. These

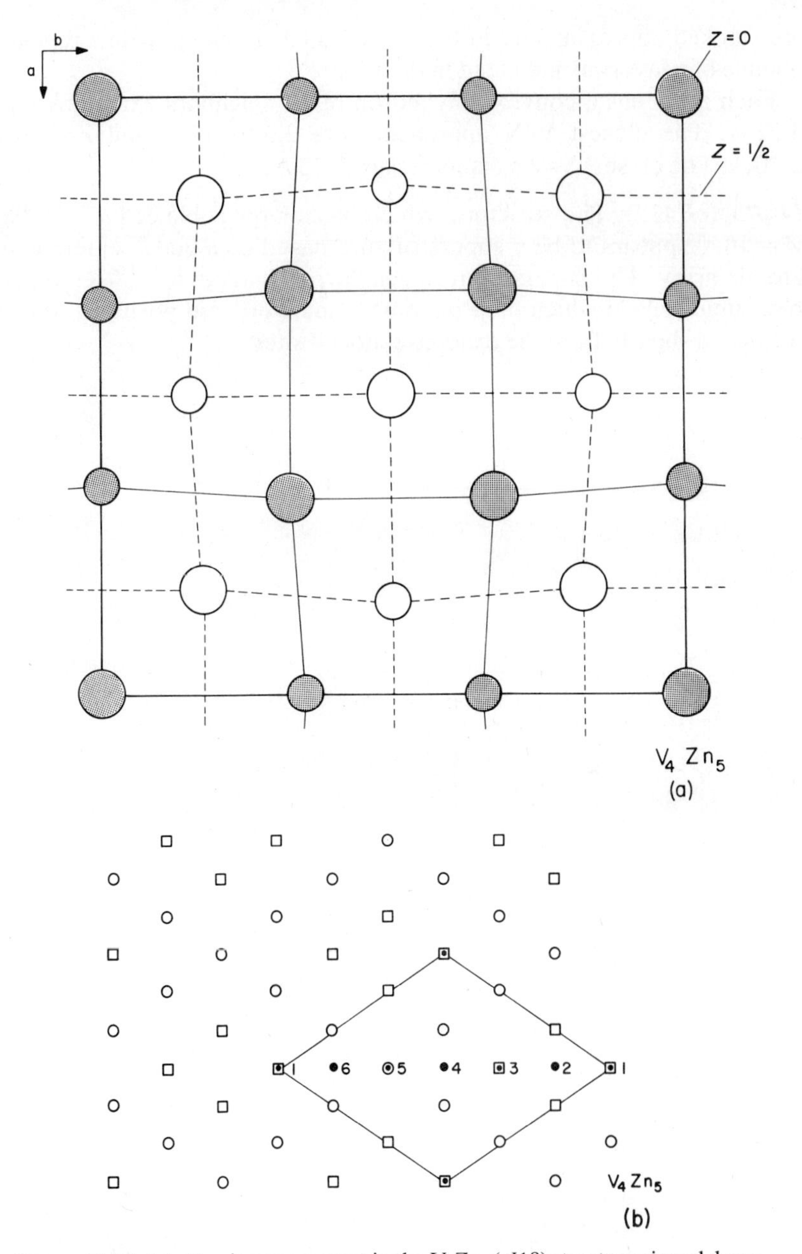

Figure 10-20. (*a*) Atomic arrangement in the V_4Zn_5 (*tI*18) structure viewed down [001]: large circles, Zn. (*b*) Arrangement of V and Zn atoms on planes parallel to (031), showing a repeat cell and the 6 stacking positions for the layers: circles, Zn; squares, V.

are stacked approximately in b.c. cubic {110} stacking, with a repeat sequence of 6 layers as indicated in the Figure.

Each atom has a convex polyhedron of 14 neighbors extending out to 3.22 Å. The closest V–V approaches are 2.50 Å (4×) and Zn–Zn are 2.70 Å. The closest V–Zn distances are 2.72 Å.

$Li_{22}Pb_5$ (*cF*432): The structure, which has a large cubic cell $a = 20.08$ Å, $Z = 16$,[30] appears to be a superstructure based on a body centered cubic atomic array. The superstructure cell corresponds to a $6 \times 6 \times 6$ block of b.c. cubic cells in which the Pb atoms assume ordered positions; they are displaced slightly from the exact pseudocell sites.

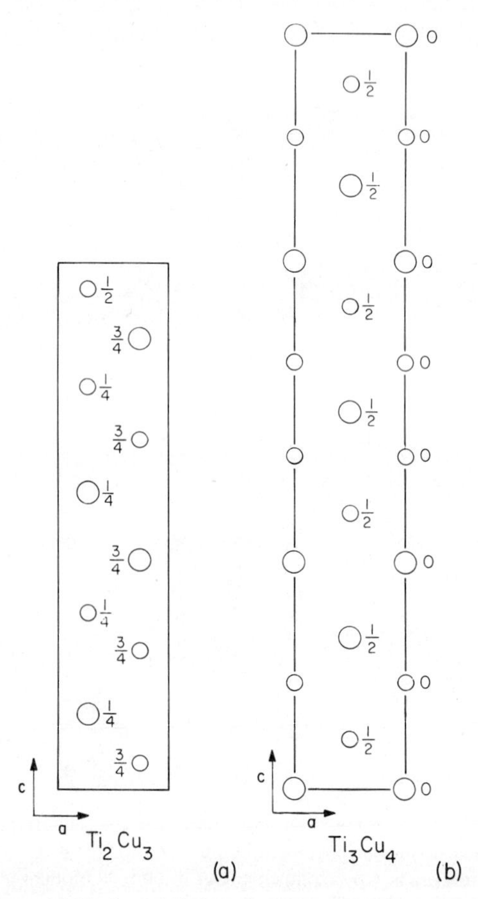

Figure 10-21. (*a* and *b*) Atomic arrangement in the Ti_2Cu_3 (*tP*10) and Ti_3Cu_4 (*tI* 14) structures viewed along [010]: Ti, large circles.

Ti_2Cu_3 (*tP*10): Ti_2Cu_3 which has cell dimensions $a = 3.13$, $c = 13.95$ Å, $Z = 2$,[31] is a superstructure of a b.c. cubic array of atoms. The unit cell contains 5 squashed b.c. cubic pseudocells stacked along *c*. The ordering of the atoms is shown in Figure 10-21*a*. Both this and the Ti_3Cu_4 structure require confirmation by single-crystal methods.

Ti_3Cu_4 (*tI*14): Ti_3Cu_4 has cell dimensions $a = 3.13$, $c = 19.94$ Å, $Z = 2$.[31] The structure is a superstructure based on a b.c. cubic array of atoms with 7 squashed b.c.c. pseudocells stacked along *c*. The ordered arrangement is shown in Figure 10-21*b*.

DEFECT SUPERSTRUCTURES BASED ON THE B.C. CUBIC STRUCTURE

$FeSi_2$ (*tP*3): The $FeSi_2$ structure with cell dimensions $a = 2.684$, $c = 5.128$ Å, $Z = 1$,[32] is a defect CsCl type with alternate layers of Cs (Fe) atoms in the *c* direction omitted. As a result the axial ratio $c/2a$ is slightly less than unity and the Si layers are displaced slightly towards each other. Thus in the [001] direction the structure is made up of a 4^4 Fe layer surrounded on either side by 4^4 Si layers displaced $(a + b)/2$ relative to the Fe layer, giving three-layer units. Fe has 8 Si neighbors at a distance of 2.3 Å and Si has 4 Fe neighbors. The ϕ Cu–Ga phase has a disordered $FeSi_2$ type structure.

$Cu_5Zn_8Fe_3Zn_{10}$ (*cI*52): The Cu_5Zn_8 γ-brass structure is usually described as a distorted defect superstructure of the b c. cube in which 27 pseudocells are stacked together, with the two sites at the supercell corners and body centers being left vacant. The atoms are considerably displaced from their ideal sites.[33,37] Actually, there are no vacant sites in the structure and although the nearest-neighbor coordinations of the 4 independent sites in Cu_5Zn_8 are 12, 12, 13, and 11, the structure can well be described as built up of interpenetrating distorted icosahedra, each atom being surrounded by 12 neighbors which have a surface coordination number of 5, giving polyhedra with 20 triangular faces. Since this description applies equally well to the Al_4Cu_9 structure, γ-brasses are tetrahedrally close packed structures, each atom forming a corner of 20 distorted tetrahedra. Single-crystal studies by the author and collaborators confirm the Cu_5Zn_8 and Cu_5Cd_8 structures and show that Fe_3Zn_{10} is isotypic (at least as regards the 52 atom pseudocell; there may be a further superstructure).

Cr_5Al_8 (*hR*26) is said to be a rhombohedral distortion of the γ-brass structure,[34] and a further superstructure type, Pu_5Hg_{21}, said to be made up of $6 \times 6 \times 6$ pseudocells and 16 vacant sites, has been attributed to several phases on the basis of X-ray powder photographs only.[35]

Al_4Cu_9 (cP52): The Al_4Cu_9 structure is a defect superstructure based on the b.c. cubic structure. The large Al_4Cu_9 cell ($a = 8.704$ Å) contains 27 CsCl type pseudocells with two sites vacant.[36] One vacant site occurs on each sublattice and the atomic distribution can be described as ($Al_{16}Cu_{10}\square$) ($Cu_{26}\square$). Two distinct clusters of 26 atoms can be recognized in the structure. In one, a tetrahedron of Cu is surrounded by another tetrahedron of Cu. This is followed by an octahedron of Cu and a cubo-octahedron of Al atoms. The other cluster is similarly constituted except that the initial tetrahedron is Al instead of Cu, and the cubo-octahedron is Cu instead of Al (Figure 10-22).

The Cu atoms occupy 6 sitesets and the Al atoms two sitesets, consistent with the description of the two 26 atom clusters. Interatomic distances vary from 2.49 to 2.79 Å. Cu(1), (2), (3), (4), and (5), and Al(1) are surrounded by convex polyhedra of 13 atoms although, with the exception of Cu(1) and (2), these include one atom at a distance of 3.3 or 3.4 Å. Cu (6) and Al(2) are surrounded by convex polyhedra of 15 atoms although four of these are at 3.3 to 3.4 Å.

δ-Sn_6Cu_{20} (hP26): The Sn_6Cu_{20} structure ($a = 7.331$, $c = 7.870$ Å, $Z = 1$)[57] determined only by X-ray powder photographs and requiring confirmation, can be regarded as a defect superstructure based on the b.c. cubic structure.

BBe_{4-5} (tP10): The cell dimensions of BBe_{4-5}, $a = 3.369$, $c = 7.050$ Å, $Z = 2$,[38] indicate that on the average the structure should be regarded as

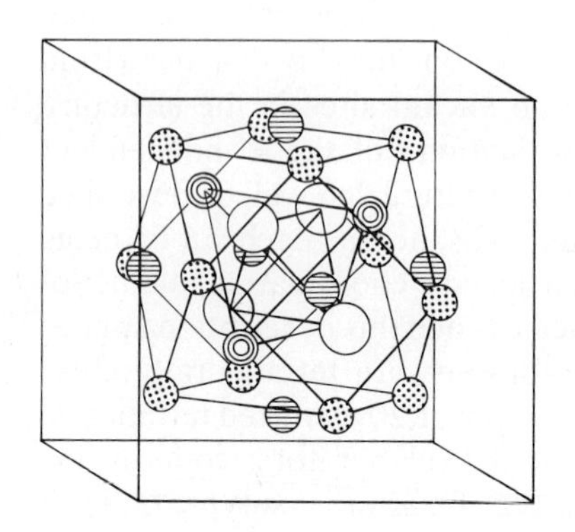

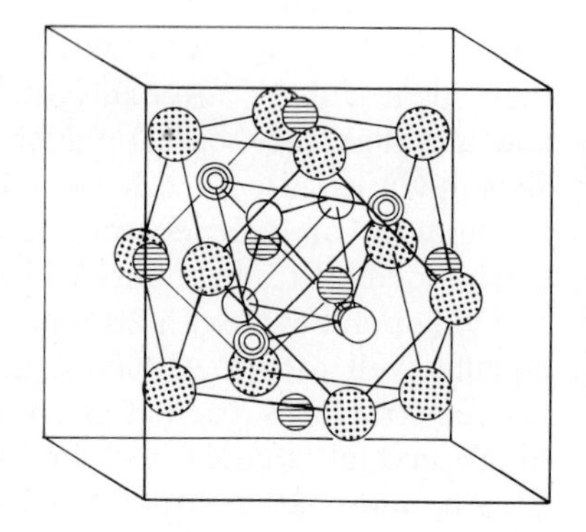

Al_4Cu_9

Figure 10-22. Pictorial views of atomic arrangements in the Al_4Cu_9 (*cP*52) structure. (Left) Cluster of atoms about the origin 0, 0, 0. (Right) Cluster of atoms about the body center $\frac{1}{2}, \frac{1}{2}, \frac{1}{2}$: large circles, Al. (Westman [36].)

a defect superstructure of the b.c. cube (average pseudocube $a = b = 2.38$, $c = 2.35$ Å) with three pseudocubes along the c direction. The first and third layers of cubes formed by Be atoms are either centered by B or uncentered, whereas the second layer of Be cubes are all centered by Be. Alternately the structure can be regarded as composed of a layer of squashed Be cubes half of which are centered by B, a layer of slightly squashed f.c. cubes of Be followed by a further layer of squashed Be cubes half centered by B (Figure 10-23). Be(1) has 4 B and 6 Be near neighbors; Be(2) has 1 B and 12 Be neighbors; Be(3) has 2 B and 10 Be neighbors and B has 9 close Be neighbors.

Al–Cu–Ni: A series of superstructures derived from a slightly rhombohedrally distorted cell has been reported from X-ray powder studies.[39] The hexagonal cells for these have 5, 11, 6, 13, 7, 15, 8, or 17 subcells stacked along c. Al atoms occupy the corners of CsCl-like rhombohedra and (Ni,Cu) or vacancies occupy the centers in ordered array, vacancies occurring along *all* triad axes. Of the hexagonal supercells those with 5, 13, and 8 stacked subcells have rhombohedral symmetry. Formulas for the superstructures range from $Al_5(Cu,Ni)_3$ to $Al_{17}(Cu,Ni)_{12}$.

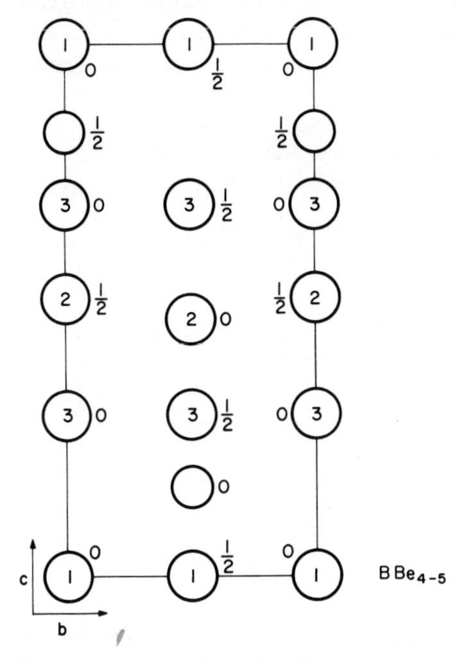

Figure 10-23. The BBe_{4-5} ($tP10$) structure projected down [100].

MIXED B.C. CUBIC {110} AND CLOSE PACKING

AuCuI (*cP*4): When c/a is less than unity in the four-atom cell, the AuCu structure provides an example of mixed b.c. cubic {110} and close packing (see p. 313).

T-CuZnAu$_2$ (*oP*8): In the *T*-$CuZnAu_2$ structure with cell dimensions $a = 4.547$, $b = 9.007$, $c = 2.915$ Å, $Z = 2$,[40] the Cu and Zn atoms are disordered on one siteset, whereas Au occupies the other set. In projection on the (001) plane the unit cell can be seen to contain 4 distorted b.c. cubic pseudocells. The atoms form distorted triangular nets in planes parallel to (110) of the pseudocells or parallel to the (010) plane of the superstructure cell. These nets are not exactly close packed; the triangles have angles between those of close packing (60°) and of b.c. cubic {110} layers (55°, 55°, 70°). The 3^6 net does not have threefold symmetry, both because of the angles, and because of the distribution of the components which give the rectangular repeat unit shown in Figure 10-24. The stacking of the two layers at $y \sim 0.13$ and 0.38, is A°B°, exactly as in a b.c. cubic structure, with the nodes of one 3^6 net over the midpoint of the edge of the triangles of the nets above and below. However, the next layer at $y \sim 0.63$ is stacked approximately in the close packed position with nodes of its 3^6 net well within the triangles of the 3^6 net at $y \sim 0.38$. The next 3^6 net at $y \sim 0.88$ is again in b.c. cubic [110]

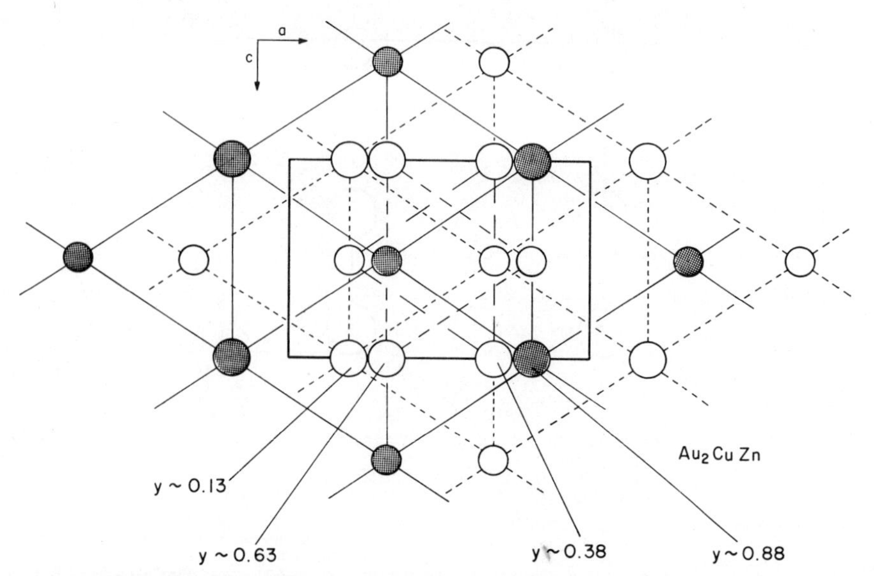

Figure 10-24. Atomic arrangement in the $CuZnAu_2$ (*oP*8) structure, seen in projection down [010]: Au, large circles; disordered Cu and Zn, small circles.

stacking relative to that at $y \sim 0.63$ and in approximately close packed stacking relative to the layer at $y \sim 0.13$ as the cycle is completed. Thus the layers are stacked in the sequence of b.c. cubic array and approximately close packing alternately.

The structure which requires confirmation by single-crystal study, should be regarded as a distorted superstructure intermediate between the b.c. cubic and c.p. arrangements and thus it is hybrid of the CsCl and AuCuI types.

SiTi (*oP*8): In the SiTi structure with cell dimensions, $a = 3.618$, $b = 6.492$, $c = 4.973$ Å, $Z = 4$,[41] the Ti atoms form triangular nets parallel to the (010) plane at $y = 0$ and $\frac{1}{2}$, and the Si atoms form rumpled triangular nets at $y \sim \frac{1}{4}$ and $\frac{3}{4}$. The angles of the triangles of these nets are much closer to those of b.c. cubic {110} nets than of close packed nets. These nets are, however, essentially stacked in close packed AB sequence along [010] (Figure 10-25). The superstructure therefore belongs to those of an intermediate class between close packing and b.c. cubic packing—a distorted hybrid of the ideal AuCuI and CsCl types. The closest Si–Ti distances are 2.31 Å.

VAu_2; *$TaPt_2$* (*oC*12): These two structures, although described in different space groups, appear to be identical (VAu_2: $a = 4.684$, $b = 8.482$, $c = 4.810$ Å, $Z = 4$, *Amm*2.[42] $TaPt_2$: $a = 8.403$, $b = 4.785$, $c = 4.744$ Å, $Z = 4$, *Cmcm*).[43] The structures can be regarded as examples of those containing close packed 3^6 nets of atoms which are stacked in A°B°, b.c.

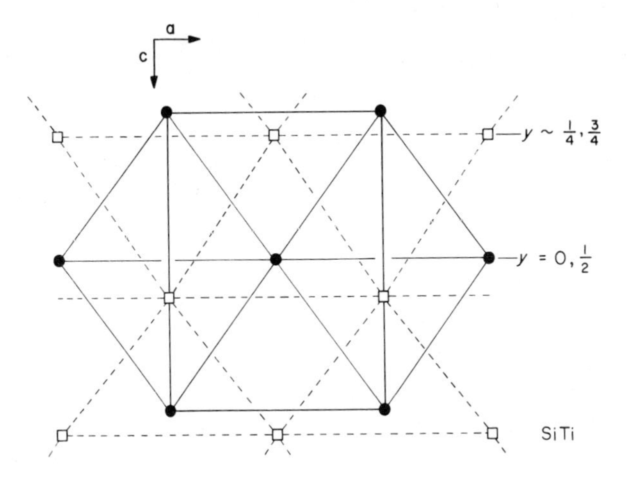

Figure 10-25. Arrangement of the atoms on 3^6 nets parallel to the (010) plane in the SiTi (*oP*8) structure: Ti, circles; Si, squares.

cubic [110] stacking (Figure 10-26), and as such they are distorted forms of the $MoSi_2$ structure with $c/a = 2.4$.

In the $TaPt_2$ description, close packed 3^6 layers of atoms occur parallel to (001) planes at $z = \pm\frac{1}{4}$, the Pt atoms forming a 6^3 graphite-like net with the Ta atoms in 3^6 nets centering the hexagons of the Pt array. Stoichiometry is thus satisfied within each layer. These layers are stacked in two positions which approximate to A°B° stacking, although the nodes of one net are displaced slightly from the midpoints of the edges of the triangles above and below, toward the triangle centers as Figure 10-26 shows. A similar situation is found in VAu_2 where, in that description, the close packed layers are parallel to the (100) plane (Figure 10-26). The structures require confirmation by single-crystal methods.

In $TaPt_2$, Ta is surrounded by 10 Pt at 2.77 to 2.80 Å and by 2 Ta at 3.17 Å and two at 3.58 Å. Pt also has CN 14 with 5 Ta neighbors and 9 Pt at distances from 2.77 to 3.58 Å; the two polyhedra appear to be almost identical in shape.

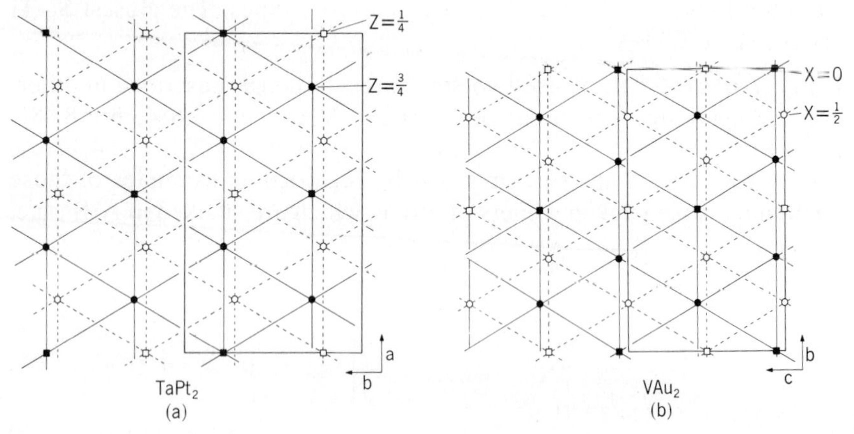

Figure 10-26. Arrangement of the atoms on 3^6 nets in the $TaPt_2$ and VAu_2 (*oC*12) structures: squares, Ta or V atoms.

MN_2: 3^6 Close Packed Layers in b.c. Cubic [110] Stacking. M on 3^6 Subnet, N on 6^3 Subnet.

Family of Polytypic Structures MN_2 with c.p. Layers Stacked in b.c. Cubic Sequence. Close packed triangular layers of atoms with two components in proportion MN_2, arranged in 3^6 and 6^3 nets respectively (Figure 10-27*a*) can not be stacked together in any close packed sequence without the M components being interlayer close neighbors. If, however, these close packed layers are stacked in the b.c. cubic [110] stacking sequence with the M atoms of one net lying over the midpoints of the N–N sides of the triangles of the nets above and below, then the M components are separated by considerably greater distances. With 6^3 and 3^6 arrangements of the two components on the close packed 3^6 nets, giving a hexagonal net repeat unit, there are four equivalent stacking positions, A°B°C°D° for b.c. cubic [110] layer stacking as shown in Figure 10-27*a*. Subsequent layers can follow each other in any one of these positions so that there is a family of polytypic structures which is based on b.c. cubic [110] stacking of close packed MN_2 layers with 6^3 and 3^6 arrangement of the components. Three members of this family of structures are recognized, $MoSi_2$ (with $c/a \sim 2.4$), $CrSi_2$, and $TiSi_2$ (Figure 10-27*b*).

$MoSi_2$ (tI6): $MoSi_2$ itself has cell dimensions $a = 3.203$, $c = 7.855$ Å, $Z = 2$ and an axial ratio c/a of 2.45.[44] Other phases with the structure mainly have axial ratios of 2.4 to 2.5 or in the range from 3.2 to 3.7 (Figure 10-28). The description of the structure depends on the recognition of two ideal axial ratios: $3 \times \sqrt{2}/\sqrt{3} = 2.38$, when exact close packed layers are stacked in b.c. cubic [110] packing, and $3 \times \sqrt{2} = 4.42$ when close packed layers are stacked in close packing. In the first case, represented well by $MoSi_2$, the structure is seen to be made up of three b.c. cubic pseudocells stacked and somewhat squashed in the c direction (Figure 10-29). Stoichiometry is satisfied in the 3^6 nets parallel to the (110) planes of the pseudocells or the superstructure; the Si and Mo atoms

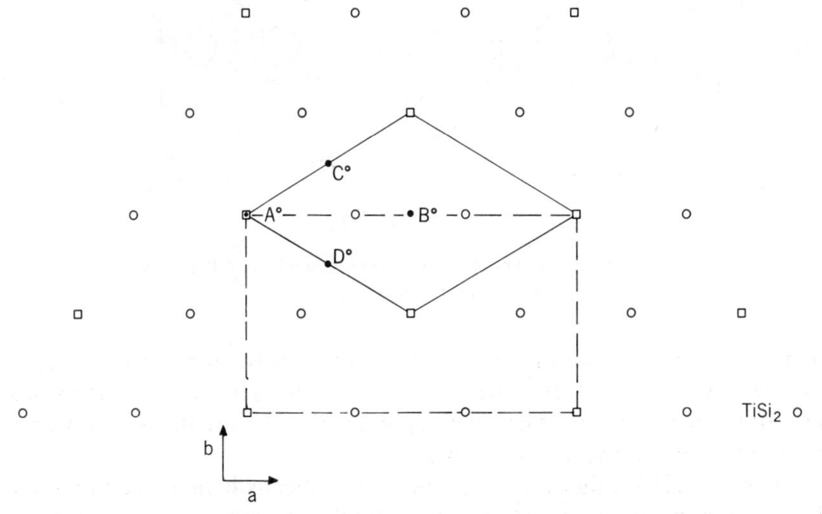

Figure 10-27. (*a*) Arrangement of atoms in close packed planes of the $TiSi_2$ (*oF*24) structure, indicating a repeat cell and the four b.c. cubic [110] type stacking positions of the layers: squares Ti atoms; broken lines, $TiSi_2$ cell viewed down [001].

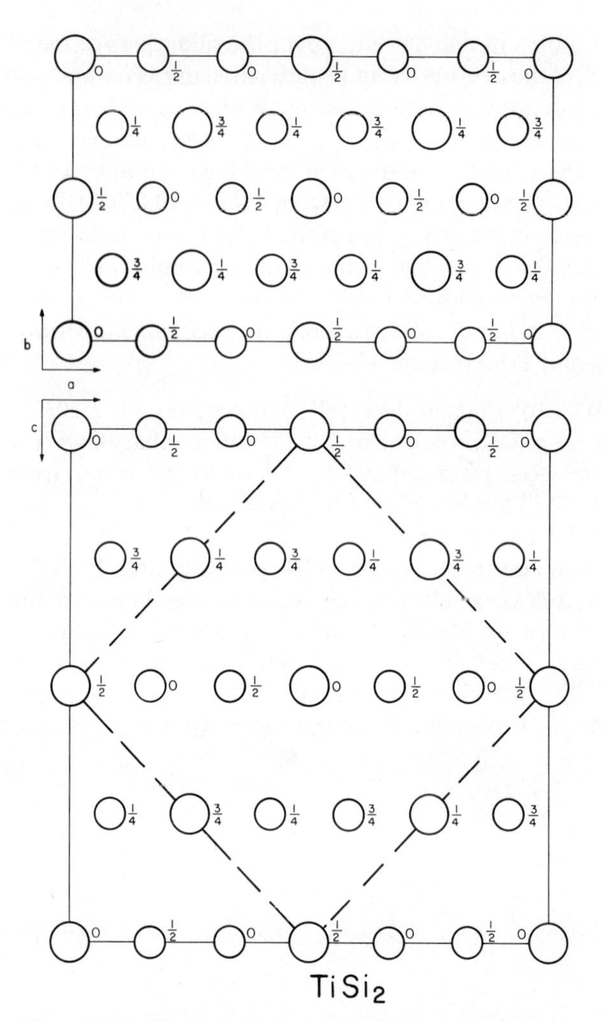

Figure 10-27. (*b*) The $TiSi_2$ structure viewed down [010] and [001]: broken lines, pseudotetragonal cell.

being arranged in 6^3 and larger 3^6 nets respectively, and the stacking sequence of the nets is A°B°. The reason for adopting b.c. cubic [110] stacking rather than close packed stacking is clearly the greater separation of the Mo atoms that the former affords.

When the axial ratio of the b.c. cubic pseudocells is increased to $\sqrt{2}$ so that c/a for the supercell has a value of 4.42, the structure is made up of f.c. cubic subcells with edge $\sqrt{2}a_0$ so that the atoms now occupy close packed 3^6 nets on {111} planes of the f.c. pseudocell. These planes are

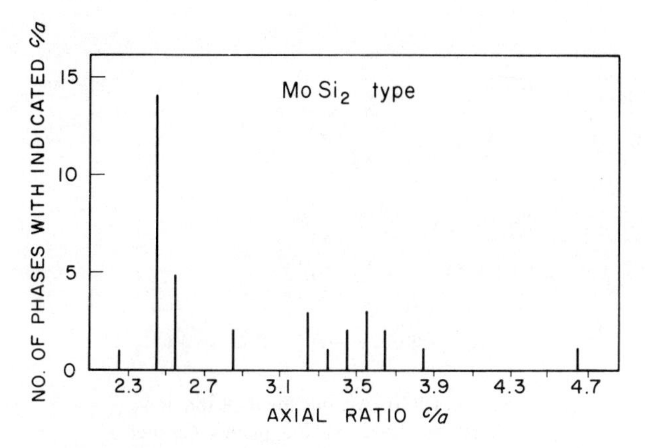

Figure 10-28. Histogram for the axial ratio, c/a, of phases with the $MoSi_2$ ($tI6$) structure.

parallel to the (013) planes of the supercell and they are stacked in close packing in a repeat sequence of nine layers in this direction as indicated in Figure 10-30. The arrangement of the atoms on these close packed planes is in rows: one of the M component to two of the N component, as shown together with the repeat cell in Figure 10-30. Although no phases with the $MoSi_2$ structure have the ideal axial ratio of 4.42 for close

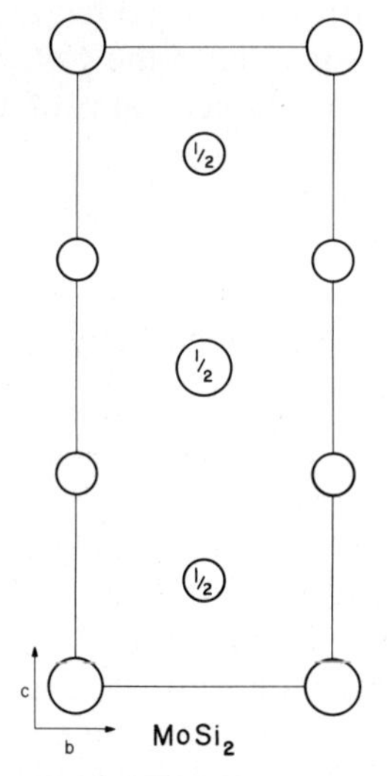

Figure 10-29. Projection of the $MoSi_2$ structure down [100].

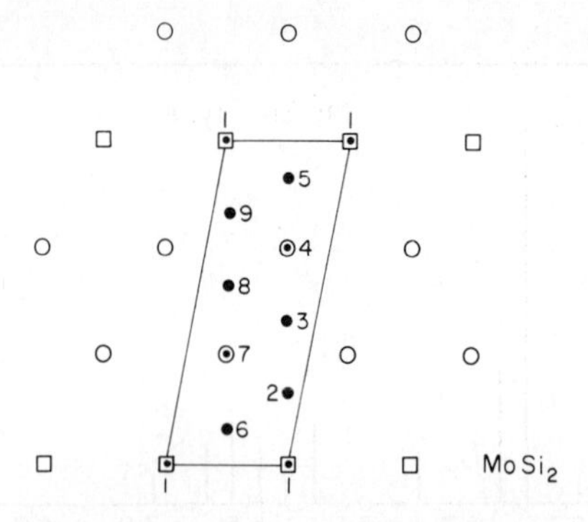

Figure 10-30. Arrangement of the MN_2 atoms on close packed planes parallel to the (013) plane of the $MoSi_2$ type structure with $c/a = 4.42$, showing a repeat cell and the 9 close packed stacking positions of the superstructure.

packing, $CdTi_2$ is said to have a larger axial ratio of 4.68, and the structures of phases with axial ratios of 3.5 or 3.6 must be considered as slight distortions of close packed layers stacked in close packing, rather than as b.c. cubic packing, since the (110) layers of the b.c. pseudo-cell have already become slightly distorted 4^4 nets of atoms rather than 3^6 nets, as indicated in Figure 10-31 which shows the (110) net for an axial ratio of 3.6.

The $MoSi_2$ structure is isopuntal to the CaC_2 type in which the larger atomic parameter results in distinct pairs of C atoms with their axis along [001].

The near-neighbor diagram (p. 53) for phases with the $MoSi_2$ structure suggests control of the structure by the geometrical factor. Phases are found for a wide range of radius ratios of the component atoms from < 0.80 to 1.25 and they are grouped between lines of atomic contacts giving 10–10 coordination.

$CrSi_2$ (hP9): Some 10 phases are known to have the $CrSi_2$ type structure; $CrSi_2$ itself has unit cell dimensions $a = 4.431$, $c = 6.364$ Å, $Z = 3$.[45] The structure is made up of three close packed 3^6 layers of atoms stacked one above the other along [001] in the unit cell. The 3^6 net is subdivided into a 6^3 net occupied by Si and a new larger 3^6 net occupied by Cr giving $CrSi_2$ stoichiometry. The close packed layers are stacked one above the other in b.c. cubic [110] stacking positions A°B°C°. The stacking of close packed layers in the b.c. cubic [110] stacking sequence results in the separation of the Cr atoms as far as possible from each other. This arrangement leads to a coordination of 14 for Cr which is surrounded by

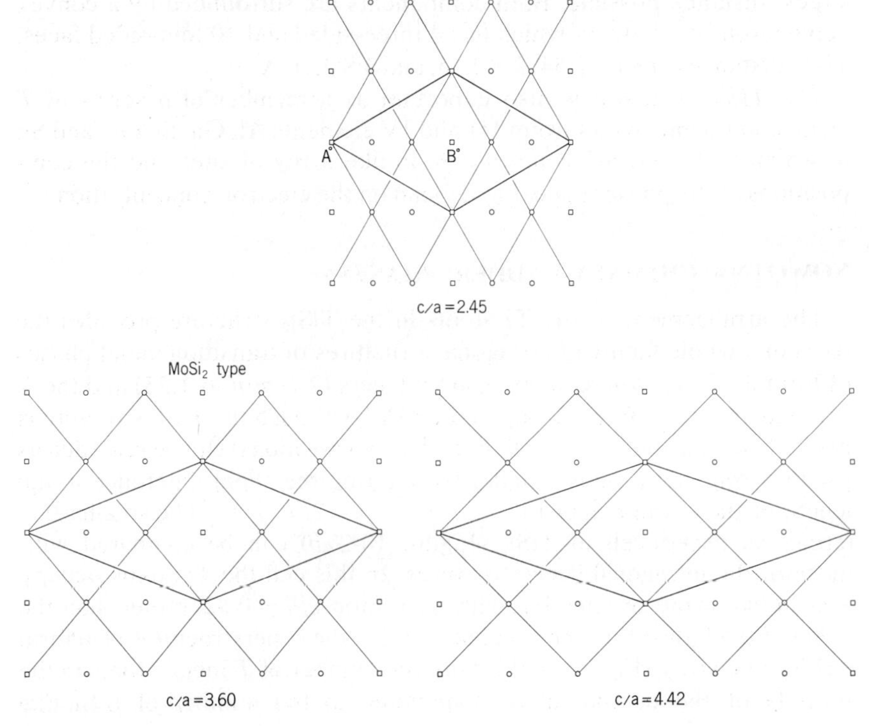

Figure 10-31. Arrangement of Mo and Si atoms parallel to the (110) plane of the b.c. cubic pseudocell in the $MoSi_2$ structure, shown for three values of the axial ratio of the superstructure. When $c/a = 2.45$ the arrangement corresponds approximately to b.c. cubic [110] stacking of close packed layers. The case for $c/a = 4.42$, where the net on the tetragonally distorted b.c. cubic (110) planes is square, corresponds to close packing of close packed layers on the distorted b.c. cubic (101) or (011) planes, which are of course (111) planes of the pseudocell which has now become f.c. cubic. The intermediate case with $c/a = 3.60$ shows the b.c. cubic (110) planes for intermediate b.c. cubic and close packing.

4+6 Si at 2.48 and 2.56 Å and 4 Cr at 3.07 Å in a polyhedron having 4 triangular and 10 four-sided faces. Si has 2+3 close Cr neighbors and 2+3 Si at 2.48 Å and 2.56 Å. The structure should be confirmed by a single-crystal study.

$TiSi_2$ (oF24): $TiSi_2$ has cell dimensions $a = 8.252$, $b = 4.783$, $c = 8.540$ Å, $Z = 8$.[46] In the $TiSi_2$ structure the atoms form close packed 3^6 layers parallel to the (001) plane (Figure 10-27). These layers have composition $TiSi_2$ and are subdivided so that Si atoms occupy 6^3 nets and Ti new larger 3^6 nets. The close packed layers are stacked in the b.c. cubic [110] stacking sequence A°C°B°D°, so that the Ti atoms are separated by the

largest distance possible. Both components are surrounded by a convex polyhedron of 14 atoms which has 4 three-sided and 10 four-sided faces. Ti–Si distances are 4×2.54, 2×2.75, and 4×2.76 Å.

The $TiSi_2$ structure is also important as a member of a series of T metal compounds with Group III and IV elements Al, Ga, Si, Ge, and Sn in which the T metal occupies a β-Sn-like array of sites and the compositions of the phases appear to depend on the electron concentration.

NOWOTNY CHIMNEY-LADDER PHASES

The arrangement of the Ti atoms in the $TiSi_2$ structure provides the basis of a whole family of tetragonal structures of transition metal phases of formula T_nX_m where m and n are integers ($2 > m/n \geq 1.25$) and the X component comes from Groups III or IV. Although the $TiSi_2$ structure is itself f.c. orthorhombic, its cell (but not atom positions) can be regarded as pseudotetragonal (b axis unique) by ignoring the slight difference in the length of the a and c axes ($a' \sim a/\sqrt{2} \sim c/\sqrt{2}$, $c' = b$). The smaller b.c. tetragonal pseudocell of $TiSi_2$ (Figure 10-27b) can be compared with those of the tetragonal superstructures. In this cell the Ti atoms occupy a β-Sn-like array of sites but with axial ratio $c/a = 0.82$, rather than the value of 0.54 for β-Sn. The T metals in all of the superstructures also form a β-Sn-like array (Figure 10-32) with the number of T metal atoms in the formula of the compound corresponding to the number of β-Sn-like

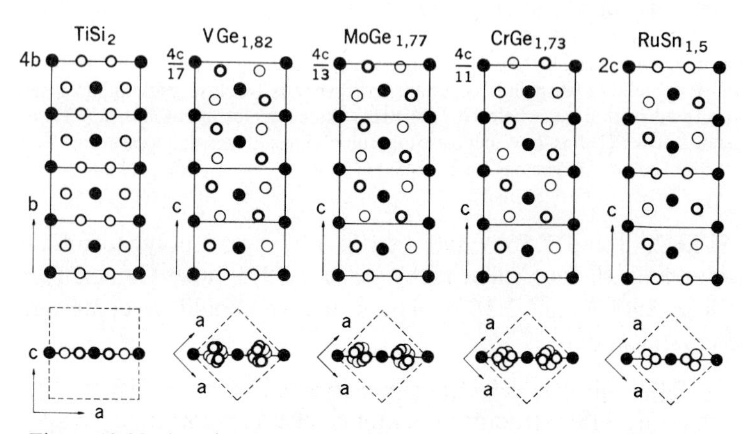

Figure 10-32. Atomic arrangements about the (110) plane in the tetragonal chimney-ladder type superstructures of $V_{17}Ge_{31}$, Mo_9Ge_{16}, $Cr_{11}Ge_{19}$, and Ru_2Sn_3, and comparison with the arrangement in the $TiSi_2$ structure. a, b and c for $TiSi_2$ refer to its orthorhombic cell: black circles, T metal positions. (Völlenkle and co-workers [49].)

pseudocells that are stacked in the c direction of the superstructures. The fourfold stacking sequence of Ti atoms arranged on equilateral triangular nets in the $TiSi_2$ structure is consistent with the tetragonal symmetry and four "molecules" per cell of all of the structures of the T_nX_m phases, but the arrangement of the Si atoms on 6^3 hexagonal nets in the $TiSi_2$ structure is not, and hence its orthorhombic symmetry. Looking down [001] of the tetragonal structures, the X atoms are seen to be distributed about the centers of the Si sites in accordance with the tetragonal symmetry (Figure 3-29 p. 103), and since $m/n < 2$ in all of the structures, their distribution is also stretched up along the [001] direction (*vide infra*).

The β-Sn-like array of Ti has a much greater axial ratio than that of β-Sn itself so as to make equilateral triangles of 3^6 subnets of Ti atoms parallel to the (001) planes of the orthorhombic cell. This occurs ideally at a value of $1/\sqrt{3} = 0.577$, whereas the relevant parameter b/a for $TiSi_2$ has a value of 0.580; for β-Sn itself the corresponding value of $c/a\sqrt{2} =$ 0.386, indicates that the triangles of the 3^6 nets of Sn in (110) planes are far distorted from equilateral. In all of these superstructures based on $TiSi_2$ type stacking of the T metals, the value of the parameter $c/na\sqrt{2}$ corresponding to $1/\sqrt{3} = 0.577$ for equilateral triangles in the 3^6 nets of T atoms, is close to the ideal value and far removed from the β-Sn value, as indicated in Table 10-1. Furthermore, the rather constant values about 0.577 do not show any dependence on composition ratio n/m, as might be expected if they depended on the relative degree of filling of the β-Sn-like

Table 10-1

Phase T_nX_m		Atomic Ratio n/m	Ratio $c/na\sqrt{2}$
β-Sn structure			0.386
Ideal $TiSi_2$ structure (close packed nets)			$1/\sqrt{3} = 0.577$
$TiSi_2$ (pseudotetragonal)	($oF24$)	2.0	0.580
$V_{17}Cr_{31}$	($tP192$)	1.82	0.590
Mo_9Ge_{16}	($tP100$)	1.78	0.577
$Mo_{13}Ge_{23}$	($tP144$)	1.77	0.577
Tc_4Si_7	($tP44$)	1.75	0.558
$Cr_{11}Ge_{19}$	($tP120$)	1.73	0.580
$Mn_{11}Si_{19}$	($tP120$)	1.73	0.561
$Mn_{15}Si_{26}$	($tP164$)	1.73	0.557
$Rh_{10}Ga_{17}$	($tP108$)	1.70	0.577
Ir_3Ga_5	($tP32$)	1.67	0.575
Ru_2Sn_3	($tP20$)	1.5	0.568
$Rh_{17}Ge_{22}$	($tI156$)	1.29	0.582
Ir_4Ge_5	($tP36$)	1.25	0.572

T metal array with the other component. Thus the stability of this family of superstructures appears to depend primarily on the 3^6 arrangement of T atoms parallel to (110) planes of their tetragonal cells, and their stacking in the $TiSi_2$ sequence A°C°B°D°, rather than on the β-Sn-like arrangement of T metal atoms, which appears to be an incidental feature of the fourfold multiplicity of the stacking sequence. The packing of the T metal atoms in sites where they can best avoid each other when the composition is 2:1 or less, is the important structural feature, and the X atoms fill up the rest of the space as best they can. The X atoms form an approximately f.c. cubic array so that four X atoms, arranged in pairs at the same height in the c direction (Figures 10-32 and 10-33), can themselves be regarded as forming a pseudocell.[47] Since both the T and X atom pseudocells contain 4 atoms, the superstructure for the phases T_nX_m contains both n T-pseudocells and m X-pseudocells, and the height of the supercell, c, is an integral multiple n or m of the c_0 dimensions of each of the pseudocells. Therefore from measurements of the subcell parameters, $c_{o(X)}$ and $c_{o(T)}$, and the supercell parameter c, the precise composition ratio n/m of phases following this structural principle can be determined, provided that all sitesets are fully occupied.[47] Since the equilateral triangles formed by the T atoms parallel to (110) and $(1\bar{1}0)$ superstructure planes are a rigid feature of the structures, giving the n β-Sn-like pseudocells of T atoms, and the atomic ratio, m/n, varies from 1.25 to < 2, the m pseudocells of X atoms (the "ladders") have to be drawn up to equal the height (c) of the n pseudocells of T atoms (the "chimneys"). For this reason this family of structures is called the "chimney-ladder" structures.

Electron concentration appears to play some role in control of this family of structures (p. 102), since the total d, s, p outer electron concentration is 14.0 ± 0.1 per transition metal atom for binary phases formed by the Groups VII, VIII_1, and VIII_2 T metals, 13.0 ± 0.1 for Group VI metals, 12.3 for Group V metals, and 12.0 for Group IV T metals, regardless of whether the X component comes from Group III or IV. The decrease in electron concentration to 12 per T metal for Group IV T metal compounds, results from the ratio m/n having a maximum value of 2. An electron concentration of 14 per T metal can only be maintained with a Group IV B component as far as the Group VI T metals, but TX_2 phases with this electron concentration are already found to take different structural arrangements. Thus $CrSi_2$ has the structure polytypic to $TiSi_2$ with A°B°C° stacking, and $MoSi_2$ ($c/a \sim 2.4$) has the structure with A°B° stacking of the 3^6 layers. When the ratio m/n exceeds the value of 2, phases are found to take structures based on other principles as in Mo_3Al_8, even though the electron concentration is 14.0 per T metal atom. With a Group III B component, 14 electrons per T atom can only be maintained

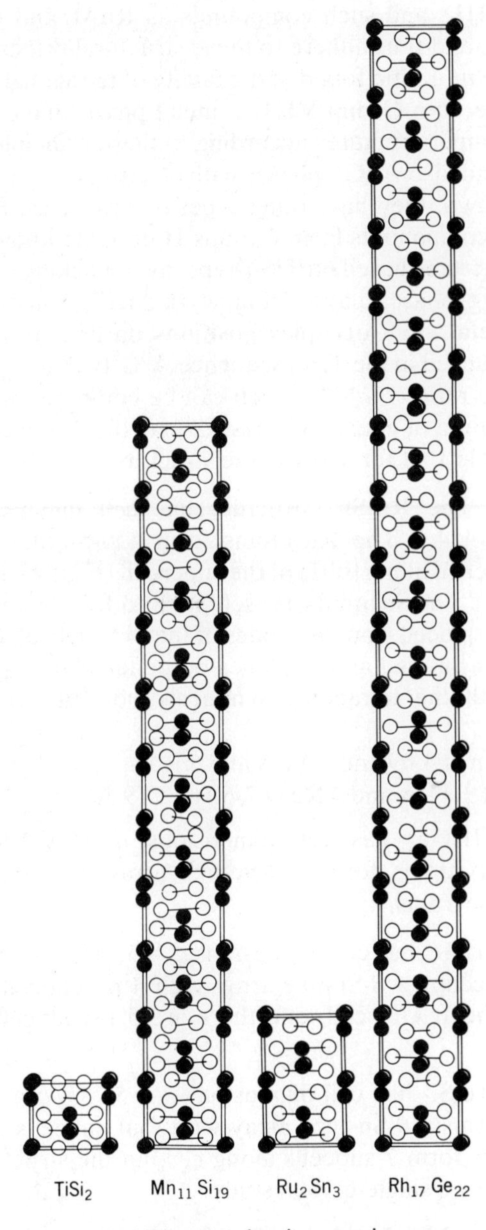

Fig. 10-33. Diagram showing atomic arrangements in $TiSi_2$, $Mn_{11}Si_{19}$, Ru_2Sn_3, and $Rh_{17}Ge_{22}$ structures. *T* metal atoms are black circles. (Jeitschko and Parthé [47].)

until Group VIII_1, and such compounds as $RuAl_2$ and $RuGa_2$ have the $TiSi_2$ structure, and thus adhere to these structural principles. The smallest value of the m/n ratio found in the family of tetragonal superstructures is 1.25 in Ir_4Ge_5. No Group VIII_3 T metal phases have at present been found to form superstructures according to these principles, and although $PtHg_2$, PtSn, and also AuGa phases with electron concentration of 14 per T atom are known, they have other types of structures. Furthermore, no phases with X components from Groups II or V are known to form tetragonal superstructures based on $TiSi_2$-type layer stacking.

The following phases have "chimney-ladder" structures. In each of them the T metal atom occupies positions on 3^6 nets parallel to (110) planes and is stacked in the $TiSi_2$ sequence A°C°B°D°.

Values of the ratio $c/na\sqrt{2}$, which can be compared with the value of 0.577 for formation of equilateral triangles by the T atoms on the 3^6 nets, are given in Table 10-1 for many of the structures.

Ru_2Sn_3 (tP20): The Ru_2Sn_3 structure has cell dimensions $a = 6.172$, $c = 9.915$ Å, $Z = 4$.[48] The Ru atoms form a β-Sn-like array with two pseudocells stacked along [001] of the supercell (Figures 10-32 and 10-33, pp. 594 and 597). The Sn atoms, present in ratio 1.5 to 1 compared to 2 to 1 for $TiSi_2$, are displaced from the nodes of the Si 6^3 nets of $TiSi_2$, both along [001] of the Ru_2Sn_3 cell and in planes perpendicular to this direction, in accordance with the tetragonal symmentry of Ru_2Sn_3 (Figures 10-32 and 10-34).

Sn(1) has 3 Sn at 3.09 and 3.11 Å and 5 Ru at 2.59 (2×) to 2.92 Å. Sn(2) has one Sn(1) at 3.11 Å and 4 Ru at 2.60 to 2.79 Å.

Ir_3Ga_5 (tP32): Ir_3Ga_5 has cell dimensions $a = 5.823$ Å, $c = 14.20$ Å, $Z = 4$.[49] The Ir atoms form a β-Sn-like array with three pseudocells stacked along the c axis.

Ir_4Ge_5 (tP36): Ir_4Ge_5 has cell dimensions $a = 5.615$, $c = 18.31$ Å, $Z = 4$.[50] The Ir atoms occupy a β-Sn-like array with 4 pseudocells stacked along the c axis of the Ir_4Ge_5 cell, and there are 5 pseudocells of Ge atoms along c.

Tc_4Si_7 (tP44): Tc_4Si_7 has cell dimensions $a = 5.737$, $c = 18.10$ Å, $Z = 4$. The T metal forms a β-Sn-type array with four subcells stacked along c, and the Si atoms form 7 subcells along c,[51] but the structure has not yet been confirmed by a single-crystal study.

Mo_9Ge_{16} (tP100): Mo_9Ge_{16} has cell dimensions $a = 5.994$, $c = 43.995$ Å, $Z = 4$.[47,52] The Mo atoms form a β-Sn-like array and the unit cell contains 9 β-Sn-like pseudocells of Mo and 16 Ge pseudocells stacked along c.

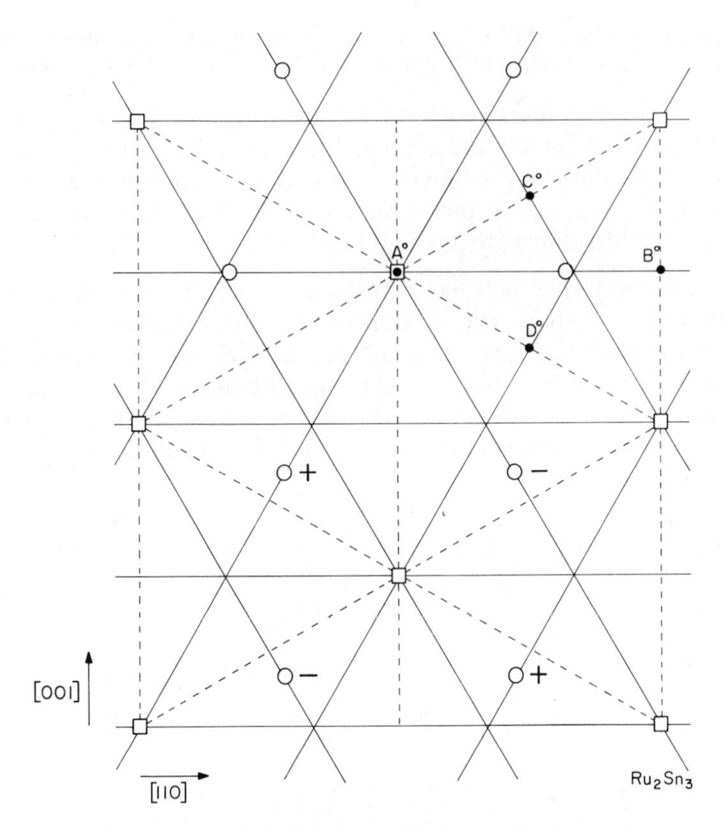

Figure 10-34. Arrangement of the Ru and Sn atoms on planes parallel to (110) of the Ru_2Sn_3 (*tP*20) structure. Ru (squares) form a 3^6 net. The Sn atoms are displaced from the sites of a 6^3 net, both in the (110) and ($1\bar{1}0$) planes and above and below them (+ and −) as shown in Figure 10-32.

$Rh_{10}Ga_{17}$ (*tP*108): $Rh_{10}Ga_{17}$ has cell dimensions $a = 5.813$, $c = 47.46$ Å, $Z = 4$.[49] The Rh atoms form a β-Sn-like array with 10 pseudocells stacked along c.

$Mn_{11}Si_{19}$ (*tP*120): $Mn_{11}Si_{19}$ has cell dimensions $a = 5.518$, $c = 48.136$ Å, $Z = 4$.[53,54] The Mn atoms form a β-Sn-like array and the supercell contains 11 β-Sn pseudocells of Mn atoms stacked in the c direction, and 19 Si-atom pseudocells. The 12 independent Mn atoms each have 8 Si neighbors at distances from 2.27 to 2.74 Å (Mn(12) has only 6). The closest Si–Si approach in the structure is 2.39 Å. $Cr_{11}Ge_{19}$ has the same structure. See also Figures 10-32 and 10-33, pp. 594 and 597. Figure 3-29, p. 103 shows the interesting variation of Cr–Ge interatomic distances as a function of the z parameter of the Cr atom, and also shows the Ge positions viewed down [001] which have a nearly circular projection.

$Mo_{13}Ge_{23}$ (*tP*144): $Mo_{13}Ge_{23}$ has cell dimensions $a = 5.987$, $c = 63.54$ Å, $Z = 4$.[54] The Mo atoms are arranged in a β-Sn-like array and the cell contains 13 β-Sn-like pseudocells and 23 Ge pseudocells stacked along c. Figure 10-35*a* shows the variation in the Mo–Ge interatomic distances about Mo atoms proceeding along [001].

$Mn_{15}Si_{26}$ (*tI*164): The cell has dimensions $a = 5.531$, $c = 65.311$ Å, $Z = 4$.[55] There are 15 β-Sn-type subcells of Mn atoms stacked in the c direction of the unit cell and 26 Si pseudocells. Some very short Mn–Si distances occur in the structure, the closest being 2.27 Å. Coordination numbers of the eight independent Mn atoms are either 12 or 14. Each of the seven independent Si atoms is surrounded by a convex polyhedron of 14 atoms.

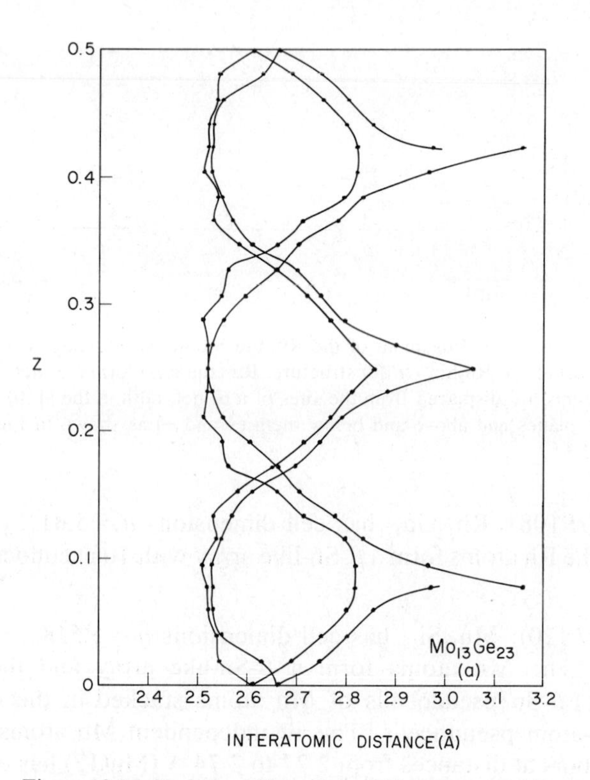

Figure 10-35. Variation of Mo–Ge and V–Ge interatomic distances about Mo and V atoms respectively, on proceeding up the c axis in the (*a*) $Mo_{13}Ge_{23}$ (*tP*144).

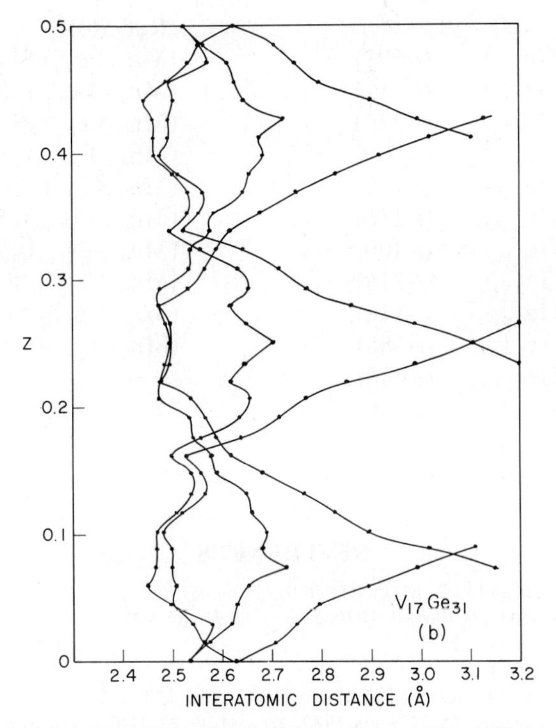

Figure 10-35. (*b*) $V_{17}Ge_{31}$ (*tP*192) structures. The ordinate is the *z* parameter of the Mo or V atoms.

$Rh_{17}Ge_{22}$ (*tI*156): $Rh_{17}Ge_{22}$ has cell dimensions $a = 5.604$, $c = 78.45$ Å, $Z = 4$.[47] The Rh atoms form a β-Sn-like array with 17 pseudocells stacked along the *c* direction and the Ge atoms form 22 pseudocells (Figure 10-33, p. 597). The Rh atoms (1) to (8) have 6 Ge neighbors at 2.37 to 2.88 Å. Rh(9) has 8 Ge at 2.47 and 2.83 Å. In addition, each Rh has 4 Rh neighbors at 2.92 to 3.16 Å. The Ge atoms have 4 or 5 Rh neighbors. The closest Ge–Ge distance is 2.77 Å.

$V_{17}Ge_{31}$ (*tP*192): $V_{17}Ge_{31}$ with cell dimensions $a = 5.91$, $c = 83.65$ Å, $Z = 4$,[54] has the V atoms in β-Sn-like array, 17 pseudocells being stacked along the *c* axis and 31 Ge pseudocells. See Figure 10-32. Figure 10-35*b* shows the variation of the V–Ge interatomic distances, about the V atoms on proceeding along [001].

Many more Nowotny chimney-ladder type structures, similar to those described above, have been discovered. In some cases single crystals of the structures have been examined.

(Ref. 54)
$Ru_{69}(Ga_{0.05}Ge_{0.95})_{104}$ (t-692)
$Ru_{11}(Ga_{0.15}Ge_{0.85})_{17}$ (t-112)
$Ru_{23}(Ga_{0.25}Ge_{0.75})_{36}$ (t-236)
$Ru_{19}(Ga_{0.35}Ge_{0.65})_{31}$ (t-200)
$Ru_{13}(Ga_{0.50}Ge_{0.50})_{22}$ (t-150)
$Ru_{23}(Ga_{0.75}Ge_{0.25})_{41}$ (t-256)
$Rh_{43}(Ga_{0.10}Ge_{0.90})_{57}$ (t-400)
$Rh_{23}(Ga_{0.25}Ge_{0.75})_{31}$ (t-216)
$Rh_{12}(Ga_{0.35}Ge_{0.65})_{17}$ (t-116)
$Rh_{39}(Ga_{0.50}Ge_{0.50})_{58}$ (t-388)
$Rh_{43}(Ga_{0.75}Ge_{0.25})_{69}$ (t-448)
$Ir_{17}(Ga_{0.15}Ge_{0.85})_{22}$ (t-156)
$Ir_{11}(Ga_{0.35}Ge_{0.65})_{15}$ (t-104)
$Ir_{19}(Ga_{0.80}Ge_{0.20})_{30}$ (t-196)

(Ref. 56)
$(Mn_{0.9}Fe_{0.1})_7Si_{12}$ (t-76)
$(Mn_{0.85}Fe_{0.15})_{17}Si_{29}$ (t-184)
$(Mn_{0.8}Fe_{0.2})_{23}Si_{39}$ (t-248)
$(Mn_{0.75}Fe_{0.25})_{29}Si_{49}$ (t-312)
$(Mn_{0.7}Fe_{0.3})_{22}Si_{37}$ (t-236)
$(Mn_{0.95}Co_{0.05})_{25}Si_{43}$ (t-272)
$(Mn_{0.95}Cr_{0.05})_{19}Si_{33}$ (t-208)
$(Mn_{0.9}Cr_{0.1})_{31}Si_{54}$ (t-340)
$(Mn_{0.8}Cr_{0.2})_{29}Si_{51}$ (t-320)
$(Mn_{0.75}Cr_{0.25})_{17}Si_{30}$ (t-188)

REFERENCES

1. W. H. Beamer and L. R. Maxwell, 1949, *J. Chem. Phys.*, **17**, 1293.
2. R. F. Mehl and C. S. Barrett, 1930, *Trans. AIME*, **89**, 575.
3. *Strukturbericht*, **1**, 28.
4. *Strukturbericht*, **1**, 25.
5. J. Böhm and O. Hassel, 1927 *Z. anorg. Chem.*, **160**, 152.
6. F. A. Bannister and M. H. Hey, 1932, *Min. Mag.*, **23**, 188; F. Grønvold, H. Haraldsen, and A. Kjekshus, 1960, *Acta Chem. Scand.*, **14**, 1879.
7. W. S. Miller and A. J. King, 1936, *Z. Kristallogr.*, **A94**, 439.
8. W. H. Zachariasen, 1952, *Acta Cryst.*, **5**, 19.
9. W. H. Zachariasen, 1952, *Acta Cryst.*, **5**, 644, 660.
10. H. Lipson and A. M. B. Parker, 1944, *J. Iron & Steel Inst.*, **149**, 123.
11. M. C. Neuburger, 1931, *Z. Kristallogr.*, **A77**, 169.
12. D. Lundqvist, 1947, *Ark. Kemi Min. Geol.*, **24A**, No. 21.
13. A. Westgren, 1938, *Z. anorg. Chem.*, **239**, 82.
14. E. Parthé, D. Hohnke, and F. Hulliger, 1967, *Acta Cryst.*, **23**, 832.
15. W. H. Zachariasen, 1952, *Acta Cryst.*, **5**, 17.
16. E. Bauer, H. Nowotny, and A. Stempfl, 1953, *Mh. Chem.*, **84**, 211, 692.
17. F. Grønvold and E. Røst, 1956, *Acta Chem. Scand.*, **10**, 1620.
18. A. J. Frueh, 1959, *Amer. Min.*, **44**, 693.
19. B. C. Giessen and N. J. Grant, 1965, *Acta Cryst.*, **18**, 1080.
20. N. C. Baenziger, R. E. Rundle, A. I. Snow, and A. S. Wilson, 1950, *Acta Cryst.*, **3**, 34.
21. N. Karlsson, 1951, *J. Inst. Met.*, **79**, 391.
22. E. Zintl and G. Brauer, 1935, *Z. Elektrochem.*, **41**, 297.
23. F. Laves, private communication.
24. E. Zintl and W. Dullenkopf, 1932, *Z. phys. Chem.*, **B16**, 195.
25. O. Heusler, 1934, *Ann. Phys. Lpz.*, **19**, 155; A. J. Bradley and J. W. Rodgers, 1934, *Proc. Roy. Soc.*, **A144**, 340.
26. C. Crevecoeur, 1964, *Acta Cryst.*, **17**, 757.
27. F. R. Morral and A. Westgren, 1934, *Svensk. Kemi Tidskr.*, **46**, 153.

28. A. Zalkin and W. J. Ramsey, 1956, *J. Phys. Chem.*, **60**, 234; A. Zalkin, W. J. Ramsey, and D. H. Templeton, 1956, *Ibid.*, **60**, 1275.
29. W. Rossteutscher and K. Schubert, 1964, *Z. Metallk.*, **55**, 617.
30. A. Zalkin and W. J. Ramsey, 1958, *J. Phys. Chem.*, **62**, 689; A. Zalkin, 1957, *Acta Cryst.*, **10**, 791.
31. K. Schubert, 1965, *Z. Metallk.*, **56**, 197.
32. B. Aronsson, 1960, *Acta Chem. Scand.*, **14**, 1414.
33. A. J. Bradley and C. H. Gregory, 1931, *Phil. Mag.*, **12**, 143.
34. A. J. Bradley and S. S. Lu, 1937, *Z. Kristallogr.*, **A96**, 20.
35. A. F. Brandt, 1966, *J. Less-Common Metals*, **11**, 216; 1967, *Ibid.*, **13**, 366.
36. S. Westman, 1965, *Acta Chem. Scand.*, **19**, 1411.
37. *Strukturbericht*, **1**, 497.
38. G. S. Markevič, Ju. D. Kondrašev, and L. Ja. Markovskij, 1960, *Ž. Neorg. Khim.*, **5**, 1783.
39. S. S. Lu and T. Chang, 1957, *Scientia Sinica*, **6**, 431.
40. M. Wilkens and K. Schubert, 1958, *Z. Metallk.*, **49**, 633; 1957. *Ibid.*, **48**, 550.
41. N. V. Ageev and V. P. Samsonov, 1959, *Ž. Neorg. Khim.*, **4**, 1590 [*Russ. J. Inorg. Chem.*, **4**, 716]. See also *Structure Reports*, 1959, **23**, 255, 256.
42. E. Stolz and K. Schubert, 1962, *Z. Metallk.*, **53**, 433.
43. B. C. Giessen and N. J. Grant, 1964, *Acta Cryst.*, **17**, 615.
44. *Strukturbericht*, **1**, 740.
45. B. Borén, 1933–35, *Ark. Kemi Min. Geol.*, **11A**, No. 10.
46. F. Laves and H. J. Wallbaum, 1939, *Z. Kristallogr.*, **A101**, 78.
47. W. Jeitschko and E. Parthé, 1967, *Acta Cryst.*, **22**, 417.
48. O. Schwomma, H. Nowotny, and A. Wittmann, 1964, *Mh. Chem.*, **95**, 1538.
49. H. Völlenkle, A. Wittmann, and H. Nowotny, 1966, *Mh. Chem.*, **97**, 506.
50. G. Flieher, H. Völlenkle, and H. Nowotny, 1968, *Mh. Chem.*, **99**, 877.
51. A. Wittmann and H. Nowotny, 1965, *J. Less-Common Metals*, **9**, 303.
52. A. Brown, 1965, *Nature, Lond.*, **206**, 502.
53. O. Schwomma, A. Preisinger, H. Nowotny, and A. Wittmann, 1964, *Mh. Chem.*, **95**, 1527.
54. H. Völlenkle, A. Preisinger, H. Nowotny, and A. Wittmann, 1967, *Z. Kristallogr.*, **124**, 9.
55. H. W. Knott, M. H. Mueller, and L. Heaton, 1967, *Acta Cryst.*, **23**, 549.
56. G. Flieher, H. Völlenkle, and H. Nowotny, 1968, *Mh. Chem.*, **99**, 2408.
A. E. Austin and J. R. Doig, 1957, *J. Metals, N.Y.*, **9**, 27.
57. O. Carlsson and G. Hägg, 1932, *Z. Kristallogr.* **A83**, 308.

11

Further Structures Generated by Square-Triangle Nets of Atoms: Cubes and Cubic Antiprisms

STRUCTURES WITH SQUARE-TRIANGLE, 3^2434 NETS OF ATOMS

The following structures are built up of 3^2434 nets of atoms which are generally stacked so that the squares lie either over the corners and center of the base of the cell, or over the midpoints of the cell edges. Nets which overlie each other directly form (distorted) cubes; those which overlie each other antisymmetrically give cubic antiprisms. When a 3^2434 net having squares over the basal cell corners is followed by a 3^2434 net with squares over the mid-points of the basal cell edges, irregular eight-cornered polyhedra are formed. Frequently the coordination polyhedra are centered by atoms forming interleaving 4^4 nets. A centered cube or anticube is generally regarded as a CU 10 polyhedron by including the two polar atoms, if 4^4 nets also lie outside of the planes of the two 3^2434 nets which create the CN 8 polyhedron. When 4^4 nets lie in the same planes as the 3^2434 nets centering the squares thereof, they create an overall 5^3+5^4 (3:2) pentagonal net as shown in Figure 11-1. Some structures contain 4^4 nets lying parallel to the basal cell edges and with the nodes over $\frac{1}{4}, \frac{1}{4}; \frac{1}{4}, \frac{3}{4}; \frac{3}{4}, \frac{1}{4}; \frac{3}{4}, \frac{3}{4}$ of the basal cell.

A simple code for describing the main features of such structures has been described on p. 48.

SeTl (*tI*16): In the SeTl structure ($a=8.036$, $c=7.014$ Å, $Z=8$)[1] 3^2434 nets of Se atoms at $z=0$ and $\frac{1}{2}$ are arranged antisymmetrically with

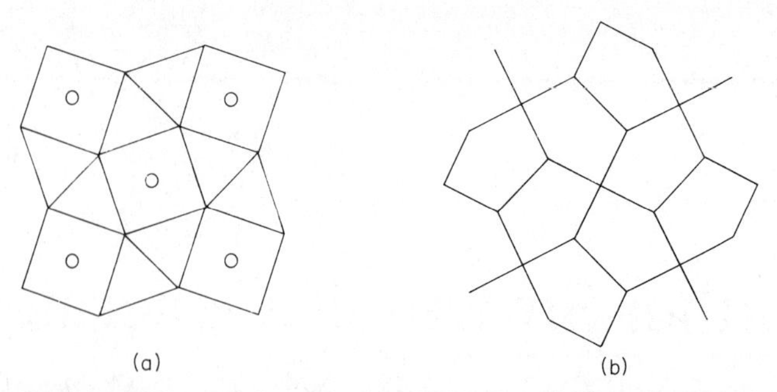

Figure 11-1. (*a*) 3^2434 net of atoms with squares centered by atoms in the same plane (4^4 net). (*b*) 5^3+5^4 (3:2) net created by the atomic arrangement described in (*a*).

respect to each other. At $z=\frac{1}{4}$ and $\frac{3}{4}$, $\frac{1}{2}.4^4$ nets of Tl(1) atoms center the squares in the Se nets which are located over the cell corners and basal plane center. Other $\frac{1}{2}.4^4$ nets of Tl(2) atoms over the midpoints of the cell edges at $z=\frac{1}{4}$ and $\frac{3}{4}$, center the diamonds of the Se 3^2434 nets as shown in Figure 11-2. The structure can therefore be regarded as a filled up $AlCu_2$ (*tI*12) type.

This structural arrangement results in

```
        Se
  \   /    \   /
   Tl        Tl
  /   \    /   \
        Se
```

chains running in the *c* direction of the crystal, each Se having two Tl(2) neighbors at 2.87 Å, and each Tl(2) four Se neighbors distributed tetrahedrally (Figure 5-19, p. 230). Tl(1) is surrounded only by rather distant neighbors; 8 Se at 3.44 Å, 2 Tl(1) at 3.51 Å and 4 Tl(2) at 4.02 Å. The structural arrangement suggests that Tl(2) is trivalent and Tl(1) is univalent and present only as an ion which provides the extra electron required for the formation of the Tl(2)–Se chains.

This structure provides another example (cf. PdS, p. 732) where a rather perfect geometrical arrangement of stacked layer networks of atoms, has apparently nothing in particular to do with the near-neighbor atomic arrangement that seems to be controlled by chemical bonds, giving chains of atoms variously oriented as they run throughout the crystal.

$CuAl_2$ (*tI*12): The $CuAl_2$ (MN_2) structure with cell dimensions, $a=6.066$, $c=4.874$ Å, $Z=4$,[2] is made up of 3^2434 nets of Al atoms at $z=0$ and $\frac{1}{2}$ which are oriented antisymmetrically relative to each other. The squares

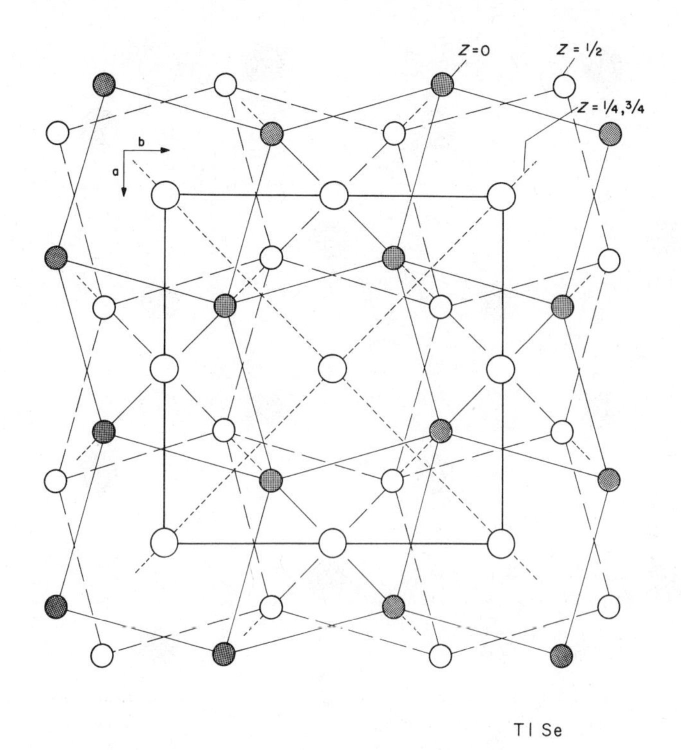

Figure 11-2. Atomic arrangement in the TlSe (*tI*16) structure projected down [001]: Tl, large circles.

in these Al layers which lie over the cell corners and basal face center, are centered by a $\frac{1}{2}.4^4$ net of Cu atoms at $z=\frac{1}{4}$ and $\frac{3}{4}$ as shown in Figure 11-3. The structure should be redetermined by modern methods.

Al has 4 Cu neighbors at 2.585 Å and one close Al at 2.745 Å ($d_{N\text{I}}$). Other Al neighbors completing a convex CN 15 polyhedron are 2×2.885 ($d_{N\text{II}}$), 4×3.115 ($d_{N\text{III}}$), and 4×3.22 Å ($d_{N\text{IV}}$). Cu has 8 Al at 2.585 in an archimedian antiprism with two polar Cu at 2.44 Å. Al neighbors with distances $d_{N\text{I}}$ and $d_{N\text{IV}}$ lie within the same net, whereas $d_{N\text{II}}$ and $d_{N\text{III}}$ refer to Al atoms of nets above and below. Figure 11-3 shows that the N atoms of one net do not exactly center the triangles of the nets above and below so that $d_{N\text{II}} \neq d_{N\text{III}}$. The condition for making $d_{N\text{II}}$ and $d_{N\text{III}}$ equal is for the x parameter of the N atoms to have a value of 0.147, whereas to make the N–N distances within the nets equal, $d_{N\text{I}} = d_{N\text{IV}}$, requires a value of 0.183. The observed values of x from 0.158 to 0.167 in compounds with the $CuAl_2$ type of structure, appear to represent a compromise to equalize as nearly as possible the 11 N–N distances. Although a

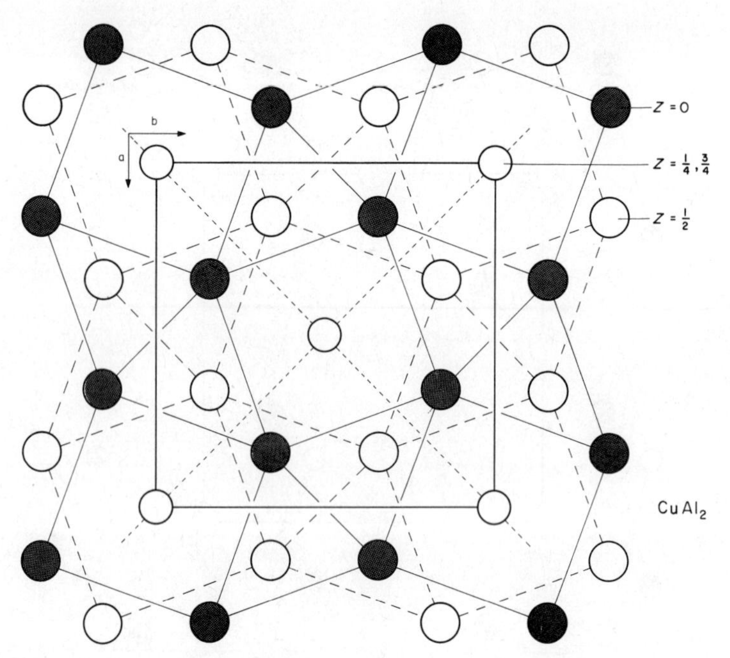

Figure 11-3. Atomic arrangement in the $CuAl_2$ ($tI12$) structure projected down [001]: Al atoms, large circles.

change of the x parameter notably influences the N–N distances, it has a relatively small influence on the M–N distances.

The average radius ratio of the components in more than 40 binary phases with the $CuAl_2$ structure is close to 0.8, and it can be seen from the near-neighbor diagram (p. 53) that this is very close to the intersection of lines for N–N_I, M–M, and M–N contacts at an x value of 0.160 (Figure 11-4). The axial ratio has a considerable influence on the near-neighbor diagram and a change to higher c/a value moves the lines for M–N and N–N_I contacts to lower $(D_M - d_M)/D_N$ values (Figure 11-4). The distribution on the near-neighbor diagram of phases with the $CuAl_2$ structure over a range of D_M/D_N values from 0.66 to 0.93 between the intersecting lines for 8–4 M–N contacts and the $1+1+4+4$ N–N contacts, suggests that the structural dimensions are controlled by a geometrical factor. This is further confirmed by the distribution of c/a values (between 0.74 and 0.88) of phases on the near-neighbor diagram which follows the movement of lines for M–N and N–N contact with c/a, so as to achieve high coordination with minimum compression of the atoms. The 2 M–M contacts have little influence on the cell dimensions of phases with the $CuAl_2$ structure.

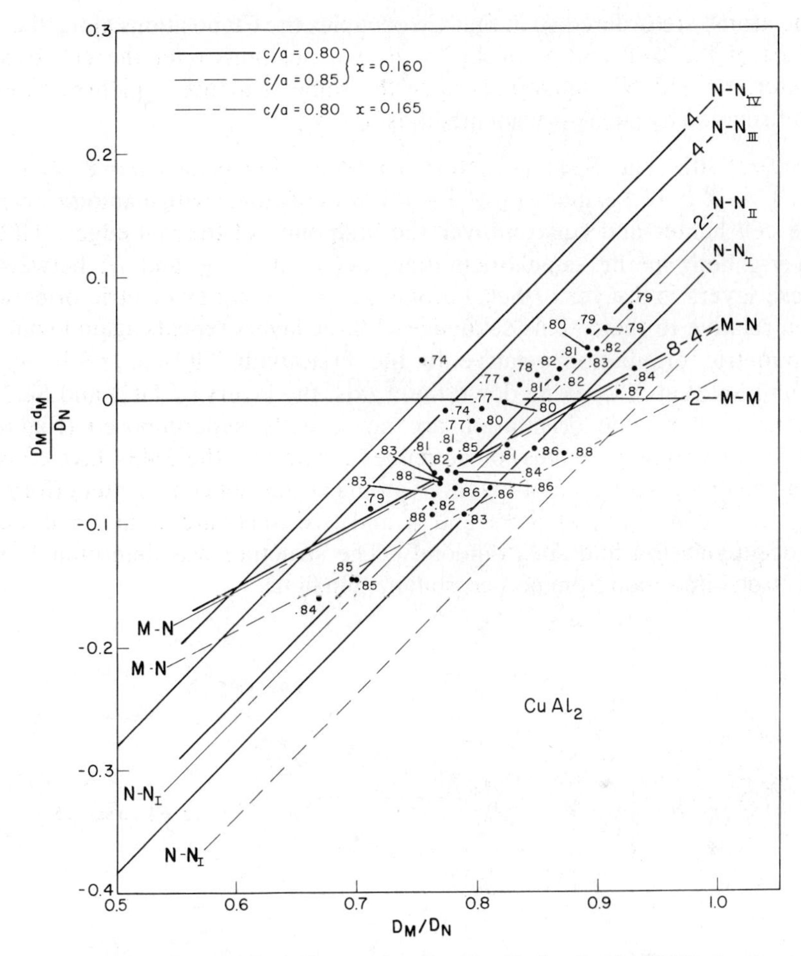

Figure 11-4. Near-neighbor diagram for phases with the $CuAl_2$ (MN_2) structure calculated for the parameters indicated on the figure. Numbers along lines indicate the number of neighbors for the contacts N–N, M–N, etc. Small numbers give the axial ratios, c/a, of phases with the $CuAl_2$ structure.

The ratio of the unit cell volume to that of the sums of the elemental volumes, $(U/4(V_M + 2V_N)$, is close to unity for many phases with the $CuAl_2$ structure; all lie between about one and 0.85, with the exception of $NiHf_2$ which apparently has a large expansion (1.14).

Nb_4CoSi (tP12): Nb_4CoSi with cell dimensions $a = 6.189$, $c = 5.053$ Å, $Z = 2$ is an ordered ternary superstructure based on the $CuAl_2$ structure, according to weak superstructure lines observed on diffraction patterns.[3]

The atoms are ordered such that Co occupies the Cu positions along the *c* edges of the cell, and Si occupies the Cu positions over the cell base center. Fe and Ni compounds have the same structure, which requires confirmation by single-crystal methods.

$SeTl_2$ (*tP*30): The $SeTl_2$ structure with cell dimensions $a = 8.54$, $c = 12.70$ Å, $Z = 10$,[4] is made up of 3^2434 layers of atoms with diamonds over the cell center and squares over the midpoints of the cell edges. Tl(2) layers, both in the same orientation, occur at $z \sim \frac{1}{12}$ and $\frac{5}{12}$; between these layers is a layer of Se(2) atoms at $z = \frac{1}{4}$ in antisymmetric orientation relative to them. This sequence of three layers repeats again in antisymmetric orientation relative to the first, with Tl(2) at $z \sim \frac{7}{12}$ and $\frac{11}{12}$ and Se(2) at $\frac{3}{4}$. Looking down the *c* axis, the layers of Tl(2) and Se(2) atoms in the same orientation are not exactly superimposed (Figure 11-5). The squares, cubes, or anticubes formed by the 3^2434 layers are centered by a sequence of layers of atoms at the cell edge centers (forming $\frac{1}{2}.4^4$ nets): Tl(1) at $z = \frac{1}{4}$ and $\frac{3}{4}$, and two Se(1) atoms at $z = 0$ and $\frac{1}{2}$, occupying fourfold sites randomly. The structure was determined by electron diffraction from polycrystalline thin films.

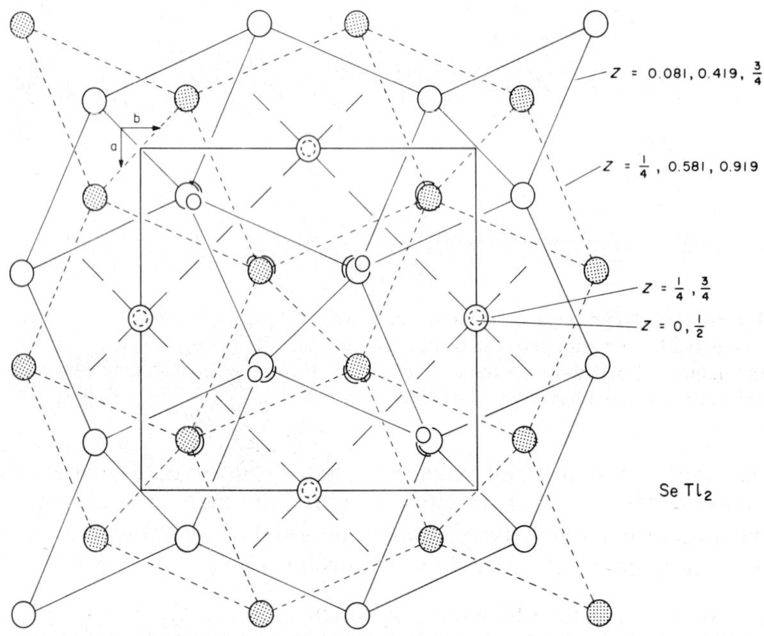

Figure 11-5. Structure of $SeTl_2$ (*tP*30) projected down [001]: Tl atoms, large circles.

$AuSn_2$ (*oP*24): $AuSn_2$ has cell dimensions $a = 6.909$, $b = 7.037$, $c = 11.789$ Å, $Z = 8$.[5] The structure is isotypic with the brookite form of TiO_2. Viewed down [001] the structure contains very rumpled 3^2434 nets with the layers about $z \sim \frac{1}{8}$ and $\frac{7}{8}$ having squares approximately over the cell base center and corners, and those about $z \sim \frac{3}{8}$ and $\frac{5}{8}$ having squares approximately over the midpoints of the cell edges, but all nets are displaced relative to exact centerings over the cell base (Figure 11-6). The squares in the 3^2434 nets are approximately centered by Au atoms. The structure can also be regarded as made up of continuous slabs of portions of pyrite-like structure parallel to the (001) plane, with Au

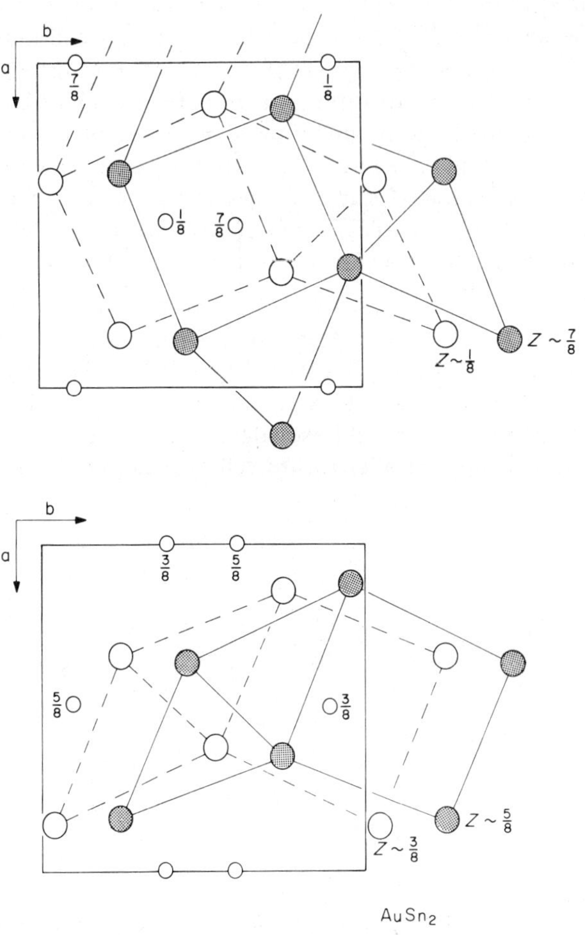

Figure 11-6. Atomic arrangement in the structure of $AuSn_2$ (*oP*24) projected down [001]: Sn, large circles.

occupying the Fe sites. After two layers of pyrite-like structure a layer of Au and Sn atoms is removed, and the next two layers of pyrite-like arrangement are displaced in the b direction as shown in Figure 11-7.

Au has 6 Sn at 2.68 to 2.86 Å and one Au at 2.99 Å. Sn(1) has 3 Au at 2.68 to 2.77 Å and Sn at 3.06 and 3.13 Å. Sn(2) has 3 Au at 2.76 and 2.86 Å and 2 close Sn neighbors.

PdP_2 ($mC12$): In the PdP_2 structure ($a = 6.207$, $b = 5.587$, $c = 5{,}874$ Å, $\beta = 111.80°$, $Z = 4$ in I setting),[6] the P atoms form zigzag chains running along [100] (P–2P = 2.20 and 2.22 Å) and the 4 P surrounding Pd form a slightly distorted square (2×3.275, 2×3.338 Å) as shown in Figure 5-10, p. 222. P has 2 P + 2 Pd neighbors in a distorted tetrahedral arrangement.

NiP_2, examined by X-ray powder diffraction, has cell dimensions $a = 6.366$, $b = 5.615$, $c = 6.072$ Å, $\beta = 126.22°$, $Z = 4$ in C setting,[7] appearing to be isostructural with PdP_2. Ni is surrounded by 4 P (2.21 Å) in a very nearly square configuration almost parallel to $20\overline{2}$ planes. P has two Ni neighbors and 2 P at 2.20 and 2.22 Å forming a considerably distorted tetrahedron. The P atoms form a rumpled 3^2434 net shown parallel to the $(10\overline{1})$ plane in Figure 11-8. The Ni atoms approximately center the squares of this net, and the nets are interconnected by P–P bonds.

$CoGe_2$ ($oC23$): In the $CoGe_2$ structure with cell dimensions $a = 5.681$, $b = 5.681$, $c = 10.818$ Å, $Z = 4$, (setting $Aba2$), 7 Co atoms randomly occupy 8 sites.[8] Ge(1) form 3^2434 nets parallel to the (001) plane with squares over cell corners and base center at $z = \frac{1}{8}$, and with squares over the midpoints of the cell base edges at $z = \frac{5}{8}$. At $z = \frac{3}{8}$ and $\frac{7}{8}$, Ge(2) form 4^4 nets oriented parallel to the cell edges, so as to locate slightly

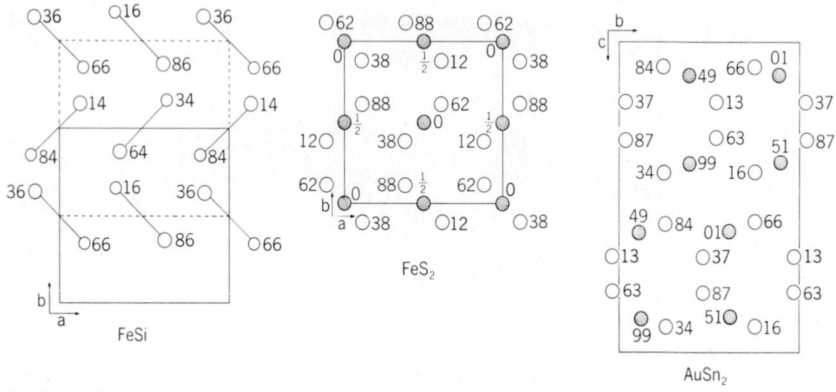

Figure 11-7. Comparison of similar atomic arrangements in the FeSi ($cP8$) and FeS_2 ($cP12$) structures, and the $AuSn_2$ ($oP24$), and FeS_2 structures. (Left) Large circles, Fe. (Center and right) Fe or Au atoms, shaded.

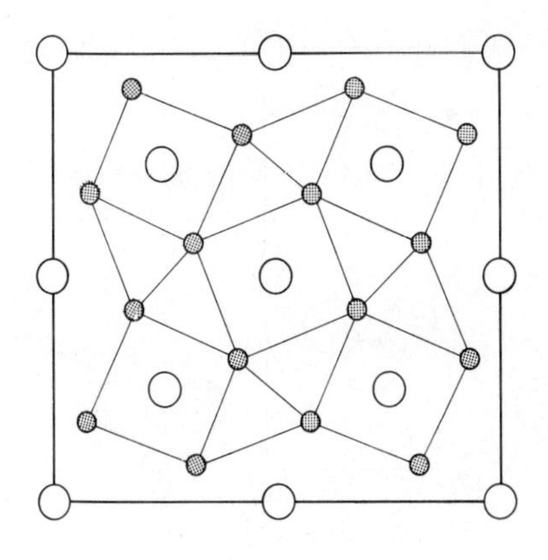

Figure 11-8. Rumpled 3^2434 net in the NiP_2 ($mC12$) structure parallel to the $(10\bar{1})$ plane: large circles, Ni atoms.

distorted squares of the net over the cell corners, base center and midpoints of the basal cell edges (Figure 11-9). $\frac{1}{2}.4^4$ nets of Co at $z = 0.012$ and 0.238 center the distorted cubic antiprisms formed between the nets at $z = 0$ and $\frac{1}{2}$, and $z = \frac{1}{4}$ and $\frac{3}{4}$ are respectively antisymmetrically arranged midpoints of the basal cell edges center the distorted cubic antiprisms made from the squares of the nets at $z = \frac{5}{8}$ and $z = \frac{3}{8}$ or $\frac{7}{8}$. The structure is related to the $PdSn_2$ type.

$PdSn_2$ (L.T.) ($tI48$): The $PdSn_2$ structure with cell dimensions $a = 6.490$, $c = 24.39$ Å, $Z = 16$,[9] has 3^2434 nets of Sn(2) atoms at heights of $z = 0$, $\frac{1}{4}$, $\frac{1}{2}$, and $\frac{3}{4}$. At $z = 0$ and $\frac{1}{2}$ the nets have squares over the cell corners and the base center, whereas at $z = \frac{1}{4}$ and $\frac{3}{4}$ diamonds lie over these locations and the squares lie over the midpoints of the cell edges. The nets at $z = 0$ and $\frac{1}{2}$, and $z = \frac{1}{4}$ and $\frac{3}{4}$ are respectively antisymmetrically arranged relative to each other. Sn(1) atoms form 1.4^4 nets at heights of $z = \frac{1}{8}$, $\frac{3}{8}$, $\frac{5}{8}$, and $\frac{7}{8}$. The nets are superimposed, with Sn atoms lying over $\frac{1}{4}$, $\frac{1}{4}$; etc. Pd atoms lying over the cell corners and base center form $\frac{1}{2}.4^4$ nets at $z = 0.09$, 0.41, 0.59, and 0.91, whereas at $z = 0.16$, 0.34, 0.66, and 0.84, Pd $\frac{1}{2}.4^4$ nets are located with the atoms over the midpoints of the cell edges. (Figure 11-10). This arrangement gives a series of skewed cubes or cubic antiprisms stacked one above the other over the cell corners, midpoints of the cell edges and the base center, so that the structure is related to the $CoGe_2$ type.

Pd has 4 close Sn(1) neighbors at 2.43 Å and 4 Sn(2) at a much larger distance of 3.32 Å. Sn(1) has 4 close Pd neighbors and 2+4 Sn at 3.16

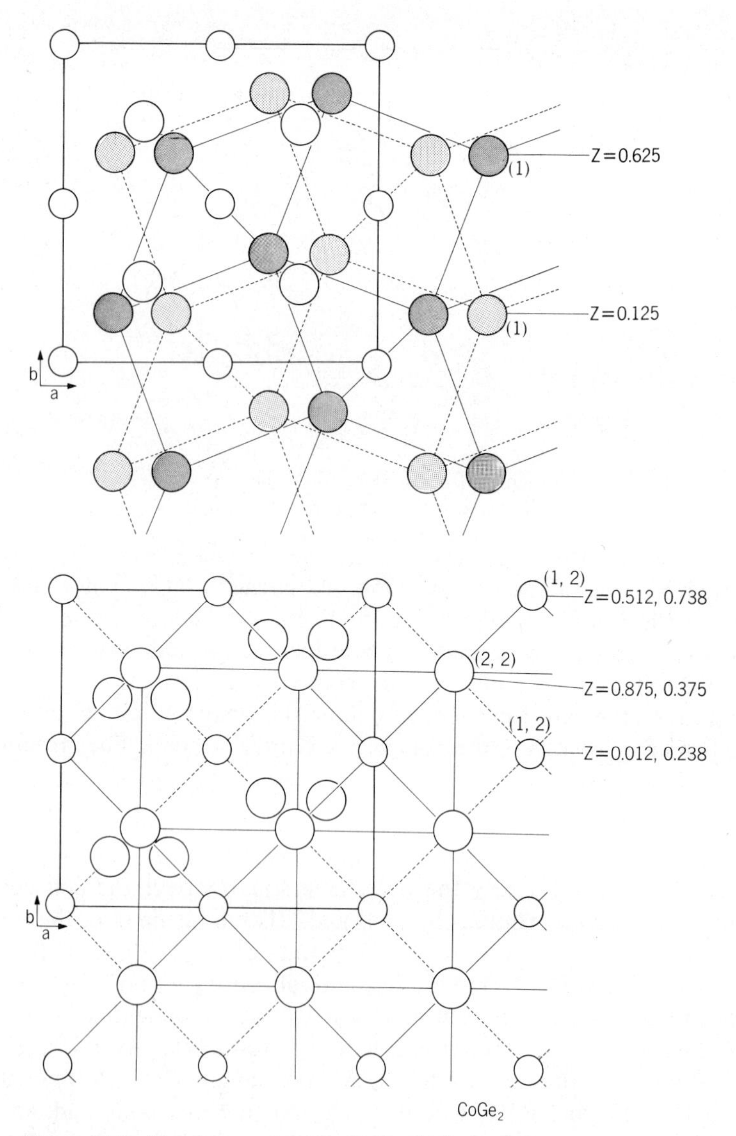

Figure 11-9. Atomic arrangements in the $CoGe_2$ ($oC23$) structure viewed down [001]: Ge, large circles.

and 3.245 Å. Sn(2) has 4 Pd neighbors, two Sn(1) at 3.16 Å, one very close Sn(2) neighbor at 2.90 Å and four more at 3.46 Å.

$PdSn_3$ ($oC32$): The structure of $PdSn_3$ ($a = 17.20$, $b = 6.47$, $c = 6.50$ Å, $z = 8$)[10] is made up of 3^2434 layers of Sn atoms and $\frac{1}{2}.4^4$ layers of Pd

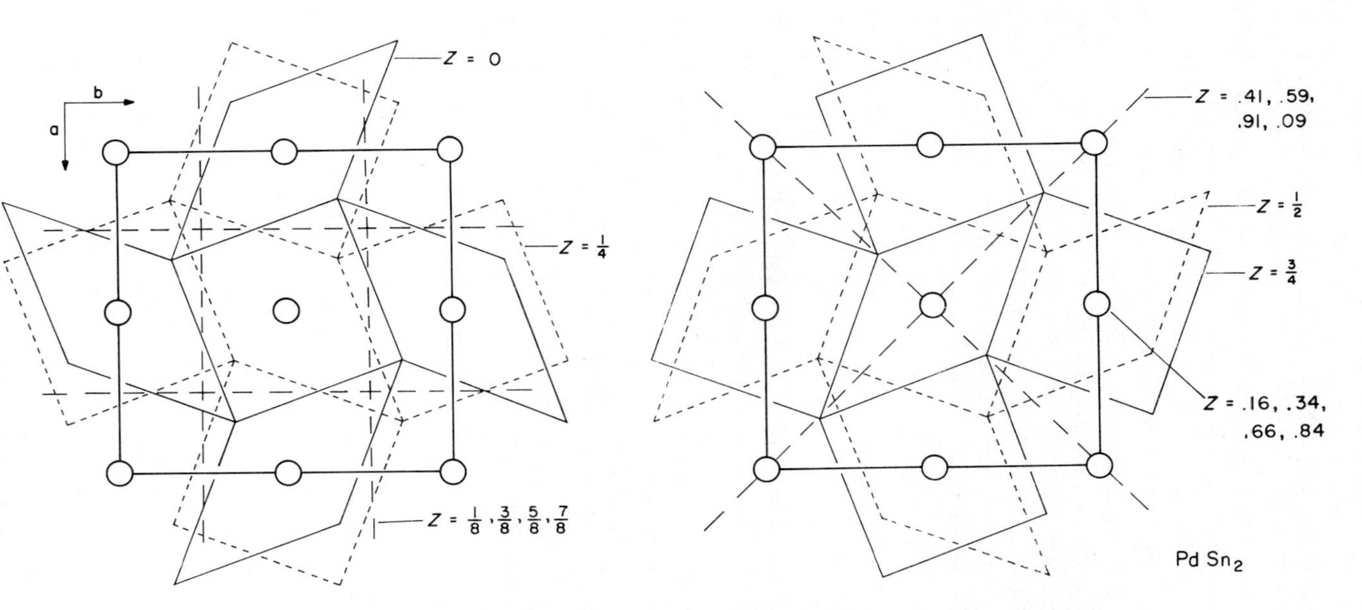

Figure 11-10. Arrangements of atoms on nets in the $PdSn_2$ (*tI*48) structure: circles, Pd atoms.

atoms. At $x=0$ the 3^2434 layer of Sn(1) has its squares over the cell corners and cell base center. At $x=\pm 0.168$, 3^2434 nets at Sn(2) are similarly arranged with squares over the cell base center, but antisymmetrically in relation to the net at $x=0$. A $\frac{1}{2}.4^4$ net of Pd at $x=\pm 0.084$ centers the cubic antiprisms of Sn so formed. Another 3^2434 net of Sn(1) is located at $x=\frac{1}{2}$, but the squares of the net are now over the midpoints of the basal cell edges. Similar nets of Sn(2) occur at $x=\frac{1}{2}\pm 0.168$, but oriented antisymmetrically to that at $x=\frac{1}{2}$, and the cubic antiprisms of Sn so formed are centered by $\frac{1}{2}.4^4$ nets of Pd at $x=\frac{1}{2}\pm 0.084$ (Figure 11-11). Pd has 8 Sn at 2.81 Å and one Pd at 2.89 Å. The structure was examined by single-crystal methods, but the atom positions were only approximately determined.

$SiPt_3$ (L.T.) (*mC*16): $SiPt_3$ in setting $F2/m$ has cell dimensions $a=7.702$, $b=7.765$, $c=7.765$ Å, $\beta=88.11°$.[11] Pt(1) and (3) atoms form 3^2434 nets parallel to the *bc* plane at $x=0$ and $\frac{1}{2}$, the nets being arranged antisymmetrically to each other with the diamonds centered over cell corners and the cell base center. The Pt(2) and Si atoms form $\frac{1}{2}.4^4$ nets at $x=\frac{1}{4}$ and $\frac{3}{4}$ with the Pt(2) atoms lying over the centers of the distorted squares of the 3^2434 net, and the Si atoms over the centers of the diamonds (Figure 11-12). The structure can therefore be regarded as further distortion of the SiU_3 type structure (p. 324) of the high-temperature form of $SiPt_3$.

The arrangement is such that Pt(2) has 4 Si at 2.745 Å as well as 4 Pt(1) (2.72 and 2.81 Å) and 4 Pt(3) (2.75 and 2.77 Å) neighbors, with two polar

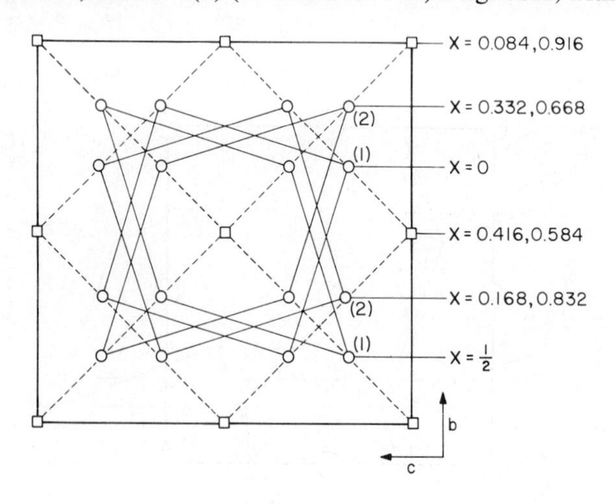

Figure 11-11. Atomic arrangements in the $PdSn_3$ (*oC*32) structure viewed down [100].

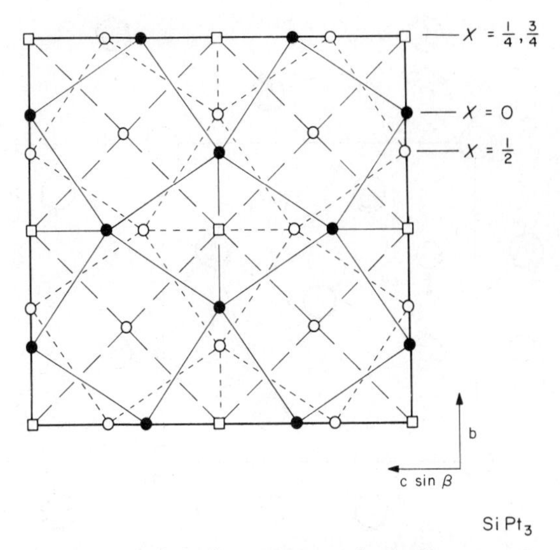

Figure 11-12. Atomic arrangement in the low temperature $SiPt_3$ ($mC16$) structure projected on to a plane normal to the [100] direction: squares, Si atoms.

Pt(2) atoms (3.85 Å). The CN 14 polyhedron has 12 four-sided faces. Pt(1) and (3) each have 13 neighbors out to 3.11 Å, and Si has 12 Pt neighbors in a distorted cubo-octahedron with 6 four and 8 three-sided faces. The structure requires confirmation by single-crystal methods.

$CoGa_3$ ($tP16$): In the $CoGa_3$ structure which has cell dimensions $a = 6.26$, $c = 6.48$ Å, $Z = 4$,[10] 3^2434 nets of Ga(2) atoms at $z = 0$ and $\frac{1}{2}$ are stacked directly above each other. Ga(1) atoms at $z = \frac{1}{4}$ and $\frac{3}{4}$ in a $\frac{1}{2}.4^4$ net center the squares of the Ga(2) nets. Co atoms at $z = \frac{1}{4}$ center one of the two triangles in half of the diamonds of the 3^2434 Ga(2) net. Co at $z = \frac{3}{4}$ center one of the two triangles in the remaining diamonds of the 3^2434 net as shown in Figure 11-13. Co has 9 neighbors (1 Co + 8 Ga); Ga(1) has 12 neighbors (2 Co + 10 Ga), one being significantly further away than the rest, and Ga(2) has 14 neighbors (3 Co + 11 Ga), 6 of which are at significantly greater distances than the others.

$ZnAu_3$ [$R2$] ($oC32$): The R2 structure which is pseudo-tetragonal has cell dimensions $a = 5.585$, $b = 5.594$, $c = 16.65$ Å, $Z = 8$.[12] The structure is made up of planar 4^4 and 3^2434 nets of atoms parallel to the (001) plane. At $z = 0$ a 3^2434 net of Au(3) has its squares over the cell corners and base center. A similar net of Au(3) at $z = \frac{1}{2}$ has its squares over the midpoints of the basal cell edges. 4^4 nets of Au(2) at $z = \frac{1}{4}$ and $\frac{3}{4}$ are arranged so as to place squares over cell corners and midpoints of the

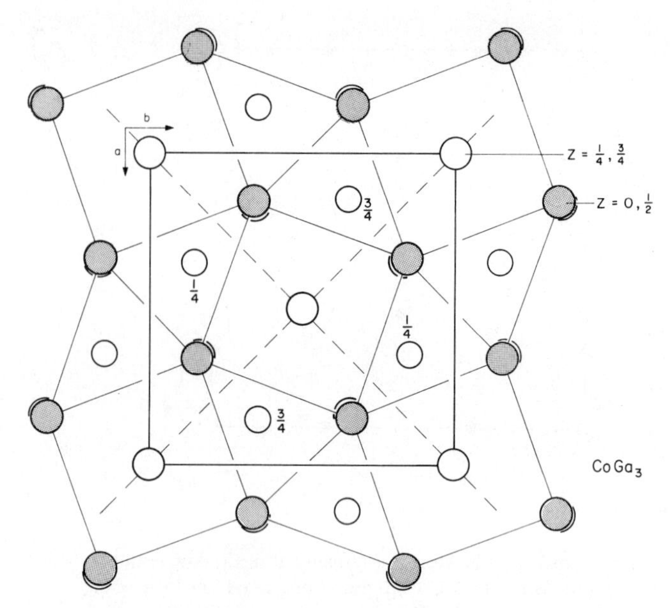

Figure 11-13. Projection of the structure of $CoGa_3$ (*tP*16) on to the (001) plane: Ga, large circles.

edges and center of the basal cell plane. Between these nets, Au(1) atoms on 4^4 nets at $z = 0.384$ and 0.616 and Zn atoms on 4^4 nets at $z = 0.136$ and 0.864 lie over the midpoints of the basal cell edges, and Au(1) atoms at $z = 0.116$ and 0.884 and Zn at $z = 0.364$ and 0.636, also on 4^4 nets, lie over the cell corners and cell base center, as shown in Figure 11-14.

$ZnAu_3$ [*R*1] (*tI*64): All atoms in the R1 structure lie on 3^2434 or 4^4 nets parallel to the (001) plane. The structure resembles that of the R2 phase, but the two 3^2434 nets of Au(3) atoms repeat in antisymmetric orientation requiring a cell with double the height, as indicated in Figure 11-14. The cell dimensions are $a = 5.586$, $c = 33.4$ Å, $Z = 16$.[12]

$FeGa_3$ (*tP*16): $FeGa_3$ has cell dimensions $a = 6.263$, $c = 6.556$ Å, $Z = 4$.[13] The Ga(2) atoms form rumpled 3^2434 nets at $z \sim \frac{1}{4}$ and $\frac{3}{4}$ which overlie each other and which have squares over the midpoints of the cell base edges. Ga(1) at $z = 0$ and $\frac{1}{2}$ center the cubes formed by the 3^2343 nets and Fe, also at $z = 0$ and $\frac{1}{2}$, center half of the triangular prisms (Figure 11-15). Fe has 8 Ga at 2.36–2.50 Å and one Fe neighbor. Ga(1) has 10 Ga + 2 Fe neighbors and Ga(2) has 3 Fe + 6 Ga out to 3.09 Å. The structure of the isotypic $CoGa_3$ was examined by single-crystal methods.

Fe_3C (*oP*16): Cementite has cell dimensions $a = 5.089$, $b = 6.743$, $c = 4.524$ Å, $Z = 4$. Various structural data are reassessed and refined by

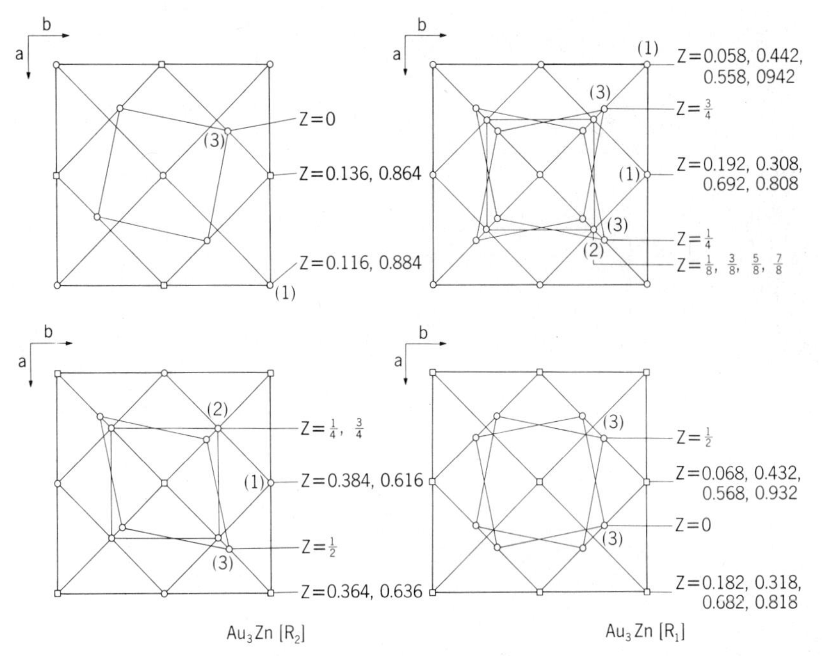

Figure 11-14. Atomic arrangements in the $ZnAu_3$ *oC*32 and *tI*64 structures viewed down [001]: squares, Zn atoms.

Herbstein and Smuts.[14] The structure can be described variously in terms of rumpled nets of atoms. For instance, there are two possible descriptions involving very rumpled 3^2434 nets of Fe atoms, one of which is parallel to (010) plane (Figure 11-16). Figure 11-16 also shows the arrangement of Fe atoms on $3^2434 + 3^34^2$ (1:2) nets parallel to (031) planes. Perhaps the most useful is that in terms of very rumpled $3^34^2 + 3^6$ (2:1) square-triangle nets of Fe atoms parallel to the (001) plane (Figure 11-17). The C atoms are surrounded by 4 Fe(2) atoms lying on one side and by 2 Fe(1) atoms lying on the other side at distances from 2.01 to 2.025 Å.

$Pd_{4.8}P$ (*mP*20): $Pd_{4.8}P$ has cell dimensions $a = 5.004$, $b = 7.606$, $c = 8.416$ Å, $\beta = 95.63°$, $Z = 18$ Pd atoms.[15] The structure has similarities to that of Fe_3C as shown in Figure 11-16. The Pd atoms form square-triangle nets parallel to the (130) plane, or rumpled square-triangle nets parallel to (010) at $y \sim 0$ and $\frac{1}{2}$. Pd has 11 or 12 neighbors and P is surrounded by 9 Pd, six forming a distorted triangular prism and the other three being situated outside the rectangular faces. The P sites are only about 60% filled randomly.

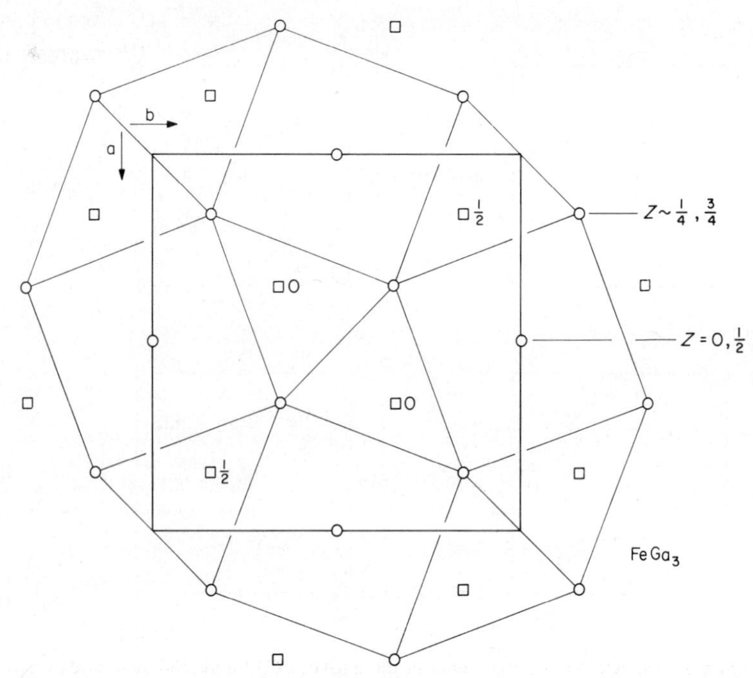

Figure 11-15. Atomic arrangements in the $FeGa_3$ (*tP*16) structure projected down [001]: squares, Fe atoms.

$PtPb_4$ (*tP*10): In the $PtPb_4$ structure ($a = 6.666$, $c = 5.978$ Å, $Z = 2$),[16] 3^2434 nets of Pb atoms at $z \sim \frac{1}{4}$ and $\sim \frac{3}{4}$ are arranged antisymmetrically with respect to each other. The squares of these nets are centered by Pt atoms at $z = 0$ which form a $\frac{1}{2}.4^4$ net, as shown in Figure 11-18. The Pb and Pt nets form a unit of three layers which is well separated from adjacent three-layer units. The structure can be compared with that of $NbTe_4$.

$NbTe_4$, sub-cell (*tP*10): The $NbTe_4$ subcell has dimensions $a = 6.499$, $c = 6.837$ Å, $Z = 2$.[17] The structure is built up of 3^2434 layers of Te atoms at $z = 0$ and $\frac{1}{2}$ which cover each other antisymmetrically (Figure 11-19). Te squares which cover the cell corners are centered by Nb at $z = \frac{1}{4}$ and $\frac{3}{4}$, whereas the Te squares about the cell center are uncentered.

$PtSn_4$ (*oC*20): $PtSn_4$ has cell dimensions $a = 6.397$, $b = 6.426$, $c = 11.38$ Å, $Z = 4$.[18] The structure is made up of 3^2434 layers of Sn atoms and 4^4 layers of Pt atoms parallel to the (001) plane. At $z = \frac{1}{8}$ and $\frac{7}{8}$ the 3^2434 nets of Sn(2) atoms have the rectangles over the cell corners and basal plane center, whereas at $z = \frac{3}{8}$ and $\frac{5}{8}$, 3^2434 nets of Sn(1) atoms have

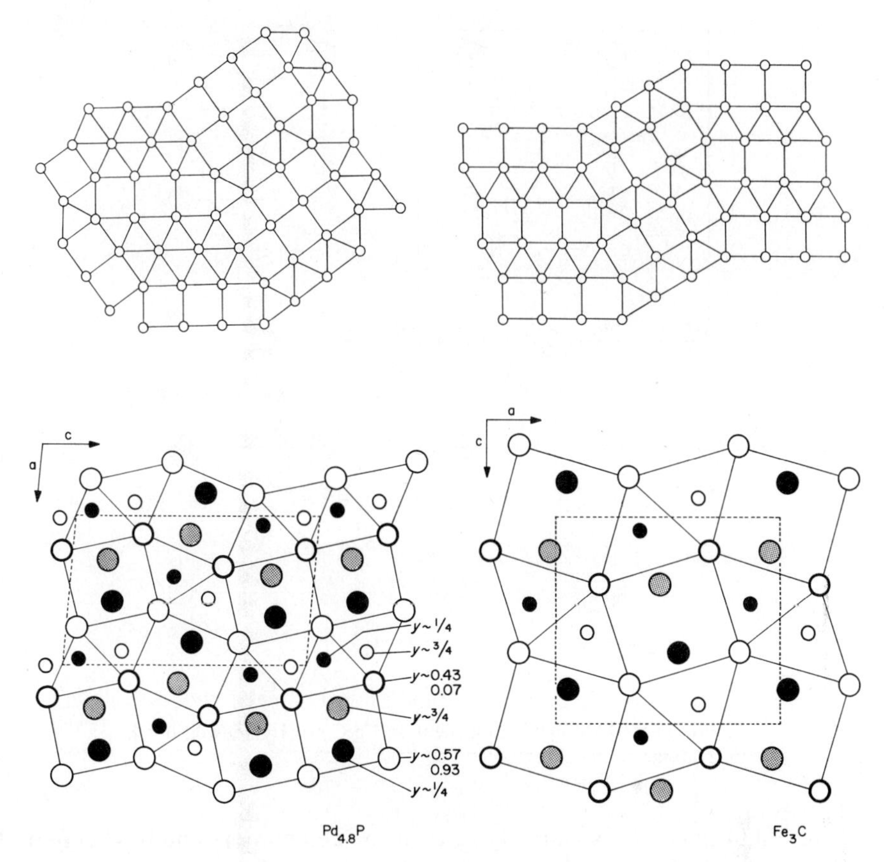

Figure 11-16. (Upper left) Arrangement of Pd atoms parallel to the (130) plane in the $Pd_{4.8}P$ (*mP*20) structure. (Upper right) Arrangement of the Fe atoms parallel to the (031) plane in the cementite Fe_3C (*oP*16) structure. (Lower left) The $Pd_{4.8}P$ structure viewed along [010]. (Lower right) The Fe_3C structure viewed along [010]. Large circles are Pd or Fe. Diagram from Sellberg [15].

diamonds over cell corners and the basal plane center and rectangles over the midpoints of the basal cell edges. The nets at $z = \frac{1}{8}$ and $\frac{7}{8}$ are arranged antisymmetrically with respect to each other and the cubic antiprisms are centered by Pt which form a $\frac{1}{2}.4^4$ net at $z = 0$. Likewise the nets at $z = \frac{3}{8}$ and $\frac{5}{8}$ are arranged antisymmetrically with respect to each other and the cubic antiprisms are centered by a 4^4 net of Pt atoms at $z = \frac{1}{2}$ which are situated over the midpoints of the basal cell edges (Figure 11-20).

$PdGa_5$ (*tI*24): The unit cell of the $PdGa_5$ structure with dimensions $a = 6.448$, $c = 10.003$ Å, $Z = 4$,[10] contains four layers of Ga(2) atoms which

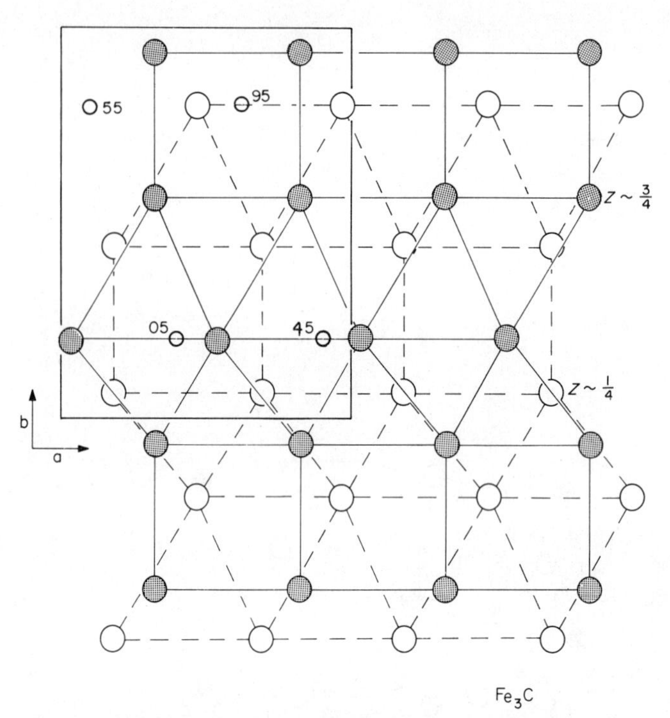

Figure 11-17. Atomic arrangement in Fe_3C ($oP16$) viewed down [001]: large circles, Fe.

form 3^2434 nets with squares lying over the cell corners and base center. Nets at $z = 0.14$ and 0.86 are of one orientation and those at $z = 0.36$ and 0.64 are arranged antisymmetrically in relation to them as shown in Figure 11-21. The Ga(2) distorted cubes or cubic antiprisms are centered at $z = 0$ and $\frac{1}{2}$ by $\frac{1}{2}.4^4$ nets of Ga(1) atoms and at $z = \frac{1}{4}$ and $\frac{3}{4}$ by similar nets of Pd atoms. Pd therefore centers the cubic antiprisms and Ga(1) the cubes of Ga(2) atoms. Ga(1) and Pd are each surrounded by a convex polyhedron of 10 atoms, and Ga(2) has 9 neighbors at distances less than 3.0 Å.

Si_2U_3 ($tP10$): Si_2U_3 has cell dimensions $a = 7.330$, $c = 3.900$ Å, $Z = 2$.[19] A 3^2434 net of U(2) atoms at $z = \frac{1}{2}$ covers a similar net of Si atoms at $z = 0$ antisymmetrically. U(1) atoms at $z = 0$ center the squares in the 3^2434 nets; they thus form a $\frac{1}{2}.4^4$ net (Figure 11-22). The Si and U(1) atoms at $z = 0$ together form a $5^3 + 5^4$ net. (See Figure 11-1, p. 606.) The Si atoms center the triangular prisms formed by U(2), having a close Si neighbor out through the rectangular face shared by two prisms. U(2) has

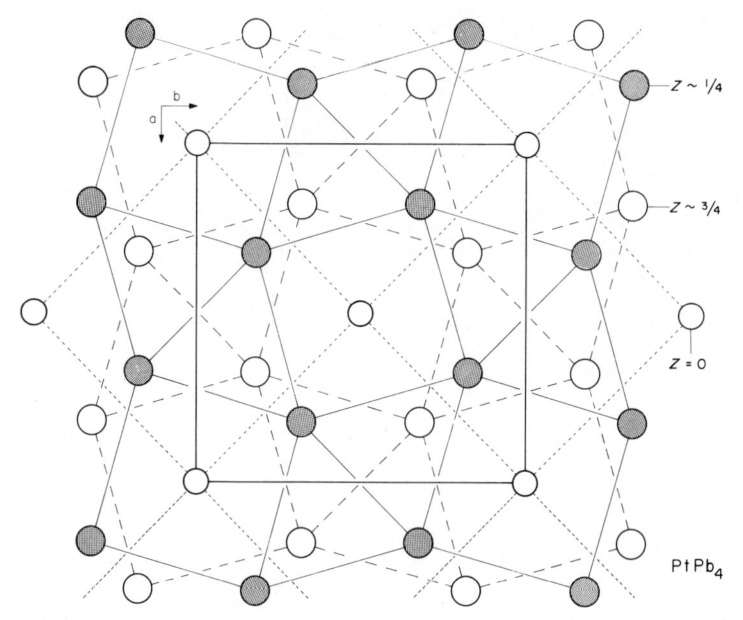

Figure 11-18. The $PtPb_4$ ($tP10$) structure shown in projection down [001]: Pb, large circles.

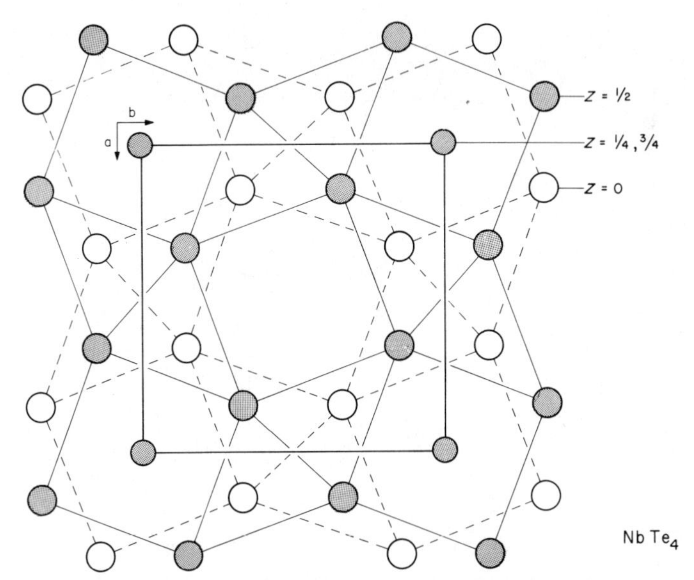

Figure 11-19. Atomic arrangement in the subcell ($tP10$) of the $NbTe_4$ structure projected on (001): Te, large circles.

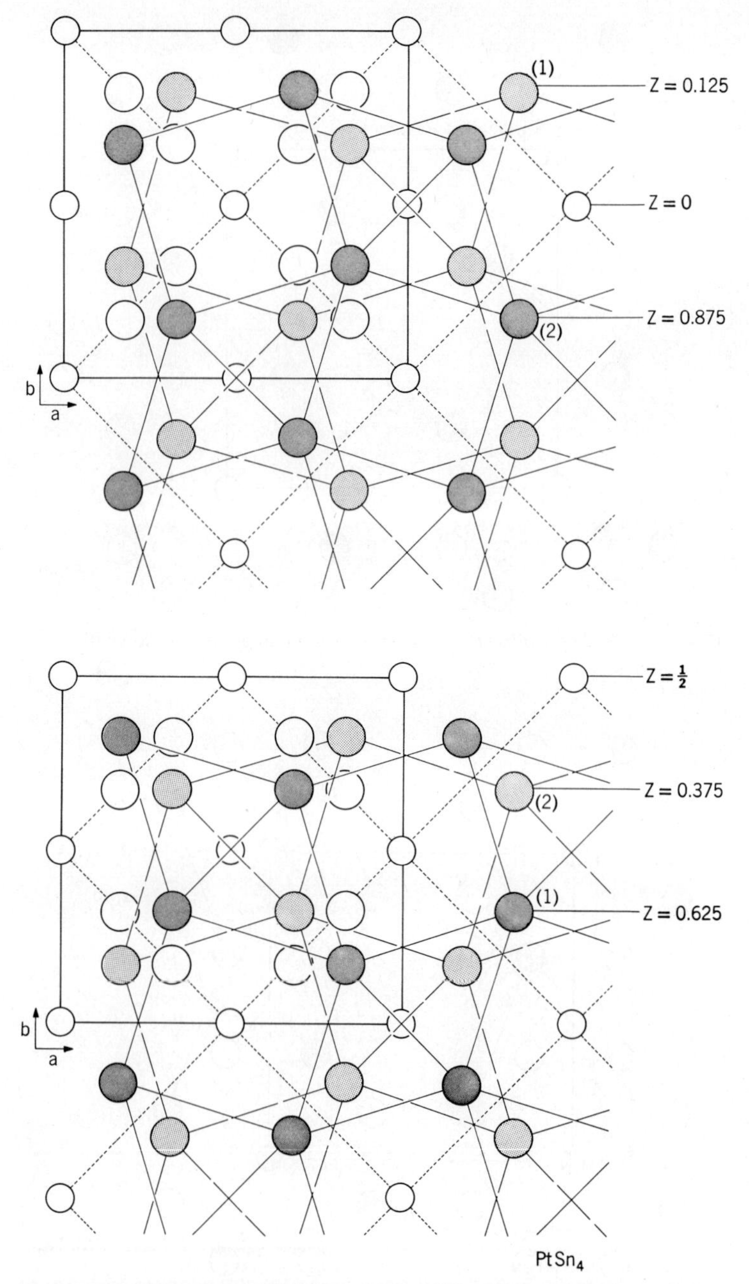

Figure 11-20. Atomic arrangements in the $PtSn_4$ (*oC*20) structure projected down [001]: Sn, large circles.

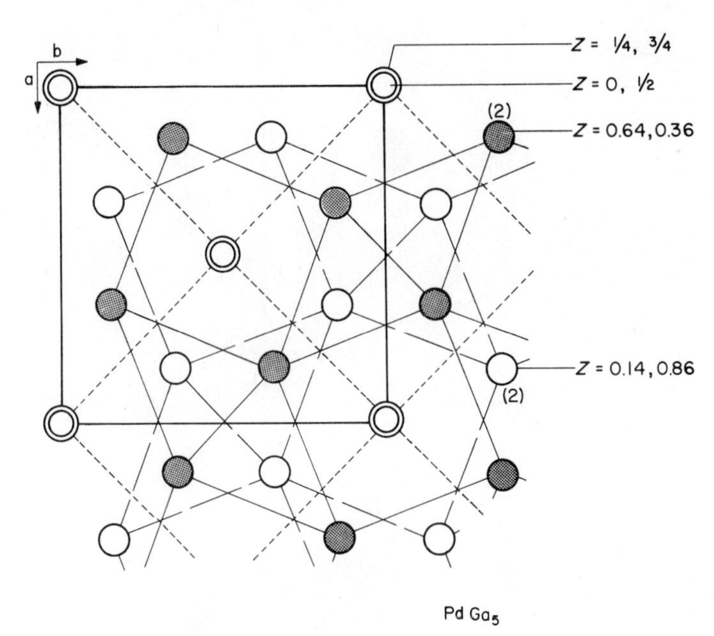

Figure 11-21. The structure of $PdGa_5$ (*tI*24) projected down [001]: Ga, large circles.

5 U(2) neighbors in its plane and is surrounded by a very narrow triangular prism of Si atoms. The structure needs to be confirmed by single-crystal methods.

Al_2Gd_3 (*Al_2Zr_3?*) (*tP*20): In the Al_2Gd_3 structure ($a = 8.344$, $c = 7.656$ Å, $Z = 4$)[20] Gd(1) and (2) atoms form 3^2434 nets at $z \sim \frac{1}{4}$ and $\frac{3}{4}$. The nets are similarly oriented at each level, but the atom positions are not exactly superimposed. Diamonds are centered over the cell base center and corners, and the rectangles are centered at the midpoints of the basal edges of the cell (Figure 11-23). Al(1) and (2), very approximately at $z = 0$ and $\frac{1}{2}$, center all of the triangles in the Gd(1)+(2) net. In so doing they also form a rumpled 3^2434 net with rectangles centered over the midpoints of the basal edges of the cell, but oriented antisymmetrically with respect to the Gd net. Gd(3) atoms form a $\frac{1}{2}.4^4$ net at $z \sim 0$ and $\frac{1}{2}$ and center the squares in the 3^2434 Gd net. Thus the Al and Gd atoms at both $z = 0$ and $\frac{1}{2}$ together form $5^3 + 5^4$ nets. (See Figure 11-1, p. 606.) The structure gives rise to a very close Al(2)–Al(2) approach of 2.45 Å; there is also one Al(1)–Al(1) distance of 2.90 Å.

Al_2Zr_3 probably has the same structure type, although it was originally described in a different space group. The closest Al–Al distance given for Al_2Zr_3 was 2.70 Å.

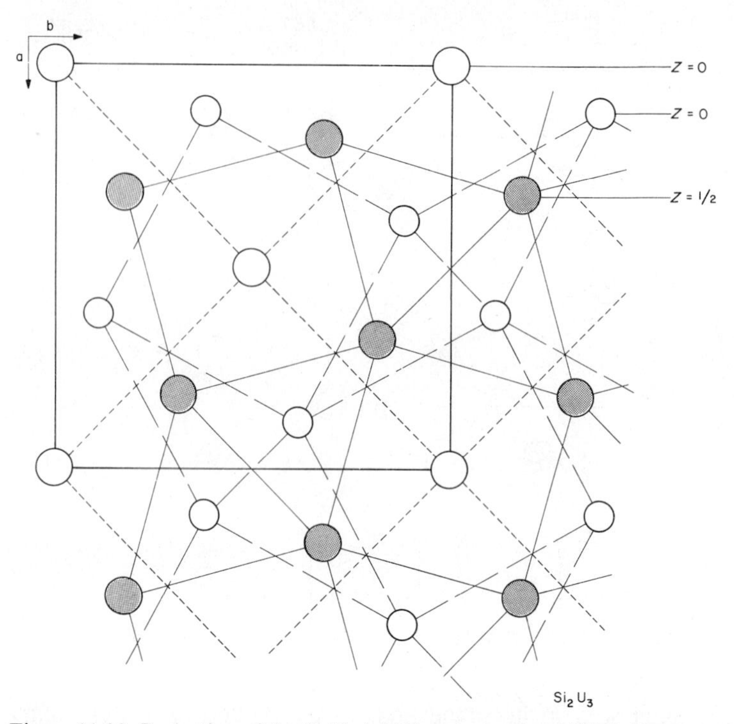

Figure 11-22. Projection of the Si_2U_3 (*tP*10) structure down [001]: U, large circles.

Co_2Al_9 (*mP*22): The dimensions of the unit cell of the Co_2Al_9 structure are $a = 6.213$, $b = 6.290$, $c = 8.557$ Å, $\beta = 94.76°$, $Z = 2$.[21] The structure can be regarded as a layer structure with Al layers parallel to (001), but the layers are not exactly planar. Al(1), (4), and (5) form a $\frac{1}{4}.4^4$ net at $z \sim 0$ and a $\frac{3}{4}.4^4$ net at $z \sim \frac{1}{2}$ (Figure 11-24*a*). At $z \sim \frac{1}{4}$ and $\frac{3}{4}$ Al(2) and (3) form 3^2434 nets in different orientations (Figure 11-24*b*). A Co atom at $z = 0.17$ approximately centers half of the squares of the Al net at $z \sim \frac{1}{4}$ and another Co at $z = 0.33$ approximately centers the other squares in the same net. Similarly Co atoms at $z = 0.67$ and 0.83 lie above and below the approximate center of the squares of the 3^2434 Al net at $z \sim \frac{3}{4}$. Co thus has 9 Al neighbors, the closest being Co–Al(1) at 2.375 Å. The Al atoms are irregularly coordinated, although Al(1) is surrounded by a distorted icosahedron of 10Al + 2Co atoms with 8 Al at 2.74 ± 4 Å, 2 Co at 2.375 Å, and 2 Al at 3.27 Å.

B_3Cr_5 (*tI*32): In the B_3Cr_5 structure ($a = 5.46$, $c = 10.64$ Å, $Z = 4$)[22] Cr(2) atoms form 3^2434 nets in the same orientation at $z = 0.85$ and 0.15, and in antisymmetric orientation at $z = 0.35$ and 0.65. The squares of these

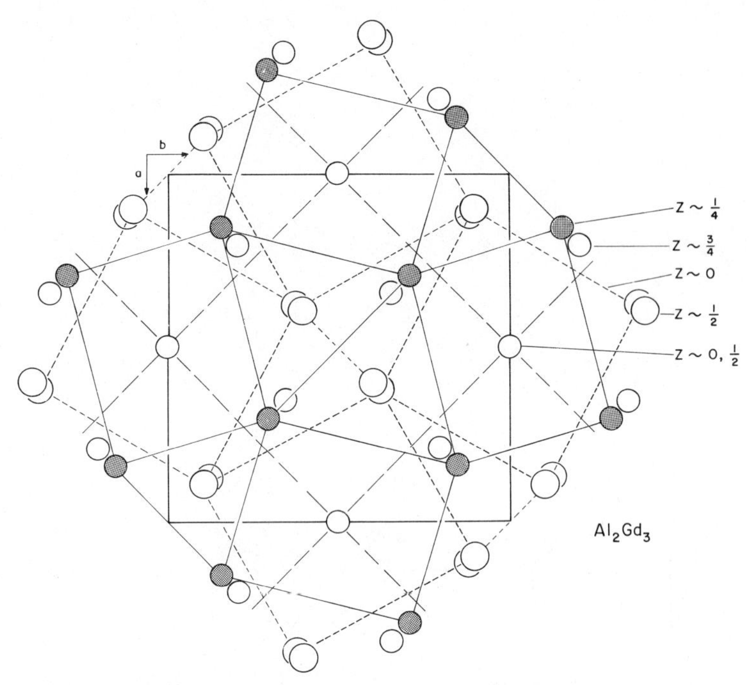

Figure 11-23. Atomic arrangement in the Al_2Gd_3 (*tP*20) structure viewed down [001]: Al, large circles.

nets lie over the cell corners and the base center of the cell. $\frac{1}{2}.4^4$ nets of Cr(1) atoms located over the cell corners and base center at $z = 0$ and $\frac{1}{2}$, center the slightly elongated cubes formed between the Cr(2) 3^2434 nets. Similar $\frac{1}{2}.4^4$ nets of B(1) atoms at $z = \frac{1}{4}$ and $\frac{3}{4}$ center the cubic antiprisms formed between the Cr(2) nets (Figure 11-25). This arrangement gives B(1) 8 Cr(2) neighbors at 2.30 Å. Finally, the structure also contains 3^2434 nets of B(2) atoms at $z = 0$ antisymmetrically oriented relative to the Cr(2) 3^2434 nets at $z = 0.85$ and 0.15, and at $z = \frac{1}{2}$ antisymmetrically oriented relative to the Cr(2) 3^2434 nets at $z = 0.35$ and 0.65. The squashed cubic antiprisms formed by Cr(2) and B(2) atoms are not centered. Triangular prisms of Cr at $z = 0.35$ to 0.65 and $z = 0.85$ to 0.15 are centered by B(2) atoms, which have a close B neighbor through the rectangular face shared between two prisms.

Mo_5SiB_2 is an ordered ternary form of the B_3Cr_5 structure. The Pb_3Ba_5 structure with cell dimensions $a = 9.038$, $c = 16.843$ Å, $Z = 4$,[23] is essentially the same as the B_3Cr_5 structure. Ba(1) is surrounded by 8 Ba(2) at 4.21 Å. Pb(1) is surrounded by 8 Ba(2) at 3.76 Å and 2 Ba(1). Ba(1) has 15 neighbors and Pb(2) nine.

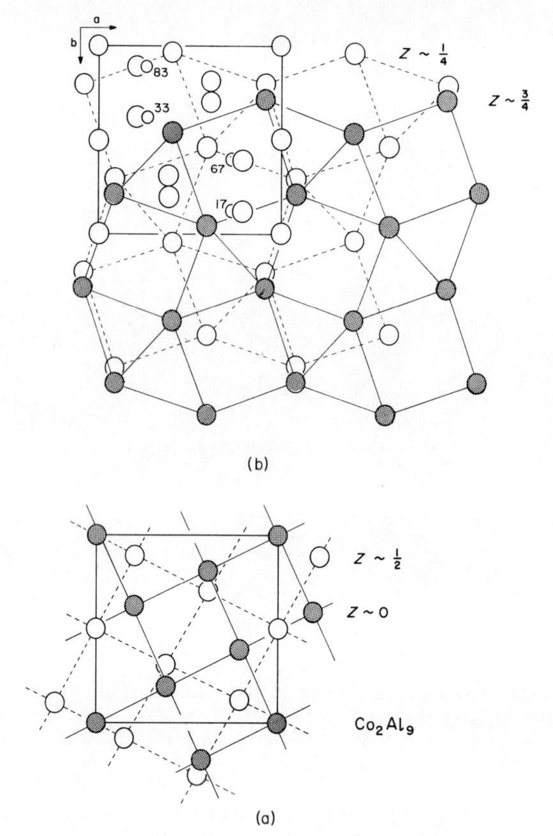

Figure 11-24. Arrangement of the atoms in the Co_2Al_9 (*mP*22) structure projected along [001]: Al, large circles.

Ge_4Sm_5 (oP36): Ge_4Sm_5 has cell dimensions $a = 7.75$, $b = 14.94$, $c = 7.84$ Å, $Z = 4$.[24] The structure is made up of layers parallel to the (010) plane. A 4^4 net of Ge(3) at $y = 0.04$ centers the squares of a slightly rumpled 3^2434 net of Sm(2) and (3) at $y \sim 0.11$. This is followed by a 3^2434 net of Ge(1) and (2) at $y = \frac{1}{4}$ which lies antisymmetrically over the Sm net at $y \sim 0.11$. A 4^4 net of Sm(3) also at $y = \frac{1}{4}$ centers the squares of the Ge net. At $y \sim 0.39$ a 3^2434 Sm net lies over that at $y \sim 0.11$ and at $y = 0.46$ another 4^4 Ge net lies over the centers of the squares of the net at $y \sim 0.39$. This sequence is repeated between $y = \frac{1}{2}$ and 1.0, with the nets in different positions so that the 4^4 nets of Ge(3) at $y = 0.545$ and 0.955 center the squares of the Ge(3) nets at $y = 0.04$ and 0.46 as shown in Figure 11-26. It should be noted that the "squares" of the 4^4 and

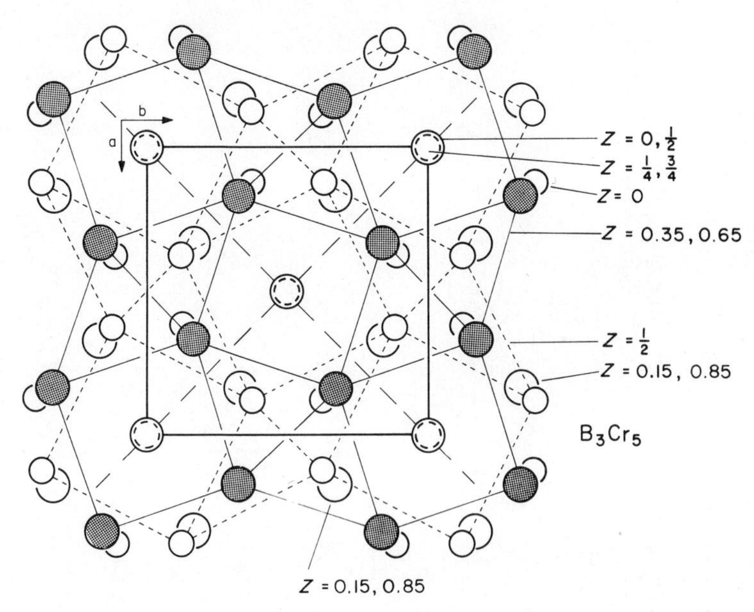

Figure 11-25. Atomic arrangement in the B_3Cr_5 (*tI*32) structure seen down [001]: Cr, large circles.

3^2434 nets are somewhat distorted. The Gd and Sm nets at both $y = \frac{1}{4}$ and $\frac{3}{4}$ together form $5^3 + 5^4$ nets. (See Figure 11-1, p. 606).

Al_7Cu_2Fe (*tP*40): The Al_7Cu_2Fe structure with cell dimensions $a = 6.336$, $c = 14.87$ Å, $Z = 4$,[25] can be considered as built up of layers of atoms stacked normal to the c axis, although some of the atoms are rather drastically displaced from the layers. The first layer is a $\frac{3}{4}.4^4$ net of Cu at $z = 0$ which has vacant Cu sites (Figure 11-27*a*). This is followed by a $\frac{3}{4}.4^4$ net of Al(3) at $z = 0.1$ with the Al(1) atoms at the cell corners displaced up from the net to $z = 0.134$ (Figure 11-27*b*). This net is distorted so that the Al atoms lie over the centers of the squares of the Cu net at $z = 0$. At $z = \frac{1}{4}$ there is a 3^2434 net of Al(2) with squares over the cell corners and center (Figure 11-27*c*). These are centered by Fe at $z = 0.2$ at the cell center and $z = 0.30$ at the cell corners. A $\frac{1}{4}.4^4$ net of Al(3) atoms follows at $z = 0.4$ with the Al(1) atoms that sit over the cell centers displaced out of the net to $z = 0.366$ (Figure 11-27*d*).

This sequence now repeats, starting with a $\frac{1}{4}.4^4$ Cu net with vacant sites at $z = \frac{1}{2}$ in different orientation (Figure 11-27*e*), and followed by a $\frac{1}{4}.4^4$ net of Al(3) and (1) at $z = 0.6$ and 0.634, a 3^2434 net of Al(2) at $z = \frac{3}{4}$ with orientation identical to that at $z = \frac{1}{4}$, and Fe at $z = 0.7$ (over cell

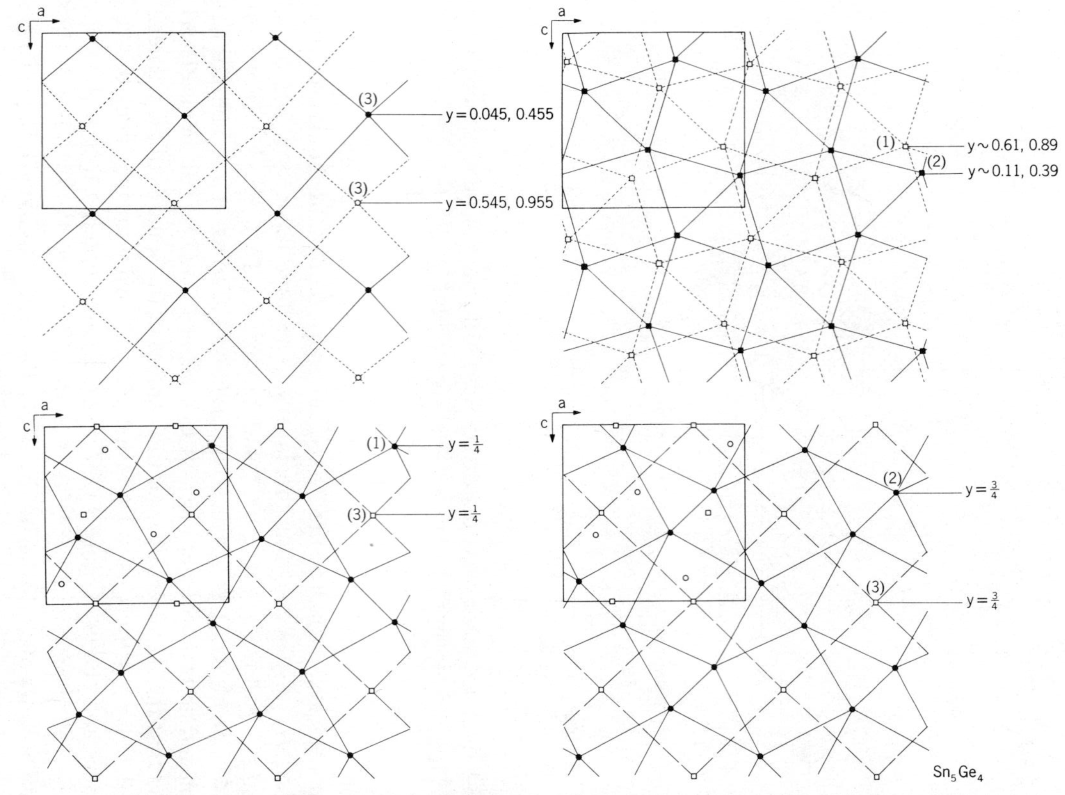

Figure 11-26. Arrangement of the atoms in nets in the Sm_5Ge_4 (*oP*36) structure viewed down [010]: Ge, circles; Sm, squares.

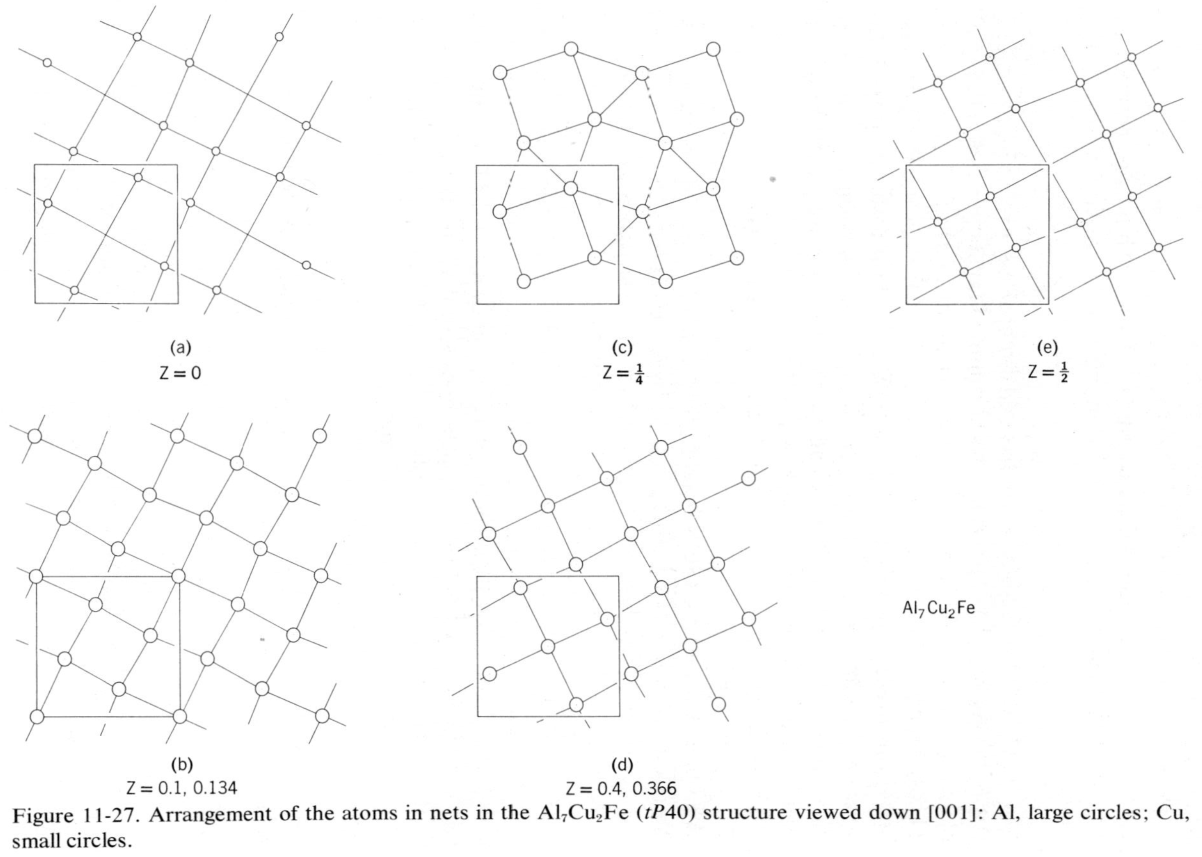

Figure 11-27. Arrangement of the atoms in nets in the Al_7Cu_2Fe (*tP*40) structure viewed down [001]: Al, large circles; Cu, small circles.

corners) and $z = 0.8$ (over cell center); finally a $\frac{3}{4}.4^4$ Al(1) and (3) net occurs at $z = 0.866$ and 0.9.

Fe has 9 neighbors all very similarly spaced. Cu has 10 neighbors at 2.51 to 2.72 Å, and one at 3.03 Å. Al(1), (2), and (3) have respectively 13, 15, and 12 neighbors (out to 3.35 Å) in the convex polyhedra surrounding them.

$CuPbAsS_3$ (*oP*24): Seligmannite has cell dimensions $a = 7.636$, $b = 8.081$, $c = 8.747$ Å, $Z = 4$.[26] The atoms form rumpled layers parallel to the (100) plane. At $x = 0$ and $\frac{1}{2}$ the S(1) and (2) atoms form $\frac{1}{2}.4^4$ nets (Figure 11-28) which are practically superimposed, and the As and Pb atoms also lie within these layers. At $x \sim \frac{1}{4}$ and $\frac{3}{4}$, S(3) and (4) form rumpled 3^2434 layers which are exactly superimposed, and the Cu atoms also lie within these layers (Figure 11-28). Some of the architectural features of this structure are reminiscent of these of Co_2Al_9.

These arrangements give Pb(1) four close S neighbors at 2.91, 2.96, 2×2.99 Å, and 2 at 3.23 Å. Pb(2) has 1 S at 2.89 Å, 2 at 2.93 Å, and 2 at 3.00 Å. Cu has 3 close S neighbors at 2.25–2.28 Å and one at 2.65 Å. As (1) has 1 S at 2.16 Å and 2 S at 2.36 Å. As(2) has 2 S at 2.37 Å and one at 2.77 Å. The closest S–S approaches in the structure are 3.31 Å.

Bournonite, $CuPbSbS_3$, has a similar structure.

Si III (H.P.) (*cI*16): The Si III phase which is formed at pressures of 160 to 200 kb, is retained on release of the pressure.[27] The atoms form very distorted tetrahedra with 3 atoms at 2.39 Å and one at 2.31 Å. They also form 3^2434 nets at fractional heights $x = 0.10$, 0.40, 0.60, and 0.90. The distorted squares at 0.10 and 0.60 lie over each other antisymmetrically and the cubic antiprisms so formed contain atoms at either 0.90 or 0.40. Similarly the distorted squares at 0.40 and 0.90 cover each other antisymmetrically, but their centers are displaced from those at 0.10 and 0.60, as indicated in Figure 11-29. The structure was determined from X-ray powder data.

CdSb (*oP*16): CdSb has cell dimensions $a = 6.471$, $b = 8.253$, $c = 8.526$ Å, $Z = 8$.[28] Sb has one Sb neighbor at 2.81 Å and two close Cd neighbors at 2.81 Å and two slightly further away at 2.91 and 3.08 Å (Figure 5-12, p. 225). Cd has 4 Sb neighbors; the closest Cd–Cd approach is 2.99 Å.

The plan of the structure on (100) resembles closely that of the Si III structure (*cI*16) and if the origin is transposed by $c/2$, then the coordinates only differ from those of Si III by $x = \pm 0.04$ or ± 0.06. Parallel to (100), the Cd and Sb atoms form rumpled distorted 3^2434 nets, or the Cd atoms alone form less distorted 3^2434 nets, that at $x \sim 0$ having squares over the cell corners and base center, and that at $x \sim \frac{1}{2}$ with squares lying over

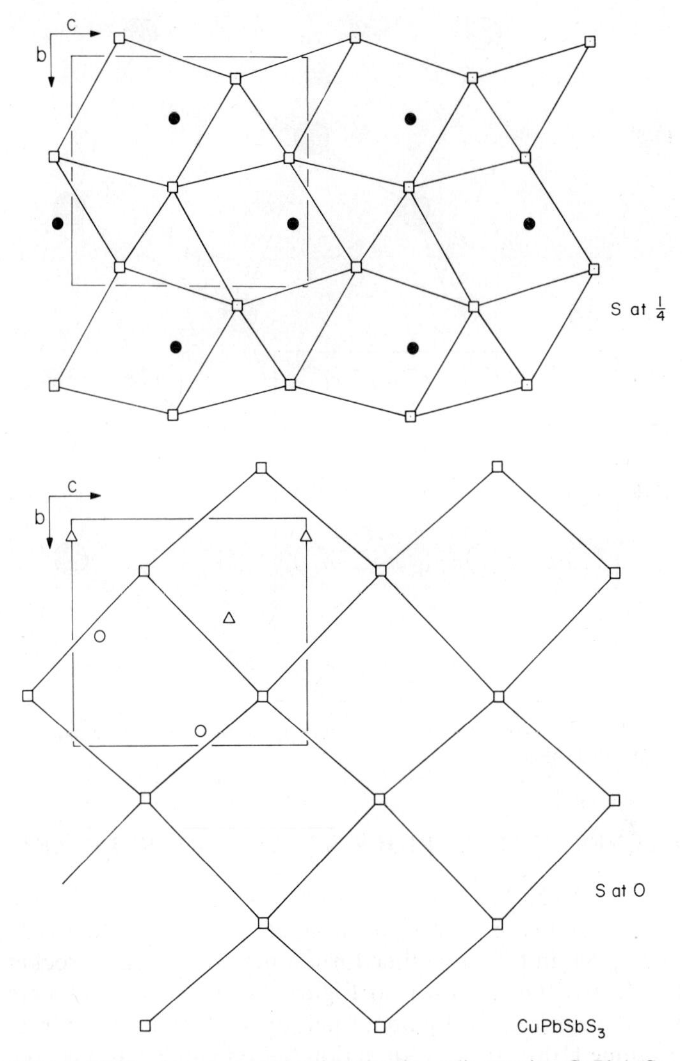

Figure 11-28. Atomic arrangement in the seligmannite $CuPbAsS_3$ ($oP24$) structure viewed down [100]: □, S; ●, Cu; ○, Sb; △, Pb.

the midpoints of the cell base edges. In CdSb $b \sim c$ but a is compressed considerably. The structure can nevertheless be regarded as a further distortion from the Si III type, which itself is a distortion from the diamond type.

The following structures also contain 3^2434 nets of atoms:

$$Mg_2Hg_5\ (tP14),\ PdS\ (tP16),\ ThB_4\ (tP20).$$

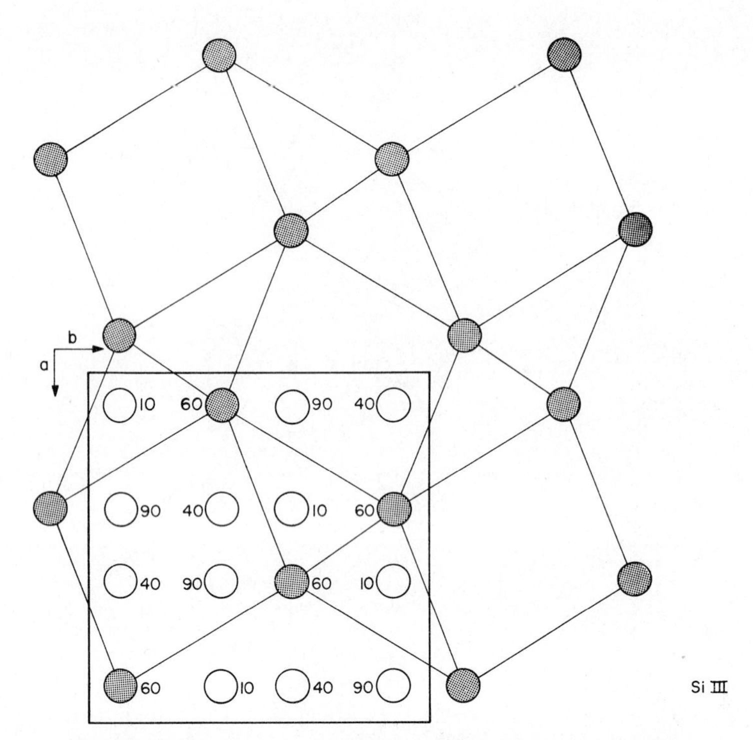

Fig. 11-29. Showing the 3^2434 nets in the Si III structure parallel to (001) planes.

STRUCTURES WITH VARIOUS SQUARE-TRIANGLE NETS

NiTh (*oP*16): NiTh has cell dimensions $a = 14.15$, $b = 4.31$, $c = 5.73$ Å, $Z = 8$.[29] The atoms are arranged in layers parallel to the (010) plane at $y = \frac{1}{4}$ and $\frac{3}{4}$. Ni and Th together form a net composed of rectangles and triangles of two types, shown in Figure 11-30. Successive nets are displaced in the a direction by about half of the length of the triangle edge that lies along [100]. In the a direction the Ni and Th atoms are arranged in rows alternately, and a Ni row at $y = \frac{1}{4}$ has a Th row above and below it at $y = \frac{3}{4}$ and $-\frac{1}{4}$.

This arrangement gives Ni(1) 6 Th neighbors at 2.68 Å to 3.11 Å; Ni(2) has 2 Ni at 2.67 Å and 7 Th at 2.91 to 3.11 Å. Th(1) has 6 Ni at 2.91 to 3.11 Å and Th(2) has 7 Ni at 2.68 to 3.11 Å. The closest Th–Th contacts are 3.54 Å.

UAl_4 (*oI*20): UAl_4 has cell dimensions $a = 4.41$, $b = 6.27$, $c = 13.71$ Å, $Z = 4$.[30] In planes parallel to (100) at $x = 0$ and $\frac{1}{2}$, the Al and U atoms

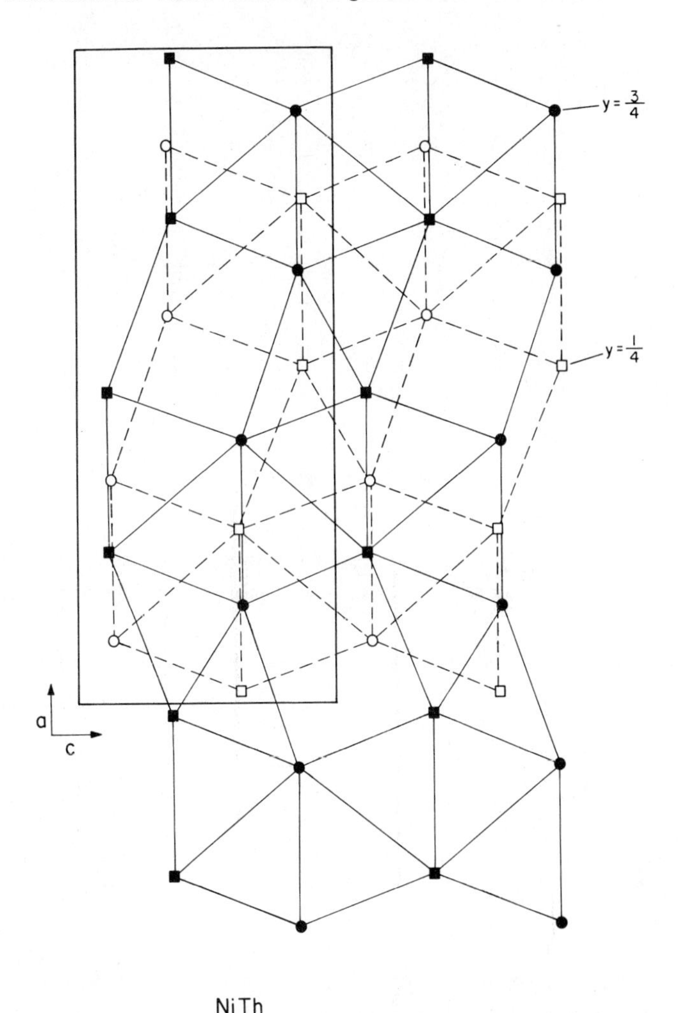

Figure 11-30. Arrangement of atoms in nets parallel to (010) in the NiTh (*oP*16) structure: circles, Th.

form slightly distorted $3^3 4^2 + 4^4$ (4:1) triangle-square nets which are stacked one above the other in two positions, so that the nodes of one net approximately center the triangles and squares of the nets above and below (Figure 11-31). The nets are stacked so that the U atoms avoid each other, although not by the greatest possible interlayer distances. U is surrounded by 13 Al atoms at 3.04 to 3.30 Å. The Al atoms have 10 to 13 neighbors at distances less than 3.30 Å; the closest Al–Al approaches are Al(2)–Al(3) 2.56 Å, Al(3)–Al(3) 2.72 Å and Al(1)–Al(3) 2.79 Å.

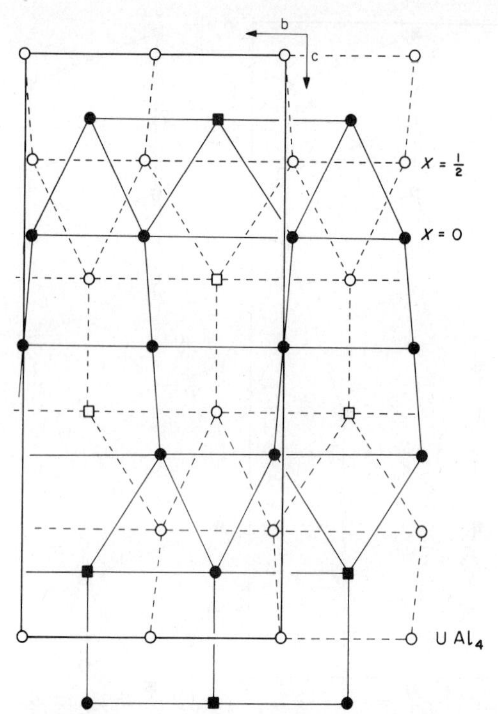

Figure 11-31. Arrangement of atoms in nets parallel to the (100) plane in the UAl_4 ($oI20$) structure.

MnU_6 ($tI28$): The MnU_6 structure with cell dimensions $a = 10.29$, $c = 5.24$ Å, $Z = 4$,[31] is made up of $3^2434 + 3^34^2$ (2 : 1) nets of U(2) and U(1) atoms at $z = 0$ and $\frac{1}{2}$. U(2) is at the 3^2434 corners and U(1) at the 3^34^2 corners, and the net at $z = \frac{1}{2}$ is antisymmetrically disposed relative to that at $z = 0$ as shown in Figure 11-32. Mn atoms at $z = \frac{1}{4}$ and $\frac{3}{4}$ center the anticubes located over the cell corners and cell center. As the c edge of the cell is only 5.24 Å, Mn has two close Mn neighbors (2.62 Å) as well as 8 U(2) neighbors (2.77 Å). U(2) has 2 Mn neighbors, 1 U(2) very close at 2.68 Å, 1 U(1) at 2.82 Å and 9 other U atoms at 3.09 and 3.45 Å. U(1) has one close U(1) neighbor at 2.74 Å, 2 U(2) at 2.82 Å, and 10 other U at 3.26–3.36 Å.

As_2Te_3 ($mC20$): In the As_2Te_3 structure, $a = 14.34$, $b = 4.006$, $c = 9.873$ Å, $\beta = 95.0°$, $Z = 4$,[32] the atoms are arranged in layers parallel to the ac plane at $y = 0$ and $\frac{1}{2}$. Parallel to the $(20\bar{1})$ plane the Te atoms form the rectangle-triangle arrangement shown in Figure 11-33 with the As atoms centering the rectangles. These layers are stacked in a tenfold sequence

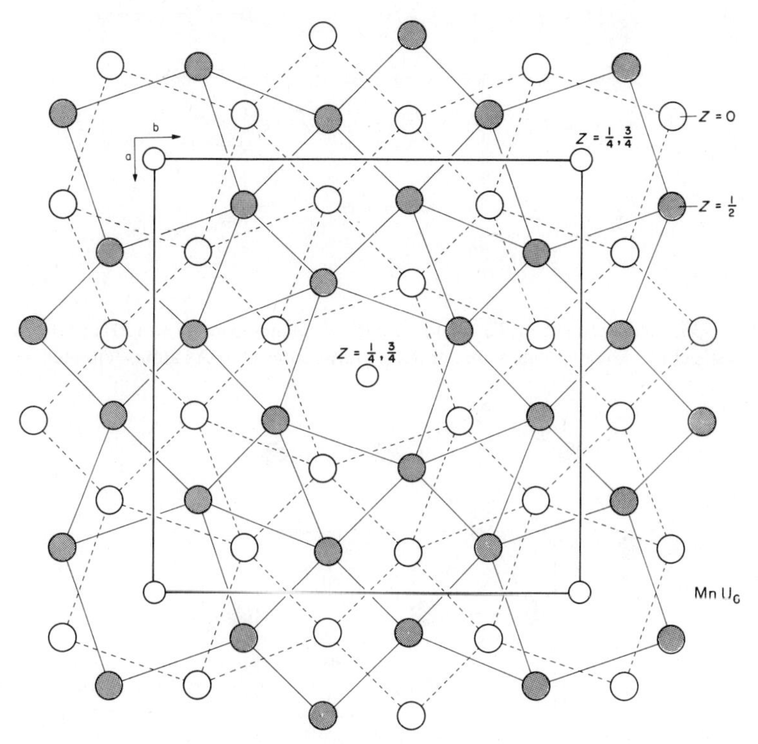

Figure 11-32. View of the MnU_6 (*tI*28) structure down [001] showing the arrangement of the atoms in nets.

with the atoms of one layer lying over the atoms of layers above and below, except when this is prevented by the triangles which introduce a different spacing between the rows. The As(2) atoms are surrounded octahedrally by Te (2.76 to 2.93 Å); the octahedra joined by edges form columns running along the *b* direction. As(1) has 3 Te neighbors (2.68 and 2.77 Å) located on a plane lying well to one side of it. Each Te has three close As neighbors.

Ge_2Pt_3 (*oC*20): With cell dimensions $a = 6.846$, $b = 12.236$, $c = 7.544$ Å, $Z = 4$,[33] the Ge_2Pt_3 structure gives the impression of being very open. The structure contains bands of Pt and Ge(2) atoms cut from a 3^2434 net and running along a line of squares (Figure 11-34). The band at $x = \frac{1}{2}$ runs through the middle of the cell parallel to the (100) plane and the rectangular prisms formed between two such layers are centered by a line of Ge(1) atoms at $x = 0$. Similar bands at the ends of the cell at $x = 0$ have the rectangular prisms formed by them centered by a line of Ge(1) atoms at $x = \frac{1}{2}$.

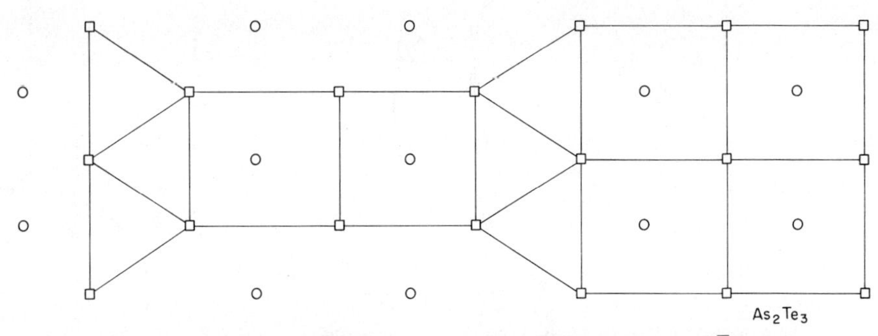

Figure 11-33. Arrangement of Te atoms in nets on planes parallel to $(20\bar{1})$ in the As_2Te_3 ($mC20$) structure. The rectangles are approximately centered by As atoms (circles).

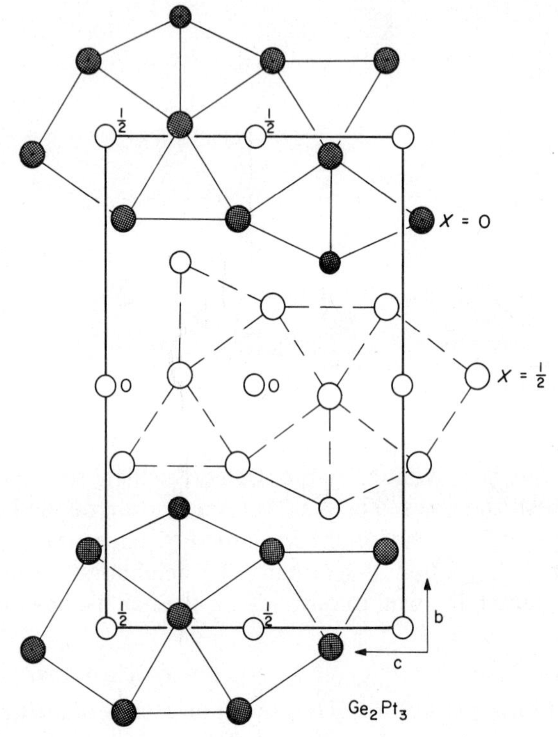

Figure 11-34. Atomic arrangement in the Ge_2Pt_3 ($oC20$) structure viewed down [100]: Pt, large circles.

Pd_5B_2 ($mC28$): In the Pd_5B_2 (Mn_5C_2) structure with cell dimensions $a = 12.79$, $b = 4.955$, $c = 5.472$ Å, $\beta = 97.03°$, $Z = 4$,[34] the Pd atoms are arranged in somewhat rumpled $3^6 + 3^34^2 + 4^4$ (2:2:1) square-triangle layers parallel to the (010) plane at $y \sim 0$ and $\frac{1}{2}$ as shown in Figure

11-35. Boron atoms are in planes at $y \sim \frac{1}{4}$ and $\frac{3}{4}$ such that they center skewed triangular prisms formed by Pd atoms. Within the prisms Pd–B distances are 2.18 and 2.19 Å; there are three other Pd neighbors out through the centers of the rectangular faces of the prisms at distances 2.60, 2.80, and 3.06 Å. There are no B–B contacts and the closest Pd–Pd

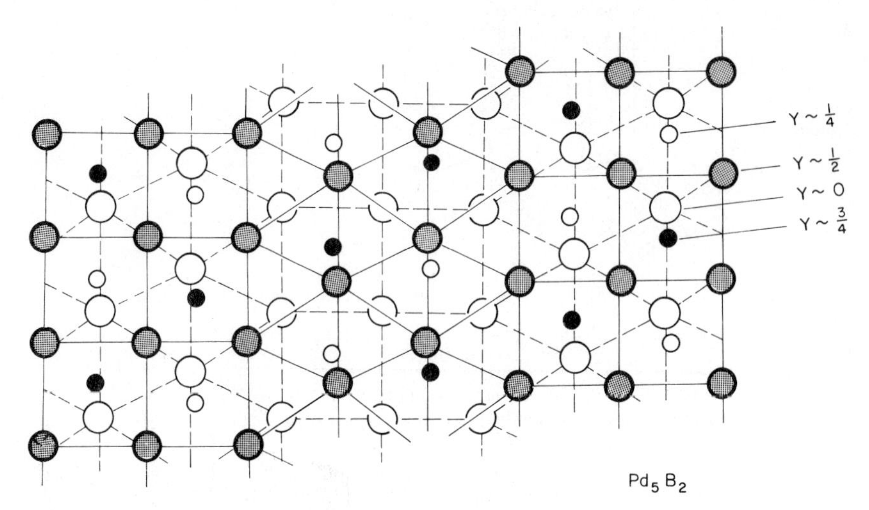

Figure 11-35. View of the Pd_5B_2 or Mn_5C_2 ($mC28$) structure showing the arrangement of the Pd atoms in nets parallel to the (010) plane. (Stenberg[34].)

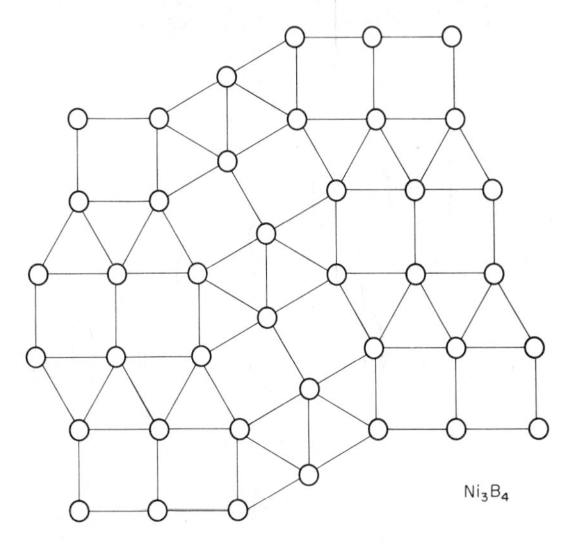

Figure 11-36. The arrangement of Ni atoms on nets parallel to the (202) plane in the Ni_4B_3 ($mC28$) structure.

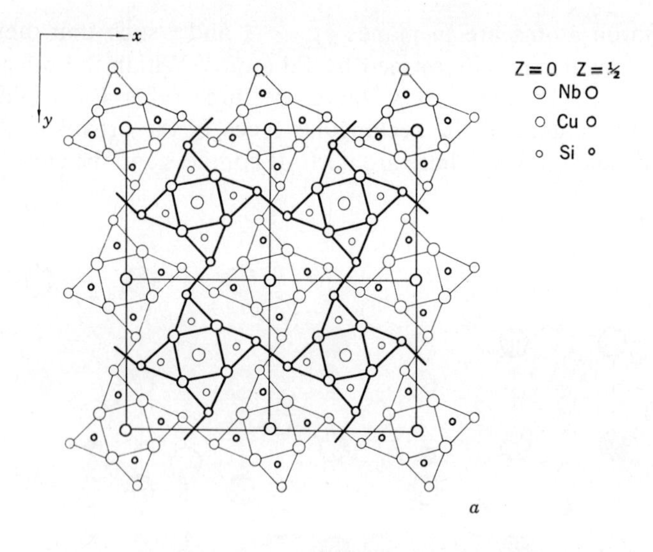

Figure 11-37. Atomic arrangement in the $Nb_5Cu_4Si_4$ (*tI*26) structure viewed along [001]. (Ganglberger[36].)

approaches are 2.70 Å. Pd(1) has 4 close B neighbors and Pd(2) and (3) have two each.

m-Ni_4B_3 (*mC*28): *m*-Ni_4B_3 has cell dimensions $a = 6.428$, $b = 4.880$, $c = 7.819$ Å, $\beta = 103.32°$, $Z = 4$.[35] The Ni atoms form rumpled $3^2434 + 3^24^2$ (1 : 1) square-triangle layers parallel to the (202) plane (Figure 11-36). The B atoms run in branched chains through the metal atom nets such that B(1) is surrounded by 2 B(2) at 1.85 Å, and B(2) has one B(1) and one B(2) (1.92 Å) neighbors. B(1) has 8 Ni at 2.10 to 2.345 Å and B(2) has 7 Ni at 2.04 to 2.17 Å. The closest Ni–Ni approaches are 2.51 Å.

$Nb_5Cu_4Si_4$ (*tI*26): The atoms in $Nb_5Cu_4Si_4$ ($a = 10.191$, $c = 3.600$ Å, $Z = 2$) form layers at $z = 0$ and $\frac{1}{2}$ as shown in Figure 11-37.[36] Nb(2) centers cubes of Nb(1) (2.99 Å), and Si centers triangular prisms of 4 Nb (2.60 and 2.70 Å) and 2 Cu (2.34 Å). In addition Si has one Nb (2.64 Å) and one Cu (2.37 Å) out through rectangular prism faces.

REFERENCES

1. J. A. A. Ketelaar, W. H. Hart, M. Moerel, and D. Polder, 1939, *Z. Kristallogr.*, **A101**, 396.
2. A. J. Bradley and P. Jones, 1933, *J. Inst. Met.*, **51**, 131.

3. E. I. Gladyševskij and Ju. B. Kuz'ma, 1965, *Ž. Strukt. Khim.*, **6**, 70 [*J. Struct. Chem.*, **6**, 60].
4. M. M. Stasova and B. K. Vajnštejn, 1958, *Kristallografija*, **3**, 141 [*Soviet Physics–Crystallography*, **3**, 140].
5. K. Schubert, H. Breimer, and R. Gohle, 1959, *Z. Metallk.*, **50**, 146.
6. W. H. Zachariasen, 1963, *Acta Cryst.*, **16**, 1253.
7. S. Rundqvist, 1961, *Acta Chem. Scand.*, **15**, 451.
8. K. Schubert and H. Pfisterer, 1950, *Z. Metallk.*, **41**, 433.
9. E. Hellner, 1956, *Z. Kristallogr.*, **107**, 99.
10. K. Schubert, H. L. Lukas, H. G. Meissner, and S. Bhan, 1959, *Z. Metallk.*, **50**, 534.
11. R. Gohle and K. Schubert, 1964, *Z. Metallk.*, **55**, 503.
12. H. Iwasaki, 1962, *J. Phys. Soc. Japan*, **17**, 1621.
13. S. S. Lu and Liang Chung-Kwei, 1965, *Chin. J. Phys.*, **21**, 1079.
14. F. H. Herbstein and J. Smuts, 1964, *Acta Cryst.*, **17**, 1331.
15. B. Sellberg, 1966, *Acta Chem. Scand.*, **20**, 2179.
16. U. Rösler and K. Schubert, 1951, *Z. Metallk.*, **42**, 395.
17. K. Selte and A. Kjekshus, 1964, *Acta Chem. Scand.*, **18**, 690.
18. K. Schubert and U. Rösler, 1950, *Z. Metallk.*, **41**, 298.
19. W. H. Zachariasen, 1948, *Acta Cryst.*, **1**, 265; 1949, *Ibid.*, **2**, 94.
20. N. C. Baenziger and J. J. Hegenbarth, 1964, *Acta Cryst.*, **17**, 620.
21. A. M. B. Douglas, 1950, *Acta Cryst.*, **3**, 19.
22. F. Bertaut and P. Blum, 1953, *C. R. Acad. Sci. Paris*, **236**, 1055.
23. D. E. Sands, D. H. Wood, and W. J. Ramsey, 1964, *Acta Cryst.*, **17**, 986.
24. G. S. Smith, Q. Johnson, and A. G. Tharp, 1966, *Acta Cryst.*, **22**, 269.
25. A Westgren in G. Phragmén, 1950, *J. Inst. Met.*, **77**, 489; H. Wiehr, 1940, *Aluminum Archiv.*, **31**, 14 pp.
26. E. Hellner and G. Leineweber, 1956, *Z. Kristallogr.*, **107**, 150; G. Leineweber, *Ibid.*, **108**, 161.
27. R. H. Wentorf and J. S. Kasper, 1963, *Science*, **139**, 338; J. S. Kasper and S. M. Richards, 1964, *Acta Cryst.*, **17**, 752.
28. K. E. Almin, 1948, *Acta Chem. Scand.*, **2**, 400.
29. J. V. Florio, N. C. Baenziger, and R. E. Rundle, 1956, *Acta Cryst.*, **9**, 367.
30. B. S. Borie, 1951, *J. Metals*, **3**, 800.
31. N. C. Baenziger, R. E. Rundle, A. I. Snow, and A. S. Wilson, 1950, *Acta Cryst.*, **3**, 34.
32. G. J. Carron, 1963, *Acta Cryst.*, **16**, 338.
33. S. Bhan and K. Schubert, 1960, *Z. Metallk.*, **51**, 327.
34. E. Stenberg, 1961, *Acta Chem. Scand.*, **15**, 861.
35. S. Rundqvist, 1959, *Acta Chem. Scand.*, **13**, 1193; S. Rundqvist and S. Pramatus, 1967, *Ibid.*, **21**, 191.
36. E. Ganglberger, 1968, *Mh. Chem.*, **99**, 549.

12

Structures Generated by Alternate Stacking of Triangular and Kagomé Nets

STRUCTURES WITH TRIANGULAR AND KAGOMÉ NETS OF ATOMS

MN_5: M and N on a 3^6 Net (Subdivided with M on 3^6, N on 6^3). N on Interleaved 3636 Kagomé Net

The $CaCu_5$ Structural Arrangement and Derived Structures. A family of structures with large coordination polyhedra is obtained by alternate stacking of 3^6 and 3636 kagomé nets of atoms with the same net period and on the same stacking sites. If the 3^6 net is subdivided into 6^3 and 3^6 nets of larger size, the hexagons of the 6^3 subnet lie antisymmetrically over the hexagons of the kagomé net and the atoms on the 3^6 subnet lie over the centers of the hexagons of both nets. The atoms of the 3^6 subnet are therefore surrounded by 18 atoms in hexagons stacked antisymmetrically and also by two polar atoms of the same type. The structure which has been described is the $CaCu_5$ type which is shown together with the CN 20 polyhedron in Figure 12-1.

Other structures can be derived from this arrangement by replacing various proportions of the Ca atoms at the centers of two fused hexagonal antiprisms, by two atoms with their common axis along the axis of the antiprisms, so that each atom centers one of the hexagonal antiprisms. Known derivatives and the fraction of Ca atoms replaced in $CaCu_5$ type pseudocells are listed in Table 12-1.

Figure 12-2 taken in part from Florio *et al.*[1] indicates the relationship

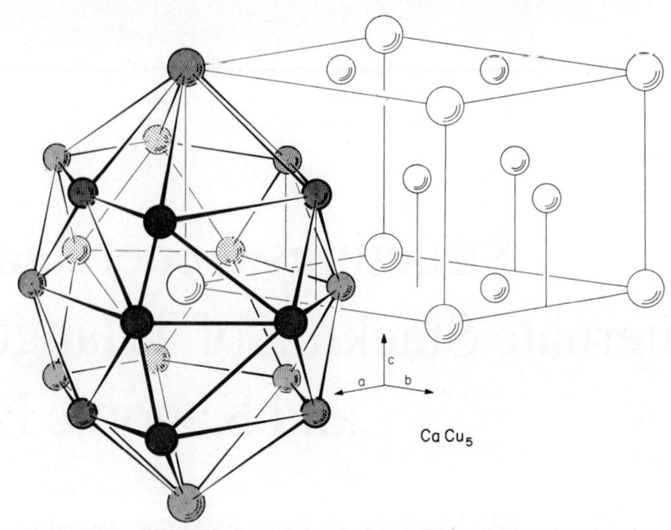

Figure 12-1. Pictorial view of the $CaCu_5$ (*hP*6) structure showing the coordination polyhedron surrounding Ca (large circles).

Table 12-1

Structure Type	Fraction of Large (Ca) Atoms Replaced in $CaCu_5$ Cell in Deriving Structure
$ThMn_{12}$	$\frac{1}{2}$
Th_2Ni_{17}	$\frac{1}{3}$
Th_2Zn_{17}	$\frac{1}{3}$
Pu_3Zn_{22}	$\frac{1}{4}$

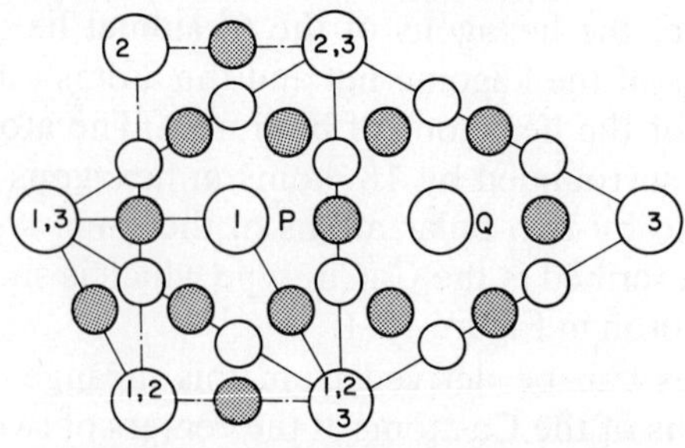

Figure 12-2. Diagram showing the relationship of the unit cells of the $CaCu_5$ (1), $ThMn_{12}$ (2), Th_2Zn_{17} (3), and Th_2Ni_{17} (3) structures. The Th or Ca atoms are large circles. Atoms represented by open circles are at one level and those by filled circles are at another level.

between the $CaCu_5$, Th_2Ni_{17}, Th_2Zn_{17}, and $ThMn_{12}$ structure types. Derived structures are also obtained by combining layer stacking sequences of the $CaCu_5$ $[\underline{A}a]\alpha$ type with those of the Laves phase type $\gamma(\underline{C}A\underline{B})$, or by removing certain of the atoms from the $CaCu_5$ arrangement as in the $PuNi_4$ structure.

Structures of the combined $CaCu_5$ and Laves type can also be regarded as derived directly from a stack of $CaCu_5$ (MN_5) type cells by replacing half of the N atoms in the $[\underline{A}a]$ type layers by M atoms, thus giving $[\underline{A}B\underline{C}]$ (M underlined) type layers which are then spread out slightly in the c direction to give $(\underline{A}B\underline{C})$. When this process is performed regularly every two $CaCu_5$ type cells, the $NbBe_3$ family of structures is obtained and when it is done every three cells along c the Er_2Co_7 family is obtained. Furthermore, the replacement of the Ca atoms by a pair of atoms on the same stacking sites converts the $CaCu_5$ type layers, $[Aa]\alpha$ to Al_3Zr_4 ($hP7$) type layers, $AaA\alpha$. The μ phase structure is made up of a combination of $AaA\alpha$ (Al_3Zr_4 type) and $\alpha(ABC)$ (Laves type) groups of layers (see p. 664).

$CaCu_5$ ($hP6$): The atomic arrangement in the $CaCu_5$ structure with unit cell dimensions $a = 5.092$, $c = 4.086$ Å, $Z = 1$,[2] is described above. The large Ca atoms center the hexagons of both the 6^3 Cu net at $z = 0$ and the 3636 kagomé nets of Cu at $z = \pm\frac{1}{2}$. Ca is thus surrounded by a coordination polyhedron of 18 Cu (6×2.94 Å, 12×3.26 Å) and 2 Ca (4.09 Å) atoms which has 24 three-sided faces and 6 four-sided faces (Figure 12-1). The Cu atoms each have 12 neighbors, but the arrangements are not icosahedral. In the notation given on p. 4, the structural arrangement is described as $[Aa]\alpha$.

Structures Derived from $CaCu_5$ by Replacing Some Ca by M_2

$ThMn_{12}$ ($tI26$): The $ThMn_{12}$ structure ($a = 8.74$, $c = 4.95$ Å, $Z = 2$)[3] can be derived directly from the $CaCu_5$ type by substitution of Mn pairs for half of the Ca (Th) atoms. This is seen in Figure 12-2, where plans of the $CaCu_5$ structure down [001] and of the $ThMn_{12}$ structure on the bc plane are superimposed. Ca (Th) atoms in alternate layers (i.e., at the $ThMn_{12}$ cell base center and halfway up the a edges of the cell) are replaced by pairs of Mn(2) atoms whose ligand lies along the direction of the a cell edges. $ThMn_{12}$ also belongs to an interesting group of structures that contain the characteristic layer net of atoms made up of squares, quadrilaterals, and triangles which is shown in Figure 12-3. Mn(2) and (3) atoms at $z = 0$ form this net with its small square lying over the cell base center; at $z = \frac{1}{2}$ they again form the net, but with the large square lying over the cell base center and the small squares lying over the cell corners

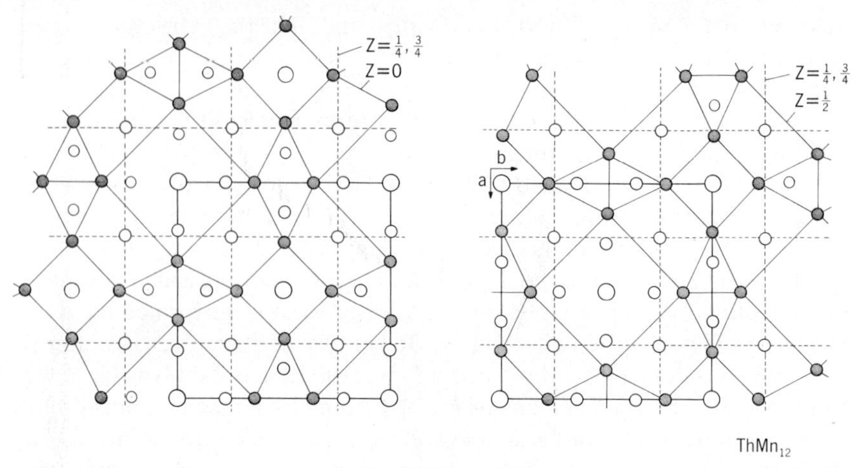

Figure 12-3. Atomic arrangement in the $ThMn_{12}$ ($tI26$) structure viewed down [001]: Th, large circles.

(Figure 12-3). Th atoms at $z = 0$ center the large squares at the cell corners and Th at $z = \frac{1}{2}$ centers the large square at the body center. The Mn(1) atoms at $z = \frac{1}{4}$ and $\frac{3}{4}$ form a 1.4^4 net; they are so placed that they center the polyhedrons formed by the quadrilaterals at $z = 0$, $\frac{1}{2}$, and 1. This arrangement surrounds Th with 20 Mn atoms at distances of 3.155 Å (12) and 3.33 Å (8), the polyhedron having four six-connected vertices; the remainder are five-connected. Mn(1) has 10 Mn and 2 Th neighbors (two pairs of Mn(1) atoms are very close at 2.475 Å). Mn(2) has 13 Mn and 1 Th neighbors (one Mn(2)–Mn(2) is only 2.43 Å) and Mn(3) has 10 Mn and 2 Th neighbors.

The near-neighbor diagram (p. 53) for phases with the $ThMn_{12}$ structure is shown in Figure 12-4.

$Th_2Zn_{15\text{-}17}$ ($hR19$): The dimensions of the equivalent hexagonal cell of the Th_2Zn_{17} structure are $a = 9.03$, $c = 13.20$ Å, $Z = 3$.[4] The structure can be derived from the $CaCu_5$ type by replacing one-third of the Ca (Th) atoms by pairs of Zn atoms whose ligand lies along the [001] direction as shown in Figure 12-2, p. 644. The base of the hexagonal Th_2Zn_{17} cell has three times the area of the base of the $CaCu_5$ cell and its height corresponds to a stack of three $CaCu_5$ cells, so that the overall volume corresponds to that of 9 $CaCu_5$ cells. The atoms are arranged in planes parallel to (001). In terms of the notation of p. 4, the structural arrangement can be described (Th underlined) as: $a'\underline{B})A(\alpha\beta\gamma)C(\underline{B}a'\underline{A})C(\gamma\alpha\beta)$ $B(\underline{A}a'\underline{C})B(\beta\gamma\alpha)A(\underline{C}$. The 6^3 and kagomé nets are planar, but there are slight displacements of the atoms from the exact net sites in the planes.

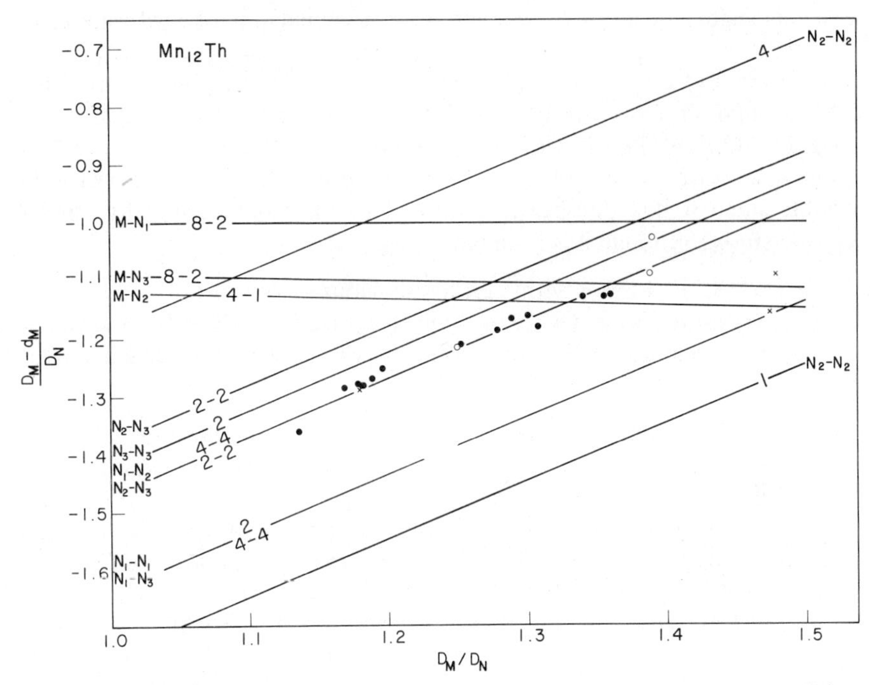

Figure 12-4. Near-neighbor diagram for phases with the $ThMn_{12}$ (MN_{12}) structure. Calculated for $c/a = 0.58$, $x_{Mn(2)} = 0.361$, $x_{Mn(3)} = 0.277$. Numbers on lines indicate the number of neighbors for contacts M–N, N–N, etc.: ○, Mn compounds taking Mn valency 2; ×, Mn compounds taking Mn valency 6.

The Th layers occur together in pairs less than 0.3 Å apart and they lie just above or below 2/3 of the hexagons in the 6^3 nets of Zn. Another feature of the structural arrangement is the close spacing (about 1.13 Å apart) of three successive kagomé layers of Zn atoms which combined give a rumpled kagomé layer with smaller net period and stacking positions of the α' type. Combining the symbols for the closely spaced layers and ignoring the rumpling of the layers that results, the structural description becomes: $[a'\underline{a}]A\alpha'C[a'\underline{c}]C\alpha'B[a'\underline{b}]B\alpha'A$.

Th is surrounded by 20 atoms. The Zn atoms are surrounded by 12, 13, or 14 atoms; Zn(2) is surrounded by a distorted icosahedron of 10 Zn and 2 Th atoms. In Pr_2Fe_{17} with this structure, the distances of 19 Fe surrounding Pr lie between 3.08 and 3.32 Å. The single Pr–Pr distance is 3.90 Å.

One of the most interesting features of phases with the Th_2Zn_{17} structure is the constancy of the axial ratio, c/a, which has a value of 1.46 ± 0.01 in more than 40 phases known to have the structure. This value exceeds slightly $\sqrt{2} = 1.41$, expected for close packing of hard spheres in the structure.

Johnson and co-workers[5] have shown that the structures reported for Th_2Fe_{17} and Th_2Co_{17}[6] are incorrect; they have the Th_2Zn_{17} type of structure. In addition, the structure reported for U_2Zn_{17}[4] is incorrect; it results from a mixture of two phases. The structure of Th_2Ni_{17} is perhaps in doubt, since it has not been confirmed at this composition by single-crystal study, although it is found in $CeMg_{10.3}$.

$PrFe_7$ (hR18.7): $PrFe_7$ with cell dimensions $a = 8.582$, $c = 12.462$ Å, $Z = 56$ atoms, is a defect Th_2Zn_{17} type structure.[7] The defects are ordered on one Pr site and one Fe site. $CeFe_7$, $NdFe_7$, and $SmFe_7$ have a similar structure.

Th_2Ni_{17} (hP38): The Th_2Ni_{17} structure with cell dimensions $a = 8.37$, $c = 8.14$ Å, $Z = 2$,[1] is a layer structure which can be described as α'B[a'b]Bα'C[a'c]C in the notation of p. 4. At $z = 0$ and $\frac{1}{2}$, Ni(2) and (4) form a kagomé net. At $z = \frac{1}{4}$ the Th(1) and (2) and Ni(3) atoms together occupy a 3^6 net; the Ni atoms form a 6^3 subnet and the Th atoms form a larger 3^6 subnet with the Th sites at position B vacant (Figure 12-5), similarly at $z = \frac{3}{4}$ with the Th sites at position C vacant.

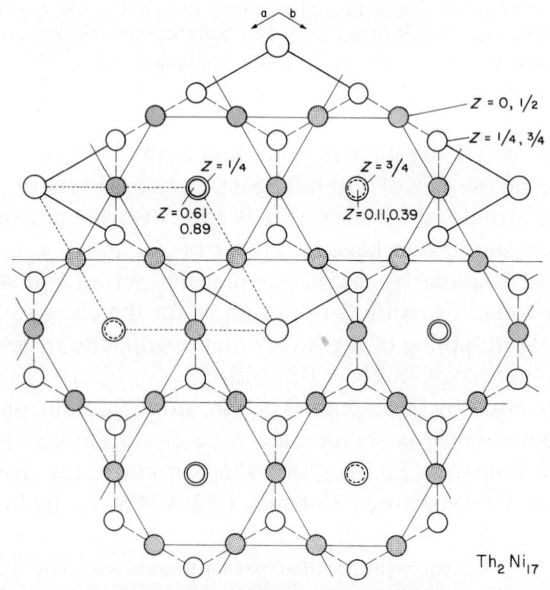

Fig. 12-5. View of the structure of Th_2Ni_{17} (*hP*38) looking down [001]: Th, large circles.

The Th atoms lie over the hexagons of the kagomé nets and two Ni(1) atoms at $z = \pm 0.11$ and $\frac{1}{2} \pm 0.11$ lie between the hexagons where the vacant Th sites are.

The structure can be derived directly from the $CaCu_5$ type as indicated in Figure 12-2, p. 644, where the two hexagonal cells are superimposed in plan down [001]. The Th_2Ni_{17} structure is obtained by replacing in alternate layers, the Ca (Th) atoms at P and Q by pairs of Ni(1) atoms whose ligand is oriented along [001]. Thus one-third of the Ca atoms in the $CaCu_5$ array are replaced by smaller atoms in deriving the Th_2Ni_{17} type structure. The base of the Th_2Ni_{17} cell has three times the area of the base of the $CaCu_5$ cell and its height corresponds to stacking two $CaCu_5$ cells along [001]. The volume of the Th_2Ni_{17} cell is therefore equivalent to that of 6 $CaCu_5$ cells.

Both Th(1) and (2) have CN 20, the former with 18 Ni at 2.79 and 3.16 Å and 2 Th at 4.07 Å, the latter with 20 Ni at 2.79 to 3.16 Å. Ni(1) at the center of a hexagonal antiprism with polar atoms has CN 14 (12 Ni at 2.58 and 3.015 Å, one close Ni(1) at 2.28 Å, and one Th at 2.93 Å). Ni(2) has icosahedral coordination by 10 Ni and 2 Th. Ni(4) also has CN 12, but not icosahedral, and Ni(3) has CN 13. The structure contains numerous short Ni–Ni distances at 2.42 Å.

The exact stoichiometry of 8.5:1 can be varied in phases with this structure and, for example, the structure of a $CeMg_{10.3}$ phase has been described[8] in which essentially Ce and Mg are disordered on one of the Th atom sites.

Pu_3Zn_{22} (tI100): Pu_3Zn_{22} which has a unit cell of dimensions, $a = 8.85$, $c = 21.18$ Å, $Z = 4$,[9] belongs to the class of structures which are derived from the $CaCu_5$ type by replacing the larger Ca (Pu) atoms by pairs of Zn atoms with ligand along [001] of the $CaCu_5$ structure. In the case of Pu_3Zn_{22} one-fourth of the Ca (Pu) atoms are effectively replaced by pairs of Zn, but the arrangement of $CaCu_5$ subcells in the structure is somewhat confused and tubes of fused hexagons containing Pu(1) run alternately along the a and b directions on proceeding up the c axis of the cell (Figure 12-6). Pu(2) lie in spaces between these tubes of hexagons. If the two polar Pu(1) atoms neighboring Pu(1) at a distance of 4.43 Å are included, both Pu(1) and Pu(2) have CN 20. Both polyhedra have 12 atoms with surface coordination 5, and 8 with surface coordination 6; both polyhedra have 36 triangular faces. Pu(2) is surrounded by Zn atoms only, at distances from 3.125 to 3.45 Å. Zn(1) and Zn(2) are surrounded by distorted icosahedra of Zn and Pu atoms; Zn(3), (4), and (5) each have CN 13.

La_3Zn_{22}, Ce_3Zn_{22}, and Pr_3Zn_{22} have a similar structure.

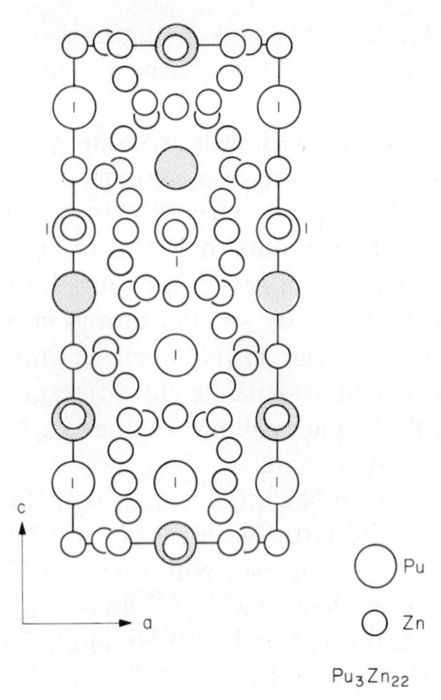

Figure 12-6. Atomic arrangement in the Pu_3Zn_{22} (*tI*100) structure viewed along [010]. Pu (2) atoms are shaded. (Johnson and co-workers [9].)

Defect $CaCu_5$ Type Structures

$PuNi_4$ (*mC*30): The $PuNi_4$ structure with cell dimensions $a = 4.87$, $b = 8.46$, $c = 10.27$ Å, $\beta = 100°$, $Z = 6$,[10] is related to that of $PuNi_5$ ($CaCu_5$, *hP*6 type) from which it is derived by removing the Ni atoms at $z = \frac{1}{2}$ from every third unit cell of $PuNi_5$, adjusting the positions of the remaining atoms slightly and changing the angle β to 100° as indicated in Figure 12-7. Pu(1) retains coordination 20, essentially identical with that in $PuNi_5$, with 6 Ni(1) at 2.815 Å, 4 Ni(2) at 3.15 Å, 8 Ni(3) at 3.165 Å, and 2 Pu at 3.59 Å. Pu(2) has 16 neighbors (14 Ni at 2.84 to 2.91 Å) and 2 Pu (3.41, 3.59 Å). The Ni atoms each have 12 close neighbors, the closest Ni–Ni approaches being 2.43 Å.

Structures Related to the $CaCu_5$ Type

$BaZn_5$ (*oC*24): $BaZn_5$ has cell dimensions $a = 10.78$, $b = 8.44$, $c = 5.32$ Å, $Z = 4$.[11] A feature of the structure is lines of Zn atoms at $z = 0$ and $\frac{1}{2}$ which mark out hexagons, emphasizing the similarity to the $SrZn_5$ structure (Figure 12-8) and a somewhat more distant relationship to the

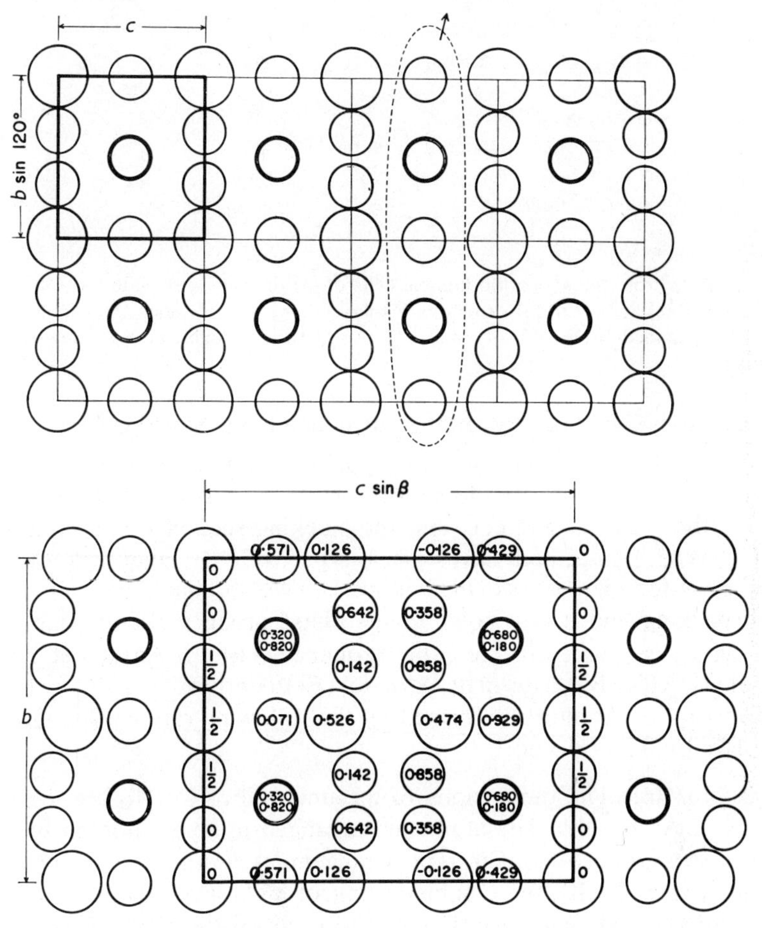

Figure 12-7. (Lower) Diagram of the $PuNi_4$ ($mC30$) structure projected down [100]. (Upper) Indicating how the $PuNi_4$ structure is derived from the $CaCu_5$ type of $PuNi_5$ by removing a row of Ni atoms from every third unit cell of $PuNi_5$. (Cromer and Larson [10].)

$CaCu_5$ structure. Ba is surrounded by 19 Zn at 3.40 to 3.89 Å and 2 Ba at 3.85 Å.

$SrZn_5$ ($oP24$): $SrZn_5$ has cell dimensions $a = 13.15$, $b = 5.32$, $c = 6.72$ Å, $Z = 4$.[11] Zn atoms at $y = 0$ and $\frac{1}{2}$ form nets of hexagons which can be compared to those formed in $BaZn_5$ (Figure 12-8). Sr is surrounded by 19 Zn at 3.28 to 3.92 Å and 2 Sr at 4.02 Å. Zn have 11 or 12 Zn and Sr neighbors.

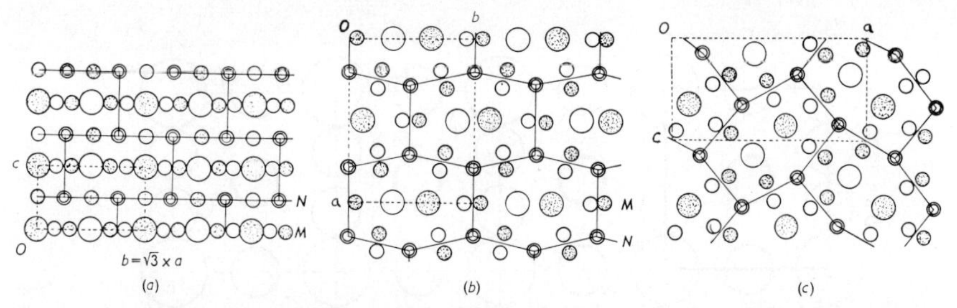

Figure 12-8. Projections of orthorhombic cells of (*a*) the $CaCu_5$ structure, (*b*) the $BaZn_5$ (*oC*24), and (c) $SrZn_5$ (*oP*24) structures. Atoms at *x*, *y*, or $z = \frac{1}{4}$, open circles; atoms at *x*, *y* or $z = \frac{3}{4}$, shaded circles; Cu or Zn at *x*, *y*, or $z = 0$ and $\frac{1}{2}$, double circles. (Baenziger and Conant[11].)

Structures Generated by Mixing $CaCu_5$ and Laves Stacking Units

Er_2Co_7 (*hR*18): The equivalent hexagonal cell has dimensions $a = 4.973$, $c = 36.11$ Å, $Z = 6$.[12] The atoms form 3^6, 6^3, or kagomé 3636 layers normal to the *c* axis of the cell, and the stacking sequence is a mixture of $CaCu_5$ type, $[Aa]\alpha$, and Laves phase type, $\gamma(CAB)$, layer arrangements (see p. 4 for significance of symbols). Two double layers of $CaCu_5$ type packing and a four-layer Laves phase group make an eight-layer unit which is stacked in the cubic sequence ABC: $\alpha[\underline{A}a]\alpha(\underline{A}B\underline{C})\gamma[\underline{C}c]$-$\gamma[\underline{C}c]\ \gamma(\underline{C}A\underline{B})\beta[\underline{B}b]\beta[\underline{B}b]\beta(\underline{B}C\underline{A})\alpha[\underline{A}a]$ (Er underlined).

The Er_2Co_7 structure is formed by the following rare earths; Gd, Tb, Dy, Ho, Er, Tm, Lu, and Y.

Ce_2Ni_7 (*hP*36): The dimensions of the unit cell of Ce_2Ni_7 are $a = 4.98$, $c = 24.52$ Å, $Z = 4$.[13] The atoms are arranged in layers normal to the *c* axis in two groups of $CaCu_5$ type arrangement, $[Aa]\alpha$, and one of Laves phase type, $\gamma(CAB)$, the combined unit of eight layers being stacked in the hexagonal AB sequence (Figure 12-9): $A\underline{B})\beta[\underline{B}b]\beta[\underline{B}b]\beta(\underline{B}A\underline{C})\gamma[\underline{C}c]$-$\gamma[\underline{C}c]\gamma(\underline{C}$ (Ce underlined). The structure is a polytype of Er_2Co_7.

The arrangement results in CN 20 for Ce(1) with 18 Ni neighbors at 2.88 to 3.32 Å and 2 Ce at 3.54 and 3.69 Å. Ce(2) has CN 16 with 12 Ni at 2.97 to 3.18 Å and 4 Ce at 3.23 and 3.54 Å in a Friauf polyhedron. Ni(1) is icosahedrally surrounded by 6 Ni (2.54 Å) and 6 Ce (2.97 Å), similarly Ni(4) with 8 Ni at 2.48 to 2.52 Å and 4 Ce at 3.10 Å. Ni(2) and (3) have 9 Ni and 3 Ce neighbors (2.88 Å) in polyhedra with surface coordination numbers 3×4, 6×5, and 3×6. Ni(5) also has CN 12 with 7 Ni and 5 Ce neighbors, including two close Ce neighbors at 2.83 Å. The closest Ni–Ni approach is 2.46 Å.

$NbBe_3$ (*hR*12): The dimensions of the $NbBe_3$ rhombohedral cell are $a = 7.495$ Å, $\alpha = 35.43°$, $Z = 3$,[14] and for the hexagonal cell $a = 4.561$,

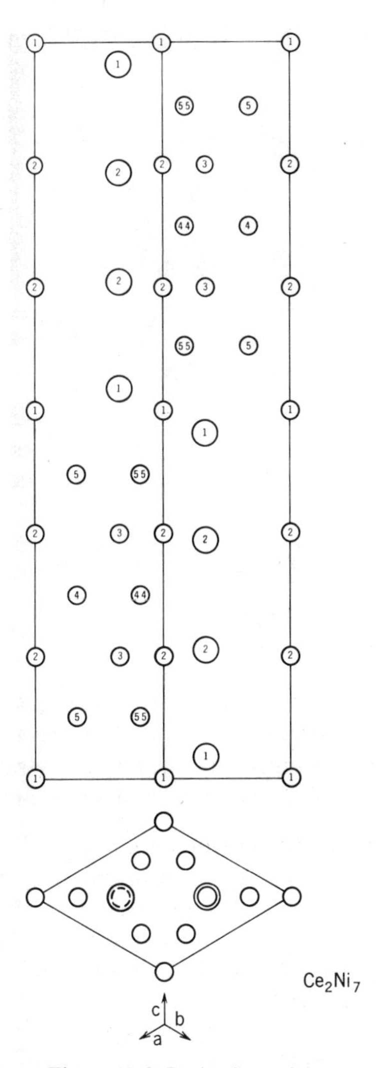

Figure 12-9. Projections of the Ce_2Ni_7 (*hP*36) structure down [100] and [110]: Ce, large circles.

$c = 21.05$ Å. The structure consists of layers of atoms normal to the c axis of the hexagonal cell. One group of $CaCu_5$ type layers ([Aa]α) and one of Laves phase type (β(BAC)) combine to give units of six layers which are stacked in ACB sequence: $[\underline{A}a]\alpha(\underline{A}B\underline{C})\gamma[\underline{C}c]\gamma(\underline{C}A\underline{B})\beta[\underline{B}b]$-$\beta(\underline{B}A\underline{C})\alpha$ (Nb atom layers underlined).

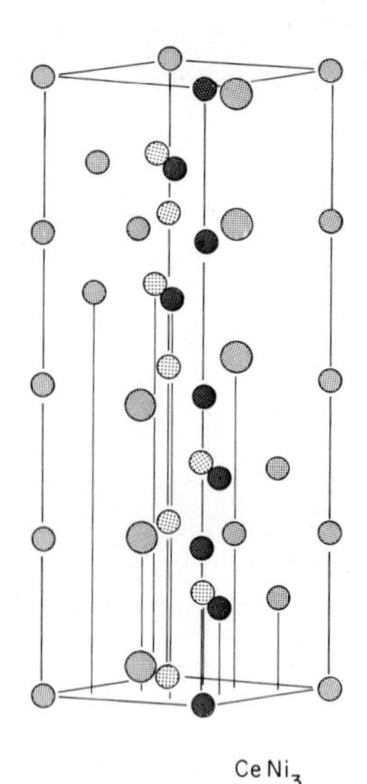

Figure 12-10. Pictorial view of the $CeNi_3$ ($hP24$) structure: Ce, large circles.

Nb(1) has CN 20 made up of 2 Nb at 2.95 Å and 18 Be. Nb(2) has CN 16 made up of 3 Nb(2) neighbors at 2.86 Å, 1 Nb(1) at 2.95 Å and 12 Be. The 12 Be atoms are arranged in a truncated tetrahedron, the resulting polyhedron having 4 regular hexagonal and 4 triangular faces.

$CeNi_3$ ($hP24$): $CeNi_3$ cell dimensions are $a = 4.98$, $c = 16.54$ Å, $Z = 6$.[15] The structure is made up of layers of atoms normal to c, one group of $CaCu_5$ type, [Bb]β, and one of Laves phase type, β(BAC), combining to give six-layer units which are stacked in the hexagonal sequence BC (Figure 12-10): A$\underline{B}$)β[$\underline{B}$b]β($\underline{B}$A$\underline{C}$)γ[$\underline{C}$c]γ($\underline{C}$ (Ce layers underlined). The structure is thus a polytype of $NbBe_3$.

Ce(1) has CN 20 being surrounded by 6 Ni at 2.875 Å, 12 Ni at 3.21 Å and 2 Ce at 3.44 Å. Ce(2) has CN 16 including 12 Ni at 2.86 to 3.14 Å, 3 Ce(2) at 3.19 Å and 1 Ce(1) at 3.44 Å. Each of the four different Ni atoms has CN 12, that about Ni(1) being a distorted icoashedron, with 6 Ni at 2.55 and 6 Ce at 2.96 Å.

MN_2: *M* on 3^6 Nets Alternately Interleaved with *N* on 3^6 or 3636 Kagomé Nets

Laves Phases: Family of Polytypic Structures. The Laves phases form a family of polytypic structures in which three closely spaced 3^6 nets of atoms are followed by a 3636 kagomé net parallel to the (001) plane when the structures are described in terms of a hexagonal cell. The 3^6 nets are stacked on the same site as the kagomé nets which they surround (e.g. β(BAC)γ(CAB)). However, the Laves phases are better described as Frank–Kasper structures (p. 664) in which, parallel to (110) planes of the hexagonal cell, pentagon-triangle main layers of atoms are stacked alternately with secondary 3^6 triangular layers whose atoms center the pentagons of the main layers. The arrangement of successive basal rows of pentagons in the pentagon-triangle nets is described by movements left (L) and right (R) along the directions of two of the edges of the 3^6 secondary nets (p. 43). $MgCu_2$ with cubic structure and all movements L (or alternately R) is the prototype for this family of structures. The structures of all other polytypes are either hexagonal or rhomobhedral, the latter when the sequence of layer movements is divisible by three.

$MgCu_2$ (*cF*24): The $MgCu_2$ structure is one of the so-called Laves phases, although its structure was determined earlier by Friauf.[16] The structure can be regarded as made up by stacking in succession, three triangular layers and a kagomé layer of atoms which lie in planes normal to [111] of the cubic cell. The layer stacking sequence is then $\underline{A})\alpha(\underline{A}C\underline{B})\beta(\underline{B}A\underline{C})\gamma(\underline{C}B$ (Mg underlined) in the notation of p. 4. The Laves phases are Frank–Kasper tetrahedrally close packed structures and the $MgCu_2$ structure can also be considered as made up of alternately stacked main pentagon-triangle $3535+35^3$ (2:3) and secondary 3^6 layers which lie parallel to the $(1\bar{1}0)$ plane (Figure 2-8*a*, p. 42). In the notation of pp. 36–45 the structure is then described as $P\underset{0}{X}P\underset{0}{/}$; L^3. The Mg atoms are surrounded by a Friauf CN 16 polyhedron of 12 Cu at 2.92 Å and 4 Mg at 3.05 Å. Cu is surrounded icosahedrally by 6 Mg atoms (3.05 Å) and 6 Cu atoms at 2.49 Å. The unit cell edge is $a = 7.048$ Å. The structure and coordination polyhedra are shown in Figure 12-11.

The near-neighbor diagram (p. 53) for phases with the $MgCu_2$ (MN_2) structure (Figure 3–5, p. 58) shows that they are formed between components with a wide range of radius ratios from 1.08 to 1.67, and are distributed between lines for 12–6 M–N and 6 N–N contacts. Together with the 4 M–M contacts that are compressed at radius ratios above about 1.2, these contacts result in high 16–12 coordination, showing that the phases owe their stability over the wide range of radius ratios of the component atoms to the geometrical factor (p. 58).

Figure 12-12*a* shows the distribution of the three common Laves phase structures as a function of the atomic volume ratio, V_M/V_N. It is interesting to note the effect of high electronegativity. Figure 3–5, p. 58,

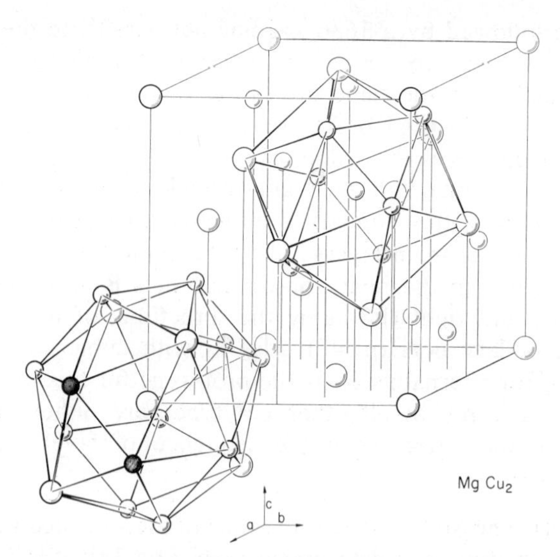

Figure 12-11. Pictorial view of the $MgCu_2$ ($cF24$) structure showing the Friauf CN 16 and icosahedral coordination polyhedra.

indicates those phases with Δx values of unity or greater. They are phases with high radius ratios and the 12–6 M–N contacts are apparently compressed in addition to the M–M contacts, indicating the strong contraction of the structural dimensions that occurs as the electrochemical factor is increased. This is shown in the histogram in Figure 12-12*b*, which gives the number of phases at various ratios of the unit cell volume to the sum of the elemental volumes of the atoms in the unit cell. Not only does the contraction of the unit cell volume compared to the sum of the elemental volumes tend to be greater for phases with the $MgCu_2$ structure than the $MgZn_2$ structure, but some of the phases show a contraction of more than 30% in volume! Some 200 phases have been reported with the $MgCu_2$ structure.

$MgSnCu_4$ ($cF24$): The $MgSnCu_4$ structure is derived from the $MgCu_2$ structure by dividing the Mg sites into two sublattices which are occupied in ordered manner by Mg at 0, 0, 0 and Sn at $\frac{1}{4}, \frac{1}{4}, \frac{1}{4}$.[17] Cu is surrounded icosahedrally by 3 Mg and 3 Sn at 2.92 Å and 6 Cu at 2.49 Å. Mg is surrounded by a CN 16 Friauf polyhedron of 12 Cu at 2.92 Å and 4 Sn at 3.05 Å. Sn is similarly surrounded by 12 Cu and 4 Mg at 3.05 Å.

$AuBe_5$ ($cF24$): The $AuBe_5$ structure is derived from the $MgCu_2$ structure by subdividing the Mg sites into two arrays.[18] One of these is occupied

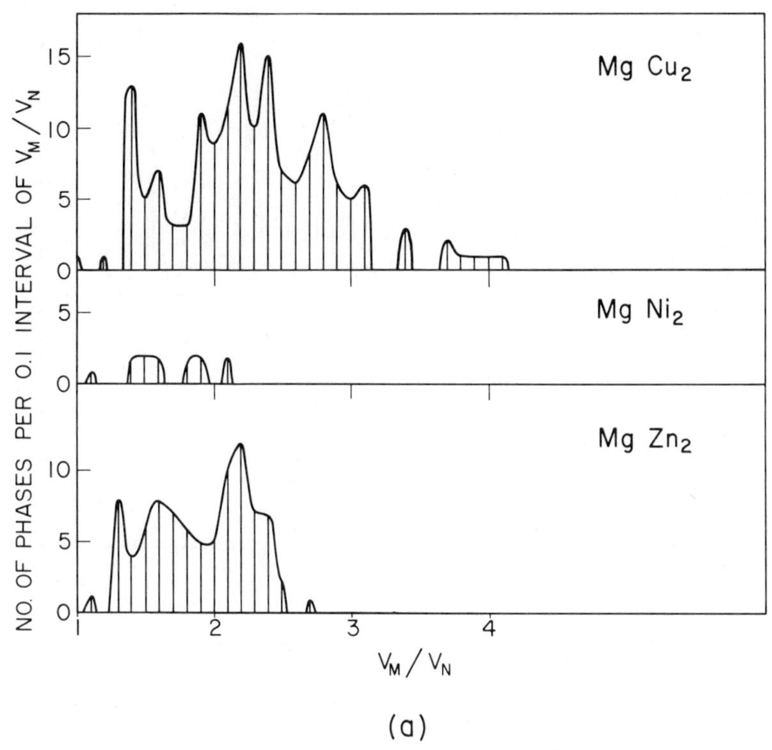

Figure 12-12. (*a*) Histogram showing the distribution of the $MgCu_2$, $MgNi_2$, and $MgZn_2$ Laves phases, MN_2, as a function of elemental atomic volume ratio, V_M/V_N.

by Au and the other by Be(1); Be(2) occupy the Cu sites. Au is surrounded by a Friauf CN 16 polyhedron of Be (12 × 2.78 Å, 4 × 2.90 Å). Be(1) is similarly surrounded by 12 Be(2) at 2.78 Å and 4 Au at 2.90 Å. Be(2) is icosahedrally surrounded by 3 Au, 3 Be(1), and 6 Be(2) (2.37 Å).

$MgZn_2$ (*hP*12): $MgZn_2$ is a Laves phase with unit cell dimensions $a = 5.18$, $c = 8.52$ Å, $Z = 4$.[19] The structure is generated by stacking together pairs of main pentagon-triangle $3535 + 35^3$ (2:3) layers and secondary 3^6 layers parallel to the (110) plane (Figure 2-8*b*, p. 43). In the notation of pp. 36–45, the structure can be described as $P\underset{0}{X}P\underset{0}{/}$; LR. Alternatively the structure can be considered as made of kagomé and 3^6 layers of atoms parallel to (001) planes, the layer sequence being AB)β(BAC)γ(C in the notation of p. 4.

Mg is surrounded by 4 Mg at 3.17 and 3.20 Å and 12 Zn at about 3.04 Å which form a Friauf polyhedron. Zn(1) and Zn(2) are both surrounded

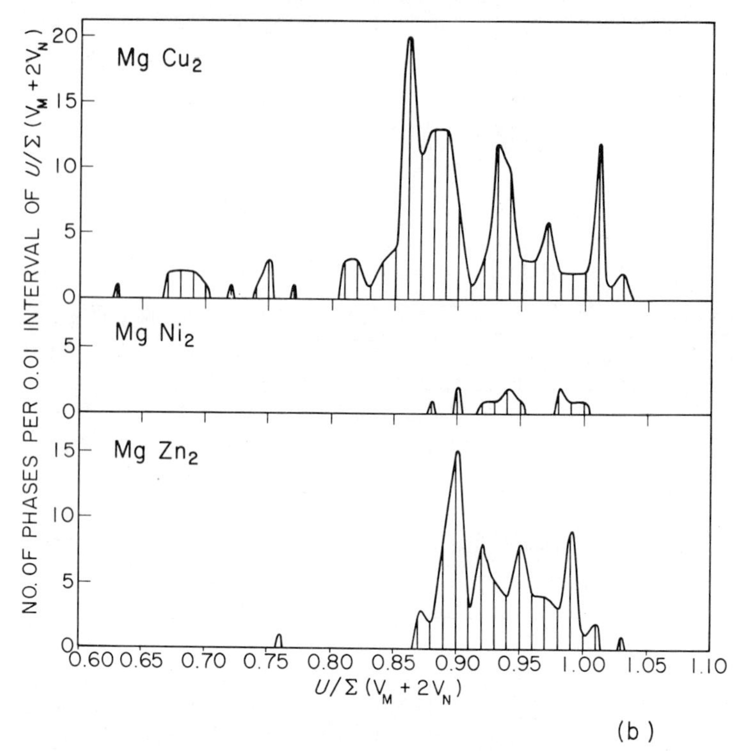

Figure 12-12. (*b*) Histogram showing the number of phases per 0.01 interval of $U/\Sigma(V_M + 2V_N)$ for Laves phases with the $MgCu_2$, $MgNi_2$, and $MgZn_2$ structures. The ratio $U/\Sigma(V_M + 2V_N)$ compares the unit cell volume to the sums of the elemental atomic volumes of all of the atoms in the unit cell.

icosahedrally by 6 Mg and 6 Zn atoms. The Zn–Zn distances are 2.54, 2.62, or 2.64 Å.

The near-neighbor diagram (p. 53) for the $MgZn_2$ structure (Figure 3-1, p. 54), which has been constructed for the ideal axial ratio and atomic parameter, shows the importance of the geometrical factor in giving stability to the structure over a range of radius ratios for the component atoms. At radius ratios above 1.225, 16–12 coordination is established at the expense of compressing four *M–M* contacts. For a more detailed discussion see $MgCu_2$, p. 59. The phases do not extend to such high radius ratios as those with the $MgCu_2$ structure, and the compression of the unit cell volume compared to the sums of the elemental volumes of the atoms in the unit cell, is not as great as in some phases with the $MgCu_2$ structure. See Figure 12-12*b*. Some 170 phases have been reported with the $MgZn_2$ structure.

URe_2 (L.T.) (*oC*24): The low-temperature form of URe_2 has cell dimensions $a = 5.600$, $b = 9.180$, $c = 8.460$ Å, $Z = 8$.[20] The structure is a distorted form of the $MgZn_2$ type structure, as is seen by comparing the orthohexagonal $MgZn_2$ cell with the URe_2 cell. U is surrounded by 16 neighbors in a Friauf polyhedron and the Re atoms are icosahedrally

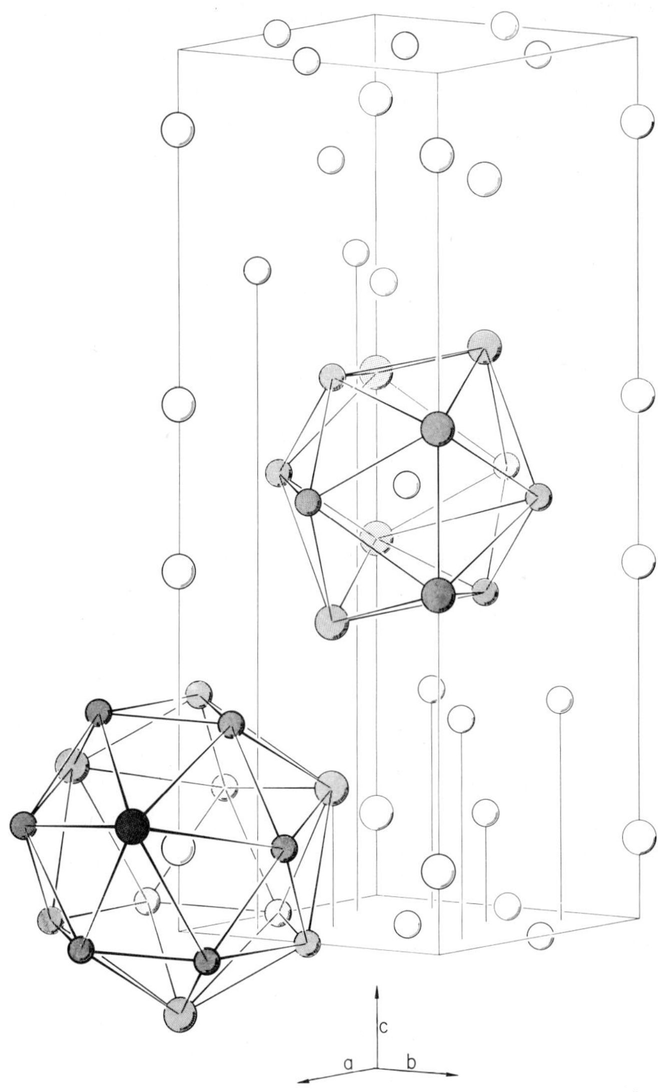

Figure 12-13. Pictorial view of the $MgNi_2$ (*hP*24) structure showing the Friauf CN 16 and icosahedral coordination polyhedra.

surrounded. The shortest Re–Re distance is 2.61 Å and Re–U distance is 3.055 Å. The structure requires confirmation by single crystal methods.

$LuMn_5$ (*hP*12): The $LuMn_5$ structure with cell dimensions $a = 5.186$, $c = 8.566$ Å, $Z = 2$,[21] is related to the $MgZn_2$ structure like the $AuBe_5$ structure is to $MgCu_2$. Thus Mn occupies half of the Mg sites in the structure and Lu occupies the other half.

$MgNi_2$ (*hP*24): $MgNi_2$ with cell dimensions $a = 4.815$, $c = 15.80$ Å, $Z = 8$,[22] is a Laves phase polytype which can be described as $P_0XP_0/$; LLRR in the notation of pp. 36–45. Alternatively, the structure can be described by the stacking of 3^6 and 3636 kagomé layers parallel to the (001) plane, the sequence being α(ABC)γ(CBA)α(ACB)β(BCA) in the notation of p. 4. The structure and coordination polyhedra are shown in Figure 12-13. Both Mg(1) and (2) atoms are surrounded by the Friauf CN 16 polyhedron of 4 Mg (2.95 and 2.97 Å) and 12 Ni (2.82 to 2.83 Å). Ni (1), (2), and (3) are each surrounded icosahedrally by 6 Mg and 6 Ni.

MgAlCu (*hR*18): This MgAlCu structure with hexagonal cell dimensions $a = 5.14$, $c = 37.89$ Å, $Z = 18$,[23] is a Laves phase polytype with a repeat sequence of 9, the description in the notation of pp. 36–45 being $P_0XP_0/$; $(LLR)^3$ in layers of atoms parallel to the (110) plane of the hexagonal cell.

$Mg(Cu,Zn)_2$: Komura[24] reported an eight compound-layer Laves phase polytype in this system at an electron concentration of about 1.98 *e*/*a*. In the notation of pp. 36–45, the structure is described as $P_0XP_0/$; LR^2LRL^2R.

$Mg(Ni,Cu)_2$: Komura[24] reported a six compound-layer Laves phase polytype described as $P_0XP_0/$; L^3R^3 in the notation of pp. 36–45.

$Mg(Zn_{0.925}Ag_{0.075})_2$ (*hP*60): With cell dimensions $a = 5.225$, $c = 42.95$ Å, $Z = 20$,[25] $Mg(Zn_{0.925}Ag_{0.075})_2$ is a Laves phase polytype having a 10 compound-layer repeat sequence. The arrangement is described in the notation of pp. 36–45 as $L^2RL^2R^2LR^2$ in planes of atoms parallel to (110). This structure replaces a five compound-layer type described earlier by Komura.[23]

REFERENCES

1. J. V. Florio, N. C. Baenziger, and R. E. Rundle, 1956, *Acta Cryst.*, **9**, 367.
2. W. Haucke, 1940, *Z. anorg. Chem.*, **244**, 17.
3. J. V. Florio, R. E. Rundle, and A. I. Snow, 1952, *Acta. Cryst.*, **5**, 449.

4. E. S. Makarov and S. I. Vinogradov, 1956, *Kristallografija*, **1**, 634.
5. Q. Johnson, G. S. Smith, and D. H. Wood, 1969, *Acta Cryst.*, **B25**, 464.
6. J. V. Florio and R. E. Rundle, 1952, *USAEC Publ. ISC*, 273.
7. A. E. Ray, 1966, *Acta Cryst.*, **21**, 426.
8. Q. Johnson and G. S. Smith, 1967, *Acta Cryst.*, **23**, 327.
9. Q. Johnson, D. H. Wood, and G. S. Smith, 1968, *Acta Cryst.*, **B24**, 480.
10. D. T. Cromer and A. C. Larson, 1960, *Acta Cryst.*, **13**, 909.
11. N. C. Baenziger and J. W. Conant, 1956, *Acta Cryst.*, **9**, 361.
12. W. Ostertag, 1967, *J. Less-Common Metals*, **13**, 385.
13. D. T. Cromer and A. C. Larson, 1959, *Acta Cryst.*, **12**, 855.
14. D. E. Sands, A. Zalkin, and O. H. Krikorian, 1959, *Acta Cryst.*, **12**, 461.
15. D. T. Cromer and C. E. Olsen, 1959, *Acta Cryst.*, **12**, 689.
16. J. B. Friauf, 1927, *J. Amer. Chem. Soc.*, **49**, 3107.
17. E. I. Gladyševskij, P. I. Kripjakevič, and M. Ju. Tesljuk, 1952, *Dokl. Akad. Nauk, SSSR*, **85**, 81.
18. N. C. Baenziger, R. E. Rundle, A. I. Snow, and A. S. Wilson, 1950, *Acta Cryst.*, **3**, 34.
19. J. Friauf, 1927, *Phys. Rev.*, **29**, 34; L. Tarschisch, A. T. Titov, and F. K. Garjanov, 1934, *Phys. Z. Sowjetunion*, **5**, 503.
20. B. A. Hatt, 1961, *Acta Cryst.*, **14**, 119.
21. F. E. Wang and J. V. Gilfrich, 1966, *Acta Cryst.*, **21**, 476.
22. F. Laves and H. Witte, 1935, *Metallwirt.*, **14**, 645.
23. Y. Komura, 1962, *Acta Cryst.*, **15**, 770.
24. Y. Komura, 1969, Eighth International Congress of Crystallography, Stony Brook, N.Y. August; Y. Komura *et al.*, 1970, *Acta Cryst.*, **B26**, 666.
25. Y. Komura, E. Kishida, and M. Inoue, 1967, *J. Phys. Soc. Japan*, **23**, 398.

13

Structures in Which Icosahedra and CN 14, 15, and 16 Polyhedra Play a Dominant Role

Structures containing Friauf CN 16 polyhedra (or in some cases CN 15 μ polyhedra) and associated interpenetrating icosahedra (or CN 13 and 14 polyhedra) which give rise to a distorted tetrahedral close packing of the atoms, and therefore good space-filling, can loosely be divided into two groups:

(i) The Frank–Kasper-type structures, in which the atoms form planar networks stacked one above the other in such a way that all atoms have either 12 (icosahedral) 14, 15, or 16 coordination (see p. 31).

(ii) Structures in which the regularity of the Frank–Kasper arrangement is lost, although the same four coordination polyhedra predominate. Many of these structures, like those examined by Samson, are based on frameworks of fused truncated tetrahedra (Friauf polyhedra minus the 4 tetrahedrally disposed atoms with surface coordination 6) which join by sharing hexagons. Arrangements of fused truncated tetrahedra also create icosahedra and each tetrahedron is interpenetrated generally by 12 icosahedra, or by 9 or 10 icosahedra and by 3 or 2 CN 13 or 14 polyhedra. Hexagonal faces of truncated tetrahedra not mutually shared are fused to hexagonal prisms, hexagonal antiprisms, or CN 15 polyhedra.

TETRAHEDRALLY CLOSE PACKED FRANK–KASPER STRUCTURES

CN 12, 14, 15, and 16 Polyhedra. Pentagon-Triangle Main Layers

Laves phases: The polytypic Laves phase structures described on pp. 654–660 are Frank–Kasper tetrahedrally close packed structures built of pentagon-triangle main layers interpenetrated by 3^6 secondary nets of atoms.

μ-Fe_7W_6 (hR13): The dimensions of the hexagonal cell are $a = 4.806$, $c = 25.84$ Å, $Z = 3$.[1] The atoms form 3^6, 6^3, or kagomé 3636 nets parallel to the (001) plane with stacking sequence in the notation of p. 4, $\mathrm{A\underline{B}\beta\underline{B}\underline{b}\underline{B}\beta\underline{B}C\underline{A}\alpha\underline{A}\underline{a}\underline{A}\alpha\underline{A}B\underline{C}\gamma\underline{C}\underline{c}\underline{C}\gamma\underline{C}}$ (W underlined). Thus the stacking sequence characteristic of the μ-phase, βBbBβBCA, is seen to be composed of a four-layer Al_3Zr_4 unit, βBbB, and a four-layer Laves unit, βBCA. However, the structure is best considered as a Frank–Kasper phase containing primary pentagon-triangle nets of W and Fe atoms parallel to the (110) plane which are interleaved with secondary 3^34^2 nets of Fe atoms (Figure 13-1). In the notation of pp. 36–45 the structure is described as $P\underset{0}{X}\underset{0}{P}/;(VL)^3$. Fe(1) is surrounded icosahedrally by 6 Fe at 2.41 Å and 6 W at 2.765 Å; Fe(2) is similarly surrounded by 5 Fe (at 2.41 and 4×2.38 Å), and 7 W at 2.57 to 2.90 Å. W(1) is surrounded by the characteristic CN 15 μ-phase polyhedron (Figure 2-2, p. 34) of 6 Fe at 2.69 and 2.71 Å, and 9 W at 2.75 to 3.06 Å. W(2) is surrounded by a Friauf CN 16 polyhedron of 12 Fe (2.69–2.90 Å) and 4 W at 2.635 and 3×2.82 Å. W(3) has the CN 14 polyhedron of 6 Fe at 2.57 Å and 8 W at 2.635, 2.69, 3×3.05, and 3×3.06 Å. The structure has been confirmed by a single-crystal study of Co_7Mo_6.[2]

In terms of Samson's[3] description of fused polyhedra, the μ structure is made up of alternate layers of truncated tetrahedra and double layers of hexagonal antiprisms (Figure 13-2) with each truncated tetrahedron being turned 60° relative to the preceding one. Since each double layer of fused hexagonal antiprisms can also be described as a layer of CN 15 μ-phase polyhedra, the structure itself can be described in terms of alternating layers of truncated tetrahedra fused to layers of CN 15 polyhedra as indicated in Figure 13-2. The sequence of centered Frank–Kasper polyhedra formed along the body diagonal of the rhombohedral cell is CN 16, CN 14, CN 14, CN 16, CN 15, CN 12 (icosahedron), CN 15; the eighth layer then repeats the first.

Al_3Zr_4 (hP7): Al_3Zr_4 has unit cell dimensions $a = 5.433$, $c = 5.390$ Å, $Z = 1$.[4] The structure can be regarded as composed of layers of atoms

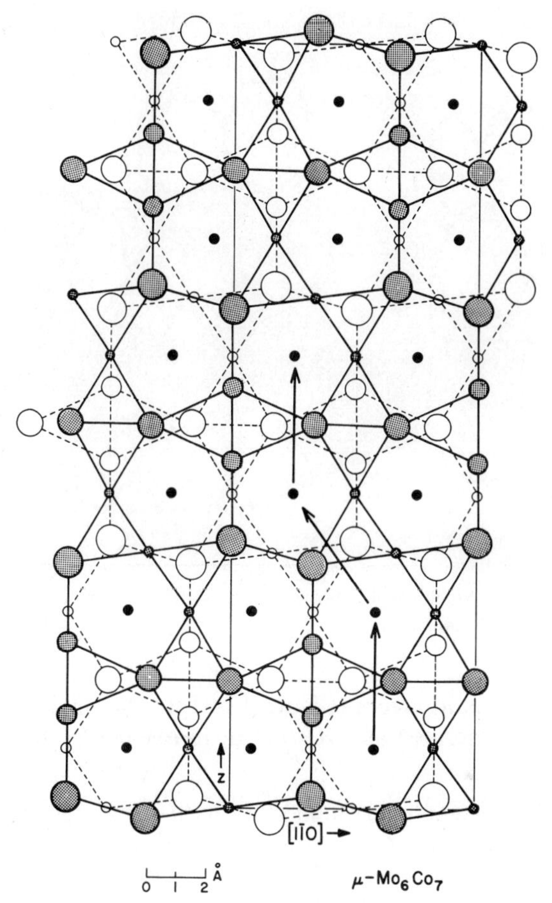

Figure 13-1. Diagram showing the development of the μ-phase structure (*hR*13) by stacking together alternately pentagon-triangle and $3^3 4^2$ nets of atoms parallel to the (110) planes of the hexagonal cell. The circles indicate in decreasing order of size, sites surrounded by 16, 15, 14, and 12 neighbors in the coordination polyhedra. (Shoemaker and Shoemaker [5].)

normal to [001] which are described as βBbB in the notation of p. 4, so that 3636 kagomé layers of Al atoms are separated successively by a 3^6 net of Zr(3) atoms, a 6^3 net of Zr (1 and 2) atoms, and another 3^6 net of Zr(3) atoms. Al_3Zr_4 is a Frank–Kasper-type structure; in the (110) plane it can be regarded as composed of main pentagon-triangle layers which are stacked antisymmetrically with respect to each other and of interleaved 4^4 rectangular secondary layers whose atoms lie over the

Figure 13-2. Stacking of coordination polyhedra in the μ-phase structure. See text. (Samson[3].)

pentagons of the main layers (Figure 13-3). In the notation of pp. 36–45 the structure is described as $P\underset{0}{X}\underset{0}{P}/;V$, or in respect to the layers parallel to (001), $H\underset{0}{X}$; L,R.

Zr(1) and (2) are each surrounded by 6 Al + 9 Zr (3.12 Å and 3.14 or 3.41 Å, respectively) in a CN 15 μ-phase polyhedron. Zr(3) is surrounded by a CN 14 Frank–Kasper polyhedron composed of 6 Al at 3.03 Å, 2 Zr at 2.695 Å and 6 at 3.41 Å. Al is surrounded icosahedrally by 4 Al (2.72 Å) and 8 Zr (3.03 and 3.12 Å). The structure has not been examined by single-crystal methods.

M Phase ($Nb_{48}Ni_{39}Al_{15}$) (*oP*52): The *M* phase with cell dimensions, $a = 9.304$, $b = 16.27$, $c = 4.933$ Å, 52 atoms per cell,[5] is one of the Frank–Kasper tetrahedrally close packed structures, containing CN 12, 14, 15, and 16 polyhedra. The structural arrangement is shown in Figure 13-4. Main pentagon-triangle layers parallel to the (001) plane are interleaved with square-triangle $3^2434 + 3^34^2$ (1 : 1) secondary layers. In the notation of pp. 36–45, the structure is described as $P/\underset{+}{P}/\underset{-}{P}X\underset{-}{P}\underset{+}{X}$; LR. Interatomic distances in the coordination polyhedra range from 2.42 to 3.15 Å, the closest approach in the two CN 16 polyhedra being 2.76 Å, and 2.83 Å in the two CN 15 polyhedra.

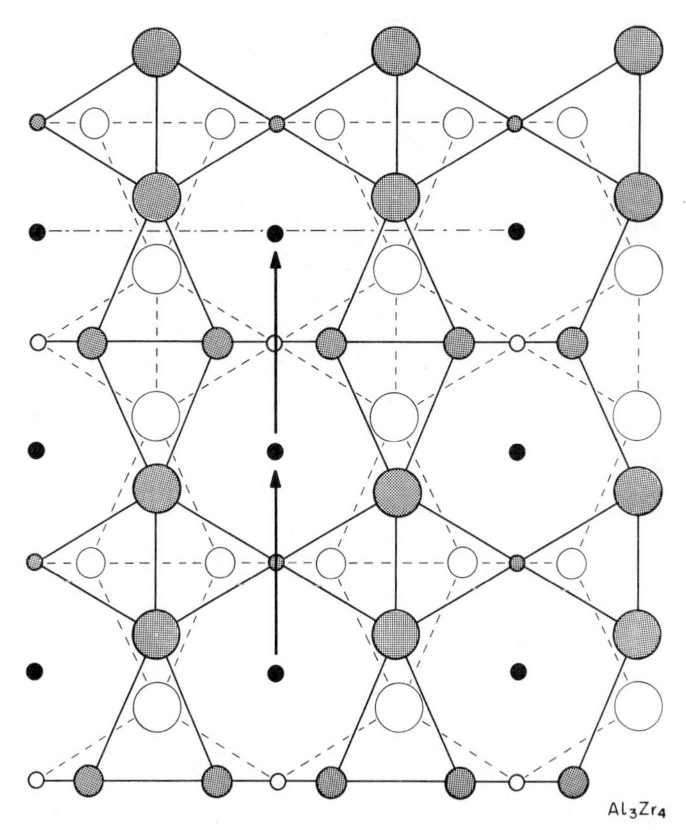

Figure 13-3. Diagram showing the development of the Al_3Zr_4 (*hP*7) structure by stacking together alternately pentagon-triangle and rectangular 4^4 nets of atoms parallel to the (110) plane. The circles indicate, in decreasing order of size, sites surrounded by 15, 14, and 12 neighbors in the coordination polyhedra. (Shoemaker and Shoemaker [5].)

Pentagon-Hexagon-Triangle Main Layers:

P phase ($Cr_{18}Mo_{42}Ni_{40}$) (*oP*56): The *P* phase has cell dimensions $a = 16.98$, $b = 4.752$, $c = 9.070$ Å and contains 56 atoms[6] in partial substitutional disorder. The structure is a Frank–Kasper tetrahedrally close packed type with atoms in CN 12, 14, 15, and 16 polyhedra only. It is made up of hexagon-pentagon-triangle primary layers parallel to the (001) plane, which are interleaved with $3^3434 + 3^34^2$ (1 : 1) secondary layers of atoms, and it is described as $\overset{+}{H}\overset{+}{X}P/\overset{-}{H}\overset{-}{X}P/$; LR in the notation of pp. 36–45 (Figure 13-5). According to Samson's description,[3] the structure is made up of rows of truncated tetrahedra (mutually sharing two hexagons)

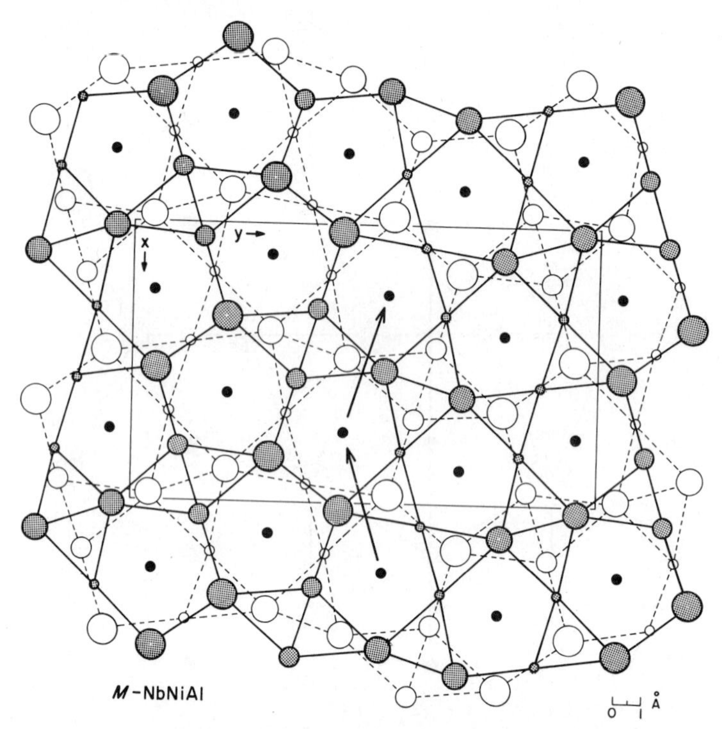

Figure 13-4. Diagram showing the development of the *M*-phase structure (*oP*52) by stacking alternately pentagon-triangle and $3^2434+3^34^2$ (1 : 1) layers of atoms parallel to the (001) plane. The circles indicate, in decreasing order of size, sites with 16, 15, 14, and 12 neighbors in their coordination polyhedra. (Shoemaker and Shoemaker[5].)

which share a further hexagon above and below with double rows of hexagonal antiprisms, the remaining hexagon being shared with μ-phase polyhedra in the same plane, as shown in Figure 13-6. The result of this arrangement is a slight reduction in the usual number of icosahedra interpenetrating the CN 16 Friauf polyhedra; only 10 vertices of the truncated tetrahedra become centers of icosahedra; the remaining two vertices center hexagonal antiprisms which have extended poles giving CN 14 polyhedra.

R–*Mo–Co–Cr* (*hR*53): The *R* phase has hexagonal cell dimensions $a = 10.80$, $c = 19.34$ Å, $Z = 159$ atoms.[7] An idealized form of the *R*-phase structure is a tetrahedrally close packed Frank–Kasper structure made up of slightly rumpled primary hexagon-pentagon-triangle layers interleaved with 3^34^2 square-triangle secondary layers and described by the

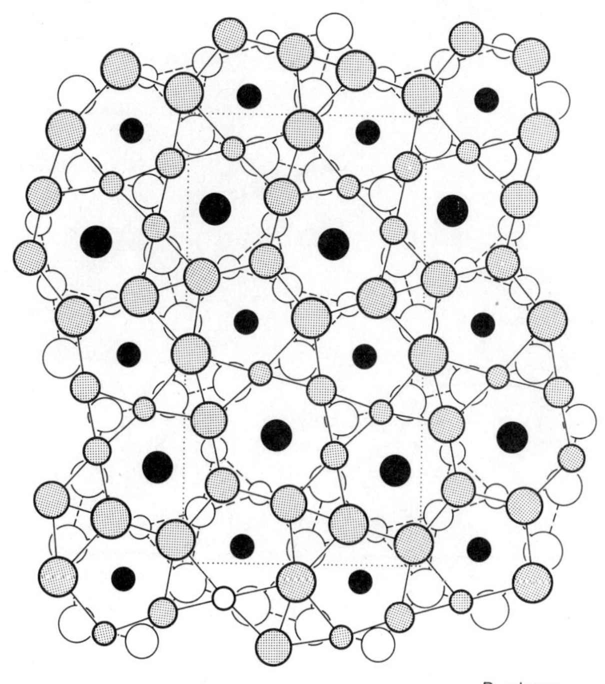

Figure 13-5. Diagram showing the development of the *P*-phase structure (*oP*56) by stacking together alternately hexagon-pentagon-triangle and $3^2434+3^34^2$ (1 : 1) nets of atoms parallel to the (001) plane. (Shoemaker and co-workers[6].)

symbols $PX_0HX_0P/_0H/_0$; $(VL)^3$ in the notation of pp. 36–45. This arrangement is indicated in Figure 13-7 giving a projection of the structure on $(\overline{13}5)$, but the figure also shows the disruption of the layer sequence in the transition regions that are found in the actual *R*-phase structure. The structure contains CN 12, 14, 15, and 16 polyhedra, the largest being occupied by Mo atoms. See also ϵ-$Mg_{23}Al_{30}$, p. 691.

Hexagon-Triangle Main Layers:

β-W or *Cr_3O* (*cP*8): In the β-W (MN_3) structure the *M* atoms form a b.c. cubic array and lines of *N* atoms run throughout the structure parallel to the edges of the b.c. cell formed by the *M* atoms (Figure 13-8).[8] The structure belongs to the Frank–Kasper type and it can be considered to be formed by the alternate stacking of main triangle-hexagon 3^26^2+3636 (2 : 1) layers and secondary 4^4 layers[9] (Figure 13-9), with the result that

Figure 13-6. Stacking of coordination polyhedra in the *P*-phase structure. See text. (Samson [3].)

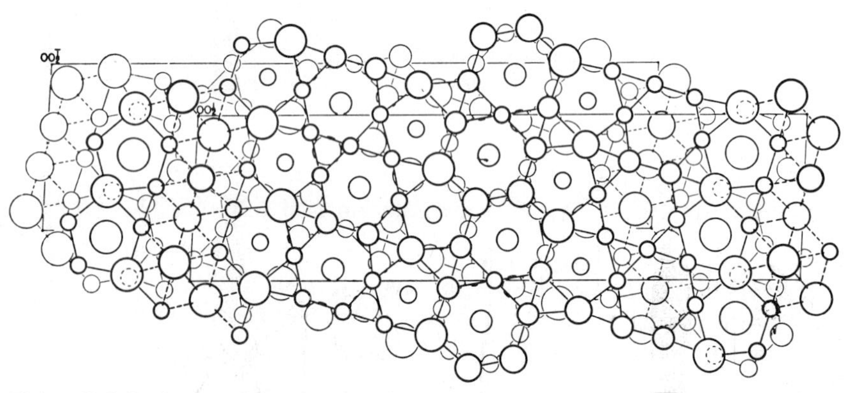

Figure 13-7. Projection of the structure of the *R* phase (*hR*53) on the $(\overline{1}\overline{3}5)$ plane. (Komura and co-workers [7].)

the *M* atoms are surrounded icosahedrally by 12 *N* atoms, and the *N* atoms are surrounded by 4 *M* and 8 + 2 *N* in a CN 14 polyhedron with triangulated faces. In $SnNb_3$ with the β-W structure, the interatomic distances are Sn–Nb, 2.96 Å, and Nb–Nb, 2 × 2.64 and 8 × 3.24 Å.

Some 70 phases known to take the β-W structure all contain transition metals as one or both components; many of the structures may be dis-

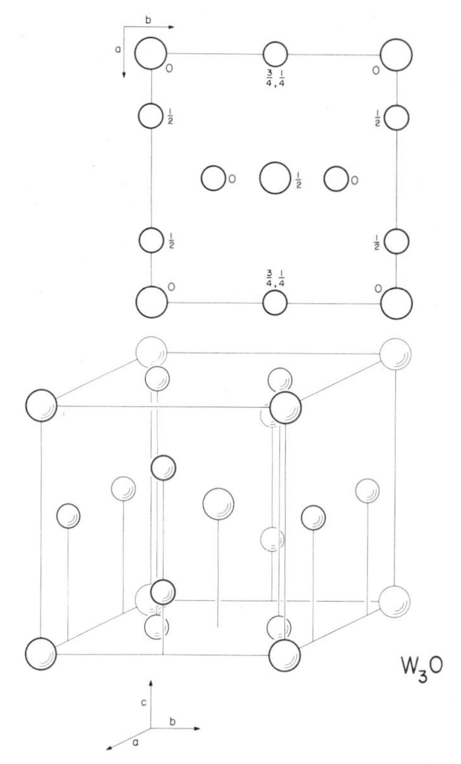

Figure 13-8. Views of the β-W or Cr_3O ($cP8$) structure.

ordered. The great interest shown in phases with this structure has been due to the very high superconducting transition temperatures of several of them.

The near-neighbor diagram (p. 53) taken for CN 12 radii, shows that the main body of phases (radius ratios 0.9 to 1.1) are located about the position where the M–N contact line crosses those for N–N contacts, and 12–14 coordination is achieved. This shows quite clearly that the coordination factor controls the occurrence of the main body of phases with the β-W structure, as would be expected for a Frank–Kasper type phase. Since many of the phases are probably disordered, it is interesting to note in Figure 3-3 that the main group of β-W phases also lie close to and parallel to the line for M–N contacts calculated for a disordered array of atoms (i.e., $d_{MN} = \frac{1}{4}(D_M + 3D_N)$, assuming the same condition for zero, i.e., $d_M = a\sqrt{3}/2$.

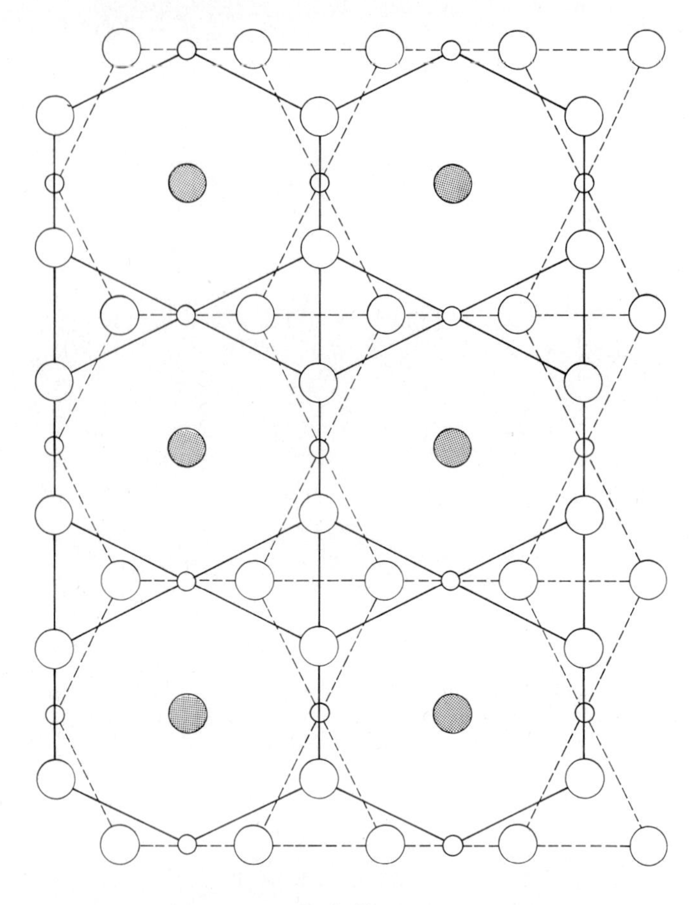

Figure 13-9. Diagram showing the β-W ($cP8$) structure made up of 3^26^2+3636 (2 : 1) main layers and 4^4 secondary layers of atoms. Large circles represent CN 14 sites and small circles CN 12 sites. (Shoemaker and Shoemaker [9].)

Figure 3-3 also compares the volume of the unit cell of β-W phases with the sums of the elemental volumes of the components. The majority of the phases show a small volume contraction which generally increases with radius ratio of the components particularly above about 1.1.

Several reviews or discussions of the β-W phases[10] have been concerned with the relationship of the unit cell edge and the radii of the M and N components. Geller,[11] for example, has derived a self-consistent set of atomic radii in the phases (based on the M–N distance) which reproduce quite well the observed unit cell edges of known or predicted phases with the β-W structure, whereas Nevitt[12] demonstrated that Geller's set of radii is not unique and other self-consistent sets, equally effective in reproducing unit cell edges, can be derived. In addition he shows that for a given *N* element, the unit cell edge is a linear function of the Goldschmidt CN 12 radius of the *M* element (or of the *M* elements of a given Group) and it is a more or less linear function of the radius of *N* elements regardless of the *M* elements.

SiV_3(L.T.)(*tP*8): A tetragonal form of β-W structure has been obtained by martensitic transformation at low temperatures.[13] For V_3Si at 4.2°K, $a = 4.715$, $c = 4.725$ Å, $c/a = 1.0022$. For $SnNb_3$ at 4.2°K, $a = 5.300$, $c = 5.252$ Å, $c/a = 0.9909$.

β-UH_3, $AuZn_3$ (*cP*32): The U atoms in β-UH_3 form a β-W array which is filled up with hydrogen atoms[14] (Figure 13-10). U(1) is surrounded by 12 H at 2.32 Å in a distorted icosahedron and U(2) by 12 H (2.31, 2.32 Å) at the corners of a truncated tetrahedron, and of these, four groups of 3 H form parts of different icosahedra about U(1). H is surrounded by 1 U(1) and 3 U(2) at the corners of a distorted tetrahedron. The hydrogen atom positions were established by powder neutron diffraction.

$AuZn_3$ has the same structural arrangement:[15] Au(1)–12 Zn = 2.705 Å, Au(2)–4 Zn = 2.68 Å, Au(2)–8 Zn = 2.85 Å. Zn has 1 Au(1) + 3 Au(2) neighbors as well as 3 Zn at 2.61 Å, 4 at 2.91 Å and one at 3.16 Å.

σ *Phase, FeCr* (*tP*30): The σ phase structure (Figure 13-11*a*) belongs to the Frank–Kasper class of tetrahedrally close packed structures. Its structure, which can be described as $H\underset{+}{X}H/;\underset{-}{L}R$ in the notation given on pp. 36–45, is made up of primary hexagon-triangle layers, $3636 + 3^26^2 + 6^3$ (3:2:1), of atoms *A*, *B*, *C*, and *D* at $z \sim 0$ and $\frac{1}{2}$. These are separated by secondary 3^2434 layers of *E* atoms at $z \sim \frac{1}{4}$ and $\frac{3}{4}$ that lie over the centers of the hexagons of the primary layers[16] (Figure 13-11*b*). This arrangement gives the following slightly distorted Frank–Kasper coordination polyhedra.

Atom	*CN*	
A	12	icosahedron
B	15	
C	14	
D	12	icosahedron
E	14	

The sigma phase structure is important in alloys of the transition metals. Evidence of ordering has been found in some alloys with the larger atoms

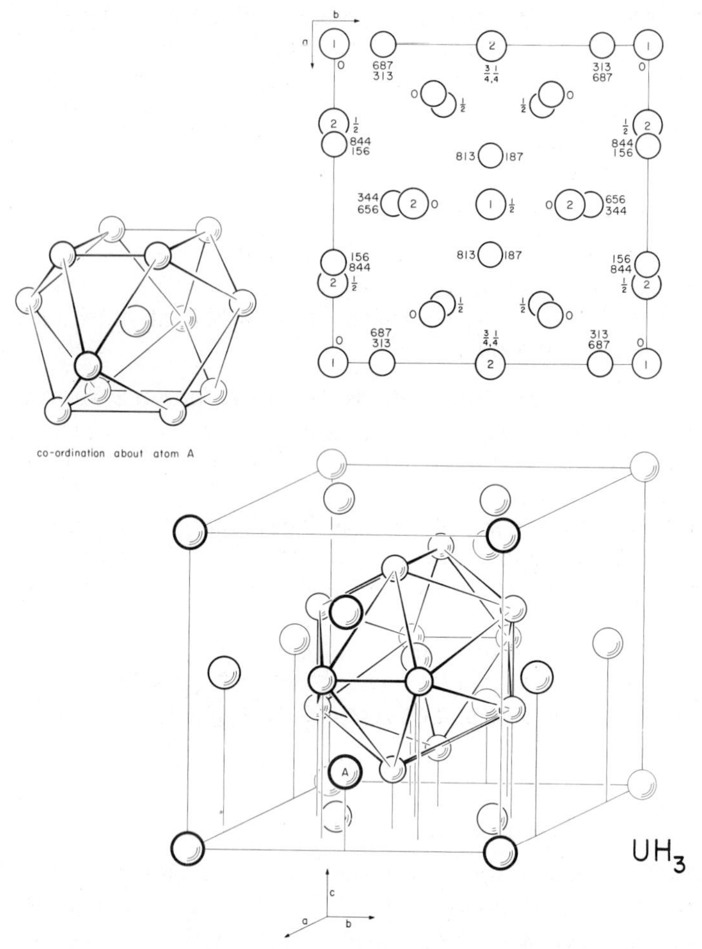

Figure 13-10. Views of the structure of UH_3 (*cP*32) showing coordination of U(1) and U(2) (A) atoms.

preferring to be located at the centers of the CN 15 and 14 polyhedra. The near-neighbor diagram (p. 53) for the σ phase structure, which is shown in Figure 3-2, p. 55, provides a good example of a structure that is controlled by the coordination factor, since all known phases are grouped closely around the intersection of lines giving the high coordination numbers found in the structure, the most favorable radius ratio for the component atoms being about 1.05 Å. It is probable that electron concentration also plays some role in controlling the composition of phases with the σ structure, since they are grouped particularly in the

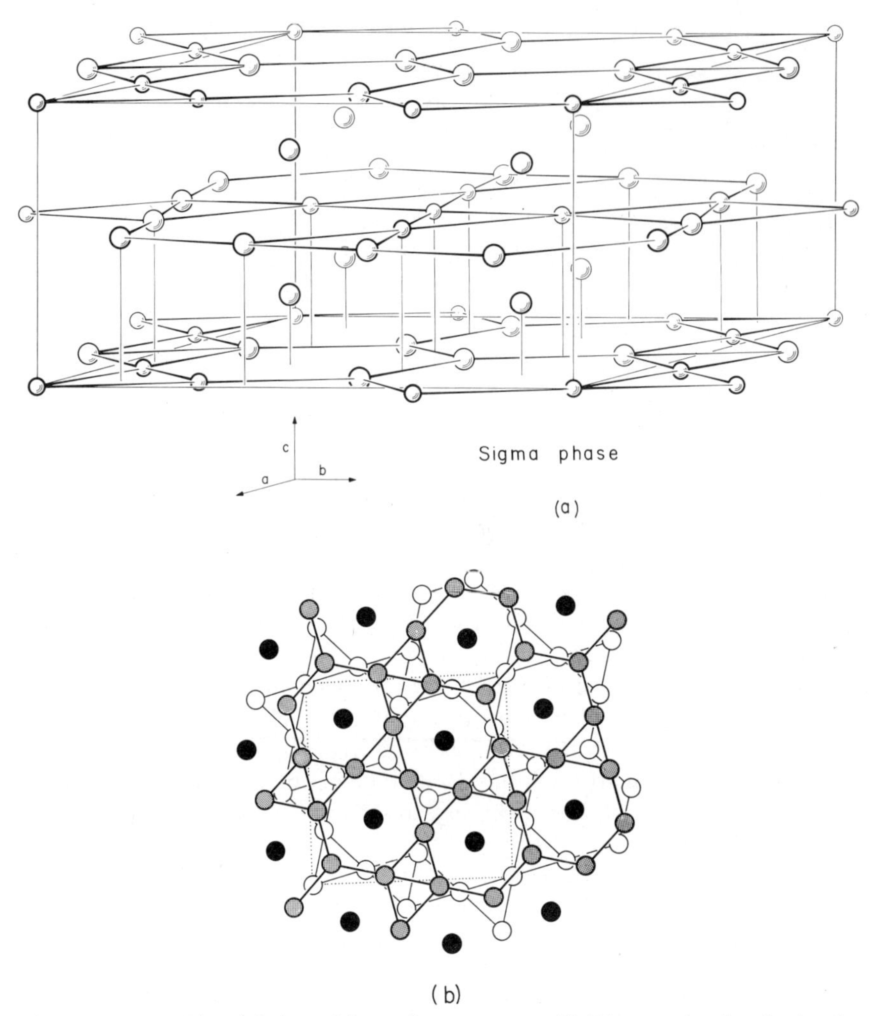

Figure 13-11. (*a*) Pictorial view of the σ-phase structure. (*b*) Diagram showing the development of the σ-phase structure (*tP*30) by stacking together alternately hexagon-triangle and 3^2434 layer networks of atoms parallel to the (001) plane. (Frank and Kasper[16].)

range from 6.2 to 7 electrons per atom (Figure 13-12), although as observed on p. 109, the effective concentration of electrons in partly filled bands at the Fermi level is likely to be 1 to 2 electrons per atom, the remainder being in filled energy bands below the Fermi level.

σ-FeCr has cell dimensions $a = 8.799$, $c = 4.544$ Å, $Z = 30$ atoms.[17] σ-phases commonly have axial ratios in the range from 0.51 to 0.52. β-U has essentially the same structure.

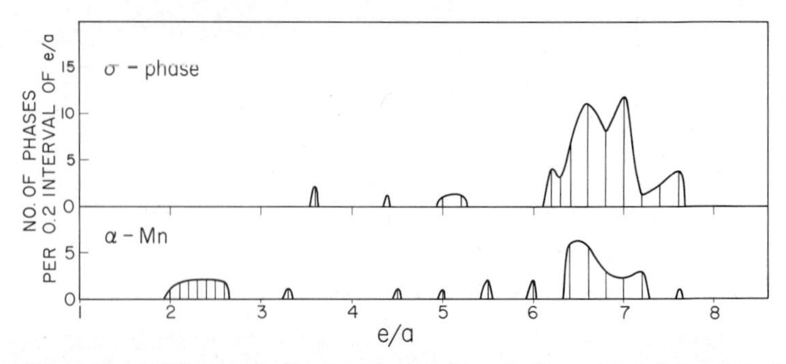

Figure 13-12. Histogram showing the number of phases per 0.2 interval of outer *s*, *p*, and *d* electron concentration per atom which have the σ phase and α-Mn structures.

INTERMETALLIC COMPOUNDS WITH GIANT CELLS

These structures which have been studied particularly by Samson,[3, 18, 19, 20, 21] can be regarded as a complex development of Frank–Kasper structures (p. 663), since the main atomic coordination is 12, 14, 15, and 16, although other polyhedra also occur. The polyhedra, instead of being all arranged uniformly in layers as in the Frank–Kasper structures, are generally fused together in groups which are built around each other successively until the unit cell content is achieved. The structures of three compounds with giant cells, β-Mg_2Al_3, $NaCd_2$, and Cu_4Cd_3, have been determined. Each has a cubic cell containing more than 1100 atoms and solution of their structures by the stochastic method is a great triumph for Samson.

Whereas the Frank–Kasper structures are regarded as generated by stacking particular layers of atoms which have the property of giving only interpenetrating CN 12, 14, 15, and 16 polyhedra, it is more convenient to consider giant cell structures as an arrangement of fused polyhedra rather than the full interpenetrating polyhedra. The fused polyhedra that result when the six-connected surface atoms which center neighboring polyhedra are ignored, are listed in Table 13-1.

Table 13-1

Polyhedron	Fused Polyhedron
CN 16	Truncated tetrahedron
CN 16	Truncated trigonal prism
CN 14	Hexagonal antiprism
CN 12	Pentagonal antiprism

Several other coordination polyhedra occur in giant cells in addition to the four Frank–Kasper polyhedra, the most important being, CN 11 formed by omitting one atom from an icosahedron, CN 13 formed by the introduction of an extra atom on the surface of an icosahedron, and CN 14 formed by the addition of an atom outside two of the rectangular faces of a CN 12 pentagonal prism with polar atoms.

Removing four spheres located at the corners of a tetrahedron from a Friauf CN 16 polyhedron composed of spheres of equal sizes leaves a truncated tetrahedron of 12 spheres. Spheres 1.35 times the radius of the standard sphere can be accommodated at the center of the polyhedron and out from the center of the four hexagons comprising the truncated tetrahedron. The central sphere is thus surrounded by 12 small spheres and 4 large ones giving the CN 16 Friauf polyhedron, and the 4 outer large atoms arranged tetrahedrally would center neighboring interpenetrating CN 16 polyhedra, the fused polyhedra being the truncated tetrahedra. Truncated tetrahedra can be arranged in close packed layers sharing three hexagonal faces within the layers and exposing the fourth hexagonal face either above or below, so that the layers can be stacked one over the other, hexagon upon hexagon (Figure 13-13). Within such an arrangement composed of contiguous N atoms of diameter a at the 12

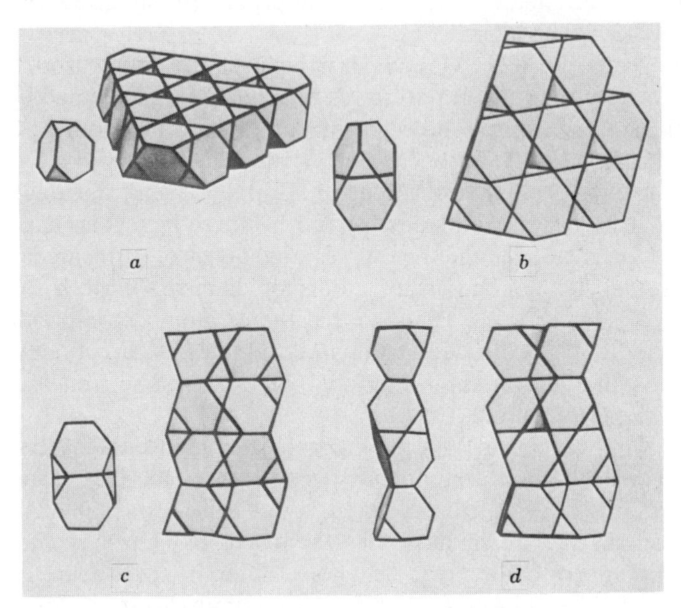

Figure 13-13. Arrangements of truncated tetrahedra (*a*) in close packed plane, (*b*) in the $MgCu_2$ structure, (*c*) in the $MgZn_2$ structure, and (*d*) in the $MgNi_2$ structure. (Samson [3].)

corners, the distance between the larger M atoms centering the polyhedra is $1.23a$, and the M–N distance is $1.17a$. Hence the radius of the M atom is 8% smaller ($0.615a$) in the M–M direction than in the M–N direction ($0.67a$) so that, as noted in the Laves phases where CN 16 polyhedra interpenetrate, the M atoms are compressed in the M–M direction (p. 59). Indeed the Laves phases provide an example of the close packing of these polyhedra as shown in Figure 13-13. In the $MgCu_2$ structure the truncated tetrahedra forming the close packed layers are rotated 60° in the second layer (Figure 13-13*b*), whereas in the $MgZn_2$ structure the second layer truncated tetrahedra are a mirror image (Figure 13-13*c*) of those in the first layer. In the $MgNi_2$ structure (Figure 13-13*d*) successive layers repeat first the $MgCu_2$ arrangement and then the $MgZn_2$ arrangement. In the other Laves phase polytypes successive stacking of truncated tetrahedra follow either the $MgCu_2$ or the $MgZn_2$ arrangements (i.e., rotation 60° or mirror image). The larger (Mg) atoms center the truncated tetrahedra and the smaller atoms occupy the vertices of the truncated tetrahedra, each of which is shared between 6 truncated tetrahedra. The large atoms are surrounded by 12 small and 4 large atoms (CN 16 Friauf polyhedron) and the small atoms, by 6 small and 6 large atoms in an icosahedral arrangement. Thus a contiguous close packing of truncated tetrahedra leads to an interpenetrating array of CN 16 Friauf polyhedra and icosahedra.

Samson calls the CN 15 polyhedra the μ-phase polyhedron, since it was first observed in the structure of this phase. When formed by atoms of equal size the central hole can accommodate an atom 1.31 times larger. The fused polyhedron resulting from the layer packing of interpenetrating CN 15 polyhedra (Figure 13-2, p. 666) is a truncated trigonal prism with 12 corners and sides formed by three hexagons having four triangular faces above and below. In the layer of truncated trigonal prisms formed by contiguous N atoms of diameter a, both the M–M and M–N distances are $1.155a$, so that the M atom is compressed 12% ($0.578a$) in the M–M direction compared to the M–N direction ($0.655a$). In this case the deformation is trigonal, whereas in the truncated tetrahedron it was tetrahedral.

The Frank–Kasper CN 14 polyhedron is composed of a hexagonal antiprism with two extra atoms outside the hexagonal faces. Such polyhedra pack together in close packed layers which stack one upon the other, hexagon over hexagon, so that the fused polyhedron is a hexagonal antiprism. Generally the two hexagons differ in size, being occupied by atoms of different sizes. When the antiprisms form a close packed layer, the size of the larger hexagon is $2/\sqrt{3} = 1.155$ times that of the smaller one. Generally hexagonal antiprisms are fused to truncated tetrahedra, truncated trigonal prisms, or, by sharing the larger hexagon with another hexagonal antiprism, forming bihexagonal antiprisms (Figure 13-2, p. 666). Samson[22] draws attention to an interesting reason why hexagonal antiprisms are common yet hexagonal prisms are uncommon features in the structural arrangements of metals: the triangular faces of centered hexagonal antiprisms give rise to tetrahedral interstices, whereas the square faces of centered hexagonal prisms create octahedral interstices and hence greater interstitial space and less close packing of the atoms. Similar reasoning applies to the preference also for pentagonal antiprisms over pentagonal prisms.

The relationship between the CN 14, 15, and 16 polyhedra is further shown in Figure 2-2, p. 34, where it is seen that the CN 15 polyhedron is derived from CN 14 by replacing the atom with surface coordination 6 outside the larger hexagon by two atoms, and the CN 16 polyhedron by a triangle of 3 atoms parallel to the plane of the hexagons. The CN 12 icosahedron is similar to the CN 14 polyhedron except that the fused polyhedron is a pentagonal antiprism rather than a hexagonal antiprism and all surface atoms therefore have 5 surface neighbors only. With 12 corner atoms, the icosahedron can also be related to the cubo-octahedron as follows: the 12 atoms forming a cubo-octahedron form three mutually orthogonal interpenetrating squares (Figure 13-14) of edge $\sqrt{2a}$, where a is the distance between atoms forming the cubo-octahedron, whereas the 12 atoms forming the icosahedron are located at the corners of three orthogonal interpenetrating rectangles whose edges are a and $1.618a$ (Figure 13-14). The center to vertex distance in a cubo-octahedron of edge a is of course also a, whereas it is only $0.951a$ in an icosahedron of edge a, and so with contiguous surface atoms of equal size, a centered icosahedron occupies a smaller volume and should be more stable than a cubo-octahedron. If all atoms are of identical size this, of course, implies a compression of the surface to center atom distance, M–M.

β-Mg_2Al_3 and $NaCd_2$: The main features of these two structures are the same, although they are partially disordered in different ways consistent with their different stoichiometric ratios, as shown by Samson.[18,20] In the ideal structures five truncated polyhedra fused about an approximate fivefold axis of symmetry form the fundamental building block of 47 atoms which contains three crystallographically different Friauf polyhedra, F_1, F_2, and F_3. Six of these groups are arranged about the vertices of an octahedron producing four more similar polyhedra (F_4) and giving a complex with T_d symmetry containing 34 polyhedra and 234 atoms (Figure 13-15). Each of these T_d complexes is connected to four others arranged at the corners of a regular tetrahedron (Figure 13-16).

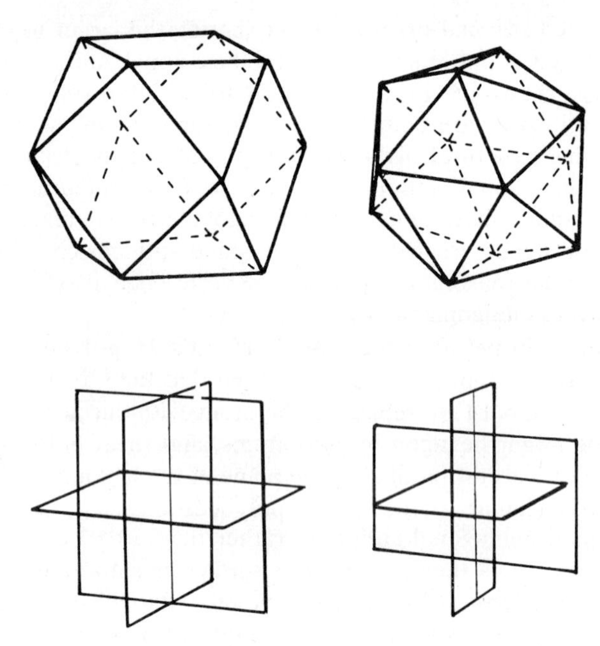

Figure 13-14. Comparison of cubo-octahedron and icosahedron. See text. (Samson[3].)

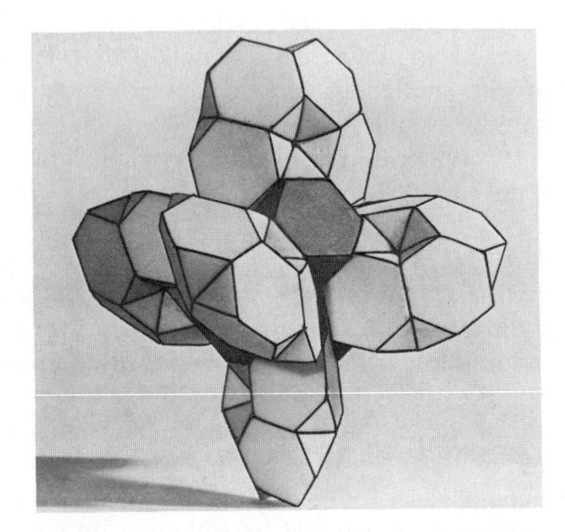

Figure 13-15. Complex of 34 truncated tetrahedra found in the β-Mg_2Al_3 and $NaCd_2$ structures. (Samson[3].)

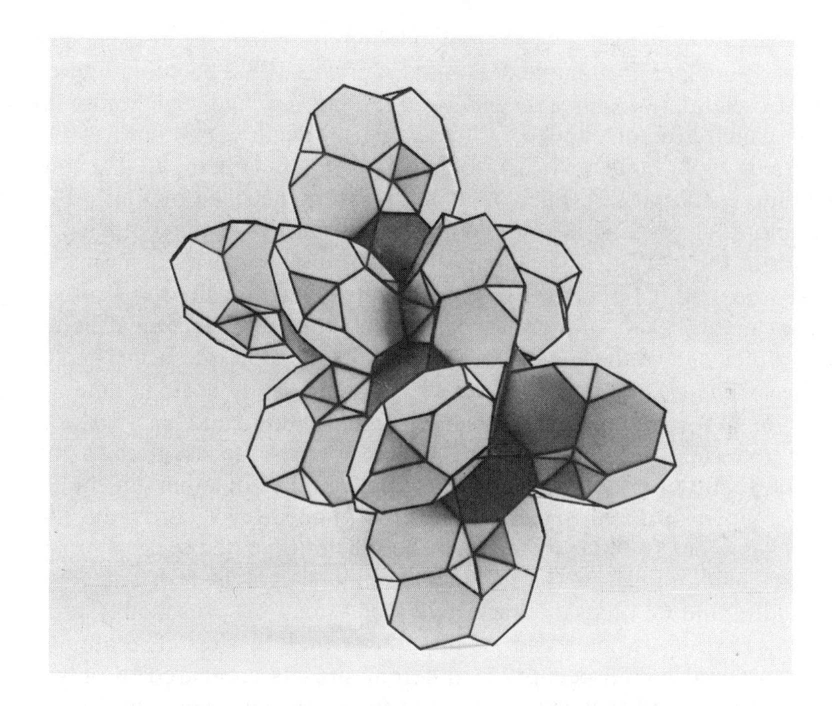

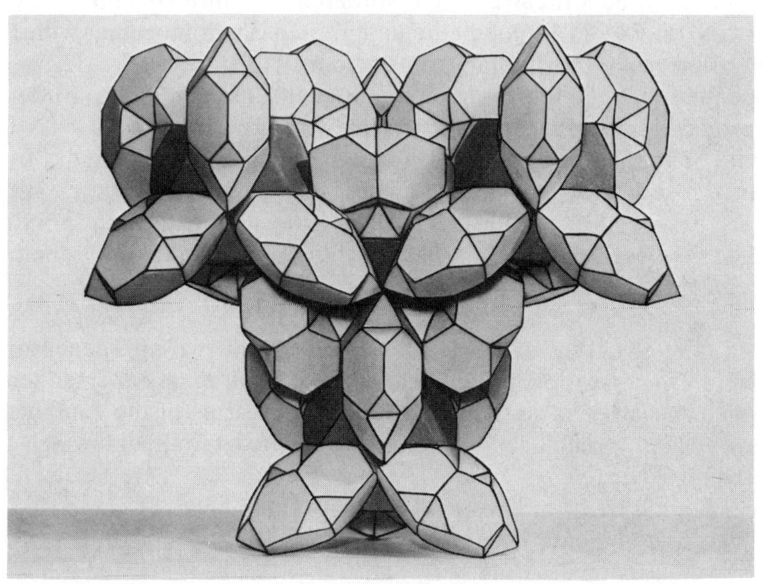

Figure 13-16. Two (upper) and 4 (lower) complexes of 34 truncated tetrahedra assembled together in the β-Mg_2Al_3 structure. (Samson [3].)

This stacking sequence is continued indefinitely so that the average numbers of atoms per T_d complex is 144 (Figure 13-17). The cubic unit cell contains eight T_d complexes giving 1152 atoms, and eight more Mg atoms which are surrounded by truncated tetrahedra (F_5) that lie at the centers of eight "spheres" of associated polyhedra (Figure 13-17) which pack into each unit cell. Each "sphere" is interpenetrated by four others. Four more Al atoms are located out from the centers of the four triangles of each of the truncated tetrahedra (F_5) surrounding these Mg atoms, the 32 Al atoms bringing the total complement of the unit cell to 1192 atoms; its edge is 28.239 Å. In the disorder that is found in the actual structure, every other one of these eight Mg-centered truncated tetrahedra and the four associated Al atoms are replaced randomly in any of six orientations by a CN 10 fused pentagonal prism which is centered and has two atoms at the poles and two atoms outside two of the prism faces, giving a complex of 15 atoms, thus reducing the total number of atoms in the unit cell of the disordered structure to 1168. Samson[20] records the ideal unit cell content as 280 truncated tetrahedra, 96 truncated trigonal prisms, 64 hexagonal antiprisms (i.e., CN 16, 15, and 14 polyhedra), 128 CN 13 polyhedra, and 624 icosahedra. The larger Mg atoms presumably prefer CN 14, 15, and 16 sites and Al the CN 12 and 13 sites. The atoms in the disordered model occupy 23 different sitesets compared to only 17 sets in the ordered structure. The disordered structure contains 252 CN 16, 24 CN 15, 48 CN 14 polyhedra and 672 icosahedra together with 170 somewhat irregular polyhedra with CN from 10 to 16.

Since the truncated tetrahedra are contiguous throughout the structure, they provide the most convenient framework on which to discuss the structure. The disorder which increases the number of icosahedra from 624 to 672, indicates that the building principle upon which such structures are based, is the establishment of as many icosahedra as is possible (tetrahedral close packing) consistent with satisfying the coordination requirements of the large atoms.

Cu_4Cd_3: The structure of Cu_4Cd_3[21] is made up of two interpenetrating diamond-like frameworks of polyhedra; one is built up of truncated tetrahedra and the other of icosahedra. The relative sizes of the component atoms are appropriate for Cu forming truncated tetrahedra which are centered by Cd.

The diamond-like framework of truncated tetrahedra is composed of three types of complex (Figure 13-18*a*, *b*, *c*). The octahedron (Figure 13-18*a*) which is made up of 4 F_1 and 6 F_2 truncated tetrahedra has T_d symmetry. The tetrahedron (Figure 13-18*c*) consists of five polyhedra F_5+4F_6 and the two groups of four polyhedra added to this (Figure

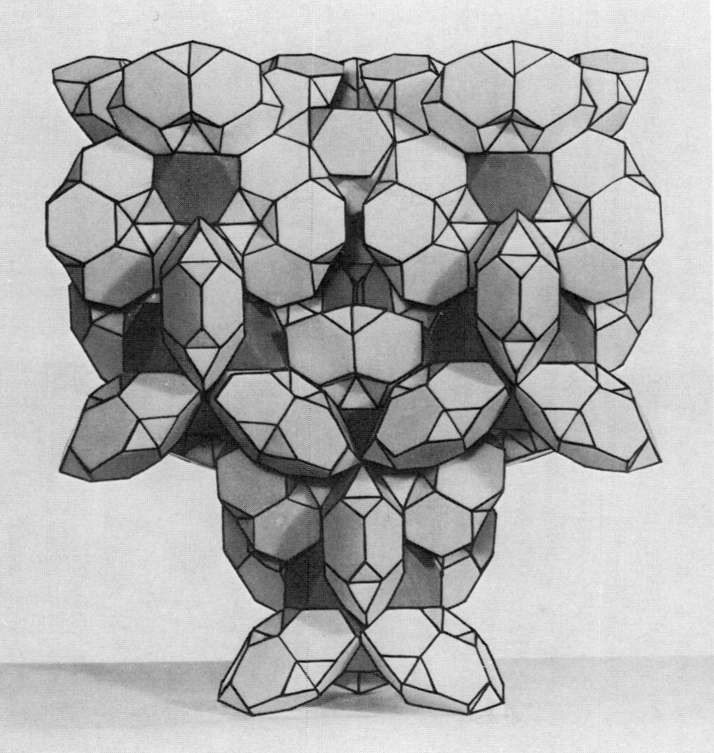

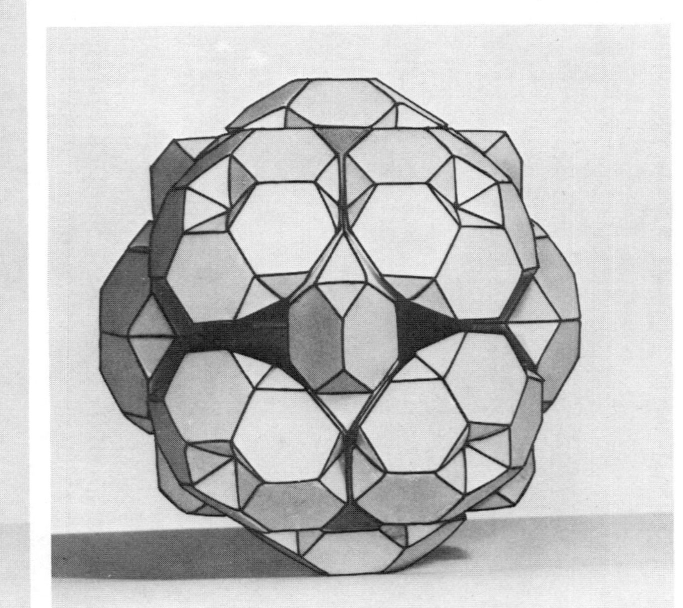

Figure 13-17. Assembling of complexes of truncated tetrahedra in the β-Mg_2Al_3 structure. On the right are shown "spheres" of polyhedra which are centered by F_5. See text. (Samson[3]).

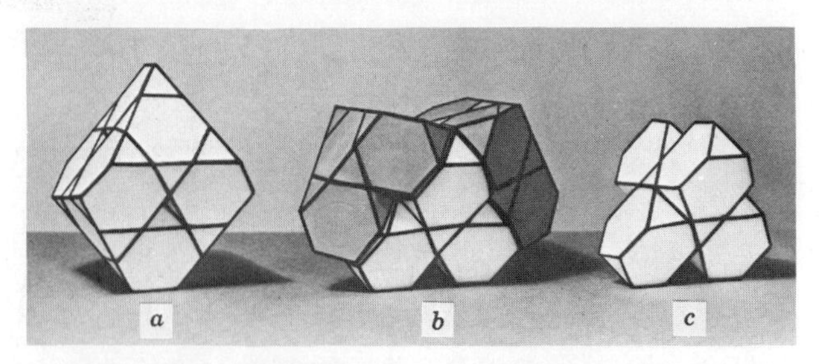

Figure 13-18. Three complexes of truncated tetrahedra found in the Cu_4Cd_3 structure. See text. (Samson[21].)

13-18*b*) are F_3 and $3F_4$. These building blocks are packed together in the unit cube (space group $F\bar{4}3m$) to give an infinite three-dimensional framework, four octahedra being arranged about sites 4(*b*) ($\frac{1}{2}, \frac{1}{2}, \frac{1}{2}; \frac{1}{2}, 0, 0; 0, \frac{1}{2}, 0; 0, 0, \frac{1}{2}$) and four tetrahedra about sites 4(*c*) ($\frac{1}{4}, \frac{1}{4}, \frac{1}{4}; \frac{1}{4}, \frac{3}{4}, \frac{3}{4}; \frac{3}{4}, \frac{1}{4}, \frac{3}{4}; \frac{3}{4}, \frac{3}{4}, \frac{1}{4}$); the groups of four (F_3+3F_4) polyhedra shown dark in Figures 13-18 and 13-19 link the octahedral and tetrahedral blocks together as shown in Figure 13-19.

The diamond-like icosahedral framework is made up of two types of building block which are now described in terms of the actual coordination polyhedra, rather than the fused polyhedra. In one, five icosahedra sharing vertices are arranged at the vertices of a pentagon so as to enclose a pentagonal prism (Figure 13-20*a*). Six of these groups can be arranged at the vertices of an octahedron of T_d symmetry (Figure 13-20*c*) in such a way that they interpenetrate, sharing icosahedra and giving a complex made up of 14 icosahedra which encloses six pentagonal prisms. Figure 13-20*b* shows two groups of five icosahedra (Figure 13-20*a*) interpenetrating at right angles so that the pentagonal prism is shared by two icosahedra (above and below) which have a common vertex centering the prism. The centers of these two icosahedra are therefore poles of the fused pentagonal prism giving a CN 12 polyhedron. Indeed, each shared pair of icosahedra vertices in the complex shown in Figure 13-20 center pentagonal prisms. 36 more pentagonal prisms are thus created and in 12 of these, two prism faces are deformed by the introduction of two more atoms giving a CN 14 fused polyhedron. The building unit shown in Figure 13-20*c* therefore represents 14 icosahedra, 30 pentagonal prisms with two atoms at the poles giving CN 12 polyhedra, and 12 more having two atoms penetrating the prism faces giving CN 14 polyhedra.

The second complex of icosahedra is composed of pairs of icosahedra Figure 13-21*a* interpenetrating so that a vertex of one forms the center of the other. Six of these pairs are arranged at the vertices of an octahedron

Figure 13-19. Three-dimensional framework in the Cu_4Cd_3 structure, which is created from the 3 complexes of truncated tetrahedra shown in the previous figure. (Samson[21].)

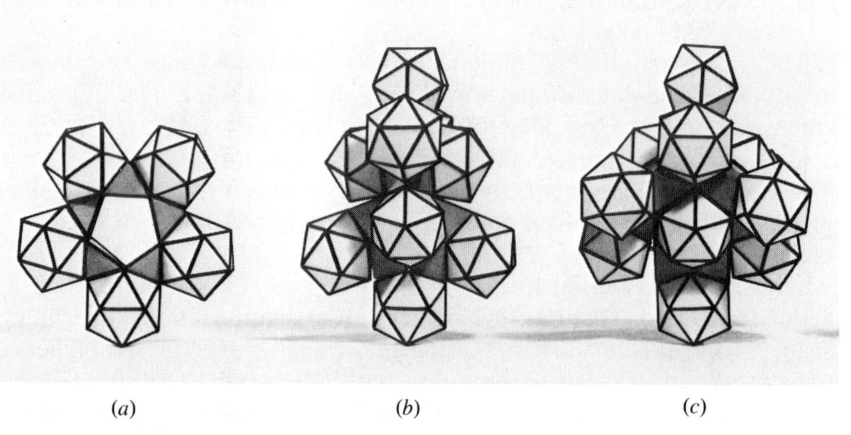

(*a*) (*b*) (*c*)

Figure 13-20. Arrangement of icosahedra in the Cu_4Cd_3 structure. See text. (Samson[21].)

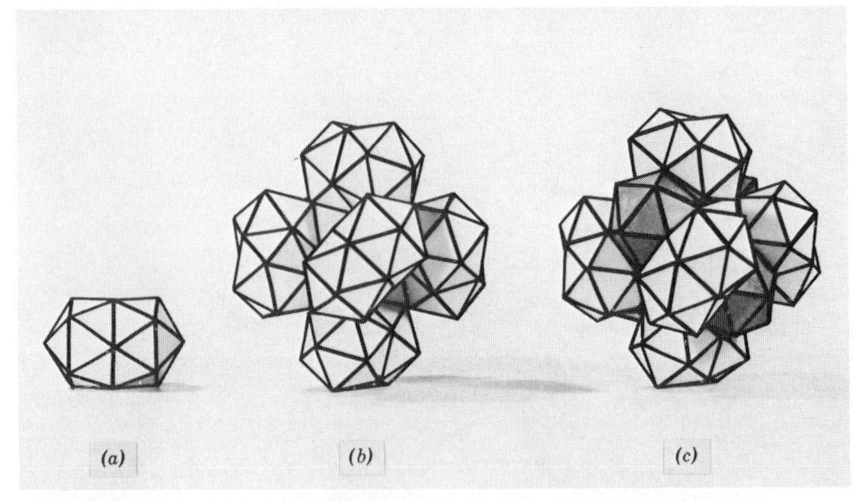

Figure 13-21. Arrangements of pairs of interpenetrating icosahedra in the Cu_4Cd_3 structure. See text. (Samson[21].)

of T_d symmetry so that the axes of diametrically opposed icosahedron pairs are orthogonal (Figure 13-21*b*). The insertion of four more icosahedra centered at the vertices of a regular tetrahedron completes the building block shown in Figure 13-21*c*. Each of these icosahedra share 6 triangles with the three groups of paired icosahedra surrounding it and therefore each vertex is shared between two paired icosahedra. This common vertex is the center of a pentagonal prism having two atoms at the poles and two atoms inserted at two of the prism faces giving a CN 14 polyhedron. The complete complex contains 16 icosahedra and 18 of the CN 14 polyhedra.

These two icosahedral building blocks are packed together to give an infinite three-dimensional framework (Figure 13-22). The unit cube contains four of the first kind located about sites 4(*d*) ($\frac{3}{4}, \frac{3}{4}, \frac{3}{4}; \frac{3}{4}, \frac{1}{4}, \frac{1}{4}; \frac{1}{4}, \frac{3}{4}, \frac{1}{4}; \frac{1}{4}, \frac{1}{4}, \frac{3}{4}$) and four of the second kind located about 4(*a*) ($0, 0, 0; 0, \frac{1}{2}, \frac{1}{2}$ ∩). These are joined together at vertices which create a further 12 pentagonal prisms with atoms at the poles (CN 12 polyhedra) between each two complexes.

The truncated tetrahedron framework (Figure 13-19) and the icosahedra framework (Figure 13-22) themselves interpenetrate, one filling space in the other, and the sharing of vertices creates new polyhedra (mostly icosahedra) which penetrate both complexes. The structure so assembled contains 124 CN 16 Friauf polyhedra, 144 CN 15 μ-phase polyhedra, 120 centered CN 14 polyhedra (pentagonal prisms with atoms

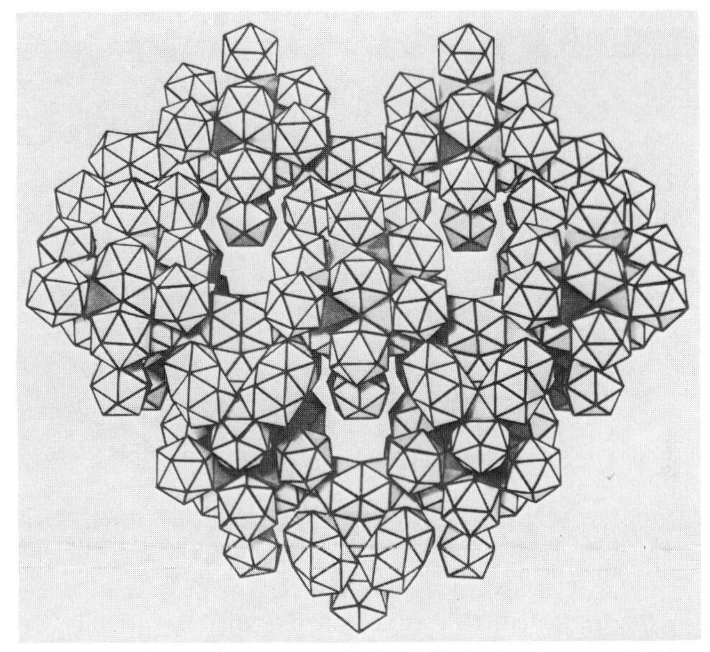

Figure 13-22. Icosahedral framework of the Cd_4Cu_3 structure which fits into cavities in the framework of truncated tetrahedra shown in Figure 11-19. (Samson[21].)

at the two poles and two inserted in prism faces), 168 centered CN 12 pentagonal prisms with atoms at the two poles, and 568 centered icosahedra. The cell has an edge $a = 25.871$ Å, and contains 1124 atoms ($Cu_{640}Cd_{484}$) distributed on 29 sitesets according to Samson.[21]

Related Structure Types in which Truncated Tetrahedra Play a Prominent Role

E Phase, $Mg_3Cr_2Al_{18}$, $ZrZn_{22}$ and α-VAl_{10} (*cF*184): $Mg_3Cr_2Al_{18}$ has a large cubic cell containing 184 atoms.[23] The structure can be regarded as composed of truncated tetrahedra of Al fused to each other by hexagonal prisms covering the four hexagonal faces and therefore with axes directed tetrahedrally as shown in Figure 13-23 taken from Samson.[3] The truncated tetrahedra are centered by Mg atoms and the hexagonal prisms (that accommodate spheres some 20% larger in radius than the atoms at the apexes), are centered by Mg and Al in disordered array. Al atoms are located outside each of the rectangular faces of the hexagonal prisms, so that they form regular octahedra centered in the cavities

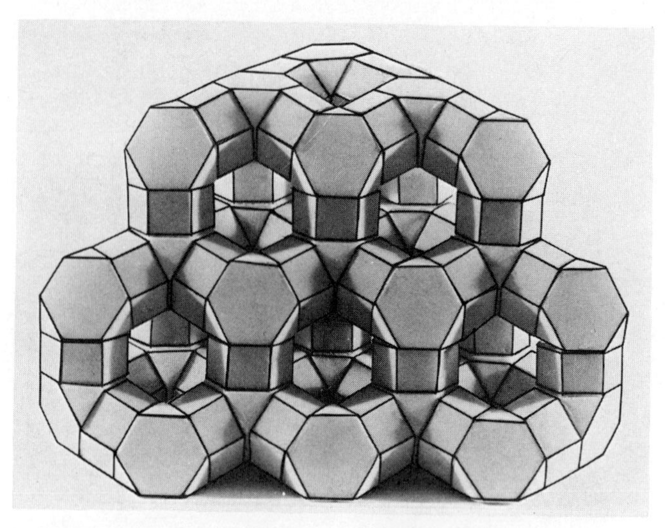

Figure 13-23. The $Mg_3Cr_2Al_{18}$ (*cF* 184) structure: showing truncated tetrahedra of Al interconnected by hexagonal prisms. (Samson [3].)

formed by the truncated tetrahedra and hexagonal prisms. The Cr atoms are located between two cavities at the center of a distorted icosahedral arrangement of Al atoms. This arrangement of centered truncated tetrahedra, hexagonal prisms, and icosahedra (CN 16, 14, and 12 polyhedra) also generates pentagonal prisms of Al atoms around the vertices of the CN 16 polyhedra. The pentagonal prisms have two atoms at the extended poles making CN 12 polyhedra that are centered by Al atoms.

In the $ZrZn_{22}$ arrangement of the *E* phase structure,[24] Zr replaces Mg at the centers of the CN 16 polyhedra and Zn replaces the Mg atoms in the $\frac{2}{3}$ of the sites at the centers of the hexagonal prisms, as well as occupying the remaining structural sites. It would thus seem that the structure contains Zn atoms of three different sizes—those centering hexagonal prisms, those forming CN 12 pentagonal prisms with two polar atoms, and those centering the icosahedra, giving a formula $ZrZn'_2Zn''_2$-Zn'''_{18}. Indeed, it is remarkable that the introduction of 4.3 at. % Zr should have this effect on an array of Zn atoms that normally cling so tenaciously to their hexagonal structure, even under pressures as high as 180 kb![25]

The α-VAl_{10} structure[26] also resembles that of $Mg_3Cr_2Al_{18}$, except that the CN 16 polyhedra are uncentered and the hexagonal prisms are centered by Al atoms alone, so that the formula could be written $Al_2\square V_2Al_{18}$ where $\square$ represents the vacant site at the center of the Friauf polyhedra.

A feature of these and other structures containing "isolated" truncated

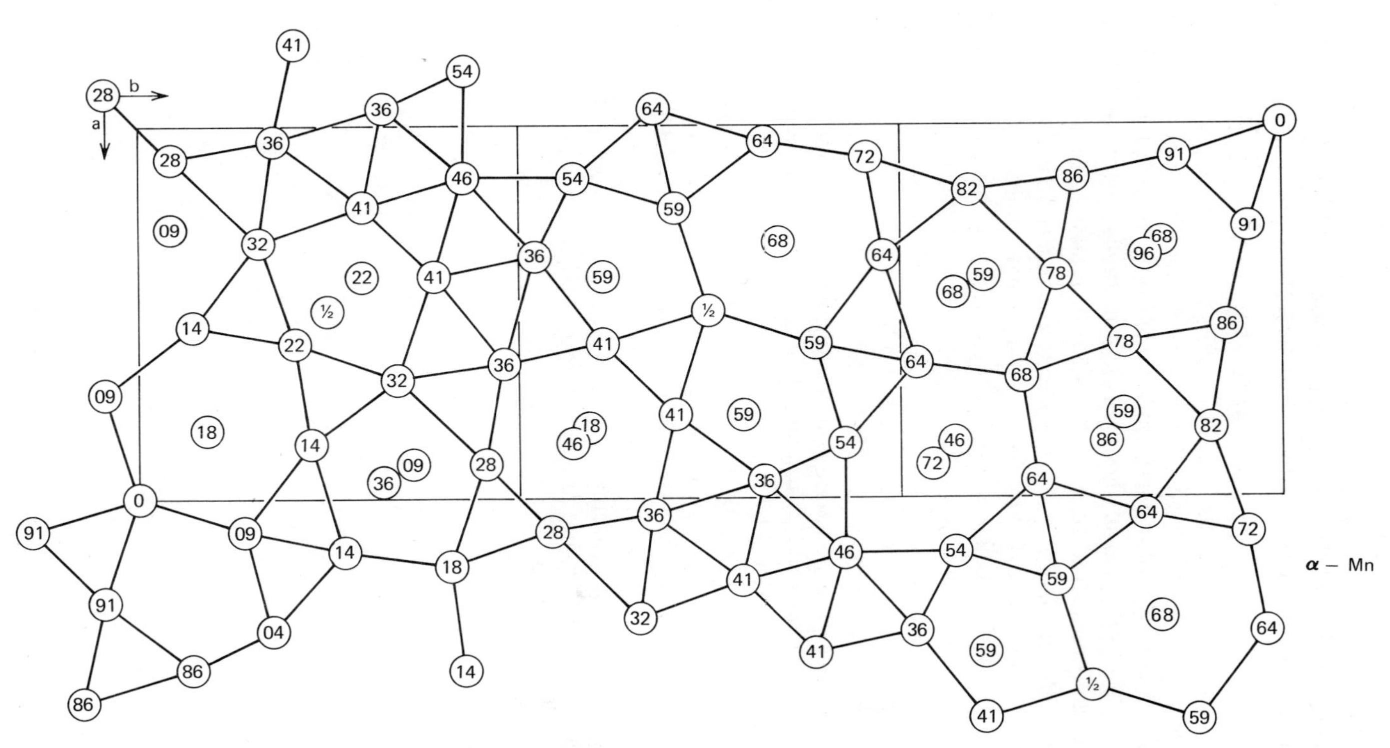

Figure 13-24. Arrangement of atoms close to the $(\overline{1}1\overline{4})$ plane in the α-Mn ($cI58$) structure on passing through three cells from one corner to that diagonally opposite.

tetrahedra (those not joined hexagon to hexagon) is that icosahedra, if present, do not interpenetrate the CN 16 polyhedra.

α-Mn, χ Phase (Ti_5Re_{24}), γ-$Mg_{17}Al_{12}$ (*cI*58): Mn occupies four different sitesets in the cubic cell ($a = 8.914$ Å) of its low-temperature form which is stable below 727°C. Two of these are similar, being surrounded by CN 16 Friauf polyhedra; one is surrounded by a very distorted Frank–Kasper CN 14 polyhedron and one is icosahedrally surrounded.[27] Mn(3) is generally considered to have CN 13 rather than 14 since one of its ligands is at a distance of 3.63 Å, whereas the others lie in the range from 2.50 to 2.96 Å. The atoms form a somewhat rumpled and irregular pentagon-hexagon-triangle net on $\{\bar{1}1\bar{4}\}$ type planes as shown in Figure 13-24. In planes parallel to {100} the atoms form rumpled irregular hexagon-triangle nets.

Mn(1) at the cell corners and body center has 4 Mn(2) at 2.825 Å and 12 Mn(4) at 2.72 Å (Figure 13-25). Mn(2) has 1 Mn(1) neighbor, 3 Mn(3) at 2.50 Å (these four neighbors having surface coordination 6), 3 Mn(3) at 2.96 Å, 6 Mn(4) at 2.70 Å and 3 at 2.895 Å. Mn(3) has 2 Mn(2) neighbors, 7 Mn(3) at 2.67 Å and 4 Mn(4) at 2.45 to 2.67 Å. Mn(4) is icosahedrally surrounded by 1 Mn(1), 3 Mn(2), 5 Mn(3) and 3 Mn(4) neighbors, the latter at very close distances of 2.24 Å and 2×2.38 Å.

χ phase alloys have the same structure. M_5N_{24} is a preferred stoichiometry for binary phases with the M atoms in the Mn(1) and Mn(2) sites centering the largest coordination polyhedra. In the binary χ phases such as Ti_5Re_{24} with the atomic ratio 5:24, the 10 large atoms per cell occupy

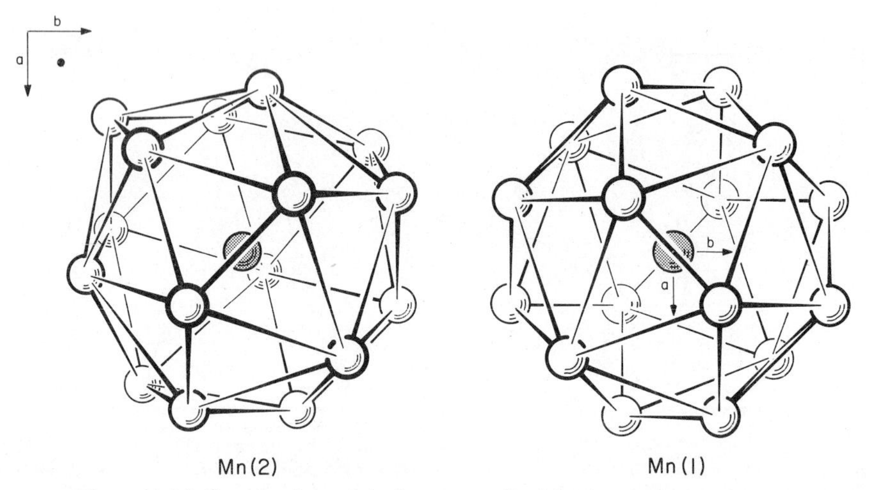

Figure 13-25. Coordination polyhedra surrounding Mn(1) and (2) atoms in α-Mn.

the centers of the 10 truncated tetrahedra, and the other atoms occupy the centers of the smaller CN 13 and CN 12 polyhedra. In α-Mn itself, the three sites of different size indicate several different valence states of Mn, and this is confirmed by the determination of the magnetic moments on the Mn atoms. The structure of α-Mn itself has only been determined by powder diffraction. Phases with the α-Mn structure occur at a preferred electron concentration of about 6.4 to 7.2 electrons per atom (Figure 13-12, p. 676).

Samson[3] describes the $Mg_{17}Al_{12}$ structure[28] which is isotypic with α-Mn, as built up of units composed of truncated tetrahedra (F_1 and F_2). F_1 shares its four hexagons with F_2, which are thus arranged at the vertices of a regular tetrahedron and centered by the 4 atoms completing the CN 16 Friauf polyhedron of F_1. These groups of 5 truncated tetrahedra are arranged about the nodes of a b.c. cubic lattice (Figure 13-26) in such a way that the 4 triangles of F_1 are shared with F_2 of other groups. The polar atoms out from the three nonshared hexagonal faces of each F_2, are vertices of other surrounding F_2 polyhedra (Figure 13-26), being themselves surrounded by 13 neighbors. Since the unit cell contains 10 CN 16 polyhedra ($2F_1$ and $8F_2$), there are 24 such vertices which are occupied by Mg, as are the centers of F_1 and F_2, giving the 34 Mg atoms of the ideal formula. Samson has found the range of the phase to extend at least to $Mg_{13}Al_{16}$, in which case Al replaces Mg at corners common to two F_2 Friauf polyhedra.

The arrangement of centered F_1 and F_2 truncated tetrahedra also generates 24 icosahedra made up of 9 Mg + 3 Al and centered by Al, as well as the 24 CN 13 polyhedra centered by Mg or Mg + Al randomly according to composition. F_1 truncated tetrahedra are penetrated by 12 icosahedra and F_2 are penetrated by 9 icosahedra and three CN 13 polyhedra.

δ-Mo-Ni (*oP*56): The δ-phase has cell dimensions $a = 9.108$, $b = 9.108$, $c = 8.852$ Å, $Z = 56$.[29] The structure contains only CN 12, 14, 15, and 16 polyhedra, and is based on tetrahedral close packing. Almost planar layers of atoms parallel to (041) and (401) planes are very similar and they have the general characteristics of hexagon-pentagon-triangle layers of Frank–Kasper structures, but there are irregularities as shown in Figure 13-27. The Mo atoms order preferentially at the centers of the larger CN 16 and 15 polyhedra.

ϵ-$Mg_{23}Al_{30}$ (*hR*53): ϵ-$Mg_{23}Al_{30}$ has a rhombohedral unit cell, $a = 10.33$ Å, $\alpha = 76.4$ Å, the equivalent hexagonal cell having the dimensions $a = 12.83$, $c = 21.75$ Å and containing 3×53 atoms.[30] Samsom describes the structure in terms of groups of truncated tetrahedra of two kinds F_1 and F_2 shown in Figure 13-28*a* and *b*, which are arranged along the

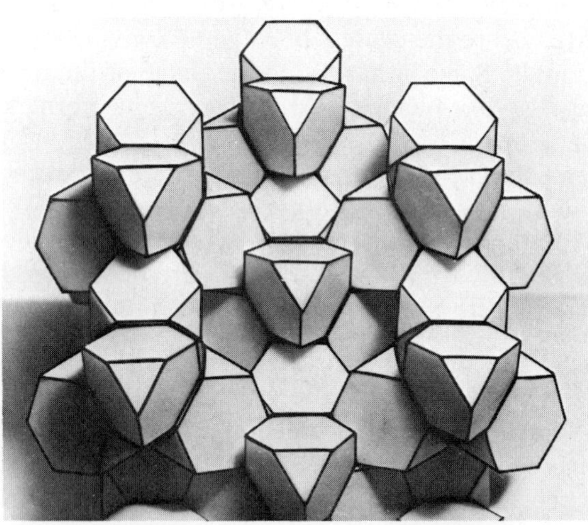

Figure 13-26. Arrangements of the truncated tetrahedra in the γ-$Mg_{17}Al_{12}$ (*cI*58) structure. (Samson[3].)

edges of the rhombohedral cell in such a way that an infinite three-dimensional framework is obtained. This contains cavities (Figure 13-28*c*) that are filled with other polyhedra including icosahedra. The rhombohedral cell contains 8 Friauf CN 16 polyhedra, 24 icosahedra, and 21 irregular polyhedra, six of which resemble the CN 15 μ-phase

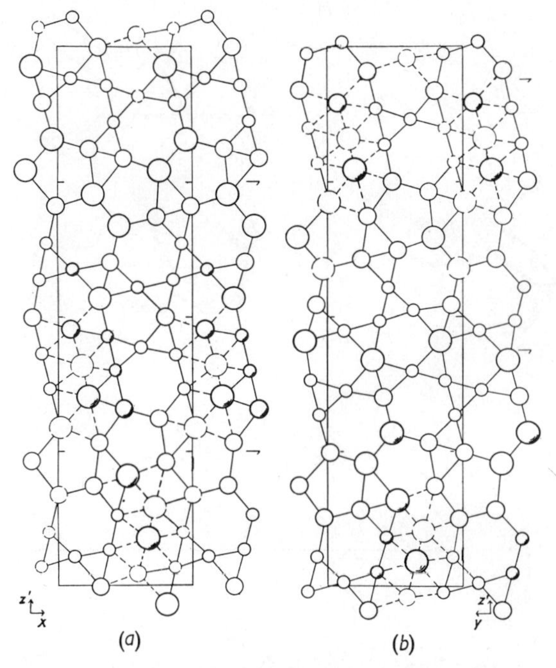

Figure 13-27. The δ-Mo-Ni (*oP*56) structure showing (*a*) projection of nearby atoms (component distance normal to the plane <0.65 Å) onto the (04$\bar{1}$) plane passing through the origin, (*b*) similarly for a (40$\bar{1}$) plane passing through (0, 0, $-\frac{1}{4}$). Broken lines connect atoms whose component interatomic distance normal to the planes is greater than 0.35 Å. (Shoemaker and Shoemaker[29].)

polyhedra. F_1 truncated tetrahedra are penetrated by 12 icosahedra, F_2 by 10 icosahedra and 2 CN 14 polyhedra. On the average, icosahedra have about half of the vertices occupied by the larger Mg atoms.

The ε-$Mg_{23}Al_{30}$ structure resembles closely that of the *R* phase (p. 668), although differences in structural parameters of the latter, which in an idealized form is a Frank–Kasper type structure (p. 31), change the polyhedra filling the cavities in the arrangement of truncated tetrahedra. Along the body diagonal of the rhombohedral cell of the *R* phase, the arrangement of contiguous centered polyhedra (Figure 13-28*d*) is a truncated tetrahedron followed by three icosahedra and another truncated tetrahedron, which then repeats. In ε-$Mg_{23}Al_{30}$, however, this arrangement is modified by widening the central icosahedron so that the atoms centering the first and third icosahedra are also ligands of that centering the middle icosahedron. The resulting sequence of coordination

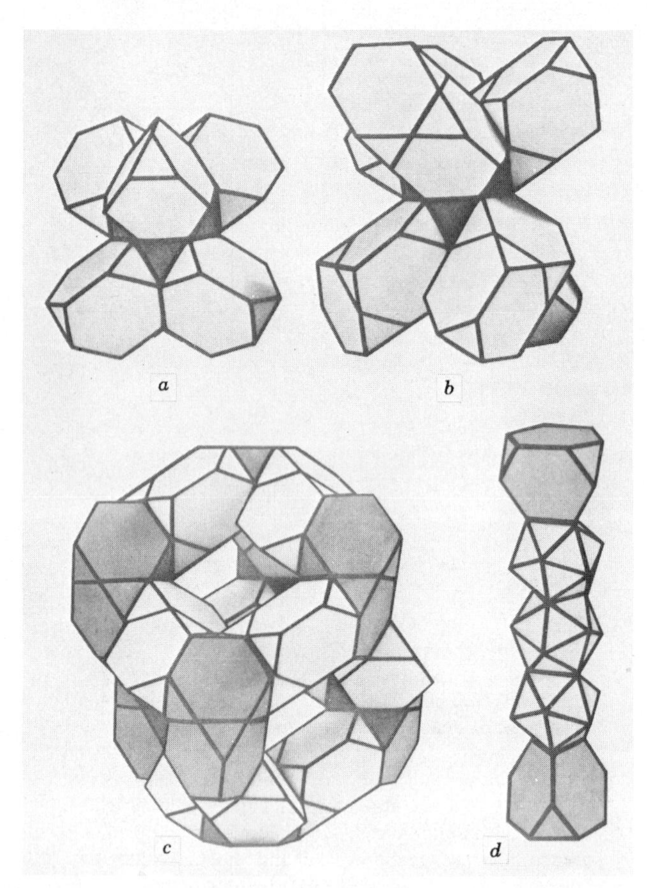

Figure 13-28. (*a*, *b*, and *c*) Arrangements of truncated tetrahedra in the ϵ-$Mg_{23}Al_{30}$ (*hR*53) structure. See text. (*d*) Arrangement of coordination polyhedra along the body diagonal of the rhombohedral cell of the *R* phase. (Samson and Gordon [30].)

polyhedra then becomes Friauf CN 16 polyhedron, CN 13, CN 14, CN 13 polyhedra, followed by another Friauf polyhedron.

$Mg_{32}(Zn,Al)_{49}$ (*cI*162): $Mg_{32}(Zn,Al)_{49}$ is one of the most interesting structures containing CN 16 polyhedra.[31] It is based on complexes of 20 Friauf polyhedra of two kinds, F_1 and F_2, which are arranged with their centers at the vertices of a pentagonal dodecahedron, giving the large almost spherical truncated icosahedron containing 113 atoms shown in Figure 13-29. These groups are centered about the lattice points of a b.c. cube as shown in Figure 13-29, and since each shares vertices with 8 surrounding complexes, the number of atoms per complex is reduced to

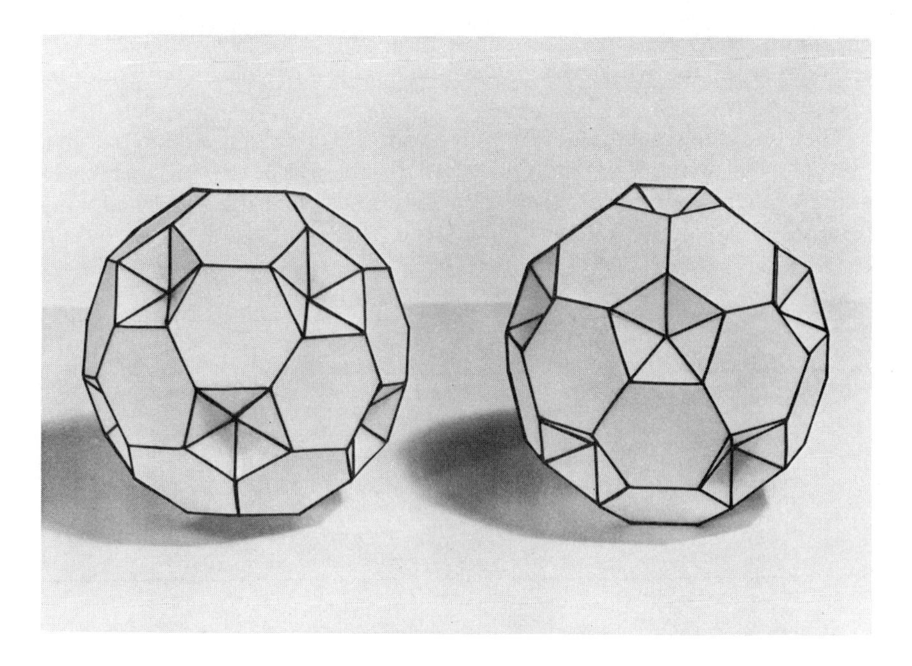

Figure 13-29. (Upper) Complexes of 20 Friauf polyhedra of two kinds F_1 and F_2 forming truncated icosahedra which occur in the $Mg_{32}(Zr,Al)_{49}$ (*cI*162) structure. (Lower) Body centered cubic stacking of the truncated icosahedra in $Mg_{32}(Zn,Al)_{49}$. (Samson [3].)

81, giving 162 atoms per unit cell with edge $a = 14.16$ Å. Figure 13-29 is taken from the work of Samson,[3] whose description of the structure is given here.

The interpenetrating CN 16 polyhedra generate icosahedra so that the unit cell contains in addition to 40 Friauf polyhedra, 98 icosahedra and 24 other irregular CN 14 and 15 polyhedra. F_1 Friauf polyhedra are penetrated by 12 icosahedra and F_2 by 10 icosahedra and two CN 14

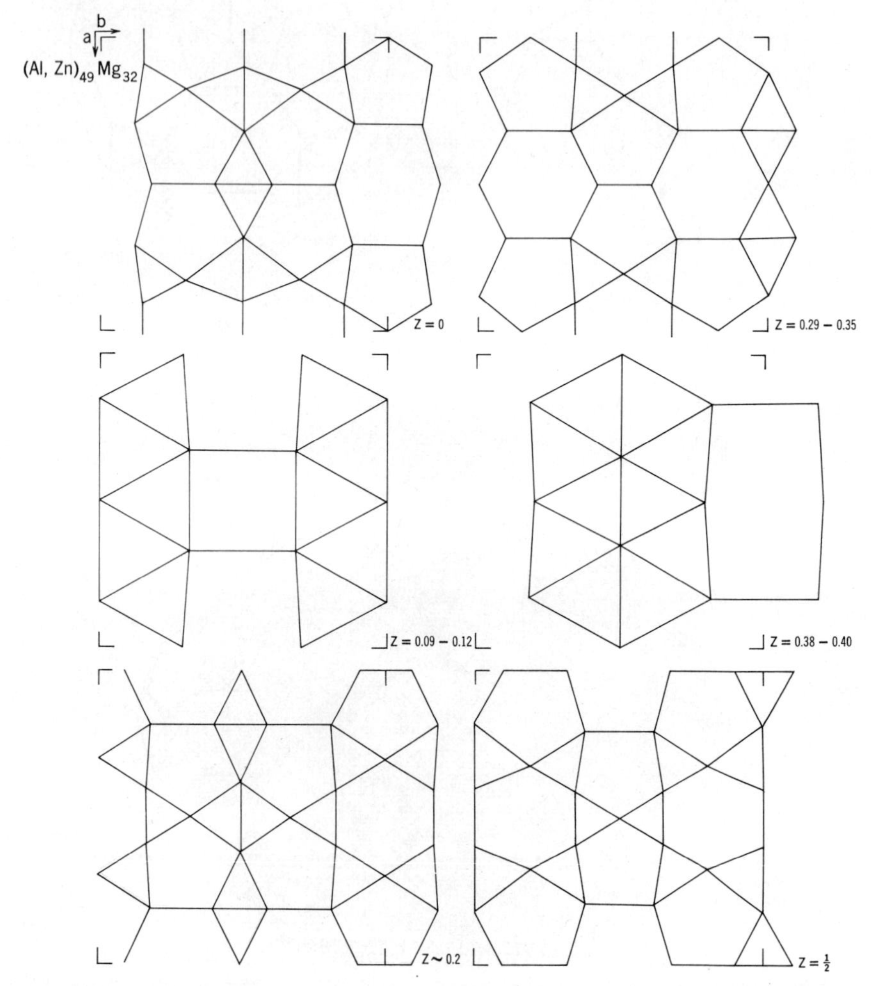

Figure 13-30. Arrangement of atoms on (rumpled) nets in the $Mg_{32}(Zn,Al)_{49}$ structure. Intersections of lines indicate atom sites. The corner brackets indicate corners of the unit cell.

polyhedra. Most of the icosahedra are made up of about half Mg atoms and half Zn + Al atoms in disordered array. The structure can also be described in terms of the icosahedral framework where, starting with an icosahedron, atoms are added in successive shells so as always to center the triangles of previous shells until the almost spherical polyhedra centered at the b.c. cubic lattice points are obtained.[31]

The $Mg_{32}(Zn,Al)_{49}$ structure in part meets the conditions for description as a Frank–Kasper layer structure, illustrating the situation where the secondary layer cannot be described by a series of parallel (zigzag) lines of atoms (p. 36). Thus Figure 13-30 shows the hexagon-pentagon-triangle layer that occurs at $z = 0$ (and at $z = \frac{1}{2}$ with b.c. translation). A $3^6 + 3^3 4^2$ subsidiary layer of atoms centering the pentagons and hexagons lies at approximately $z = 0.1$ and 0.9. Another very approximately planar hexagon-pentagon-triangle layer (Figure 13-30) occurs at $z =$ 0.2 and 0.8, and with b.c. translation at $z = 0.3$ and 0.7. The subsidiary layer at $z \sim 0.1$ also centers the hexagons and pentagons in the layer at $z = 0.2$. This sequence of layers at z coordinates of 0.8 primary, 0.9 secondary, 0 primary, 0.1 secondary, and 0.2 primary, which repeats with $\frac{1}{2}$, $\frac{1}{2}$, 0 translation at z values of 0.3, 0.4, 0.5, 0.6, and 0.7, leaves primary layers at $z = 0.2$ and 0.3, also at $z = 0.7$ and 0.8, which are not separated by secondary layers of atoms. The Ru_3Be_{17} structure (*cI*160) has a rather similar arrangement of layers as shown in Figure 13-32, and

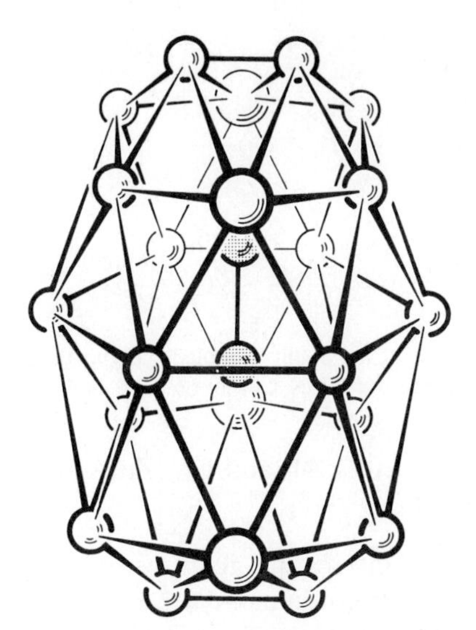

Figure 13-31. Pairs of Be(6) atoms in the Ru_3Be_{17} structure which are surrounded by 22 atoms.

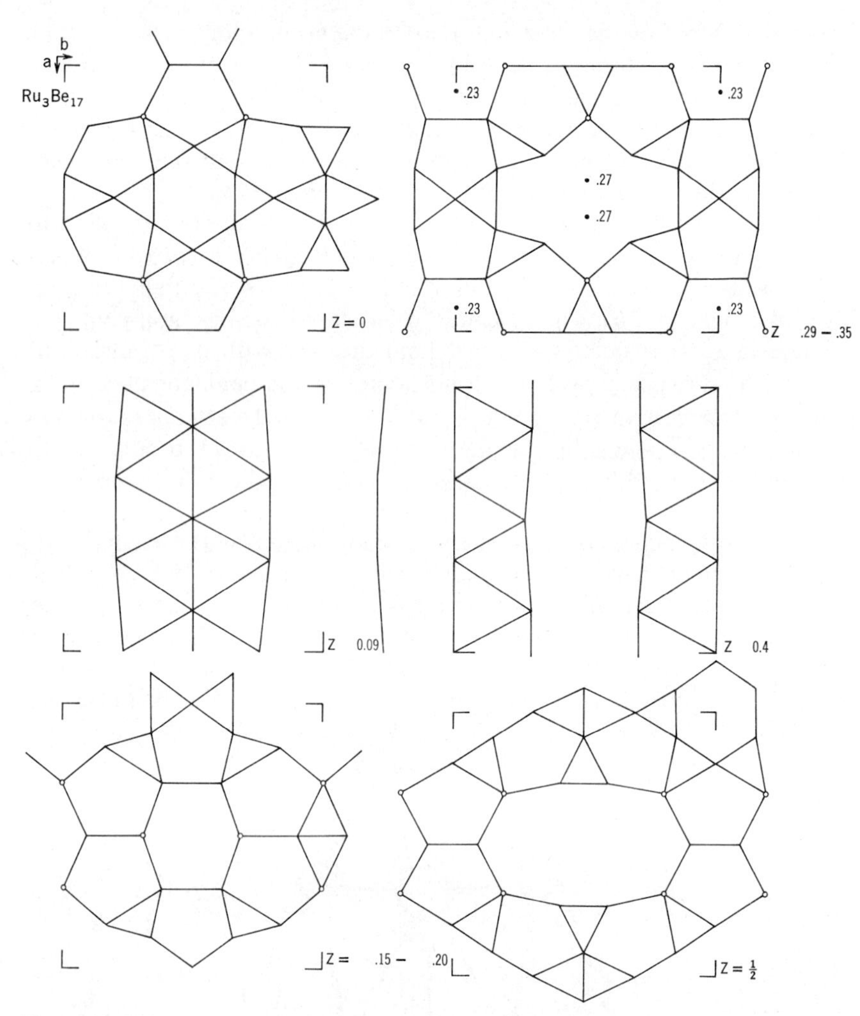

Figure 13-32. Arrangement of atoms on (rumpled) nets in the Ru_3Be_{17} ($cI160$) structure. Intersections of lines indicate atom sites. Corner brackets indicate corners of the unit cell.

the phases Mg_4CuAl_6, Li_3CuAl_5 and $Li_{32}(Zn,Al)_{49}$ are probably icostructural with $Mg_{32}(Zn,Al)_{49}$.

Ru_3Be_{17} ($cI160$): Two interesting features of the Ru_3Be_{17} structure[32] are i) the occurrence of Be(6) in pairs at 2.12 Å which are surrounded by the cage of 22 atoms shown in Figure 13-31, each Be(6) atom itself being surrounded by a μ phase CN 15 polyhedron, and ii) a large hole that occurs about the origin of the structure and which is surrounded by

12 Be(3) atoms at a distance of 2.81 Å. The Be atoms occupy six sitesets and Ru one. Ru is surrounded by a CN 16 Friauf-type polyhedron of Be with Ru–Be distances of 2.38 to 2.61 Å. Be(2) to (5) have 3 Ru neighbors and Be(1) and (6) have two. The closest Be–Be approaches are 2.04 Å. The arrangement of atoms in layers in the Ru_3Be_{17} and $Mg_{32}(Al,Zn)_{49}$ structures is rather similar as can be seen by comparing Figures 13-30 and 13-32.

OTHER STRUCTURES IN WHICH ICOSAHEDRA (AND CN 14, 15, OR 16 POLYHEDRA) ARE IMPORTANT

$CoAs_3$ (*cI*32): In the skutterudite structure the As atoms form almost perfect icosahedra (Figure 13-33). The Co atoms lie between the icosahedra being surrounded octahedrally by 6 As at 2.35 Å. Each As has 2 Co neighbors and 2 As at 2.46 Å; four other As atoms occur at 3.12 Å.[33]

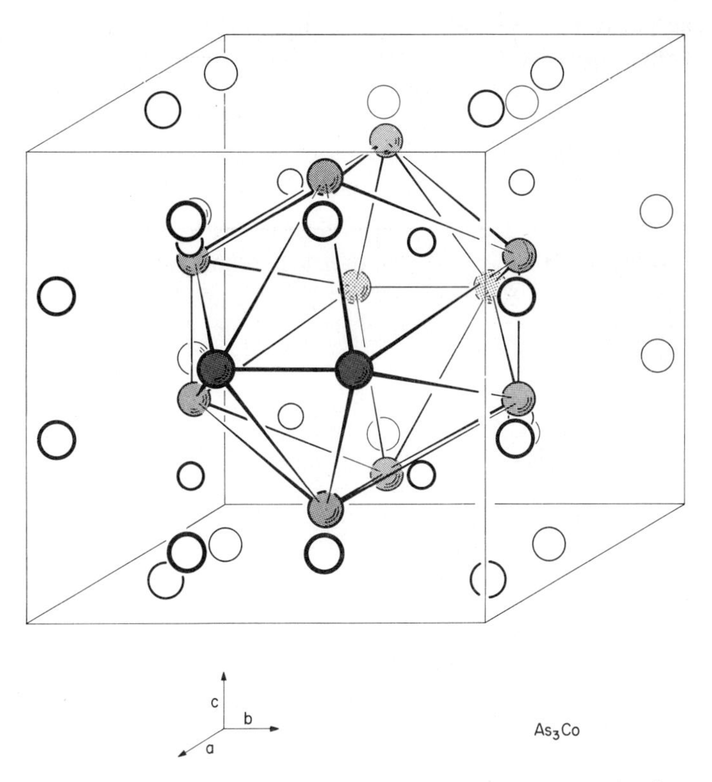

Figure 13-33. Pictorial view of the $CoAs_3$ (*cI*32) structure showing icosahedra of As atoms.

The structure can also be regarded as being made up of planes of Co atoms with squares of As atoms (As–2As = 2.46 Å) arranged between the planes as shown in Figure 5-11, p. 223.

ζ-CoZn$_{13}$ (*mC* 28): ζ-$CoZn_{13}$ has cell dimensions $a = 13.306$, $b = 7.535$, $c = 4.992$ Å, $\beta = 126.78°$, $Z = 2$.[34] The Co atoms are presumably located at 0, 0, $\frac{1}{2}$; $\frac{1}{2}$, $\frac{1}{2}$, $\frac{1}{2}$, since the neighbors to these sites are significantly closer than the distances between other sites. They are thus surrounded by slightly distorted icosahedra of 12 Zn (2.50 to 2.61 Å). These icosahedra share vertices linking them together in chains parallel to the c axis, and the chains of icosahedra are packed together in an approximately hexagonal array (Figure 13-34). Zn(1) atoms are located in interstices between the chains. The building principle of the structure is thus similar to that of the boron-icosahedra interstitial structures.

B (*tP*50): With cell dimensions $a \sim 8.75$, $c \sim 5.06$ Å, the B atoms in tetragonal boron form uncentered icosahedra which are arranged in f.c. cubic packing.[35] Four of these icosahedra are so arranged in the unit cell that there are also B–B contacts between icosahedra (Figure 13-35). Up to two atoms randomly occupy sites at the midpoints of the cell edges at $z = \frac{1}{2}$ and in the center of the basal plane at $z = 0$. These atoms (B(5)) have 4 B(1) neighbors. The other B atoms (1–4) each have 6 B neighbors.

α-B (*hR* 12): The hexagonal cell has dimensions $a = 4.908$, $c = 12.57$ Å, $Z = 36$.[36] The structure is made up of nearly regular icosahedra in slightly

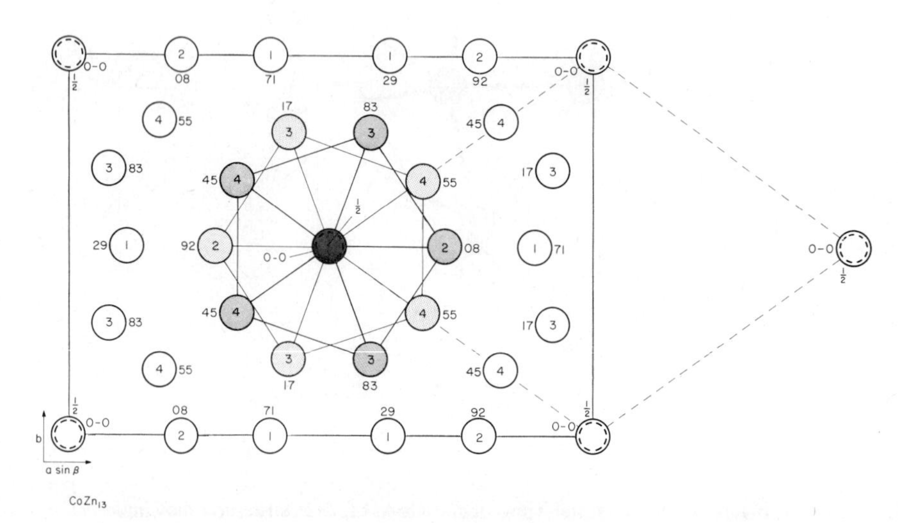

Figure 13-34. View of ζ-$CoZn_{13}$ structure projected onto a plane normal to [001].

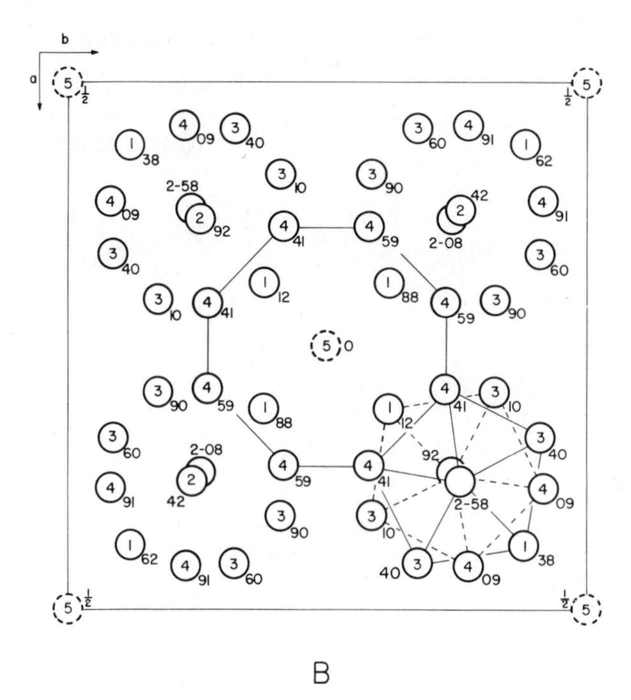

Figure 13-35. The structure of tetragonal B ($tP50$) viewed down [001].

deformed f.c. cubic close packing. Figure 13-36 shows the arrangement of icosahedra at the hexagonal cell corners; they are also centered about $\pm(\frac{1}{3}, \frac{2}{3}, \frac{2}{3})$. In addition to intra-icosahedral B–B contacts of 1.73–1.79 Å, half of the B atoms form intericosahedral contacts of 1.71 Å. The other six B per icosahedra have a contact at 2.03 Å to a B atom in each of two neighboring icosahedra. If these three B atoms at the vertices of an equilateral triangle share 2 electrons, the interatomic distances are in exact conformity with theoretical treatment of B icosahedra[37] which leads to the conclusion that 26 electrons are required for internal bonding, leaving 10 electrons free for external connections.

The boron arrangement is similar to that found in B_4C; in α-B omission of C from the octahedral holes permits the icosahedra to approach each other more closely and form the three-center B bonds.

B_4C, etc. ($hR15$): The structures of B_4C[38] ($a = 5.61$, $c = 12.14$ Å for hexagonal cell) and other "compounds" such as B_4Si, B_7O, or $B_{13}O_2$ and $B_{13}As_2$[39] have the icosahedral arrangement of B atoms found in α-B, in which six of the icosahedral atoms are directly bonded to B atoms in

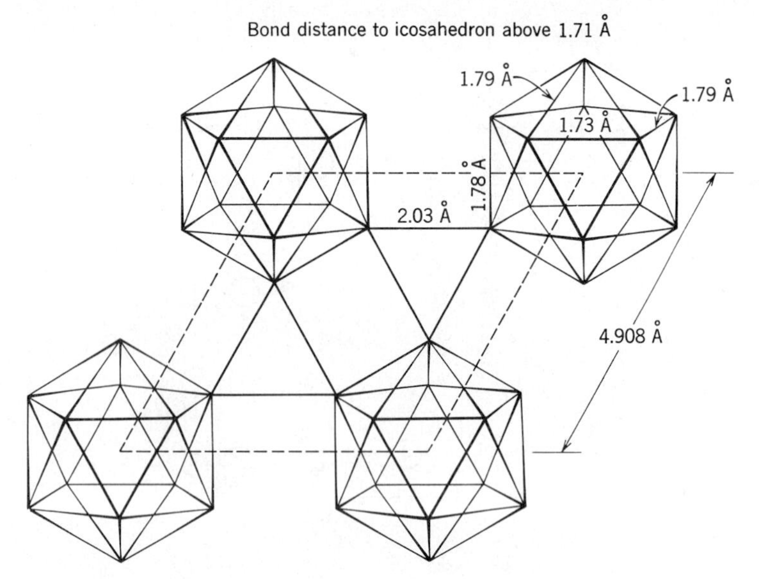

Figure 13-36. Arrangement of the icosahedra in the α-B ($hR12$) structure. (Decker and Kasper[36].)

neighboring icosahedra. The other 6 B atoms per icosahedra form bonds to carbon (or other) atoms in the large octahedral holes of the structure. Three carbon atoms form chains within these holes. The end carbons of the chains are each bonded to 4 B atoms of two icosahedra (B–C = 1.64 Å) and the middle carbon of the chain is bonded only to the end members of the chain (C–C = 1.39 Å).

AlC_4B_{24} ($oC58$): AlC_4B_{24}, which has cell dimensions $a = 5.69$, $b = 8.88$, $c = 9.10$ Å, $Z = 2$, is said to be the phase previously called AlB_{10},[40] although Will[41] later published a structure of AlB_{10}. The gross features of both structures are similar, but the AlC_4B_{24} determination putting C in the bridges between icosahedra is more attractive than the uncertain situation advanced by Will. Further attempts at the structure, now described as $AlCa_4B_{24}$ have been made by Will,[42] and by Perrotta, *et al.*[43] who describe the phase as $Al_{2.1}C_8B_{51}$.

Boron atoms form icosahedra in an approximately close packed hexagonal arrangement. The icosahedra are joined together, both by direct B–B bonds (1.77 Å) and by nonlinear bridges involving a carbon atom (B–C = 1.62 Å[43]). This much seems certain. Perrotta *et al.* also maintain that interstitial B creates linear C–B–C chains (B–C = 1.47 Å) while giving C a tetrahedral surrounding. The Al atoms are also located in the interstitial space.

$MgAlB_{14}$(oI~62): $MgAlB_{14}$ has unit cell dimensions $a = 10.313$, $b = 8.115$, $c = 5.848$ Å, $Z = 4$.[44] The structure contains icosahedral B_{12} groups which are approximately in b.c. cubic {110} arrangement on (100) planes ($b/c = 1.39 \approx \sqrt{2}$). These layers of icosahedra are stacked directly above each other along [100]. The icosahedra are interconnected by direct B–B bonds (1.75 Å) and by bridges through interstitial Al, Mg, and B(5) atoms. Al and Mg sites are partially occupied; Al is surrounded by 12 and Mg by 14 B neighbors.

YB_{66} (cF1608): The structure[45] is built up of groups of 156 B atoms composed of 12 B_{12} icosahedra arranged icosahedrally about a central B_{12} icosahedron, which is smaller and more regular than the others. These groups in two orientations 90° to each other are packed in a NaCl type arrangement, giving a large cubic cell with edge 23.44 Å that contains 1248 B atoms. The structure contains empty channels that run parallel to the fourfold axes at $\frac{1}{4}, \frac{1}{4}, x$. Some 336 more B atoms and 24 Y atoms are located interstitially in these channels. The B atoms on partially occupied sites form cages, and the Y atoms are located so that they have B neighbors both from the cages and from the icosahedral framework. The Y sites actually occur in pairs less than 2 Å apart, so that only one can be occupied at a time.

$Cu_{16}Mg_6Si_7$, Th_6Mn_{23} (cF116): The $Cu_{16}Mg_6Si_7$ and Th_6Mn_{23} structures are very similar. Mn(1) and (2) have identical positions to Si(1) and (2) and Mn(3) positions are almost identical to those of Cu(2). The atomic coordinates of Mn(4) and Cu(1) and of Th and Mg differ somewhat.

In the Th_6Mn_{23} structure[46] octahedra of Th atoms arranged in face centered array are surrounded by 44 Mn(2), (3), and (4) atoms to give the polyhedron shown in Figure 9-63 p. 546. These polyhedra pack together sharing faces and leaving 4 holes per unit cell at $\frac{1}{2}, \frac{1}{2}, \frac{1}{2}$; etc., which are filled by Mn(1) atoms. Mn(1) is thus surrounded by a cube of Mn(3) at 2.65 Å. Mn(2) is surrounded icosahedrally by 4 Mn(3) at 2.73 Å, 4 Mn(4) at 2.57 Å, and 4 Th at 3.185 Å. Mn(3) has 10 Mn neighbors at 2.65 to 3.06 Å and 3 Th at 3.08 Å. Mn(4) is also surrounded icosahedrally by 3 Mn(2), 3 Mn(3) at 2.69 Å, 3 Mn(4) at 2.55 Å, and 3 Th at 3.17 Å. Th is surrounded by 12 Mn at 3.07 to 3.185 Å and 4 Th at 3.595 Å.

In the $Cu_{16}Mg_6Si_7$ structure[47] Mg octahedra (corresponding to those of Th) are arranged in face centered array. Each Mg has 16 neighbors (4 Mg at 2.005 Å, 4 Si(2) at 3.02 Å, 4 Cu(1) at 2.78 Å, and 4 Cu(2) at 3.04 Å). Si(1) has 8 Cu(2) arranged at the corners of a cube at 2.48 Å. Si(2) is surrounded icosahedrally by 4 Mg, 4 Cu(1) at 2.38 Å, and 4 Cu(2) at 2.54 Å, as also is Cu(1) with 3 Mg, 3 Si(2), 3 Cu(1) (2.69 Å), and 3 Cu(2)

(2.54 Å) neighbors. Cu(2), like Mn(3), has 13 neighbors. Hence it can be seen that the slight differences in atomic parameters do not change the atomic coordination in the two structures and they should therefore be regarded as isostructural.

In these structures the Cu(1) and (2) (Mn(3) and (4)) atoms form 48^2 nets like the Cr(4) atoms in the $Cr_{23}C_6$ structure (see Figure 14-38 p. 747).

$SrMg_4$ (hP90): $SrMg_4$ has cell dimensions $a = 10.51$, $c = 28.36$ Å, $Z = 18$.[48] The Sr atoms are grouped in octahedral clusters about the cell corners and the midpoints of the *c* edges of the cell, and triangular clusters at heights $z = \pm\frac{1}{4}$. The arrangement of 50 Mg atoms about the octahedral Sr clusters is the same as that found in Sr_6Mg_{23} which has the Th_6Mn_{23} or $Cu_{16}Mg_6Si_7$ structure (*cF*116), as indicated in Figure 9-63 p. 546. The triangular groups of Sr atoms are surrounded by 33 Mg atoms as shown in Figure 9-63. Sr(2) atoms are surrounded by 12 Mg + 4 Sr neighbors in a convex polyhedron with 24 three-sided and two four-sided faces, each atom having a surface coordination of 5.

β-Mn (cP20): There are two types of Mn atoms in the β-Mn structure[49] which is stable in the temperature range from 1095° to 727°C. The 8 Mn(1) atoms in the unit cell are each surrounded by 3 Mn(1) at 2.37 Å and 9 Mn(2) (3×2.53 Å and 6×2.69 Å) in a distorted icosahedron. Mn(2) has 6 Mn(1) neighbors (2×2.53 Å and 4×2.69 Å) and 8 Mn(2) (2×2.61 Å 4×2.67 Å, and 2×3.25 Å) in a very distorted Frank–Kasper CN 14 type polyhedron. A few alloy phases take the β-Mn type structure, which is shown in Figure 13-37. It was first determined by single-crystal methods[49] and later confirmed by powder neutron diffraction.[50]

Al_2Mo_3C (cP24): The Al_2Mo_3C structure is a filled up β-Mn type with the carbon atoms added in positions 4 (*a*) ($\frac{3}{8}, \frac{3}{8}, \frac{3}{8}$), of space group $P4_132$.[51] Al occupies the Mn(1) and Mo the Mn(2) positions. Carbon is at the center of a considerably distorted octahedron of Mo atoms (Figure 13-38). Several ternary carbides have this structure. The carbon atom positions need to be confirmed by a neutron diffraction study.

Fe_3W_3C (cF112): In the Fe_3W_3C structure[52] the W atoms form an array of octahedra joined by sharing faces. The octahedra are alternately regular and uncentered, or slightly distorted and centered by carbon. The Fe atoms form regular tetrahedra distributed between the W octahedra. Thus C has 6 W at 2.10 Å and W has 2 C and 4 W neighbors at 2.90 Å. In addition, it has 6 Fe at 2.75, 2.79, and 2.80 Å. The 12 atoms surround W in a distorted icosahedron. Fe(1) is surrounded icosahedrally by 6 Fe(2) at 2.35 and 6 W at 2.79 Å. Fe(2) is also surrounded by a distorted icosahedron of 6 Fe at 2.35 and 6 W at 2.75 and 2.80 Å. The

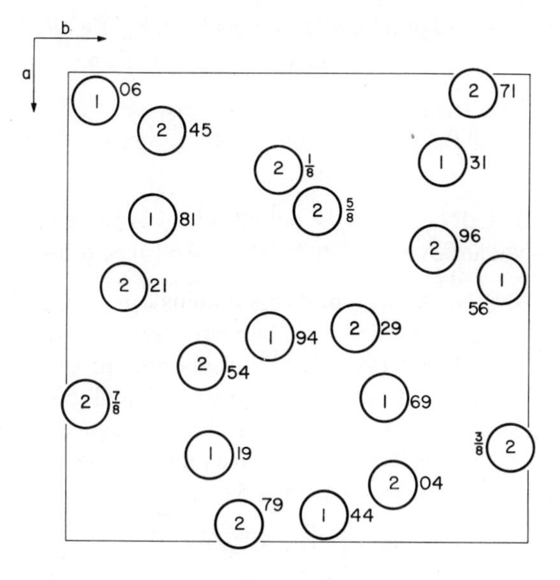

Figure 13-37. Arrangement of the atoms in the β-Mn ($cP20$) structure.

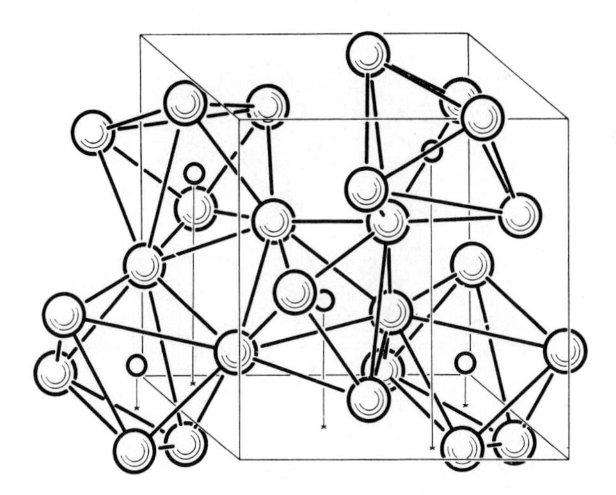

Figure 13-38. View showing the octahedral surrounding of C by Mo in the Al_2Mo_3C ($cP24$) structure, which is a filled up form of the β-Mn structure. (Jeitschko and co-workers [51].)

structure can be regarded as a filled up $NiTi_2$ type. Fe_6W_6C and Ti_4Cu_2O forms of the structure have been examined by neutron diffraction.

$NiTi_2$ (cF96): The structure examined by powder neutron diffraction,[53] is similar to that of Fe_3W_3C except that there are no C atoms centering Ti(2) octahedra.

$CdNi_{1-x}$ (cF92): $CdNi_{1-x}$ has a defect Ti_2Ni type structure with 4 Ni sites randomly vacant over the Ni 32(*e*) and 16 (*c*) positions.[54]

α'-V_7Al_{45} (mC 104): α'-V_7Al_{45} has cell dimensions $a = 25.60$, $b = 7.621$, $c = 11.08$ Å, $\beta = 128.92°$, $Z = 2$.[55] The atoms occur in layers parallel to the (010) plane. The V(1) and (2) atoms are surrounded by slightly distorted icosahedra of Al. V(3) are also surrounded icosahedrally by 11 Al and one V(3) at 2.64 Å so that two V(3) icosahedra interpenetrate each other; otherwise the icosahedra share faces or edges. Al–V distances in the icosahedra lie in the range from 2.51 to 2.93 Å.

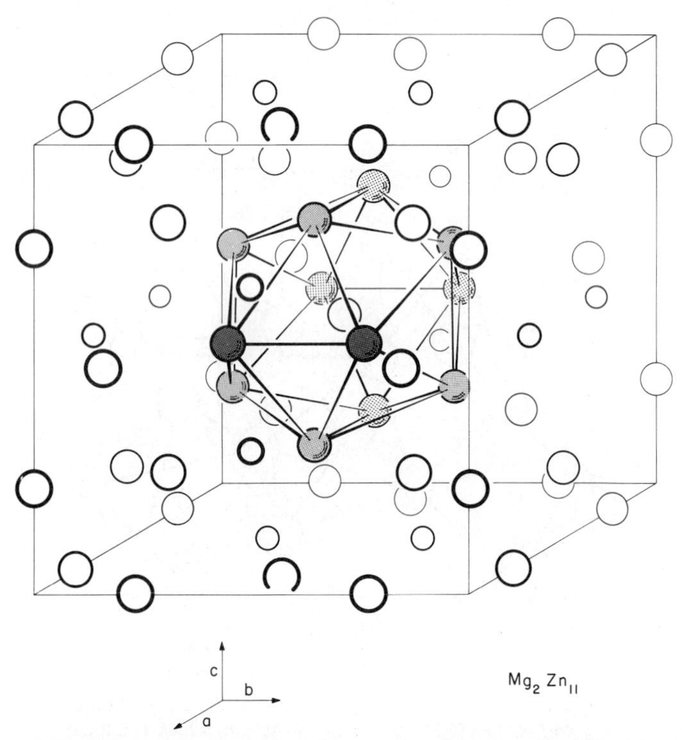

Figure 13-39. Pictorial view of the structure of Mg_2Zn_{11} (*cP*39), showing an icosahedron of Zn(5) surrounding Zn(1): large circles, Mg.

Mg_2Zn_{11}, $Mg_2Al_5Cu_6$ (*cP*39): The atoms can be regarded as arranged in layers in the Mg_2Zn_{11} structure, although there appears to be little significance in this. In the ternary alloy, $Mg_2Al_5Cu_6$,[56] Al(1), (2), and (3) occupy the Zn(1), (3), and (4) positions, Cu(1) and (2) the Zn(2) and (5) positions of the Mg_2Zn_{11} structure. Cu(1) atoms form regular octahedra about the unit cell corners with Cu–4Cu = 2.64 Å. Al(3) atoms are situated outside each triangular face of the octahedra and equidistant (2.54 Å) from each of the three Cu(1) atoms of the faces. The Al(1) atom at the cell body center is surrounded by a regular icosahedron of 12 Cu(2) atoms at a distance of 2.54 Å (Cu(2)–Cu(2) = 2.685 and 2.71 Å). It is strange to find the larger atom at the center of an icosahedron of 12 smaller atoms, since the central hole is smaller than the surface atoms if they are in contact. In this case the Al–Cu distance is compressed and the Cu–Cu separation is larger than in metallic Cu. An array of Al(2) and Mg atoms links the 14 Cu(1) and Al(3) atoms at the cell corners and the Cu(2) icosahedra, although there are also direct contacts from Cu(2) to Al(3).

In the Mg_2Zn_{11}[57] structure Mg is surrounded by a convex polyhedron of 16 Zn (2.91–3.55 Å) and 1 Mg (3.08 Å). Zn(1) is surrounded icosahedrally by 12 Zn(5) at 2.575 Å (Figure 13-39) and Zn(5) is also surrounded icosahedrally by 3 Mg and 9 Zn. Zn(4) has an irregular coordination polyhedron of 3 Mg and 9 or 12 Zn, and Zn(2), which form the octahedra

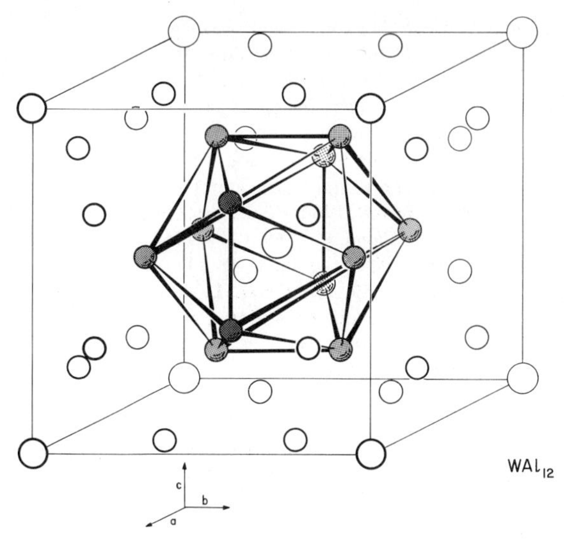

Figure 13-40. Pictorial view of the structure of WAl_{12} showing an icosahedron of Al surrounding W.

about the cell corners (4 Zn(2) at 2.84 Å), also have 2 Zn(3) at 2.65 Å and 4 Zn(4) at 2.69 Å, together with 2 Mg at 3.55 Å.

In plan many of the structure's features are similar to those found in the β-UH_3 or $AuZn_3$ structure (p. 673).

WAl_{12} (*cI*26): The W atoms lie at the cell corners and body center and the Al atoms are arranged about each W atom in almost perfect icosahedra (Figure 13-40). W–12Al = 2.73 Å. Each Al has 1 W and 10 Al neighbors at 2.79 to 2.90 Å. The icosahedra do not interpenetrate, but there are Al–Al connections between them as close as 2.80 Å.[58] The structure requires confirmation by single-crystal methods.

REFERENCES

1. H. Arnfelt and A. Westgren, 1935, *Jernkont. Ann.*, **119**, 185.
2. J. B. Forsyth and L. M. d'Alte da Veiga, 1962, *Acta Cryst.*, **15**, 543.
3. S. Samson, 1968, in *Structural Chemistry and Molecular Biology*, San Francisco: W. H. Freeman & Co., p. 687.
4. C. G. Wilson, D. K. Thomas, and F. J. Spooner, 1960, *Acta Cryst.*, **13**, 56.
5. C. B. Shoemaker and D. P. Shoemaker, 1967, *Acta Cryst.*, **23**, 231.
6. D. P. Shoemaker, C. B. Shoemaker, and F. C. Wilson, 1967, *Acta Cryst.*, **10**, 1.
7. Y. Komura, W. G. Sly, and D. P. Shoemaker, 1960, *Acta Cryst.*, **13**, 575.
8. H. Hartmann, F. Ebert, and O. Bretschneider, 1931, *Z. anorg. Chem.*, **198**, 116; W. G. Burgers and J. A. M. van Liempt, 1931, *Rec. Trav. Chim. Pays-Bas*, **50**, 1050; G. Hägg and N. Schönberg, 1954, *Acta Cryst.*, **7**, 351; N. Schönberg, 1954, *Acta Chem. Scand.*, **8**, 221.
9. C. B. Shoemaker and D. P. Shoemaker, 1969, in *Developments in the Structural Chemistry of Alloy Phases*, ed, B. C. Giessen, New York: Plenum Press, p. 107.
10. P. Greenfield and P. A. Beck, 1956, *J. Metals*, **8**, 265; F. Laves, 1956, *Theory of Alloy Phases*, Cleveland: American Society for Metals, p. 124.
11. S. Geller, 1956, *Acta Cryst.*, **9**, 885.
12. M. V. Nevitt, 1963, in *Electronic Structure and Alloy Chemistry of the Transition Elements*. ed, P. A. Beck, New York: Interscience, p. 101.
13. B. W. Batterman and C. S. Barrett, 1964, *Phys. Rev. Letters*, **13**, 390.
14. R. E. Rundle, 1947, *J. Amer. Chem. Soc.*, **69**, 1719; 1951, *Ibid.*, **73**, 4172.
15. E. Günzel and K. Schubert, 1958, *Z. Metallk.*, **49**, 234.
16. F. C. Frank and J. S. Kasper, 1959, *Acta Cryst.*, **12**, 483.
17. B. G. Bergman and D. P. Shoemaker, 1954, *Acta Cryst.*, **7**, 857.
18. S. Samson, 1962, *Nature, Lond.*, **195**, 259.
19. S. Samson, 1964, *Acta Cryst.*, **17**, 491.
20. S. Samson, 1965, *Acta Cryst.*, **19**, 401.
21. S. Samson, 1967, *Acta Cryst.*, **23**, 586.
22. S. Samson, 1969, in *Structural Developments in Alloy Phases*, ed, B. C. Giessen, New York: Plenum Press, p. 65.
23. S. Samson, 1958, *Acta Cryst.*, **11**, 851.
24. S. Samson, 1961, *Acta Cryst.*, **14**, 1229.
25. R. W. Lynch and H. G. Drickamer, 1965, *J. Phys. Chem. Solids*, **26**, 63.
26. P. J. Brown, 1957, *Acta Cryst.*, **10**, 133.

27. A. J. Bradley and J. Thewlis, 1927, *Proc. Roy. Soc.*, **A115**, 456.
28. F. Laves, K. Löhberg, and P. Rahlfs, 1934, *Nachr. Ges. Wiss. Göttingen, Math-Phys. Kl. Neue Folge*, **1**, 67.
29. C. B. Shoemaker and D. P. Shoemaker, 1963, *Acta Cryst.*, **16**, 997.
30. S. Samson and E. K. Gordon, 1968, *Acta Cryst.*, **B24**, 1004.
31. G. Bergman, J. L. P. Waugh, and L. Pauling, 1957, *Acta Cryst.*, **10**, 254.
32. D. E. Sands, Q. C. Johnson, A. Zalkin, O. H. Krikorian, and K. L. Kromholtz, 1962, *Acta Cryst.*, **15**, 832.
33. I. Oftedal, 1928, *Z. Kristallogr.*, **A66**, 517; U. Ventriglia, 1957, *Period Miner.*, **26**, 345, also 147.
34. P. J. Brown, 1962, *Acta Cryst.*, **15**, 608 (Note the x_1 values on p. 610 are misprinted; they should read 0.6110, 0.6120, 0.6155, and 0.60).
35. J. L. Hoard, R. E. Hughes, and D. E. Sands, 1958, *J. Amer. Chem. Soc.*, **80**, 4507.
36. B. F. Decker and J. S. Kasper, 1959, *Acta Cryst.*, **12**, 503.
37. H. C. Longuet-Higgins and M. de V. Roberts, 1955, *Proc. Roy. Soc.*, **A230**, 110.
38. H. K. Clark and J. L. Hoard, 1943, *J. Amer. Chem. Soc.*, **65**, 2115.
39. S. LaPlaca and B. Post, 1961, *Planseeber. Pulvermet.*, **9**, 109.
40. V. I. Matkovich, J. Economy, and R. F. Giese, 1964, *J. Amer. Chem. Soc.*, **86**, 2337.
41. G. Will, 1967, *Acta Cryst.*, **23**, 1071.
42. G. Will, 1969, *Acta Cryst.*, **B25**, 1219.
43. A. J. Perrotta, W. D. Townes, and J. A. Potenza, 1969, *Acta Cryst.*, **B25**, 1223.
44. V. I. Matkovich and J. Economy, 1970, *Acta Cryst.*, **B26**, 616.
45. S. M. Richards and J. S. Kasper, 1969, *Acta Cryst.*, **B25**, 237.
46. J. V. Florio, R. E. Rundle, and A. I. Snow, 1952, *Acta Cryst.*, **5**, 449.
47. G. Bergman and J. L. T. Waugh, 1956, *Acta Cryst.*, **9**, 214.
48. F. E. Wang, F. A. Kanda, C. F. Miskell, and A. J. King, 1965, *Acta Cryst.*, **18**, 24.
49. G. D. Preston, 1928, *Phil. Mag.*, **5**, 1207.
50. J. S. Kasper and B. W. Roberts, 1956, *Phys. Rev.*, **101**, 537.
51. W. Jeitschko, H. Nowotny, and F. Benesovsky, 1963, *Mh. Chem.*, **94**, 247.
52. A. Westgren, 1933, *Jernkont. Ann.*, **117**, 1; J. Leciejewicz, 1964, *J. Less-Common Metals*, **7**, 318; M. H. Mueller and H. W. Knott, 1963, *Trans. Met. Soc. AIME*, **227**, 674.
53. M. H. Mueller and H. W. Knott, 1963, *Trans. Met. Soc. AIME*, **227**, 674; See *Structure Reports*, **22**, 888.
54. J. K. Critchley and J. W. Jeffery, 1965, *Acta Cryst.*, **19**, 674.
55. P. J. Brown, 1959, *Acta Cryst.*, **12**, 995.
56. S. Samson, 1949, *Acta Chem. Scand.*, **3**, 809.
57. S. Samson, 1949, *Acta Chem. Scand.*, **3**, 835.
58. J. Adam and J. B. Rich, 1954, *Acta Cryst.*, **7**, 813.

14

Further Structures Generated by Stacking Nets Containing Squares, Pentagons, Hexagons, etc.

PENTAGON-TRIANGLE NETS

CeAl (*oC*16): CeAl has cell dimensions $a = 9.270$, $b = 7.680$, $c = 5.760$ Å, $Z = 8$.[1] In planes parallel to (001), Ce and Al(1) atoms form pentagon-triangle nets at $z = \frac{1}{4}$ and $\frac{3}{4}$ (Figure 14-1), the atomic arrangement being the same as that in the A_1 layers of the δ-Mn_4Al_{11}, $MnAl_6$, and WAl_4 structures. The pentagons at one height cover those at the other antisymmetrically and the pentagonal antiprisms so formed are centered by Al(2) at $z = 0$ and $\frac{1}{2}$, so that the structure contains columns of interpenetrating distorted icosahedra along [001] which are centered by Al(1). Al(2) are also surrounded by 12 atoms giving a polyhedron with 20 triangular faces and surface coordination 5 for all ligands. Ce have 15 neighbors at distances out to 4.06 Å, and the closest Al–Al and Ce–Ce distances are 2.65 Å and 3.32 Å, respectively. The structure requires confirmation by single-crystal methods.

DyAl (*oP*16): DyAl has cell dimensions $a = 5.822$, $b = 11.369$, $c = 5.604$ Å, $Z = 8$.[1] The Al(1) and Dy atoms exist in layers parallel to the (001) plane at heights $z = \frac{1}{4}$ and $\frac{3}{4}$ where they form somewhat distorted pentagon-triangle nets. The pentagons of these nets overlap each other more or less antisymmetrically and the pentagonal antiprisms so formed are centered by Al(2) atoms at $z = 0$ and $\frac{1}{2}$. The structure can be compared with that of CeAl (*oC*16). Both Dy(1) and (2) are similarly coordinated

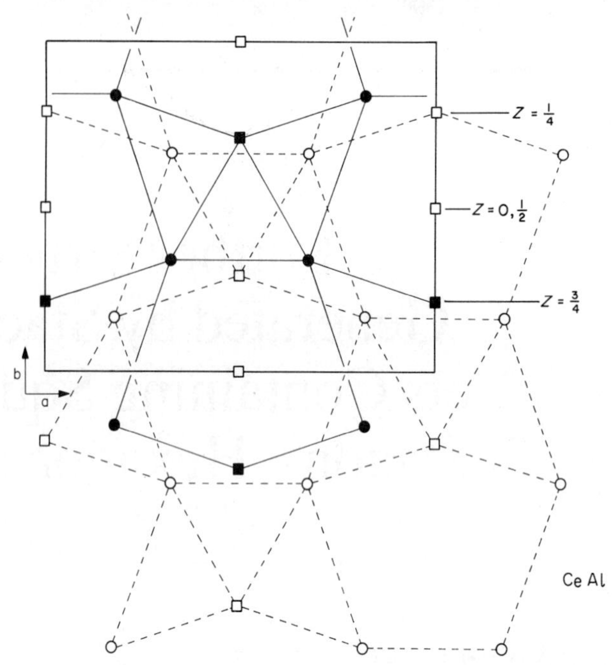

Figure 14-1. Pentagon-triangle nets of atoms parallel to the (001) plane in the CeAl ($oC16$) structure: squares, Al atoms.

with 8 Al neighbors at 3.04 to 3.51 Å and 7 Dy at 3.28 to 3.87 Å. The closest Al–Al approaches are 2.67 Å.

δ-Mn_4Al_{11} ($aP15$): δ-Mn_4Al_{11} has a triclinic cell with dimensions $a = 5.092$, $b = 8.862$, $c = 5.047$ Å, $\alpha = 85°19'$, $\beta = 100°24'$, $\gamma = 105°20'$, $Z = 1$, although it is described in terms of a more convenient C-centered (pseudomonoclinic) cell with $b = 17.1$ Å.[2] The structure contains planes of atoms parallel to the (010) plane of this cell. These are of two types, A_1 and A_2, shown in Figure 14-2*a* and *b*. The Al(5) atom in Figure 14-2 is about 0.4 Å from the mean level of the A_2 plane. The stacking sequence in the structure is $XA_1A_2A_2A_1XA_1A_2A_2A_1$ where X represents an Al atom at 0, 0, 0 or 0, 0, $\frac{1}{2}$. The A layers are spaced at $y \sim \frac{1}{16}, \frac{3}{16}, \frac{5}{16}, \frac{7}{16}, \frac{9}{16}, \frac{11}{16}, \frac{13}{16}$, and $\frac{15}{16}$. It is apparent from Figure 14-2 that the rumpled A_2 net is a distorted form of the 3^2434 net found in the $CuAl_2$ structure, and the A_1 net which is obtained from A_2 by removing the Al(5) atoms marked, is a pentagon-triangle net similar to those found in CeAl ($oC16$) and DyAl ($oP16$). The $MnAl_6$, WAl_4, and Fe_2Al_5 ($oC14$)[3] structures also contain A_1 type layers.

Both Mn(1) and (2) have 10 neighbors out to 2.82 Å, the environ-

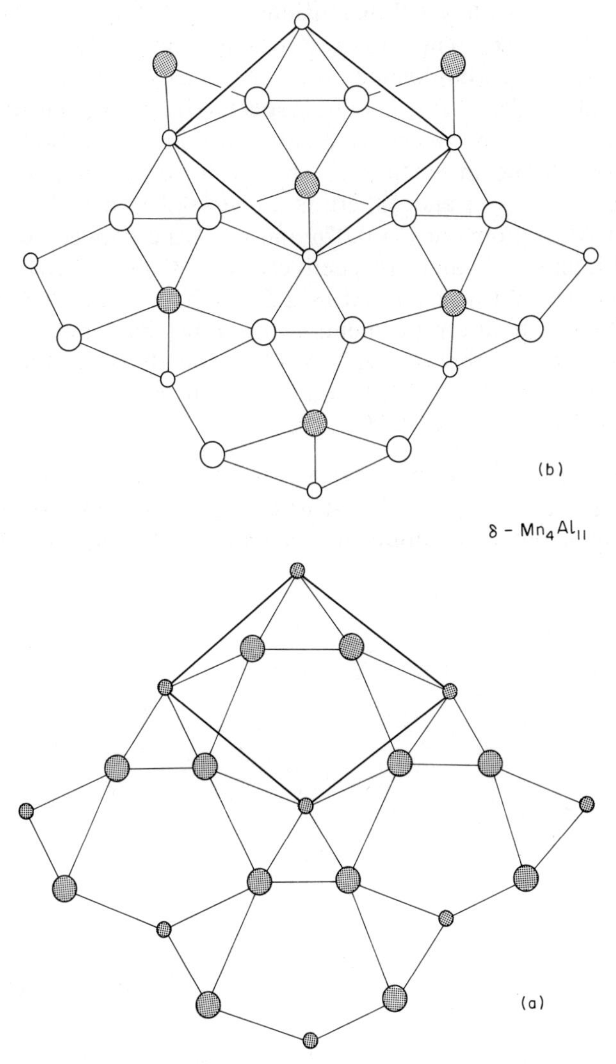

Figure 14-2. Diagram showing the two types of layers of atoms, A_1 and A_2, which lie parallel to the (010) plane in the δ-Mn_4Al_{11} (*aP*15) structure.

ment of Mn(1) resembling that of Mn in $MnAl_6$. The CN 10 polyhedron can be regarded as an icosahedron with two vertices removed. Mn(1) has two close Al neighbors at 2.42 and 2.50 Å and Mn(2) also has two at 2.45 and 2.48 Å. Al(0) has 12 neighbors and the other Al have 11, including either 3 or 4 Mn.

WAl_4 (mC30): WAl_4 has cell dimensions $a = 5.272$, $b = 17.77$, $c = 5.218$ Å, $\beta = 100.2°$, $Z = 6$.[4] The structure, which is related to those of $MnAl_6$ and δ-Mn_4Al_{11}, has well-defined layers of atoms parallel to the *ac* plane. At $y = 0$ and $\frac{1}{2}$ and $y \sim \frac{1}{8}, \frac{3}{8}, \frac{5}{8}$, and $\frac{7}{8}$ respectively, planar and slightly non-planar layers occur in the arrangement shown in Figure 14-3. These are the A_1 layers found in δ-Mn_4Al_{11} which can be regarded as pentagon-triangle nets. At $y \sim \frac{1}{4}$ and $\frac{3}{4}$ there is a distorted 3^6 net of Al(6) and (7) atoms and Al(3) atoms at $y = 0.076$ and 0.576 lie between the layers at $y \sim 0$ and $\frac{1}{8}$, and $y \sim \frac{1}{2}$ and $\frac{5}{8}$ respectively. The W atoms avoid each other in the structure. W(1) has 10 Al at 2.53 to 2.86 Å and W(2) has 11 at 2.575 to 2.86 Å, both arrangements being somewhat similar, and that of W(1) is the same as that of Mn in $MnAl_6$. The 10 Al neighbors of W(1) together with 2 W(1) at 3.31 Å complete a distorted icosahedron. The closest Al–Al approach is 2.575 Å.

$MnAl_6$, α-$CuFe_4Al_{23}$ (oC28): $MnAl_6$ has cell dimensions $a = 7.552$, $b = 6.498$, $c = 8.870$ Å, $Z = 4$.[5] α-$CuFe_4Al_{23}$ has a similar structure with Cu on Al sites, but slight atomic displacements lower the symmetry from

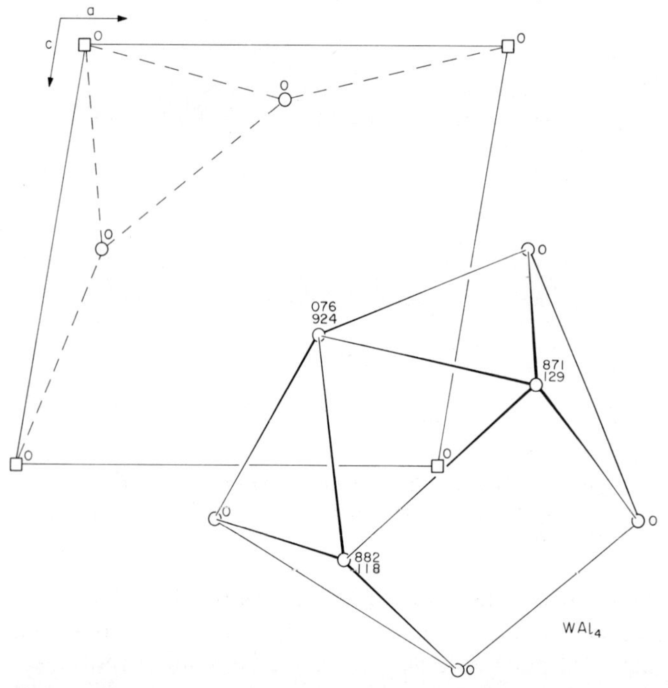

Figure 14-3. Layer arrangement in WAl_4 (*mC*30) viewed down [010]: squares, W atoms.

Cmcm to $Ccm2_1$ and the same may also be true to a lesser extent for $MnAl_6$.[6] The arrangement of the atoms is shown in Figure 14-4, Mn(Fe) being located at $z = \pm\frac{1}{4}$. At $z = \frac{1}{4}$, and at $\frac{3}{4}$ in different orientation, Mn and Al form the characteristic A_1 arrangement found in the Mn_4Al_{11} and other aluminum-transition-metal alloy structures, which can be regarded as a pentagon-triangle net. At $z = 0$ and $\frac{1}{2}$, Al atoms form a 6^3 net similar to that formed by the Zn atoms in the $BaZn_5$ (*oC*24) structure. Other Al atoms are located between these four layers of atoms. 10 Al surround each Mn in a complex polyhedron (2.435 to 2.64 Å), the closest approach being Mn–2Al(2). In the centrosymmetric description, Al atoms have 11 Al and Mn neighbors. The short Al(2)–Al(2) distances of 2.57 and 2.62 Å are notable; also Al(1)–Al(1) at 2.64 Å.

CoGe (*mC*16): In the CoGe structure which is a filled up Ni_3Sn_4 type with cell dimensions, $a = 11.65$, $b = 3.81$, $c = 4.95$ Å, $\beta = 101.10°$, $Z = 8$,[7] the atoms form pentagon-triangle layers at $y = 0$ and $\frac{1}{2}$. The layers are made up of alternate bands of triangles and bands of pentagons separated by triangles, which run in the *c* direction. Double triangle bands at $y = 0$ lie over pentagon-triangle bands at $y = \frac{1}{2}$ and *vice versa* as shown in Figure 14-5. The pentagons are formed by 1 Co and 4 Ge atoms and the pentagons of one layer are off-centered by Ge(1) atoms of the layers above and below. Ge(2) atoms do not center pentagons.

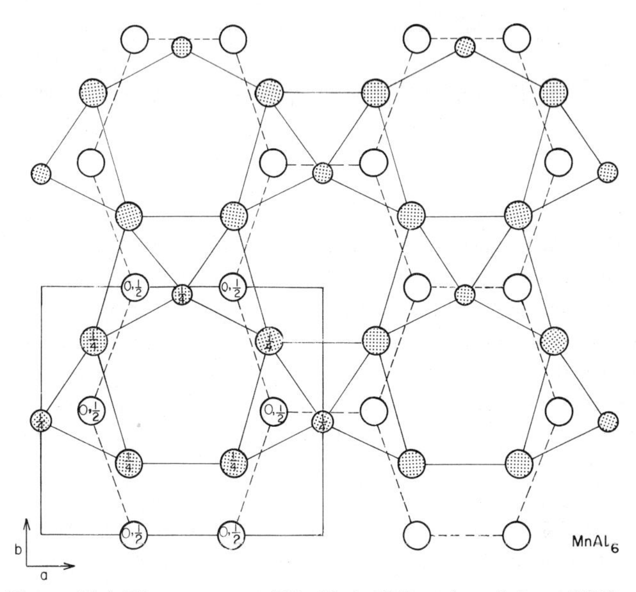

Figure 14-4. The structure of $MnAl_6$ (*oC*28) projected down [001]: small circles, Mn atoms. (Nicol [5].)

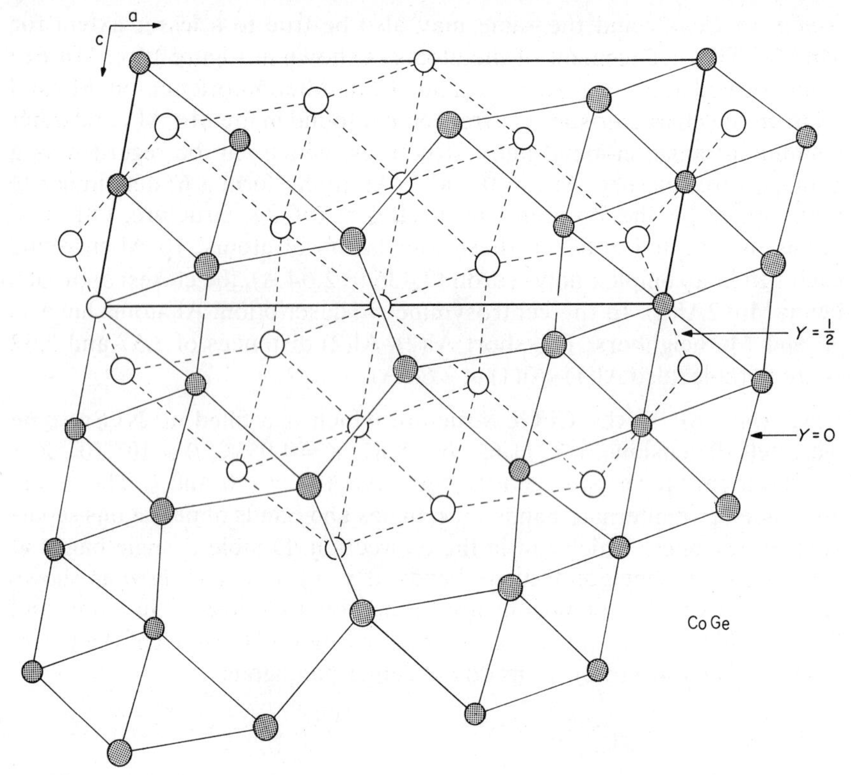

Figure 14-5. Pentagon-triangle nets of atoms in the CoGe structure seen down [010].

Ge(1) has 6 Co neighbors at 2.34 to 2.63 Å and 5 Ge neighbors at 2.79 to 3.27 Å. Ge(2) has 6 Co at 2.345 to 2.45 Å and one at 3.08 Å, 1 Ge(2) at 2.74 Å, and 4 other Ge at 3.08 and 3.27 Å. The Co(1) and (2) atoms each have 10 neighbors in a characteristic coordination polyhedron shown in Figure 14-6, which has 16 three-sided faces and ligand surface coordination numbers: 4×4, 4×5, and 2×6. Similar coordination is found about the Co atoms in CoSn, *hP*6 (p. 536). Co(3) also has 10 neighbors at 2.34–2.71 Å and one at 3.08 Å, but they are arranged differently. The structure was approximately determined by a single-crystal study.

δ-Ni_3Sn_4 (*mC*14): The δ-Ni_3Sn_4 structure with $a = 12.22$, $b = 4.061$, $c = 5.187$ Å, $\beta = 103°47'$, $Z = 2$,[8] is a defect CoGe type, the Ni (Co(2)) sites at 2(*c*) 0, 0, $\frac{1}{2}$ being vacant; otherwise, the atomic arrangements are very similar. This leaves half of the characteristic CN 10 polyhedra

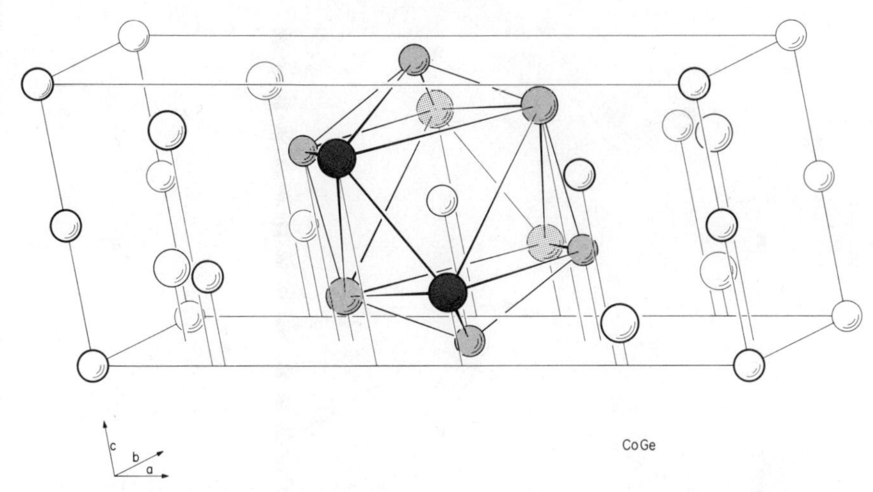

Figure 14-6. Pictorial view of the CoGe (*mC*16) structure showing the coordination about Co (small circle).

(Figure 14-6) uncentered and converts the other half to CN 8 polyhedra. Ni(2) (Co(3)) now has 10 neighbors.

Th_3Pd_5 (hP8): Th_3Pd_5 has cell dimensions $a = 7.149$, $c = 3.899$ Å, $Z = 1$.[9] The building principle of this structure is similar to that of Mn_5Si_3 (compare Figures 14-7 and 14-9), except that in Th_3Pd_5 the major polygons are pentagons instead of hexagons and they are composed of one component only. The Pd atoms at $z = 0$ lie on an interesting $35^3 + 5^3$ (3:2) network giving pentagonal prisms that are centered by the Th atoms which form a slightly distorted 3^6 net at $z = \frac{1}{2}$. Th has 10 Pd neighbors (4×3.035, 4×2.93, 2×3.64 Å) and 6 Th equatorially at 3.90 and 4.03 Å. Pd(1) has 6 Th at 3.035 Å. Pd(2) has 4 Th at 2.93, 2 at 3.64 Å, 2 Pd at 2.72, and 2 at 2.875 Å. The structure requires confirmation by single-crystal methods.

It is interesting to note that there is another pentagon-triangle net conjugate to the one existing in this structure so that pentagons of one net cover those of the other antisymmetrically. A structure made up of these two nets with an interleaving 3^6 secondary net would resemble the Frank–Kasper structures but it is unknown at present.

Pt_5P_2 (mC28): Pt_5P_2 has cell dimensions $a = 10.764$, $b = 5.385$, $c = 7.438$ Å, $\beta = 99.17°$, $Z = 4$.[10] Parallel to the (100) plane, Pt(2) and (3) and P form characteristic rumpled pentagon-triangle nets at $x \sim 0.14$, 0.36, 0.64, and 0.86 (Figure 14-8*b*). Alternatively, parallel to the (010) plane, Pt(2) and (3) atoms form rumpled octagon-"square" nets at $y \sim \frac{1}{4}$ and $\frac{3}{4}$.

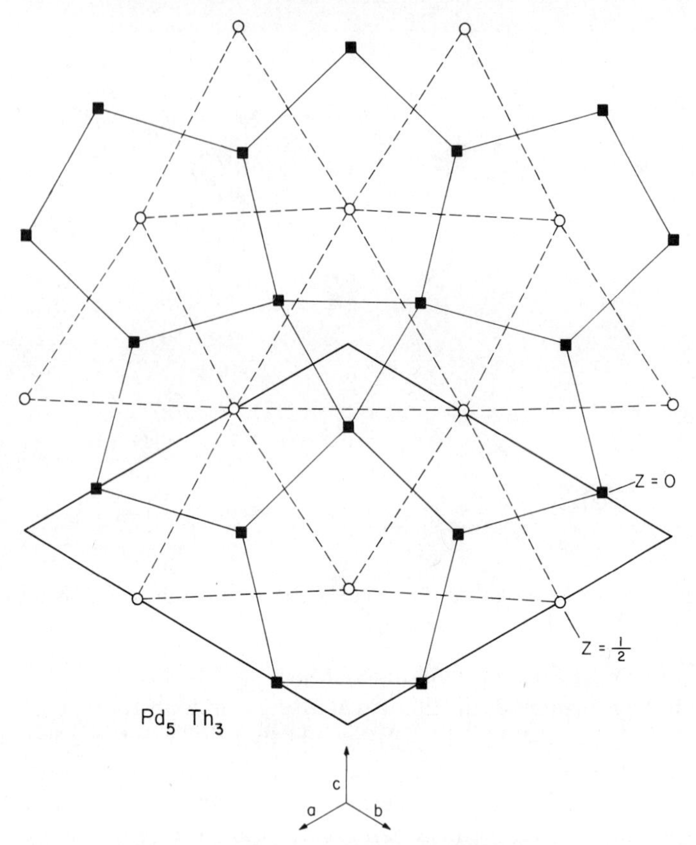

Figure 14-7. Atomic arrangement in the Th_3Pd_5 (*hP*8) structure seen down [001]: squares, Pd atoms.

The "squares" of one layer center the octagons of the next. At $y \sim 0$ and $\frac{1}{2}$, Pt(1) and P together form much more rumpled and distorted 4^4 nets; that at $y \sim \frac{1}{2}$ is displaced by half the net period along a from the other (Figure 14-8*a*). The nodes of these nets lie along lines centering the octagons and squares of the other nets.

P has 5 close Pt neighbors at 2.31–2.37 Å and one at 2.61 Å which are arranged in an irregular polyhedron. Pt(1) has 8 Pt and 2 P neighbors and Pt(2) and (3) each have 7 Pt and 2 and 3 P neighbors respectively. The shortest Pt–Pt distance is 2.76 Å.

Si_3Mn_5 (*hP*16): The dimensions of the hexagonal cell are $a = 6.910$, $c = 4.814$ Å, $Z = 2$. Mn(2) and Si atoms form layers at $z = \frac{1}{4}$ and $\frac{3}{4}$, being arranged in six-sided polygons centered about the sites $\frac{2}{3}$, $\frac{1}{3}$ and

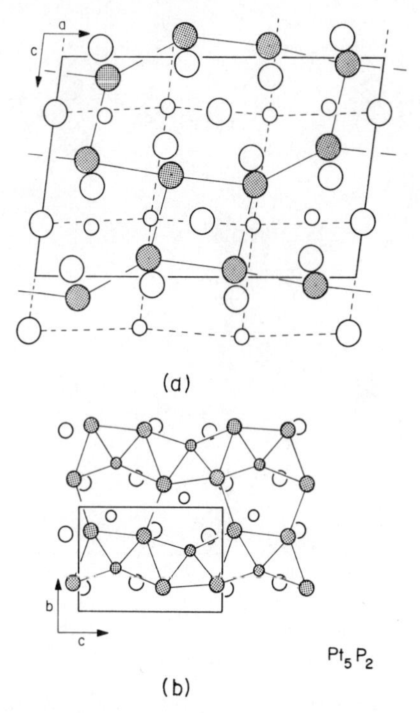

Figure 14-8. (*a*) The structure of P_2Pt_5 (*mC*28) projected down [010]. Pt are large circles. Shaded atoms are at $y \sim \frac{3}{4}$. Open atoms connected by broken lines at $y \sim 0$. (*b*) Atomic arrangement in P_2Pt_5 parallel to the (100) plane: shaded atoms, at $x \sim 0.14$; open atoms, at $x \sim 0.36$.

$\frac{1}{3}$, $\frac{2}{3}$. Polygons at $z = \frac{1}{4}$ and $\frac{3}{4}$ are not superimposed but related more or less antisymmetrically, forming polyhedra which are centered by Mn(1) atoms at $z = 0$ and $\frac{1}{2}$ (Figure 14-9). Mn(1) therefore has CN 14.[11] The structure is frequently stabilized, often in preference to other M_3N_5 types, by the presence of a small amount of interstitial carbon, although B and N can also impart stability to a lesser extent. X-ray and neutron diffraction studies together with chemical analysis and density measurements on the alloy $Mo_{4.8}Si_3C_{0.6}$, have shown[12] that the C atoms occupy the octahedral voids at 0, 0, 0; 0, 0, $\frac{1}{2}$ and that some of the Mo(1) sites (4*d*) remain unoccupied, the formula being $Mo_6Mo_{4-x}Si_6C_{2-y}$.

The near-neighbor diagram (p. 53) for phases with the Si_3Mn_5 structure indicates that the structure forms mainly for geometrical reasons,

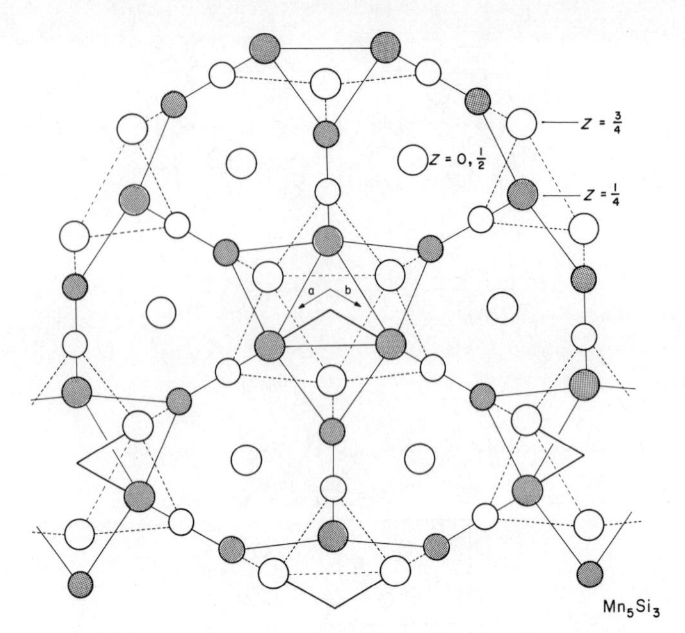

Figure 14-9. Atomic arrangement in the Mn_5Si_3 (*hP*16) structure viewed down [001]: Mn atoms, large circles.

since the phases are distributed in a band lying between lines for *M–N* and *N–N* contacts, corresponding to 9–14 or 9–15 coordination.

Pb_3Ca_5 (*hP*48): With cell dimension $a = 16.23$, $c = 7.04$ Å, $Z = 6$, Pb_3Ca_5 is a distorted superstructure based on the Si_3Mn_5 type.[13] The mixed layers at $z \sim \frac{1}{4}$ and $\frac{3}{4}$ are not planar as in Si_3Mn_5 and this accounts for the larger cell of Pb_3Ca_5 which has three times the volume of the Si_3Mn_5 cell.

Ga_4Ti_5 (*hP*18): The Ga_4Ti_5 structure with cell dimensions $a = 7.861$, $c = 5.452$ Å, $Z = 2$, is a filled up Si_3Mn_5 (*tI*18) structure.[14] Ga(2), Ti(1) and (2) atoms occupy positions of the Si_3Mn_5 structure, and Ga(1) occupy the cell corners and midpoints of the *c* edges of the cell. They are thus surrounded octahedrally by 6 Ti at 2.656 Å and by 2 Ga(1) at 2.726 Å. There are a further 6 Ga(2) atoms at 3.283 A. The structure requires confirmation by single-crystal methods.

$ZnSb_3Hf_5$ with cell dimensions $a = 8.514$, $c = 5.747$ Å, $Z = 2$,[15] is also a filled up Si_3Mn_5 type, with the Zn atoms occupying the same positions as Ga(1) in Ga_4Ti_5.

Laves and μ phases: Laves, μ, and *M* phases and the Al_3Zr_4 structures described on pp. 654–667 are also built up of pentagon-triangle nets of atoms.

HEXAGON-PENTAGON-TRIANGLE NETS

Si_2Ni_3 (oC80): Si_2Ni_3 has cell dimensions $a = 12.23$, $b = 10.81$, $c = 9.24$ Å $Z = 16$.[16] The atoms form layers parallel to the (100) plane. At $x = 0$ and $\frac{1}{2}$, Ni and Si together form the hexagon-pentagon-triangle nets shown in Figure 14-10. The nets at one level are displaced along b compared to those at the other, so that the pentagons of one lie over the hexagons of the other. At $x \sim \pm\frac{1}{6}$ and $x \sim \pm\frac{1}{3}$ there are rumpled Ni–Si nets made up of triangles and distorted rectangles as shown in Figure 14-10. The nets at $x \sim \pm\frac{1}{6}$ have the same positions and these are displaced along b relative to those at $x \sim \pm\frac{1}{3}$. Atomic coordinations are generally irregular and dissimilar. Ni–Ni distances vary from 2.47 to 2.76 Å and Ni–Si from 2.20 to 2.55 Å, those about 2.31 Å being the most common.

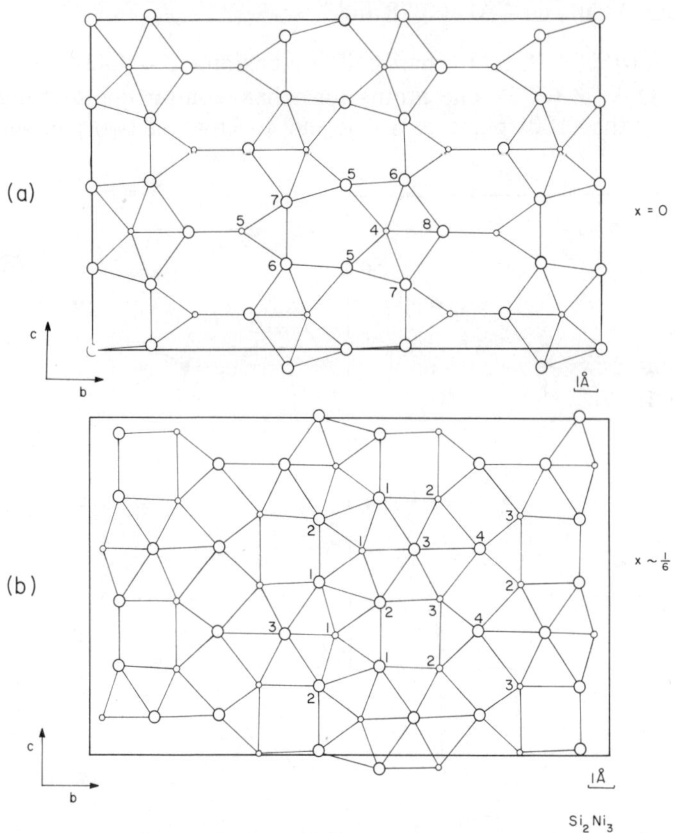

Figure 14-10. (*a*) Showing the atomic arrangement in the Ni_3Si_2 (*oC*80) structure in planes parallel to (100) at $x = 0$: Ni atoms, large circles. (*b*) Atomic arrangement in the rumpled layer about $x = \frac{1}{6}$.

δ-CuMgAl$_2$ (*oC*16): δ-$CuMgAl_2$ has cell dimensions $a = 4.01$, $b = 9.25$, $c = 7.15$ Å, $Z = 4$.[17] The atoms form hexagon-pentagon-triangle layers parallel to the (100) plane at heights $z = 0$ and $\frac{1}{2}$. The pentagons formed by Al and Cu are joined together in rows running in the [001] direction and are separated from each other by rows of hexagons and triangles (Figure 14-11). The layers at $z = 0$ and $\frac{1}{2}$ are so placed over each other that Cu atoms of one layer center the triangular prisms formed by the layers above and below. Mg of one layer centers the pentagonal prisms formed by the layers above and below, and the Al of one layer lie over the hexagons of the other, close to the Cu atoms.

These arrangements result in Cu having 3 Mg neighbors at 2.72 and 2.765 Å and 6 Al at 2.515 and 2.54 Å. Mg has 10 Al neighbors at 2.97, 3.04, and 3.15 Å, together with 3 Cu at 2.72 and 2.765 Å. Al has 5 Mg neighbors, 3 Cu, and 4 Al at 2.78 and 2.92 Å.

α-La$_3$Al$_{11}$ (*oI*28): La_3Al_{11} has cell dimensions $a = 4.431$, $b = 13.142$, $c = 10.132$ Å, $Z = 2$.[18] The atoms form hexagon-pentagon-triangle nets parallel to the (100) plane at $x = 0$ and $\frac{1}{2}$. Rows of two pentagons fol-

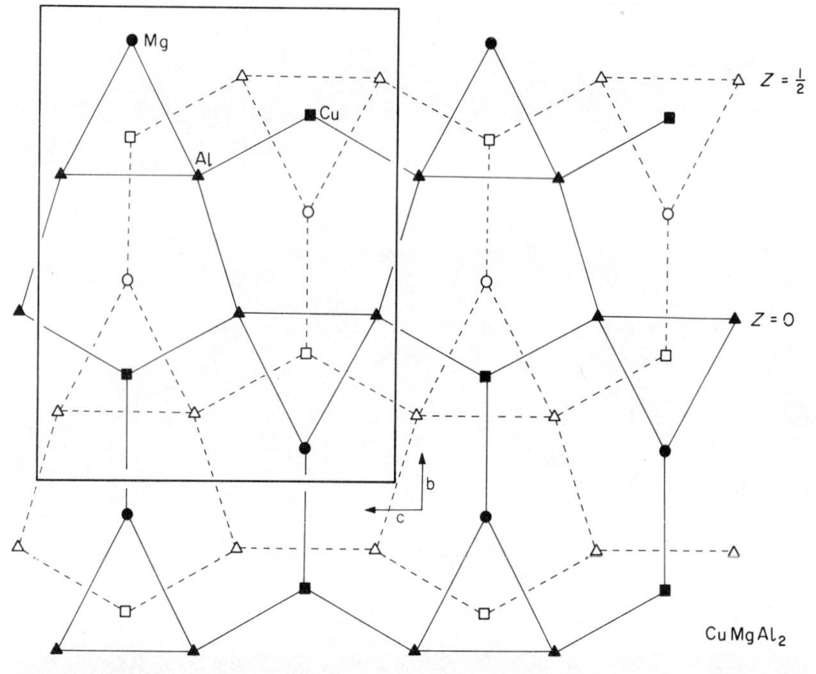

Figure 14-11. View showing the atomic arrangement in the δ-$CuMgAl_2$ (*oC*16) structure projected down [100].

lowed by a hexagon of Al atoms run in the [010] direction. La(1) atoms center the hexagonal prisms and La(2) the pentagonal prisms formed by the layers above and below (Figure 14-12). Within the same plane La(1) are surrounded by a hexagon of 4 Al + 2 La(2) which, together with the Al hexagons of the layers above and below and the 2 polar La atoms, give La(1) CN 20. La(2) are surrounded by 6 Al, 1 La(1) and 1 La(2) within their plane, and the 5 Al above and below, plus two polar La atoms, so that they also have CN 20.

One single-crystal study[18] maintained that the phase is La_3Al_{11} (space group *Immm*) whereas another single-crystal study carried out at the same time[19] maintained that it is $LaAl_4$ (*oI*30) (space group *Imm*2).

Os_4Al_{13} (*mC*34): The Os_4Al_{13} structure, with cell dimensions $a = 17.64$, $b = 4.228$ $c = 7.773$ Å, $\beta = 115.15°$, $Z = 2$,[20] has the atoms on somewhat irregular hexagon-pentagon-triangle nets at $y = 0$ and $\frac{1}{2}$ (Figure 14-13). The shortest Os–Al distance is 2.57 Å.

P and R Phases: P and *R* phase structures described on pp. 667–669, are built up of hexagon-pentagon-triangle layers of atoms.

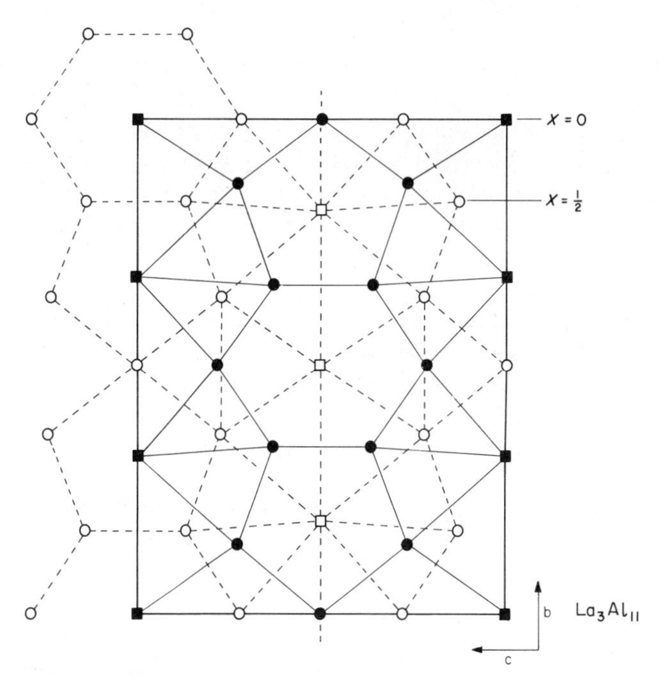

Figure 14-12. Atomic arrangement in the La_3Al_{11} (*oI*28) structure viewed down [100]: circles, Al atoms.

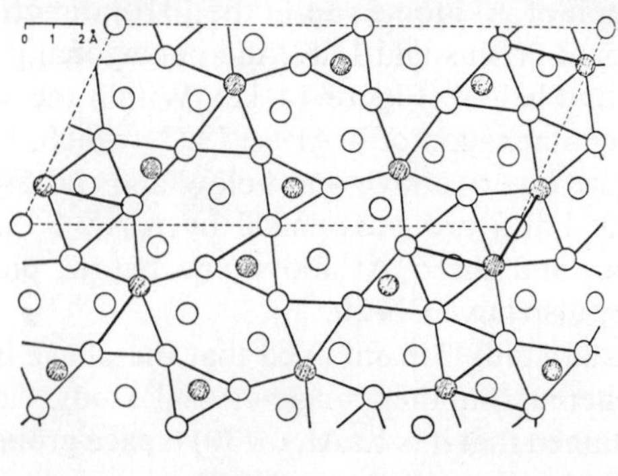

Figure 14-13. Atomic arrangement in the Os_4Al_{13} ($mC34$) structure viewed down [010]: Al atoms, open circles. Connected atoms are at $y = 0$, unconnected at $y = \frac{1}{2}$. (Edshammar[20].)

NETS WITH HEXAGONS, PENTAGONS, SQUARES (QUADRILATERALS), AND/OR TRIANGLES

Ir_3Ge_7 ($cI40$): The Ir_3Ge_7 structure[21] contains characteristic layer nets of the type shown in Figure 14-14, which occur in several structures (e.g., p. 646). These are made up of squares of two sizes, quadrilaterals and triangles. In Ir_3Ge_7, Ir and Ge(1) atoms at $z = 0$ form this array with the small squares at the cell center and the large squares at the cell corners. The array occurs again at $z = \frac{1}{2}$ with the large square at the cell center. Squares of Ge(2) atoms are located over the cell base center at $z = 0.34$ and 0.66, and over the cell corners at $z = 0.16$ and 0.84, and Ir atoms are located over the cell corners and cell base center at the same heights respectively. Other Ge(1) atoms at the midpoints of the cell edges at $z = \frac{1}{4}$ and $\frac{3}{4}$ center the diamonds of the arrays of atoms at $z = 0$ and $\frac{1}{2}$. Since the structure is cubic, this description of layers perpendicular to the c axis applies also to layers that occur perpendicular to the a and b axes.

These arrangements surround Ir with a convex polyhedron of 9 atoms (8 Ge at 2.52 and 2.58 + Ir at 2.76 Å). Ge(1) has 4 Ge(1) at 3.09 Å and 4 Ir at 2.58 Å, whereas Ge(2) has 3 Ge(2) at 2.725 Å, one at 2.84 Å and 3 Ir at 2.52 Å. Single-crystal studies have been made of various phases with this structure.[22]

α-SV_3 (H.T.) ($tI32$): α-SV_3 has cell dimensions $a = 9.470$, $c = 4.589$ Å, $Z = 8$.[23] V(1) and S atoms at $z = 0$ and $\frac{1}{2}$ form the characteristic

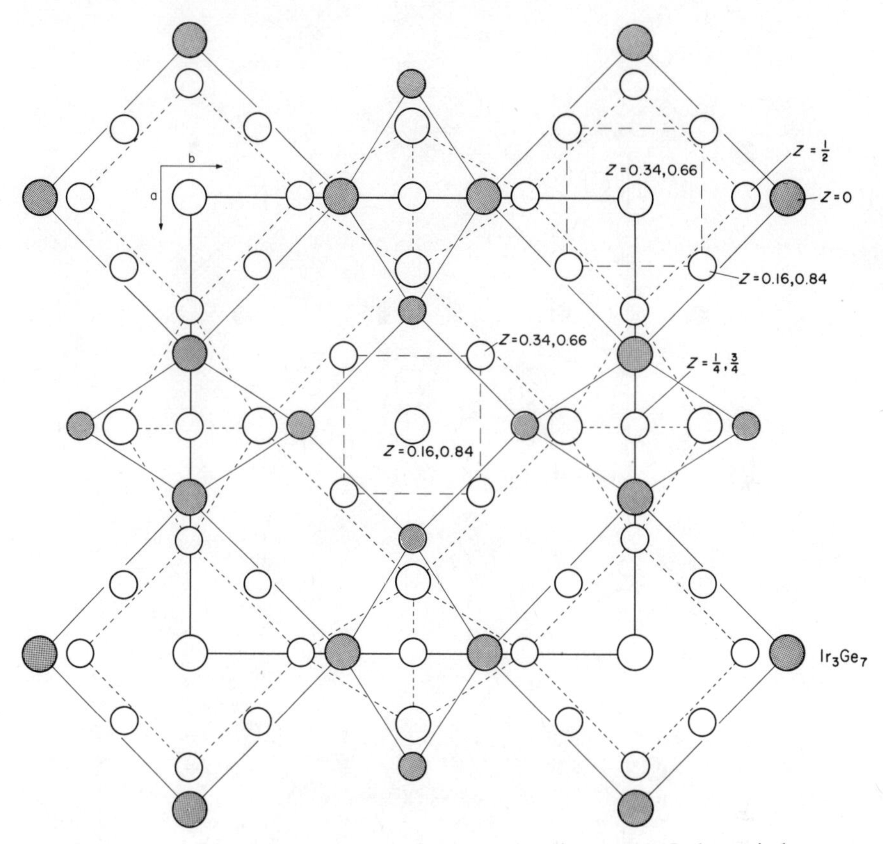

Figure 14-14. Atomic arrangement in the Ir_3Ge_7 ($cI40$) structure: Ir, large circles.

square-quadrilateral-triangle net (Figure 14-15) which is also found in the Ir_3Ge_7 and $ThMn_{12}$ structures. The squares are of two sizes, the larger one occurring at the cell corners and the smaller one at the cell base center in the layer at $z = 0$ and *vice versa* in the layer at $z = \frac{1}{2}$ as shown in Figure 14-15. V(2) and (3) atoms form $3^28^2 + 38^2$ (1:1) nets at $z = \frac{1}{4}$ and $\frac{3}{4}$. In these nets pairs of triangles form diamonds which are located over the cell corners and base centers. The diamonds in one layer are oriented at 90° to those in the layers above and below. The diamonds of the square-quadrilateral-triangle nets at $z = 0$ and $\frac{1}{2}$ lie over the centers of the octagons of the $3^28^2 + 38^2$ nets.

These arrangements give S eight close V neighbors. V(1) has 2 S + 12 V neighbors, V(2) has 2 S + 9 V, and V(3) has 4 S + 9 V neighbors at distances of 3 Å and less.

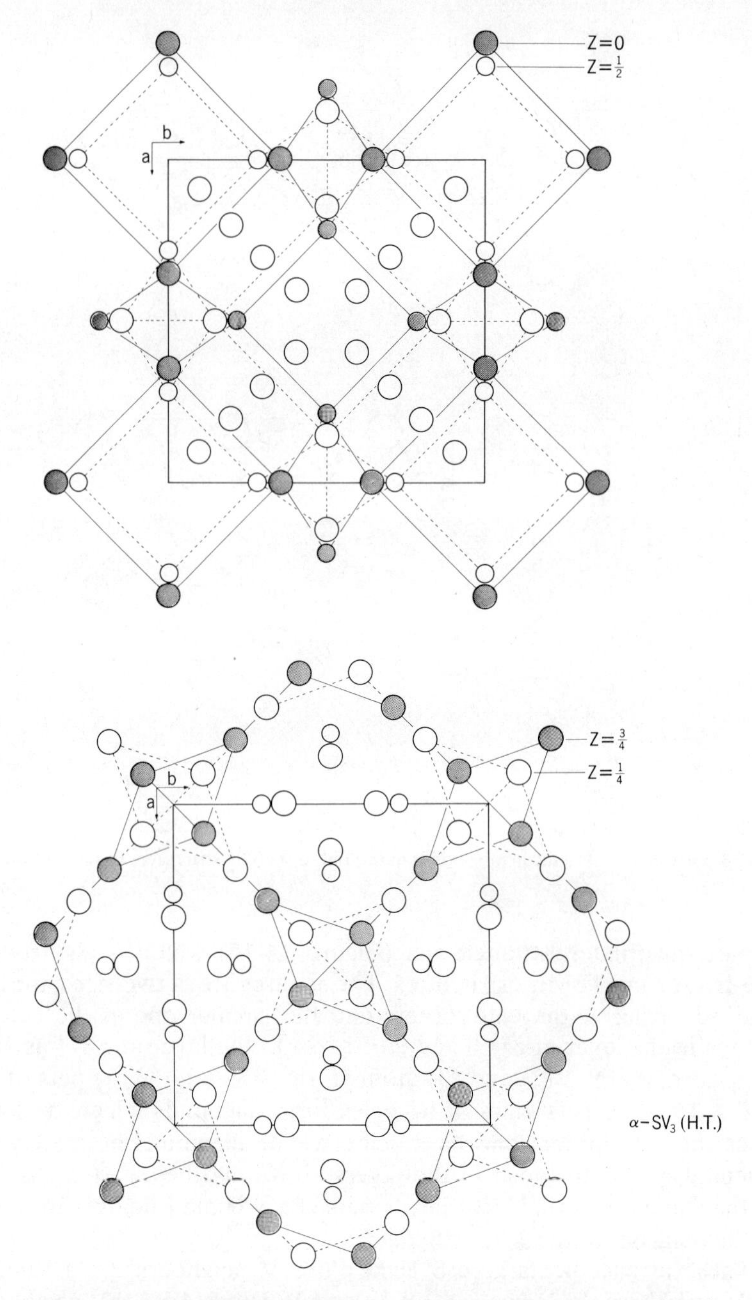

Figure 14-15. Atomic arrangement in the high-temperature α-SV_3 ($tI32$) structure viewed down [001]: V, large circles.

Mn_2Hg_5 (*tP*14): Mn_2Hg_5 with cell dimensions $a = 9.758$, $c = 2.998$ Å, $Z = 2$,[24] has an interesting structure in which $3545 + 3535$ (4 : 1) pentagon-square-triangle nets of Hg atoms at $z = 0$ are stacked one above the other in the [001] direction. The Mn atoms at $z = \frac{1}{2}$ form a 3^3434 net which centers the pentagons of the Hg net (Figure 14-16). Both Hg(1), on the cell edges, and Hg(2) are surrounded by 4Mn + 6Hg, whereas Mn is surrounded by a pentagonal prism of 2Hg(1) + 8Hg(2) atoms; two polar Mn neighbors complete a CN 12 polyhedron.

V-phase. $Zr_4Co_4Ge_7$ (*tI*60): $Zr_4Co_4Ge_7$ has cell dimensions $a = 13.228$, $c = 5.229$ Å, $Z = 4$.[25] The structure is made up of layers of atoms parallel to the (001) plane. At $z = 0$ and $\frac{1}{2}$ there are pentagon-square-triangle nets of Zr and Ge(1) to (3) atoms with the squares over the cell base center and corners. The nets are oriented antisymmetrically relative to

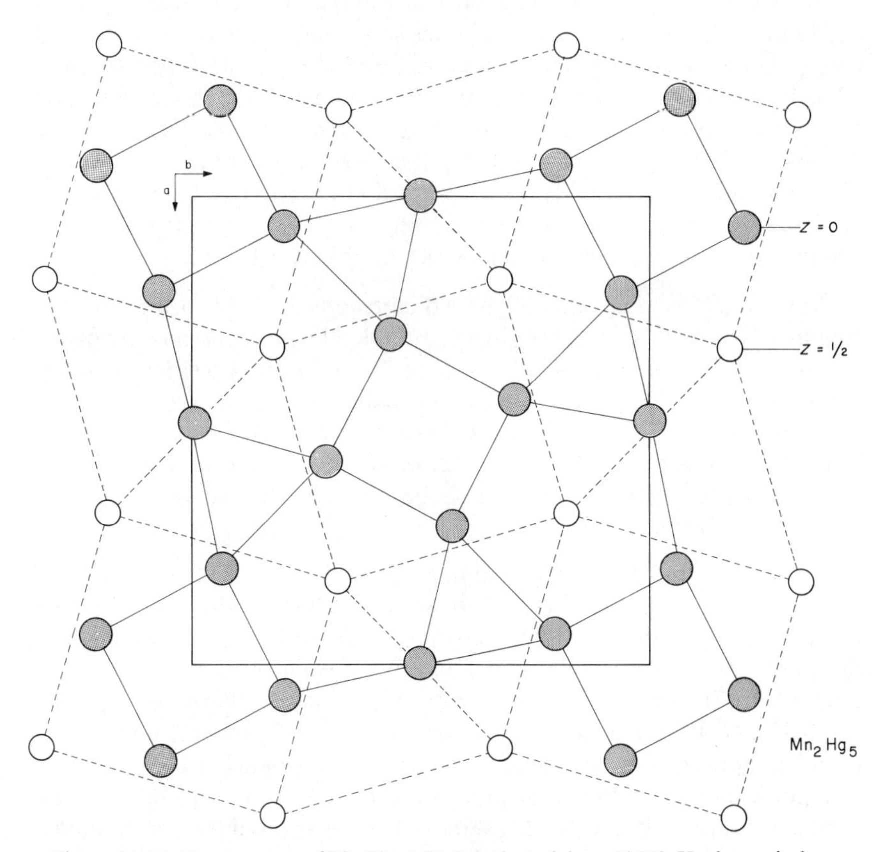

Figure 14-16. The structure of Mn_2Hg_5 (*tP*14) projected down [001]: Hg, large circles.

the nets above and below. The Co atoms at $z = \pm\frac{1}{4}$ form a 48^2 net, each Co lying over the center of a pentagon of the other nets, and the Ge (4) atoms at $z = \pm\frac{1}{4}$ form a large $\frac{1}{2}\,.\,4^4$ net lying over the centers of the squares of the nets at $z = 0$ and $\frac{1}{2}$.

The Zr atoms are surrounded by a CN 17 polyhedron which is derived from the CN 15 μ-phase polyhedron by replacing two atoms by pairs of atoms. Similar coordination occurs in the Ti_5Ga_4 structure. Co and Ge(1) are surrounded by distorted icosahedra. In the Ge(2) CN 11 polyhedron, two Ge atoms forming what would otherwise have been an icosahedron, are replaced by Zr; similar coordination occurs in the Mn_5Si_3 (*hP*16) structure. Ge(3) is surrounded by a CN 14 polyhedron with one of the neighboring Ge(3) atoms very close at 2.46 Å, so that the pair of Ge(3) atoms can be considered as surrounded by a CN 20 polyhedron resembling the CN 22 polyhedra around the Be(6) pairs in the Ru_3Be_{17} structure. Ge(4) is surrounded by 10 neighbors in a cubic antiprism with two polar neighbors. The structure thus combines some features of the tetrahedrally close packed Frank–Kasper structures and the $CuAl_2$ type structure. This is readily seen from Figure 14-17 where main layer pentagon-triangle regions resembling those of the Laves phases, separate cubic antiprism regions such as are found in the $CuAl_2$ structure.

Zr–Ge distances vary from 2.73 to 2.91 Å, Co–Ge distances from 2.45 to 2.49. The closest Zr–Zr approach is 3.29 Å, Co–Co is 2.62 Å, and in addition to the Ge(3) pairs, Ge(4)–Ge(4) is 2.615 Å.

~Mo_3CoSi (tI56): The unit cell, with dimensions $a = 12.646$, $c = 4.889$ Å, contains 56 atoms arranged in slightly rumpled hexagon-pentagon-square-triangle nets parallel to the (001) plane at $z \sim 0$ and $\frac{1}{2}$[26] (Figure 14-18). Nets at $z \sim \frac{1}{2}$ are disposed antisymmetrically to those at $z \sim 0$. 4^4 nets at $z = \frac{1}{4}$ and $\frac{3}{4}$ center the hexagons, pentagons, and squares of the other nets. Thus, the atoms have CN 15, 14, 12, or 10. The structure resembles the Frank–Kasper types, but cannot so be classed because of the squares occurring in the main layers.

β-Pu (*mC*34): In β-Pu with cell dimensions at 93°C, $a = 9.227$, $b = 10.45$, $c = 7.824$ Å, $\beta = 92.54°$, $Z = 34$,[27] in setting $I2/m$, the Pu(1) to (4) atoms form a pentagon-distorted-square-triangle network at $y = 0$ and $\frac{1}{2}$, with the pentagons at each level partly overlying each other (Figure 14-19). Pairs of Pu(7) atoms at 2.59 Å are located within the overlying regions at $y = 0.15$ and 0.35 and at $y = 0.65$ and 0.85. Pu(5) and (6) atoms form rumpled hexagon-triangle nets at $y \sim \frac{1}{4}$ and $\frac{3}{4}$ (Figure 14-19) with the hexagons overlying the pentagons of the nets at $y = 0$ and $\frac{1}{2}$. These arrangements result in CN 12, 13, or 14 for the seven different Pu atoms. Interatomic distances correspond to a CN 12 radius of 1.59 Å for Pu at

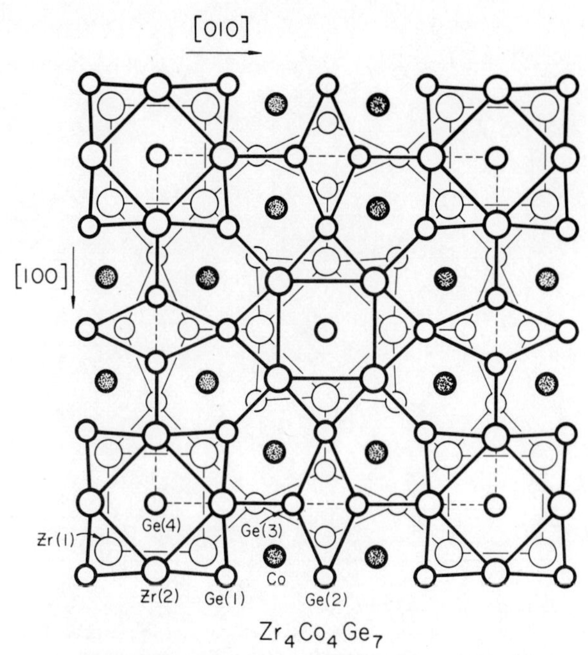

Figure 14-17. Atomic arrangement in the *V* phase Zr_4Co_4-Ge_7 (*tI*60) viewed down [001]. Atoms at $z = \frac{1}{2}$ are connected by thick lines, those at $z = 0$ by thin lines. Other atoms are at $z = \frac{1}{4}$ and $\frac{3}{4}$: Zr, large circles; Ge, small circles; Co, small shaded circles. (Jeitschko [25].)

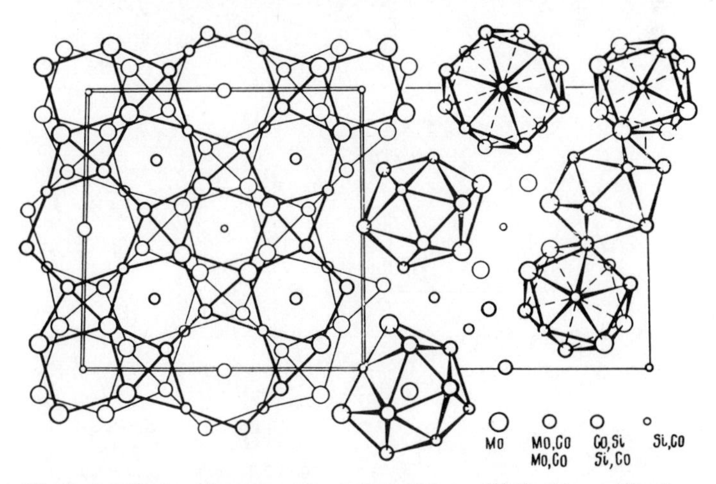

Figure 14-18. Atomic arrangement viewed down [001] and coordination in the ~ Mo_3CoSi (*tI*56) structure. (Gladysevskij and co-workers [26].)

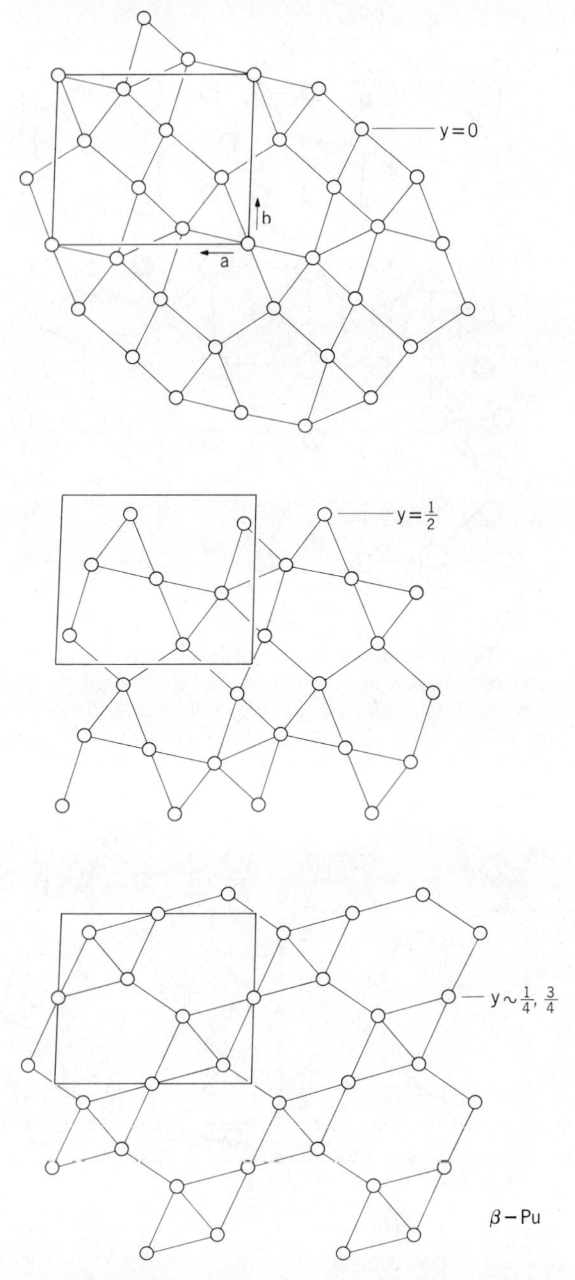

Figure 14-19. Arrangement of the atoms in layers in the β-Pu (*mC*34) structure.

room temperature. In addition to the very short Pu–Pu distance noted above, there are numerous contacts at a distance of about 2.80 Å. The structure was determined from X-ray powder data taken at high temperature.

α-US₂ ($tI30$), *Si₃W₅* ($tI32$), *β-SbTi₃* ($tI32$): The structures of these three substances resemble each other closely, differing only in the stoichiometry of the components contained in the unit cells. In the case of α-US_2, only two of the sites in positions 4(a) of space group $I4/mcm$ are occupied (randomly). In the general structure type, atoms of two kinds (sitesets 16(k) and 8(h)) form 3^2634+3^36 (2:1) nets at $z=0$ and $\frac{1}{2}$, which are oriented antisymmetrically relative to each other (Figure 14-20). Atoms of position 4(a) at $z=\frac{1}{4}$ and $\frac{3}{4}$ over the cell corners and over the base center, center the cubic antiprisms of the 3^2634+3^36 array of atoms. Atoms of position 4(b) also at $z=\frac{1}{4}$ and $\frac{3}{4}$ over the midpoints of the cell edges, center the hexagons of the 3^2634+3^36 nets. The distributions of atoms in the three structures is summarized in Table 14-1.

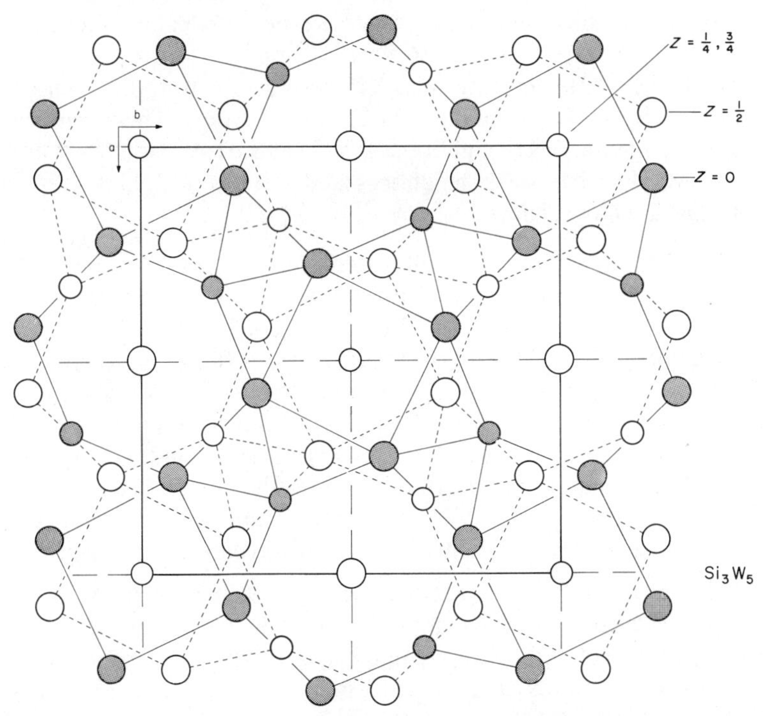

Figure 14-20. The structure of Si_3W_5 ($tI32$) projected down [001]: W, large circles.

Table 14-1

Structure	3^2634 Corners Position $16(k)$	3^36 Corners Position $8(h)$	Position $4(a)$ Centering Cubic Antiprisms	Position $4(b)$ Centering Hexagonal Antiprisms
α-US_2[28]	S(2)	U(2)	U(1)	S(1)
Si_3W_5[29]	W(2)	Si(2)	(2 atoms randomly) Si(1)	W(1)
β-$SbTi_3$[30]	Ti(3)	Sb	Ti(1)	Ti(2)

In the Si_3W_5 structure ($a = 9.604$, $c = 4.970$ Å, $Z = 4$) Si(1) has 2 Si (2.485 Å) and 8 W neighbors at 2.58 Å; Si(2) has 10 W neighbors at distances from 2.52 to 2.85 Å. W(1) has 4 Si(2) at 2.63 Å, 2 W(1) at 2.485 Å and 8 W(2) at 3.03 Å. W(2) has 6 Si at 2.58–2.85 Å and 9 W neighbors, the closest being 2.77 Å and 2×2.87 Å.

Te_4Ti_5 (tI18): Te_4Ti_5 with cell dimensions $a = 10.16$, $c = 3.772$ Å, $Z = 2$,[31] is a layer structure with $3436 + 3646$ (1:1) nets of atoms at $z = 0$ and $\frac{1}{2}$ arranged antisymmetrically with respect to each other (Figure 14-21). Te atoms are at the 3436 corners and Ti(2) atoms at the 3646 corners. Ti(1) atoms forming large 4^4 nets at $z = 0$ and $\frac{1}{2}$, center the squares formed by Te in the Te + Ti(2) nets of atoms. This arrangement results in Te having 6 Ti neighbors: 1 Ti(1) at 2.91 Å, 3 + 2 Ti(2) at 2.77 and 2.82 Å. Ti(1) has 4 Te neighbors and 8 Ti(2) at 2.95 Å. Ti(2) has 3 + 2 Te and 2 Ti(1) neighbors.

Mg_2Ga_5 (tI28): Mg_2Ga_5 has cell dimensions $a = 8.627$, $c = 7.111$ Å, $Z = 4$.[32] Ga atoms form very rumpled $46^2 + 6^4$ (1:1) nets about $z \sim \frac{1}{4}$ and $\frac{3}{4}$ which are interconnected by Ga(1)–Ga(1) links (2.57 Å) along [001] (Figure 14-22). Mg atoms located at $z = 0$ or $\frac{1}{2}$ lie in the space between the rumpled Ga hexagons, but displaced toward one Ga(2) end, being surrounded by 2 Ga(2) (2.87 Å) and 8 Ga(1) (4×2.89, 4×2.97 Å). Ga(2) is surrounded by 4 Ga(1) (2.68 Å) and 4 Mg (2.87 Å) in a squashed cubic antiprism. In addition to the Ga(1)–Ga(1) neighbors cited above, Ga(1) also has 2 Ga(1) at 2.65 Å and Ga(2) at 2.68 Å completing a very distorted tetrahedron; it is also surrounded by 4 Mg.

PdS (tP16): In the PdS structure ($a = 6.429$, $c = 6.608$ Å, $Z = 8$)[33] Pd(1) and (3) atoms form hexagon-triangle $3636 + 3^26^2$ nets at $z = 0$ and $\frac{1}{2}$ which cover each other antisymmetrically. The hexagon centers together with Pd(2) atoms at $z = \frac{1}{4}$ and $\frac{3}{4}$ lie over the cell corners as shown in Figure 14-23, the Pd atoms forming a distorted β-W arrangement. The S atoms form rumpled 3^2434 nets at $z \sim \frac{1}{4}$ and $\frac{3}{4}$ which are superimposed.

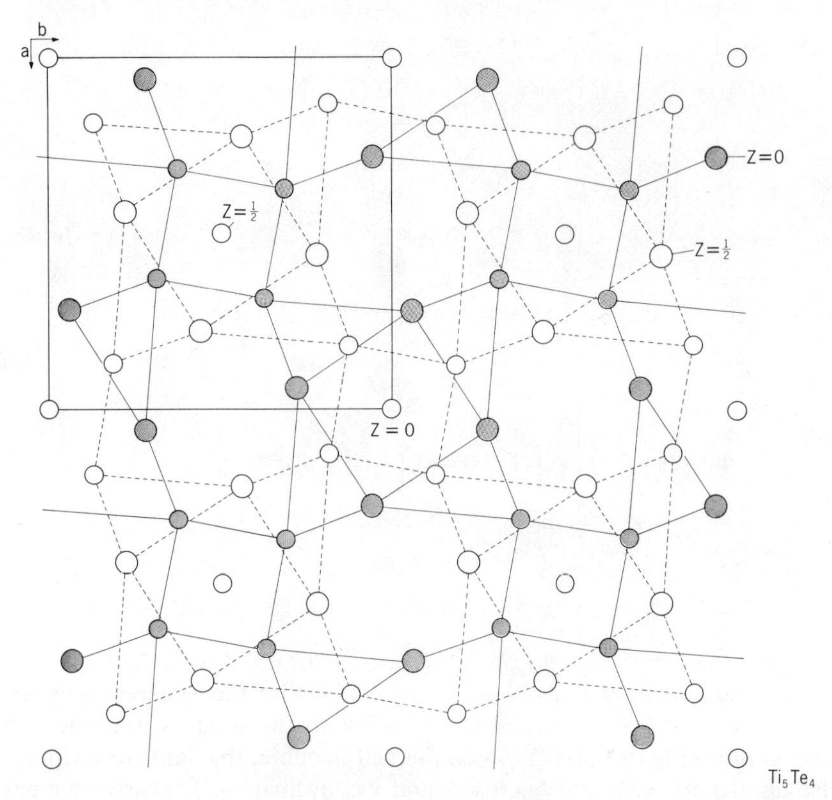

Figure 14-21. Atomic arrangement in the Te_4Ti_5 structure (*tI*18) viewed down [001]: Te, large circles.

The squares of these nets lie over the cell center and corners. S has 4 close Pd neighbors at distances varying from 2.26 to 2.43 Å, but the arrangement is considerably distorted from a tetrahedron. Each Pd atom has 4 close S neighbors in a planar arrangement. The structure thus appears to be one which is determined by the chemical bonds, and the arrangement of the Pd atoms in $3636+3^26^2$ and 4^4 layer nets, suggestive of a structure controlled by high coordination, is probably an unimportant feature of the structural architecture.

Ca_3Ag_8 (*cI*44): The cubic structure[34] can be regarded as made up of hexagon-square-triangle nets of Ca and Ag(2) atoms at $z=0$ with the squares over the cell corners and base center and a similar net in anti-symmetric orientation at $z=\frac{1}{2}$. Ag(1) atoms form 1.4^4 nets at $z=\frac{1}{4}$ and $\frac{3}{4}$ and Ca form $\frac{1}{2}.4^4$ nets at $z=0.245$ and 0.755. Ag(2) atoms also form

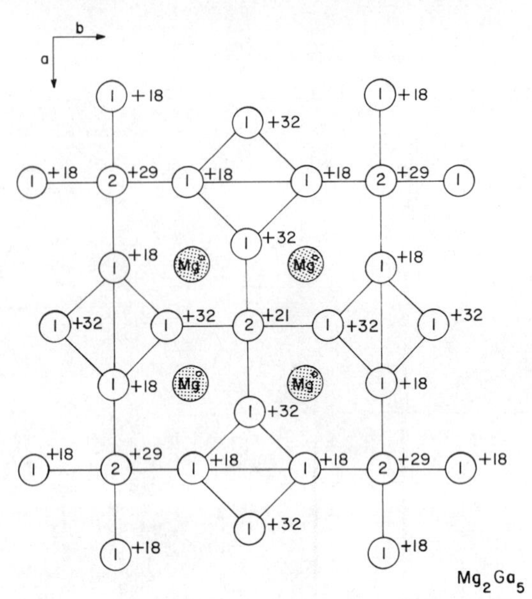

Figure 14-22. Structure of Mg_2Ga_5 (*tI*28) viewed down [001].

octagon-square nets at $z = 0.347$ and 0.653 with the squares over the cell corners, and at $z = 0.153$ and 0.847 with the squares over the cell base center (Figure 14-24). Since the cell is cubic, the same description also applies to the *a* and *b* axes (*x* and *y* coordinates). This arrangement surrounds Ca with 4 Ca (3.54 Å), 4 Ag(1) (3.47 Å) and 8 Ag(2) (4×3.21, 4×3.52 Å). Ag(1) has 6 Ca and 6 Ag(2) surrounding it in a slightly puckered hexagon at 2.80 Å. Ag(2) also has CN 12 (4 Ca, 8 Ag neighbors).

$Ce_3Ni_6Si_2$ (cI44): $Ce_3Ni_6Si_2$ is an ordered ternary form of the Ca_3Ag_8 type structure.[35] Ce is surrounded by 4 Ni at 2.89 Å, 4 at 2.96 Å, 4 Si at 3.14 Å, and 4 Ce at 3.51 Å. Si has 6 Ni neighbors at 2.43 and 6 Ce at 3.14 Å which form a distorted icosahedron. Ni is surrounded by 10 neighbors. Thirteen $Ln_3Ni_6Si_2$ phases are known. The structure requires confirmation by single-crystal methods.

Some Additional Structures Generated by Stacking Hexagonal (6^3) Nets

$AgTlSe_2$ (hP24): $AgTlSe_2$ has cell dimensions $a = 9.70$, $c = 8.25$ Å, $Z = 6$.[36] The Se atoms form hexagons about the cell corners at $z = 0$ and $\frac{1}{2}$ (Figure 14-25). Tl and Ag severally form triangular 3^6 nets at $z = \pm 0.258$. Taken together, Ag and Tl form 6^3 nets at $z = \pm 0.258$ which are stacked in hexagonal sequence $c'b'$ (notation of p. 4). Se–Se

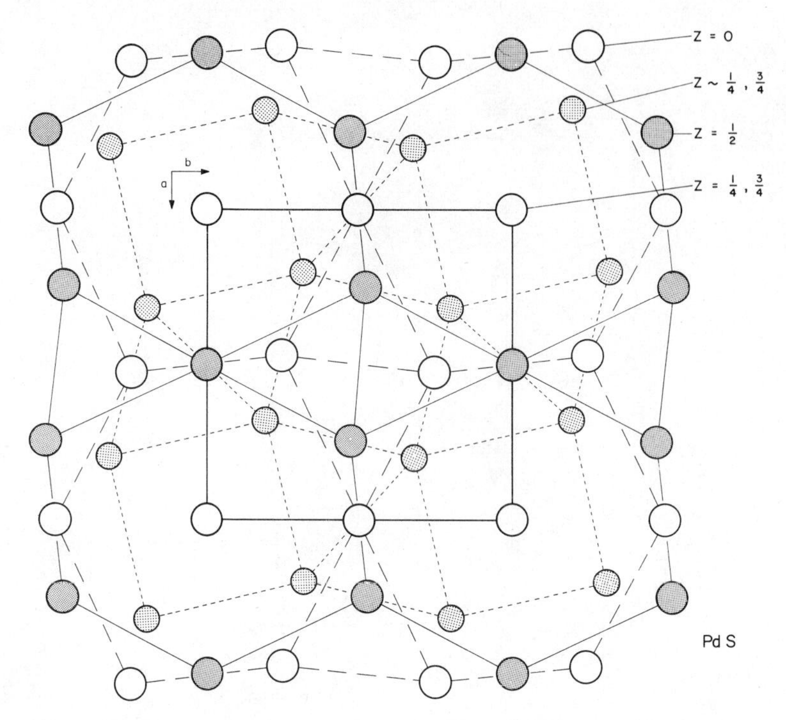

Figure 14.23. Arrangement of the atoms in the PdS (*tP*16) structure viewed down [001]: Pd, large circles.

distances within the hexagons are 2.38 Å. The shortest Tl–Se distances are 3.11 and 3.20 Å; Ag–Se distances are 2.68 and 2.78 Å and Ag–Tl distances are 3.15 and 3.39 Å. The structure was found in thin evaporated films and determined from oblique textured electron diffraction photographs.

Ga (*oC*8): The stable Ga structure has cell dimensions $a = 4.520$, $b = 7.663$, $c = 4.526$ Å, $Z = 8$.[37] The Ga atoms form a 6^3 network of distorted hexagons parallel to the (100) plane at heights $x = 0$ and $\frac{1}{2}$ (Figure 14-26). Every Ga has one short contact (2.485 Å) and two at 2.69 A in the same plane and four more contacts at 2.73 and 2.79 Å in planes above and below. The shortest Ga–Ga bonds join rumpled layers of Ga atoms which lie parallel to the (010) plane.

Structures Generated by Nets Containing Heptagons and Larger Polygons

ScB_2C_2 (*oP*20): ScB_2C_2 has cell dimensions $a = 5.175$, $b = 10.075$, $c = 3.440$ Å, $Z = 4$.[38] The C and B atoms form a heptagon-pentagon net at

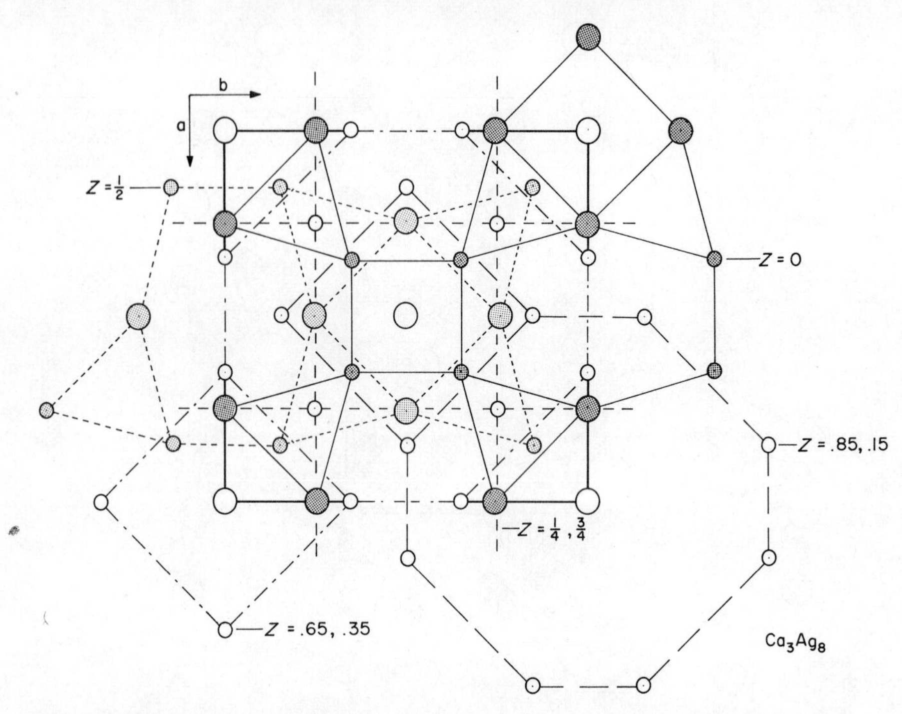

Figure 14-24. Atomic arrangement in the Ca_3Ag_8 ($cI44$) structure viewed down [001]: Ca, large circles.

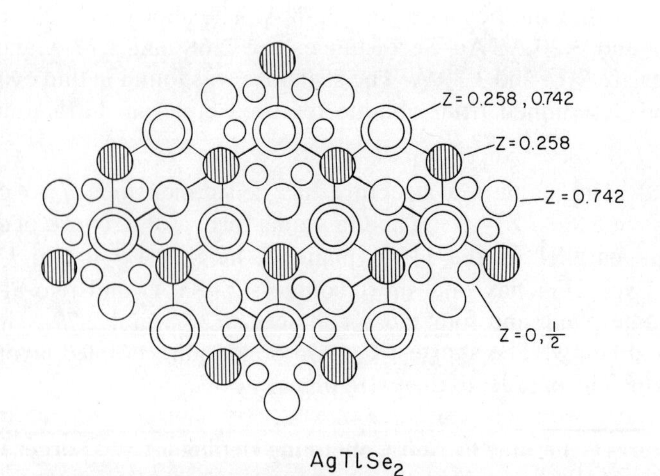

Figure 14-25. Structure of $AgTlSe_2$ ($hP24$) viewed down [001]: Ag, shaded circles; Se, small circles.

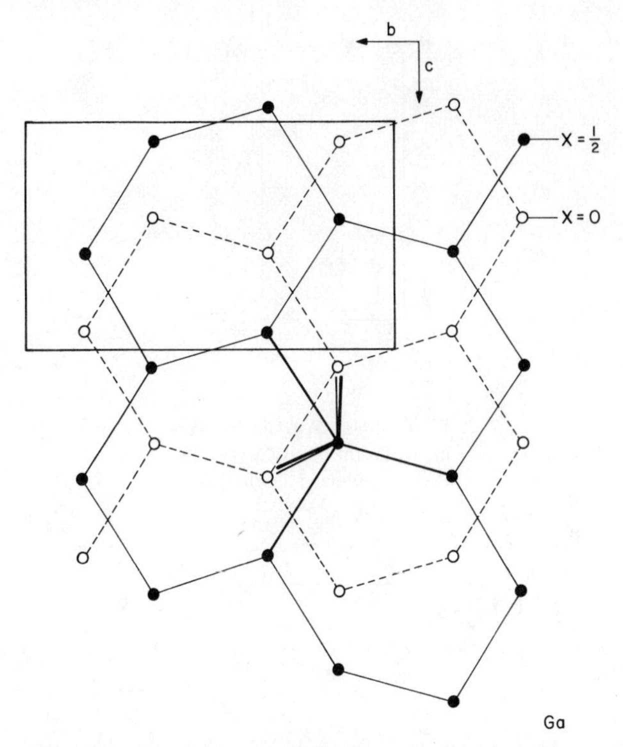

Figure 14-26. Atomic arrangement in the Ga structure seen down [100].

$z = \frac{1}{2}$ and Sc atoms at $z = 0$ center the heptagonal prisms (Figure 14-27). Sc also has 2 Sc at 3.295 Å and one at 3.319 Å in the same plane making, with two polar Sc neighbors at 3.44 Å, a CN 19 polyhedron. Sc–B and Sc–C distances vary from 2.40 to 2.52 Å. C–C = 1.447 Å, B–B = 1.589 Å, C–B distances vary from 1.52 to 1.61 Å. The structure was determined by single-crystal X-ray diffraction.

$IrB_{1.35}$ (mC19): $IrB_{1.35}$ has cell dimensions $a = 10.525$, $b = 2.910$, $c = 6.099$ Å, $\beta = 91°4'$, $Z = 8$.[39] The atoms lie in layers in the *ac* plane at $y = 0$ and $\frac{1}{2}$, where they form an arrangement of triangles and large polygons that leaves large voids over the midpoints of the cell edges at $y = 0$, and over the cell corners and base centers at $y = \frac{1}{2}$. The Ir atoms of one layer lie in the voids of the layers above and below (Figure 14-28).

The closest Ir–Ir approaches are 2.66 and 2.85 Å. The closest Ir–B distances are 1.99 Å and the closest B–B approaches 1.57 Å.

Ni_3P (tI32): In the Ni_3P structure ($a = 8.954$, $c = 4.386$ Å, $Z = 8$),[40] Ni(1)

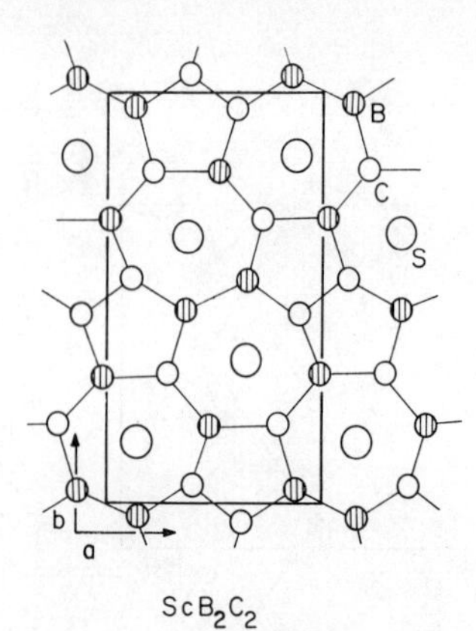

Figure 14-27. Atomic arrangement in the ScB_2C_2 (*oP*20) structure viewed down [001].

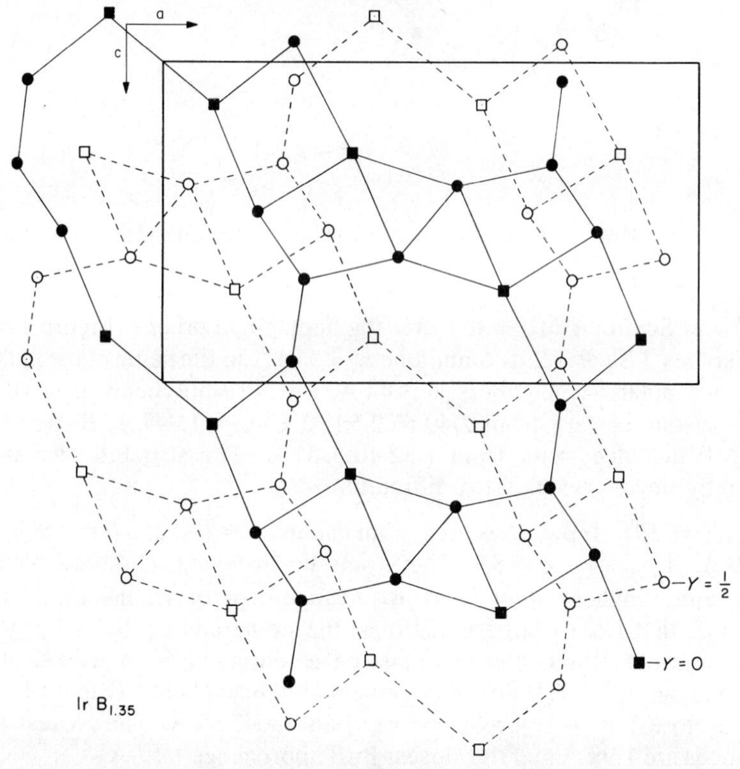

Figure 14-28. Atomic arrangement in the $IrB_{1.35}$ (*mC*19) structure viewed down [010]: squares, Ir atoms.

and (3) atoms at $z \sim \frac{1}{4}$ and $\frac{3}{4}$ form $3^2 8^2 + 38^2$ (1 : 1) networks which contain large octagons, large squares with atoms at the midpoints of the sides, and pairs of triangles forming diamonds as shown in Figure 14-29. The diamonds lie over the cell center and cell corners. Neighboring layers are so arranged that the octagons of one layer cover the large squares of the other. The space between these pairs of octagons and squares is occupied by diamonds formed by two Ni(2) and two P atoms at heights $z \sim 0$ and $\sim \frac{1}{2}$.

These arrangements give Ni(1) 2P + 12Ni neighbors, Ni(2) 4P + 10 Ni neighbors, and Ni(3) 3P + 10 Ni neighbors, all at distances less than 3 Å. P is surrounded by 9 Ni at distances from 2.22 to 2.34 Å.

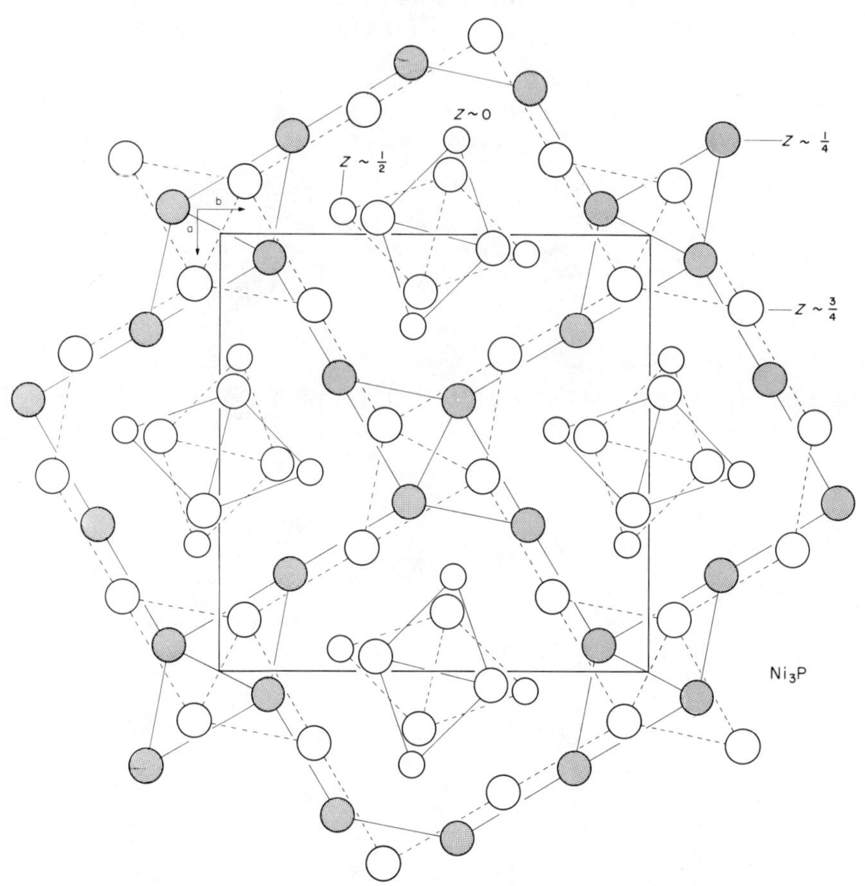

Figure 14-29. Arrangement of the atoms in the Ni_3P structure ($tI32$) viewed down [001]: Ni, large circles.

β-SV$_3$ (L.T.) (*tP*32): β-SV_3 with cell dimensions $a = 9.381$, $c = 4.663$ Å, $Z = 8$,[41] forms a layer structure. V(1) and (2) atoms at $z = 0$ and $\frac{1}{2}$ form $3^28^2 + 38^2$ (1:1) nets, one being oriented antisymmetrically with respect to the other (Figure 14-30). The octagons are centered over the cell corners and base center, and the diamonds lie over the centers of the cell edges. The S and V(3) atoms together form diamonds at $z = \frac{1}{4}$ and $\frac{3}{4}$ which center the octagons in the V(1) and (2) nets. The diamonds at $z = \frac{3}{4}$ are oriented 90° to those at $z = \frac{1}{4}$. S has 8 close V neighbors (2.30 to 2.47 Å) and two further away at 3.025 Å. V(1), (2), and (3) are surrounded by 14, 15, and 15 neighbors, respectively, at distances less than 3.06 Å.

Ni$_{12}$P$_5$ (*tI*34): The $Ni_{12}P_5$ structure ($a = 8.646$, $c = 5.070$ Å, $Z = 2$)[42] is a layer structure with 48^2 nets of Ni(2) atoms at $z = \frac{1}{4}$ and $\frac{3}{4}$. The nets

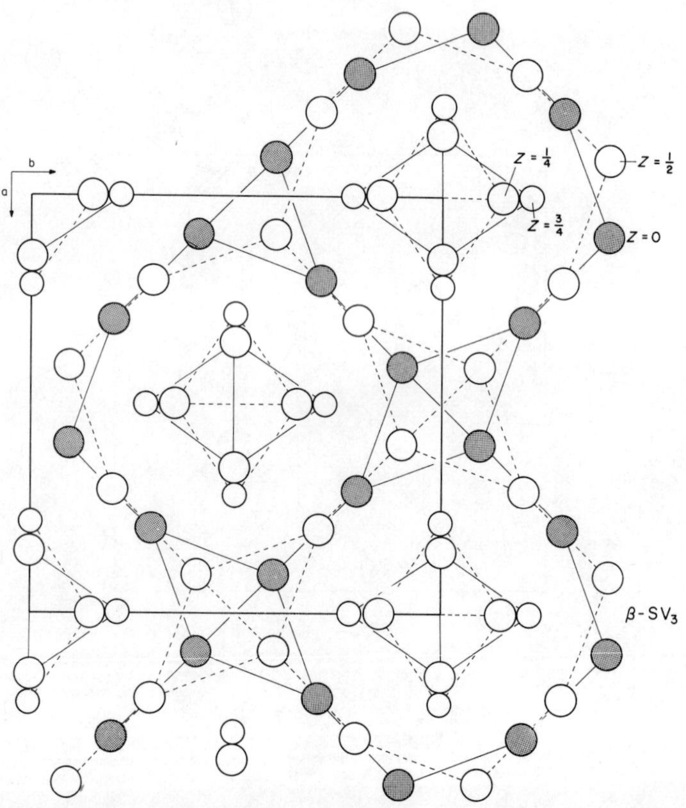

Figure 14-30. Atomic arrangement in the low temperature β-SV_3 (*tP*32) structure viewed down [001]: V, large circles.

at $z = \frac{1}{4}$ and $\frac{3}{4}$ are exactly superimposed and they are arranged with squares over the cell corners and base center. The octagons have their centers over the midpoints of the basal cell edges, where there also are located at $z = 0$ and $\frac{1}{2}$, diamonds made up of two Ni(1) and two P(2) atoms (Figure 14-31). In addition, the P(2) atoms together with P(1) form $\frac{3}{4}.4^4$ nets at $z = 0$ and $\frac{1}{2}$ as indicated in Figure 14-31. Ni(1) has two close P neighbors (2.21, 2.24 Å) and two at 2.60Å; it also has 7 close Ni neighbors and 4 rather farther away. Ni(2) has 4P + 8Ni neighbors. P(1) is surrounded by 8 Ni at 2.25 Å and P(2) has 10 Ni neighbors at various distances up to 2.60 Å.

$Pd_{17}Se_{15}$ (cP64): $Pd_{17}Se_{15}$ has a cell with edge, $a = 10.606$ Å, $Z = 2$.[43] The structure can be regarded as made up of layers of atoms. At $z = 0$ Pd(2) and (3) and Se(3) form a net of squares and crosses (Figure 14-32*a*). At $z \sim 0.16$ and at ~ 0.84 a square and four triangles formed by Pd(4)

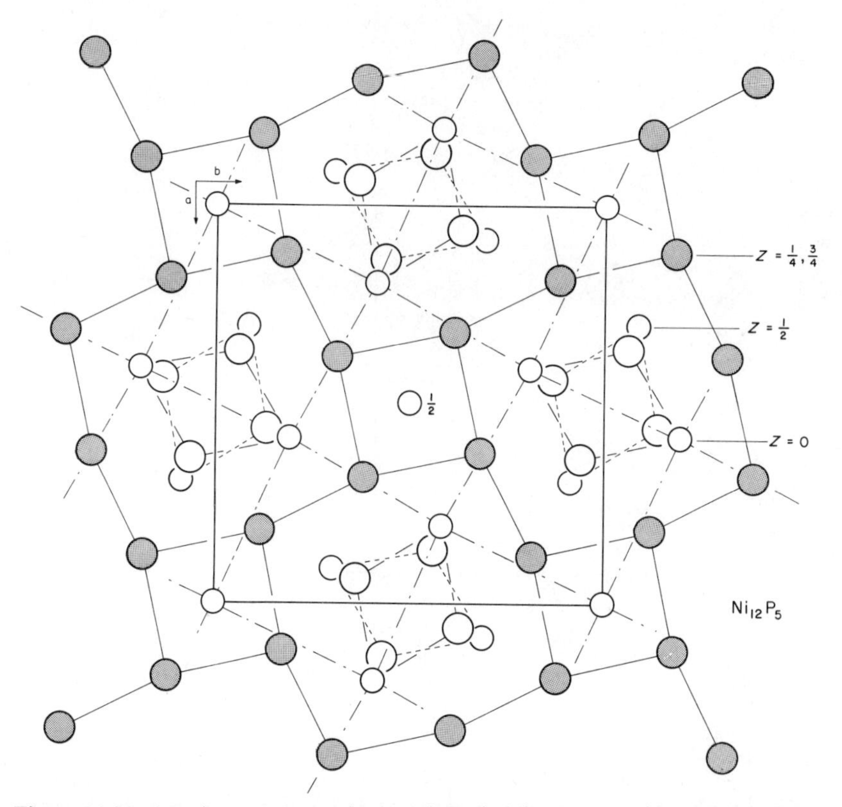

Figure 14-31. Atomic arrangement in the $Ni_{12}P_5$ (*tI*34) structure viewed down [001]: Ni, large circles.

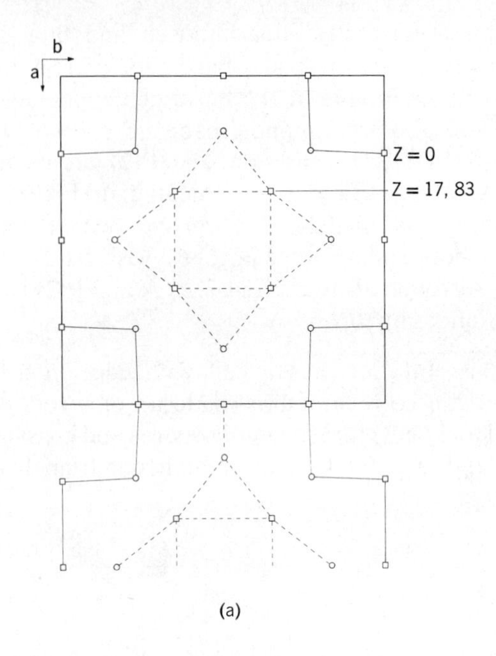

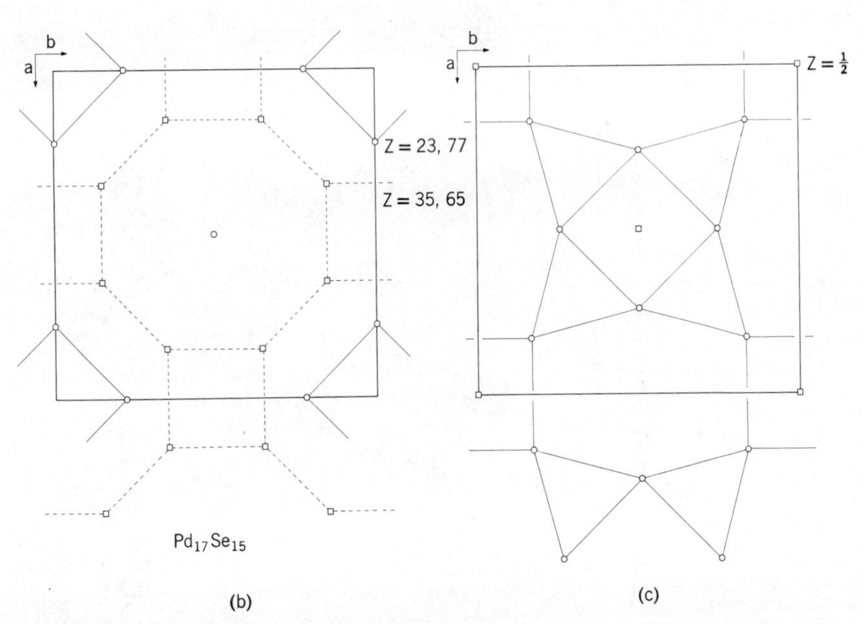

Figure 14-32. Arrangement of the atoms in layers in the $Pd_{17}Se_{15}$ (*cP*64) structure: circles, Se atoms.

and Se(3) atoms lie over the centers of the crosses (Figure 14-32*a*). At $z \sim \frac{1}{4}$ and $\sim \frac{3}{4}$, Se(2) form squares about the cell corners which are centered by Pd(3), and Se(1) occurs over the cell base center (Figure 14-32*b*). These Se atoms lie over the centers of the octagons in 48^2 nets formed by Pd(4) at $z = 0.35$ and 0.65 (Figure 14-32*b*). Finally, at $z = \frac{1}{2}$, Se(1) and (3) form a 3436 + 3646 net with the hexagons centered over the midpoints of the cell edges and the squares over the cell corners and base center (Figure 14-32*c*). The squares are centered by Pd(2) over the cell corners and Pd(1) over the cell center at $z = \frac{1}{2}$. Since the structure is cubic, this description applies down each of the crystal axes *a*, *b*, and *c*.

Pd(1) at the cell body center is surrounded by a regular octahedron of Se(1) at 2.58 Å. Pd(2), (3), and (4) have square or approximately square coordination by Se. Pd(2)–Se(3) = 2.53 Å, Pd(3)–Se(2) = 2.44 Å, and Pd(4)–Se = 2.43–2.51 Å. Pd(2) has 2 close Pd(3) neighbors and Pd(3) has one Pd(2) neighbor at 2.78 Å. Se(1) has 5 Pd neighbors; Se(2) has four and Se(3) five. Large voids occur in the structure about the three sites $0, \frac{1}{2}, \frac{1}{2} \curvearrowright$.

Pd_2Hg_5 (*tP*14): Pd_2Hg_5 has cell dimensions $a = 9.463$, $c = 3.031$ Å, $Z = 2$.[44] The atoms for layers at $z = 0$ and $\frac{1}{2}$. At $z = 0$, Hg(2) atoms form a network of squares and eight-sided figures, and at $z = \frac{1}{2}$, Pd form a net of squares and diamonds, the latter being centered by Hg(1) at $z = \frac{1}{2}$. The Hg(1) atom and 2 Pd atoms lie over the eight-sided polygon at $z = 0$ (Figure 14-33). Hg has 12 Hg at 3.00 to 3.08 Å and 1 Hg at 3.21 Å, together with 2 Pd at 2.81 Å and 2 Pd at 2.89 Å. Pd has 9 Hg at 2.81 to 2.89 Å.

γ-Ni_3In_{7-x} ($0.4 \leq x \leq 0.8$) (*cI*38): The Ni_3In_{7-x} structure was determined by electron diffraction from polycrystalline and textured thin films.[45] In(1) and (2) atoms form octagon-square-triangle nets at $z = \frac{1}{3}$ and $\frac{2}{3}$, the octagons being centered by Ni lying over the cell corners. Between these at $z = \frac{1}{2}$ is a square-quadrilateral net of In(2) and Ni atoms. The larger Ni squares over the cell center form very distorted cubic antiprisms with the In(1) squares at $z = \pm\frac{1}{3}$, and the smaller In(2) squares form cubic antiprisms with In(1) squares in layers at $z = \frac{1}{6}$ and $\frac{5}{6}$. These are centered by Ni atoms lying over the cubic corners at $z = \frac{1}{3}$ and $\frac{2}{3}$ (In–Ni = 2.62 and 2.75 Å). The layers at $z = \frac{1}{6}$ and $\frac{5}{6}$ also contain a Ni atom over the cell center which is surrounded by a distorted anticube of In formed by the In squares at $z = \pm\frac{1}{3}$ and In squares of another square-quadrilateral net of In(2) and Ni at $z = 0$ (In–Ni = 2.62 and 2.75 Å). The larger Ni squares of this net form very distorted anticubes over the cell corners with the In(1) atoms of the layers at $z = \pm\frac{1}{6}$.

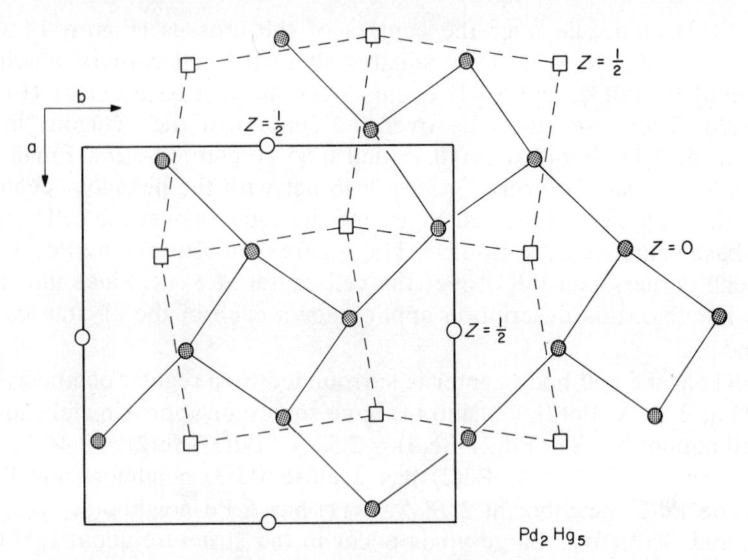

Figure 14-33. Atomic arrangement in the Pd_2Hg_5 (*tP*14) structure viewed down [001]: circles, Hg atoms.

Since the structure is cubic, this description holds for each of the three crystal axes and it results in cubes of In(1), whose positions are only 85% occupied, about the cell center and corners (In–In = 2.92 Å), octahedra of Ni about the cell centers and corners (Ni–Ni = 3.03 Å) and squares of In(2) on the cell faces (In–In = 3.24 Å) as shown in the pictorial view of Figure 14-34.

MgGa$_2$ (*oP*24): $MgGa_2$ has cell dimensions $a = 6.802$, $b = 16.35$, $c = 4.111$ Å, $Z = 8$.[46] The atoms form planar layers at $z = 0$ and $\frac{1}{2}$ as shown in Figure 14-35, the Mg atoms being confined to the layer at $z = 0$. At $z = \frac{1}{2}$ the Ga atoms form a pentagon-octagon net, the octagons arising through the omission of a Ga atoms that would have created two more pentagons and two triangles. The Mg atoms of the net at $z = 0$ lie over the centers of the pentagons at $z = \frac{1}{2}$ and also over the centers of the two unformed pentagons that create part of the octagon. The Ga(1) atoms form pairs (2.56 Å) at $z = 0$ which lie across the rectangle comprizing the center of the octagons at $z = \frac{1}{2}$. Each Ga(1) is surrounded by 3 Mg in approximately equilateral triangles which share a common edge. These 6 atom units lie over the open spaces in the net at $z = \frac{1}{2}$ where the Ga–Ga distances are 2.61 to 2.64 Å and 2.835 Å, and 3.045 Å between Ga(3) and (4). Coordination numbers of Mg(1), Mg(2), Ga(1), Ga(2), Ga(3), and Ga(4) are respectively 13, 15, 8, 9, 10, and 10. The structure contains holes of about 5 Å diameter at 0, 0, $\frac{1}{2}$ and $\frac{1}{2}, \frac{1}{2}, \frac{1}{2}$.

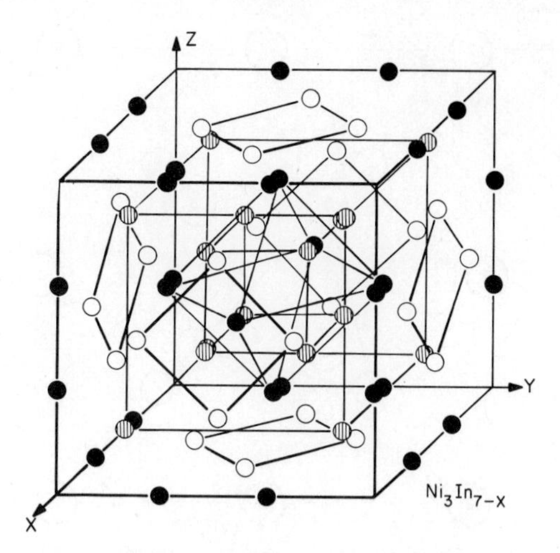

Figure 14-34. Pictorial view of the γ-Ni_3In_{7-x} (*cI*38) structure: open and shaded circles, In sites; black circles, Ni sites.

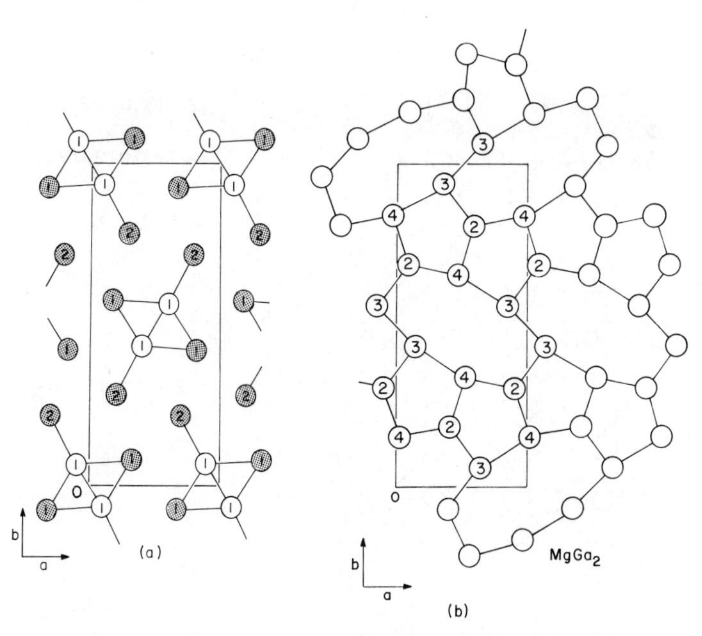

Figure 14-35. The $MgGa_2$ (*oP*24) structure viewed down [001]: shaded circles, Mg.

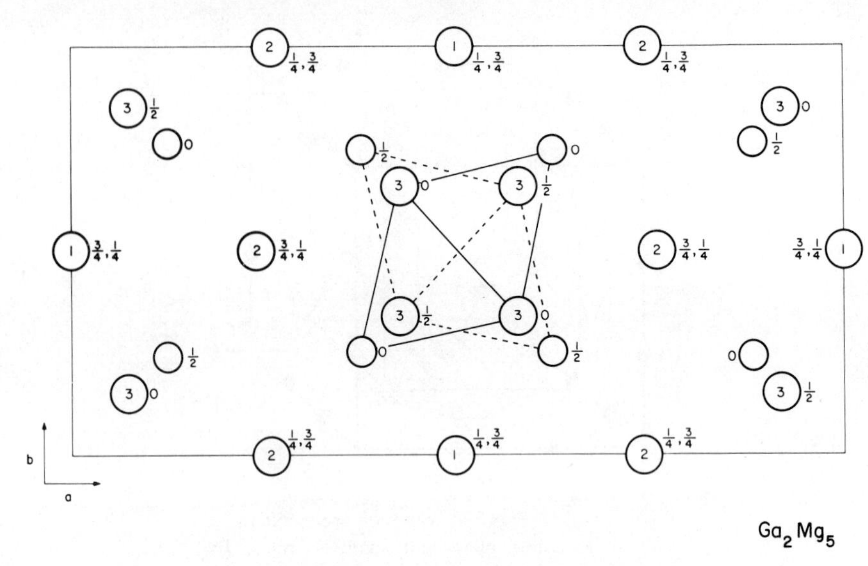

Figure 14-36. The Ga_2Mg_5 ($oI28$) structure viewed down [001]: Mg, large circles.

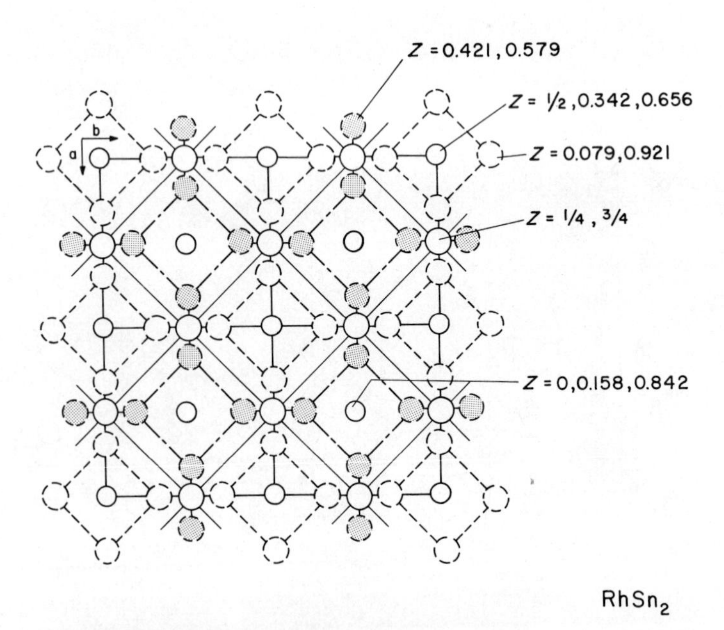

Figure 14-37. Atomic arrangement in the $RhSn_2$ ($tI18$) structure viewed down [001]. Four unit cells are shown: Sn, large circles.

Ga_2Mg_5 (oI28): Ga_2Mg_5 has cell dimensions $a = 13.71$, $b = 7.017$, $c = 6.020$ Å, $Z = 4$.[47] The Mg(1) and (2) atoms at $z = \frac{1}{4}$ and $\frac{3}{4}$ form an $8^3 + 8^2$ (3:1) net, where the octagons are approximately squares with atoms at the midpoints of the edges (Figure 14-36). Between the centers of the octagons at $z = 0$ and $\frac{1}{2}$ there are diamonds made up of 2 Mg(3) and 2 Ga atoms. The $8^3 + 8^2$ nets result from an array of distorted cubes of Mg atoms, each of which have 2 corners missing along the [001] direction.

$RhSn_2$ (tI18): $RhSn_2$ has cell dimensions $a = 4.487$, $c = 17.72$ Å. $Z = 6$.[48] The structure is made up of half-occupied 48^2 nets of Sn(2) atoms at $z = \frac{1}{12}, \frac{5}{12}, \frac{7}{12}$, and $\frac{11}{12}$. Those at $\frac{1}{12}$ and $\frac{11}{12}$ have squares covering the cell corners and those at $\frac{5}{12}$ and $\frac{7}{12}$ have squares covering the cell base center

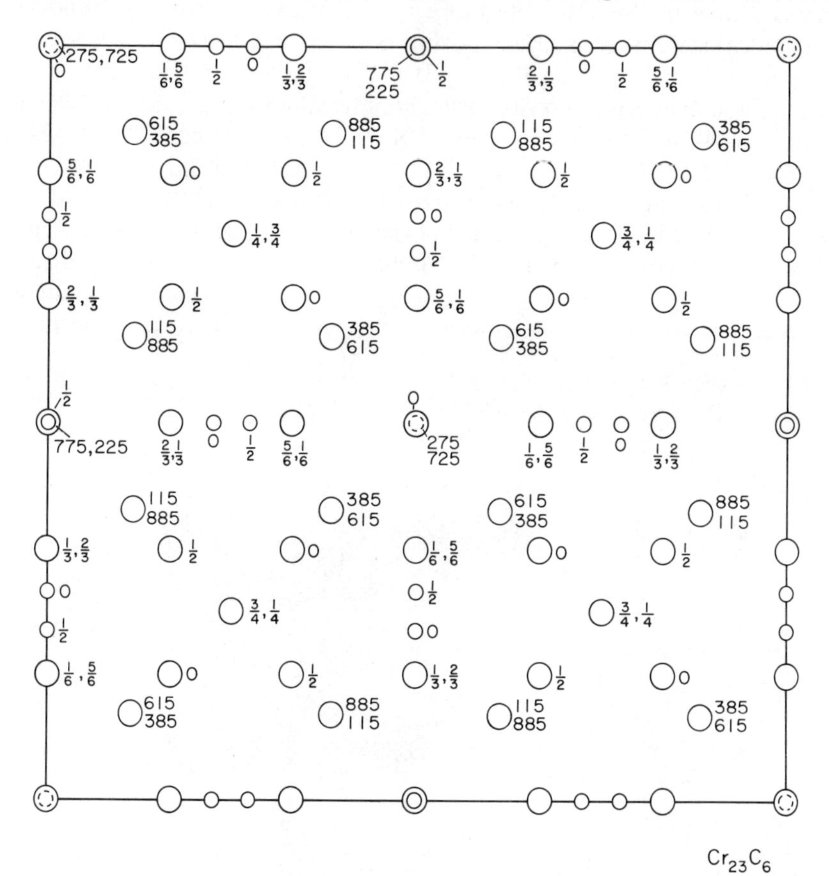

Figure 14-38. Atomic arrangement in the $Cr_{23}C_6$ (*cF*116) structure: large circles, Cr atoms.

and octagons covering the cell corners as shown in Figure 14-37. 4^4 Sn(1) layers occur at $z = \frac{1}{4}$ and $\frac{3}{4}$ with the atoms located above the midpoints of the cell edges. Rh(1) atoms at $\frac{1}{2}, \frac{1}{2}, 0$, lie over the centers of the Sn octagons at $z = \frac{11}{12}$ and $\frac{1}{12}$, and at $0, 0, \frac{1}{2}$ over the centers of the octagons at $z = \frac{5}{12}$ and $\frac{7}{12}$. They also have 2 Rh neighbors along [001]. Rh(2) atoms at $z \sim \frac{1}{3}$ and $\frac{2}{3}$ are located above the cell corners and at $z \sim \frac{1}{6}$ and $\frac{5}{6}$ are located above the basal plane center. They have one Rh and 8 Sn(2) neighbors on one side and 4 Sn(1) on the other along [001]. The structure is described as being made up of two $CuAl_2$ type layers which alternate with a CaF_2 type layer, there being six layers in the unit cell along [001].

$Cr_{23}C_6$ (cF116): The Cr(3) atoms form the same space filling array of cubo-octahedra, large truncated octahedra and truncated tetrahedra as the B atoms in the UB_{12} structure (p. 757); except that the truncated octahedra are centered about positions $\frac{1}{2}, \frac{1}{2}, \frac{1}{2}$; etc., the cubo-octahedra about the 0, 0, 0; etc. positions and the truncated tetrahedra about $\frac{1}{4}, \frac{1}{4}, \frac{1}{4}$; etc.[49] Cr(2) atoms center the truncated tetrahedra (12 Cr(3) at 2.96 Å), with 4 Cr(4) at 2.49 Å completing CN 16 Friauf polyhedra. Cr(1) center the cubo-octahedra (12 Cr(3) at 2.49 Å). The Cr(4) atoms form 48^2 nets at $z = 0.115$ and 0.885 (squares over midpoints of cell edges) and $z = 0.385$ and 0.615 (squares over cell corners and base center) so that a cube of 8 Cr(4) is situated about the center of the large truncated octahedra. Finally, the carbon atoms are also located within the truncated octahedra outside the cubes of Cr(4) atoms, lying on lines which pass normally

Figure 14-39. Arrangements of coordination polyhedra in the $Cr_{23}C_6$ structure. See text. (Samson [50].)

through the midpoints of the cube faces. They are thus surrounded by a distorted cubic antiprism of 4 Cr(4) at 2.09 Å and 4 Cr(3) at 2.11 Å. Cr(3) has 2 C neighbors, 2 Cr(2) at 2.96 Å, 4 Cr(3) at 2.49 Å and one at 2.56 Å, and 4 Cr(4) at 2.70 Å. Cr(4) has 3 C neighbors, one Cr(2) at 2.49 Å, 6 Cr(3) at 2.70 Å and 3 Cr(4) at 2.45 Å. The atomic arrangement is shown in Figure 14-38.

Figure 14-39 taken from Samson[50] shows the arrangement of cubo-octahedra and large truncated octahedra along ⟨001⟩ directions and a collection of cubo-octahedra, truncated octahedra, and truncated tetrahedra formed by the Cr(3) atoms. Although the structure was originally determined by single-crystal methods, it requires confirmation and the carbon atom positions to be established by neutron diffraction.

Several other structures, such as the CaB_6 (p. 756), UB_{12} (p. 757) and $AlTh_2H_4$ (p. 760) types are also built up of nets of atoms containing octagons.

REFERENCES

1. C. Bècle and R. Lemaire, 1967, *Acta Cryst.*, **23**, 840.
2. J. A. Bland, 1958, *Acta Cryst.*, **11**, 236.
3. K. Schubert, 1964, *Kristallstrukturen zweikomponentiger Phasen*, Berlin: Springer, p. 296.
4. J. A. Bland and D. Clark, 1958, *Acta Cryst.*, **11**, 231.
5. A. D. I. Nicol, 1953, *Acta Cryst.*, **6**, 285.
6. P. J. Black, O. S. Edwards, and J. B. Forsyth, 1961, *Acta Cryst.*, **14**, 993.
7. S. Bhan and K. Schubert, 1960, *Z. Metallk.*, **51**, 327.
8. H. Nowotny and K. Schubert, 1946, *Z. Metallk.*, **37**, 23.
9. J. R. Thomson, 1963, *Acta Cryst.*, **16**, 320.
10. E. Dahl, 1967, *Acta Chem. Scand.*, **21**, 1131.
11. K. Åmark, B. Borén, and A. Westgren, 1936, *Svensk. Kem. Tid.*, **48**, 273; *Metallwirt.*, **15**, 835.
12. E. Parthé, W. Jeitschko, and V. Sadagopan, 1965, *Acta Cryst.*, **19**, 1031.
13. O. Helleis, H. Kandler, E. Leicht, W. Quiring, and E. Wölfel, 1961, *Z. anorg. Chem.*, **320**, 86.
14. M. Pötzschke and K. Schubert, 1962, *Z. Metallk.*, **53**, 474.
15. W. Rieger and E. Parthé, 1968, *Acta Cryst.*, **B24**, 456.
16. G. Pilström, 1961, *Acta Chem. Scand.*, **15**, 893.
17. H. Perlitz and A. Westgren, 1943, *Ark. Kemi Min. Geol.*, **16B**, No. 13.
18. A. H. Gomes de Mesquita and K. H. J. Buschow, 1967, *Acta Cryst.*, **22**, 497.
19. I. I. Zalucky and P. I. Kripjakevič, 1967, *Drop. Akad. Nauk, Ukr. RSR*, No. 4, 362.
20. L.-E. Edshammar, 1964, *Acta Chem. Scand.*, **18**, 2294.
21. O. Nial, 1947, *Svensk. Kem. Tidskr.*, **59**, 172.
22. P. Jensen and A. Kjekshus, 1966, *Acta Chem. Scand.*, **20**, 417; *Idem*, 1967, *J. Less-Common Metals*, **13**, 357; P. Jensen, A. Kjekshus, and T. Skansen, 1969, *Ibid.*, **17**, 455.
23. B. Pedersen and F. Grønvold, 1959, *Acta Cryst.*, **12**, 1022.
24. J. F. deWet, 1961, *Acta Cryst.*, **14**, 733.

25. W. Jeitschko, 1969, *Acta Cryst.*, **B25**, 557.
26. E. I. Gladyševskij, P. I. Kripjakevič, and R. V. Skolozdra, 1968, *Dokl. Akad. Nauk, SSSR*, **175**, 1047 [*Soviet Physics–Doklady*, **12**, 755].
27. W. H. Zachariasen and F. H. Ellinger, 1963, *Acta Cryst.*, **16**, 369.
28. R. C. L. Mooney Slater, 1964, *Z. Kristallogr.*, **120**, 278.
29. B. Aronsson, 1955, *Acta Chem. Scand.*, **9**, 1107.
30. A. Kjekshus, F. Grønvold, and J. Thorbjørnsen, 1962, *Acta Chem. Scand.*, **16**, 1493.
31. F. Grønvold, A. Kjekshus, and F. Raaum, 1961, *Acta Cryst.*, **14**, 930.
32. G. S. Smith, Q. Johnson, and D. H. Wood, 1969, *Acta Cryst.*, **B25**, 554.
33. T. F. Gaskell, 1937, *Z. Kristallogr.*, **A96**, 203.
34. L. D. Calvert and C. Rand, 1964, *Acta Cryst.*, **17**, 1175.
35. E. I. Hladyschewskyj, P. I. Krypiakewytsch, and O. I. Bodak, 1966, *Z. anorg. Chem.*, **344**, 95.
36. R. M. Imamov and Z. G. Pinsker, 1965, *Kristallografija*, **10**, 199, [*Soviet Physics–Crystallography*, **10**, 148].
37. C. S. Barrett, 1962, *Adv. in X-ray Analysis*, Vol. 5, New York: Plenum Press, p. 33; B. D. Sharma and J. Donohue, 1962, *Z. Kristallogr.*, **117**, 293.
38. G. S. Smith, Q. Johnson, and P. C. Nordine, 1965, *Acta Cryst.*, **19**, 668.
39. B. Aronsson, 1963, *Acta Chem. Scand.*, **17**, 2036.
40. S. Rundqvist, E. Hassler, and L. Lundvik, 1962, *Acta Chem. Scand.*, **16**, 242.
41. B. Pedersen and F. Grønvold, 1959, *Acta Cryst.*, **12**, 1022.
42. S. Rundqvist and E. Larsson, 1959, *Acta Chem. Scand.*, **13**, 551.
43. S. Geller, 1962, *Acta Cryst.*, **15**, 713.
44. P. Ettmayer, 1965, *Mh. Chem.*, **96**, 884.
45. R. V. Baranova and Z. G. Pinsker, 1965, *Kristallografija*, **10**, 614 [*Soviet Physics–Crystallography*, **10**, 523].
46. G. S. Smith, K. F. Mucker, Q. Johnson, and D. H. Wood, 1969, *Acta Cryst.*, **B25**, 549.
47. K. Schubert, F. Gauzzi, and K. Frank, 1963, *Z. Metallk.*, **54**, 422.
48. E. Hellner, 1956, *Z. Kristallogr.*, **107**, 99; H. Jagodzinski and E. Hellner, 1956, *Ibid.*, **107**, 124.
49. A. Westgren, 1933, *Jernkont. Ann.*, **117**, 501.
50. S. Samson, 1968, in *Structural Chemistry and Molecular Biology*, San Francisco: W. H. Freeman & Co., p. 687.

15

Structures with Large Coordination Polyhedra

In this Chapter the structures of phases with large coordination polyhedra are considered, although several of these such as $CaCu_5$ (p. 645), $ThMn_{12}$, Th_2Zn_{17}, Th_2Ni_{17}, Pu_3Zn_{22}, $PuNi_4$ (pp. 645–650), Er_2Co_7, Ce_2Ni_7, $NbBe_3$, $CeNi_3$ (pp. 652–654), and α-La_3Al_{11} (p. 722) with CN 20 polyhedra have already been described. The relative sizes of the component atoms are generally very disparate in phases with large coordination polyhedra and Figure 3-14, p. 75, for example, indicates how the relative sizes of the two atoms in binary phases influence which MN_x ($x > 10$) structure a phase adopts.

BaHg$_{11}$ (*cP*36): Ba occupies the midpoints of the edges of the large cubic cell, $a = 9.60$ Å, $Z = 3$.[1] The Hg atoms form a series of squares and octagons at eight different levels which are centered over the base center of the cell (Figure 15-1). Hg(4), the closest of these, surround Hg(1) at the body center of the cell in a cubo-octahedron (Hg–Hg = 3.055 Å). The Ba atoms are surrounded by 20 Hg (8 Hg(2) at 3.92 Å, 8 Hg(3) at 3.63 Å, and 4 Hg(4) at 3.73 Å). There are no polar atoms closing the polyhedron, so the end faces are four-sided. Other polygon faces are triangular. Hg(2) and (3) have surface CN 5 and Hg(4) has surface CN 6 (Figure 15-1). Hg(2) is surrounded by 3 Ba and 9 Hg in a distorted icosahedron. Hg(3) has 2 Ba and 8 Hg neighbors and Hg(4) has 1 Ba and 11 Hg neighbors in an irregular convex polyhedron.

A number of Hg and Cd phases with Group I or II A metals or rare earths take the $BaHg_{11}$ structure. The near-neighbor diagram (p. 53) indicates that the coordination factor is important in determining which phases can adopt the structure, since known phases with the $BaHg_{11}$ struc-

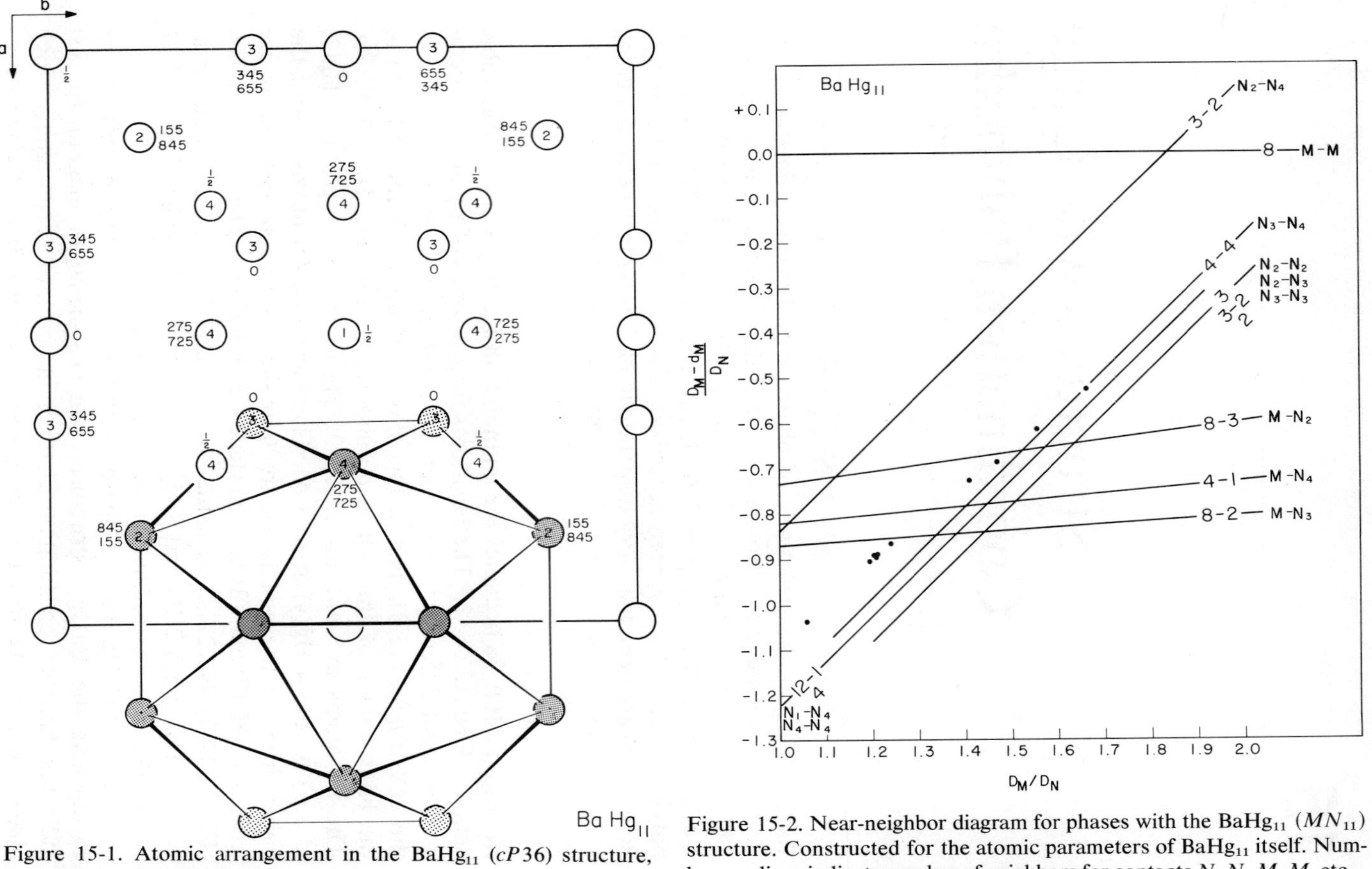

Figure 15-1. Atomic arrangement in the $BaHg_{11}$ (*cP*36) structure, showing coordination about Ba.

Figure 15-2. Near-neighbor diagram for phases with the $BaHg_{11}$ (MN_{11}) structure. Constructed for the atomic parameters of $BaHg_{11}$ itself. Numbers on lines indicate number of neighbors for contacts *N–N*, *M–M*, etc.

ture are distributed about the region of the diagram where the lines for 20–6 M–N contacts cross those for N–N contacts (Figure 15-2). Naturally the N atoms control the overall dimensions of the structure since they constitute over 90% of the atoms present in it. The near-neighbor diagram shows that the most favored radius ratios of the component atoms for achieving CN 20 about the M atoms are in the range from 1.35 to 1.60 Å.

Ce_5Mg_{41} (*tI*92): Ce_5Mg_{41} has cell dimensions $a = 14.78$, $c = 10.43$ Å, $Z = 2$.[2] Ce(1) is surrounded by 20 Mg in a polyhedron similar to that surrounding Th in the $ThMn_{12}$ structure, and Ce(2) is surrounded by 18 Mg, both polyhedra being shown in Figure 15-3*a*. Four Ce(2) polyhedra surround a Ce(1) polyhedron as shown in Figure 15-3*b*. Four of these clusters centered at 0, 0, 0 surround further clusters centered at $\frac{1}{2}, \frac{1}{2}, \bar{\frac{1}{2}}$ and $\frac{1}{2}, \frac{1}{2}, \frac{1}{2}$ (Figure 15-3*c*) and thus the structure of Ce_5Mg_{41} is built up. Mg(4), (5), and (7) are surrounded icosahedrally by 10 Mg

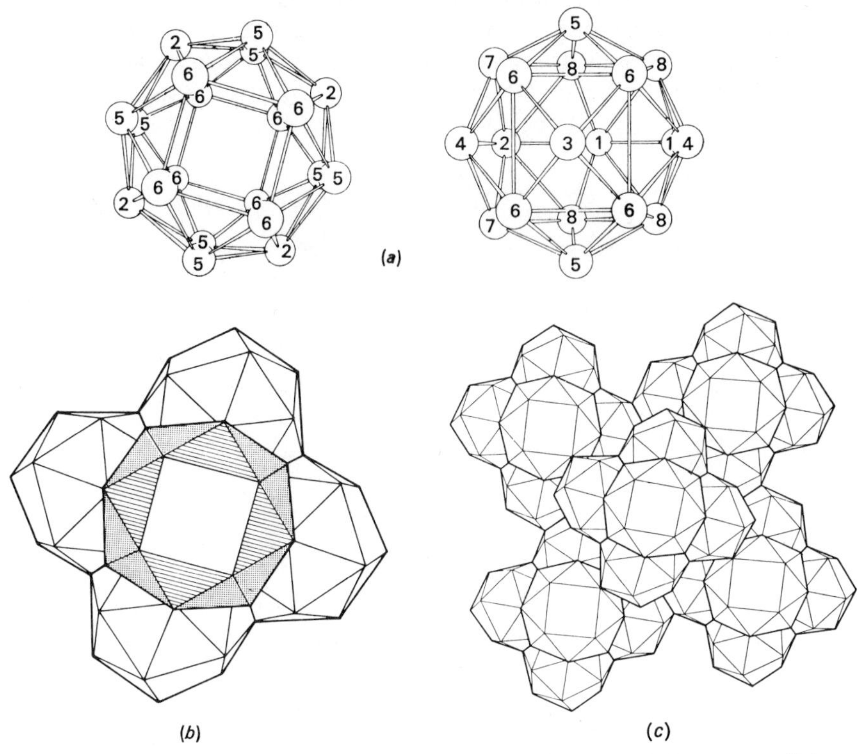

Figure 15-3. Coordination polyhedra and buildup of the Ce_5Mg_{41} (*tI*92) structure. (Johnson and Smith [2].)

(2.98–3.25 Å) and 2 Ce (3.42–3.80 Å). Mg(2) (CN 14), Mg(3) (CN 12) and Mg(6) (CN 13) have polyhedra observed in the Th_2Ni_{17} and Mg_2Zn_{11} structures. Mg(1) and (8) can be regarded as together centering a dumbell-shaped polyhedron.

$BaCd_{11}$ (*tI*48): $BaCd_{11}$ has cell dimensions, $a = 12.02$, $c = 7.74$ Å, $Z = 4$.[3] Ba has 22 Cd neighbors at distances from 3.73 to 4.18 Å. With the exception of the two polar atoms, all members of the coordination polyhedron have a surface coordination number of 5. The arrangement of Ba atoms in the cell is the same as that of the Sn atoms in the β-Sn structure. Cd(1) is surrounded by 14 neighbors which form a Frank–Kasper polyhedron and Cd(2) is surrounded icosahedrally by 12 neighbors. Cd(3) is also surrounded by 12 neighbors, although the polyhedron is not an icosahedron as two of the surface atoms only have four surface neighbors. The structure is shown in plan in Figure 15-4 and Figure 15-5 shows the near-neighbor diagram (p. 53) for phases with the $BaCd_{11}$ structure.

$NaZn_{13}$ (*cF*112): In the $NaZn_{13}$ structure ($a = 12.284$ Å, $Z = 8$)[4] the Na and Zn(1) atoms are in a CsCl type arrangement (8 cells). Zn(2) atoms surround Zn(1) icosahedrally at 2.66 Å, and one triangular face of each of the 8 icosahedra facing Na make up a slightly distorted snub cube of

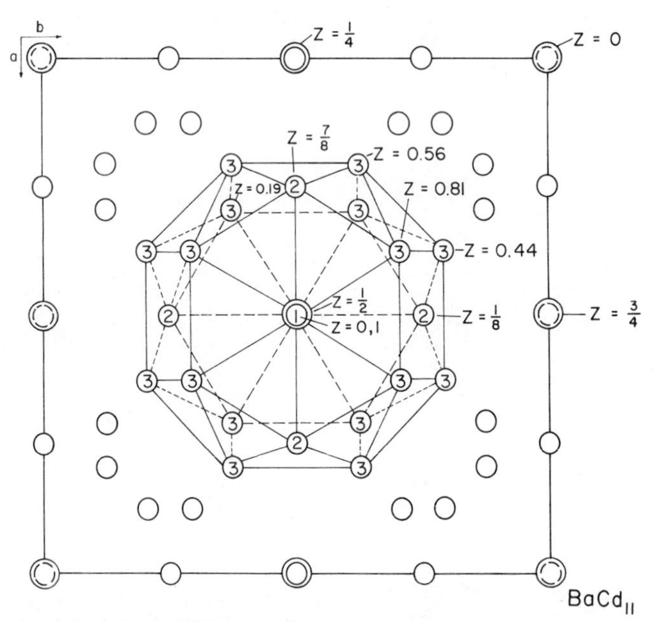

Figure 15-4. The $BaCd_{11}$ (*tI*48) structure viewed down [001], showing the coordination about Ba.

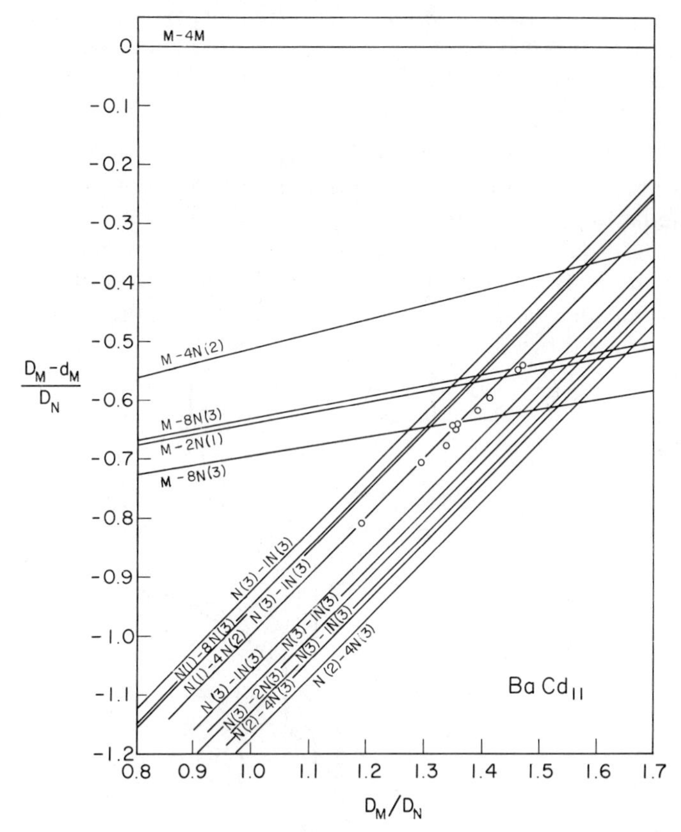

Figure 15-5. Near-neighbor diagram for the $BaCd_{11}$ ($tI48$) (MN_{11}) structure. Drawn for the axial ratio and atomic parameters of $BaCd_{11}$ itself. Numbers indicate number of M–N, etc. neighbors. Numbers in parentheses indicate N(1), (2), or (3) atoms.

24 Zn(2) atoms which surround Na at 3.57 Å. The Zn(2) icosahedra pack in alternate orientations differing by 90°. Zn(2) is surrounded by 2 Na and 10 Zn at 2.57 to 2.92 Å.

The near-neighbor diagram (p. 53) indicates that the $NaZn_{13}$ structure is particularly expected to occur when the radius ratio R_M/R_N of the two components has a value of 1.6 to 1.7 where the line for 24 M–N contacts crosses those for 12 and 10 N–N contacts. The group of phases with this radius ratio lying at the intersection of these lines, confirms the importance of the coordination factor in giving the structure stability. The distribution of the phases generally parallel to the lines for N–N contacts is only to be expected since N atoms comprise some 93% of the atoms in the structure. Nevertheless, it is interesting to observe that Zn and Cd phases severally appear to be distributed on a single line parallel to those of N–N contacts over quite a wide range of radius ratios (Figure 15-6).

CaB_6 (cP7): In the CaB_6 structure[5] Ca atoms and regular octahedra of B atoms are in a CsCl type of arrangement with the octahedra centered about the body centers of the cells and their axes parallel to the cell axes (Figure 15-7). The value of the x parameter of the boron atoms, 0.207, results in the B–B distances within the octahedra being equal to the B–B distances between octahedra, so that the B atoms form 48^2 nets at $x = \frac{1}{2}$. These nets and the B atoms at $x = 0.207$ and 0.793, result in Ca being

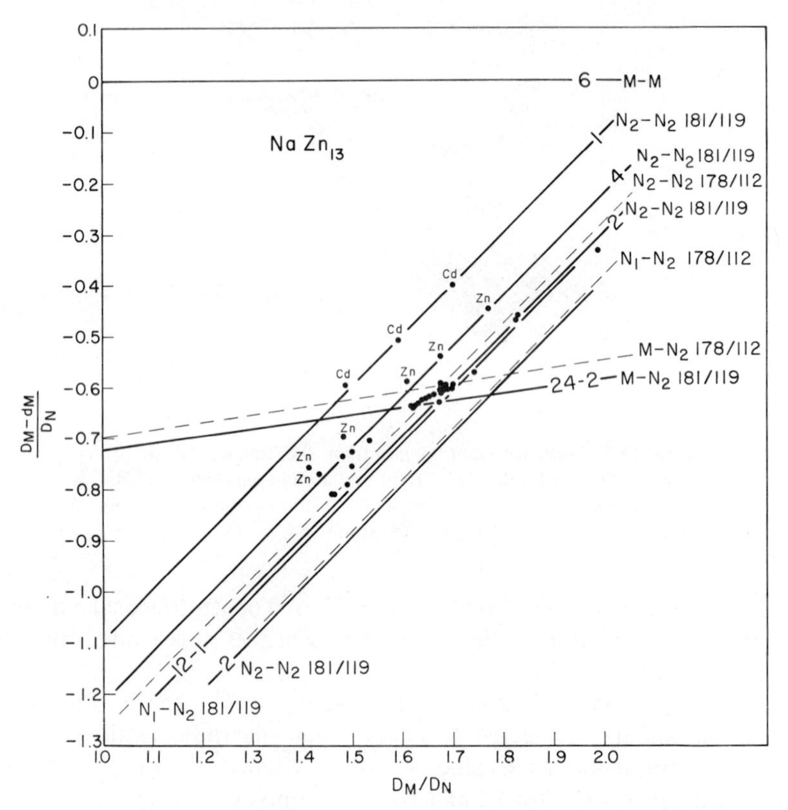

Figure 15-6. Near-neighbor diagram for phases with the $NaZn_{13}$ ($cF112$) (MN_{13}) structure. Constructed for the atomic parameters of $NaZn_{13}$ ($y = 0.181$, $z = 0.119$) full lines, and for parameter values $y = 0.178$, $z = 0.112$, broken lines. Numbers on lines indicate the number of neighbors for M–M, N–N, etc. contacts. Phases formed by Cd and by Zn are indicated.

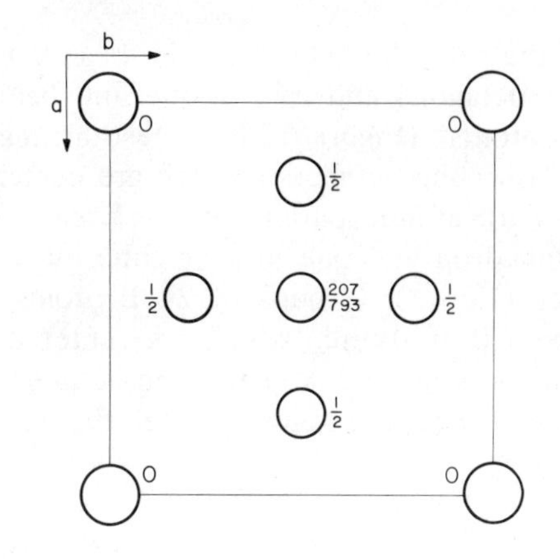

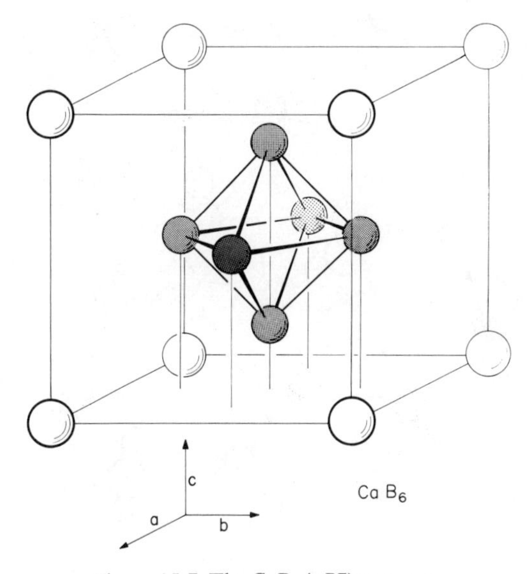

Figure 15-7. The CaB_6 (*cP*7) structure.

surrounded by 24 B (at 3.054 Å) in a regular truncated cube, and the octahedra and truncated cubes fill space. Boron has 4 Ca neighbors and 5 B at 1.72 Å.

UB_{12} (*cF*52): The unit cell edge of UB_{12} is $a = 7.477$ Å, $Z = 4$.[6] B atoms at $z = 0$ and $\frac{1}{2}$ form 48^2 nets; at $z = 0$ the octagons are centered about the cell base center and corners, and at $z = \frac{1}{2}$ they are centered over the

midpoints of the cell edges. U atoms in 4^4 nets at $z = 0$ and $\frac{1}{2}$ lie at the center of the octagons and $\pm\frac{1}{3}z$ above and below each U, there are squares of B atoms† (Figure 15-8). These arrangements are such that 12 B atoms form cubo-octahedra which are centered on the F lattice at $\frac{1}{2}$, 0, 0; etc. With atomic parameter $x = \frac{1}{6}$, each B has 4 neighbors in the cubo-octahedron and one in a neighboring cubo-octahedron at the same distance (1.76 Å). Groups of 24 B atoms form larger truncated octahedra about 0, 0, 0; etc., which are centered by U so that each U has 24 B neighbors at 2.79 Å. The cubo-octahedra and truncated octahedra do not fill space; the voids between them are truncated tetrahedra

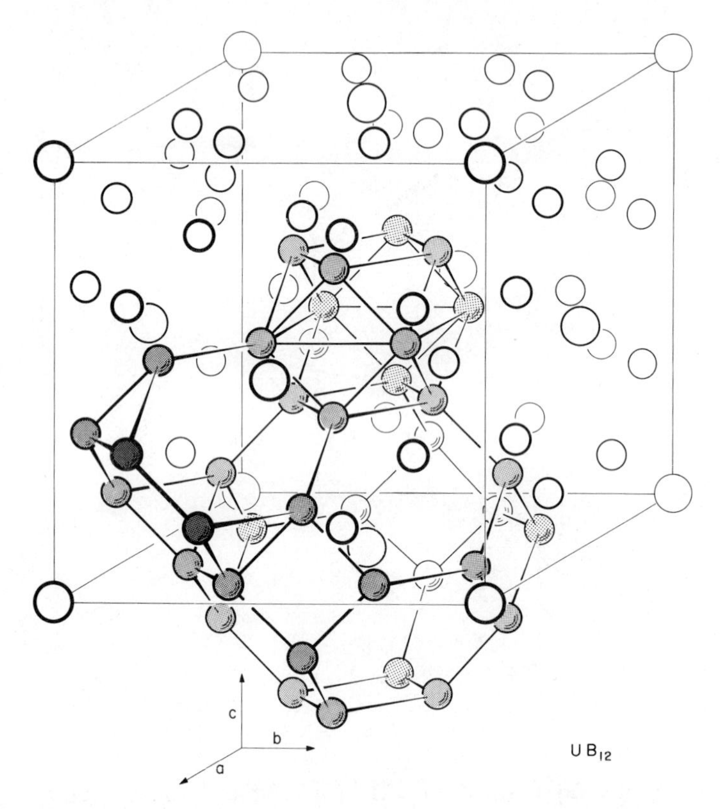

Figure 15-8. Pictorial view of the UB_{12} ($cF52$) structure showing cubo-octahedra of 12 B atoms, truncated octahedra of 24 B atoms centered by U, and uncentered truncated tetrahedra of B atoms.

†Since the structure is cubic, these descriptions hold equally well in terms of the x and y parameters.

located about $\frac{1}{4}$, $\frac{1}{4}$, $\frac{1}{4}$; etc. See Figure 15-8. In addition to 5 B neighbors, each B also has 2 U neighbors.

ThB_4 (tP20): ThB_4 has cell dimensions $a = 7.256$, $c = 4.113$ Å, $Z = 4$.[7] The Th atoms form a 3^2434 net at $z = 0$, with the squares centered over the cell corners and cell base center. The B(2) and (3) atoms at $z = \frac{1}{2}$ form a $47^2 + 7^3$ (4:3) net with the small squares centered over the squares in the Th net and oriented antisymmetrically with respect to them. B(1) atoms at $z \sim 0.2$ and 0.8 lie over the centers of the squares of the Th and B(2)+(3) nets and themselves form a $\frac{1}{2}.4^4$ net (Figure 15-9). A more realistic way of considering the boron arrangement in the structure is to recognize that they form octahedra located about the centers of the distorted cubes of Th atoms. These octahedra of B(1) and B(3) atoms (B(1)–4B(3) = 1.74 Å, B(3)–2B(3) = 1.80 Å) are linked together via B(2) atoms (B(2)–2B(3) = 1.79 Å), and a three-dimensional array of B atoms is completed by further B(1)–B(1) (1.74 Å) and B(2)–B(2) (1.79 Å) links.

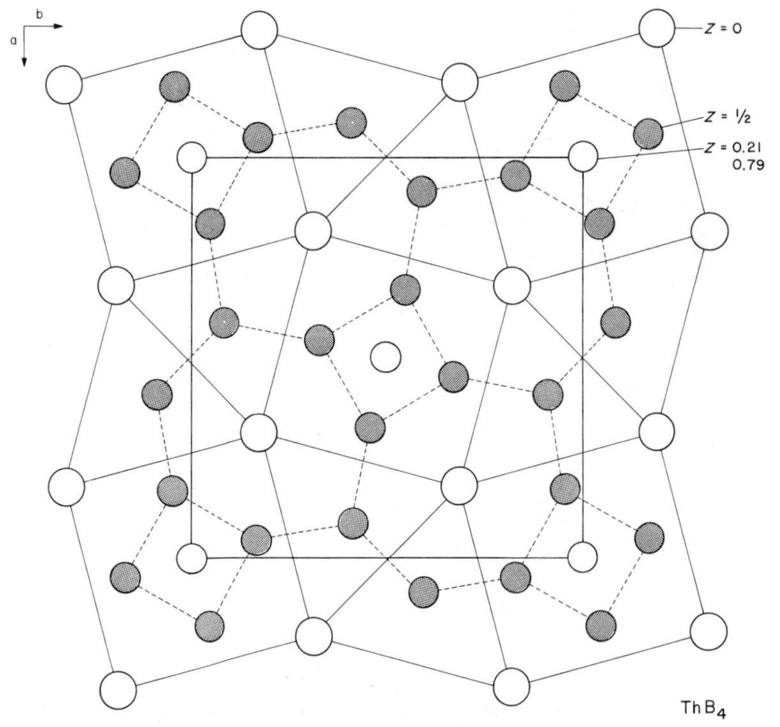

Figure 15-9. Atomic arrangement in the ThB_4 (*tP*20) structure viewed down [001]: Th, large circles.

Thus the Th atoms sit in a cage of 18 B atoms (4 B(1) at 2.78 Å, 4 B(2) at 2.96 Å, 2 B(2) at 3.10 Å and 8 B(3) at 2.84 Å). In addition Th has 5 Th neighbors in the same plane (4 × 3.74 Å, 1 × 3.85 Å) and two neighbors above and below along [001] at 4.11 Å, giving a total coordination of 25.

~*$AlTh_2H_4$* (*tI*28): The $AlTh_2H_4$ structure with cell dimensions $a = 7.632$, $c = 6.531$ Å, $Z = 4$,[8] contains layers of atoms normal to the c axis. Al atoms at $z = 0$ and $\frac{1}{2}$ form 48^2 nets as shown in Figure 15-10. Between these layers 3^3434 nets of hydrogen atoms with squares over cell corners and the base center occur at $z = 0.137$, 0.363, 0.637, and 0.863. These nets alternate in orientation so that those above and below any given net are disposed antisymmetrically relative to it. Th atoms at $z = \frac{1}{4}$ and $\frac{3}{4}$ over the cell corners and base center, lie at the centers of the octagonal prisms of Al atoms and cubic antiprisms of H atoms, achieving, together with two polar Th atoms, a coordination number of 26. The structure and hydrogen positions were determined by powder neutron diffraction.

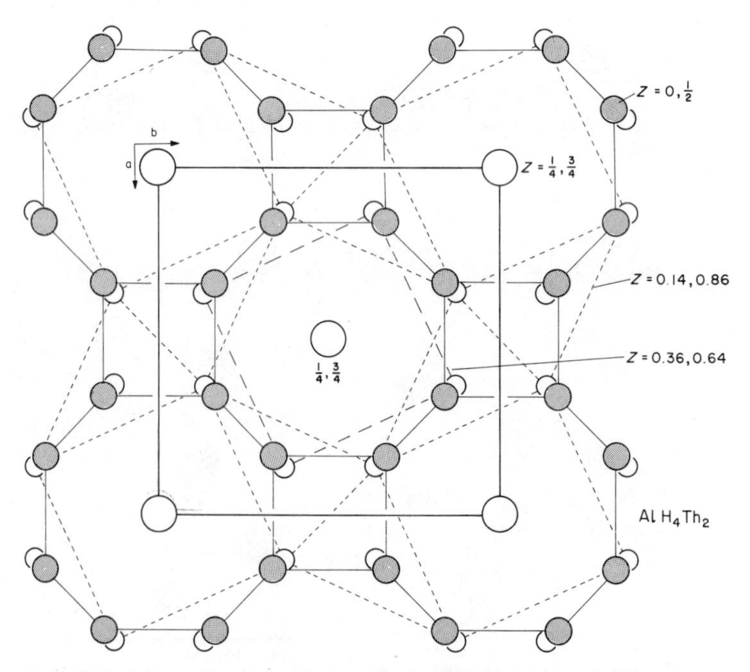

Figure 15-10. Atomic arrangement in the $AlTh_2H_4$ (*tI*28) structure viewed down [001].

REFERENCES

1. G. Peyronel, 1952, *Gazz. Chim. Ital.*, **82**, 679.
2. Q. Johnson and G. S. Smith, 1967, *Acta Cryst.*, **22**, 360.
3. M. J. Sanderson and N. C. Baenziger, 1953, *Acta Cryst.*, **6**, 627.
4. D. P. Shoemaker, R. E. Marsh, F. J. Ewing, and L. J. Pauling, 1952, *Acta Cryst.*, **5**, 637.
5. P. Blum and F. Bertaut, 1952, *Acta Cryst.*, **7**, 81; G. Allard, 1932, *Bull. Soc. Chim. Fr.*, **51**, 1213.
6. F. Bertaut and P. Blum, 1949, *C. R. Acad. Sci. Paris,* **229**, 666; 1954, *Acta Cryst.*, **7**, 81.
7. A. Zalkin and D. H. Templeton, 1953, *Acta Cryst.*, **6**, 269.
8. J. H. N. van Vucht, 1963, *Philips Res. Repts.*, **18**, 21, 35.

16

Idiosyncratic Structures

It has not been possible to fit some 50 structure types conveniently into the classification given in Chapters 7 to 15; these structures are described in this chapter.

STRUCTURES OF METALLIC ALLOYS

α-Np (*oP*8): α-Np has cell dimensions $a = 6.663$, $b = 4.723$, $c = 4.887$ Å, $Z = 8$.[1] The atoms form very distorted 3^6 nets parallel to the (010) plane at $y = \frac{1}{4}$ and $\frac{3}{4}$. The nets are arranged one above the other (Figure 16-1) so that the Np of one net lies close to the acute-angled apex of half of the triangles of the nets above and below. Each Np has 4 neighbors at 2.60 to 2.64 Å and one at 3.06 Å. The arrangement cannot be considered as distorted close packing as the distribution of centered triangles in one net by the nodes of the nets above and below, is incorrect. The structure was determined by powder photographs.

α-Pu (*mP*16): The monoclinic cell of the α-Pu structure has dimensions $a = 6.183$, $b = 4.822$, $c = 10.963$ Å, $\beta = 101.79°$, $Z = 16$.[2] The atoms are arranged in layers at $y = \frac{1}{4}$ and $\frac{3}{4}$ where they form a network of irregular quadrilaterals and/or triangles (Figure 16-2). The eight independent atoms are surrounded by irregular coordination polyhedra, six of them having CN 14(4), one CN 12(5), and one CN 16(3). The numbers in parenthesis give the numbers of close Pu–Pu approaches of 2.57–2.79 Å that are included in each polyhedron. The structure was determined by X-ray powder diffraction data.

γ-Pu (*oF*8): γ-Pu has cell dimensions $a = 3.159$, $b = 5.768$, $c = 10.162$ Å at 235°C, $Z = 8$.[3] The atoms lie on two interpenetrating f.c. orthorhombic lattices, one displaced by $\frac{1}{4}$ of the body diagonal from the other (Figure 16-3). Each Pu has 10 neighbors: 4 at 3.026 Å, 2 at 3.159 Å and 4 at 3.288 Å. Six of these surround each Pu in a plane parallel to (001) and

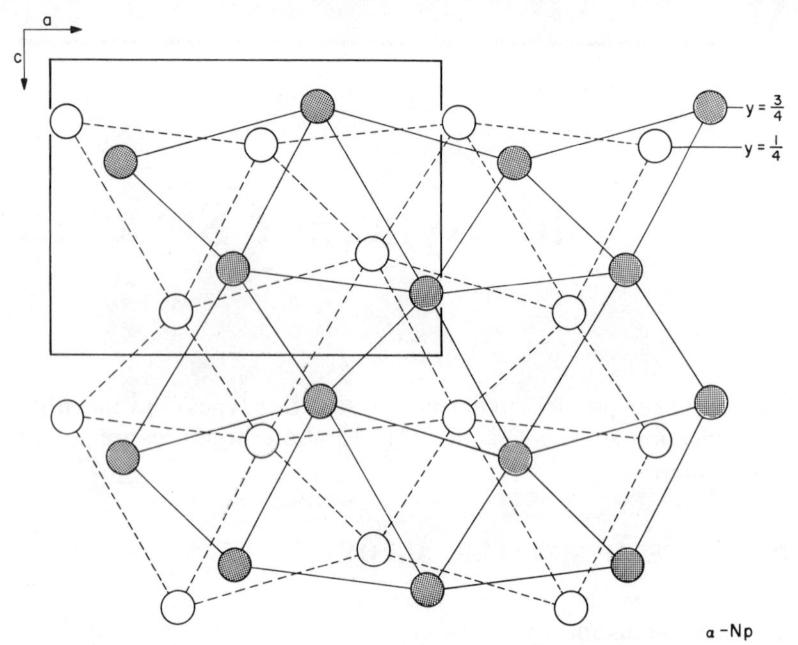

Figure 16-1. Atomic arrangement in the α-Np ($oP8$) structure viewed down [010].

two others lie above and below it. The structure was determined from X-ray powder data.

Bi II ($mC4$) (H.P. formed at 25.4 kb): Bi II has cell dimensions, $a = 6.674$, $b = 6.117$, $c = 3.304$ Å, $\beta = 110.33°$, $Z = 4$.[4] The positions give a two-layer structure parallel to the ab plane (Figure 16-4). Bi has 3 neighbors in the double layer (3.15 and 2×3.17 Å) and three neighbors in neighboring double layers (2×3.30 Å, 3.40 Å). Other neighbors are at distances greater than 3.70 Å. X-ray powder data.

α-U ($oC4$): α-U has cell dimensions $a = 2.854$, $b = 5.870$, $c = 4.955$ Å, $Z = 4$.[5] Each U has two neighbors at 2.75 Å which form zigzag chains running along the [001] direction and two more neighbors at 2.85 Å forming lines running along the [100] direction. Other neighbors are at distances greater than 3.25 Å. The atomic arrangement is similar to that of metastable Ga ($oC4$) (see Figure 16-5) but the cell proportions differ.

Ga (metastable) ($oC4$): This metastable modification with cell dimensions, $a = 2.90$, $b = 8.13$, $c = 3.17$ Å, $Z = 4$,[6] can be prepared by crystallization of a supercooled melt at -16.3°C. The atomic arrangement is the same as in α-U, but the cell proportions differ considerably. The nearest Ga neighbors (2×2.68 Å) form zigzag chains along the [001]

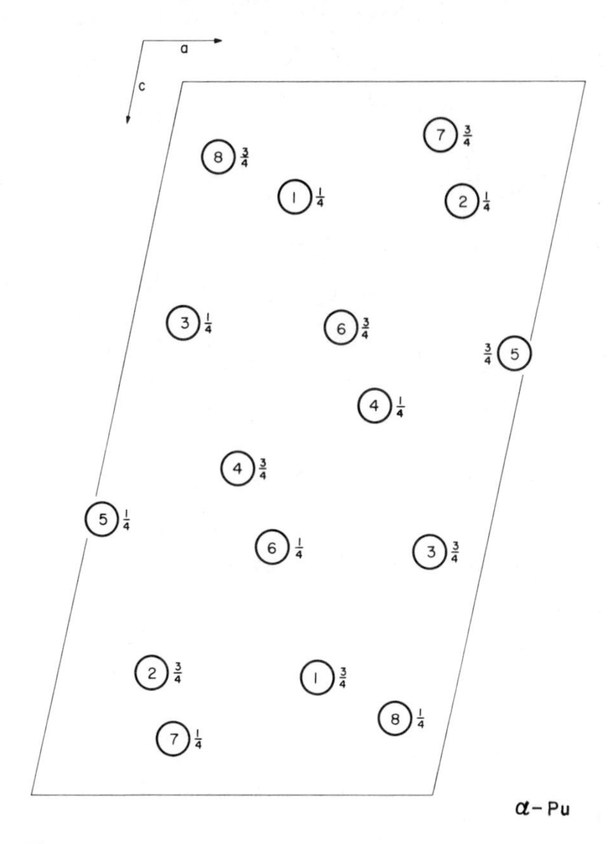

Figure 16-2. The structure of α-Pu ($mP16$) viewed down [010].

direction (Figure 16-5). Otherwise Ga has CN 10 (4×2.87, 2×2.90, 2×3.17 Å).

GaU ($oC32$): In the GaU structure ($a = 9.40$, $b = 7.60$, $c = 9.42$ Å, $Z = 16$) the U and Ga(2) atoms are arranged in layers parallel to the (001) plane.[7] At $z = 0$ and $\frac{1}{2}$, U atoms form a distorted 3^6 net in the same positions and at $z = \frac{1}{4}$ and $\frac{3}{4}$ they occupy a much larger 3^6 net and Ga(2) form a distorted rectangular 4^4 net. These nets are displaced relative to each other, and between them are inserted pairs of Ga(1) atoms (2.49 Å) with their ligand direction parallel to [001] (Figure 16-6). The U atoms form chains so that the structure maintains some of the features of both the α-U and Ga structures. Ga(1) and (2) each have two neighbors of the other kind at 2.86 Å and Ga(1) has 3 U at 2.61 Å, Ga(2) having 3 U at 2.66 and 2.67 Å. U have 2 or 4 close Ga neighbors and 1, 2, or 4 U at 2.82 to 2.85 Å.

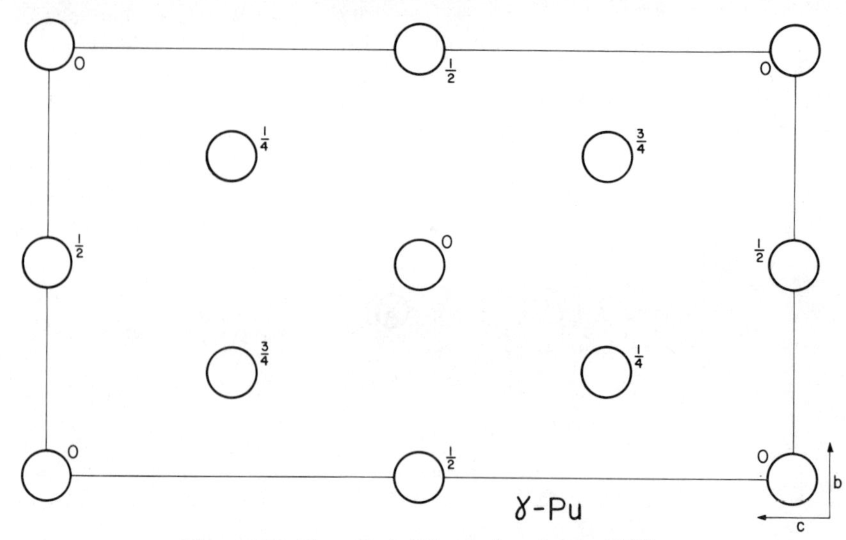

Figure 16-3. The γ-Pu ($oF8$) cell viewed down [100].

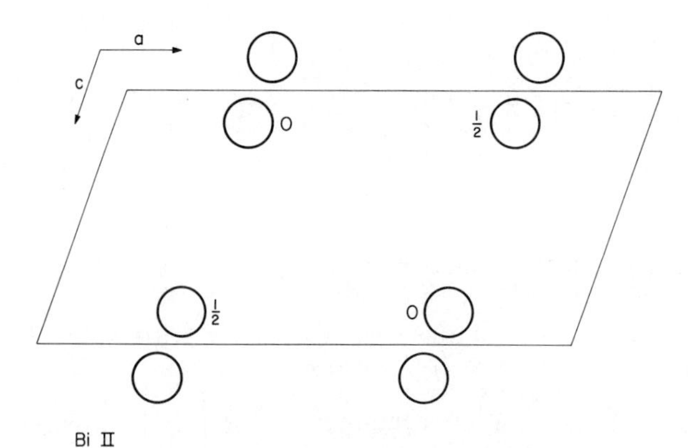

Figure 16-4. The structure of the Bi II H.P. phase ($mC4$) viewed down [010].

GaMg ($tI32$): The GaMg structure ($a = 10.53$, $c = 5.53$ Å, $Z = 16$)[8] contains diamonds formed by 2 Ga and 2 Mg atoms which lie parallel to the basal plane, although they are not quite planar. These diamonds are stacked one above the other in alternating orientations. The Mg atoms which occupy the inner sites (Figure 16-7) form strings of tetrahedra along c. The tetrahedra are joined by sharing edges parallel to the (001) plane.

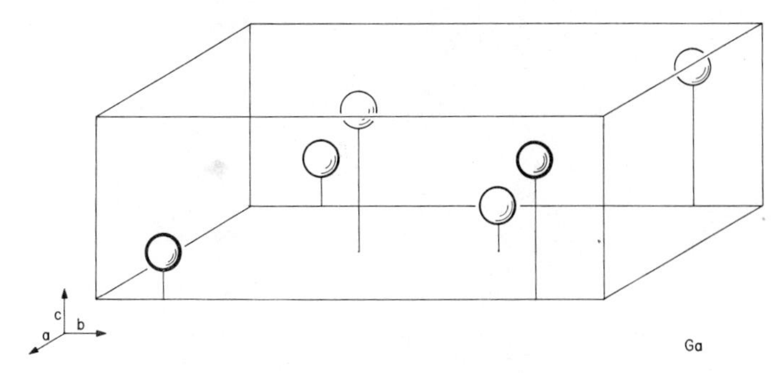

Figure 16-5. Pictorial view of the structure of metastable Ga ($oC4$).

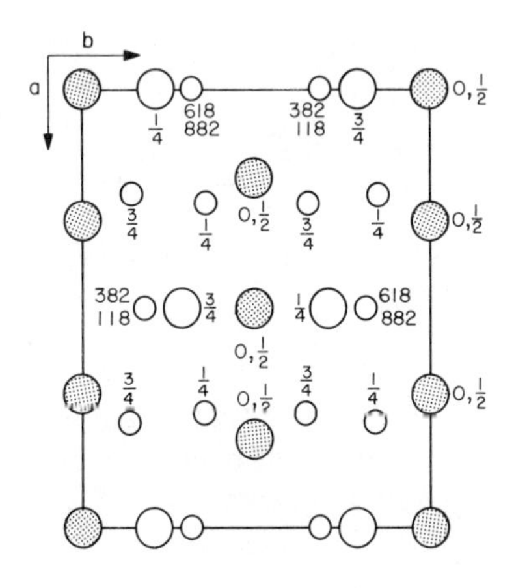

Figure 16-6. The structure of GaU ($oC32$) viewed down [001]: U, large circles.

HgK ($aP8$): HgK with cell dimensions $a = 6.59$, $b = 6.76$, $c = 7.06$ Å, $\alpha = 106°5'$, $\beta = 101°52'$, $\gamma = 92°47'$, $Z = 4$,[9] contains slightly distorted square planar groups of Hg (3.02 and 3.04 Å). These are connected together with a Hg–Hg link (3.36 Å) between parallelograms which makes an angle of 157° with the parallelogram planes. K–Hg distances vary from 3.56 to 3.75 Å.

ζ-AgZn ($hP9$): The ζ-AgZn cell dimensions are $a = 7.6360$, $c = 2.8179$ Å, $Z = 4.5$.[10] Ag + Zn (4.5 + 1.5 atoms respectively) form distorted triangular layers at $z = \frac{1}{4}$ and $\frac{3}{4}$ which are stacked in the sequence CB. Layers at $z = \frac{\bar{1}}{4}$ and $\frac{3}{4}$ and those at $z = \frac{1}{4}$ and $\frac{5}{4}$ form interpenetrating triangular prisms,

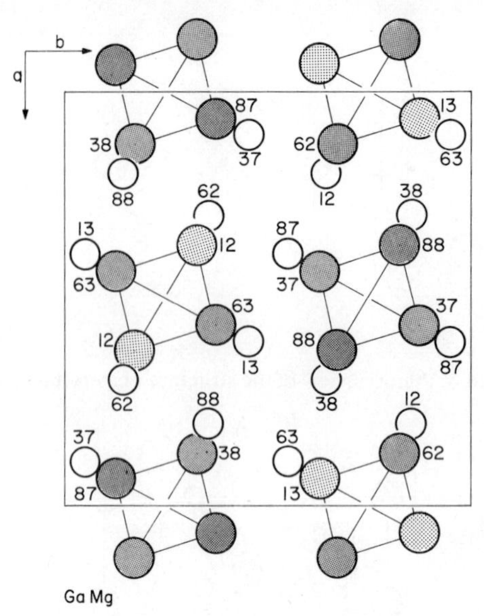

Figure 16-7. The structure of GaMg ($tI32$) viewed down [001].

atoms of one lot centering some of the prisms of the other lot and *vice versa*. Zn (2) atoms also at $z = \frac{1}{4}$ and $\frac{3}{4}$ center some of these prisms, having also three (Ag + Zn) neighbors in the same plane out through the rectangular faces of their surrounding prisms. Zn(1) at the cell corners occupy the centers of what would have been octahedral holes in the Ag + Zn array, if the layers had not been too close along [001]. The structure requires confirmation by single-crystal methods.

$SiLi_2$ ($mC12$): $SiLi_2$ has cell dimensions $a = 7.70$, $b = 4.41$, $c = 6.56$ Å, $\beta = 113.4°$, $Z = 4$.[11] The structure contains Si–Si pairs at 2.37 Å and each Si atom has 7 Li neighbors at 2.59 to 2.77 Å. The atoms form layers at $y = 0$ and $\frac{1}{2}$ (Figure 16-8).

RuB_2 ($oP6$): RuB_2 has cell dimensions $a = 4.644$, $b = 2.867$, $c = 4.045$ Å, $Z = 2$.[12] The atoms form

```
         B         B
       /   \     /   \
    Ru   |   Ru   |   Ru
       \   /     \   /
         B         B
```

chains (B–Ru = 2.165 and 2.25 Å; B–B = 1.77 Å) running in the c direction at heights $y = \frac{1}{4}$ and $\frac{3}{4}$. The chains are positioned so that

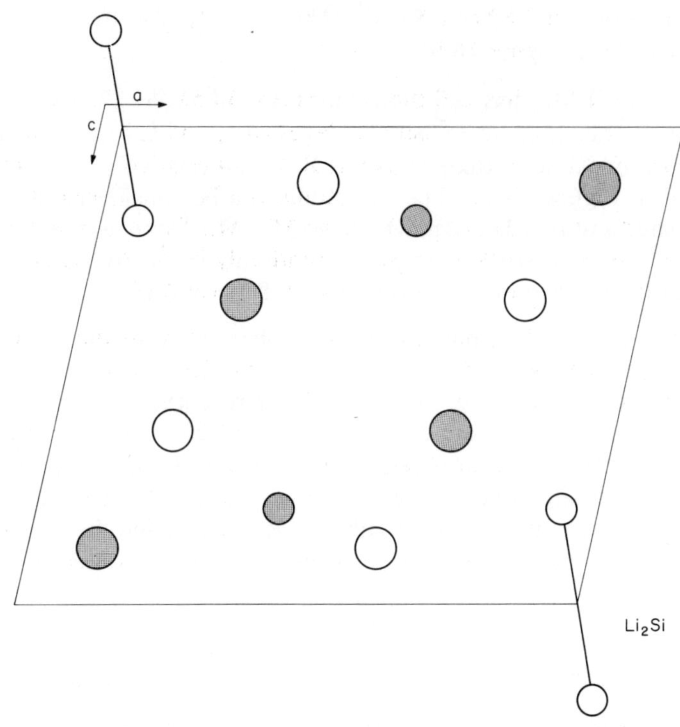

Figure 16-8. The structure of $SiLi_2$ ($mC12$) viewed down [010], showing Si–Si pairs: Li atoms, large circles; atoms at $y = \frac{1}{2}$ are shaded.

a chain at $y = \frac{1}{4}$ lies between two chains at $y = \frac{3}{4}$ and *vice versa,* but the chains approach each other so closely that they are interconnected by B–B links (2 × 1.90 Å) and by B–Ru links (2 × 2.20 Å). Each Ru has 8 B neighbors at distances referred to above, and 6 Ru neighbors (2 × 2.865, 4 × 2.99 Å) all on one side of it.

$InMg_3$ (*$In_{21}Mg_{79}$*) ($hR16$): $InMg_3$ has hexagonal cell dimensions $a = 6.323$, $c = 31.06$ Å, $Z = 12$.[8] The atoms are disposed in mixed layers at every twelfth of the cell edge along the [001] direction, the In atoms forming 3^6 nets and the Mg atoms kagomé nets. Although stoichiometry is satisfied in each layer, the relative arrangement of the two subnetworks is not always the same. The In atoms alternately lie over the centers of the hexagons of the kagomé networks in two layers, and then the triangles in the next two layers. In the notation of p. 4 the stacking arrangement is (In underlined) $[\underline{A}\alpha][\underline{B}\gamma][\underline{C}\beta][\underline{A}\alpha][\underline{C}\gamma][\underline{A}\beta][\underline{B}\alpha][\underline{C}\gamma][\underline{B}\beta][\underline{C}\alpha][\underline{A}\gamma][\underline{B}\beta]$. The structure requires confirmation by single-crystal methods.

$RhBi_4$ ($cI120$): In the $RhBi_4$ structure[13] Rh is surrounded by a distorted

cube of Bi atoms at 2.82 to 2.83 Å and Bi is surrounded by 2 Rh and 9 Bi at 3.14 to 3.89 Å (Figure 16-9).

BMn_4 (oF40): BMn_4 has cell dimensions $a = 14.53$, $b = 7.293$, $c = 4.209$ Å, $Z = 8$.[14] The atoms form identical layers at $z = 0, \frac{1}{4}, \frac{1}{2}$, and $\frac{3}{4}$ which are stacked in four different positions (A, B, C, D) relative to each other as indicated in Figure 16-10. The Mn atoms can be considered as forming parallel sheets of tetrahedra in which the Mn–Mn distances are 2.41–2.44 or 2.71 Å. The 8 B atoms were placed randomly in the 16 largest holes in the structure where they are surrounded by 8 Mn at 2.19 Å.

Te_2Rh_3 (oC20): Te_2Rh_3 has an orthorhombic cell with dimensions $a = 3.697$, $b = 12.446$, $c = 7.694$ Å, $Z = 4$.[15] The Rh atoms form slightly distorted squares at $x = 0$ and $\frac{1}{2}$ which share corners so as to form a chain running in the c direction. The shared corners create triangles between neighboring squares (Figure 16-11). The Te atoms form a chain of triangles running in the c direction at $x = \frac{1}{2}$ and 0, so that Te(1) atoms center the tetragonally extended cubes of Rh atoms, and the Te(2) atoms located at $x = \frac{1}{2}$, lie outside the rectangles formed by the outer faces of the

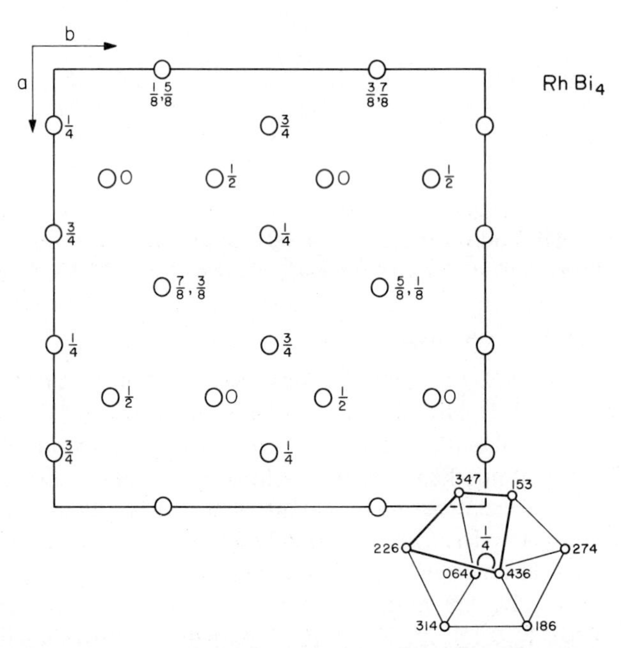

Figure 16-9. Diagram showing the coordination of Rh by B in the $RhBi_4$ (*cI*120) structure. In the view of the unit cell down [001] only the sites of the Rh atoms are shown.

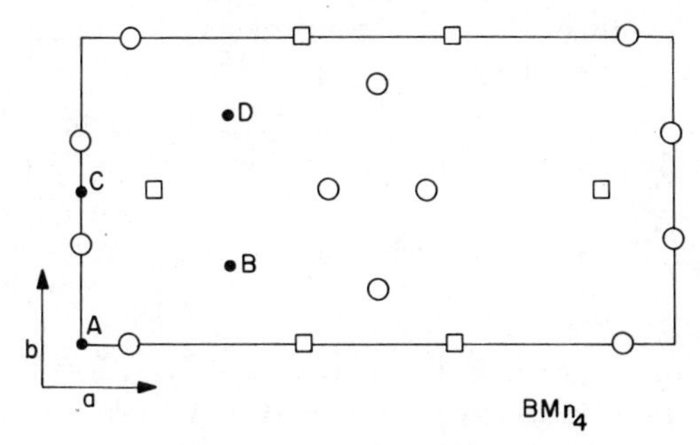

Figure 16-10. View down [001] showing the atomic arrangement in the unit cell of BMn_4 ($oF40$) at $z = 0$, and the four stacking positions for the layers at $z = 0, \frac{1}{4}, \frac{1}{2}$, and $\frac{3}{4}$. Circles are Mn atoms, squares are boron atoms.

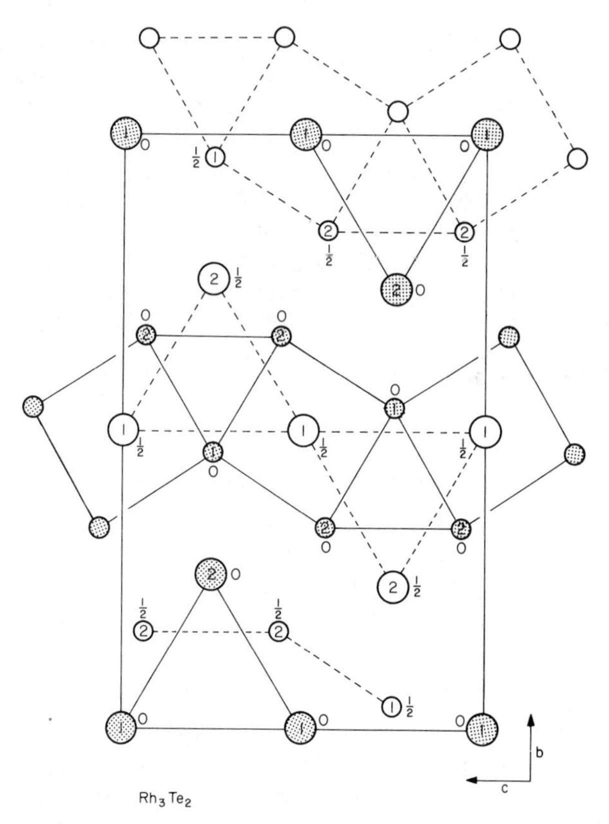

Figure 16-11. The structure of Te_2Rh_3 ($oC20$) viewed down [100].

Rh triangles at $x = 0$ and 1. Te(2) atoms at $x = 0$ lie outside the rectangles formed by Rh at $x = \pm\frac{1}{2}$ (Figure 16-11).

The Rh "squares" have edges 2.82 and 2.86 Å and each Rh has 5 Te neighbors in a distorted square pyramid at 2.61 to 2.76 Å. Te(1) has 4 Rh(1) at 2.70 Å and 4 Rh(2) at 2.76 Å. Te(2) has 4 Rh(2) at 2.63 Å above and below, and 2 Rh(2) at 2.61 Å and 1 Rh(1) at 2.64 Å at the same height.

Si_2Li_7 (oP36): Si_2Li_7 has cell dimensions $a = 7.99$, $b = 15.21$, $c = 4.43$ Å, $Z = 4$.[16] The atoms form layers at $z = 0$ and $\frac{1}{2}$. At $z = 0$ their distribution has some resemblance to a 3^6 net. The Si atoms at $z = \frac{1}{2}$ form pairs at 2.38 Å; those at $z = 0$ are separated by distances of more than 4.4 Å. The paired Si have 5 Li at 2.50 to 2.66 Å and three others at 2.84 to 2.94 Å. The isolated Si atoms have 7 Li at 2.31 to 2.73 Å and 4 more at 2.92 to 3.00 Å. No details of the method of structure determination are given.[16]

Th_3P_4 (cI28): The Th_3P_4 structure[17] is made up of short planar strips of atoms which lie normal to {001} planes and zigzag alternately + and − about ⟨100⟩ directions, giving the effect of boards joined together along their length to create a staircase. Such staircases run along each of the ⟨100⟩ directions and interpenetrate each other. This results in the P atoms being surrounded by 6 Th at 2.98 Å in a greatly distorted octahedron and Th being surrounded by 8 P in an arrangement which is not a cube (Figure 16-12). Th also has 8 Th neighbors at 4.03 Å. Th_3P_4 is a popular structure type which cannot at present be placed within the systematic organization of structures.

Si_4Zr_5 (tP36): Si_4Zr_5 has cell dimensions $a = 7.122$, $c = 13.00$ Å, $Z = 4$.[18] The arrangement of atoms is shown in Figure 16-13. Si(1) has one Si at 2.47 Å and 7 Zr at 2.60 to 3.06 Å. Si(2) has one Si and 8 Zr at 2.67 to 3.07 Å. The closest Zr–Zr approaches are 3.015 and 3.11 Å.

P_4Ni_5 (hP36): P_4Ni_5 has cell dimensions $a = 6.789$, $c = 10.986$ Å, $Z = 4$.[19] The atomic arrangements are somewhat irregular. Ni(1) and (3) atoms each have 10 Ni and P neighbors and Ni(2) and (4) have 11 neighbors. P(3) is surrounded closely by a slightly distorted tetrahedron of 3 Ni (2.186 Å) and one P(2) atom (2.189 Å). P(1) has 7 Ni neighbors, P(2) has 5 Ni and 1 P neighbor and P(4) has 9 Ni neighbors.

$FeAl_3$ (Fe_4Al_{13})(mC101): $FeAl_3$ has cell dimensions $a = 15.49$, $b = 8.083$, $c = 12.48$ Å, $\beta = 107.7°$, $Z = 76.8Al + 24Fe$, and atoms which occupy 20 different sitesets.[20] This complex structure can be discussed in terms of alternate flat and puckered layers of atoms (Figure 16-14) perpendicular to the *b* axis. The Fe atoms have 10 or 11 Al neighbors. Fe(3) has an

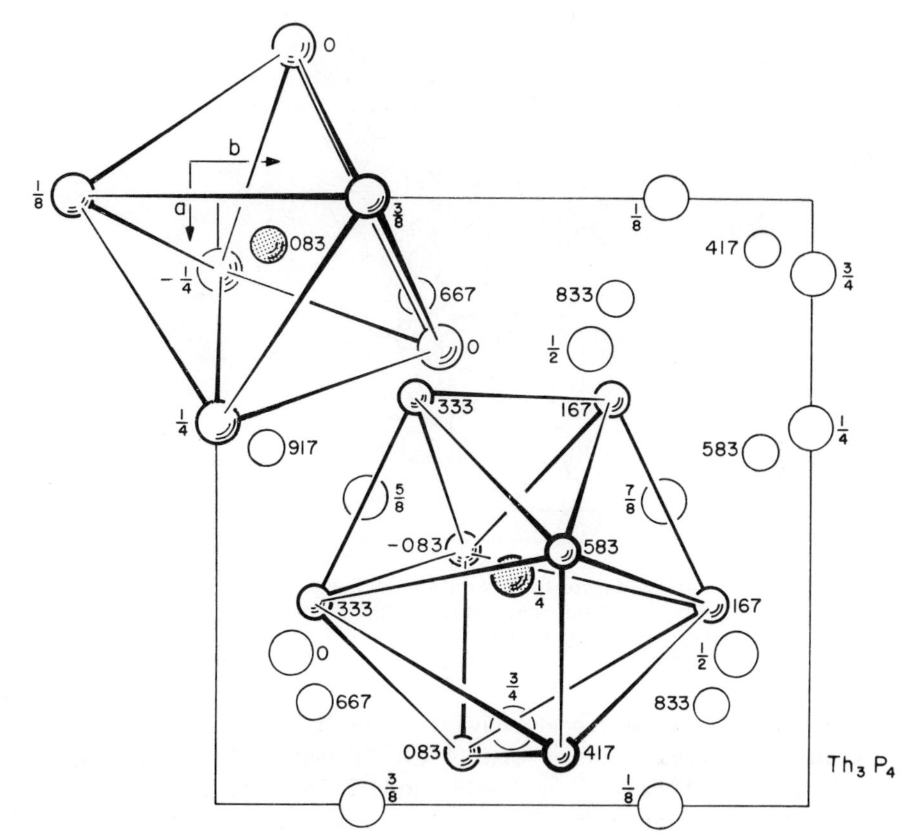

Figure 16-12. The Th_3P_4 (*cI*28) structure viewed down [001] showing the coordination of the P (small circles) and Th atoms.

Fe(3) neighbor at 2.905 Å. The closest Al–Fe distances are 2.26 and 2.31 Å.

Co_4Al_{13}(mC93): The structure of Co_4Al_{13} with cell dimensions $a = 15.18$, $b = 8.12$, $c = 12.34$ Å, $\beta = 107.9°$, $Z = 24.4Co + 68.3Al$, is very similar to that of $FeAl_3$.[21]

Si_4Cu_{15} (Th_4H_{15}) (cI76): Si_4Cu_{15} has a cubic cell with edge $a = 9.714$ Å, $Z = 4$.[22] Cu(1) is surrounded by 4 Si (2.62 Å) and 8 Cu(2) (2.47 and 2.505 Å). Cu(2) has 3 Si neighbors (2.38, 2.515 and 2.60 Å), 2 Cu(1) and 8 Cu(2) at 2.52 to 2.99 Å. Si is surrounded by a very distorted icosahedron of 3 Cu(1) and 9 Cu(2) at 2.38 to 2.62 Å. The atomic arrangement is shown in Figure 16-15.

In the Th_4H_{15} structure[23] the H(1) and Th positions are identical to

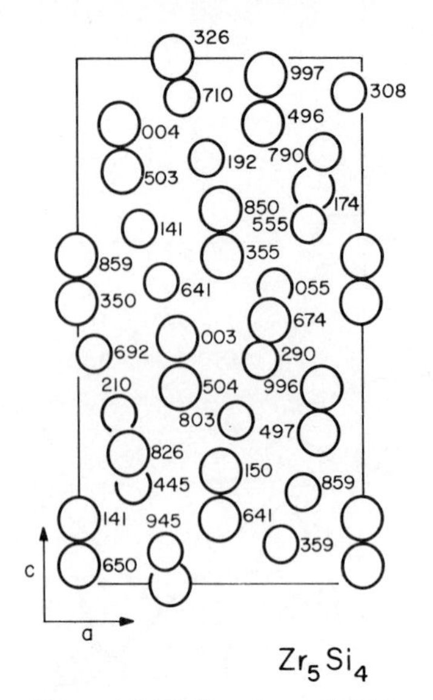

Figure 16-13. Structure of Si_4Zr_5 (*tP*36) viewed along [010]: large circles, Zr atoms.

those of Cu(1) and Si respectively; those of H(2) differ slightly. The resulting structure resembles that of $Cu_{15}Si_4$ rather closely, although details differ. H(1) has 4 Th at 2.46 Å and H(2) has 3 at 2.28, 2.285, and 2.31 Å. Th is surrounded by 12 H, but the arrangement is not icosahedral. The structures require appropriate confirmation by single-crystal and neutron-diffraction studies, respectively.

Sn_5Nb_6 (oI44): Sn_5Nb_6 has cell dimensions $a = 5.656$, $b = 9.199$, $c = 16.843$ Å, $Z = 4$.[24] The atoms show pronounced layering parallel to the (100) plane and some of the atomic arrangements are similar to those found in the Sn_5Ti_6 structure parallel to the (001) plane, although the atomic coordination in the two structures differs. Chains of Nb atoms occur only along [100] in Sn_5Nb_6 and the Nb–Nb separation, 2.83 Å, is larger than that found in $SnNb_3$. In Sn_5Nb_6 there are also chains of Sn atoms running along [100] with spacings alternately 2.76 and 2.90 Å. The Nb atoms have 11 or 12 close neighbors.

Zr_7Ni_{10} (oC68), $Zr_{15}Ni_{19}$ (oP68): Stoichiometric Zr_7Ni_{10} with cell dimensions $a = 12.39$, $b = 9.156$, $c = 9.211$ Å, $Z = 4$ takes Zr into solid solution with attendant distortion and change of space group in $Ni_{19}Zr_{15}$.[25] However, the cell shape and overall atomic arrangement remains the

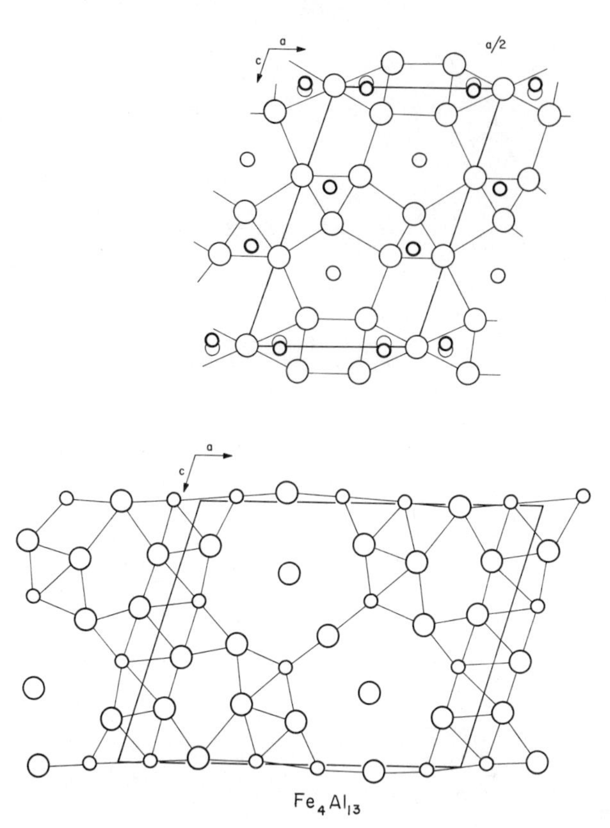

Figure 16-14. Atomic arrangements in the Fe_4Al_{13} (*mC*101) structure viewed down [010]: (upper) Flat layers; (lower) Puckered layers; large circles, Al atoms.

same in both structures. Zr_7Ni_{10} is said to resemble the CrB (*oC*8) structure of NiZr. Both structures consist of alternate layers of Ni and Zr atoms; in Zr_7Ni_{10} these lie parallel to (011) and $(01\bar{1})$ planes and in NiZr parallel to (021) and $(02\bar{1})$ planes, although the Zr layers in the former do contain some Ni atoms, and the layers are rumpled.

N_8Ca_{11} (*tP*38): N_8Ca_{11} has cell dimensions $a = 14.45$, $c = 3.60$ Å.[26] The atoms form layers at $z = 0$ and $\frac{1}{2}$. The structure contains $(N_2Ca_3)_n$ and $(N_3Ca_4)_n$ chains running in the [001] direction. Ca–N distances lie between 2.31 and 2.50 Å, and there is a short Ca–Ca distance of 3.11 Å.

$Ge_{10}Ho_{11}$ (*tI*84): The large tetragonal cell of $Ge_{10}Ho_{11}$ ($a = 10.79$, $c = 16.23$ Å) contains 5 independent Ge atoms and 4 Ho atoms. The structure contains some Ge–Ge bonds but the Ge chains which are a feature of

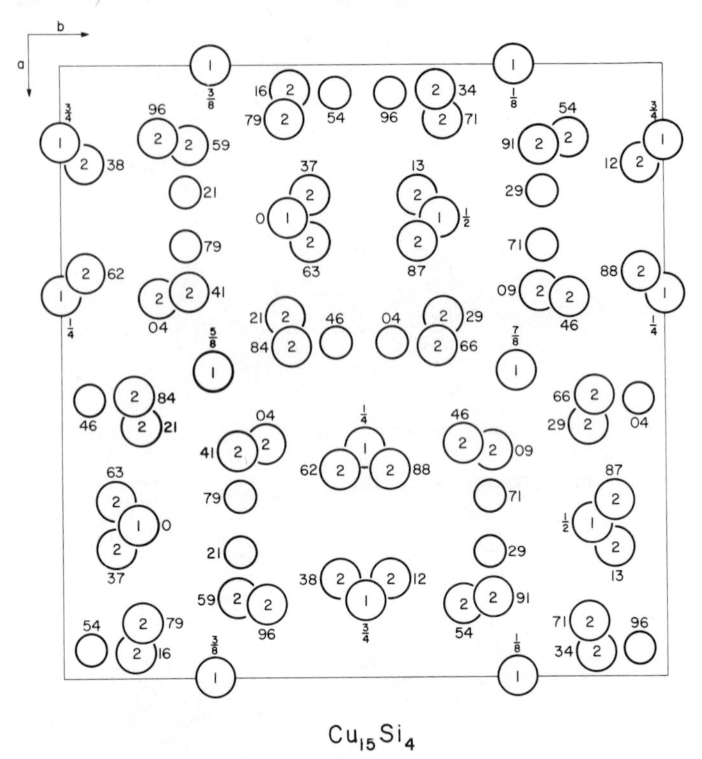

Figure 16-15. The atomic arrangement in the Si_4Cu_{15} (*cI*76) structure viewed down [001]: large circles, Cu atoms.

GeHo in the CrB structure are absent.[27] Ge(1) and Ge(5) severally form Ge–Ge pairs at 2.96 and 2.54 Å respectively, and Ge(3) form square clusters with separation 2.58 Å. The CN of Ge is generally 8 or 9 and of Ho 15–17.

$Co_{11}Ce_{24}$ (*hP*70): The $Co_{11}Ce_{24}$ structure with cell dimensions $a = 9.587$, $c = 21.825$ Å, $Z = 2$, is made up by stacking large triangular 3^6 and kagomé layers in the sequence $\gamma(A\alpha\beta B)\gamma(A\beta)\gamma\alpha(\beta C)(\alpha A\gamma)\beta(A\alpha C\gamma)$-$\beta(A\gamma)\beta\alpha(\alpha B)(\underline{\alpha A}\beta)$ in the notation of p. 4, with the Co layers underlined. The main feature of the structure is that the larger Ce atoms always avoid each other in successive layers so that, for instance, the atoms of one layer lie over the hexagons of kagomé layers above or below. The Co atoms generally occur as close neighbors to Ce atoms. Apart from these observations the structure does not appear to have features relating it closely to other types.

There are 10 different Ce atoms and 5 different Co atoms in the structure. The Ce atoms are surrounded by convex polyhedra of 14 to 16 atoms, the closest Co–Ce and Ce–Ce distances being 2.59 Å and 3.29 Å respectively. The Co atoms are surrounded by convex polyhedra of 9 or 10 Ce atoms.

$Gd_{13}Zn_{58}$ ($Pu_{13}Zn_{58}$) (hP142): $Gd_{13}Zn_{58}$ has cell dimensions $a = 14.35$, $c = 14.21$ Å, $Z = 2$.[29] Figure 16-16 taken from the paper by Wang, shows the arrangement of Zn and Gd atoms at various levels. From this it can be seen that the Gd atoms are so arranged that there are no contacts between them in three dimensions. The CN of the Gd atoms is 16 to 18, the smallest CN of Zn is 9. The structure contains several short interatomic distances, notably: Zn–Zn, 2.20, 2.35, and 2.40 Å, and Zn–Gd, 2.74 and 2.76 Å.

The structure of $Pu_{13}Zn_{58}$ investigated by Larson and Cromer[30] may be of the same type as $Gd_{13}Zn_{58}$ as the results are very similar. $Pu_{13}Zn_{58}$ has a hexagonal subcell with dimensions $a = 14.43$, $c = 14.14$ Å, $Z = 2$.

$CrUC_2$ (oP16): $CrUC_2$ has cell dimensions $a = 5.433$, $b = 3.232$, $c = 10.637$ Å, $Z = 4$.[31] The atoms are confined to planes parallel to (010) at heights $y = \frac{1}{4}$ and $\frac{3}{4}$. Bands of U and Cr atoms alternately lie parallel to the (001) plane. The atoms form distorted tetrahedra sharing edges that separate zigzag ladders of C atoms distributed in planes parallel to (001) and running in the [010] direction (Figure 16-17). Thus C(1) has 1 C(2) atom at 1.27 Å and 2 C(1) at 1.83 Å; C(2) has 1 C(1) neighbor and 2 C(2) at 2.05 Å. Cr has 3 C(1) at 2.08–2.10 Å, 2 C(2) at 2.23 Å, and 4 U neighbors (2.84 and 2.85 Å). U has 1 C(1) at 2.52 Å, 3 C(2) at 2.46, and 2.49 Å and 4 Cr neighbors. The structure was determined by single-crystal X-ray diffraction.

T_3-$MnZnAl_4$, $Ni_4Mn_{11}Al_{60}$ (oC152): The atomic positions in the complex structure T_3-$MnZnAl_4$ ($a = 7.78$, $b = 23.8$, $c = 12.6$ Å) have been determined together with the probable locations of the component atoms.[32] The structure is related to that of $Ni_4Mn_{11}Al_{60}$ in that all positions, with the exception of Al(10) in 4(*c*), are occupied in T_3.

α-$Mn_{12}(Al,Si)_{57}$, α-$Si_7Fe_{12}Al_{50}$ (cP138): α-MnAlSi has cell dimensions $a = 12.68$ Å, $Z = 2$.[33] The structure is a three-dimensional array of coordination polyhedra. Mn(1) is surrounded by 10 Al at 2.43 to 2.84 Å, and Mn(2) by 9 Al at 2.53 to 2.62 Å with one very short Mn(2)–Al(5) distance of 2.27 Å. Al–Al distances range from 2.44 Å to 3.08 Å and the 9 independent Al atoms have CN from 11 to 15. There are large holes in the structure at the cell corners and at the cell center. These are surrounded by Al(4) and (5) atoms at distances about 2.43 Å.

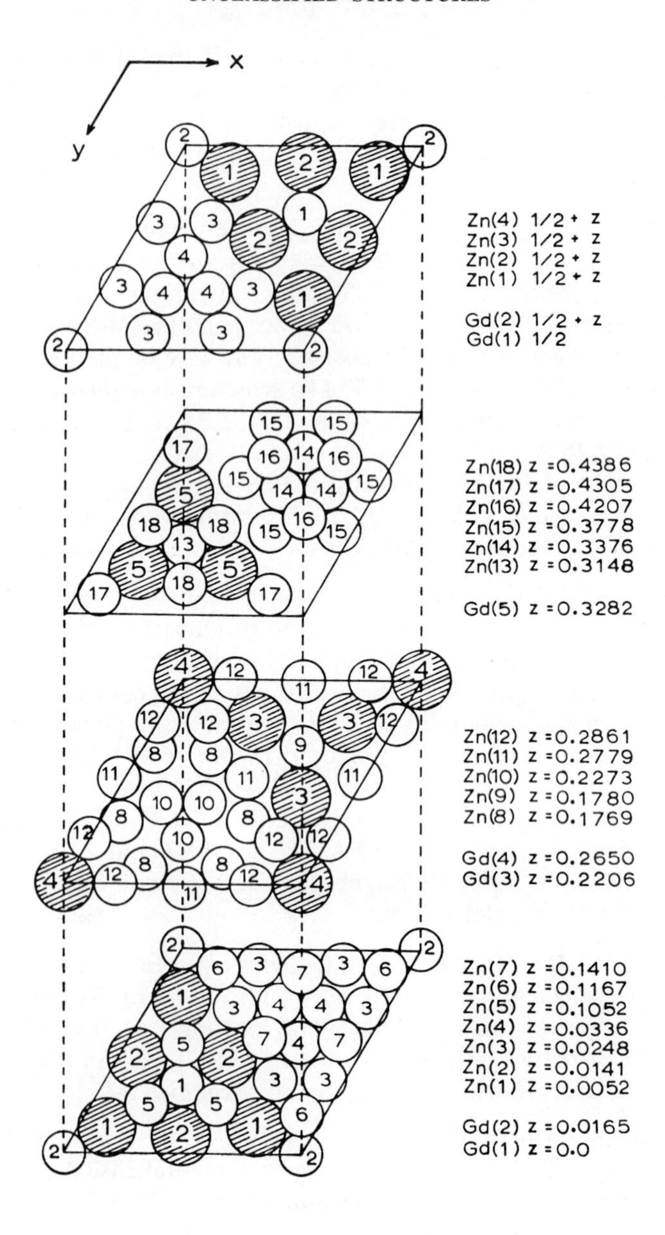

Figure 16-16. Schematic view of the atomic arrangement in the $Gd_{13}Zn_{58}$ (*hP*142) structure. (Wang[29].)

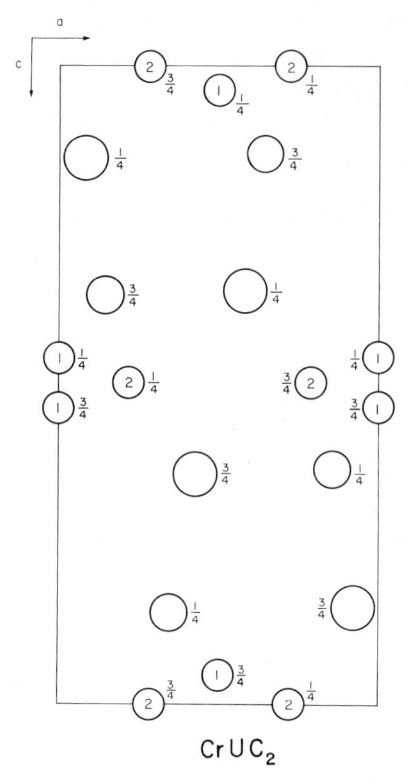

Figure 16-17. The structure of $CrUC_2$ (*oP*16) viewed down [010]: U, largest circles; C, smallest circles.

In α-SiFeAl[34] the short Fe–Al bond is 2.34 Å for both Fe(1) and Fe(2) atoms.

STRUCTURES OF VALENCE COMPOUNDS

Black Phosphorus (*oC*8): With cell dimensions $a = 3.314$, $b = 10.48$, $c = 4.376$ Å, $Z = 8$,[35] black phosphorus forms the double layer structure shown in Figure 5-21, p. 237. Each P atom forms two bonds at 2.22 Å within its layer and one at 2.24 Å linking the double layers.[35]

α-Se, β-Se (*mP*32): Se forms eight-membered nonplanar rings in the α and β-Se structures which both have monoclinic symmetry and 32 atoms per unit cell.[36]

AsLi (*mP*16): The AsLi cell has dimensions $a = 5.79$, $b = 5.24$, $c = 10.70$

Å, $\beta = 117.4°$, $Z = 8$.[37] A feature of the structure is As–As contacts giving As chains that form spirals running throughout the structure in the *b* direction (Figure 5-8, p. 220). Thus As(1) has 2 As(2) neighbors at 2.45 and 2.47 Å and As(2) has 2 As(1) neighbors. Both As(1) and As(2) have 6 Li neighbors at 2.63 to 2.88 Å in a considerably distorted octahedral arrangement. Li(1) and (2) each have 6 As neighbors and there are close Li(2)–Li(2) and Li(1)–Li(2) approaches of 2.98 and 3.00 Å respectively.

AsS (*mP*32): The dimensions of the unit cell of realgar in standard $P2_1/c$ setting are $a = 6.56$, $b = 13.50$, $c = 9.705$ A, $\beta = 113.75°$, $Z = 16$.[38] The structure contains discrete As_4S_4 molecules separated from each other by distances corresponding to van der Waals bonds. Each of the four independent S atoms forms two bonds to As at distances from 2.18 to 2.28 Å. Each of the four independent As atoms forms two bonds to S and one to another As atom (2.59 or 2.60 Å).

HgS (*hP*6): The dimensions of the HgS cell are $a = 4.146$, $c = 9.497$ Å, $Z = 3$.[39] Unlike most other simple structures with hexagonal symmetry, the cinnabar structure must be regarded as containing spirals of atoms twined about the *c* edges of the unit cell, rather than planes of atoms lying normal to the *c* axis of the crystal (Figure16-18). The spirals have a pitch of 6, Hg and S atoms alternating along the chains. Hg–S distances along the chains are 2.39 Å. Other Hg–S distances are 2×3.10 Å and 2×3.30 Å.

NiS (*hR*6): The dimensions of the hexagonal cell of millerite are $a = 9.610$, $c = 3.148$ Å, $Z = 9$.[40] The structure is made up of triangles of Ni about 0, 0, 0.088; $\frac{2}{3}$, $\frac{1}{3}$, 0.421 and $\frac{1}{3}$, $\frac{2}{3}$, 0.755, and slightly larger triangles of S about $\frac{1}{3}$, $\frac{2}{3}$, 0.263; 0, 0, 0.596 and $\frac{2}{3}$, $\frac{1}{3}$, 0.929 (Figure 16-19). Ni has 1 S at 2.18 Å, 5 at 2.35 to 2.40 Å and 2 Ni at 2.53 Å. S has 6 Ni neighbors. The structure should be confirmed by modern single-crystal methods.

CuO (*mC*8): In the tenorite structure, which has cell dimensions $a = 4.66$, $b = 3.42$, $c = 5.12$ Å, $\beta = 99.48°$, $Z = 4$,[41] Cu is surrounded by 4 O approximately in a square, and O is surrounded by 4 Cu approximately tetrahedrally (Figure 16-20).

$SCu_{1.96}$ (*tP*12): The metastable $SCu_{1.96}$ structure with cell dimensions $a = 3.996$, $c = 11.29$ Å, $Z = 4$, is shown in Figure 16-21.[42] S has 6 irregularly arranged equidistant neighbors. Cu has three close S neighbors, one Cu at 2.715 Å and 6 more at 2.965 Å. The structure requires confirmation by single-crystal methods.

SCu_2 II (H.T.) (*hP*6): Chalcosite has lattice parameters $a = 3.89$, $c =$

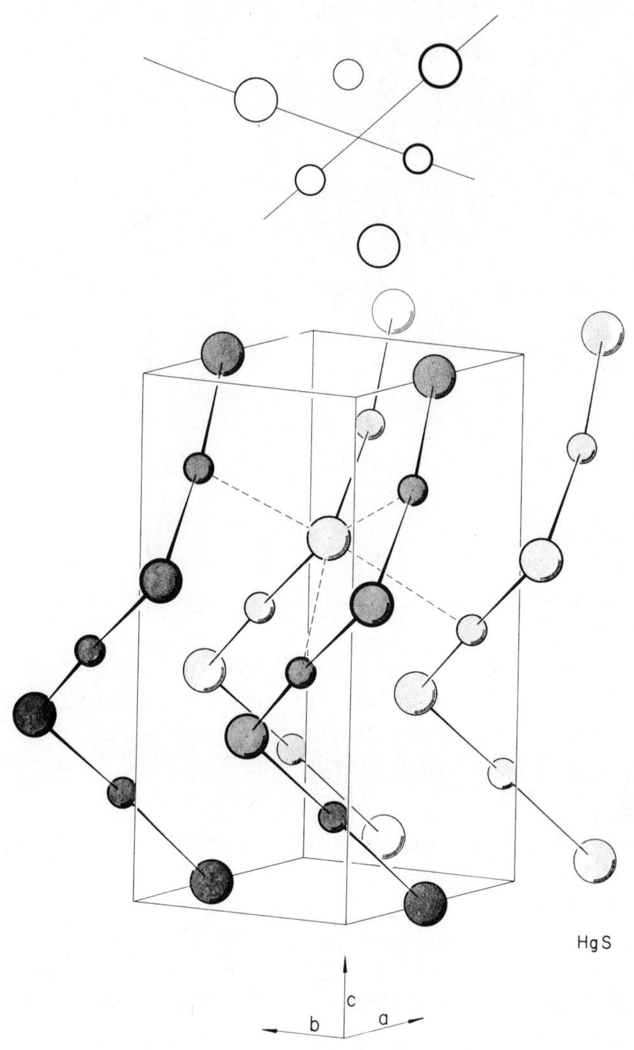

Figure 16-18. Pictorial view of the atomic arrangement in the HgS (*hP*6) structure.

6.88 Å, $Z = 2$,[43] at 125°C. The S atoms form close packed layers stacked in the sequence AB ($c/a = 1.77$) and the Cu atoms partially occupy three sitesets.

$TeAg_2$ (L.T.) (*mP*12): The mineral hessite has unit cell dimensions $a = 8.09$, $b = 4.48$, $c = 8.96$ Å, $\beta = 123.33°$, $Z = 4$.[44] Within the unit cell, atoms lie on four lines in the *ac* plane at heights $z \sim 0.16, 0.34, 0.66,$

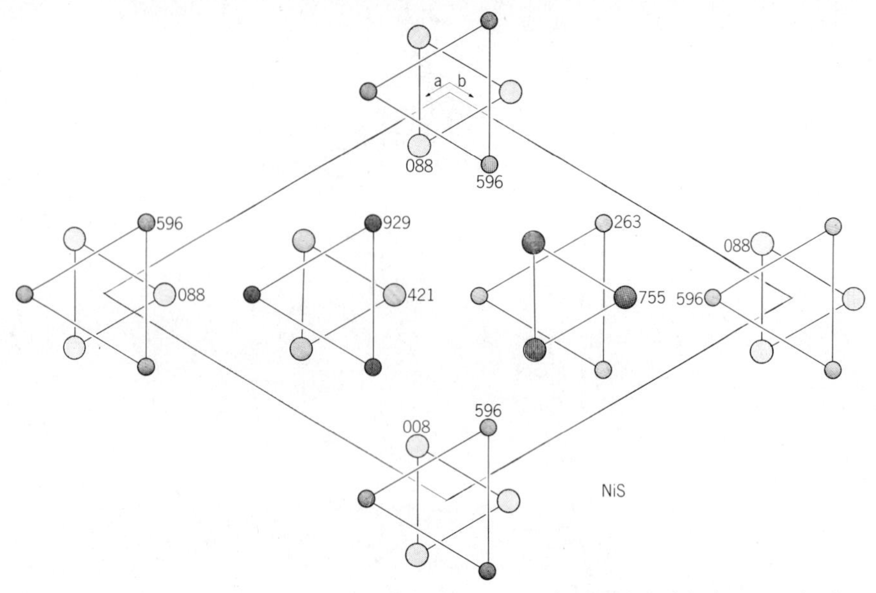

Figure 16-19. View of the millerite NiS ($hR6$) structure down [001] of the hexagonal cell.

0.84, but the lines of atoms do not run continuously throughout the crystal (Figure. 16-22). This arrangement surrounds Ag(1) with 10 neighbors at distances closer than 3.14 Å, and Ag(2) and Te are surrounded by 8 atoms.

α-SeAg₂ ($oP12$): Low-temperature α-$SeAg_2$ obtained in evaporated and annealed thin films, has cell dimensions $a = 7.05$, $b = 7.85$, $c = 4.33$ Å, $Z = 4$.[45] The Ag atoms have two close Se neighbors at 2.53 Å which form zigzag Ag–Se–Ag–Se chains parallel to [001]. The structure contains very short Ag–Ag distances of 2.61, 2.62, and 2.64 Å. The structure was obtained from oblique-texture electron diffraction photographs.

$SeIn_2$ ($oP24$): In_2Se has cell dimensions $a = 15.24$, $b = 12.32$, $c = 4.075$ Å, $Z = 8$.[46] The atoms lie in planes at $z = 0$ and $\frac{1}{2}$. There are Se–In–Se–In–Se chains (In–Se = 2.72 Å) which are linked together by Se–Se bonds (2.20 Å). See Figure 16-23. Close In–Se distances lie between 2.65 and 2.74 Å and close In–In distances between 2.86 and 2.99 Å. The structure was determined by oblique-texture electron diffraction photographs taken of thin evaporated films.

$SrSi_2$ ($cP12$): Si atoms form a three-dimensional array each having 3 Si neighbors at 2.41 Å.[47] Sr is surrounded by 6 Si at 3.21 Å and has two more

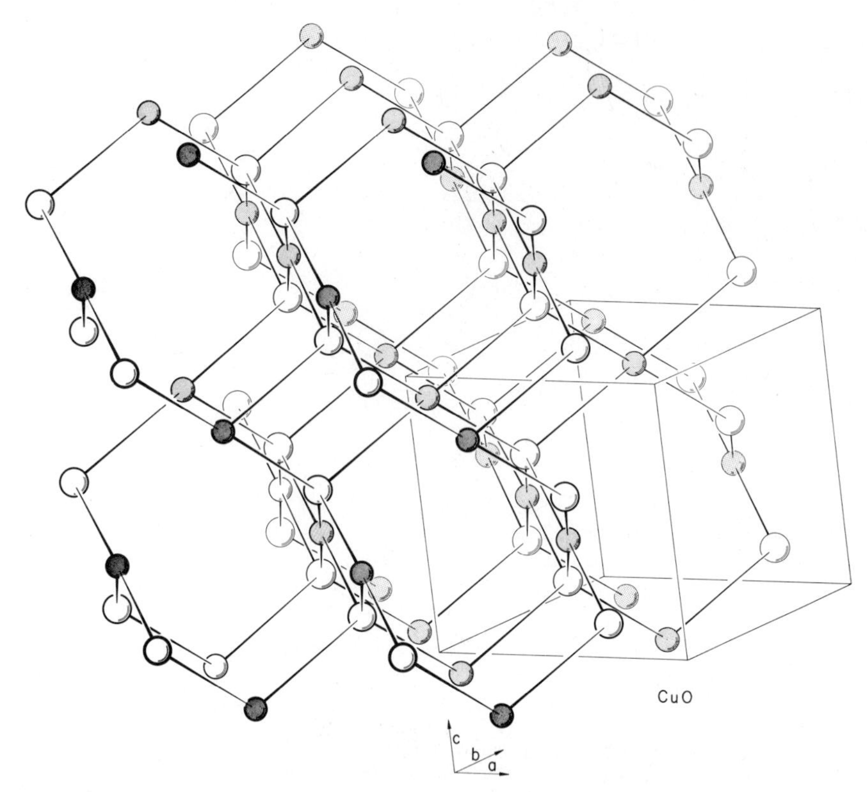

Figure 16-20. Pictorial view of the structure of CuO (*mC*8): large circles, oxygen.

neighbors at 3.43 Å (Figure 16-24). It appears that the structure requires refinement by single-crystal methods.

α-CdP$_2$ (*oP*12): α-CdP_2 has an orthorhombic cell, $a = 9.90$, $b = 5.408$, $c = 5.171$ Å, $Z = 4$.[48] The structure contains spiral chains of P atoms (P–P = 2.05 and 2.39 Å alternately) which run in the *c* direction. Each P also forms bonds to two Cd atoms which are surrounded by 4 P approximately tetrahedrally (Cd–P = 2.50 to 2.63 Å) linking the chains together.

Mo$_2$*As*$_3$ (*mC*20): Mo_2As_3 has cell dimensions $a = 16.061$, $b = 3.2349$, $c = 9.643$ Å, $\beta = 136.74°$, $Z = 4$.[49] The atoms lie in layers at $y = 0$ and $\frac{1}{2}$ without any significant net pattern. Mo atoms are surrounded by distorted octahedra of 6 As atoms, which are joined together by sharing common edges. The Mo atoms are displaced from the centers of the octahedra toward one face, resulting in the formation of zigzag chains of Mo running in the [010] direction (Mo(1)–Mo(1) = 2.940 Å, Mo(2)–Mo(2) =

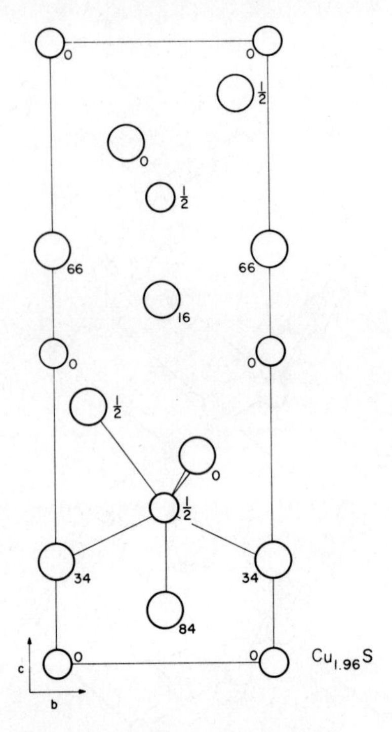

Figure 16-21. The structure of $SCu_{1.96}$ ($tP12$) viewed down [100].

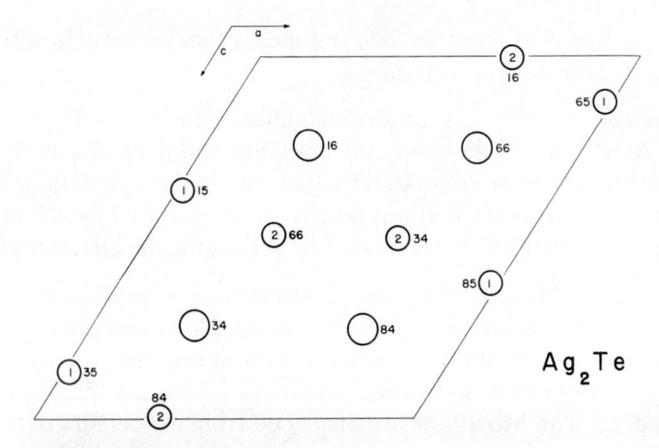

Figure 16-22. The structure of the low-temperature form of $TeAg_2$ ($mP12$) viewed down [010].

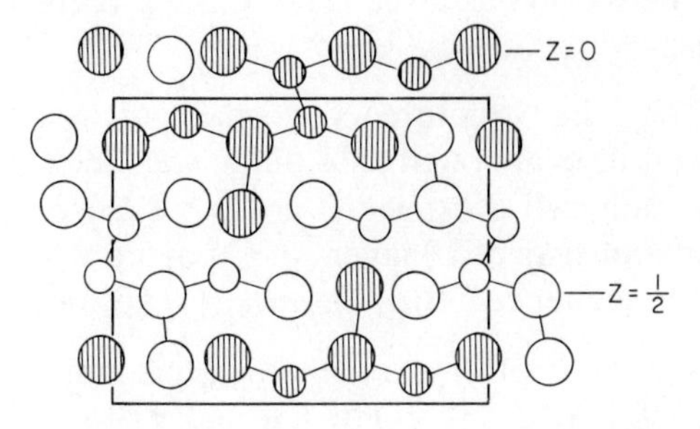

Figure 16-23. The atomic arrangement in the $SeIn_2$ (*oP*24) structure viewed down [001]: In, large circles.

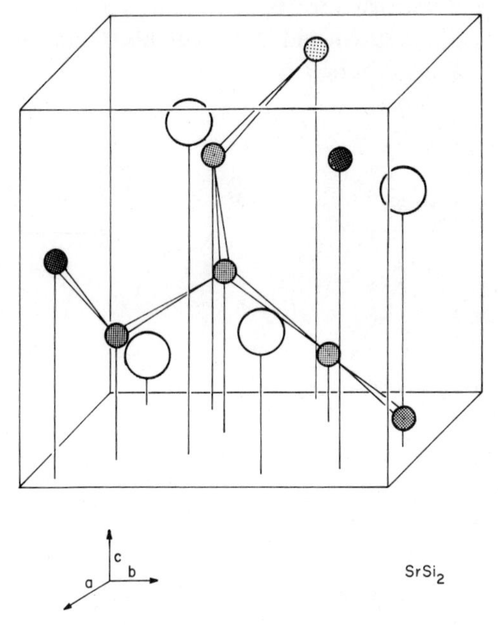

Figure 16-24. Pictorial view of the $SrSi_2$ (*cP*12) structure.

2.955 Å). The structure also contains close pairs of As(3) atoms (2.445 Å). Mo–As distances are 2.51 to 2.60 Å.

As_2S_3 (*mP*20): The dimensions of the unit cell of orpiment are $a = 4.25$, $b = 9.59$, $c = 12.22$ Å, $\beta = 109.9°$, $Z = 4$.[50] The structure is made up of three-layer units S, As, S parallel to (010). Within the three-layer groups

each As has three S neighbors and each S has two close As neighbors. S–S distances between the three-layer groups correspond to van der Waals distances.

β^1-*Te_4Cu_7* (*hP*22): The structure of a phase of assumed composition Te_4Cu_7, observed in evaporated thin films, was determined by electron diffraction. The unit cell dimensions are $a = 8.28$, $c = 7.22$ Å, $Z = 2$,[51] and the structure consists of 10 layers of Cu or Te atoms parallel to (001) and stacked in the sequence CB($\underline{\alpha}$A)$\beta\gamma\underline{\alpha}\beta$($\gamma\underline{A}$) (Te underlined).

γ-*$Ag_{0.93}Cu_{1.07}S$* (*oC*12): Stromeyerite has cell dimensions $a = 4.066$, $b = 6.628$, $c = 7.972$ Å, $Z = 4$.[52] The Cu and S atoms form 6^3 graphite-like nets at $z = \frac{1}{4}$ and $\frac{3}{4}$, but the nodes of the nets do not overlie each other (Figure 16-25). The Ag atoms form overlying 3^6 nets at $z = 0$ and $\frac{1}{2}$, and are so located in relation to the Cu + S hexagons above and below, that they are surrounded by 3 Cu and 3 S atoms (2.40 Å). Each S has 3 Cu (2.26, 2 × 2.29 Å) and 2 Ag neighbors and each Cu has 3 S neighbors and 4 Ag neighbors.

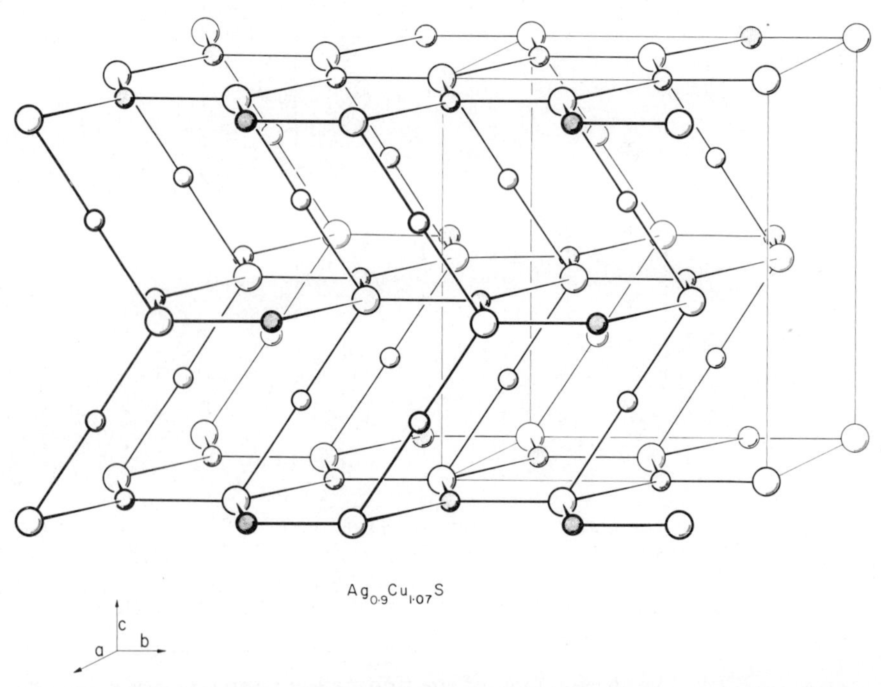

Figure 16-25. Pictorial view of the atomic arrangement in the stromeyerite, $Ag_{0.9}Cu_{1.07}S$ (*oC*12) structure: S, largest circles; Cu, small shaded circles; Ag, small open circles. The origin was transposed in the drawing to place S at the cell corners.

$AsTlS_2$ (*mP*32): The dimensions of the lorandite unit cell are $a = 6.11$, $b = 11.33$, $c = 12.27$ Å, $\beta = 104.2°$, $Z = 8$.[53] The structure contains spiral chains, running in a direction parallel to [010].

```
2.05 Å        2.40      2.15         2.35      2.05
———As(1)———S(4)———As(2)———S(3)———
     | 2.25               | 2.16
     S(1)                 S(2)
```

The Tl atoms are located without the spirals, and the structure could in principle be accounted for if Tl were present as Tl^{I} ions, supplying the extra electron satisfying the valence requirements of the AsS_2 chains However, interatomic distances indicate that the structure is not so simple, and that Tl atoms supply the needed electron to satisfy valence octets in the As and S atom chains by covalent bonding.

Thus S(1), one of the atoms attached to the As–S(3)–As–S(4) chains has a Tl(1) neighbor at a distance of only 2.87 Å, and a Tl(2) neighbor at 3.00 Å. On the other hand S(2), the other atom attached to the chains, has no Tl neighbor closer than Tl(2) at 2.99 Å, although Tl(2), itself, approaches S(4), one of the chain atoms, rather closely at 2.83 Å. The other chain atom S(3) has no Tl neighbors closer than 3.30 Å. In addition, Tl(2)–Tl(2) at 3.51 Å is a rather close contact. This arrangement of S–Tl contacts results in an uneven distribution of As–S distances along the chains as indicated in the sketch above.

$AsAg_3S_3$ (*hR*14): In the structure of proustite (hexagonal cell $a = 10.74$, $c = 8.658$ Å, $Z = 6$) the atoms are arranged in planes parallel to (001).[54] The S and Ag atoms form spirals with a pitch of c along [001]. The spirals are connected together by triangular pyramids of S with As as apex. However, there are two quite independent unconnected arrays of spirals, one rotating right, the other left. One set of spirals is connected by As atoms at $z = 0$, $\frac{1}{3}$, and $\frac{2}{3}$, the other by As atoms at $z = \frac{1}{6}$, $\frac{1}{2}$, $\frac{5}{6}$. Ag–S = 2.40 Å, As–S = 2.25 Å. Ag_3SbS_3 is isotypic.

$AsAg_3S_3$ (*mC*56): Xanthokon with cell dimensions $a = 12.00$, $b = 6.26$, $c = 17.08$ Å, $\beta = 110.0°$, $Z = 8$,[55] is composed of double sheets of atoms parallel to (001) which are connected by Ag–S bonds. Figure 16-26 shows the structure projected down the b axis. Ag(1) and (3) have planar coordination by 3 S at 2.45 to2.60 Å, whereas the 3 S neighbors of Ag(2) (2.48 to 2.78 Å) are not coplanar with it. Ag(2) also has 3 Ag neighbors at 2.99 to 3.07 Å and another S at 2.965 Å. Ag(1) and (2) also have one Ag(2) neighbor. As atoms have 3 S neighbors at 2.22 to 2.27 Å disposed

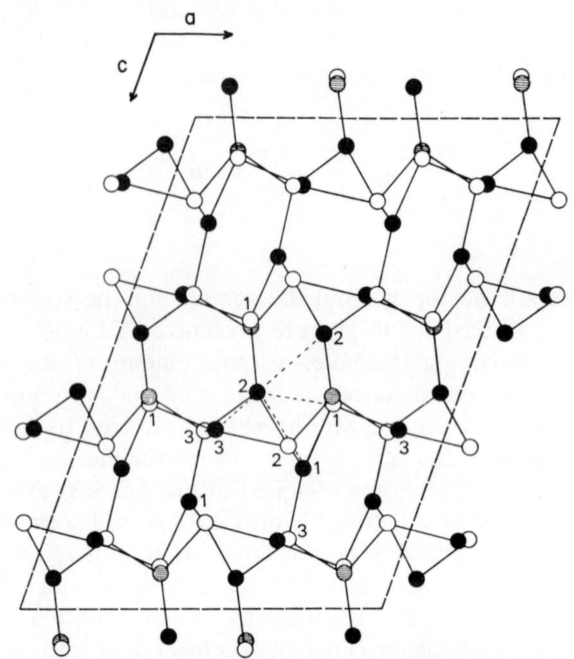

Figure 16-26. The structure of xanthokon, Ag_3AsS_3 ($mC56$) projected down [010]: sulphur, open circles; As, shaded circles; Ag, black circles. (Engel and Nowacki [55].)

pyramidally. Each S atom has 1 As and 3 Ag neighbors, which are disposed tetrahedrally in the case of S(3) and (with strong distortion) S(1).

$AgSbS_2$ ($mC32$): The structure of miargyrite has cell dimensions $a = 12.86$, $b = 4.411$, $c = 13.22$ Å, $\beta = 98.63°$, $Z = 8$.[56] Although the Ag and Sb atoms individually form distorted 3^6 nets parallel to the *ac* plane at $y \sim 0$ and $\frac{1}{2}$, the structure can probably best be understood by consideration of atomic arrangements in (102) planes. Here an ideal structure can be imagined in which the atoms form 3^6 nets made up of alternate rows of S and of Ag + Sb atoms. The layers are then distorted by moving one S row towards a neighboring Ag + Sb row (Figure 16-27). The direction of this displacement alternates between successive layers, and within layers after every fourth S atom in a row.

These arrangements result in Sb having 3 S neighbors at 2.48 to 2.58 Å lying in a plane that does not contain Sb. Ag(1) has S neighbors at 2.44, 2.50, 2.58, and 2.72 Å, whereas Ag(2) has two close S neighbors at 2.36

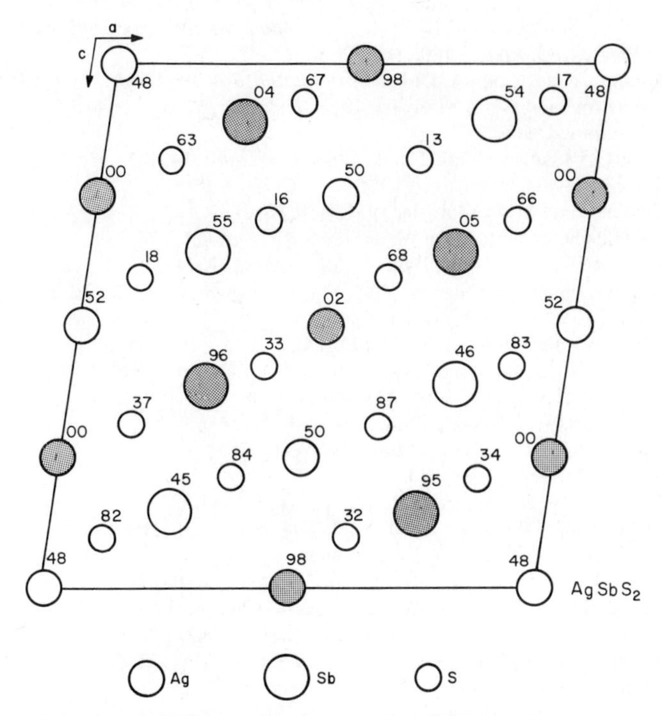

Figure 16-27. The structure of miargyrite, $AgSbS_2$ ($mC32$) viewed down [010].[56]

and 2.47 Å; the next nearest S atom occurs at 2.89 Å. The S atoms have three Ag and Sb neighbors.

$Tl_2Sb_2Se_4$ (oP8): $Tl_2Sb_2Se_4$ has cell dimensions $a = 4.18$, $b = 4.50$, $c = 12.00$ Å, $Z = 1$.[57] The structure is made up of $\frac{1}{2} \cdot 4^4$ nets of Sb and Tl atoms at $z = 0$ and $\frac{1}{2}$, separated by zigzag chains of Se atoms at $z \sim \frac{1}{4}$ and $\frac{3}{4}$ which run in the a direction. Sb has 4 Tl neighbors at 3.07 Å. Tl has 4 Sb neighbors and 2 Se at 2.74 Å, and Se has 1 Tl neighbor and 2 Se neighbors along the chains at 2.16 Å. The structure was determined by electron diffraction from thin films.

REFERENCES

1. W. H. Zachariasen, 1952, *Acta Cryst.*, **5**, 660.
2. W. H. Zachariasen, 1963, *Acta Cryst.*, **16**, 784; W. H. Zachariasen and F. H. Ellinger, 1963, *Ibid.*, **16**, 777.
3. W. H. Zachariasen and F. H. Ellinger, 1955, *Acta Cryst.*, **8**, 431.
4. R. M. Brugger, R. B. Bennion, and T. G. Worlton, 1967, *Phys. Letters,* **24A**, 714.

5. See e.g. E. F. Sturcken and B. Post, 1961, *Advances in X-ray Analysis*, New York: Plenum Press, **4**, 85; *Idem*, 1960, *Acta Cryst.*, **13**, 852.
6. H. Curien, A. Rimsky, and A. DeFrain, 1961, *Bull. Soc. Fr. Minér. Crist.*, **84**, 260.
7. E. S. Makarov and V. A. Levdik, 1956, *Kristallografija*, **1**, 644 [*Soviet Physics–Crystallography*, **1** **506**].
8. K. Schubert, F. Gauzzi, and K. Frank, 1963, *Z. Metallk.*, **54**, 422.
9. E. J. Duwell and N. C. Baenziger, 1955, *Acta Cryst.*, **8**, 705.
10. I. G. Edmunds and M. M. Qurashi, 1951, *Acta Cryst.*, **4**, 417.
11. H. Axel, H. Schäfer, and Armin Weiss, 1965, *Angew Chem.*, **77**, 379.
12. R. B. Roof and C. P. Kempter, 1962, *J. Chem. Phys.*, **37**, 1473.
13. N. N. Žuravlev and G. S. Ždanov, 1955, *Ž. Éksper. Teoret. Fiz. SSSR*, **28**, 237.
14. R. Kiessling, 1950, *Acta Chem. Scand.*, **4**, 146.
15. W. H. Zachariasen, 1966, *Acta Cryst.*, **20**, 334.
16. H. Schäfer, H. Axel, and Armin Weiss, 1965, *Z. Naturforsch.*, **B20**, 1010.
17. K. Meisel, 1939, *Z. anorg. Chem.*, **240**, 300.
18. H.-U. Pfeifer and K. Schubert, 1966, *Z. Metallk.*, **57**, 884.
19. M. Elfström, 1965, *Acta Chem. Scand.*, **19**, 1694.
20. P. J. Black, 1955, *Acta Cryst.*, **8**, 43.
21. R. C. Hudd and W. H. Taylor, 1962, *Acta Cryst.*, **15**, 441.
22. F. R. Morral and A. Westgren, 1934, *Ark. Kemi, Min. Geol.*, **11B**, No. 37.
23. W. H. Zachariasen, 1953, *Acta Cryst.*, **6**, 393.
24. J. R. Ogren, T. G. Ellis, and J. F. Smith, 1965, *Acta Cryst.*, **18**, 968.
25. M. E. Kirkpatrick, J. F. Smith, and W. L. Larsen, 1962, *Acta Cryst.*, **15**, 894.
26. Y. Laurent, J. Lang, and M. Th. LeBihan, 1969, *Acta Cryst.*, **B25**, 199.
27. G. S. Smith, Q. Johnson, and A. B. Tharp, 1967, *Acta Cryst.*, **23**, 640.
28. A. C. Larson and D. T. Cromer, 1962, *Acta Cryst.*, **15**, 1224.
29. F. E. Wang, 1967, *Acta Cryst.*, **22**, 579.
30. A. C. Larson and D. T. Cromer, 1967, *Acta Cryst.*, **23**, 70.
31. H. Nowotny, R. Kieffer, F. Benesovsky, and E. Laube, 1958, *Mh. Chem.*, **89**, 692.
32. A. Damjanovic, 1961, *Acta Cryst.*, **14**, 982.
33. M. Cooper and K. Robinson, 1966, *Acta Cryst.*, **20**, 614.
34. M. Cooper, 1967, *Ibid.*, **23**, 1106.
35. A. Brown and S. Rundqvist, 1965, *Acta Cryst.*, **19**, 684.
36. R. D. Burbank, 1951, *Acta Cryst.*, **4**, 140; R. E. Marsh, L. Pauling, and J. D. McCullough, 1953, *Ibid.*, **6**, 71.
37. D. T. Cromer, 1959, *Acta Cryst.*, **12**, 36.
38. T. Ito, N. Morimoto, and R. Sadanaga, 1952, *Acta Cryst.*, **5**, 775.
39. K. L. Aurivillius, 1950, *Acta Chem. Scand.*, **4**, 1413.
40. See e.g. V. G. Kuznecov *et al.* 1961 [*Proc. 4th All Union Conf. on Semiconductor Materials*. Translated by Consultants Bureau, New York, p. 128].
41. G. Tunell, E. Posnjak, and C. J. Ksanda, 1935, *Z. Kristallogr.*, **A90**, 120.
42. A. Janosi, 1964, *Acta Cryst.*, **17**, 311.
43. M. J. Buerger and B. J. Wuensch, 1963, *Science*, **141**, 276.
44. A. J. Frueh, 1959, *Z. Kristallogr.*, **112**, 44.
45. Z. G. Pinsker, Chou Ching-liang, R. M. Imamov, and E. L. Lapidus, 1965, *Kristallografija*, **10**, 275 [*Soviet Physics–Crystallography*, **10**, 225].
46. L. I. Man and S. A. Semiletov, 1965, *Kristallografija*, **10**, 407 [*Soviet Physics–Crystallography*, **10**, 328].
47. K. Janzon, H. Schaefer, and Armin Weiss, 1965, *Angew. Chem.*, **77**, 258.

48. J. Goodyear and G. A. Steigmann, 1969, *Acta Cryst.*, **B25**, 2371; O. Olofsson and J. Gullman, 1970, *Ibid.*, **B26**, 1883.
49. P. Jensen, A. Kjekshus, and T. Skansen, 1966, *Acta Chem. Scand.*, **20**, 1003.
50. M. J. Buerger, 1942, *Amer. Min.*, **27**, 301; N. Morimoto, 1949, *X-sen Kondanka, Osaka, Univ.*, **5**, 115.
51. R. V. Baranova, 1967, *Kristallografija*, **12**, 266. [*Soviet Physics–Crystallography*, **12**, 221].
52. A. J. Frueh, 1955, *Z. Kristallogr.*, **106**, 299.
53. A. Zemann and J. Zemann, 1959, *Acta Cryst.*, **12**, 1002.
54. D. Harker, 1936, *J. Chem. Phys.*, **4**, 381.
55. P. Engel and W. Nowacki, 1968, *Acta Cryst.*, **B24**, 77.
56. C. R. Knowles, 1964, *Acta Cryst.*, **17**, 847.
57. G. Pinsker, S. A. Semelitov, and E. N. Belova, 1956, *Dokl. Akad. Nauk, SSSR*, **106**, 1003.

Subject Index

In order to find a specific alloy or crystal structure, reference must be made to the *Formula Index*, since chemical names are not listed in the *Subject Index*. Trivial names (e.g. NaCl: rocksalt) may, however, be included in this index. A page reference given in bold face indicates the page where the description of the crystal structure type is to be found. Authors are not indexed.

Formula Index

A page reference given in bold face indicates the place where a description of the crystal structure type is to be found. I have attempted to ensure that all references to alloys and compounds in this index give some significant information about them: that is to say, pages on which they are referred to trivially in the text, are not listed.

Indexing is alphabetical by chemical symbols, except where one component of a solid solution (indicated by parentheses) cannot be placed in alphabetical order and therefore must be ignored. Thus $(Cu,Zn)_2Mg$ follows Cu_2Mg, yet precedes Cu_3P. Alloys of indefinite composition are indicated by dashes separating the components, e.g. Cu-Mn.